Fungal Biotechnology

Prospects and Avenues

Editors:

Sunil K. Deshmukh
Nano Biotechnology Centre
The Energy and Resources Institute, New Delhi, India

Kandikere R. Sridhar
Department of Biosciences
Mangalore University, Mangalore, India

Susanna M. Badalyan
Laboratory of Fungal Biology and Biotechnology
Institute of Pharmacy, Department of Biomedicine
Yerevan State University, Yerevan, Armenia

CRC Press
Taylor & Francis Group
Boca Raton London New York

CRC Press is an imprint of the
Taylor & Francis Group, an **informa** business
A SCIENCE PUBLISHERS BOOK

First edition published 2022
by CRC Press
6000 Broken Sound Parkway NW, Suite 300, Boca Raton, FL 33487-2742

and by CRC Press
4 Park Square, Milton Park, Abingdon, Oxon, OX14 4RN

CRC Press is an imprint of Taylor & Francis Group, LLC

Library of Congress Cataloging-in-Publication Data (applied for)

ISBN: 978-1-032-16385-7 (hbk)
ISBN: 978-1-032-16386-4 (pbk)
ISBN: 978-1-003-24831-6 (ebk)

DOI: 10.1201/9781003248316

Typeset in Times New Roman
by Radiant Productions

Preface

As the first phase of development, traditional mycological studies mainly addressed the damages and devastation caused by fungi on plants, animals, air-borne pathogens, wood decomposition, and mycotoxins. Advances in applied mycology in the second phase enhanced our knowledge on their significance in fermentation technology (flours, bakery goods, and cheese) and production of fermented products (wine, beer, and spirit) (Hyde et al., 2019; Vendukand Veljović, 2021). Further, value-added approaches of mycology revealed the importance of fungi in the production of bioactive metabolites, pharmaceuticals, and cosmeceuticals useful in treatment of human lifestyle diseases and plant diseases (e.g., pest control) (Sridhar and Deshmukh, 2021). Owing to heavy dependence on fossil resources, production of renewable merchandise projected the significance of fungi as potential candidates towards a sustainable global economy (Meyer et al., 2020). As the third phase, recent developments in mycology revealed their significance in the fields of advanced research (e.g., building materials, packaging resources, electronic devices, and leather-like merchandise) (Ghazvinian et al., 2019). In addition, fungi possess several qualities to cleanup non-biodegradable components and xenobiotics in the ecosystem to promote sustainable bioremediation. The fifty different ways of industrial utilization of fungi that have been documented of latesignifies their importance in human affairs (Hyde et al., 2019).

The book **Fungal Biotechnology: Prospects and Avenues** documents current biotechnological advances and avenues for bioprospects of fungi. Fungal biopolymers have many applications like prebiotics, therapeutics, immunoceuticals, drug delivery, and oil retrieval. Fungal mycelial biocomposites are useful in the fields of civil engineering, electronics, insulations, packaging, wearables, arts, and designs. Fungal bioactive metabolites have several implications beyond antibiotics such as volatiles, biofuels, nematicides, pigments, and agriculturally compatible metabolites. They also serve to produce nanoparticles of medicinal, nutritional, and industrial significance. In view of environmental protection, fungal activity and their products support bioremediation via degradation of xenobiotics and recalcitrant materials. This book introduces not only the value of fungi but it serves as a knowledge bank to the graduates, post-graduates, and researchers contemplating fungal biotechnology, particularly in the areas of industrial, environmental, medical, molecular, and environment mycology.

This book deals with (1) application of genomics in the field of fungal biotechnology; (2) production of fungal biomaterials like chitosan, pigments, and biopolymers of industrial and therapeutic significance; (3) generation of value-added metabolites of medicinal significance (broad spectrum antifungal agents like echinocandins and their derivatives; metabolites of *Ganoderma* and *Lignosus*), and biofuels using fungi (yeasts and wood rot fungi); (4) potential of fungi in the production of metabolites of nutritional significance in food, feed (coenzyme Q_{10}), fungal probiotics and prebiotics, beverages, and mycotoxins; (5) significance of fungi biofuel production (e.g., *Pichia* and wood rot fungi); (6) nematicidal effect of nematophagous fungi in plant protection. It also addresses uses of fungal metabolites as medicine as well as nutraceuticals including therapeutic biopolymers.

We are grateful to all the contributors for their timely submission/revision of stimulating chapters in the field of fungal biotechnology. We are indebted to the reviewers for value addition to the chapter by meticulous examination. The CRC Press has extended all kind of encouragement and cooperation during the tenure of this book despite heavy backlog, owing to the pandemic,to present this book as scheduled.

New Delhi, India — **Sunil K. Deshmukh**
Mangalore, India — **Kandikere R. Sridhar**
Yerevan, Armenia — **Susanna M. Badalyan**

References

Ghazvinian, A., Farrokhsiar, P., Vieiru, F., Pecchia, J. and Gursoy, B. (2019). Mycelium-based bio-composites for architecture: Assessing the effects of cultivation factors on compressive strength. pp. 505–514. *In*: Sousa, J.P., Henriques, G.C. and Xavier, J.P. (Eds.). *Material Studies and Innovation.* Volume 2, eCAADe 37/SigraDI.

Hyde, K.D., Xu, J., Rapior, S., Jeewon, R., Lumyong, S. et al. (2019). The amazing potential of fungi: 50 ways we can exploit fungi industrially. *Fungal Diversity*, 97: 1–136.

Meyer, V., Basenko, E., Benz, J.P., Braus, G.H., Caddick, M.X. et al. (2020). Growing a circular economy with fungal biotechnology: A white paper. *Fungal Biol. Biotechnol.*, 7: 5. https://doi.org/10.1186/s40694-020-00095-z.

Sridhar, K.R. and Deshmukh, S.K. (2021). *Advances in Macrofungi: Pharmaceuticals and cosmeceuticals.* Boca Raton, USA:CRC Press, Taylor and Francis Group, p. 319.

Venduk, J. and Veljović, S. 2021. Macrofungi in the production of alcoholic beverages: Beer, wine, and spirits. pp. 108–141. *In*: Sridhar, K.R. and Deshmukh, S.K. (Eds.). *Advances in Macrofungi: Industrial avenues and prospects.* Boca Raton, USA:CRC Press.

Contents

Food, Feed and Nutrition

Biofuels

Plant Protection

List of Contributors

Asmaa Abbas
Department of Life Technologies, Faculty of Technology, University of Turku, FI-20014, Turku, Finland.
Department of Chemistry, Faculty of Science, Sohag University, Sohag, 82524, Egypt.

C Contigli
Serviço de Biologia Celular, Diretoria de Pesquisa e Desenvolvimento, Fundação Ezequiel Dias, Belo Horizonte, MG 30510-010, Brazil.

Shin Yee Fung
Medicinal Mushroom Research Group (MMRG), Department of Molecular Medicine, Faculty of Medicine, University of Malaya, Kuala Lumpur, Malaysia.
Centre for Natural Products Research and Drug Discovery (CENAR), University of Malaya, Kuala Lumpur, Malaysia.
University of Malaya Centre for Proteomics Research (UMCPR), University of Malaya, Kuala Lumpur, Malaysia.

Saji George
KRIBS-BIONEST (RGCB Campus -3), Kerala Technology Innovation Zone, BTIC Building, KINFRA Hi-Tech Park, Kalamessary, Kochi, India, Pin 683503.

Puja Ghosh
Department ofBiotechnology, Pondicherry University, Kalapet, Pondicherry-605014, India.

Ichiro Kamei
Faculty of Agriculture, University of Miyazaki, 1-1 Gakuen-kibanadai-nishi, Miyazaki 889-2192, Japan.

Le Duy Khuong
Faculty of Environment, Ha Long University, 258 Bach Dang Street, Uong Bi District, Quang, Ninh Province, Vietnam.

Boon Hong Kong
Medicinal Mushroom Research Group (MMRG), Department of Molecular Medicine, Faculty of Medicine, University of Malaya, Kuala Lumpur, Malaysia.

Agata Krakowska
Jagiellonian University Medical College, Faculty of Pharmacy, Department of Inorganic and Analytical Chemistry, 9 Medyczna Street, 30–688 Kraków, Poland.

Paweł Kudrys
University of Opole, Faculty of Natural Sciences and Technology Institute of Environmental Engineering and Biotechnology, Poland.

Bikash Kumar
Bioprocess and Bioenergy Laboratory, Department of Microbiology, Central University of Rajasthan, NH-8, Bandarsindri, Kishangarh, Ajmer 305817, Rajasthan, India.

Fernanda Cortez Lopes
Graduate Program in Cell and Molecular Biology, Center for Biotechnology, Universidade Federal do Rio Grande do Sul, Porto Alegre, Brazil.

MTNS Lima
Departamento de Ciência de Alimentos, Faculdade de Farmácia, Universidade Federal de Minas Gerais, Belo Horizonte, MG 31270-901, Brazil.

ShivanNegowda Mahadevakumar
Department of Studies in Botany, University of Mysore, Mysore, Karnataka, India.

BA Martins
Departamento de Química, Instituto de Ciências Exatas, Universidade Federal de Minas Gerais, Belo Horizonte, MG 31270-901, Brazil.

Dunja Miletić
Institute of Food Technology and Biochemistry, Faculty of Agriculture, University of Belgrade, 11000 Belgrade, Serbia.

Ewa B Moliszewska
University of Opole, Faculty of Natural Sciences and Technology Institute of Environmental Engineering and Biotechnology, Poland.

Sesha Subramanian Murugan
Biomaterials Research Laboratory, Yenepoya Research Centre, Yenepoya (Deemed to be University), Deralakatte, Mangalore, Karnataka 575018, India.

Bożena Muszyńska
Jagiellonian University Medical College, Faculty of Pharmacy, Department of Pharmaceutical Botany, 9 Medyczna Street, 30–688 Kraków, Poland.

Małgorzata Nabrdalik
University of Opole, Faculty of Natural Sciences and Technology Institute of Environmental Engineering and Biotechnology, Poland.

Revanth Babu Pallam
Department of Biotechnology, Pondicherry University, Kalapet, Pondicherry 605014, India.

Navnit Kumar Ramamoorthy
Centre for Biotechnology, Anna University, Chennai-600025, India.

Prafull Ranadive
Organica Biotech Pvt. Ltd., 36, Ujagar Industrial Estate, W.T. Patil Marg, Govandi, Mumbai, India, Pin 400 088.

S Renganathan
Centre for Biotechnology, Anna University, Chennai-600025, India.

Vemuri V Sarma
Department of Biotechnology, Pondicherry University, Kalapet, Pondicherry-605014, India.

Gi Hun Seong
Department of Bionano Engineering, Center for Bionano Intelligence Education and Research, Hanyang University, Ansan 426-791, South Korea.

Aleksandra Sknepnek
Institute of Food Technology and Biochemistry, Faculty of Agriculture, University of Belgrade, 11000 Belgrade, Serbia.

Kandikere R Sridhar
Department of Biosciences, Mangalore University, Mangalore, Karnataka, India
Centre for Environmental Studies, Yenepoya (Deemed to be University), Mangalore, Karnataka, India.

Katarzyna Sułkowska-Ziaja
Jagiellonian University Medical College, Faculty of Pharmacy, Department of Pharmaceutical Botany, 9 Medyczna Street, 30–688 Kraków, Poland.

JA Takahashi
Departamento de Química, Instituto de Ciências Exatas, Universidade Federal de Minas Gerais, Belo Horizonte, MG 31270-901, Brazil.

Tuğba Teker
Institute of Graduate Studies in Sciences, Programme of Molecular Biotechnology and Genetics, Istanbul University, 34116, Istanbul, Turkey.

Pradipta Tokdar
KRIBS-BIONEST (RGCB Campus -3), Kerala Technology Innovation Zone, BTIC Building, KINFRA Hi-Tech Park, Kalamessary, Kochi, India, Pin 683503.

Chu Luong Tri
Faculty of Agriculture, University of Miyazaki, 1-1 Gakuen-kibanadai-nishi, Miyazaki 889-2192, Japan.

Jayachandran Venkatesan
Biomaterials Research Laboratory, Yenepoya Research Centre, Yenepoya (Deemed to be University), Deralakatte, Mangalore, Karnataka 575018, India.
Department of Bionano Engineering, Center for Bionano Intelligence Education and Research, Hanyang University, Ansan 426-791, South Korea.

Pradeep Verma
Bioprocess and Bioenergy Laboratory, Department of Microbiology, Central University of Rajasthan, NH-8, Bandarsindri, Kishangarh, Ajmer 305817, Rajasthan, India.

Hui-Yeng Yeannie Yap
Department of Oral Biology and Biomedical Sciences, Faculty of Dentistry, MAHSA University, Bandar Saujana Putra, Selangor, Malaysia.

Tapani Yli-Mattila
Department of Life Technologies, Faculty of Technology, University of Turku, FI-20014, Turku, Finland.

Emre Yörük
Department of Molecular Biology and Genetics, Faculty of Arts and Sciences, Istanbul Yeni Yuzyil University, 34010, Istanbul, Turkey.

Genomics

1

Applications of Genomics in Fungal Biotechnology

Shivannegowda Mahadevakumar[1,2,*] and *Kandikere R. Sridhar*[3]

1. Introduction

The kingdom of fungi is ranked alongside the green plants and animals within the domain Eukarya. The fungal eukaryotes are heterotrophic with wall containing chitin but no plastids in their cytoplasm. Similarly, Oomycetes, Slime moulds and Plasmodiophoromycetes were treated under the group of fungi (fungi-like organisms as they possess the ability to produce hyphae or resting spores). The members of Oomycota relatively resemble the members of the Stramenopile algae and the Mycetozoans closely resemble the Amoebozoa. According to the recent estimate, the predicted global number fungal species is 2.2–3.8 million (Hawksworth and Lücking, 2017; Hawksworth, 2019), Hawksworth (2019) predicts occurrence of 96,000 species of fungi in the Indian subcontinent. The recent taxonomic literature related to fungi is increasing exponentially mainly due to the wide usage of genomic resources, and shortly a better understanding could be visualized about the total fungal biodiversity estimates.

The genome sequences offer a first look at the genetic basis of biodiversity of filamentous fungi as well as yeasts. *Saccharomyces cerevisiae*, a budding unicellular yeast with a small genome, has a long history of genetic and molecular basis owing to its genome sequence during the 1990s. The discovery of fission in yeasts (*S. chizosaccharomyces pombe*) and filamentous fungus (*Neurospora crassa*) revolutionized the fungal genomics. Many of the fungal sequenced genomes are accessible in the public domain and represent the largest sampling of any eukaryotic

[1] Department of Studies in Botany, University of Mysore, Mysore, Karnataka, India.
[2] Department of Studies in Microbiology, Karnataka State Open University, Mukthagangotri, Mysore, Karnataka, India.
[3] Department of Biosciences, Mangalore University, Mangalore, Karnataka, India.
* Corresponding author: mahadevakumars@gmail.com

kingdom so far. The genome-sequencing programme improves the research amplitude in the fields of agriculture, bioenergy, biotechnology, bioremediation, ecology, and medicine. Fungal genomics has greater potential to improve the human, environmental, and planetary health. More fungal genomes sequenced means deciphering more genes, encoding antibiotics, enzymes, organic acids, and pathways. The genomics, proteomics, transcriptomics, and other OMICS-related data enhance the knowledge on biology, functional and evolutionary domains of the host fungus in an environment and its interaction with other organisms (Mahadevakumar and Sridhar, 2020). From the OMICS platforms, it is possible to dissect the mechanisms of parasitism, symbiosis, and other related processes during development. With minimum expenditure, the high-throughput DNA sequence technology allowed the disciplines like microbial ecology, microbial biotechnology, and plant pathology to enter the era of genomics. In the molecular era, meta-genomics provide high-quality references and have begun to enhance the knowledge on the fungal diversity, evolution and interaction with environment and other organisms. Currently, although over 100 genome sequences of fungi are available, however, there are no comprehensive reviews.

The pace of sequencing of fungal genomes is slow. However, in the last decade, scientists working with Whitehead Institute and MIT Center for Genome Research accelerated sequencing study (presently the Broad Institute). The fungal Genome Initiative was launched (www.broad.mit.edu/annotation/fgi/); and as a result many databases are regularly updated, greatly assisting many fields of fungal biology (Fig. 1.1). The website www.genomeonline.org/ has a mirror site in Greece: www.gold.imbb.forth.gr/ (Liolios et al., 2007). The fungal genome helps in a better understanding of the eukaryotic cell processes. In addition to health, it has addressed the issues concerned to food safety and environmental diversity. Fungal genomics, as a scientific discipline, decipher the genome as well as the hereditary information of fungi. For example, fungal genomics could be used to assess the fungal evolution and the outbreaks of fungal diseases. Towards basic and applied research, this review provides a road map for the fungal genome-sequencing programme and application in the fields of taxonomy, pharmacy, and biotechnological progress.

2. Fungal Genomes and the Evolutionary Relationships

2.1 Critics of the Genome Quality Published in the Beginning

The complete genome sequences of *Haemophilus influenzae* and *Saccharomyces cerevisiae* were published during 1995 and 1996, respectively, which were followed by the human genome sequences during 2001 (Fleischmann et al., 1995; Goffeau et al., 1996; Venter et al., 2001; Lander et al., 2001). They have been used as 'reference genomes' to study genetic variation in other individuals of the same species (Weigel and Mott, 2009; Aflitos et al., 2014; www.nlgenome.nl/; www.ebi.ac.uk/ega/; www.rug.nl/cit/; www.rug.nl/target/; www.biggrid.nl/; www.surfsara.nl/; www.molgenis.org/). Sequencing has become one of the most popular modern genome analysis workflows since 2006. It is reasonable to expect that this is only the beginning of the discovery of genomic variation within the populations.

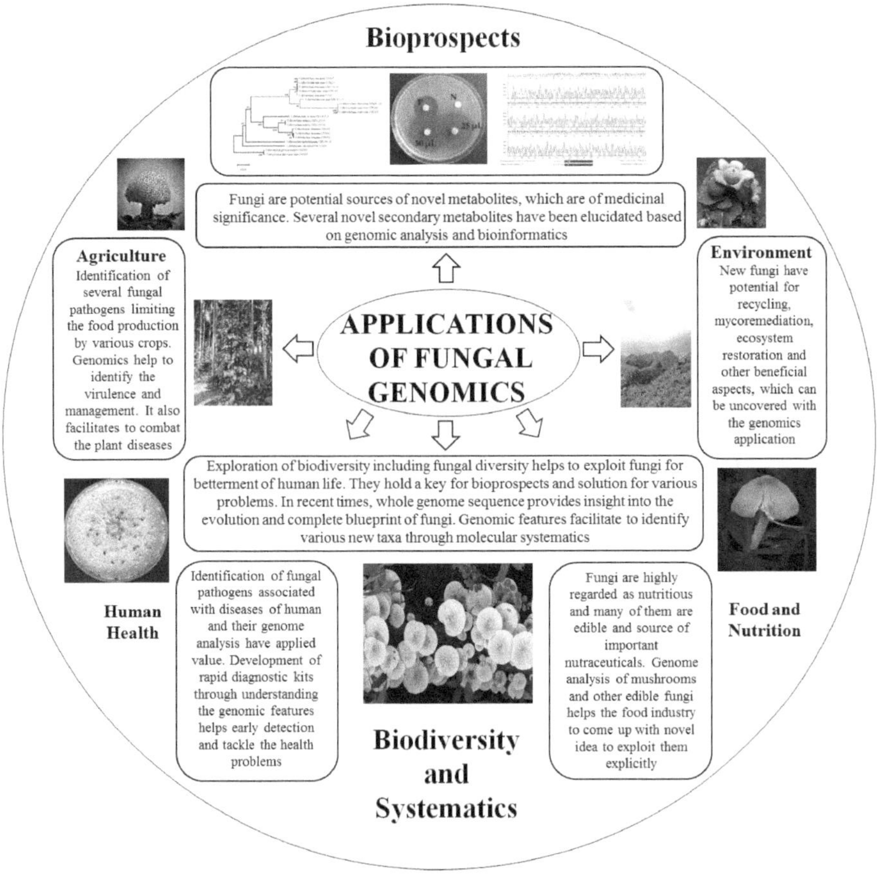

Fig. 1.1. Various facets of application of genomics in the field of fungal biotechnology.

The complete genome of a bacterium (*Haemophilus influenzae*) was first published in 1995 and for the eukaryotic organism (*Saccharomyces cerevisiae*) in 1996. Many complete genome sequences followed there after served as 'reference genomes' for comparative studies. The advancement in sequencing platforms helped to produce a large number of high quality genome resources. Therefore, a reference genome has become obsolete as more genomes are sequenced (Tettelin et al., 2005; Morgante et al., 2007). A reference sequence can be the genome of a single individual, a population consensus, a 'functional' genome (no gene-disablement mutations), or a maximal genome (every sequence ever detected). Although they may be valid in their own context, they are not represented in many of the early reference sequences. Because the sequence patches derived from the experimental material were available, it has become a random mix from different sources. Several individual genome sequences of populations, taxonomic units (Rogers and Gibbs, 2014), or environments have only recently been determined. However, Marschall (2018) opined that, a reliable 'reference genome' should have competencies beyond those genomes listed above to fully exploit the data.

Hence, instead of just one reference genome, it is apt to use "*pan-genome*"—a representation of whole genomic content in one species or a clade. The word pan-genome was coined by Sigaux (2000) to refer a public database of genome as well as transcriptome alterations in experimental models, tissues, and tumours. For our deeper understanding of genomics, Tettelin et al. (2005) demarcated the pan-genome consisting of two genomes: a 'core' genome that exists in all strains, and a 'dispensable' genome (flexible or accessory genome) missing from one or more of the strains. It could include, besides genes, other variations found in the genome collection or database.

3. 'Next Generation' Sequencing

From the time of the first fungal genome sequences, there have been technological advancements in sequencing platforms. Several cell-free as well as electrophoresis-free sequence tools are referred to as 'next' or "second generation", and have recently become cost-effective. Sanger sequencing was the dominant technology when large-scale sequencing projects were initiated. This usually involved the PCR and electrophoresis to create bands that could be used to decode the sequence. Although some parts of the DNA could not be cloned in *Escherichia coli*, the next generation sequencing (NGS) technology/platforms employ electrophoresis-free schemes to decode the DNA sequence through monitoring every sequencing step in situ. With a large quantity of sequence data generated by a single instrument run, all the NGS technologies have gained profound influence in modern sequence platforms (SanMiguel, 2011). Various sequencing platforms explored for fungal genome research and their specifications are presented in Table 1.1. The NGS can

Table 1.1. Comparative account of commercially available next-generation sequences.

	Sequences Generated (billion bases)		Flow Cells/ Run	Regions (Lanes)/ Flow Cell	
	Per run	**Per day**		**Max.**	**Min.**
Life Technologies (Applied Biosystems) 3730XL	0.0001	0.001	NA	NA	NA
Roche GS-FLX (454)	0.5	0.5	1	16	2
Illumina Genome Sequencer (GA-2x)	25	2	1	8	8
Life Technologies (Applied Biosystems) SOLiD	60	4	2	8	1
Helicos HeliScope	20	2.5	2	25	25
	Run time	**Max. output**	**Max. reads/run**	**Max. read Length**	
Illumina iSeq 100	9.5–19 hrs	1.2Gb	4 million	2×150 bp	
Illumina MiniSeq	4–24	7.5	25 million	2×150 bp	
Illumina MiSeq Series+	4–55	15	25+ million	2×300 bp	
Illumina NextSeq 550 Series+	12–30	120	400 million	2×150 bp	
Illumina NextSeq 1000 & 2000	11–48	330	1.1 billion	2×150 bp	

Notes: SanMiguel, 2011; NA, not applicable.

rapidly produce huge numbers of DNA sequences and the cost per base is decreasing. Depending on the project scope, one NGS platform may be better than others and the following sections provide their major uses in fungal genomics (https://sapac.illumina.com/systems/sequencing-platforms.html).

3.1 The De novo Assembly of Fungal Genomes

The discrepancies of *de novo* sequencing are determined by the genome's size and the repetitive fraction. Sequencing large, highly repetitive genomes like mammals and plants takes hundreds of times more resources than resequencing the same genome. But fungi have compact genomes with little repetitive DNA. Various techniques have been developed to reassemble a full genome from sequence reads. Notables are assembly programmes (or "assembly engines") and their algorithms for overlapping short reads into contigs. Sequencing "paired end" as well as "mate pair" DNA fragments is vital for extending the sequence contigs and linking adjacent contigs those devoid of sequence bridges into scaffolds. Other methods for genome assembly include optical mapping, restriction enzyme mapping and fluorescent microscopy.

Use of a reference sequence reduces the necessity for long sequence reads in the programmes of resequencing. Because fungal genomes are small and simple, all the NGS technologies are suitable for resequencing. The fungal pathogens affect plants as well as animals and some are model/reference species for major taxonomic units. Excluding a few, most of such sequenced fungal species are too distant to be used for comparative genomics (Kellis et al., 2004). Numerous strains could be sequenced using a single instrument run (may be SOLiD or Solexa sequence platforms). Using SOLiD 3.5 chemistry, at an approximate cost of U$1,600, can generate up to 320 Mb of raw sequence of each strain of fungi and at the Purdue Genome Core Facility sequencing eight strains were reported. It means, 30 coverage for each strain by presuming the genome size as 40 Mb.

Resequencing the closely related species or the strains of single species (or races or subspecies) can reveal genetic mechanisms underlying species or strain differentiation. Comparative assessment of genome sequences of narrowly linked fungi can identify genes involved in mycotoxin or phytotoxin production, regulation of sexual and or asexual reproduction, race specificity determination, and colonization of specific hosts or ecological niches. The data generated by NGS may be overwhelming for fungi biologists. While sequencing costs continue to fall, fungal biologists will likely spend more time towards handling and analysing sequence data. If the sequence reads approach short transcript length, simplifies assembly (Sanger et al., 1977). Because transcript abundance indicates gene activity, measuring relative transcript abundance across time, tissue, or environmental conditions is a common first step in decoding biological mechanisms. The NGS can decipher the sequences of transcribed genes as well as the relative abundance of each transcript type of a specific fungal tissue. However, the latter may not necessitate the expensive long-read methodologies. Compared to the microarrays, the NGS is better at identifying and profiling low abundance transcripts (Metzker, 2010). The NGS is expected to replace the microarray assays and other traditional gene expression profiling techniques as sequencing costs fall and software becomes available.

4. Functional Genomics

Fungal genomics is a branch of biology that studies the identification and characterization of all the sequences in a fungal genome. Determination of the genome sequence, on the other hand, is only the beginning of genomics. Subsequently, one can time to look at the function of many genes (called functional genomics), comparison of genes with genes from other organisms (called comparative genomics), assessment of expression profile (called transcriptome), and lastly study the protein "expression profiling" (called proteome). The new "OMICS" science led to metabolomics, as well as referred to the whole study of low molecular weight of compounds (Silva, 2016). As on August 31, 2021, over 2068 fungal genomes are published and they are available for public (www.genome.jgi.doe.gov/fungi/fungi.info.html). Fungal genomics, as well as other 'omics' such as transcriptomics, metabolomics, proteomics, and lipidomics are increasingly routinely employed by biologists to better understand basic biology and investigate related applications on a larger scale (biotechnology, medicine, and agriculture) (Xiao et al., 2017). Functional genomics examines the expression of gene and protein function at a global scale (genome-wide), with an emphasis on gene transcription, translation, protein-protein interactions, and frequently employing high-throughput methods. This approach facilitates to investigate the functions as well as interactions of genes or their products known as functional genomics (Liu et al., 2010a,b; Buza and McCarthy, 2013). The entire genome must be sequenced to accomplish functional genomics. Subsequent publication of the first complete yeast genome in 1992 (*Saccharomyces cerevisiae*), sequencing techniques have progressed significantly (Goffeau et al., 1996). Currently, a combination of recently emerging methods (e.g., the PacBio with Illumina providing fully closed genomes) resulted in a plethora of fungal genome databases (Haas et al., 2011), which contain more than 100 fungal genomes or FungiDB and Ensembl Fungi (Kersey et al., 2010; Stajich et al., 2012). There are numerous examples to target various product classes, however, a few examples are given below on currently feasible for secondary metabolite biosynthesis by the OMICS and bioinformatics tools.

Among the many genes and gene clusters found in the genomes of fungi are those that encode valuable beneficial products, such as industrially vital enzymes and secondary metabolites. A significant improvement has been made in bioinformatics tools for genome annotation as well. Nowadays, in addition to the well-recognized antiSMASH fungal-specific algorithms for identification of clusters of biosynthetic gene such as FunGeneClusterS are available (Vesth et al., 2016; Blin et al., 2017). Researchers may express heterologously the gene clusters in many familiar hosts (e.g., *Aspergillus niger*) by the development of these tools by Boecker et al. (2018). Such an approach could be employed to make the process easier to produce a known metabolite at large scale. However, to express the previously silent gene clusters discovered during analysing the genome, using the "genome mining" technique opens up a new avenue towards the discovery of innovative anti-infecting agents. The heterologous expression of precise gene clusters in different hosts permits and leads towards investigation of biosynthetic pathways for known antibiotics as well as natural products. Biosynthesis of the mycotoxins, trichothecene, was one of the first fungal biosynthesis studies carried out in this pattern (Tokai et al., 2007). Other studies on the Ascomycota followed over the years; however, more recently, researchers have turned their attention towards Basidiomycota to understand the synthesis of secondary metabolites (Lin et al., 2019). Biosynthesis of the antifungal strobilurins in *Aspergillus oryzae* has been studied through expression (Nofiani et al., 2018). A current research trend demonstrated the possibilities to modify the biosynthesis of fungal metabolite towards increased production at a large-scale, or to produce drugs that are more bioavailable and compatible to the patients. Regulation of secondary metabolite biosynthesis has also been studied in model organisms; such results are also available to employ across the fungal kingdom (Brakhage and Schroeckh, 2011). This discovery of genome mining soon led to complete elucidation of new classes of metabolites as demonstrated for enzymes in the past (Dilokpimol et al., 2018). An extremely high quantity of bioactive metabolite was obtained in a relatively short period because of using of modern bioprocess technology, and systems biology methods.

5. Genomics in Fungal Taxonomy

The fungal taxonomy is an evolving field, which has been influenced by technological and research advancements. It was not until the invention of the microscope that many fungal micro-morphological characteristics could be observed, and it was this invention that greatly aided in the recognition of new species beginning in the 17th century. Since the 19th century, application of physiological and biochemical characteristics (e.g., colour reaction, nutrient utilization assays, and thin layer chromatography) of secondary metabolites has improved the reliability of fungal classification (Zhang et al., 2017). In addition, the development of sequencing of the gene as a method of barcoding as well as phylogenetic reconstructions at various taxonomic levels significantly improved understanding of the delimitation of fungal species. Increased accessibility of gene sequences obtained from different genetic loci like internally transcribed spacer region 1/2 (ITS1/2), 18s rDNA, 5.8S rDNA, 28S rDNA (ribosomal DNA), or beta-tubulin (*tub*), translation elongation factor (*tef*-1) 1 alpha, DNA-directed RNA polymerase II subunit (RPB2) and others (protein-coding genes), phylogenetic inferences have transitioned from single-gene to multigene phylogeny and now whole genome sequence assembly (Zhang et al., 2017; Dornburg et al., 2017). The cost of development of extremely rapid methods to generate complete produce of sequences of many organisms is currently driving biological research in to the 21st century. Such new approaches are being used to construct strong taxonomic frameworks are known as phylogenomics (Nagy and Szöllisi, 2017). Despite their importance in natural product chemistry, fungal identification remains a daunting task for researchers owing to the complexity in morphological features, cultural characters, and cryptic behaviour shared by many of the closely related organisms. Therefore, molecular identification of fungi has gained significance and it has travelled a long way.

In the previous sections, the advancements from morphology to cultural to molecular identification of fungi are presented. However, molecular analysis also evolved ever since the discovery of PCR invention and sequencing platforms and have become a major service provider to explore the identity based on multi-locus sequence analysis. There are several genes targeted for identification of *Fusarium* species employing several primer combinations along with PCR conditions are provided for *Diaporthe*, *Colletotrichum*, and *Pestalotiopsis*-like genera. At this point, to illustrate the advancement of events in identification to molecular genome analysis, *Fusarium* species is considered as an example and the various events in taxonomy of the *Fusarium* species is presented in Box 1.1.

Based on the 1000 Fungal Genomes Project (https://1000.fungalgenomes.org/home/), until now, studies on the phylogenome have been carried out only in a minor number of taxonomic groups within the fungal kingdom (Grigoriev et al., 2014). However, to date, over 1300 genomes of fungi are sequenced (Grigoriev et al., 2014). Consequence of the huge variety and abundance of fungi (2.2–3.8 million estimated species) (Blackwell, 2011; Hawksworth and Lücking,

Box 1.1 Landmarks taxonomy and identification of *Fusarium* (after Crous et al., 2021).

1809	*Fusarium* Genus first described by Link
1821	Sanctioning work (Fries)
1910	First *Fusarium* monograph by Appel and Wollenweber
1912	Separate naming of asexual fungi
1913	Use of pure cultures, introduction of morphological sections
1916	Fusaria autographice eptoria first published; subsections introduced
1918	Identity of *F. roseum* sensu Link. Was questioned; Series introduced
1935	Major taxonomic revision, study of fungarium and pure cultures
1940	Single spore culturing in *Fusarium*
1958	Type designation mandatory
1971	Conidiogenesis as main criterion in *Fusarium* classification
1977	*Fusarium* morphology not a reflection of genealogical relationship
1983	Abandonment of Snyder and Hansen's taxonomic system
1989	Broad generic definition to include microconidial isolates Application of molecular characters in *Fusarium* systematics
1990	Description of Mesoconidia and their importance in classification First Universal fungal rDNA PCR primers are published
1993	Use of additional genes (multi-locus phylogenies for *Fusarium*) Induction of phylogenetic species complexes, morphological sections are abandoned
1996	First fungal genome published
1997	Proposal to lectotypify *Fusarium* with a conserved type
1999	Lectotypification of *Fusarium* accepted by the Nomenclature Committee for Fungi
2000	Phylogenetic species recognition concept
2006	Era of phylogenomics
2007	First *Fusarium* genome is published
2010	Mobile pathogenicity chromosomes discovered in *Fusarium*
2011	Molecular evidence for segregation of *Fusarium sensu lato* Dual nomenclature for pleomorphic fungi is abandoned
2013	Fusarium to be conserved over *Gibberella* Proposal of broad cladistics definition of *Fusarium*
2015	Fusarium confined to the species with *Gibberella* sexual morphs
2021	Phylogenomic arguments to retain broad *Fusarium* concept.

2017), only a few or solitary members in the families are pursued, making it difficult to address interspecific relationships at the taxonomic levels below ordinal rank using the public repository. Because of their economic and scientific significance, a few groups of fungi have their extensively sequenced genomes, which have provided the basis for phylogenomic analyses of these fungal groups. Many Hypocreales (Ascomycota) members, including nine *Trichoderma* species, have recently been studied (Druzhinina et al., 2018). The members of the Aspergillaceae have been studied considering a dataset of *Aspergillus* (45 spp.) and *Penicillium* spp. (33 spp.) (Druzhinina et al., 2018; Steenwyk et al., 2019). Based on the use of 1669 gene matrix, the latter study provides the largest phylogenomic reconstruction at the family level that has been completed till date. For the first time, Pizarro et al. (2018) produced numerous sequences of genome solely for the purpose of resolving relationships among 51 lichenized fungi (*Parmeliaceae*), which was previously unresolved. The researchers also obtained up to 2556 orthologous single-copy genes for the purpose of developing a tree even though majority of the strains have low genome quality (mean, N50 of 20,000 bp). This has resulted in a very stable tree topology consisting of monophyletic subclades (Pizarro et al., 2018). There were no other noteworthy taxonomic groups in the Sordariomycetes, which is one of the largest classes of the Ascomycota consisting of 37 orders served towards the comparative genomic studies except for the families such as Magnaporthales and Hypocreales (Wijayawardene et al., 2018; Zhang et al., 2018).

The first application of MLST (multi-locus sequence typing) in fungi was to recognize the species. Traditional techniques of recognition of species of fungi were through phenotype or by mating tests, which are superseded by the phylogenetic methods by nucleotide sequence of multiple genealogies. Identifying the species as clades of individuals that are apart genetically has exposed the new and hidden species in many plants, pathogenic as well as toxigenic fungi (O'Donnell et al., 2000). Some of the taxa earlier recognized by the phenotypic studies are misleading.

Recently identified species have significant differences in phenotype, especially the virulence. Despite having 2.2–3.8 million species, fungi are one of the least studied groups in biodiversity under the multicellular eukaryotes (Hawksworth and Lücking, 2017). Despite their importance in ecology, only a quarter of the estimated species (~ 120,000 spp.) have been formally described, and many of them which are devoid of sequences of DNA have been placed in the databases (~ 85,000 spp.) (Fisher et al., 2012; Hawksworth and Lücking, 2017). Fungal metagenome sequencing has provided the scientists a huge number of datasets consisting of sequences from non-cultivable or environmentally derived fungi with meagre taxonomic annotation. Hence, connecting the curated DNA sequence data against the expertly identified voucher specimen is one of the critical steps in patching up the current gap between described and sequenced fungi.

The ITS regions of the gene cluster of nuclear ribosomal RNA were identified as the chief fungal barcode marker (Schoch et al., 2012; Consortium for the Barcode of Life in 2012). The highly different non-coding ITS1 as well as ITS2 regions are housed between the small subunit (SSU 18S) and the large subunit (LSU 28S) coding genes, separated by the 5.8S coding gene. The ITS1 and ITS2 employed in fungal species identification of fungi for over 20 years owing to their rapid evolutionary rate (Nilsson et al., 2008). The genes 18S, 5.8S, and 28S (highly conserved) have permitted the universal primers to be developed for PCR amplification and subsequent sequencing of the whole ITS region or the ITS1 as well as ITS2 regions (White et al., 1990; Gardes and Bruns, 1993; Martin and Rygiewicz, 2005; Toju et al., 2012). As the fungal genome consists of gene cluster with many tandemly repeated copies of the ribosomal RNA (comprising the ITS), amplification of such region is possible from a small quantity of DNA (Xu, 2016). Because sequencing of such marker regions could be followed easily in living organisms, hence, it is mainly useful for fungal assemblage of DNA barcoding (Osmundson et al., 2013).

5. Fungal Genomics in Plant Pathology

Crop losses due to fungi exceeded U$200 billion annually. Thus, controlling the plant diseases is critical for sustainable food production in limited agricultural land, water, fertilizer, and fuel. Lack of knowledge about the genetic basis, the biochemical basis of pathogenicity, infection mechanism and the resistance of host species has hampered the disease control. The 10 most economically vital fungal pathogens include: *Blumeria graminis* (powdery mildew on Hordeum vulgare), *Botrytis cinerea* (fruit rot on more than 200 hosts), *Colletotrichum* spp. (known to cause anthracnose disease on a large number of hosts), *Fusarium graminearum* (head blight of maize), *F. oxysporum* (vascular wilt on a wide host range), *Magnaporthe oryzae* (blast disease in rice), *Melampsora lini* (rust in flax), *Mycosphaerella graminicola* (eptoria leaf blotch), *Puccinia* spp. (wheat rust), and *Ustilago maydis* (maize smut) (Dean et al., 2012; Mahadevakumar and Sridhar 2021). Genomics continues to provide insights into the rapidly expanding areas in plant pathology such as causes of pathogenesis, host-pathogen interactions, host defence responses and changes in the genome structure. The genome-based knowledge on the pathogenic fungi like *Pythium* spp. and *Rhizoctonia* spp. associated with wheat, potato and rice are essential to identify the precise genes, specific pathways, and molecular markers of phenotypes as well as pathosystems. Advances in sequencing as well as technology of sequence assembly enabled comparative genomics at the laboratory and the field levels.

5.1 Molecular Genetics to Target the Diversity of Pathogen Traits

Plant pathogen genome sequencing enables deeper understanding of the mechanism of pathogenesis and documentation of key effector genes augment infection (Moller and Stukenbrock, 2017; Plissonneau et al., 2017). Finding superior genome reads is very critical to understand the biology, their role and comparative analysis with known genome resources from public databases. The Broad Institute assembled the genome of *F. oxysporum* (f. sp. *Lycopersici* strain) for the first time during 2007 using 6 × Sanger sequencing. This genome has 15 chromosomes with 62 Mb long (Ma et al., 2013). Lately, Ayhan et al. (2018) assembled the genome of *F. oxysporum* f. sp. *Lycopersici* with the help of the Illumina platform. The benchmark genome resources assembly will be utilized by the NCBI Genome Database as a defaulting genome for *F. oxysporum*. Assembly of genomes and resources are also differed significantly. The gene copy number, gene absence or presence, repeat structure, and single nucleotide polymorphisms are known to vary significantly between the genotypes of the same species (Schatz et al., 2014; Thudi et al., 2016; Bayer et al., 2017).

Availability of quality genome assembly is very important to understand the pathogenesis. However, not all the major plant pathogens are having quality genome resources. For example, for the flax (*Linum usitatissimum*) pathogen *F. oxysporum* f. sp. *Lini*, there were no sequenced genomes. So far, only the rusts associated with linseed, i.e., *Melampsora lini* has its genome and this assembly is not at the chromosome level (21130 scaffolds with N50 = 31 kb). However, the *M. lini* has paved the way to understand the pathogenesis as it is the only reliable source to evaluate the molecular mechanism of pathogenicity (Nemri et al., 2014).

Fusarium oxysporum contains many strains that exhibit a high level of genetic and functional diversities; thus it is one of the valuable research tools. These non-pathogenic as well as pathogenic strains have no morphological distinction, hence the strains of pathogens have a narrow host specificity to the plants they infect (Nelson et al., 1981; Steinberg et al., 2016). Knowledge on molecular basis of pathogenicity of the genus *Fusarium* is mainly dependent on the information gained about the source material via molecular genetic studies, which is currently at its infancy.

Fusarium oxysporum is one of the large species complexes widely distributed in the soil, indoor, and aquatic habitats (Brandt and Park, 2013; Bell and Khabbaz, 2013; Kauffman et al., 2013). The FOSC (*F. oxysporum* species complex) includes soil-borne plant pathogens are devastating and resulting in vascular wilt diseases (O'Donnell et al., 2004; Ma et al., 2013). The concept of *formae speciales* was developed mainly to categorize the plant pathogens that infect specific hosts. In the case of a pathogenic isolate of tomato (*Solanum lycopersicum*), however, the horizontal gene transfer of lineage-specific (LS) chromosomes can transmit the pathogenicity to a specific host plant species (Ma et al., 2010, 2013). Clinical isolates of the *F. oxysporum* are phylogenetically diverse as well as polyphyletic according to the global survey of genetic diversity (O'Donnell et al., 1998, 2009).

Fusarium oxysporum is known to infect plants as well as humans. The pathogenicity of *F. oxysporum* was determined by horizontal gene transfer by lineage-specific (LS) chromosomes. But the LS chromosomes are not known among the human pathogenic isolates. Zhang et al. (2020) reported four distinct LS chromosomes in the strain (NRRL 32931), which is a human pathogenic strain that affects the leukemia patient. The LS chromosomes lack housekeeping genes; however, they are enriched in metal ion as well as cation transporters. A ceruloplasmin homologue and genes involved in the expansion of the alkaline pH-responsive transcription factor (PacC/Rim1p), which is housed in the genome of the strain NRRL 47514, has been linked to the outbreak of *Fusarium keratitis*. Accordingly, this is the first clue of compartmentalization of genome in the two human pathogenic genomes of fungi.

The *F. oxysporum* f. sp. *lini* is one of the most harmful pathogens to the flax, which has been sequenced for the first time by Krasnov et al. (2020). They employed two sequencing platforms [Oxford Nanopore Technologies (MinION system) with long noisy reads as well as Illumina (HiSeq 2500 instrument)] with short accurate reads to assemble a high-quality genome. For genomic analysis, there exist several tools for genome assembly, but their results vary based on the sequencing data volume, genome complexity, and read length including the quality. In addition to

Canu, Flye, MaSuRCA, Shasta, and wtdbg2, they tested for Nanopore polishers (Medaka and Racon) as well as Illumina polishers (Pilon and POLCA). The Canu assembly with Medaka and POLCA was deemed to be the most complete and accurate. Repetitive contigs were further removed using Purge Haplotigs, leaving a 59 Mb genome with N50 of 3.3 Mb and 99.5 per cent completeness (BUSCO). They also got a 38.7 kb circular mitochondrial genome. The assembled *F. oxysporum* study expands plant-pathogen interactions in flax (Krasnov et al., 2020). Genome sequence data of selected *Colletotrichum* (11 strains) and *Fusarium* (10 strains) with their host have been given in Table 1.2.

Table 1.2 Important species of *Colletotrichum* and *Fusarium* genome sequence data.

	Isolate/Strain	**Host**	**References**
Colletotrichum			
Colletotrichum acutatum	KC05	Pepper	Han et al., 2016
Colletotrichum orbiculare	MAFF 240422	Cucurbits	Gan et al., 2013
Colletotrichum australisinense	GX1655	Rubber tree	Liu et al., 2020
Colletotrichum gloeosporioides	CgLc1	*Liriodendron*	Fu et al., 2020
Colletotrichum graminicola	M1.001	Maize	O'Connell et al., 2012
Colletotrichum higginsianum	IMI 349063	*Brassica campestris*	Zampounis et al., 2016
Colletotrichum incanum		Radish	Gan et al., 2016
Colletotrichum shisoi		*Perilla frutescens*	Gan et al., 2019
Colletotrichum siamense	HBCG01	Rubber tree	Liu et al., 2020
Colletotrichum sublineola	CgSl1	*Sorghum*	Buiate et al., 2017
Colletotrichum tanaceti	BRIP57314	*Pyrethrum*	Lelwala et al., 2019
Fusarium			
Fusarium graminearum		Cereals	Cuomo et al., 2007
Fusarium oxysporum	Strain Fo5176	Cabbage	Fokkens et al., 2020
Fusarium oxysporum f. sp.	KGSJ26F3	Potato	Xie et al., 2020
Fusarium oxysporum f. sp. *cubense*	Strain 160527	Banana	Asai et al., 2019
Fusarium oxysporum f. sp. *cubense*	Strain Fol4287	Banana	Guo et al., 2014
Fusarium equiseti	D25–1	Barley	Li et al., 2021
Fusarium oxysporum f. sp. *koae*	Fo koae	*Acacia koa*	Dobbs et al., 2020
Fusarium oxysporum f. sp. *lini*	MI39, F329, F324, F282, F287	Flax	Kanapin et al., 2020
Fusarium oxysporum f. sp. *melonis*	38 Strains	Melon	Sabahi et al., 2021
Fusarium oxysporum f. sp. *vasinfectum*	TF1 (race 1), 89-1A (race 4), LA1E, LA3B and 14-004	Cotton	Seo et al., 2020

6. Genomics of Endophytic Fungi

Endophytic fungi and pathogenic fungi frequently co-occur reliant on the plant-microbiome and virulence factors associated with it. There are several comprehensive reviews focused on comparative genomics of pathogens as well as endophytes, which unravel the complex interactions between host vs. endophytes and host vs. pathogens (Tejasvi et al., 2017). Lpez-Fernandez et al. (2015) compared the genomes of three strains of *Pantoea ananatis* (isolates from the healthy maize seeds) to recognize the functional genes and the genetic drivers of niche adaptation. Besides disease protection, many endophytes promote plant growth. Because some endophytes can induce root or shoot development, the endophytic community composition can change. These findings were depending on the 16S rRNA gene V4 and V6 region and the 454 GS-FLX platforms (Yu et al., 2015). Unlike bacteria and fungi, endophytic yeasts have received little attention. For the first time, the Sanger whole genome shotgun method was used to sequence an endophytic yeast genome (Firrincieli et al., 2015). Despite being Basidiomycete symbionts of poplar, *R. graminis* WP1 may use a different signalling pathway than *Laccaria bicolor*. In the transcriptome of the diazotrophic endophytic *Herbaspirillum seropedicae*, Tadra-Sfeir et al. (2015) used the RNA-Seq to identify the essential genes necessary for endophyte establishment and interaction. Endophytic microbiomes are gaining importance in fields like agriculture as well as forestry, and they may play a role in the climate change. Differences in the indicator species linked with legume-based agroforestry suggest that a flow from the microbiomes of neighbouring plants (Köberl et al., 2015). This revolutionizes cropping systems as beneficial traits like biocontrol, as well as nitrogen fixation, which could be acquired from the surrounding plant communities. Okubo et al. (2015) used 454 GS FLX sequencing to investigate the impact of climate change on the methane emissions by the rice microbiome. The relative abundance of methane oxidizing Methylocystaceae decreased with increased CO2. As a result, elevated CO2 slowed methane oxidation and increased methanogenesis in rice paddy fields.

As the plant-endophyte connections are multifaceted, it is difficult to comprehend. This may be owing to mutualistic, commensalistic, symbiotic, and trophobiotic interactions among the host vs. endophytes and it can change over during plant development. With the advancement of genomics data of various beneficial endophytic fungi, it is possible to decipher the molecular basis of these interactions (Orozco-Mosqueda and Santoyo, 2021). Although hundreds of fungal genomes are in databases, a few beneficial endophyte genomes exist (Harman et al., 2021; Wonglom et al., 2020; Yang et al., 2014). Some of the endophytic fugal genome resources available along with host details and accession numbers are presented in Table 1.3.

The genome sequences of *Trichoderma atroviride* and *T. virens* determined by Kubicek et al. (2011) and compared the genomic features with *T. reesei*. They mainly focused on unravelling the genes encoding for chitinolytic enzymes and 1,3-glucanases, as well as secondary metabolite genes like NRPS, polyketide synthases, and pore-forming cytolytic peptides. To explore the genomic features of various endophytic fungi, the study of mutants is crucial for gene functional research. Pachauri et al. (2020), recently reported the M7 *T. virens* mutant using gamma ray-

Table 1.3. Some of the endophytic fungal species with whole genome sequence data.

	Strain	Host	Accession #	References
Alternaria sp.	MG1	*Vitis vinefera*	QPFE00000000	Lu et al., 2019
Aspergillus versicolor	0312	*Paris polyphylla*	-	Wang et al., 2019
Cadophora sp.	DSE1049	*Salix rosmarinifolia*	PCYN00000000	Knapp et al., 2018
Calcarisporium arbuscula	NRRL 3705	Russulaceae		Cheng et al., 2020
Colletotrichum tofieldiae	-	*Arabidopsis thaliana*	-	Hacquard et al., 2016
Colletotrichum truncatum		*Glycine max*	VUJX00000000	Rogerio et al., 2020
Fusarium solani	JS-169	*Morus alba*	NGZQ00000000	Kim et al., 2017
Harpophora oryzae	R5-6-1	Wild rice	-	Xu et al., 2015
Microdochium bolleyi	J235TASD1	Beach Grass	LSSP00000000	David et al., 2016
Periconia macrospinosa	DSE2036	*Festuca vaginata*	PCYO00000000	Knapp et al., 2018
Pestalotiopsis fici	CGMCC3.15140	*Camilia sinensis*	ARNU00000000	Wang et al., 2015
Phialocephala scopiformis	DAOMC 229536	*Picea glauca*	LKNI00000000	Walker et al., 2016
Phialocephala subalpina	UAMH 11012	-	-	Schlegel et al., 2016
Sarocladium brachiariae	HND5	*Brachiaria brizantha*	RQPE00000000	Yang et al., 2019
Serendipita indica	DSM 11827	*Arabidopsis thaliana*	CAFZ00000000	Zuccaro et al., 2011
Xylona heveae	TC161	*Hevea brasiliensis*	JXCS00000000	Gazis et al., 2016

induced mutagenesis. The M7 mutant had altered the morphology and lacked plant interaction and mycoparasitism functions, including the ability to produce several volatile and nonvolatile metabolites. Further, the transcriptome analysis revealed downregulation of genes related to secondary metabolism, carbohydrate metabolism, hydrophobicity, and transportation in the mutant. The authors sequenced the entire genome and discovered a 250 kb deletion that lost 71 predicted open reading frames (ORFs). New insights into the genetics of morphogenesis, metabolite production, and mycoparasitism may lead to the discovery of plant-beneficial traits in *Trichoderma* genus. Similarly, *Aspergillus* is a fascinating genus of fungi that may also be beneficial plant endophyte (Mehmood et al., 2019). Several species of *Aspergillus* (*A. awamori*, *A. flavipes*, *A. flavus*, *A. flocculus*, *A. fumigatus*, and *A. terreus*) have been isolated and characterized as endophytes on various host plants (Gubiani et al., 2019; Mehmood et al., 2019).

Endophytic fungal research is gaining importance in the recent past mainly due to the diversity of secondary metabolites produced by the endophytes and their significance in pharmaceutical applications. Understanding the secondary metabolites and associated gene clusters of endophytes help the biologists to explore and produce the products in large quantity. Identification of putative genes responsible for secondary metabolite production has advanced rapidly. Identifying polyketide synthases as well as non-ribosomal peptide synthases encoding gene along with the functions of adjacent genes are the basis for identifying specific genes (Bergmann et al., 2007). The orphan and silent gene clusters are the terms used to describe the undiscovered metabolite and gene clusters (Chiang et al., 2011). Despite of isolating several genes from their natural environment in the laboratory, the secondary metabolite gene clusters often go silent.

Understanding the molecular mechanisms and their biosynthesis may allow precise control of production of new natural derivatives. Since there are whole genome sequence data available in the genome databases, it has offered an opportunity to perform comparative analysis of published genomes (comparative genomics), which will give a detailed insight into the gene clusters responsible for specific secondary metabolite synthesis. Kjærbølling et al. (2018) reported six whole-genome sequences representing the uncharted branches of the genus *Aspergillus*. Although *Aspergillus* species is an important plant as well as human pathogen, it is an important source of enzyme production (Kjærbølling et al., 2018). Genomics data have been exploited and reported the PacBio sequences for four species of *Aspergillus* in elucidating the biochemical pathways for secondary metabolite production through comparative genomics (Kjærbølling et al., 2018).

7. Genomics of Macrofungi

Most macrofungi could not be studied in the laboratory because they lack a reference genome. Recent advances in sequencing technology and analytical tools for molecular population genetics and evolution have accelerated the generation, release, and updating of the draft data. The availability of genome resources pertaining to the macrofungi has eased the research on exploring the diversity of gene corresponding to secondary metabolite biosynthesis (Corre and Challis, 2009; Dai, 2019). Similarly, there are many medicinally important mushrooms and they have been used as traditional medicine in many countries, which includes China, India, Japan, Korea, and others. Recently, numerousmacrofungi have been sequenced for a whole genome sequence and important ones to mention here are *Auricularia heimuer*, *Ganoderma lucidum*, and *Wolfiporia cocos* genomes (Chen et al., 2012; Floudas et al., 2012; Dai, 2019). Globally many and varied macrofungal species are a potential source of drugs.

Currently, several large fungal genome projects have been launched. There are thousands of assemblies of the microbial genome mainly contributed by the Human Microbiome Project, Examples include: Microbial Dark Matter Project (Cisse and Stajich, 2019), and 1000 Fungal Genomes Project (http://1000.fungalgenomes.org) (Grigoriev et al., 2013). Li et al. (2018) published the largest genomic dataset for macrofungal species (90 draft genome assemblies) during 2018. One such example for exploring the genomic resources of medicinally important macrofungi is Sanghuang (*Sanghuangporus sanghuang*) from China by Shao et al. (2020). The

Sanghuang is a medicinally valued fungi used in traditional Chinese, Japanese, and Korean medicines. After long-term use of this fungus, it has been linked to the extended life, removal of toxins, and improvement in digestion in humans (Zhu et al., 2008; Lee et al., 2011; Cai et al., 2019). Shao et al. (2020) observed the lack of available genomic resources for this medicinal mushroom. The NCBI database contains about 120 nucleotide sequences, where most of them have been exploited for phylogeny treatment. Therefore, the genomic resources, presented by Shao et al. (2020), is treated as one of benchmark genome resource as it gives detailed insight into diverse secondary metabolite pathways. Sanghuang possesses 34.5 Mb size, and it encodes 11,310 predicted genes. Up to 16.88% (1909) of the predicted genes are from Eukaryotic Orthologous Groups, while up to 27.23% (665) are involved in metabolism. So also, 334 CAZyme genes and their characteristics were compared to other fungi. The expression of homologous genes responsible in flavonoid polysaccharide, and triterpenoid biosynthesis was studied in four developmental stages such as mycelia of 10- and 20-days old, and fruit bodies of 1- and 3-years old (Shao et al., 2020). The absence of chalcone isomerase 1 in the pathway of flavonoid biosynthesis indicated that this mushroom used different mechanisms than plants to synthesize the flavonoids. Up to 343 transporters as well as four velvet family proteins were identified, which are involved in the regulation of secondary metabolite, uptake, and redistribution. The fungus' genome provides information on its diverse secondary metabolites, which may be useful in future medical research. Similarly, to treat Type 2 diabetes as well as obesity, (24E)-3,4-seco-cucurbita-4,24-diene-3-hydroxy-26.29-2-dioic acid was extracted from the sporocarps of *Russula lepida* (Liu, 2007). Many *Lactarius* metabolites have antiviral as well as antitumor potential (Tala et al., 2017). *Auricularia auricula* polysaccharides are potent antioxidants (Fan et al., 2007). In-depth study of most macrofungal species could not be studied in the laboratory owing to a lack of reference genomes. There have been a few publications of macrofungal genomes published; draught genome assemblies of 90 macrofungi, except for *Annulohypoxylon stygium* and *Russula foetens*, which are toxic (Martin et al., 2008, 2010; Stajich et al., 2010; Morin et al., 2012; Floudas et al., 2012; Kohler et al., 2015; Nagy et al., 2016; Li et al., 2018).

Macrocybe gigantea is another important mushroom known to produce varied antioxidants, bioactive compounds, and water-soluble polysaccharides. Recently, Kui et al. (2021) assembled the genome resources for *M. gigantea.* Its genome (41.23 Mb) was compared with 11 other macrofungal genomes. A comparative genomics insight confirmed that *M. gigantea* was in the *Macrocybe* genus, and distinct from the genus *Tricholoma*. They also discovered that the glycosyl hydrolase family 28 (GH28) had conserved motifs which were distinct from the genus *Tricholoma*. The genomic resource of *M. gigantea* will help to understand better the fungal biology, especially the differences in growth rates as well as energy metabolism.

8. Genomics in Human Health

Fungi have been linked to many acute and chronic human ailments. First, sequencing fungi that cause health risks to humans or serve as research models was prioritized. Human fungal pathogen *Candida albicans* causes skin as well as mucosal infections in healthy subjects, which will be life-threatening in immunocompromised patients. In *C. albicans*, the allelic differences may uplift the genetic diversity and the development of resistance to drug (Cowen et al., 2002). Jones et al. (2004) reported the 14 Mb genome of an environmental sensor (calmodulin signalling pathway along with a protein kinase) and adaptation of the gene set in *Candida*. Comparative genomics help managing the disease by following evolution of pathogen and pathogenesis (Jones et al., 2004). Severe bacterial infections have been linked to the use of antibiotics as have antimicrobials. In the evolution yeast species, comparative genomics has helped to identify and to characterize novel ORFs in *S. cerevisiae* (Cai et al., 2008). Towards understanding of evolution as well as natural history, Rhind et al. (2011) compared the genomes as well as transcriptomes of the *Saccharomyces japonicus, S. octosporus, S. pombe,* and *Schizosaccharomyces cryophilus*. Genomes of *S. cerevisiae* and *Cryptococcus neoformans* revealed differences in protein structure as well as function (Engel and Cherry, 2013; Ormerod et al., 2013).

A pulmonary pathogen, *Coccidioides immitis* and *C. posadasii* infect at least 150,000 people per year (Hector and Laniado-Laborin, 2005; Sharpton et al., 2009). The adaptive changes to the existing genes and acquisition of a small number of new genes have transformed them from plant-based to animal-based for nutritional support (Sharpton et al., 2009). Another ascomycete *Aspergillus fumigatus* targets patients with neutropenia, corticosteroid users, and those with T cell defects (Ronning et al., 2005). Comparative genomics of two clinical isolates of *A. fumigatus* revealed species-specific chromosomal islands that contribute to rapid adaptation to heterogeneous environments (Ronning et al., 2005; Fedorova et al., 2008).

Cryptococcus neoformans (basidiomycetous yeast) is one of the most dangerous human fungal pathogens. This basidiomycete model yeast for fungal pathogenesis has been sequenced for its genome (Loftus et al., 2005). Increased transposon abundance may cause karyotype instability, phenotypic variation, altering virulence factors, nutrition, and metabolic profiles (Loftus et al., 2005; Ormerod et al., 2013). The genomes of 118 *C. gattii* isolates were recently sequenced, revealing the PNW subtypes and global diversity of the molecular type VGII (Engelthaler et al., 2014). Population diversification in pathogenic microevolution and clonal expansion under differential selection has been seen (Ormerod et al., 2013).

Recently, the genomes of *Malassezia globosa* and *M. restricta* were sequenced (Martinez et al., 2012). *Malassezia* has several secreted lipases and hydrolases that help its adaptation to the human skin. Its close relationship to the maize pathogen *Ustilago maydis* suggests an ancestral shift from plant to animal host preference (Xu et al., 2007). Unexpectedly, *M. globosa* secreted protein clusters are found in the *U. maydis* genome, despite their apparent plant-specific functions. Its genome was sequenced and compared with other related species for skin and nail infection genes (Martinez et al., 2012). The dermatophytes' entire genome has duplicated over time and is enriched with the genes for proteases, kinases, and secondary metabolites that may contribute to cause the disease (Martinez et al., 2012).

Comparative genome sequencing studies of human fungal pathogens enable identification of genes and variants associated with virulence and drug resistance. Genomes for some important fungal pathogens were only recently assembled, revealing gene family expansions in many species and the extreme gene loss in one of the obligate species. The scale and scope of species sequenced is rapidly expanding, leveraging technological advances to assemble and annotate genomes with their precision. The whole genome approaches provide the resolution necessary for comparison of closely related isolates, e.g., in the analysis of outbreaks or sampled across time within a single host. Genome analysis of fungal pathogens associated with human has shown a great significant improvement in identification of virulence, associated drug resistance genes, and their variants. Genomic studies also demonstrated gene family (responsible for a particular disease) enlargements in many species and on the contrary, loss of gene in some species. In recent years, increased number of fungal pathogens sequenced, which allowed for assembly of genome and their annotation by comparing the reference genomes. The whole genome approaches provide the resolution needed to compare closely related isolates, such as those from an outbreak or collected over time from a single host. Cuomo (2017) has given a detailed review on the applications of genomics in medical mycology. Some of the important fungal pathogens associated with humans and their whole genome sequence data are provided in Table 1.4.

Nowadays, invasive fungal infections (IFI) have become more common, especially among immunosuppressed subjects. They die more frequently, and they are highly susceptible to fungal infections. Owing to the limitations of available antifungal therapies, the IFIs have higher morbidity and mortality than other infections. Thus, this creates difficulty in finding new treatments or new drugs. So

Table 1.4. Genome data of human pathogenic fungi with whole genome sequence resources.

	Genome Size	**Associated Disease**	**References**
Aspergillus fumigatus	29 Mb	Chronic pulmonary Aspergillosis	Fedorova et al., 2008; Nierman et al., 2005
Blastomyces dermatidis	75 Mb	Blastomycosis	Muñoz et al., 2015
Candida albicans	15 Mb	Candidiasis	Muzzey et al., 2013
Candida glabrata	15 Mb	Superficial mucosal and bloodstream infections	Muzzey et al., 2013
Coccidioides immitis	29 Mb	Coccidioidomycosis	Sharpton et al., 2009
Cryptococcus neoformans	19 Mb	Cryptococcal meningitis	Janbon et al., 2014
Cryptococcus gattii	19 Mb	Cryptococcosis	D'Souza et al., 2011; Farrer et al., 2015
Histoplasma capsulatum	41 Mb	Pulmonary and disseminated histoplasmosis	Histoplasma genome project
Paracoccidioides brasiliensis	30 Mb	Paracoccidioidomycosis	Muñoz et al., 2016;
Penicillium marneffei	29 Mb	Fatal systemic mycosis	Nierman et al., 2015; Woo et al., 2011
Pneumocystis jirovecii	8 Mb	Pneumocystis pneumonia	Ma et al., 2016; Cissé et al., 2012
Rhizopus oryzae	46 Mb	Rhinocerebral mucormycosis	Chibucos et al., 2016; Ma et al., 2009

far, no new IFI-controlling drugs have been developed using genomics platforms. The comparative genomics can help finding new drug targets that are shared by the pathogens. Abadio et al. (2011) identified fungal drug targets for human IFIs. The human pathogenic fungi have produced several drug targets. On the other hand, they chose kre2 and erg6 genes for *in silico* testing. The orthologs of these g7 genes were found in eight human fungal pathogens. Towards the fungal survival within the host, two genes (kre2 and erg6) identified have potential drug targets. The *in silico* as well as manual mining identified 57 potential drug targets, based on the 55 essential *Aspergillus fumigatus* or *Candida albicans* genes and two genes (kre2 and erg6) pertinent for survival of fungi within the host system. These 57 targets have orthologs in eight human pathogenic fungi (*A. fumigatus*, *Blastomyces dermatitidis*, *C. albicans*, *Coccidioides immitis*, *Cryptococcus neoformans*, *Histoplasma capsulatum*, *Paracoccidioides brasiliensis*, and *P. lutzii*).

9. Challenges in Fungal Genomics Studies

Different perspectives exist on fungi as eukaryotes and their evolution. Although scientists have discovered several traits using cultural characteristics, morphological traits, and biochemical features, the possibilities with genomics are endless. As a result of modern sequencing technologies being cost-effective and feasible, we now have unlimited access to numerous fungal genomes. An organism's biological functions could be explored using modern biological tools, especially the whole genome sequencing. Gabaldon (2020) outlines the grand challenges in fungal genomics as well as evolution. Many of their lineages are poorly represented, and information on their early splitting patterns is often incomplete (Naranjao-Ortiz and Gabaldon, 2019a,b). The technologies available can fix the poorly represented lineages, but it will not fix all the fTOL issues.

The first major issue is the accuracy of the genome sequenced. De Vries et al. (2017) emphasized that a publicly released genome does not need to be complete or annotated correctly. Except for a few manually curated genome sequences, the actual and perceived state of completeness and accuracy differ greatly (de Vries et al., 2017). As vast genomic resources are deposited in public databases, complete, accurate and permanent genome sequencing criteria are required. The current sequencing platforms offer low-cost, deep coverage sequencing with long sequence reads. With advanced bioinformatics tools and resources, the entire genome may be represented. However, the high-quality assembly of sequence reads is required to obtain a complete genome and chromosome structure represented by telomere repeats (de Vries et al., 2017). So far, two fungi such as *Myceliophthora thermophila* and *Thielavia terrestris* have been reported to have complete genomes (Berka et al., 2011; de Vries et al., 2017). To identify genes and assign functions, genome annotation is very critical. The number of resources (algorithms and bioinformatics resources) for genome annotation has increased dramatically. The new bioinformatics resources use RNA-seq data to annotate gene models (Reid et al., 2014; Testa et al., 2015; Hoff et al., 2016). Researchers are now seeking gold standard genome sequences to simplify the task.

10. Conclusion and Future Outlook

In combination of traditional and high-throughput sequencing platforms, a tremendous upsurge took place in the availability of genomic resources. The genetic engineering of the metabolic pathways has triggered the biosynthesis of secreted proteins and many secondary metabolites helped to employ new fungal strains towards biocontrol functions. Fungal genomics is evolving and in the recent past, many fungal species have been subjected for whole genome sequencing by various workers across the globe. Though researchers struggled to sequence a single gene in the past, it is not a problem in recent times due to advancements and evolution in the sequencing platforms. Discovery of genome sequences provides a new dimension about the plant-endophyte interactions. To confirm the discovery of these genomic determinants of plant-endophyte interaction, experiments should be accomplished.

Human fungal pathogens are increasingly being studied using genome sequencing. Currently, researchers are comparing the genomes of hundreds of pathogen isolates from a single species. Genome data provides the resolution needed to examine isolate microevolution during infection and pinpoint the outbreak source and transmission networks. Genomic approaches to diagnosis may become more common, and they may provide additional clinically actionable information like drug resistance predictions. Even prior to some treatments, metagenomic data may identify potentially harmful fungi and microbes. A high-resolution view of the specific fungal isolates causing disease is possible using microbial sequencing methods.

Obtaining a complete genome sequence of a fungus is not cumbersome and the whole genome sequencing is being treated as a starting point for various advanced studies to explore further applications. The rapid pace rate at which the new genomes or raw sequence data deposited in the public databases is growing is unbelievable. However, the difficulty is the analysis and interpretation of the whole genome sequence data. Thanks are due to rapid developments in the field of bioinformatics, which is complimenting the data analysis. In future, genomics is going to perform a very critical role in exploring the potential of each organism for the benefit of mankind.

Acknowledgements

The first author (MKS) is grateful to the Council of Scientific and Industrial Research, New Delhi for the award of Research Associateship and the support given by the Department of Studies in Botany, University of Mysore.

References

Abadio, A.K., Kioshima, E.S., Teixeira, M.M., Martins, N.F., Maigret, B. and Felipe, M.S. (2011). Comparative genomics allowed the identification of drug targets against human fungal pathogens. *BMC Genomics*, 12: 75. doi: 10.1186/1471-2164-12-75.

Aflitos, S., Schijlen, E., de Jong, H., de Ridder, D. and Smit, S. (2014). Exploring genetic variation in the tomato (Solanum section Lycopersicon) clade by whole-genome sequencing. *Plant J.*, 80(1): 136–148. https://doi.org/10.1111/tpj.12616.

Asai, S., Ayukawa, Y., Gan, P., Masuda, S., Komatsu, K. et al. (2019). High-quality draft genome sequence of *Fusarium oxysporum* f. sp. *cubense* Strain 160527, a causal agent of panama disease. *Genome Sequences*, 9(29): e00654-19. https://doi.org/10.1128/MRA.00654-19.

Ayhan, D.H., López-Díaz, C., Di Pietro, A. and Ma, L.-J. (2018). Improved assembly of reference genome *Fusarium oxysporum* f. sp. *lycopersici* Strain Fol4287. *Microbiol. Resour. Announc.*, 7: e00910–18.

Bayer, M.M., Rapazote-Flores, P., Ganal, M., Hedley, P.E., Macaulay, M. et al. (2017). Development and evaluation of a barley 50k iSelect SNP array. *Front. Plant Sci.*, 8: 1792. doi: 10.3389/fpls.2017.01792.

Bell, B.P. and Khabbaz, R.F. (2013). Responding to the outbreak of invasive fungal infections: The value of public health to Americans. *JAMA*, 309: 883–884.

Bergmann, S., Schümann, J., Scherlach, K., Lange, C., Brakhage, A.A. and Hertweck, C. (2007). Genomics-driven discovery of PKS-NRPS hybrid metabolites from *Aspergillus nidulans*. *Nat. Chem. Biol.*, 3(4): 213–217. https://doi.org/10.1038/nchembio869.

Berka, R.M., Grigoriev, I.V., Otillar, R., Salamov, A., Grimwood, J. et al. (2011). Comparative genomic analysis of the thermophilic biomass-degrading fungi *Myceliophthora thermophila* and *Thielavia terrestris*. *Nat. Biotechnol.*, 29(10): 922–927.

Blackwell, M. (2011). The fungi: 1, 2, 3 … 5.1 million species? *Am. J. Bot.*, 98: 426–438. https://doi.org/10.3732/ajb.1000298.

Blin, K., Wolf, T., Chevrette, M.G., Lu, X., Schwalen, C.J. et al. (2017). antiSMASH 4.0-improvements in chemistry prediction and gene cluster boundary identification. *Nucleic Acids Res.*, 45: W36–W41. https://doi.org/10.1093/nar/gkx319.

Boecker, S., Grätz, S., Kerwat, D., Adam, L., Schirmer, D. et al. (2018). *Aspergillus niger* is a superior expression host for the production of bioactive fungal cyclodepsipeptides. *Fungal Biol., Biotechnol.*, 5: 4. https://doi.org/10.1186/s40694-018-0048-3.

Brakhage, A.A. and Schroeckh, V. (2011). Fungal secondary metabolites—Strategies to activate silent gene clusters. *Fungal Gen. Biol.*, 48(1): 15–22. https://doi.org/10.1016/j.fgb.2010.04.004.

Brandt, M.E. and Park, B.J. (2013). Think fungus-prevention and control of fungal infections. *Emerg. Infect. Dis.*, https://doi.org/10.3201/eid1910131092.

Buiate, E.A.S., Xavier, K.V., Moore, N., Torres, M.F., Farman, M.L. et al. (2017). A comparative genomic analysis of putative pathogenicity genes in the host-specific sibling species *Colletotrichum graminicola* and *Colletotrichum sublineola*. *BMC Genomics*, 18: 67. https://doi.org/10.1186/s12864-016-3457-9.

Buza, T. and McCarthy, M. (2013). Functional genomics: Applications to production agriculture. *CAB Reviews*, 8: 054. DOI: 10.1079/PAVSNNR20130054.

Cai, C., Ma, J., Han, C., Jin, Y., Zhao, G. and He, X. (2019). Extraction and antioxidant activity of total triterpenoids in the mycelium of a medicinal fungus, *Sanghuangporus sanghuang*. *Sci. Rep.*, 9: 7418. doi: 10.1038/s41598-019-43886-43880.

Cai, J., Zhao, R., Jiang, H. and Wang, W. (2008). *De Novo* origination of a new protein coding gene in *Saccharomyces cerevisiae*. *Genetics*, 179: 487–496.

Chen, S., Xu, J., Liu, C., Zhu, Y., Nelson, D.R. et al. (2012). Genome sequence of the model medicinal mushroom *Ganoderma lucidum*. *Nat. Comm.*, 3: 913. https://doi.org/10.1038/ncomms1923.

Cheng, J.T., Cao, F., Chen, X.A., Li, Y. and Mao, X. (2020). Genomic and transcriptomic survey of an endophytic fungus *Calcarisporium arbuscula* NRRL 3705 and potential overview of its secondary metabolites. *BMC Genomics*, 21: 424. https://doi.org/10.1186/s12864-020-06813-6.

Chiang, Y., Chang, S., Oakley, B.R. and Wang, C.C.C. (2011). Recent advances in awakening silent biosynthetic gene clusters and linking orphan clusters to natural products in microorganisms. *Curr. Opinion Chem. Biol.*, 15(1): 137–143. https://doi.org/10.1016/j.cbpa.2010.10.011.

Chibucos, M.C., Soliman, S., Gebremariam, T., Lee, H., Daugherty, S. et al. (2016). An integrated genomic and transcriptomic survey of mucormycosis-causing fungi. *Nat. Commun.*, 7: 12218. https://doi.org/10.1038/ncomms12218.

Cissé, O.H., Pagni, M. and Hauser, P.M. (2012). *De novo* assembly of the *Pneumocystis jirovecii* genome from a single bronchoalveolar lavage fluid specimen from a patient. *mBio.*, 4(1): e00428–12. https://doi.org/10.1128/mBio.00428-12.

Cisse, O.H. and Stajich, J.E. (2019). FGMP: assessing fungal genome completeness. *BMC Bioinformatics*, 20: 184. https://doi.org/10.1186/s12859-019-2782-9.

Corre, C. and Challis, G.L. (2009). New natural product biosynthetic chemistry discovered by genome mining. *Nat. Prod. Rep.*, 26(8): 977–986. https://doi.org/10.1039/B713024B.

Cowen, L.E., Anderson, J.B. and Kohn, L.M. (2002). Evolution of drug resistance in *Candida albicans*. *Ann. Rev. Microbiol.*, 56: 139–165.

Crous, P.W., Lombard, L., Sandoval-Denis, M., Seifert, K.A., Schroers, H.-J. et al. (2021). *Fusarium*: more than a node or a foot-shaped basal cell. *Stud. Mycol.*, 98: 100116 (2021). https://doi.org/10.1016/j.simyco.2021.100116.

Cuomo, C.A., Güldener, U., Xu, J., Trail, F., Turgeon, B.G. et al. (2007). The *Fusarium graminearum* genome reveals a link between localized polymorphism and pathogen specialization. *Science*, 317(5843): 1400–1402. https://doi.org/10.1126/science.1143708.

D'Souza, C.A., Kronstad, J.W., Taylor, G., Warren, R., Yuen, M. et al. (2011). Genome variation in *Cryptococcus gattii*, an emerging pathogen of immunocompetent hosts. *mBio.*, 2(1): e00342. https://doi.org/10.1128/mBio.00342-10.

Dai, Y.C. (2019). Whole genome sequence of *Auricularia heimuer* (Basidiomycota. Fungi), the third most important cultivated mushroom worldwide. *Genomics*, 111: 50–58. https://doi.org/10.1016/j.ygeno.2017.12.013.

David, A.S., Haridas, S., LaButti, K., Lim, J., Lipzen, A. et al. (2016). Draft genome sequence of Microdochium bolleyi, a dark septate fungal endophyte of beach grass. *Genome Announc.*, 4: e00270–16. https://doi.org/10.1128/genomeA.00270-16.

de Vries, R.P., Riley, R., Wiebenga, A., Aguilar-Osorio, G., Amillis, S. et al. (2017). Comparative genomics reveals high biological diversity and specific adaptations in the industrially and medically important fungal genus *Aspergillus*. *Genome Biol.*, 18: 28. https://doi.org/10.1186/s13059-017-1151-0.

Dean, R., Van Kan, J.A.L., Pretorius, Z.A., Hammond-Kosack, K.E., Pietro, A.D. et al. (2012). The top 10 fungal pathogens in molecular plant pathology. *Pl. Mol. Pathol.*, 13: 414–430.

Dilokpimol, A., Mäkelä, M.R., Varriale, S., Zhou, M., Cerullo, G. et al. (2018). Fungal feruloyl esterases: Functional validation of genome mining-based enzyme discovery including uncharacterized subfamilies. *New Biotechnol.*, 41: 9–14. https://doi.org/10.1016/j.nbt.2017.11.004.

Dobbs, J.T., Kim, M., Dudley, N.S., Klopfenstein, N.B., Yeh, A. et al. (2020). Whole genome analysis of the koa wilt pathogen (*Fusarium oxysporum* f. sp. *koae*) and the development of molecular tools for early detection and monitoring. *BMC Genomics*, 21: 764. https://doi.org/10.1186/s12864-020-07156-y.

Dornburg, A., Townsend, J.P. and Wang, Z. (2017). Maximizing power in phylogenetics and phylogenomics: a perspective illuminated by fungal big data. *Adv. Genet.*, 100: 1–47. https://doi.org/10.1016/bs.adgen.2017.09.007.

Druzhinina, I.S., Chenthamara, K., Zhang, J., Atanasova, L., Yang, D. et al. (2018). Massive lateral transfer of genes encoding plant cell wall-degrading enzymes to the mycoparasitic fungus *Trichoderma* from its plant-associated hosts. *PLOS Genet.*, 14: e1007322. https://doi.org/10.1371/journal.pgen.1007322.

Engel, S.R. and Cherry, J.M. (2013). The new modern era of yeast genomics: community sequencing and the resulting annotation of multiple *Saccharomyces cerevisiae* strains at the *Saccharomyces* genome database 2013. doi: 10.1093/database/bat012.

Engelthaler, D.M., Hicks, N.D., Gillece, J.D., Roe, C.C., Schupp, J.M., Driebe, E.M. et al. (2014). Cryptococcus gattii in North American Pacific Northwest: Whole-population genome analysis provides insights into species evolution and dispersal. *mBio*, 5(4): e01464–14. https://doi.org/10.1128/mBio.01464-14.

Fan, L.S., Zhang, S.H., Yu, L. and Ma, L.-J. (2007). Evaluation of antioxidant property and quality of breads containing *Auricularia auricula* polysaccharide flour. *Food Chem.*, 101: 1158–1163.

Farrer, R.A., Desjardins, C.A., Sakthikumar, S., Gujja, S., Saif, S. et al. (2015). Genome evolution and innovation across the four major lineages of *Cryptococcus gattii*. *mBio.*, 6: e00868–e00815. https://doi.org/10.1128/mBio.00868-15.

Fedorova, N.D., Khaldi, N., Joardar, V.S., Maiti, R., Amedeo, P. et al. (2008). Genomic islands in the pathogenic filamentous fungus *Aspergillus fumigatus*. *PLOS Genet.*, 4: e1000046. https://doi.org/10.1371/journal.pgen.1000046.

Firrincieli, A., Otillar, R., Salamov, A., Schmutz, J., Khan, Z. et al. (2015). Genome sequence of the plant growth promoting endophytic yeast *Rhodotorula graminis* WP1. *Front. Microbiol.*, 6: 978. doi: 10.3389/fmicb.2015.00978.

Fleischmann, R.D., Adams, M.D., White, O., Clayton, R.A., Kirkness, E.F. et al. (1995). Whole-genome random sequencing and assembly of *Haemophilus influenzae* Rd. *Science*, 269(5223): 496–512. https://doi.org/10.1126/science.7542800.

Floudas, D., Binder, M., Riley, R., Barry, K., Blanchette, R.A., et al. (2012). The Paleozoic origin of enzymatic lignin decomposition reconstructed from 31 fungal genomes. *Science*, 336: 1715–1719. https://doi.org/10.1126/science.1221748.

Fokkens, L., Guo, L., Dora, S., Wang, B., Ye, K. et al. (2020). A chromosome-scale genome assembly for the *Fusarium oxysporum* strain fo5176 to establish a model *Arabidopsis*-fungal pathosystem. *G3: Genes, Genomes, Genetics*, 10(10): 3549–3555 https://doi.org/10.1534/g3.120.401375.

Fu, F., Hao, Z., Wang, P., Lu, Y., Xue, L. et al. (2020). Genome Sequence and comparative analysis of *Colletotrichum gloeosporioides* isolated from *Liriodendron* leaves. *Phytopathol.*, 110(7): 1260–1269. https://doi.org/10.1094/PHYTO-12-19-0452-R.

Gabaldón, T. (2020). Grand challenges in fungal genomics and evolution. *Front. Fungal Biol.*, 1: 594855. doi: 10.3389/ffunb.2020.594855.

Gan, P., Ikeda, K., Irieda, H., Narusaka, M., O'Connell, R.J. et al. (2013). Comparative genomic and transcriptomic analyses reveal the hemibiotrophic stage shift of *Colletotrichum* fungi. *New Phytol.*, 197(4): 1236–1249. https://doi.org/10.1111/nph.12085.

Gan, P., Narusaka, M., Kumakura, N., Tsushima, A., Takano, Y. et al. (2016). Genus-wide comparative genome analyses of *Colletotrichum* species reveal specific gene family losses and gains during adaptation to specific infection lifestyles. *Genome Biol. Evol.*, 8(5): 1467–1481. https://doi.org/10.1093/gbe/evw089.

Gan, P., Tsushima, A., Hiroyama, R., Narusaka, M., Takano, Y. et al. (2019). *Colletotrichum shisoi* sp. nov., an anthracnose pathogen of *Perilla frutescens* in Japan: Molecular phylogenetic, morphological, and genomic evidence. *Sci. Rep.*, 9: 13349. https://doi.org/10.1038/s41598-019-50076-5.

Gardes, M. and Bruns, T.D. (1993). ITS primers with enhanced specificity for basidiomycetes–application to the identification of mycorrhizae and rusts. *Mol. Ecol.*, 2(2): 113–118. https://doi.org/10.1111/j.1365-294X.1993.tb00005.x.

Gazis, R., Kuo, A., Riley, R., LaButti, K., Lipzen, A. et al. (2016). The genome of Xylona heveae provides a window into fungal endophytism. *Fungal Biol.*, 120(1): 26–42. https://doi.org/10.1016/j.funbio.2015.10.002.

Goffeau, A., Barrell, B.G., Bussey, H., Davis, R.W., Dujon, B. et al. (1996). Life with 6000 genes. *Science*, 274: 563–567.

Grigoriev, I.V. and Martin F.A. (2013). Changing landscape of fungal genomics. pp. 1–20. *In*: Hoboken, N.J. (Ed.). *The Ecological Genomics of Fungi*. John Wiley & Sons, Inc.

Grigoriev, I.V., Nikitin, R., Haridas, S., Kuo, A., Ohm, R.A. et al. (2014). MycoCosm portal: Gearing up for 1000 fungal genomes. *Nucleic Acids Res.*, 42: D699–D704. https://doi.org/10.1093/nar/gkt1183.

Gubiani, J.R., Oliveira, M.C., Neponuceno, R.A., Camargo, M.J., Garcez, W.S. et al. (2019). Lett. Cytotoxic prenylated indole alkaloid produced by the endophytic fungus *Aspergillus terreus* P63. *PhytochemLett*, 32: 162–167. https://doi.org/10.1016/j.phytol.2019.06.003.

Guo, L., Han, L., Yang, L., Zeng, H., Fan, D. et al. (2014). Genome and transcriptome analysis of the fungal pathogen *Fusarium oxysporum* f. sp. cubense causing banana vascular wilt disease. *PLOS ONE*, 9(4): e95543. https://doi.org/10.1371/journal.pone.0095543.

Haas, B.J., Zeng, Q., Pearson, M.D., Cuomo, C.A. and Wortman, J.R. (2011). Approaches to fungal genome annotation. *Mycology*, 2(3): 118–141. https://doi.org/10.1080/21501203.2011.606851.

Hacquard, S., Kracher, B., Hiruman, K., Munch, P.C., Garrido-Oter, R. et al. (2016). Survival trade-offs in plant roots during colonization by closely related beneficial and pathogenic fungi. *Nat. Comm.*, 7: 11362. https://doi.org/10.1038/ncomms11362.

Han, J., Chon, J., Ahn, J., Choi, I., Lee, Y. et al. (2016). Whole genome sequence and genome annotation of *Colletotrichum acutatum*, causal agent of anthracnose in pepper plants in South Korea. *Genomics Data*, 8: 45–16. https://doi.org/10.1016/j.gdata.2016.03.007.

Harman, G.E., Doni, F., Khadka, R.G. and Uphoff, N. (2021). Endophytic strains of *Trichoderma* increase plants' photosynthetic capability. *J. Appl. Microbiol.*, 130(2): 529–546. https://doi.org/10.1111/jam.14368.

Hawksworth, D.L. and Lücking, R. (2017). Fungal diversity revisited: 2.2 to 3.8 million species. *Microbiol Spectr.*, 5(4). https://doi.org/10.1128/microbiolspec. FUNK-0052-2016.

Hawksworth, D.L. (2019). The macrofungal resources: Extant, current utilization, future prospects and challenges. pp. 1–9. *In*: Sridhar, K.R. and Dsshmukh, S.K. (Eds.). *Advances in Macrofungi: Diversity, Ecology and Biotechnology*. CRC Press, Boca Raton.

Hector, R.F. and Laniado-Laborin, R. (2005). Coccidioidomycosis—A fungal disease of the Americas. *PLoS Medicine*, 2: e2. https://doi.org/10.1371/journal.pmed.0020002.

Hoff, K.J., Lange, S., Lomsadze, A., Borodovsky, M. and Stanke, M. (2016). BRAKER1: Unsupervised RNA-Seq-Based genome annotation with GeneMark-ET and AUGUSTUS. *Bioinformatics*, 32(5): 767–769. https://doi.org/10.1093/bioinformatics/btv661.

Janbon, G., Ormerod, K.L., Paulet, D., Byrnes, E.J., Yadav, V. et al. (2014). Analysis of the genome and transcriptome of *Cryptococcus neoformans* var. *grubii* reveals complex RNA expression and microevolution leading to virulence attenuation. *PLOS Genet.*, 10: e1004261. https://doi.org/10.1371/journal.pgen.1004261.

Jones, T., Federspiel, N.A., Chibana H., Dungan, J., Kalman, S. et al. (2004). The diploid genome sequence of *Candida albicans*. *Proc. Nat. Acad. Sci.*, 101: 7329–7334.

Kanapin, A., Samsonova, A., Rozhmina, T. and Bankin, M. (2020). The genome sequence if five highly pathogenic isolates of *Fusarium oxysporum* f. sp. *lini*. *Mol. Plant Microbe Inter.*, 33(9): 1112–1115. https://doi.org/10.1094/MPMI-05-20-0130-SC.

Kauffman, C.A., Pappas, P.G. and Patterson, T.F. (2013). Fungal infections associated with contaminated methylprednisolone injections. *New Eng. J. Med.*, 368: 2495–2500.

Kellis, M., Patterson, N., Endrizzi, M., Birren, B. and Lander, E.S. (2004). Sequencing and comparison of yeast species to identify genes and regulatory elements. *Nature*, 423: 241–254.

Kersey, P.J., Lawson, D., Birney, E., Derwent, P.S., Haimel, M. et al. (2010). Ensembl genomes: extending Ensembl across the taxonomic space. *Nucl. Acid Res.*, 38: D563–569. https://doi.org/10.1093/nar/gkp871.

Kim, J.A., Jeon, J., Park, S., Kim, K., Choi, G. et al. (2017). Genome sequence of an endophytic fungus, *Fusarium solani* JS-169, which has antifungal activity. *Genome Announc.*, 5: e01071–17. https://doi.org/10.1128/genomeA.01071-17.

Kjærbølling, I., Vesth, T.C., Frisvad, J.C., Nybo, J.L., Theobald, S. et al. (2018). Linking secondary metabolites to gene clusters through genome sequencing of six diverse *Aspergillus* species. *PNAS*, 115(4): E753–E761. https://doi.org/10.1073/pnas.1715954115.

Knapp, D.G., Németh, J.B., Barry, K., Hainaut, M., Henrissat, B. et al. (2018). Comparative genomics provides insights into the lifestyle and reveals functional heterogeneity of dark septate endophytic fungi. *Sci. Rep.*, 8: 6321. https://doi.org/10.1038/s41598-018-24686-4.

Köberl, M., Dita, M., Martinuz, A., Staver, C. and Berg, G. (2015). Agroforestry leads to shifts within the gammaproteobacterial microbiome of banana plants cultivated in Central America. *Front. Microbiol.*, 6: 91. https://doi.org/10.3389/fmicb.2015.00091.

Kohler, A., Kuo, A. and Martin, F. (2015). Convergent losses of decay mechanisms and rapid turnover of symbiosis genes in mycorrhizal mutualists. *Nat Genet.*, 47: 410–415.

Krasnov, G.S., Pushkova, E.N., Novakovskiy, R.O., Kudryavtseva, L.P., Rozhmina, T.A. et al. (2020). High-quality genome assembly of *Fusarium oxysporum* f. sp. *lini*. *Front. Genet.*, 11: 959. https://doi.org/10.3389/fgene.2020.00959.

Kubicek, C.P., Herrera-Estrella, A., Seidl-Seiboth, V., Martinez, D.A., Druzhinina, I.S. et al. (2011). Comparative genome sequence analysis underscores mycoparasitism as the ancestral lifestyle of Trichoderma. *Genome Biol.*, 12(4): R40. https://doi.org/10.1186/gb-2011-12-4-r40.

Kui, L., Zhang, Z., Wang, Y., Zhang, Y., Li, S. et al. (2021). Genome assembly and analyses of the macrofungus *Macrocybe gigantea*. *Hindawi BioMed Res. Int.*, 6656365. https://doi.org/10.1155/2021/6656365.

Lander, E.S., Linton, L.M., Birren, B., Nusbaum, C., Zody, M.C. et al. (2001). Initial sequencing and analysis of the human genome. *Nature*, 409: 860–921. https://doi.org/10.1038/35057062.

Lee, D., Kim, S.C., Kim, D., Kim, J.H., Park, S.P. et al. (2011). Screening of *Phellinus linteus*, a medicinal mushroom, for anti-viral activity. *J. Kor. Soc. Appl. Biol. Dhem.*, 54: 475–478. doi: 10.3839/jksabc.2011.073.

Lelwala, R.V., Korhonen, P.K., Young, N.D., Scott, J.B., Ades, P.K. et al. (2019). Comparative genome analysis indicates high evolutionary potential of pathogenicity genes in *Colletotrichum tanaceti*. *PLOS ONE*, 14(5): e0212248. https://doi.org/10.1371/journal.pone.0212248.

Li, H., Wu, S., Ma, X., Chen, W., Zhang, J. et al. (2018). The genome sequences of 90 mushrooms. *Sci. Rep.*, 8: 9982. https://doi.org/10.1038/s41598-018-28303-2.

Li, X., Xu, S., Zhang, J. and Li, M. (2021). Assembly and annotation of whole-genome sequence of *Fusarium equiseti*. *Genomics*, 113(4): 2870–2876. https://doi.org/10.1016/j.ygeno.2021.06.019.

Lin, H.C., Hewage, R.T., Lu, Y.C. and Chooi, Y.H. (2019). Biosynthesis of bioactive natural products from Basidiomycota. *Org. Biomol. Chem.*, 7: 1027–1036.

Liolios, K., Mavromatis, K., Tavernarakis, N. and Kyrpides, N.C. (2007). The genomes online database (GOLD) in 2007: status of genomic and metagenomic projects and their associated metadata. *Nucl. Acid Res.*, 36(1): D475–479. https://doi.org/10.1093/nar/gkm884.

Liu, J.K. (2007). New terpenoids from Basidiomycetes, *Russula lepida*. *Drug Disc. Ther.*, 1: 94–103.

Liu, L., Agren, R., Bordel, S. and Nielsen, J. (2010a). Use of genome-scale metabolic models for understanding microbial physiology. *FEBS Lett.*, 584: 2556–2564.

Liu, L., Gao, H., Chen, X., Cai, X., Yang, L., Guo, L. et al. (2010b). Brasilamides A-D: sesquiterpenoids from the plant endophytic fungus *Paraconiothyrium brasiliense*. *Eur. J. Org. Chem.*, 17: 3302–3306.

Liu, X., Li, B., Yang, Y., Cai, J., Shi, T. et al. (2020). Pathogenic adaptations revealed by comparative genome analyses of two *Colletotrichum* spp., the causal agent of anthracnose in rubber tree. *Front. Microbiol.*, 11: 1484. https://doi.org/10.3389/fmicb.2020.01484.

Loftus, B.J., Fung, E., Roncaglia, P., Rowley, D., Amedeo, P. et al. (2005). The genome of the basidiomycetous yeast and human pathogen *Cryptococcus neoformans*. *Science*, 307: 1321–1324. https://doi.org/10.1126/science.1103773.

Lòpez-Fernàndez, S., Sonego, P., Moretto, M., Pancher, M., Engelen, K. et al. (2015). Whole-genome comparative analysis of virulence genes unveils similarities and differences between endophytes and other symbiotic bacteria. *Front. Microbiol.*, 6: 419. doi: 10.3389/fmicb.2015.00419.

Lu, Y., Ye, C., Che, J., Xu, X., Shao, D. et al. (2019). Genomic sequencing, genome-scale metabolic network reconstruction, and *in silico* flux analysis of the grape endophytic fungus *Alternaria* sp. MG1. *Microb. Cell Factor*, 18: 13. https://doi.org/10.1186/s12934-019-1063-7.

Ma, L., Chen, Z., Huang da, W., Kutty, G., Ishihara, M. et al. (2016). Genome analysis of three *Pneumocystis* species reveals adaptation mechanisms to life exclusively in mammalian hosts. *Nat. Commun.*, 7: 10740. https://doi.org/10.1038/ncomms10740.

Ma, L.-J., Geiser, D.M., Proctor, R.H., Rooney, A.P., O'Donnell, K. et al. (2013). *Fusarium* pathogenomics. *Ann. Rev. Microbiol.*, 67: 399–416. https://doi.org/10.1146/annurev-micro-092412-155650.

Ma, L.J., Ibrahim, A.S., Skory, C., Grabherr, M.G., Burger, G. et al. (2009). Genomic analysis of the basal lineage fungus *Rhizopus oryzae* reveals a whole-genome duplication. *PLOS Genet.*, 5: e1000549. https://doi.org/10.1371/journal.pgen.1000549.

Ma, L.-J., van der Does, H.C., Borkovich, K.A., Coleman, J.J., Daboussi, M.-J. et al. (2010). Comparative genomics reveals mobile pathogenicity chromosomes in *Fusarium*. *Nature*, 464: 367–373. https://doi.org/10.1038/nature08850.

Mahadevakumar, S. and Sridhar, K.R. (2020). Diagnosis of *Pythium* by classical and molecular approaches. pp. 200–224. *In*: Rai, M.K., Abd-Elsalam, K. and Ingle, A.P. (Eds.). *The genus Pythium: Diagnosis, diseases and management*. Boca Raton, Florida: CRC Press.

Mahadevakumar, S. and Sridhar, K.R. (2021). Diversity of plant pathogenic fungi in agricultural crops. pp. 101–149. *In*: Dubey, S.K. and Verma, S.K. (Eds.). *Plant, Soil and Microbes in Tropical Ecosystem*. Singapore Springer Nature Pte Lted.

Marschall, T. (2018). Computational pan-genomic: Status, promises and challenges. Brief. *Bioinform.*, 19(1): 116–135. https://doi.org/10.1093/bib/bbw089.

Martin, F., Aerts, A., Ahrén, D., Brun, A. et al. (2008). The genome of *Laccaria bicolor* provides insights into mycorrhizal symbiosis. *Nature*, 452: 88–92.

Martin, F.M., Kohler, A., Murat, C., Balestrini, R.M., Coutinho, P.M. et al. (2010). Périgord black truffle genome uncovers evolutionary origins and mechanisms of symbiosis. *Nature*, 464: 1033–1038.

Martin, K.J. and Rygiewicz, P.T. (2005). Fungal-specific PCR primers developed for analysis of the ITS region of environmental DNA extracts. *BMC Microbiol.*, 5: 28. https://doi.org/10.1186/1471-2180-5-28.

Mehmood, A., Hussain, A., Irshad, M., Hamayun, M., Iqbal, A. et al. (2019). *In vitro* production of IAA by endophytic fungus *Aspergillus awamori* and its growth promoting activities in *Zea mays*. *Symbiosis*, 77(3): 225–235. https://doi.org/10.1007/s13199-018-0583-y.

Metzker, M.L. (2010). Sequencing technologies: The next generation. *Nature Rev.*, 11: 31–46.

Moller, M. and Stukenbrock, E.H. (2017). Evolution and genome architecture in fungal plant pathogens. *Nat. Rev. Microbiol*, 15(12): 756–771. https://doi.org/10.1038/nrmicro.2017.76.

Morgante, M., De Paoli, E. and Radovic, S. (2007). Transposable elements and the plant pan-genomes. *Curr. Opin. Plant Biol.*, 10(2): 149–155. https://doi.org/10.1016/j.pbi.2007.02.001.

Morin, E.I., Kohler, A., Baker, A.R., Foulongne-Oriol, M., Lombard, V. et al. (2012). Genome sequence of the button mushroom *Agaricus bisporus* reveals mechanisms governing adaptation to a humic-rich ecological niche. *Proc. Nat. Acad. Sci.*, 109: 17501–17506.

Muñoz, J.F., Gauthier, G.M., Desjardins, C.A., Gallo, J.E., Holder, J. et al. (2015). The dynamic genome and transcriptome of the human fungal pathogen Blastomyces and close relative Emmonsia. *PLOS Genet.*, 11: e1005493. https://doi.org/10.1371/journal.pgen.1005493.

Muñoz, J.F., Farrer, R.A., Desjardins, C.A., Gallo, J.E., Sykes, S. et al. (2016). Genome diversity, recombination, and virulence across the major lineages of paracoccidioides. *mSphere*, 1(5): e00213–16. https://doi.org/10.1128/mSphere.00213-16.

Muzzey, D., Schwartz, K., Weissman, J.S. and Sherlock, G. (2013). Assembly of a phased diploid *Candida albicans* genome facilitates allele-specific measurements and provides a simple model for repeat and indel structure. *Genome Biol.*, 14(9): R97. https://doi.org/10.1186/gb-2013-14-9-r97.

Nagy, L., Riley, R., Tritt, A., Adam, C., Daum, C. et al. (2016). Comparative genomics of early-diverging mushroom-forming fungi provides insights into the origins of lignocellulose decay capabilities. *Mol. Biol. Evol.*, 33: 959–370.

Nagy, L.G. and Szöllősi, G. (2017). Fungal phylogeny in the age of genomics: Insights into phylogenetic inference from genome-scale datasets. *Adv. Genet.*, 100: 49–72. https://doi.org/10.1016/bs.adgen.2017.09.008.

Naranjo-Ortiz, M.A. and Gabaldón, T. (2019a). Fungal evolution: Diversity, taxonomy, and phylogeny of the fungi. *Biol. Rev.*, 94: 2101–2137. doi: 10.1111/brv.12550.

Naranjo-Ortiz, M.A. and Gabaldón, T. (2019b). Fungal evolution: major ecological adaptations and evolutionary transitions. *Biol. Rev.*, 94: 1443–1476. doi: 10.1111/brv.12510.

Nelson, P.E., Toussoun, T.A. and Cook, R.J. (1981). *Fusarium: Diseases, biology, and taxonomy*. University Park, PA: Pennsylvania State University Press, p. 474.

Nemri, A., Saunders, D.G.O., Anderson, C., Upadhyaya, N.M., Win, J. et al. (2014). The genome sequence and effector complement of the flax rust pathogen *Melamspora lini*. *Front. Pl. Sci.*, 5: 98. https://doi.org/10.3389/fpls.2014.00098.

Nierman, W.C., Pain, A., Anderson, M.J., Wortman, J.R., Kim, H.S. et al. (2005). Genomic sequence of the pathogenic and allergenic filamentous fungus *Aspergillus fumigatus*. *Nature*, 438: 1151–1156. https://doi.org/10.1038/nature04332.

Nierman, W.C., Fedorova-Abrams, N.D. and Andrianopoulos, A. (2015). Genome sequence of the AIDS-associated pathogen *Penicillium marneffei* (ATCC18224) and its near taxonomic relative *Talaromyces stipitatus (ATCC10500)*. *Genome Announc.*, 3: https://doi.org/10.1128/genomeA.01559-14.

Nilsson, R.H., Kristiansson, E., Ryberg, M., Hallenberg, N. and Larsson, K. (2008). Intraspecific ITS variability in the kingdom fungi as expressed in the international sequence databases and its implications for molecular species identification. *Evol. Bioinform.* Online, 4: 193–201. https://doi.org/10.4137/ebo.s653.

Nofiani, R., de Mattos-Shipley, K., Lebe, K.E., Han, L., Iqbal, Z. et al. (2018). Strobilurin biosynthesis in Basidiomycete fungi. *Nature comm.*, 9: 3940. https://doi.org/10.1038/s41467-018-06202-4.

O'Connell, R.J., Thon, M.R., Hacquard, S., Amyotte, S.G., Kleemann, J. et al. (2012). Lifestyle transitions in plant pathogenic *Colletotrichum* fungi deciphered by genome and transcriptome analyses. *Nat. Genet.*, 44(9): 1060–1065.

O'Donnell, K., Kistler, H.C., Cigelnik, E. and Ploetz, R.C. (1998). Multiple evolutionary origins of the fungus causing Panama disease of banana: Concordant evidence from nuclear and mitochondrial gene genealogies. *Proc. Nat. Acad. Sci.*, 95: 2044–2049.

O'Donnell, K., Kistler, H.C., Tacke, B.K. and Casper, H.H. (2000). Gene genealogies reveal global phylogeographic structure and reproductive isolation among lineages of *Fusarium graminearum*, the fungus causing wheat scab. *Proc. Nat. Acad. Sci.*, 97: 7905–7910.

O'Donnell, K., Sutton, D.A., Rinaldi, M.G., Magnon, K.C., Cox, P.A. et al. (2004). Genetic diversity of human pathogenic members of the *Fusarium oxysporum* complex inferred from multilocus DNA sequence data and amplified fragment length polymorphism analyses: evidence for the recent dispersion of a geographically widespread clonal lineage and nosocomial origin. *J. Clin. Microbiol.*, 42: 5109–5120. https://doi.org/10.1128/JCM.42.11.5109-5120.2004.

O'Donnell, K., Gueidan, C., Sink, S., Johnston, P.R., Crous, P.W. et al. (2009). A two-locus DNA sequence database for typing plant and human pathogens within the *Fusarium oxysporum* species complex. *Fungal Genet. Biol.*, 46: 936–948. https://doi.org/10.1016/j.fgb.2009.08.006.

Okubo, T., Liu, D., Tsurumaru, H., Ikeda, S., Asakawa, S. et al. (2015). Elevated atmospheric CO_2 levels affect community structure of rice root-associated bacteria. *Front. Microbiol.*, 6: 136. https://doi.org/10.3389/fmicb.2015.00136.

Ormerod, K.L., Morrow, C.A., Chow, E.W.L., Lee, I.R., Arras, S.D.M. et al. (2013). Comparative genomics of serial isolates of Cryptococcus neoformans reveals gene associated with carbon utilization and virulence. *G3 Genes Genom. Genet.*, 3: 675–686.

Orozco-Mosqueda, M.C. and Santoyo, G. 2021. Plant-microbial endophytes interactions: Scrutinizing their beneficial mechanisms from genomic explorations. *Curr. Pl. Biol.*, 25: 100189. https://doi.org/10.1016/j.cpb.2020.100189.

Osmundson, T.W., Robert, V.A., Schoch, C.L., Baker, L.J., Smith, A. et al. (2013). Filling gaps in biodiversity knowledge for macrofungi: Contributions and assessment of an herbarium collection DNA Barcode Sequencing Project. *PLOS ONE*, 8(4): e62419. https://doi.org/10.1371/journal.pone.0062419.

Pachauri, S., Sherkhane, P.D., Kumar, V., and Mukherjee, P.K. (2020). Whole genome sequencing reveals major deletions in the genome of M7, a gamma ray-induced mutant of *Trichoderma virens* that is repressed in conidiation, secondary metabolism, and mycoparasitism. *Front. Microbiol.*, 11: 1030. https://doi.org/10.3389/fmicb.2020.01030.

Pizarro, D., Divakar, P.K., Grewe, F., Leavitt, S.D., Huang, J.-P. et al. (2018). Phylogenomic analysis of 2556 single-copy protein-coding genes resolves most evolutionary relationships for the major clades in the most diverse group of lichen-forming fungi. *Fungal Divers.*, 92: 31–41. https://doi.org/10.1007/s13225-018-0407-7.

Plissonneau, C., Benevenuto, J., Mohd-Assaad, N., Fouché, S., Hartmann, F.E. and Croll, D. (2017). Using population and comparative genomics to understand the genetic basis of effector-driven fungal pathogen evolution. *Front. Pl. Sci.*, 8: 119. https://doi.org/10.3389/fpls.2017.00119.

Reid, I., O'Toole, N., Zabaneh, O., Nourzadeh, R., Dahdouli, M. et al. (2014). SnowyOwl: accurate prediction of fungal genes by using RNA-Seq and homology information to select among *ab initio* models. *BMC Bioinform.*, 15: 229. https://doi.org/10.1186/1471-2105-15-229.

Rhind, N., Chen, Z., Yassour, M., Thompson, D.A., Haas, B.J. et al. (2011). Comparative functional genomics of the fission yeasts. *Science*, 332: 930–936.

Rogério, F., Boufleur, T.R., Ciampi-Guillardi, M., Sukno, S.A., Thon, M.R. et al. (2020). Genome sequence resources of *Colletotrichum truncatum, C. plurivorum, C. musicola*, and *C. sojae*: four species pathogenic to soybean (*Glycine max*). *Phytopathology*, 110: 1497–1499. https://doi.org/10.1094/PHYTO-03-20-0102-A.

Rogers, J. and Gibbs, R.A. (2014). Comparative primate genomics: emerging patterns of genome content and dynamics. *Nat. Rev. Genet.*, 15(5): 347–359. https://doi.org/10.1038/nrg3707.

Ronning, C.M., Fedorova, N.D., Bowyer, P., Coulson, R., Goldman, G., Kim, H.S. et al. (2005). Genomics of *Aspergillus fumigatus*. *Rev. Iberoam. Micol.*, 22(4): 223–228. https://doi.org/10.1016/s1130-1406(05)70047-4.

Sabahi, F., de Sain, M., Banihashemi, Z. and Rep, M. (2021). Comparative genomics of *Fusarium oxysporum* f. sp. melonis strains reveals nine lineages and a new sequence type of AvrFom2. *Environ. Microbiol.*, 23(4): 2035–2053. https://doi.org/10.1111/1462-2920.15339.

Sanger, F., Nicklen, S. and Coulson, A.R. (1977). DNA sequencing with chain-terminating inhibitors. *Proc. Nat. Acad. Sci.*, 74: 5463–5467.

SanMiguel, P. (2011). Next-generation sequencing and potential applications in fungal genomics. pp. 51–60. *In*: Xu, J.-R. and Bluhm, B.H. (Eds.). *Fungal Genomics: Methods and protocols, methods in molecular biology*, Volume # 722, Springer *Nature*.

Schatz, M.C., Maron, L.G., Stein, J.C., Wences, A.H., Gurtowski, J. et al. (2014). Whole genome *de novo* assemblies of three divergent strains of rice, *Oryza sativa*, document novel gene space of *aus* and *indica*. *Genome Biol.*, 15: 506. https://doi.org/10.1186/s13059-014-0506-z.

Schlegel, M., Münsterkötter, M., Güldener, U., Bruggmann, R., Duò, A. et al. (2016). Globally distributed root endophyte *Phialocephala subalpina* links pathogenic and saprophytic lifestyles. *BMC Genomics*, 17: 1015. https://doi.org/10.1186/s12864-016-3369-8.

Schoch, C.L., Seifert, K.A., Huhndorf, S., Rober, V., Spouge, J.L. et al. (2012). Nuclear ribosomal internal transcribed spacer (ITS) region as a universal DNA barcode marker for *Fungi*. *Proc. Nat. Acad. Sci.*, 109(16): 6241–6246. https://doi.org/10.1073/pnas.1117018109.

Seo, S., Pokhrel, A. and Coleman, J.J. (2020). The genome sequence of five genotypes of *Fusarium oxysporum* f. sp. *vasinfectum*: A resource for studies on *Fusarium* wilt of cotton. *Molecular Plant-Microbe Interactions*, 33(2): 138–140. https://doi.org/10.1094/MPMI-07-19-0197-A.

Shao, Y., Guo, H., Zhang, J., Liu, H., Wang, K. et al. (2020). The genome of the medicinal macrofungus sanghuang provides insights into the synthesis of diverse secondary metabolites. *Front. Microbiol.*, 10: 3035. https://doi.org/10.3389/fmicb.2019.03035.

Sharpton, T.J., Stajich, J.E., Rounsley, S.D., Gardner, M.J., Wortman, J.R. et al. (2009). Comparative genomic analyses of the human fungal pathogens *Coccidioides* and their relatives. *Genome Res.*, 19: 1722–1731. https://doi.org/10.1101/gr.087551.108.

Sigaux, F. (2000). Cancer genome or the development of molecular portraits of tumors [in French]. *Bull. Acad. Natl. Med.*, 184(7): 1441–1449.

Silva, R.N. (2016). Perspectives in genomics: The future of fungi in OMICS era. *Current Genomics*, 17(2): 82–84. https://doi.org/10.2174/1389202917021603031748 29.

Stajich, J.E., Harris, T., Brunk, B.P., Brestelli, J., Fisher, S. et al. (2012). FungiDB: An integrated functional genomics database for fungi. *Nucl. Acids Res.*, 40: D675–D681.

Stajich, J.E., Wilke, S.K., Ahrén, D., Au, C.H. et al. (2010). Insights into evolution of multicellular fungi from the assembled chromosomes of the mushroom *Coprinopsis cinerea* (Coprinus cinereus). *Proc. Nat. Acad. Sci.*, 107: 11889–11894.

Steenwyk, J.L., Shen, X.X., Lind, A.L., Goldman, G.H and Rokas, A. (2019). A robust phylogenomic time tree for biotechnologically and medically important fungi in the genera *Aspergillus* and *Penicillium*. *mBio*, 10: 1–25. https://doi.org/10.1128/mBio.00925-19.

Steinberg, C., Lecomte, C., Alabouvette, C. and Edel-Hermann, V. (2016). Root interactions with nonpathogenic *Fusarium oxysporum* Hey *Fusarium oxysporum*: What do you do in life when you do not infect a plant? pp. 281–299. *In*: Vos, C.M.F. (ed.). *Belowground Defence Strategies in Plants*, Springer Cham.

Tadra-Sfeir, M.Z., Faoro, H., Camilios-Neto, D., Brusamarello-Santos, L, Balsanelli, E. et al. (2015). Genome wide transcriptional profiling of *Herbaspirillum seropedicae* SmR1 grown in the presence of naringenin. *Front. Microbiol.*, 6: 491. doi: 10.3389/fmicb.2015.00491.

Tala, M.F., Qin, J., Ndongo, J.T. and Laatsch, H. (2017). New azulene-type sesquiterpenoids from the fruiting bodies of *Lactarius deliciosus*. *Nat. Prod. Bioprospect.*, 7: 269–273.

Tejesvi, M.V., Pirttilä, A.M. and Frank, A.C. (2017). Editorial: Emerging tools for emerging symbioses-using genomics applications to studying endophytes. *Front. Microbiol.*, 8: 859. doi: 10.3389/fmicb.2017.00859.

Testa, A.C., Hane, J.K., Ellwood, S.R. et al. (2015). Coding Quarry: highly accurate hidden Markov model gene prediction in fungal genomes using RNA-seq transcripts. *BMC Genomics*, 16: 170. https://doi.org/10.1186/s12864-015-1344-4.

Tettelin, H., Masignani, V., Cieslewich, M.J., Donati, C., Medini, D. et al. (2005). Genome analysis of multiple pathogenic isolates of *Streptococcus agalactiae*: Implications for the microbial "pan-genome". *Proc. Nat. Acad. Sci.*, 102(39): 13950–13955. https://doi.org/10.1073/pnas.0506758102.

Thudi, M., Khan, A.W., Kumar, V., Gaur, P.M., Katta, K. et al. (2016). Whole genome re-sequencing reveals genome-wide variations among parental lines of 16 mapping populations in chickpea (*Cicer arietinum* L.). *BMC Pl. Biol.*, 16(Suppl 1): 10. https://doi.org/10.1186/s12870-015-0690-3.

Toju, H., Tanabe, A.S., Yamamoto, S. and Sato, H. (2012). High-coverage ITS Primers for the DNA-based identification of ascomycetes and basidiomycetes in environmental samples. *PLOS ONE*, 7(7): e40863. https://doi.org/10.1371/journal.pone.0040863.

Tokai, T., Koshino, H., Takahashi-Ando, N., Sato, M., Fujimura, M. et al. (2007). *Fusarium* Tri4 encodes a key multifunctional cytochrome P450 monooxygenase for four consecutive oxygenation steps in trichothecene biosynthesis. *Biochem. Biophys. Res. Com.*, 353: 412–417.

Venter, J.C., Adams, M.D., Myers, E.W., Li, P.W., Mural, R.J. et al. (2001). The sequence of the human genome. *Science*, 291: 1304–1351. https://doi.org/10.1126/science.1058040.

Vesth, T.C., Brandl, J. and Anderson, M.R. (2016). FunGeneClusterS: Predicting fungal gene clusters from genome and transcriptome data. *Syn. Sys. Biotechnol.*, 1(2): 122–129. https://doi.org/10.1016/j.synbio.2016.01.002.

Walker, A.K., Frasz, S.L., Seifert, K.A., Miller, J.D., Mondo, S.J. et al. (2016). Full genome of Phialocephala scopiformis DAOMC 229536, a fungal endophyte of spruce producing the potent anti-insectan compound rugulosin. *Genome Announc.*, 4: e01768–15. https://doi.org/10.1128/genomeA.01768-15.

Wang, W.G., Du, L.Q., Sheng, S.L., Li, A., Li, Y.P. et al. (2019). Genome mining for fungal polyketide-diterpenoid hybrids: discovery of key terpene cyclases and multifunctional P450s for structural diversification. *Org. Chem. Front.*, 6: 571–578. https://doi.org/10.1039/c8qo01124a.

Wang, X., Zhang, X., Liu, L., Xiang, M., Wang, W. et al. (2015). Genomic and transcriptomic analysis of the endophytic fungus *Pestalotiopsis* fici reveals its lifestyle and high potential for synthesis of natural products. *BMC Genomics*, 16: 28. https://doi.org/10.1186/s12864-014-1190-9.

Weigel, D. and Mott, R. (2009). The 1001 Genomes project for *Arabidopsis thaliana*. *Genome Biol.*, 10: 107. https://doi.org/10.1186/gb-2009-10-5-107.

White, T., Bruns, T., Lee, S. and Taylor, J. (1990). Amplification and direct sequencing of fungal ribosomal RNA genes for phylogenetics. pp. 315–322. *In*: Innis, M.A., Gelfand, D.H., Sninsky, J.J. and White, T.J. (Eds.). *PCR Protocols: A Guide to Methods and Applications.* New York: Academic Press.

Wijayawardene, N.N., Hyde, K.D., Lumbsch, H.T., Liu, J.K., Maharachchikumbura, S.S.N. et al. (2018). Outline of ascomycota: 2017. *Fungal Divers.*, 88(1): 167–263. https://doi.org/10.1007/s13225-018-0394-8.

Wonglom, P., Ito, S.I. and Sunpapao, V. (2020). Volatile organic compounds emitted from endophytic fungus *Trichoderma asperellum* T1 mediate antifungal activity, defense response, and promote plant growth in lettuce (*Lactuca sativa*). *Fungal Ecol.*, 43: Article 100867. https://doi.org/10.1016/j.funeco.2019.100867.

Woo, P.C.Y., Lau, S.K.P., Liu, B., Cai, J.J., Chong, K.T.K. et al. (2011). Draft genome sequence of *Penicillium marneffei* strain PM1. *Eukaryot. Cell.*, 10: 1740–1741. https://doi.org/10.1128/EC.05255-11.

Xiao, G., Zhang, X. and Gao, Q. 2017. Bioinformatic approaches for fungal Omics. *Biomed Res. Int.*, 7270485. https://doi.org/10.1155/2017/7270485.

Xie, K., Yue, Y., Qiu, H. and Hu, X. (2020). Complete mitochondrial genome sequence of potato pathogenic fungus, *Fusarium oxysporum* f. sp. KGSJ26F3. *Mitochondrial DNA Part B Resources*, 5(3): 2408–2409. https://doi.org/10.1080/23802359.2020.1773334.

Xu, J. (2016). Fungal DNA barcoding. *Genome*, 59(11): 913–932. https://doi.org/10.1139/gen-2016-0046.

Xu, J., Saunders, C.W., Hu, P., Grant, R.A., Boekhout, T., Kuramae, E. et al. (2007). Dandruff-associated Melassezia genomes reveal convergent and divergent virulence traits shared with plant and human fungal pathogens. *Proceedings of National Academy of Science, USA*, 104(47): 18730–18735. https://doi.org/10.1073/pnas.0706756104.

Xu, X.-H., Su, Z.-Z., Wang, C., Kubicek, C.P., Feng, X.-X. et al. (2015). The rice endophyte Harpophora oryzae genome reveals evolution from a pathogen to a mutualistic endophyte. *Sci. Rep.*, 4: 5783. https://doi.org/10.1038/srep05783.

Yang, H., Wang, Y., Zhang, Z., Yan, R. and Zhu, D. (2014). Whole-genome shotgun assembly and analysis of the genome of *Shiraia* sp. strain Slf14, a novel endophytic fungus producing huperzine A and hypocrellin A. *Genome Announc.*, 2 (1): e00011–e00014, https://doi.org/10.1128/genomeA.00011-14.

Yang, Y., Liu, X., Cai, J., Chen, Y., Li, B. et al. (2019). Genomic characteristics and comparative genomics analysis of the endophytic fungus *Sarocladium brachiariae*. *BMC Genomics*, 20: 782. https://doi.org/10.1186/s12864-019-6095-1.

Yu, X., Yang, J., Wang, E., Li, B. and Yuan, H. (2015). Effects of growth stage and fulvic acid on the diversity and dynamics of endophytic bacterial community in *Stevia rebaudiana* Bertoni leaves. *Front. Microbiol.*, 6: 867. https://doi.org/10.3389/fmicb.2015.00867.

Zampounis, A., Pigné, S., Dallery, J.-F., Wittenberg, A.H.J., Zhou, S. et al. (2016). Genome sequence and annotation of *Colletotrichum higginsianum*, a causal agent of *Crucifer* anthracnose disease. *Genome Announcements*, 4(4): e00821–16. https://doi.org/10.1128/genomeA.00821-16.

Zhang, N., Cai, G., Price, D.C., Crouch, J.A., Gladieus, P. et al. (2018). Genome wide analysis of the transition to pathogenic lifestyles in Magnaporthales fungi. *Sci. Rep.*, 8: 5862. https://doi.org/10.1038/s41598-018-24301-6.

Zhang, N., Luo, J. and Bhattacharya, D. (2017). Advances in fungal phylogenomics and their impact on fungal systematics. *Adv. Genet.*, 100: 309–328. https://doi.org/10.1016/bs.adgen.2017.09.004.

Zhang, Y., Yang, H., Turra, D. Zhou, S., Ayhan, D.H. et al. (2020). The genome of opportunistic fungal pathogen *Fusarium oxysporum* carries a unique set of lineage-specific chromosomes. *Comm. Biol.*, 3: 50. https://doi.org/10.1038/s42003-020-0770-2.

Zhu, T.B., Kim, S.H., and Chen, C.Y. (2008). A medicinal mushroom: *Phellinus linteus*. *Curr. Med. Chem.*, 15: 1330–1335. doi: 10.2174/092986708784534929.

Zuccaro, A., Lahrmann, U., Güldener, U., Langen, G., Pfiffi, S. et al. (2011). Endophytic life strategies decoded by genome and transcriptome analyses of the mutualistic root symbiont *Piriformospora indica*. *PLOS Pathog.*, 7(10): e1002290. https://doi.org/10.1371/journal.ppat.1002290.

Biomaterials

2

Fungal Chitosan

Sources, Production, and Applications

Sesha Subramanian Murugan,[1] *Jayachandran Venkatesan*[1,2,*] and *Gi Hun Seong*[2,*]

1. Introduction

Chitin is a glycosidic biopolymer with at least 60% N-acetyl-D-glucosamine and glycosidic linkages between subunits (Dhillon et al., 2013). Chitosan may be found in lobster, mollusk radulum, arthropod exoskeleton, cephalopod endoskeleton, and fungal cell walls. Crustacean's cells walls are often utilized to produce the chitin and chitosan using chemical and enzymatic methods. Using crustacean's raw materials which are available abundantly and these methods are simple and produce huge volumes of chitin and chitosan. Fungal biomass which is an alternative resource for chitin and chitosan production, is also abundant in nature and commercially viable, and may be used in the production of chitosan. The recent researches were focused on chitosan extraction from the fungal biomass by using fermentation technology through solid-state and submerged fermentation methods. Solid-state fermentation was used to produce a large amount of fungal biomass, which was then used to make chitosan in large quantities. Bulk fungal biomass was used in large-scale chitosan production and produced through solid-state fermentation. In the proper medium, the fermentation was exposed to the most essential parameters of temperature, pH, and growth factors. Peptones, D-glucose, and yeast extract were the most common supplements that help fungal growth and use carbon sources in chitosan production. Fungi have numerous chitin polysaccharides and glycoproteins such as mannoproteins, galactoproteins, glucuronoproteins, and xylomannoproteins in their cell walls. The β-1,3-linked glucose, which was found in fungal cell walls

[1] Biomaterials Research Laboratory, Yenepoya Research Centre, Yenepoya (Deemed to be University), Deralakatte, Mangalore, Karnataka 575018, India.

[2] Department of Bionano Engineering, Center for Bionano Intelligence Education and Research, Hanyang University, Ansan 426-791, South Korea.

* Corresponding authors: jvenkatesan@yenepoya.edu.in; ghseong@hanyang.ac.kr

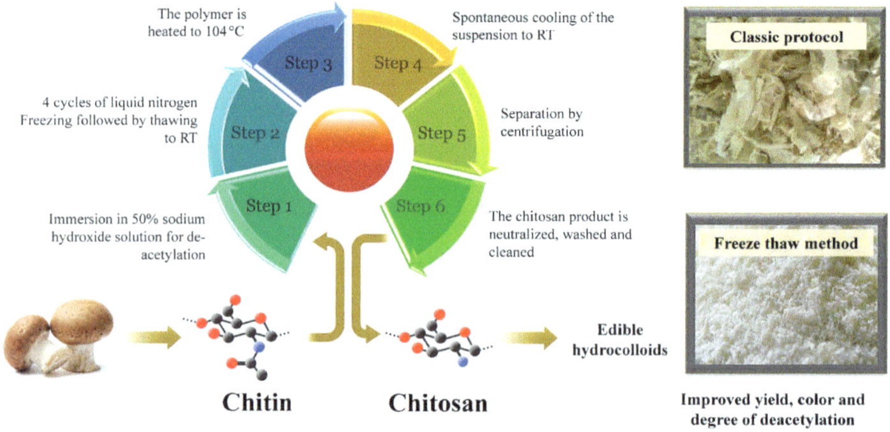

Fig. 2.1. Steps involved in the production of chitosan from fungal source (Ban et al., 2018).

Note: Chemical structure of the chitosan (Figure is redrawn from the reference research article: Sami El-banna et al., 2019).

and forms a complex network with chitin and glycoprotein components (Kannan et al., 2010). According to Dhillon et al. (2013), chitin was discovered in *Aspergillus niger*, *Mucor rouxii* (45%), and *Penicillium notatum* (20%). In the presence of acidic acid and lactic acid aqueous media, chitin can be deacetylated and converted to chitosan. The precise chemical structure (Fig. 2.1), biodegradable, biocompatible, and non-toxic character of chitosan attracts a lot of interest in cell encapsulation, food additives, wastewater treatment, enzyme immobilization, wound-healing dressings, and drug-delivery systems, etc. (Dhillon et al., 2013). The steps involved in the production of chitin from mushroom fungal wall along with structure of chitosan is shown in Fig. 2.1.

2. Sources of Chitosan

The major source of chitosan is crustacean crab shell, shrimp, and beetle's exoskeleton, but only for a season. The microbes of protists, algae, and kingdom fungal of molds and macro mushrooms have the potential growing interest from natural sources. The possible sources of chitin and chitosan produced from fungal-based sources include *Absidia glauca*, *Aspergillus niger*, *Gongronella butleri*, *Mocors rouxii*, *Rhizopus oryzae*, and *Trichoderma reesei*. The chitosan were found in the inner cellular components of fungus (Merzendorfer, 2011; Dhillon et al., 2013). The fungi have no seasonal limitations and have many protein contents and heavy metals of nickel and copper. The chitosan from the fungal sources possesses biocompatible and biodegradable character (Aranaz et al., 2009; Ghormade et al., 2017).

2.1 Production from Natural Sources

The shrimp consists of 45% exoskeletal and cephalothoraxes, which include 50–70% astathin, astaxanthin, lutein, canthaxanthin, and β-carotene, as well as 30–50% protein, 30% calcium carbonates, and 20–30% astathin. Acid and alkaline treatments were used to remove chitin from related proteins and calcium chloride (Acosta et al.,

1993). Another investigation shows that chitosan is also produced by the hydrolysis of chitin acetamide groups. The crude crustacean shells were washed, ground, and produced through the sieving procedure to fine particles of 250 μm. The crustacean shell powder has been demineralized to remove calcium carbonate from the raw materials using 0.25 M HCl. The proteins in the raw materials were removed by deproteinization by using 1 M NaOH. The chitin which was obtained was treated with 45% NaOH either by microwave radiation or at 110°C to get a fine product of chitosan. The yield of chitosan produced *Absidia, Gongronella, Mucor*, and *Rhizopus* was investigated, and *Gongronella butleri* has recorded the highest output by solid-state fermentation (SSF) (Nwe and Stevens, 2008). A comparative study of chitosan production among different species of *Aspergillus niger, Lentinus edodes, Pleurotus sajo-caju, Rhizopus oryzae*, and yeast strains of *Candida albicans* TISTR523 and *Zygosaccharomyces rouxii* TISTR5058 were investigated by using modification methods of solid-state formation. The results show that the *R. oryzae* has the highest output of chitosan production of 139 mg/g with 14% product content (Pochanavanich and Suntornsuk, 2002). Suntornsuk et al. (2002) studied the chitosan production on four different fungal strains (*A. niger* TISTR3245, *R. oryzae* TISTR3189, *Z. rouxii* TISTR5058, and *C. albicans* TISTR5239) on the soybean and mungbean residues. The finely powdered residues were treated with 1 N NaOH for 15 minutes at 121°C and were centrifuged. The centrifuged product was processed at 2% acetic acid for 8 hours at a temperature of 95°C and centrifuged at 12,000 g for 15 min before being dried at a temperature of 60°C. According to the findings, *R. oryzae* had the highest chitosan output of 4.3 g/kg on soybean residue (Suntornsuk et al., 2002). The presence of the chitosan from the *R. oryzae* was confirmed by the x-ray diffractometer in the 2 θ range 10.5° and 20.0° and fourier transform infrared spectroscopy bands at 1,598 cm^{-1} (amino deformation mode), 3,450 cm^{-1} (NH bond stretching), 1,649 cm^{-1} (amide bands) and 898 cm^{-1} (β—anomer) (Wang et al., 2008). It is reported that the glucose concentrations of 1–3% also influenced the chitosan production from the fungal biomass (Rane and Hoover, 1993). Studies by the Tajdini et al. (2010) showed yield of chitosan from the *Rhizomucor miehei* and *Mucor racemosus* up to 98.6% and 97.1%, respectively. The chitosan from both species manifests antibacterial and antifungal activities against *C. albicans, C. glabrata, Pseudomonas aeroginosa*, and *Escherichia coli* (Tajdini et al., 2010).

The study of Nwe proved that the production of the chitosan from the fungal mycelia which supplied urea at the rate of 7.2 g urea/kg and had the best optimal conditions for the solid-state fermentation method (Nwe and Stevens, 2004). The strong acid treatment was used to break down the chitin and β-glucan which usually forms the rigid cross-linked network. Also, the enzymatic digestion was performed to break down the chitin-glucan complex by using glucanase, chitinase, and amylase (Nwe et al., 2010). Usually, the two-step extraction process was carried for the production of the chitosan from the fungi cell wall through the fermentation process in acidic and basic conditions. The raw materials of mycelia were treated with NaOH alkaline conditions at 90–121°C for about 20 minutes to degrade the cell wall proteins. The chitin was released from the complex network through the deacetylation process. The alkali insoluble materials in the processed materials were treated with hydrochloric, lactic, or acetic acid at 25–95°C for 1–2 hours to remove the phosphate sources from the fungal cell wall. The resulted final product of chitosan-rich fungal biomass was centrifuged at the pH condition of 9–10 with repeated acetone and ethanol washing (Tasar et al., 2016). The filamentous fungus of *Gongronella butleri* CCT4274 was cultured in the medium of the aqueous extract of apple pomace and projected the production of chitosan at the rate of 1.19 g/L with 21% biomass content (Streit et al., 2009).

3. Applications in Biomedical Applications

The chitosan derived from the fungal biomass is extremely attractive in the biomedical and pharmaceutical fields due to its controlled molecular weight and low toxicity (Freitas et al., 2015; Ospina et al., 2015; Batista et al., 2018; Zhu et al., 2019; Rao et al., 2020). It also has the characteristics of non-antigenic action and is employed as a drug carrier in the delivery systems for non-viral genes. The biodegradable and biocompatible nature of chitosan made an impact in the wound healing process, the contact lens coating, the material for bone grafting, bandaging and was employed as a hemorrhage controller.

3.1 Wound Healing

Wound healing is the process in which the cells of the immune system get accelerated and matrix deposition of the cells takes place and tissue remodel happens. Naturally, the immune system has an accelerator of its own, but it works slowly, and a severe wound requires an external stimulator to stimulate immune cells, which chitosan will provide. Furthermore, the fungal chitosan is employed as a membrane in a variety of wound healing applications (Chung et al., 1994; Sathiyaseelan et al., 2017; Anbazhagan and Thangavelu, 2018; Jones et al., 2020; Moeini et al., 2020; Rao et al., 2020, 2021). Chitosan stimulates the dynamic environment, allowing the injured tissue to restore its integrity (Sinno and Prakash, 2013). It is reported that freeze-drying techniques were used to regulate the release of antiseptics using a mixture of polyvinyl alcohol and chitosan materials. Chitosan and polyethylene glycol composites were employed as wound healing and pore generating agents in cotton textiles. Chitosan gels were also utilized to provide the growth components necessary for wound healing. According to Okamoto et al. (2003), the use of chitosan enhances the blood coagulation process for wound healing. Also, it has several characteristic features of attracting the platelets and activating the macrophages and neutrophils for the healing process involved in tissue regeneration (Okamoto et al., 2003).

3.2 Antimicrobial Activity

The chitosan is the positively charged materials that attract the negatively charged elements in the microbial cell wall and inhibits the effectiveness of toxic components. The activity is against the many varieties of polycationic bacteria and fungi which binds with the DNA of the microbes and inhibit the mRNA synthesis and proteins required for bacterial growth. Also in the wound-dressing applications, the chitosan with the antipyretic has long-term antibacterial activity. The research study shows the effectiveness of the molecular weight of chitosan's role in antibacterial activity

(Tajdini et al., 2010; Tayel et al., 2010; Prabu and Natarajan, 2012; Hosseinnejad and Jafari, 2016; Rao et al., 2021). The results suggested that the molecular weight of Mv = 9.16×10^4 has excellent antimicrobial activity and includes algae, fungi, certain gram-positive and gram-negative bacteria. Moreover, the chitosan was used as natural preservatives due to the evaluated results against the foodborne pathogens of *Salmonella typhimurium* and *Staphylococcus aureus* (Fei Liu et al., 2001; Moussa et al., 2013). Chitosan preservatives are used on a variety of foods in the form of film, coating, solution, and powder. The antibacterial action of chitosan is mostly due to its cationic character. The antimicrobial effect is triggered by the electrostatic contact between positively charged sites and negatively charged microbial membranes, which results in cell lysis and the breakdown of peptidoglycans in the microbial walls. This leads to the death of the microorganisms (Severino et al., 2014; Goy et al., 2016; Hafdani and Sadeghinia, 2011). Another study shows that chitosan is effective against both gram-positive and gram-negative bacteria. In the gram-negative bacteria of the *Escherichia coli*, the chitosan replaces the divalent cations of Ca^{2+} and Mg^{2+} from the binding sites and minimizes the interaction of lipopolysaccharides protein bilayer in the outer membrane of the bacterial cell and this leads to cell lysis (Fig. 2.2A) (Khan et al., 2015). Meanwhile in the gram-positive bacteria (*S. aureus*), the polycationic long chain molecules of the chitosan bind to the cell wall and inhibit the cellular function in an effective manner (Fig. 2.2B) (Aranda-Martinez et al., 2016; Chao et al., 2019).

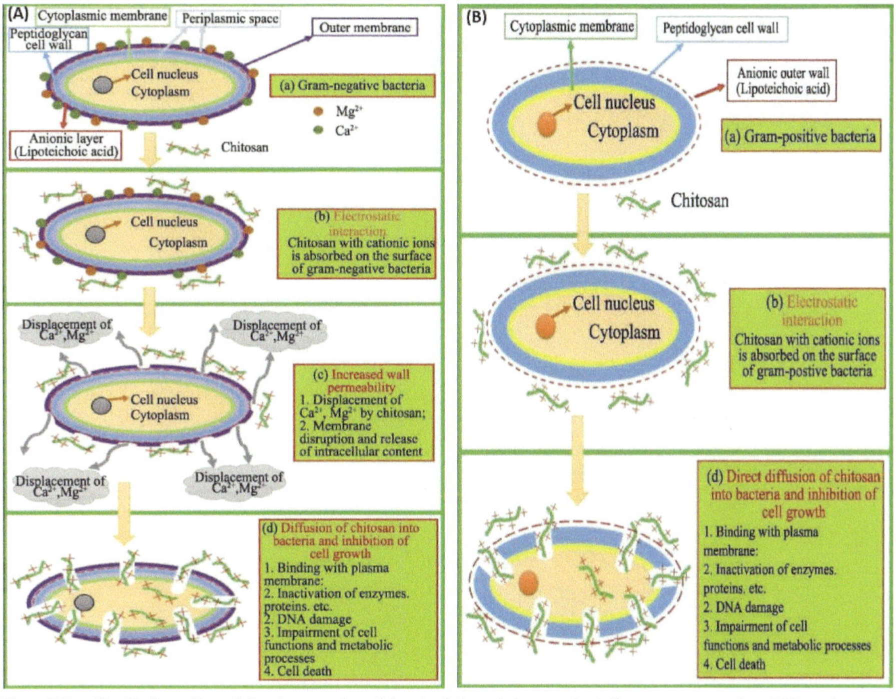

Figs. 2.2A,B. Cell wall architecture and anti-bacterial activity against the gram-negative (A) and gram-positive (B) bacteria.

Note: The figure is adopted from the reference research article (Chao et al., 2019).

3.3 Antitumor Activity and Drug Delivery

Fungal derived chitosan have been extensively used for anticancer applications (Balamurugan, 2012; Elsoud and Kady, 2019; Almutairi et al., 2020; Alalawy et al., 2020). Chitosan-based nanogels are commonly used in antitumor activity and drug delivery pharmaceutical products. The micelles of chitosan were frequently used as amphiphilic nanocarriers in tumor treatment. Chitosan anticancer activity was demonstrated by the release of interleukin-1 and 2, which leads to the development of cytolytic T cells (Tokoro et al., 1988). By injecting adenocarcinoma cells with MDA-MB-231 human cells, the efficacy of the chitosan-ZnO nanocarrier was tested subcutaneously in female-athymic BALB/c (nu+/nu+) mice aged 6–7 weeks. The ZnO nanocarrier with paclitaxel (called FCPZnO) was tested using natural paclitaxel and hollow chitosan-ZnO nanocarriers (HZnO) (HZnO). Through using an *in vivo* fluorescence imaging system with FCP ZnO IR Dye 680 was able to assess tumor particle absorption (Brown et al., 2016). Various chitosan nanoparticle encapsulated medications targeting the tumor sites and delivering specific drugs include cisplatin, paclitaxel, docetaxel, camptotecin, and doxorubicin for inhibiting the growth of tumor cells. In comparison to other nanoparticles, chitosan nanoparticles have wide blood circulation and tumor-specific delivery in cell and animal models (Jee et al., 2012). The efficacy of fungal chitosan nanoparticles as carriers for curcumin encapsulation and increasing its anticancer effect against different human cancer cells was investigated. The results showed that the curcumin-loaded fungal chitosan

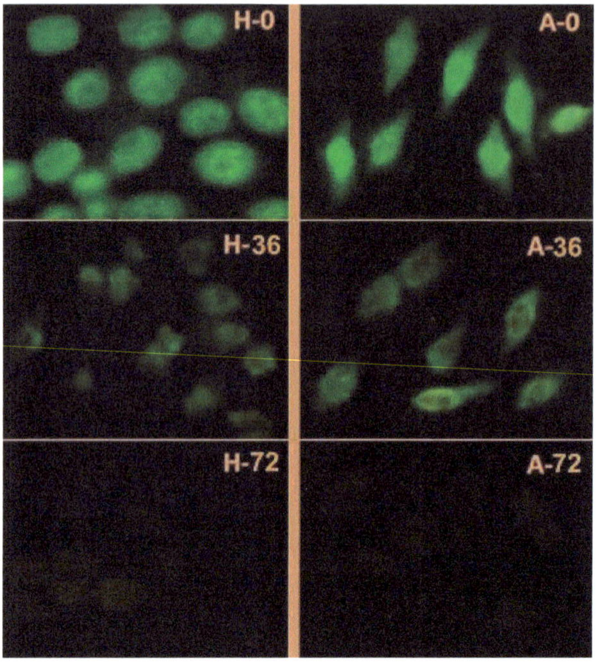

Fig. 2.3. Microscopic fluorescence imaging of cancer cell lines HCT-116 (H) and A-549 (A) as influenced by curcumin/fungal chitosan nanoparticles exposure for 0, 36, and 72 hours.

Note: The figure is adopted from the reference research article (Almutairi et al., 2020).

nanoparticle composites, which served as natural and biosafe agents, dramatically reduced the viability of A549 and HCT116 cancer cells, causing autophagy and death (Fig. 2.3) (Almutairi et al., 2020).

3.4 Gene Delivery

It has proven that chitosan was the ideal candidate for oral DNA delivery (Balamurugan, 2012; Elsoud and Kady, 2019). Chitosan from the fungal sources was used to encapsulate the plasmid DNA in a nanoparticle form resulting in improved biocompatibility and gene delivery (Plapied et al., 2010). Also, the chitosan can be formulated into various nanoparticles using multiple cross-linking methods (emulsion-droplet coalescence, ionic cross-linking, reverse micelle, covalent cross-linking, and precipitation method) for use in targeted drug delivery systems such as buccal drug delivery, vaccine delivery, mucosal drug delivery, vaginal drug delivery, and pulmonary drug delivery. Stability, reduced toxicity, and improved biocompatibility made the advantageous characteristic feature of chitosan selection for gene delivery systems (Garg et al., 2019).

4. Conclusion

Chitosan polysaccharides from the fungal source can be employed in numerous biological applications. Chitosan-based products offer a lot of economic promise in biomedical and tissue engineering applications. For this aim, several methods for altering or dealing with chitosan have been reported. The allergic sensitivity of crustaceans to chitosan has limited the usage of chitosan in commercial applications. However, further research is needed to decrease the impact of allergy diseases and to commercialize the benefits of fungal chitosan applications.

Acknowledgment

This research was supported by the Basic Science Research Program through the National Research Foundation (NRF) of Korea (2018R1A6A1A03024231 and 2021R1A2C1003566).

References

Acosta, N., Jiménez, C., Borau, V. and Heras, A. (1993). Extraction and characterization of chitin from crustaceans. *Biom. Bioener.*, 5(2): 145–153.

Alalawy, A.I., Rabey, H.A.E., Almutairi, F.M., Tayel, A., Duals, M.A.A. et al. (2020). Effectual anticancer potentiality of loaded bee venom onto fungal chitosan nanoparticles. *Int. J. Polym. Sci.*, 2020(2a): 1–9.

Almutairi, F.M., Rabey, H.A.E., Tayel, A.A., Alalawy, A.I., Al-Duais, M.A. et al. (2020). Augmented anticancer activity of curcumin loaded fungal chitosan nanoparticles. *Int. J. Biol. Macromol.*, 55: 861–867.

Anbazhagan, S. and Kalaichelvan, P.T. (2018). Application of tetracycline hydrochloride loaded-fungal chitosan and Aloe vera extract based composite sponges for wound dressing. *J. Adv. Res.*, 14: 63–71.

Aranaz, I., Mengíbar, M., Harris, R., Panos, I., Miralles, B. et al. (2009). Functional characterization of chitin and chitosan. *Curr. Chem. Biol.*, 3(2): 203–230.

Aranda-Martinez, A., Lopez-Moya, F. and Lopez-Llorca, L.V. (2016). Cell wall composition plays a key role on sensitivity of filamentous fungi to chitosan. *J. Basic Microbilol.*, 56(10): 1059–1070.

Balamurugan, M. (2012). Chitosan: A perfect polymer used in fabricating gene delivery and novel drug delivery systems. *Int. J. Pharm. Pharm. Sci.*, 4(3): 54–56.

Ban, Z., Horev, B., Rutenberg, R., Danay, O., Bilbao, C. et al. (2018). Efficient production of fungal chitosan utilizing an advanced freeze-thawing method; quality and activity studies. *Food Hydrocolloids*, 81: 380–388.

Batista, A.C.D.E., Neto, F.E.D.S. and Paiva, W.D.S. (2018). Review of fungal chitosan: Past, present and perspectives in Brazil. *Polímeros*, 28: 275–283.

Brown, D., Brunt, K. and Rehmann, N. (2016). Chitosan biopolymer from fungal fermentation for delivery of chemotherapeutic agents. *Mat. Matt.*, 11: 86–89.

Chung, Ll.Y., Schmidt, R.J., Hamlyn, P.F., Sagar, B.F., Andrews, A.M. and Turner, T.D. (1994). Biocompatibility of potential wound management products: Fungal mycelia as a source of chitin/chitosan and their effect on the proliferation of human F1000 fibroblasts in culture. *J. Biomed. Mat. Res.*, 28(4): 463–469.

Duan, C., Meng, X., Jingru, M., Khan, L., Dai, L. et al. 2019. Chitosan as a preservative for fruits and vegetables: A review on chemistry and antimicrobial properties. *J. Bioresour. Bioprod.*, 4(1): 11–21.

Dhillon, G., Dhillon, S.K., Brar, S.K. and Verma, M. (2013). Green synthesis approach: Extraction of chitosan from fungus mycelia. *Crit. Rev. Biotechnol.*, 33(4): 379–403.

El-banna Sami, F.S., Mahfouz, M.E., Leperatti, S., E.-Kemary, M. and Hanafy, N.A.N. (2019). Chitosan as a natural copolymer with unique properties for the development of hydrogels. *Appl. Sci.*, 9(11): 2193.

Elsoud, M.M.A. and Kady, E.M.E. (2019). Current trends in fungal biosynthesis of chitin and chitosan. *Bull. Nat. Res. Centre*, 43(1): 1–12.

Freitas, F., Roca, C. and Reis, M.A.M. (2015). Fungi as sources of polysaccharides for pharmaceutical and biomedical applications. pp. 61–104. *In*: Thakur, V. K. and Thakur, M.K. (Eds.). *Handbook of Polymers for Pharmaceutical Technologies*, Vol. 3.

Garg, U., Chauhan, S., Nagaich, U. and Jain, N. (2019). Current advances in chitosan nanoparticles based drug delivery and targeting. *Adv. Pharmaceut. Bull.*, 9(2)S: 195–204.

Ghormade, V., Pathan, E.K. and Deshpande, M.V. (2017). Can fungi compete with marine sources for chitosan production? *Int. J. Biol. Macromol.*, 104: 1415–1421.

Goy, R.C., Morais, S.T.B. and Assis, O.B.G. (2016). Evaluation of the antimicrobial activity of chitosan and its quaternized derivative on *E. coli* and *S. aureus* growth. *Rev. Bras. Farmacog.*, 26: 122–127.

Hafdani, F.N. and Sadeghinia, N. (2011). A review on application of chitosan as a natural antimicrobial. *World Acad. Sci. Eng. Technol.*, 50: 252–256.

Hosseinnejad, M. and Jafari, S.M. (2016). Evaluation of different factors affecting antimicrobial properties of chitosan. *Int. J. Biol. macromol.*, 85: 467–475.

Jee, J.-P., Na, J.H., Lee, S., Kim, S.H., Choi, K. et al. (2012). Cancer targeting strategies in nanomedicine: Design and application of chitosan nanoparticles. *Curr. Opin. Solid State Mat. Sci.*, 16(6): 333–342.

Jones, Mitchell, Marina Kujundzic, Sabu John and Alexander Bismarck. (2020). Crab vs. mushroom: A review of crustacean and fungal chitin in wound treatment. *Mar. Drugs*, 18(1): 64. 10.3390/md18010064.

Khan, A., Vu, K.D., Riedi, B. and Lacroix, M. (2015). Optimization of the antimicrobial activity of nisin, Na-EDTA and pH against gram-negative and gram-positive bacteria. *LWT-Food Sci. Technol.*, 61(1): 124–129.

Liu, X.F., Guan, Y.L., ang, D.Z., Li, Z. and Yao, K.D. (2001). Antibacterial action of chitosan and carboxymethylated chitosan. *J. Appl. Oplym. Sci.*, 79(7): 1324–1335.

Marikani, K., Nesakumari, M., Rajarathinam, K. and Singh, R. (2010). Production and characterization of mushroom chitosan under solid-state fermentation conditions. *Adv. Biol. Res.*, 4 (1): 10–13.

Merzendorfer, H. (2011). The cellular basis of chitin synthesis in fungi and insects: Common principles and differences. *Eur. J. Cell Biol.*, 90(9): 759–769.

Moeini, A., Pedram, P., Makvandi, P., Malinconico, M. and Gomez d'Ayala, G. (2020). Wound healing and antimicrobial effect of active secondary metabolites in chitosan-based wound dressings: A review. *Carbohydr. Polym.*, 233: 115839.

Moussa, S.H., Tayel, A.A., Al-Hassan, A.A. and Farouk, A. (2013). Tetrazolium/formazan test as an efficient method to determine fungal chitosan antimicrobial activity. *J. Mycol.*, 1–7.

Nwe, N. and Stevens, W.F. (2008). Production of chitin and chitosan and their applications in the medical and biological sector. pp. 161–176. *In*: Tamura, H. (Ed.). *Recent Research in Biomedical Aspects of Chitin and Chitosan*. Kerala, India: Research Signpost.

Nwe, N., Tetsuya, F. and Hiroshi, T. (2010). Production of fungal chitosan by enzymatic method and applications in plant tissue culture and tissue engineering: 11 years of our progress, present situation, and future prospects. *Biopolymers*, 7(135): 137–162.

Nwe, N. and Willem, F.S. (2004). Effect of urea on fungal chitosan production in solid substrate fermentation. *Proc. Biochem.*, 39(11): 1639–1642.

Okamoto, Y., Yano, R., Miyatake, K., Tomohiro, I., Shigemasa, Y. and Minami, S. (2003). Effects of chitin and chitosan on blood coagulation. *Carbohydr. Polym.*, 53(3): 337–342.

Ospina, N.M., Alvarez, S.P.O., Sierra, D.M.E., Vahos, D.F.R., Ocampo, P.A.Z. and Orozco, C.P.O. (2015). Isolation of chitosan from *Ganoderma lucidum* mushroom for biomedical applications. *J. Mat Sci. Mat. Med.*, 26(3): 135. 10.1007/s10856-015-5461-z.

Plapied, L., Vandermeulen, G., Vroman, B., Préat, V. and des Rieux, A. (2010). Bioadhesive nanoparticles of fungal chitosan for oral DNA delivery. *Int. J. Pharmaceut.*, 398(1-2): 210–218.

Pochanavanich, P. and Suntornsuk, W. (2002). Fungal chitosan production and its characterization. *Lett. Appl. Microbiol.*, 35(1): 17–21.

Prabu, K. and Natarajan, E. (2012). *In vitro* antimicrobial and antioxidant activity of chitosan isolated from *Podophthalmus vigil*. *J. Appl. Pharmaceut. Sci.*, 2(9): 75–82.

Rane, K.D. and Dallas, G.H. (1993). Production of chitosan by fungi. *Food Biotechnol.*, 7(1): 11–33.

Rao, K.M., Sudhaker, K., Suneetha, M., Won, S.Y. and Han, S.S. (2021). Fungal-derived carboxymethyl chitosan blended with polyvinyl alcohol as membranes for wound dressings. *Int. J. Biol. Macromol.*, 190: 972–800.

Rao, K.M., Suneetha, M., Park, G.T., Babu, A.G. and Han, S.S. (2020). Hemostatic, biocompatible, and antibacterial non-animal fungal mushroom-based carboxymethyl chitosan-ZnO nanocomposite for wound-healing applications. *Int. J. Biol. Macromol.*, 155: 71–80.

Sathiyaseelan, A., Shajahan, A., Kalaichelvan, P.T. and Kaviyarasan, V. (2017). Fungal chitosan based nanocomposites sponges: An alternative medicine for wound dressing. *Int. J. Biol. Macromol.*, 104: 1905–1915.

Severino, R., Vu, K.D., Densi, F., Salmieri, S., Ferrari, G. and Lacroix, M. (2014). Antibacterial and physical effects of modified chitosan based-coating containing nanoemulsion of mandarin essential oil and three non-thermal treatments against *Listeria innocua* in green beans. *Int. J. Food Microbiol.*, 191: 82–88.

Sinno, H. and Prakash, S. (2013). Complements and the wound healing cascade: An updated review. *Plastic Surg. Int.*, 2013: 146764. 10.1155/2013/146764.

Streit, F., Koch, F., Laranjeira, M.C.M. and Ninow, J. (2009). Production of fungal chitosan in liquid cultivation using apple pomace as substrate. *Br. J. Microbiol.*, 40: 20–25.

Suntornsuk, W., Pochanavanich, P. and Suntornsuk, L. (2002). Fungal chitosan production on food processing by-products. *Proc. Biochem.*, 37(7): 727–729.

Tajdini, F., Amini, M.A., Nafissi-Varcheh, N. and Faramarzi, M.A. (2010). Production, physiochemical and antimicrobial properties of fungal chitosan from Rhizomucor miehei and *Mucor racemosus*. *Int. J. Biol. Macromol.*, 47(2): 180–183.

Tasar, O.C., Erdal, S. and Taskin, M. (2016). Chitosan production by psychrotolerant *Rhizopus oryzae* in non-sterile open fermentation conditions. *Int. J. Biol. Macromol.*, 89: 428–433.

Tayel, A.A., Moussa, S., El-Tras, W.F., Knittel, D., Opwis, K. and Schollmeyer, E. (2010). Anticandidal action of fungal chitosan against *Candida albicans*. *Int. J. Biol. Macromol.*, 47(4): 454–457.

Tokoro, A., Tatewaki, N., Suzuki, K., Mikami, T., Suzuki, S. and Suzuki, M. (1988). Growth-inhibitory effect of hexa-N-acetylchitohexanse and chitohexaose against Meth-A solid tumor. *Chem. Pharmaceut. Bull.*, 36(2): 784–790.

Wang, W.P., Du, Y.-M. and Wang, X.-Y. (2008). Physical properties of fungal chitosan. *World J. Microbiol. Biotechnol.*, 24(11): 2717–2720.

Zhu, L.F., Li, J.S., Mai, J. and Chang, M.W. (2018). Ultrasound-assisted synthesis of chitosan from fungal precursors for biomedical applications. *Chem. Eng. J.*, 357: 498–507.

3

Sources and Industrial Applications of Fungal Pigments

Fernanda Cortez Lopes

1. Introduction

Color plays an important role in the acceptability of products in several industries, since consumers generally judge the quality of a product based on its color (Wrolstad and Culver, 2012). The use of synthetic colorants is widespread in all industries due to their low cost and better stability (Ranaweera et al., 2020). However, synthetic colorants are presenting many side effects, some are immunosuppressive, carcinogenic, teratogenic, and cause allergies (Babitha, 2009; Mukherjee et al., 2017; Lopes and Ligabue-Braun, 2021). The concern about these harmful effects is not only to the human but also to the environment; since these synthetic colorants are not biodegradable, it has raised growing concern in natural coloring alternatives (Narsing Rao et al., 2017; Kalra et al., 2020; Sánchez-Muñoz et al., 2020).

Natural colorants are primarily derived from the plants, insects, mineral ores, or microorganisms. The production of pigments by microorganisms was described in bacteria, filamentous fungi, yeasts, and algae. Microbial pigments have many advantages compared to animal- and plant-derived pigments, including rapid growth, easy processing, independence of weather conditions and they do not compete for limited farming land (Narsing Rao et al., 2017; Pombeiro-Sponchiado et al., 2017; Sen et al., 2019; Lopes and Ligabue-Braun, 2021). Another advantage of using microorganisms to produce pigments is the possibility of using inexpensive substrates, such as agro-industrial residues, as these are generated in high amounts by many industries as byproducts and pose disposal problems (Lopes and Ligabue-Braun, 2021).

Graduate Program in Cell and Molecular Biology, Center for Biotechnology, Universidade Federal do Rio Grande do Sul, Porto Alegre, Brazil.
Email: fernandacortezlopes@gmail.com

Filamentous fungi are promising candidates, because of their ability to produce primary and secondary metabolites such as peptides, enzymes, organic acids, heterologous proteins, antibiotics, and pigments (Radzio and Kück, 1997; Hajjaj et al., 1998). Many fungal pigments are now considered safe and economical (Mukherjee et al., 2017) and are sometimes produced due to insufficiency of nutrients. When the nutritional supply of vital nutrients decreases or there is some adverse environmental condition, fungus produces secondary metabolites, including pigments. It is considered as an adaptive advantage to the fungus, mainly because some pigments have other biological activities. Yeasts are also promising as they can produce pigments, mainly the carotenoids. They are also important hosts to the production of secondary metabolites through heterologous expression of biosynthetic pathways (Siddiqui et al., 2012; Chreptowicz et al., 2019). Lastly, even the higher fungi like mushrooms have also been reported to produce various pigments although this has not been explored industrially. Currently, mushroom pigments are used only as a taxonomical tool because industrial production of these fungi for commercial purpose is not feasible (Mukherjee et al., 2017; Kalra et al., 2020). This chapter deals with sources of fungal pigments, their biological functions, and industrial applications.

2. Sources of Fungal Pigments

An ideal candidate to produce pigments to meet the industrial needs should possess some specific characteristics: (i) organisms should be cultivable; (ii) a fast growth rate; (iii) allow towards optimization; (iv) high productivity; (v) availability and ability to produce throughout the year; (vi) non-toxigenic; (vii) nonpathogenic; (viii) ability to grow in a wide range of agro-wastes; (ix) tolerant to broad physical and chemical parameters during fermentation (Ramesh et al., 2019). Fungi, mostly ascomycetous and basidiomycetous are known to produce an enormous range of colors, including several chemical classes of pigments such as melanins, azaphilones, flavins, carotenoids, naphtoquinones, and anthraquinones. The chemical structures of the main classes of these pigments are given in Fig. 3.1. In addition, lichens as symbiont of fungus with green alga and or cyanobacterium are also capable of producing pigments (Kalra et al., 2020).

Filamentous fungi that belong to the genera *Aspergillus*, *Fusarium*, *Monascus*, *Penicillium*, and *Trichoderma* produce quinones, anthraquinones, rubropunctamines, rubropunctatin, ankaflavin, monascin, melanins, and many other pigments possessing red, purple, yellow, brown, orange, blue, and green colors. Most of these pigments are water soluble, which allows an easy extraction without using organic solvents. These are produced as secondary metabolites through one of the pathways such as polyketide, mevalonate, and shikimate. Filamentous fungi produce more remarkable and stable pigments of industrial interest than any other natural pigments (Mukherjee et al., 2017; Narsing Rao et al., 2017; Venil et al., 2020).

Regarding yeasts, mainly carotenoids are produced by these organisms such as γ-and β-carotene, lycopene, torulene, torularhodin (*Rhodotorula* spp. and *Sporobolomyces roseus*), and astaxanthin (*Phaffia rhodozyma*). Carotenoids possessing yellow, red, and orange colors are widely used as food and feed

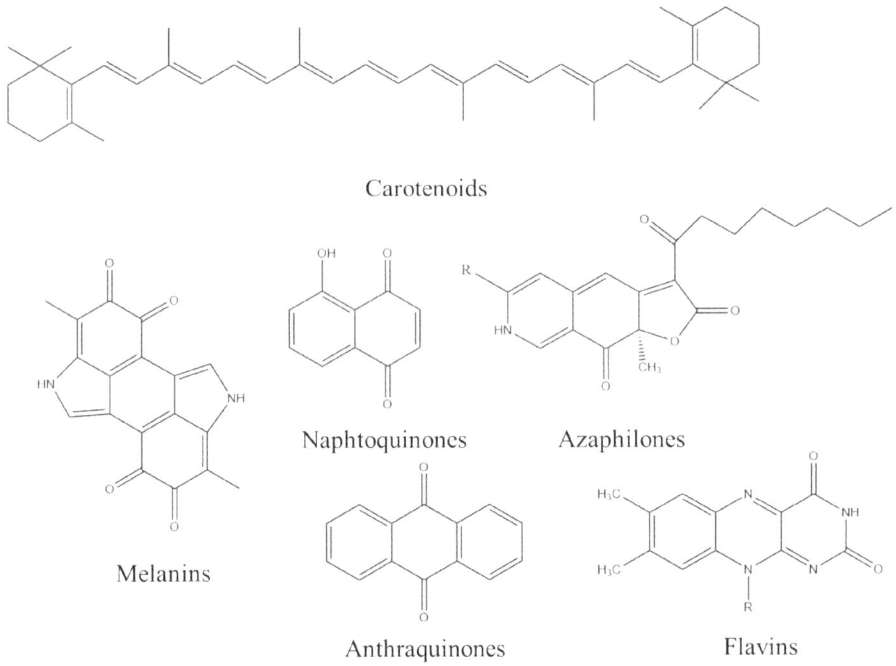

Fig. 3.1. Chemical structures of main classes of fungal pigments.

supplements, as well as antioxidants in pharmaceutical industries (Mukherjee et al., 2017). Production of carotenoids is also possible by the fungal genera such as *Aschersonia*, *Aspergillus*, *Blakeslea*, *Cercospora*, *Mucor*, *Penicillium*, *Phycomyces*, *Sclerotinia*, *Sclerotium*, and *Ustilago* (Avalos and Limón, 2015).

A complete and updated list of pigment-producing fungi and their respective pigments is given by Lagashetti et al. (2019). A few pigment-producing fungi have been documented in the Table 3.1. Some fungi are capable of creating a vast array of pigments such as *Monascus* spp. which is known to produce two yellow (monascin and ankaflavin), two orange (rubropunctatin and monascorubrin), and two red (rubropunctamine and monascorubramine) colorants. These pigments have been used for centuries in traditional oriental foods (red mold rice) in Southern China, Japan, and Southeast Asia (Dufossé et al., 2005; Feng et al., 2012). The bottleneck of *Monascus* spp. industrial widespread use is the co-production of one mycotoxin called citrinin by some strains (Dufosse et al., 2014). Many processes have been used to decrease the production of citrinin by altering some of the fermentation variables: nitrogen sources, pH, dissolved oxygen, and some important genetic alterations. *M. purpureus* SM001, an industrial strain, does not produce citrinin due to disruption of the polyketide synthase gene pksCT through *Agrobacterium tumefaciens*-mediated transformation (Jia et al., 2010). *Monascus* spp. also produce other interesting metabolites such as monacolins (cholesterol-lowering agents), γ-amino butyric acid (an antihypertensive substance), and dimerumic acid (an antioxidant) (Chen et al., 2015).

Table 3.1. Some representative producing-pigments fungi (data from: Dufossé, 2016; Sen et al., 2019; Lopes and Ligabue-Braun, 2021).

Fungi	Types of Pigments	Colors
Monascus spp.	Ankaflavine, monascin, monascorubin, rubropunctatin, monascorubramine and rubropuntamine	Yellow, orange, and red
Penicillium purpurogenum and *Talaromyces atroroseus*	Azaphilones	Red
Penicillium oxalicum var. *armeniaca* CCM 8242	Anthraquinones (ArPink Red™)	Red and other hues
Fusarium sp. and *Cordyceps unilateralis*	Nafthoquinones	Red, yellow, orange-brown, and brown
Blakeslea trispora, Fusarium sporotrichioides, Mucor circinelloides, Neurospora crassa, Phycomyces blakesleeanus and *Rhodotorula rubra*	Carotenoids (lycopene and β-carotene)	Yellow, orange, and red
Cryptococcus sp., *Saccharomyces neoformans* var. *nigricans* and *Yarrowia lipolytica*	Melanines	Dark-brown and black
Ashbya gossypi	Riboflavin	Yellow

Some species of *Talaromyces*/*Penicillium* secrete a large amount of azaphilones red pigments. Many studies have identified these pigments as *Monascus*-like pigments, namely monascorubrin, ankaflavin, monascin, monascorubramine, monascorubin, and other *Monascus*-like pigments derivatives (Mapari et al., 2008; Frisvad et al., 2013; Shah et al., 2014; Woo et al., 2014; Afshari et al., 2015). Lopes and coworkers shown in a phylogenetic tree based on ITS (intergenic transcribed spacer) region sequencing that *Penicillium* and *Monascus* genera show high similarity, having a common ancestor. This can explain the production of the *Monascus* pigments also by some strains of *Penicillium* (Lopes et al., 2013). In contrast, *Penicillium oxalicum* var. *armeniaca* CCM 8242 is used at an industrial scale for food colorant production Arpink red™ (Natural Red™) (Sardaryan, 2002). It was the first commercialized fungal red pigment; however, the company is not functioning nowadays. The red colorant is an extracellular metabolite of the anthraquinone class. Mapari et al. (2005) hypothesized that fungus identification is probably incorrect, based on the information found in the patent about fungal morphology. Anthraquinones are also produced by other filamentous fungi such as *Aspergillus*, *Eurotium* sp., *Fusarium* sp., and *Trichoderma* (Venil et al., 2020).

With similar structure, naphtoquinones are mainly produced by *Fusarium* sp., presenting yellow, orange, and brown colors (Babula et al., 2009). *Cordyceps unilateralis* BCC 1869 is capable of producing six extracellular red naphthoquinones (Unagul et al., 2005). The naphthoquinones are widespread in fungi with important biological functions such as phytotoxic, antimicrobials, insecticidal, anti-carcinogenic, and cytostatic activities (Venil et al., 2020).

Melanin production has been reported by a wide variety of microorganisms such as *Aspergillus fumigatus*, *Colletotrichum lagenarium*, *Cryptococcus neoformans*, *Magnaporthe grisea*, *Paracoccidioides brasiliensis*, and *Sporothrix schenckii*. However, as some of these fungi are human pathogens, they should be avoided for industrial melanin production (Langfelder et al., 2003).

Riboflavin (vitamin B2) is a water-soluble pigment that has a strong yellowish-green fluorescence. This pigment has several applications as a dietary supplement and food additive in dairy products, sauces, baby foods, fruit, and energy juices (Rana et al., 2021). *Ashbya gossypii* was the first organism used in industrial riboflavin production. It was isolated as a plant pathogen and characterized as a natural riboflavin producer (Kato and Park, 2012).

Using biotechnological processes, the fungal pigments can be produced industrially on a large scale. Pigments are products that are generated by fermentation; therefore they are affected by temperature, pH, carbon source, and type of fermentation used (solid or submerged) (Mukherjee et al., 2017). Use of solid substrate fermentation in many cases could be advantageous to produce pigments as this type of fermentation mimics better the physiological state of the filamentous fungi. However, use of solid substrate fermentation is still difficult considering industrial production, thus submerged fermentations are preferred for producing pigments (Lopes and Ligabue-Braun, 2021).

There are several challenges in the production and recovery of microbial pigments such as low yield, high production cost, and difficulty in downstream steps. Selection of appropriate microorganism and growth conditions, carbon and nitrogen source are the key parameters for the scale-up production. Manipulation of culture conditions and co-culturing with other organisms can also be helpful in improving the expression and increasing the yield of a particular pigment (Kalra et al., 2020; Sánchez-Muñoz et al., 2020).

3. Industrial Applications

There is a worldwide demand for colorants of natural origin in the food, cosmetic and textile industries. Natural pigments can be used as dyes for different substrates such as leather, paper, paints, and coatings, in cosmetics, and as food additives (Kalra et al., 2020). The success of pigments produced by fermentation depends on several factors such as acceptability by the consumers, regulatory approval, and the investments necessary to uplift the product to the market (Dufosse et al., 2014). Sustainability and advance of the fungal-based pigment industry depend on three important and critical factors: (i) absence of mycotoxin co-production that guarantees the safety of the product; (ii) satisfactory pigment yield; (iii) pigment stability and purity, depending on the pigment application (Kalra et al., 2020).

Some of the fungal pigments are stable against light, heat, and pH (Joshi et al., 2003). These properties of fungal pigments make them attractive for industrial application. Some microbial pigments are produced industrially, such as β-carotene (*Blakeslea trispora* and *Dunaliella salina*) and lycopene (*B. trispora*), *Monascus*-derived pigments, Natural Red™ (*Penicillium oxalicum*), *Monascus* pigments (*Monascus* spp.), riboflavin (*Ashbya gossypii*), phycocianin (*Spirulina platensis*), and astaxanthin (*Paracoccus carotinifasciens* and *Haematococcus pluvialis*) (Dufosse et al., 2014). It is important to highlight that the filamentous fungi *A. gossypii*, *B. trispora*, *Monascus* spp., and *P. oxalicum* have great importance and potential in the production of pigments. Several patents for the use of *Monascus*-like pigments in food and cosmetics were deposited by the companies such as Nestlé, Unilever, The Quaker Oat, and L'Oreal (Caro et al., 2017).

Table 3.2 describes some industrial applications of fungal pigments. Most of the studies published on fungal colorings are related to their use as food colorants. Many fungal pigments are usually used as additives, antioxidants, and color intensifiers to enhance the organoleptic properties of food. Some of the fungal dyes have already

Table 3.2. Industrial applications of fungal pigments.

Pigment(s)	Producer	Industrial Applications	References
Arpink red, riboflavin, β-carotene and *Monascus* pigments	*Ashbya gossypii, Blakeslea trispora, Monascus* spp. and *Penicillium oxalicum*	Food	Caro et al., 2017; Lagashetti et al., 2019
Melanin, patulin and pink/red/blue/ yellow/green/black pigments*	*Arthrographis cuboidea, Ceratocystis* sp., *Chlorociboria aeruginascens, Lasiodiplodia theobromae, Ophiostoma* sp., *Penicillium griseofulvum, Phialocephala* sp., *Scytalidium cuboideum, Scytalidium ganodermophthorum, Trametes versicolor, Xylaria polymorpha*	Wood dyeing	Liu et al., 2020
Monascus pigments and *Monascus*-like pigments, melanin, and carotenoids	*Alternaria alternata, Aspergillus nidulans, Blakeslea trispora, Cordyceps* spp., *Emericella* spp., *Fusarium* spp., *Monascus* sp., *Penicillium* spp., and *Phaffia rhodozyma*	Cosmetics and pharmaceuticals	Caro et al., 2017; Lagashetti et al., 2019; Meruvu and Dos Santos, 2021
Anthraquinones, asperyellone, azaphilones, carotenoids, 2, 4-di-tert-butylphenol, magenta pigment*, polyketides and quinones	*Acrostalagmus, Alternaria alternate, Alternaria* sp., *Aspergillus niger, Aspergillus* sp., *Bisporomyces* sp., *Chlorociboria aeruginosa, Cunninghamella, Curvularia lunata, Emericella nidulans, Fusarium verticillioides, Isaria farinosa, Monascus purpureus, Penicillium chrysogenum, Penicillium italicum, Penicillium murcianum, Penicillium purpurogenum, Penicillium* oxalicum, *Penicillium regulosum, Phoma herbarum, Phymatotrichum* sp., *Talaromyces australis, Talaromyces verruculosus, Thermomyces* sp., *Trichoderma* sp., *Trichoderma virens, Sclerotinia* sp., *Scytalidium cuboideum,* and *Scytalidium ganodermophthorum*	Textile dyeing	Venil et al., 2020

Note: *Some articles did not identify the chemical structure of the pigments.

entered the market as food colorants such as *Monascus* pigments, Arpink red, riboflavin and β-carotene. Carotenoids, such as lycopene and β-carotene can also act as a sunscreen to preserve the quality of food even in intense light (Caro et al., 2017; Lagashetti et al., 2019; Meruvu and Dos Santos, 2021). According to Caro et al. (2017), some fungal colors are in the development stage to be used as food colorant such as red azaphilones from *Penicillium purpurogenum* and *Talaromyces atroroseus* and yellow-orange carotenoid from *Mucor circinelloides*.

The biodegradable and sustainable production of natural pigments from fungal sources can be considered advantageous to textile dyeing, because they have colorfastness and interesting staining properties (Venil et al., 2020). The applications of *Monascus*-like pigments from *Penicillium* sp. on textile, cotton, wool, leather, paper, paint, and others have been patented by Mapari et al. (2009), representing a promising alternative against the synthetic dyes. Fuck et al. (2018) used *M. purpureus* extract to leather dyeing. These authors observed that the dyed leather showed good penetration, color homogeneity and heat fastness. It was proposed that this extract could be used in leather accessories, clothing, and footwear (Fuck et al., 2018). This study agrees with other works that include *M. purpureus* pigments with applications in the textile and tannery industries. Other strains those can be used with these aims are: *Emericella nidulans*, *Fusarium verticillioides*, *Isaria farinosa*, *M. ruber*, *Penicillium marneffei*, and *Thermomyces* (Ogbonna, 2016; Lagashetti et al., 2019).

In the wood industry, fungal pigments are highly useful. They improve the surface quality of wood, promote decorative effect, and add value to the final product and decrease the costs of processing (Liu et al., 2020). Fungal colorants, especially melanin, carotenoids and lycopene have been reported for their application in cosmetics, sunscreens, sun lotions, sunblock lotions, face creams, anti-aging facials and lipsticks (Narsing Rao et al., 2017; Sajid and Akbar, 2018).

4. Biological Activities

Apparently, microbial pigments are not simply colors, but they are a mixture of diverse chemical components with multifaceted potential biological activities (Kim, 2013). Some pigments have many therapeutic applications like immune modulators, anticancer, antioxidant and antimicrobial, becoming interesting to the food industry, cosmetics, and pharmaceuticals. They are attractive because their biological properties increase the possibility of applications (Mukherjee et al., 2017; Narsing Rao et al., 2017; Lagashetti et al., 2019; Sánchez-Muñoz et al., 2020).

Fungal pigments protect against biotic and abiotic agents (antagonistic microbes and UV radiation) (Eisenman and Casadevall, 2012). For example, melanin has been reported as a "fungal armor", because it can protect fungi from adverse conditions. This pigment can neutralize oxidants generated by stress. In addition, it has other activities such as thermoregulatory, radio- and photo-protective, antimicrobial, antiviral, cytotoxic, anti-inflammatory, and immunomodulatory (Pombeiro-Sponchiado et al., 2017). Melanin has physicochemical properties and biological activities that support its suitability for a wide range of applications in cosmetic, pharmaceutical, electronic, and food-processing industries (Pombeiro-Sponchiado et al., 2017).

Several studies have shown that the pigments or pigment extracts of certain species of fungal genera (*Aspergillus*, *Fusarium*, *Monascus*, *Penicillium*, *Talaromyces*, and *Trichoderma*) and yeast *Rhodotorula glutinis* possess antimicrobial activity against different pathogenic bacteria including yeasts and filamentous fungi. It has been reported that fungal pigments especially carotenoids, and naphthoquinones have antioxidant potential (Lagashetti et al., 2019). All these biological activities need to be better explored towards future applications, mainly in foods, cosmetics, and pharmaceuticals, because these industries need pigments with high quality, purity, and safety.

5. Conclusions and Prospects

Fungi can serve as cell factories for production of many metabolites, including pigments in an economical and human friendly way (Mukherjee et al., 2017). While considering search of new pigment-producing fungi, an interesting approach could be enforced to explore new environments. Marine ecological niches are still mostly unexplored and some characteristics like salinity, low temperature, and darkness of this environment induce microbes to produce novel metabolites (Venil et al., 2020). Chemical diversity of fungi should be explored for identifying new promising pigments and toxicological testing must be carried out to ensure their safety to consumers, avoiding the presence of mycotoxins in the final products (Venil et al., 2020). Many biotechnological tools are now available such as new generation sequencing, to sequence complete genomes of the new fungal isolates, 'omics' technologies, that help the understanding of biosynthetic routes and genome-editing methodologies, such as CRISPR-Cas9, that can be used to delete genes responsible for mycotoxin production (Paillié-Jiménez et al., 2020). Therefore, the use of all such methodologies can provide new and safe fungal pigments to be used industrially.

References

Afshari, M., Shahidi, F., Mortazavi, S.A., Tabatabai, F. and Es' haghi, Z. (2015). Investigating the influence of pH, temperature, and agitation speed on yellow pigment production by *Penicillium aculeatum* ATCC 10409. *Nat. Prod. Res.*, 29: 1300–1306.

Avalos, J. and Limón, M.C. (2015). Biological roles of fungal carotenoids. *Curr. Genet.*, 61: 309–324.

Babitha, S. (2009). Microbial pigments. pp. 147–162. *In*: Singh, nee' Nigam P. and Pandey, A. (Eds.). *Biotechnology for Agro-Industrial Residues Utilisation*, Springer.

Babula, P., Adam, V., Havel, L. and Kizek, R. (2009). Noteworthy secondary metabolites naphthoquinones: Their occurrence, pharmacological properties, and analysis. *Curr. Pharm. Anal.*, 5: 47–68.

Caro, Y., Venkatachalam, M., Lebeau, J., Fouillaud, M. and Dufossé, L. (2017). Pigments and colorants from filamentous fungi. *Fungal Metab.*, 499–568.

Chen, W., He, Y., Zhou, Y., Shao, Y., Feng, Y., Li, M. et al. (2015). Edible filamentous fungi from the species *Monascus*: Early traditional fermentations, modern molecular biology, and future genomics. *Compr. Rev. Food Sci. Food Saf.*, 14: 555–567.

Chreptowicz, K., Mierzejewska, J., Tkáčová, J., Młynek, M. and Čertik, M. (2019). Carotenoid-producing yeasts: Identification and characteristics of environmental isolates with a valuable extracellular enzymatic activity. *Microorganisms*, 7(653): 1–18.

Dufossé, L., Galaup, P., Yaron, A., Arad, S.M., Blanc, P. et al. (2005). Microorganisms and microalgae as sources of pigments for food use: A scientific oddity or an industrial reality? Trends *Food Sci. Technol.*, 16: 389–406. doi: 10.1016/j.tifs.2005.02.006.

Dufossé, L., Fouillaud, M., Caro, Y., Mapari, S.A.S. and Sutthiwong, N. (2014). Filamentous fungi are large-scale producers of pigments and colorants for the food industry. *Curr. Opin. Biotechnol.*, 26: 56–61.

Dufossé, L. (2016). Current and potential natural pigments from microorganisms (bacteria, yeasts, fungi, microalgae). pp. 337–354. *In*: Carle, R. and Schweiggert, R. (Eds.). *Handbook on Natural Pigments in Food and Beverages*, Elsevier.

Eisenman, H.C. and Casadevall, A. (2012). Synthesis and assembly of fungal melanin. *Appl. Microbiol. Biotechnol.*, 93: 931–940.

Feng, Y., Shao, Y. and Chen, F. (2012). *Monascus* pigments. *Appl. Microbiol. Biotechnol.*, 96: 1421–1440.

Frisvad, J.C., Yilmaz, N., Thrane, U., Rasmussen, K.B., Houbraken, J. and Samson, R.A. (2013). *Talaromyces atroroseus*, a new species efficiently producing industrially relevant red pigments. *PLOS One*, 8: e84102.

Fuck, W.F., Lopes, F.C., Brandelli, A. and Gutterres, M. (2018). Screening of natural dyes from filamentous fungi and leather dyeing with *Monascus purpureus* Extract. *J. Soc. Leather Technol. Chem.*, 102: 69–74.

Hajjaj, H., Blanc, P.J., Goma, G. and François, J. (1998). Sampling techniques and comparative extraction procedures for quantitative determination of intra- and extracellular metabolites in filamentous fungi. *FEMS Microbiol. Lett.*, 164: 195–200. doi: 10.1016/S0378-1097(98)00191-8.

Jia, X.Q., Xu, Z.N., Zhou, L.P. and Sung, C.K. (2010). Elimination of the mycotoxin citrinin production in the industrial important strain *Monascus purpureus* SM001. *Metab. Eng.*, 12: 1–7. doi: 10.1016/j.ymben.2009.08.003.

Joshi, V.K., Attri, D., Bala, A. and Bhushan, S. (2003). Microbial pigments. *Indian J. Biotechnol.*, 2: 362–369.

Kalra, R., Conlan, X.A. and Goel, M. (2020). Fungi as a potential source of pigments: Harnessing filamentous fungi. *Front. Chem.*, 8: 1–23.

Kato, T. and Park, E.Y. (2012). Riboflavin production by *Ashbya gossypii*. *Biotechnol. Lett.*, 34: 611–618.

Kim, S.-K. (2013). *Marine Biomaterials: Characterization, Isolation and Applications*. CRC press.

Lagashetti, A.C., Dufossé, L., Singh, S.K. and Singh, P.N. (2019). Fungal pigments and their prospects in different industries. *Microorganisms*, 7(604): 1–36.

Langfelder, K., Streibel, M., Jahn, B., Haase, G. and Brakhage, A.A. (2003). Biosynthesis of fungal melanins and their importance for human pathogenic fungi. *Fungal Genet. Biol.*, 38: 143–158.

Liu, Y., Yu, Z., Zhang, Y. and Wang, H. (2020). Microbial dyeing for inoculation and pigment used in wood processing: opportunities and challenges. *Dye. Pigment.*, 109021.

Lopes, F.C., Tichota, D.M., Pereira, J.Q., Segalin, J., De Oliveira Rios, A. and Brandelli, A. (2013). Pigment production by filamentous fungi on agro-industrial byproducts: An eco-friendly alternative. *Appl. Biochem. Biotechnol.*, 171: 616–625.

Lopes, F.C. and Ligabue-Braun, R. (2021). Agro-Industrial residues: Eco-friendly and inexpensive substrates for microbial pigments production. *Front. Sustain. Food Syst.*, 5: 65.

Mapari, S.A.S., Nielsen, K.F., Larsen, T.O., Frisvad, J.C., Meyer, A.S. and Thrane, U. (2005). Exploring fungal biodiversity for the production of water-soluble pigments as potential natural food colorants. *Curr. Opin. Biotechnol.*, 16: 231–238. doi: 10.1016/j.copbio.2005.03.004.

Mapari, S.A.S., Hansen, M.E., Meyer, A.S. and Thrane, U. (2008). Computerized screening for novel producers of Monascus-like food pigments in *Penicillium* species. *J. Agric. Food Chem.*, 56: 9981–9989. doi: 10.1021/jf801817q.

Mapari, S.S., Thrane, U., Meyer, A.S. and Frisvad, J.C. (2009). Production of *Monascus*-like azaphilone pigment. U.S. Patent Application n. 12/674,752, 13 out. 2011.

Meruvu, H. and Dos Santos, J.C. (2021). Colors of life: A review on fungal pigments. *Crit. Rev. Biotechnol.*, 1–25.

Mukherjee, G., Mishra, T. and Deshmukh, S.K. (2017). Fungal pigments: An overview. *Dev. Fungal Biol. Appl. Mycol.*, 525–541.

Narsing Rao, M.P., Xiao, M. and Li, W.-J. (2017). Fungal and bacterial pigments: secondary metabolites with wide applications. *Front. Microbiol.*, 8: 1113.

Ogbonna, C.N. (2016). Production of food colourants by filamentous fungi. *African J. Microbiol. Res.*, 10: 960–971.

Pombeiro-Sponchiado, S.R., Sousa, G.S., Andrade, J.C.R., Lisboa, H.F. and Gonçalves, R.C.R. (2017). Production of melanin pigment by fungi and its biotechnological applications. *Melanin*, pp. 47–75.

Radzio, R. and Kück, U. (1997). Synthesis of biotechnologically relevant heterologous proteins in filamentous fungi. *Process Biochem.*, 32: 529–539.

Ramesh, C., Vinithkumar, N.V., Kirubagaran, R., Venil, C.K. and Dufossé, L. (2019). Multifaceted applications of microbial pigments: Current knowledge, challenges, and future directions for public health implications. *Microorganisms*, 7(186): 1–46.

Rana, B., Bhattacharyya, M., Patni, B., Arya, M. and Joshi, G.K. (2021). The realm of microbial pigments in the food color market. *Front. Sustain. Food Syst.*, 5: 54.

Ranaweera, S.J., Ampemohotti, A. and Arachchige, U.S.P.R. (2020). Advantages and considerations for the applications of natural food pigments in the food industry. *Int. J. Eng. Res. Technol.*, 1: 8–15.

Sajid, S. and Akbar, N. (2018). Applications of fungal pigments in biotechnology. *Pure Appl. Biol.*, 7(3): 922–930.

Sánchez-Muñoz, S., Mariano-Silva, G., Leite, M.O., Mura, F.B., Verma, M.L., da Silva, S.S. et al. (2020). Production of fungal and bacterial pigments and their applications. pp. 327–361. *In*: Verma, M. and Chandel, A. (Eds.). *Biotechnological Production of Bioactive Compounds* (Elsevier).

Sardaryan, E. (2002). Strain of the microorganism *Penicillium oxalicum* var. *Armen. its Appl. US Pat.*, 6: 586.

Sen, T., Barrow, C.J. and Deshmukh, S.K. (2019). Microbial pigments in the food industry: Challenges and the way forward. *Front. Nutr.*, 6: 7.

Shah, S.G., Shier, W.T., Tahir, N., Hameed, A., Ahmad, S. et al. (2014). *Penicillium verruculosum* SG: A source of polyketide and bioactive compounds with varying cytotoxic activities against normal and cancer lines. *Arch. Microbiol.*, 196: 267–278.

Siddiqui, M.S., Thodey, K., Trenchard, I. and Smolke, C.D. (2012). Advancing secondary metabolite biosynthesis in yeast with synthetic biology tools. *FEMS Yeast Res.*, 12: 144–170.

Unagul, P., Wongsa, P., Kittakoop, P., Intamas, S., Srikitikulchai, P. and Tanticharoen, M. (2005). Production of red pigments by the insect pathogenic fungus *Cordyceps unilateralis* BCC 1869. *J. Ind. Microbiol. Biotechnol.*, 32: 135–140.

Venil, C.K., Velmurugan, P., Dufossé, L., Devi, P.R. and Ravi, A.V. (2020). Fungal Pigments: Potential coloring compounds for wide ranging applications in textile dyeing. *J. Fungi*, 6: 68.

Woo, P.C.Y., Lam, C.-W., Tam, E.W.T., Lee, K.-C., Yung, K.K.Y. et al. (2014). The biosynthetic pathway for a thousand-year-old natural food colorant and citrinin in *Penicillium marneffei*. *Sci. Rep.*, 4: 1–8.

Wrolstad, R.E. and Culver, C.A. (2012). Alternatives to those artificial FD&C food colorants. *Annu. Rev. Food Sci. Technol.*, 3: 59–77.

4

Medicinal Mushrooms as a Source of Therapeutic Biopolymers

Bożena Muszyńska,[1,*] *Agata Krakowska*[2] and
Katarzyna Sułkowska-Ziaja[1]

1. Introduction

In the world of fungi, there are numerous possibilities of biosynthesis leading to the production of compounds with complex chemical structures showing high biological activity. As saprophytic organisms, fungi are biochemically related to the composition of the substrate on which they live. This largely determines their biosynthetic capabilities. Among major constituents of fungi, important compounds include those derived from the basic metabolism of sugars (polysaccharides); transformation of active acetate (isoprenoids, sterols), and amino acids (peptides, proteins).

Mushrooms have been used for thousands of years in traditional Chinese medicine due to their healing properties (Ayurveda, TCM, Kampo), to prevent or cure many diseases. Contemporary research proved that fruiting bodies or mushrooms are a rich source of bioactive compounds with a wide spectrum of impact on the human system. Numerous studies have shown a high therapeutic activity of these compounds. They possess antitumor, antioxidant, antibacterial, antifungal, antiviral, anti-inflammatory, and immunomodulatory activities confirmed in experimental studies. In addition, they possess anti-cholinergic, hepatoprotective, hypoglycemic, hypolipidemic, neuroprotective, anti-allergic, and anti-malarial effects, and support the beneficial intestinal microflora. Their therapeutic effect has also been confirmed

[1] Jagiellonian University Medical College, Faculty of Pharmacy, Department of Pharmaceutical Botany, 9 Medyczna Street, 30–688 Kraków, Poland.
[2] Jagiellonian University Medical College, Faculty of Pharmacy, Department of Inorganic and Analytical Chemistry, 9 Medyczna Street, 30–688 Kraków, Poland.
* Corresponding author: bozena.muszynska@uj.edu.pl, muchon@poczta.fm

in atherosclerosis, and arterial hypotension. This chapter addresses the therapeutic significance of carbohydrates, proteins, lectins, terpenes, and sterols, as well as enzymes derived from various mushrooms.

2. Carbohydrates

Carbohydrates (mainly glycogen, mannitol, sorbitol, arabinitol) constitute 1–6% of the mass of mushrooms' fruiting bodies. Apart from monosaccharides, the fruiting bodies of Basidiomycota mushrooms also contain glucose, galactose, mannose, xylose, fructose, fucose, and sedoheptulose in free form. Fructose is used due to its antibacterial effect as a component of disinfectant and anti-acne lotions and a preservative in cosmetic preparations. The alcoholic sugar, mannitol is also common. It has the ability to protect the skin and is used in skin-care cosmetics. The only free disaccharide found is trehalose, composed of two glucose molecules linked by an O-glycosidic bond, which is abundant in the fruiting bodies of *Boletus* spp., *Cortinarius* spp. and *Suillus* spp. (Wannet et al., 1999). This compound has moisturizing, protective (against temperature, UV, and chemical factors), and antibacterial properties. The occasional symptoms of poisoning may result from gastrointestinal intolerance to this compound. The dietary value and biological properties of these carbohydrates are well known. One of the functional groups of mushrooms are carbohydrates along with fiber. Due to the degree of solubility in water, dietary fiber is divided into insoluble fiber (cellulose and chitin) and soluble fiber, among which the main components are β–glucans and chitosans. The total fiber content on average ranges from 2.7 to 4 g per 100 g of the fruiting body. The need for most nutrients that affect health are also those that contain natural resources (Howlett et al., 2010; Slavin, 2013).

2.1 Chitin

Chitin is a dietary fiber that possesses a variety of chemical, physical, and physiological properties (Fig. 4.1). Chitin (Greek: Χιτών chiton = outer coat) is an organic compound from the group of biopolymers, from which the cell walls of mushrooms are built. Chitin has a similar structure to cellulose, but instead of glucose units, it has N–acetylglucosamine monomers. Physicochemical properties and molecular weight of chitosans isolated from mushrooms are identical to those isolated from crustaceans. Chitosans are derivatives of chitin, formed because of its

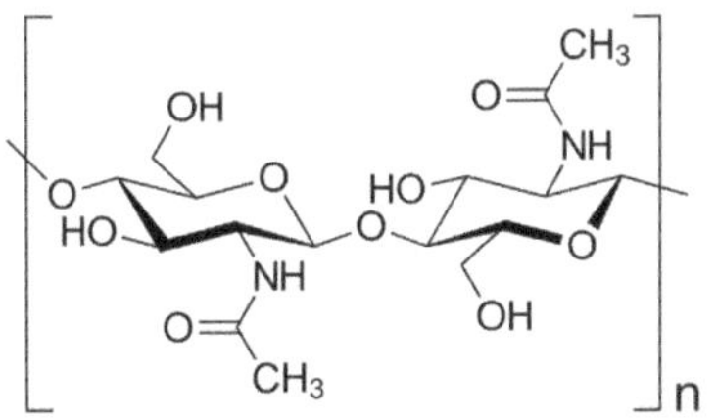

Fig. 4.1. The structural formula of chitin

partial deacetylation. Chitosans lower the concentration of LDL cholesterol in the blood and liver, as well as triacylglycerides in the serum and therefore reduce the risk of cardiovascular diseases. Chitosans also affect the absorption of cholesterol in the gastrointestinal tract, causing a decrease in its total blood concentration and levels of low-density lipoprotein (LDL) fraction without affecting the concentration of high-density lipoprotein (HDL) fraction. Due to the presence of large amounts of dietary fiber in edible mushrooms, especially glucans (increasing the viscosity of the hyphae) and chitin, the excretion of bile acids and neutral steroids increases (Marcle, 2008; Sułkowska-Ziaja et al., 2018; Kumari and Kishor, 2020). In the acidic environment of the stomach, the amino groups present in chitosan particles assume a positive charge and combine with the negatively charged residues of bile acids. Low pH makes chitosan complexes insoluble in bile acids which are excreted from the human body. Glucosamine obtained because of enzymatic breakdown of chitin is used as a precursor for the synthesis of hyaluronic acid used in the treatment of joint capsule dysfunction. In cosmetology, it is used as a moisturizing agent. In the pharmaceutical industry, chitin and chitosans are used as drug carriers and slimming aids, while in cosmetology they are mainly used in preparations for skin care (moistening properties) and hair regeneration products.

2.2 Polysaccharides

Polysaccharides represent an important group of carbohydrates with anticancer properties. The interest in cytotoxic substances contained in mushroom dates back to the beginning of 1940s. A polysaccharide complex with immunostimulatory properties was isolated from the cell wall of *Saccharomyces cerevisiae* (baker's yeast). After oral administration of this complex, macrophage activation and stimulation of the reticuloendothelial system have been observed (Lemieszek et al., 2012). The anticancer properties of extracts derived from fruiting bodies of the Agaricomycetes fungus *Boletus edulis* were reported in sarcoma-180-bearing mice (Lucas et al., 1957). The components of the extracts, however, have not been identified. The first anticancer drugs, which are mushroom-derived polysaccharides, were officially used for therapy in Japan (Blagodatski et al., 2018). The importance of edible mushrooms as agents effective in the prevention and treatment of neoplastic diseases has been confirmed in clinical trials in recent years (Xu et al., 2012). The main compounds responsible for this effect are polysaccharides, proteoglycans, or steroids (Ruthes et al., 2015; Meng et al., 2016; Singdevsachan et al., 2016; Kosanić et al., 2018).

The β-glucans are one of the best-known groups of compounds possessing anticancer activity (Patel and Goyal, 2012). The content of polysaccharides depends on the structure of mushrooms and stage of their morphogenesis (Cheung, 2013). Different mushroom species produce various types of polysaccharides that can be water soluble and insoluble. Simple sugar molecules that combine into polysaccharides through glycosidic bonds are made up of only residues of one type of monosaccharide—homopolysaccharides, while some of them belong to heteropolysaccharides and are made up of residues of different monosaccharides. These compounds can be used as prebiotics in the treatment of neoplastic or viral diseases, such as acquired immunodeficiency syndrome (AIDS) (Daba and Ezeronye, 2003). It is assumed that polysaccharides isolated from the edible mushrooms activate the immune response *in vitro* and *in vivo*, acting as biological stimulants. Heteropolysaccharides that show an anti-proliferative effect on neoplastic cells turned out to be substances that inhibit the growth of tumors, mainly after intraperitoneal or oral administration (Zong et al., 2012; Zhu et al., 2015; Singdevsachan et al., 2016). The role of mushroom polysaccharides in modulation of immune system and thereby a potential antitumor effect is of particular importance. One of the first clinically described effects of mushroom-derived polysaccharides in the treatment of cancer dates to 1957 (Byerrum et al., 1957; Meng et al., 2016).

The mechanism of action of mushroom-derived polysaccharides on the immune system has been confirmed in subsequent studies; it involves stimulating cells of the immune system, including T lymphocytes and cytotoxic T lymphocytes (CTL), B lymphocytes, granulocytes (eosinophils and neutrophils), NK (natural killer) cells or macrophages (Zhang et al., 2007; Roupas et al., 2012; Meng et al., 2016; Singdevsachan et al., 2016). This mechanism is particularly characteristic of β–1,3–glucans, but numerous studies also suggest that β–glucans may enhance specific cellular response by enhancing the secretion of IL–6, IL–8, IL–12 and IFN–γ cytokines from neutrophils, macrophages, and NK cells (Meng et al., 2016; Singdevsachan et al., 2016). In addition, β–glucans found in fruiting bodies of edible mushroom may stimulate new effector cells contributing to the formation of antibodies against tumor antigens, which is less popular than the classic cytotoxic effect that chemotherapy induces (Singdevsachan et al., 2016). The ability to bind other molecules, such as proteins and steroids is also important for the anticancer effect, which results in increased antitumor activity. In clinical practice, mushroom polysaccharides are most often used as an element of polytherapy in addition to standard chemotherapy or radiotherapy treatments. Most of the polysaccharides used are glucans, as well as homo- or heteroglycans, when combined with other protein molecules, can be converted into glycoproteins, glycopeptides, or proteoglycans. The conformation of the polysaccharide chain is also crucial for eliciting a therapeutic effect. The most active polysaccharides are often complexes with proteins with a molecular weight of 10,000 kDa. It has been reported that human macrophages possess a polysaccharide receptor highly specific for glucose and mannose molecules, from which the antitumor activity (ATA) of polysaccharides with these groups may result (Zhang et al., 2007; Patel and Goyal, 2012; Roupas et al., 2012; Meng et al., 2016; Tian et al., 2016). The antitumor effect has been demonstrated among the edible mushrooms rich in polysaccharides with the structure described above. These mushrooms include *Agaricus bisporus*, *A. blazei*, *A. campestris*, *Armillaria mellea*, *Boletus edulis*, *Cantharellus cibarius*, *Flamulina velutipes*, *Grifola frondosa*, *Hericium erinaceus*, *Imleria badia*, *Lactarius deliciosus*, *Lentinula edodes*, *Macrolepiota procera*, *Pleurotus ostreatus*, *Sparassis crispa*, and *Tremella fuciformis*.

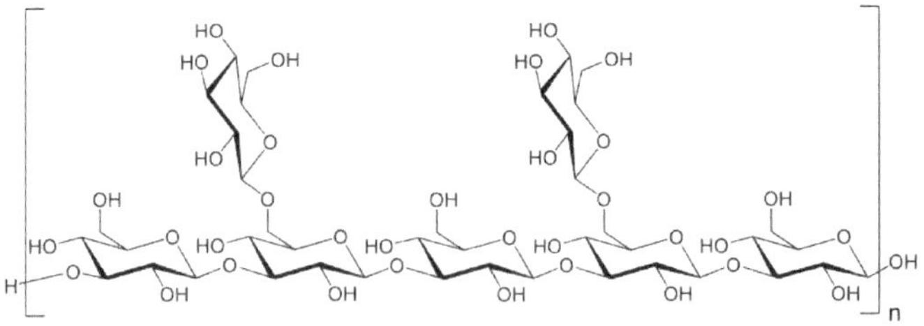

Fig. 4.2. The structural formula of lentinan

2.3 Lentinan

In modern traditional medicine, *L. edodes* is used not only as a strength-enhancing agent, but also as an anticancer substance from this species. Lentinan was isolated from fruiting bodies of *L. edodes* (Fig. 4.2). This compound is chemically a β (1 → 3) glucan with branches β (1 → 6), with a molecular weight of 500 kDa, having the structure of a right-handed helix. It is considered the most active among the known anticancer polysaccharides. It prevents the formation of neoplastic changes caused by chemical carcinogens and viruses, as well as inhibits the development of allogeneic and some syngeneic tumors and acts as a prebiotic, increasing the absorption of nutrients and stimulating the intestinal microbiota, and augmenting the bioavailability of medicinal substances. The lentinan also reduces the effects of impaired absorption and function of the intestinal microbiota during radio- and chemotherapy, not only during oncological therapy, but also during other monotherapy and polytherapy. This polysaccharide is most often used in the treatment of solid tumors of the stomach, colon, breast, lung, and malignant leukemia. It extends the average survival duration of cancer patients. It is likely to function by stimulating T lymphocytes, increasing the production of interleukin-1 and 3 and nitric oxide (NO) by cells of the immune system, stimulating the secretion of Colony Stimulating Factor (CSF), acute phase proteins and direct or indirect (by lymphocytes T) effect on macrophages. Studies on the mechanism of action of lentinan also show that it is thymic-dependent and enhances the response of helper T-cell precursors and macrophages, and thus some cytokines, produced by lymphocytes, after the diagnosis of cancer cells. The induction of IFN-Ɣ is also important for this activity.

Lentinan is most often used in the treatment of gastric cancer as an adjunct to conventional treatment, including surgical removal of tumor and chemotherapy or radiotherapy. It is a kind of synergistic action by improving the general condition of the patient. Effective treatment of hepatocellular carcinoma, colorectal, and pancreas has also been made while reducing the side effects and to improve the quality of life. Lentinan is also effective in the treatment of leukemia as a factor inhibiting the proliferation of lymphocytes. It is also important that it has a selective anti-proliferative effect on neoplastic skin cells (CH72) without affecting healthy keratinocytes (C_{50}). It has also shown that lentinan inhibits metastasis in the liver of mice in adenocarcinoma-26. This action is made possible by the activation of Kupffer cells in the liver. Lentinan has no cytotoxic activity, practically no side effects; it may lead only to local irritation after the injection or occasional fever and vomiting. Usually, these are only episodes and, most importantly, are generally well-tolerated by the patient's body. The use of lentinan as an adjuvant has been shown to improve the quality of life in cancer patients because it also eliminates the side effects of chemo- and radiotherapies, like other mushroom polysaccharides (Lindequist et al., 2005; Mantovani et al., 2008; Patel and Goyal, 2012; Roupas et al., 2012; Muszyńska et al., 2013; Meng et al., 2016; Singdevsachan et al., 2016).

2.4 Pleuran

Pleuran is a β 1,3–D–glucan isolated from *P. ostreatus* fruiting bodies. It has been shown to slow down the formation of precancerous lesions in the colon in Wistar rats. This action consisted of inhibiting the proliferation of tumor cells and inducing their apoptosis. Additionally, the new polysaccharide (POPS-1) obtained from hot water extracts of this mushroom significantly reduces the toxicity compared to the commonly used 5-fluorouracil (5-FU). A diet rich in dried *P. ostreatus* reduced toxicity in mice treated with cyclophosphamide and decreased pathological changes resulting from the appearance of dimethylhydrazine-induced colorectal cancer in rats. This effect resulted from the strong antioxidant and immunostimulatory potential of *P. ostreatus* fruiting bodies and their polysaccharide content, especially pleuran, as part of the dietary fiber.

Another example of mushroom-derived polysaccharides used in therapy of neoplastic diseases is Krestin (*Trametes versicolor*). Polysaccharide fractions isolated from mushrooms, in the form of glucuronoglucans, xyloglucans, mannoglucans, and xylomannoglucans represent a particular interest in terms of their anticancer properties (Miyazaki and Nishijima, 1981).

2.5 β–Glucans

The study of antitumor polysaccharides from edible medicinal species *A. mellea* revealed the presence of peptidoglucan (Devkota and Hammerschmidt, 2020). The authors demonstrated ATA of the polysaccharide fraction. The results of research on the chemical structure of the sugar part of peptide-glucan, the presence of glucose molecules connected by β (1 → 3) and β (1 → 6) bonds was found. Recent study research proved the presence of polysaccharides with a different chemical structure in the fruiting bodies of *A. mellea* (Yan et al., 2018). Two α–glucan polysaccharide fractions were isolated. The main fraction consisted of linear chains of α (1 → 3) – and α (1 → 4)–glucan bound to the protein, while the second fraction contained α (1 → 3)–glucan. These studies focused only on the chemistry of polysaccharides.

Another example of antitumor polysaccharide is the β–glucan isolated from *S. crispa*, which is mainly found in edible mushroom. Previous clinical study trials have been conducted in which powdered fruiting bodies of this species were administered orally to patients suffering from tumors in the amount of 300 mg/day. In most of the patients, a clear improvement was observed compared to the control group (Roupas et al., 2012).

Antitumor potential of two edible mushrooms *L. deliciosus* and *M. procera* was tested against the human epithelial carcinoma (HeLa), human colorectal carcinoma (LS174), and human lung carcinoma (A549) cells. IC_{50} values ranged from 19.01 to 74.01 μg/ml for *L. deliciosus* extract and from 25.55 to 68.49 μg/ml for *M. procera* extract depending on the type of cell lines. The activity of these extracts, however, was not as effective as in the case of cis-dichlorodiammine platinum (Cis-PP). *Macrolepiota procera* showed a stronger ATA against A549 and LS174 cell lines, while *L. deliciosus* showed greater activity against HeLa. However, it has not been proven whether this action is selective for neoplastic cells or also for other types of neoplastic cells. It is possible, however, that the presence of CD is one of the factors influencing the cytotoxic effect on the tested cell lines, but these are further studies confirming this activity. In the case of *M. procera*, aqueous extracts showed inhibitory activity against tumor metastasis of colon 26–M3.1 (Meng et al., 2016).

Edible medicinal mushroom *A. blazei* (=*A. subrufescens*) is a species commonly used in cancer prevention because it exhibits immunomodulatory and antimutagenic properties. High β–glucan polysaccharide fractions obtained from this species proved effective in the treatment of both androgen dependent and non-androgen dependent prostate cancer. The induction of apoptosis in prostate cancer cells was directly related to the activation of caspase-3 as an apoptotic factor. The polysaccharide fraction obtained from this species turned out to be effective in the treatment of neoplasms in *in vivo* studies, with no cytotoxic activity tested *in vitro*. Additionally, in combination with 5-fluorouracil (5-FU), this species protected against leukopenia as a side effect of treatment with this substance. Attempts have been made to improve β–glucan obtained from this species by modifying the incorporation of sulphate groups to improve solubility (Mantovani et al., 2008; Patel and Goyal, 2012; Roupas et al., 2012; Zong et al., 2012).

A. campestris and the glycoprotein fraction obtained from this species showed ATA against sarcoma-180 in ICR mice. Importantly, the protein group of the glycoprotein contributing to the anticancer effect consists of 17 amino acids (Singdevsachan et al., 2016). In turn, water extracts obtained at high temperature from *H. erinaceus* fruiting bodies revealed high levels of β–glucan, the administration of which reduced the tumor mass in mice with induced colorectal cancer. The reduction in tumor mass was due to the induction of tumor necrosis factor (TNF) and NK cell secretion, as well as macrophage activation and inhibition of angiogenesis (Lindequist et al., 2005; Patel and Goyal, 2012; Roupas et al., 2012; Meng et al., 2016; Singdevsachan et al., 2016).

2.6 *Grifolan*

Grifolan, obtained from medicinal fungus *Grifola frondosa* (Maitake), is also a source of β–glucan, including the active D fraction and the polysaccharide fraction obtained by further purification (both improve the effectiveness of cisplatin therapy). This fraction works by enhancing the activity of inhibiting the spread of tumors and reduction of nephrotoxicity and immunosuppression induced by treatment with cisplatin (Masuda et al., 2009). In combination therapy with cyclophosphamide, this fraction acts by inducing the production of granulocyte colony stimulating factor (G-CSF), which contributes to the promotion of granulopoiesis and mobilization of granulocytes in mice treated with cyclophosphamide. Water-insoluble polysaccharides such as, for example, chemically sulfated polysaccharide (S–GAP–P) have also been classified as being effective in the treatment of human gastric cancer (SGC–7901 cancer cells). Grifolan has been shown to be effective not only in combination therapy alongside 5-FU (5-fluorouracil), but also as monotherapy. S–GAP–P induced apoptosis of SGC–7901 cancer cells. This effect was dose-dependent. Additionally, the use of 5-FU and S–GAP–P side-by-side enhanced the anticancer effect. β–Glucan from *G. frondosa* also showed a cytotoxic effect against human prostate cancer cells (PC-3 cells). The induction of apoptosis has been confirmed in *in vitro* studies in the case of an androgen-independent tumor (Patel and Goyal, 2012; Roupas et al., 2012; Meng et al., 2016; Singdevsachan et al., 2016).

2.7 Other Polysaccharides

The studies showed that the polysaccharide isolated from other mushrooms, such as *Pleurotus pulmonarius*, is effective in the treatment of liver cancer, both *in vitro* and *in vivo*. The effect was to inhibit the growth of this tumor through an inhibitory effect on VEGF-induced (vascular endothelial growth factor) PI3K/AKT signaling pathway (Cheung, 2013). In *B. edulis*, the antitumor effect was studied in mice. This action was to inhibit Sarcoma-180 cancer cells (Daba and Ezeronye, 2003). Medical applications are mainly made of polysaccharides in the form of a triple helix, e.g., the already mentioned lentinan, although the mechanism of their action due to their structure has not been fully investigated.

Nevertheless, polysaccharides in other forms, e.g., linear, such as (1 → 3)–β–D–glucan, soluble in water and isolated from *Auricularia auricula*, also have a strong anticancer potential (Zhang et al., 2007; Ruthes et al., 2015; Meng et al., 2016; Singdevsachan et al., 2016). Depending on the structure, mushroom polysaccharides also exhibit immunomodulatory activity, strengthening and accelerating the body's defense response, thus causing an antitumor effect, which is more common than typical cytotoxic effect (Zhang et al., 2007; Meng et al., 2016; Singdevsachan et al., 2016). The effect of the use of mushroom polysaccharides may be to prevent the formation of neoplastic changes, and in the case of detected neoplasms, the arrest of the cell cycle and induction of tumor cell apoptosis, which will disturb its growth and development (inhibition of growth by up to 50%), as well as stop the formation of metastases, while extending the survival rate of patients (Zhang et al., 2007; Zong et al., 2012).

The preventive effect of polysaccharides was observed especially in the group of farmers involved in the commercial cultivation of edible mushrooms, including *F. velutipes* and *A. blazei*. As part of a daily diet result from consuming mushrooms, the cancer incidence was estimated about 40% lower than in the general incidence in the population. Studies using mice were also carried out to support these studies. They were fed regularly with these species of mushroom and then injected with tumor cells. It was clearly observed that unlike in the control group, no murine tumors developed in the research group, which confirms the strong preventive antitumor effect of polysaccharides from the species *F. velutipes* and *A. blazei* (Zhang et al., 2007). The source of immunomodulatory polysaccharides are not only the fruiting bodies of many Basidiomycota species, but also their mycelial cultures (Wasser, 2002).

3. Amino Acids and Proteins

Most proteinogenic amino acids are present in mushrooms. The favorable ratio of exogenous to endogenous amino acids makes them a source of high-quality protein. Studies on the chemical composition of edible species have shown that these proteins are characterized by high bioavailability, even reaching 90%. It was confirmed that dried mushrooms may contain up to 25% of digestible proteins. One quarter of the amino acids contained in fruiting bodies is in the free state and the mutual proportions of amino acids are akin to those that are found in plant products. The most common free amino acids in mushrooms are glutamic acid, alanine, proline, aspartic acid, arginine, glutamine, tryptophan, and tyrosine (Ribeiro et al., 2008; El Enshasy et al., 2013; Struck, 2015).

A. bisporus fruiting bodies contain large amounts of easily digestible protein (25%), the amino acid composition of which is represented, among others, by alanine, aspartic acid, and glutamic acid (Ribeiro et al., 2008). The fruiting bodies of *B. edulis* also contain 25% of digestible protein. Smaller amounts of available protein are found in the fruiting bodies of *C. cibarius* (about 4%), which contain alanine, arginine, aspartic acid, glutamic acid, leucine, lysine, phenylalanine, proline, tyrosine, valine, and tryptophan. In turn, the fruiting bodies of *Imleria badia* contain about 3% of digestible protein and are particularly rich in tryptophan, cystine, methionine, lysine, aspartic acid, and glutamic acid (Kalač, 2009). Phenylalanine, tyrosine, and tryptophan present in mushrooms are essential aromatic amino acids that act as precursors for neurotransmitters, such as serotonin and catecholamines (adrenaline, noradrenaline, dopamine) in the brain. L-Tryptophan is a precursor to vitamin B_3 (niacin) and stimulates the secretion of insulin and growth hormone (Bach et al., 2017).

Many studies have shown that mushrooms produce a new family of protein immunomodulators, called FIPs. Since the discovery of the first FIP (Ling-Zhi-8 from *Ganoderma lucidum*) in 1989, 11 different types of FIP have been isolated among others from *F. velutipes*, *G. lucidum*, *G. tsugae*, and *G. sinensis*. These proteins have been shown to inhibit autoimmune diabetic responses in an animal model and increase graft survival in allogeneic skin transplanted mice, with less nephrotoxicity compared to other immunosuppressive drugs, such as cyclosporin A (Tanaka et al., 1989; Wang et al., 2004).

Lectins are carbohydrate-binding proteins or glycoproteins. Initially, they were described in plants, but they are also found in fungi. Lectins are involved in various interactions between cells, e.g., in the regulation of cell adhesion, defense against plant infections, interactions between *Rhizobium* and a plant cell, and as receptors of pathogens that bind to animal cells. Some lectins are strong mitogens (Ho et al., 2004). They are bound to sugar residues of glycoproteins present on the surface of lymphocytes. Thus, they contribute to the clumping of cells, which often leads to their activation and, consequently, to their blastic transformation. The first mushroom lectin discovered in 1907 was the hemagglutinin derived from *Amanita muscaria*. The presence of lectins is confirmed in edible mushrooms, such as *A. bisporus*, *A. campestris*, *B. edulis*, and *C. cibarius* (Singh et al., 2010).

Lectins are most common in taxonomic groups belonging to the genus *Lactarius*, *Russula*, and the families *Boletaceae* and *Agariaceae*. Lectins synthesized in fruiting bodies and mycelium of a given species may be similar in structure and properties (*Kuehneromyces mutabilis*, *L. deliciosus*, and *Lactarius deterrimus*) or may be completely different (*Laetiporus sulphureus* and *Pholiota squarrosa*). In some species, lectins have not been found in the mycelium, although they are found in fruiting bodies (Nikitina et al., 2017). The physiological role of lectins in mushrooms is not fully understood. The differences in the function of lectins result from different developmental biology and diet of individual species. The participation of higher fungal lectins in the mobilization and transport of sugars, fusing mycelium hyphae, and in the mycorrhiza process has been proven (Singh et al., 2009). The ATA of lectins is based on inducing apoptosis of cancer cells; however, the mechanism of action is not fully understood (Konska, 2006; Singh et al., 2014; Singh et al., 2016; Zhao et al., 2020).

4. Enzymes

Mushroom enzymes are compounds with a protein structure. They present a wide spectrum of biological activities, one of the most important of which seems to be anticancer activity (e.g., tyrosinase) and the non-selective ability to decompose dead organic matter. Thanks to their activity, human waste is also decomposed in the environment. Lignolytic enzymes can even decompose wood and lignin in it. Lignin modifying enzymes (LMEs) are oxidoreductases that catalyze the flow of electrons from one substrate to another. LMEs work by generating free radicals that randomly react with the lignin polymer, breaking covalent bonds and releasing phenolic compounds. There are two main types of lignin-modifying enzymes: peroxidases and laccases (phenolic oxidases). As previously mentioned, the main LMEs are lignin peroxidase (LiP), manganese peroxidase (MnP), versatile peroxidase (VP), and laccase (laccase). In addition, these fungi secrete high molecular weight molecules, increasing a range of potentially biodegradable compounds. LMEs are secondary metabolites of WRF, because lignin oxidation does not provide energy to fungi (Wesenberg et al., 2003). These enzymes are responsible for generation of highly reactive and nonspecific free radicals. During advanced oxidative reactions, these enzymes can oxidize persistent and harmful compounds (Kersten and Cullen, 2007). Laccase is an oxidase with four copper cations in an active center. It catalyzes oxidation by transferring one electron from four substrate molecules to one oxygen molecule, which in turn is reduced to water (Wesenberg et al., 2003). Laccase has a low substrate specificity and may react with diphenols, arylamines, and aminophenols. The redox potential of laccase is between 780–800 mV. In the presence of a mediator, laccase can also oxidize nonphenolic molecules (Wesenberg et al., 2003; Wong, 2009). This enzyme has interesting properties depending on the species of mushrooms and its location in the hypha. It is believed to be involved in the processes of virulence (yeast, bacteria, pathogenic fungi), lignin degradation (white rot mushrooms), deposition (plants), pigment synthesis (fungi, bacteria) and insect molting (Wong, 2009). This enzyme also has an ATA. Recent scientific reports indicate the possibility of using this enzyme as a substance with antitumor and antiviral properties. The effect of the active form of laccase isolated from the white rot fungus *Cerrena unicolor* on human cervical cancer cells has been investigated (Matuszewska et al., 2013). Lignin peroxidase (LiP) was the first enzyme isolated from the fungus *Phanerochaete chrysosporium*, (Caramelo et al., 1999). This enzyme is a glycoprotein with a molecular weight between 38 and 47 kDa, with a low optimum pH of about 3. It has the potential to catalyze the oxidation of phenolic and aromatic compounds similar in structure to lignin. Lignin peroxidase exhibits a classic peroxidase mechanism: it can react with phenolic and aromatic substrates, causing the formation of phenyl radicals, but it is also capable of oxidizing substrates with a high redox potential (up to 1400 mV) (Wesenberg et al., 2003). This enzyme shows relative specificity to its substrates, and one of them is veratril alcohol (VA) which is a natural secondary metabolite of WRF increasing the activity of lignin peroxidase and the rate of lignin degradation (Wong, 2009). Manganese peroxidase (MnP) is an extracellular enzyme discovered in *P. chrysosporium* by Kuwahara (1984) and considered as the most common lignolytic peroxidase produced by almost all white rot fungi of the Basidiomycota (Riva, 2006; Wesenberg et al., 2003). Manganese peroxidase is a glycoprotein with a molecular weight between 32 and 62.5 kDa. This enzyme has a similar catalytic cycle to other peroxidases involving two electron oxidations, however, MnP is capable of oxidizing Mn^{2+}, resulting in the formation of diffusive oxidants (Mn^{3+}) capable of penetrating the cell wall and oxidizing phenolic substrates (Wong, 2009). Universal peroxidase combines the substrate specificity of three other fungal peroxidases: Mn, LiP and peroxidase produced by *Coprinopsis cinerea*. As a result, it is capable of oxidizing substrates with low and high redox potential, including Mn^{2+} ions, phenolic and nonphenolic lignin dimers, substituted phenolic compounds, veratril alcohol, and various types of dyes (Reactive Black 5) (Hofrichter, 2002). Universal peroxidase is only produced by fungi of the genera *Bjerkandera*, *Lepista*, and *Pleurotus*. It shows different optimal pH values for oxidation, e.g., pH 5for Mn^{2+}, and pH 3 for aromatics. These values are similar to those of lignin peroxidase and manganese peroxidase. In the universal peroxidase catalytic cycle, two electron oxidations of inactive peroxidase (containing Fe^{3+}) by hydrogen peroxide take place. The compound I is formed (C-IA containing the complex cationic radical oxo–Fe^{4+}–porphyrin), the reduction of which because of two one-electron reactions leads to the formation of the intermediate compound II (C-IIA, containing oxo–Fe^{4+}–porphyrin), Mn^{3+} and the resting form of the enzyme. The compounds C-IB and C-IIB, which are in equilibrium with C-IA and C-IIA, are involved in the oxidation of veratril alcohol and other aromatic compounds with a high redox potential (Perez Boada et al., 2005). It has been shown that the presence of moderate concentrations of Mn^{2+}, strongly inhibit the oxidation of LiP substrates, such as veratril alcohol (Mester and Field, 1998).

5. Terpenenoids

Terpenes (isoprenoids) are organic compounds with the general formula $(C_5H_8)_n$. The main skeleton of terpenes is formed by combining five-carbon isoprene units. Terpenes are formed through the mevalonic acid pathway from active acetate through mevalonic acid to “active isoprene”. Due to the size of the molecules, terpenes are divided into monoterpenes (C10 composed of 2 isoprene units), sesquiterpenes (C15 composed of 3 isoprene units), diterpenes (C20 made of 4 isoprene units,) triterpenes (C30 made of 6 isoprene units), and tetraterpenes (C40 made of 8 isoprene units). Terpenes are a widely represented group of secondary metabolites in fungi.

5.1 Monoterpenes

Monoterpenes are compounds with the formula $C_{10}H_{16}$. The Geranyl pyrophosphate is the precursor to monoterpenes (GPP). It occurs by condensation of isopentenyl pyrophosphate (IPP) or dimethylallyl (DMAPP). These compounds are volatile substances and are responsible for the characteristic smell of fruiting bodies and mycelium.

Research conducted on fresh fruiting bodies of wild-growing mushrooms (including *Gomphidius glutinosus*, *Hydnum repandum*, *Lepista nuda*, *Marasmius aliaceus, Mycena pura*, and *Suillus luteus*) confirmed the presence of monoterpenes, including α– and β–pinene, α–fenchene, campfene, β–fellandren, limonene, linalool, and terpinolene (Breheret et al., 2019). Previous study confirmed the presence of 3,7–dimethylocta–1,6–dien–3–ol (linalool) in fruiting bodies of *G. lucidum*, *H. erinaceus*, and *Fomitopsis betulina*; 1–methyl–4–(1–methylethenyl)–cyclohexene (limonene) in *G. lucidum*, *H. erinaceus*, *L. edodes*; 2–(4–methyl–1–cyclohex–3–enyl) propane–2–ol (α–terpineol) in *Antrodia camphorate*, *G. lucidum*, *F. betulina*, *Pleurotus erygii* and 1–methyl–4–(prop–1–en)–2–yl)–cyclohexan–1–ol (β–terpineol) in *Antrodia camphorata* and *H. erinaceus* (Pennerman et al., 2015).

Monoterpene-alcohols including: 3,7–dimethyl–1,6–octadiene–3,4–diol; 4–[(3R,4S)–3–hydroxy–3,7–dimethylocta–1,6–dienyl(Z)–9–octadecenoate; 4–[(3R, 4S)–3–hydroxy–3,7–dimethylocta–1,6–dienyl (9Z,12Z)–9,12–octadecadienoate; 3,7–dimethyl–1,6–octadiene–3,4,5–triol; and *bis*[6–(3,4,7–trihydroxy–3,7–dimethyloctenyl) ether were isolated from a mushroom *Dictyophora indusiate* (syn. *Phallus indusiatus*) (Ishiyama et al., 1999). In turn, 1,2–dihydroxymintlactone isolated from *Cheimonophyilum candidissimum* showed nematicidal activity (Stadler et al., 1995). In the fruiting bodies of *F. velutipes*, a monoterpene with a chemical structure of 5–hydroxymethyl–2–(1–methyl–ethenyl)–1–cyclohexanol has been identified (Cai et al., 2013). A new monoterpene, 6,7,8–trihydroxy–2,6–dimethyloctanoic acid, was also isolated from the toxic species *Trogia venenata*, which is a frequent cause of poisoning in some regions of China (Ying-Chao et al., 2018).

5.2 Sesquiterpenes

Sesquiterpenes are a class of terpenes that consist of three isoprene units with the molecular formula $C_{15}H_{24}$. The precursor in the synthesis of sesquiterpenes is farnesyl diphosphate (FPP), which occurs independently in *cis* and *trans* forms, and its isomer—nerolidyl diphosphate (NPP). FPP biosynthesis occurs by condensation of geranyl diphosphate (GPP) with IPP.

About 200 basic structures of natural sesquiterpenes are currently known. Sesquiterpenes are compounds with different degrees of oxidation, additionally merging with combinations of esters and ethers.

Sesquiterpenes are a widely represented group of secondary metabolites in mushrooms, and among others are derivatives of iludan, hirsutane, lactarate, cuparan, and bisabolans. Iludan derivatives, such as illudine M and S, were isolated from *Omphalotus illudens* and *Lampteromyces japonicus*. These compounds have antibacterial activity (ABA) and ATA properties and presented toxicity (Duru and Çayan, 2015).

The derivatives of hirsutane: hypnophilin, pleurotellic acid, and pleurotellol, isolated from mushroom *Pleurotellus hypnophilus* mycelial cultures, exhibit a broad spectrum of antibiotic activity against both G(+) and G(–) bacteria, AFA (antifungal activity), and cytotoxic activity (Giannetti et al., 1986).

Other hirsutane derivatives hypnophilin and 1–desoxy–hypnophilin, which were obtained due to *Lentinus crinitus* biofermentation, have shown ATA and cytotoxic activity against mouse L929 fibroblastoma (Abate and Abraham, 1994; Qiu et al., 2018). Two hirsutane derivatives: hirsutic acid and (–) complicatic acid with antimicrobial properties were isolated from the wood-decaying polypore *Stereum complicatum* (Perera et al., 2020). Species of the genus *Panus* contain sesquiterpene nematolone whose cytotoxicity is related to the presence of α and β unsaturated ketone groups (Thakur et al., 2013). Coriolin, an antibiotic obtained from mycelial cultures of *Coriolus (=Trametes) consors*, also has a sesquiterpene structure. This compound is active against *Staphylococcus aureus*, *Mycobacterium flavus*, *Bacillus subtilis*, *B. anthracis*, and *Trichomonas vaginalis*. Moreover, it inhibits the development of Yoshida's sarcoma. A coriolin derivative, coriolin B, isolated from the same species shows ABA, ATA, and immunostimulatory activity (Sushila et al., 2012). Several isolactarane sesquiterpenes were identified from fruiting bodies and mycelial cultures: merulidial, merulanic acid, merulactone, merulialol, and the sterpurenes tremediol and tremetriol (Abraham, 2001). A merulidial showed a cytotoxic activity, inhibiting the synthesis of DNA in ECA cells. The other compounds mentioned above lead to inter alia, apoptosis of HL60 human leukemia promyelocytes (Sułkowska-Ziaja et al., 2005). Enokipodins A-D, cuparane sesquiterpenes that were isolated from *F. velutipes* mycelial cultures showed AFA against *Cladosporium herbarum* and ABA against *B. subtilis* and *S. aureus* strains (Ishikawa et al., 2005). Flamvelutpenoids A–D, four cuparene–type sesquiterpenes, were isolated from the mycelial culture of *F. velutipes*. These compounds showed an antibacterial activity (ABA) against *E. coli*, *B. subtilis*, and *S. aureus* (Wang et al., 2012). The mycelial cultures of *Coprinus cinereus* were a source of logopodins (quinone sesquiterpenes), possessing antimicrobial properties (Agger et al., 2009). Bisabolane sesquiterpenes, cheimonophyllon A–E, and cheimonophyllal were isolated from mycelial cultures of *Cheimonophyllum candidissium*. These compounds show antibiotic and nematocidal activities (Stadler et al., 1994). The sesquiterpenes aryl esters are the main constituents of *A. mellea* fruiting bodies and mycelial cultures. The first isolated sesquiterpene aryl ester from *A. mellea* was a melloid. Since then, more than 37 compounds have been identified, including armillarin and armillaridin, armillaryin, armillarybine, armillarigine, armillarigin, and armillarikin. Another research group isolated 14 new sesquiterpenes aryl esters with ABA and AFA (Donnelly et al., 1987; Yang et al., 1989; Midland et al., 2001; Gao et al., 2009). A multitude of lactarane sesquiterpenes are isolated from the species of the genus *Lactarius*. The mono-alcohol 3-desoxylactarorufin A is known from *L. necator*; blennins A, B, C, D from *L. blennius*; lactarorufin B from *L. rufus*, *L. necator*—anhydrolactarorufin A, and lactarorufin N and 8–epi–pipertriol. The vellerolactone and pyrovellerolactone were isolated from a different *Lactarius* species: 15-Hydroxyblennin A was extracted from *L. torminosus*. *Lactarius* spp. produced piperdial, piperalol, lactarorufin D and E, and 4,8-dihydroxyfuran (Abraham, 2001).

The hallucinogenic species, *Gymnopilus junonius*, is rich in trichothecene and tremulan sesquiterpenes, were evaluated for their cytotoxic effects on human lung and prostate cancer cell lines. They showed significant immunosuppressive similarity to the cytotoxicity of the control drug, doxorubicin (Lee et al., 2020).

5.3 Diterpenes

According to the biosynthesis scheme, diterpenes are derived from geranylgeraniol, which is formed by the prenyl chain extension reaction with farnesyl diphosphate and IPP. This reaction is catalyzed by prenyltransferase. These types of transferases have been isolated from various organisms, including yeast, and numerous plants. The precursor compound in the synthesis of diterpenes is geranyl-geranyl-PP (GGPP), which transforms into geranylgeraniol by cleavage of the diphosphate moiety (with the participation of diphosphoesterase). The ability to biosynthesize diterpenes in the Basidiomycota species is limited compared to other taxonomic groups of fungi producing valuable gibberellins (Kuznetsova and Vlasenko, 2020).

Diterpene with a proven biological activity is pleuromutilin, isolated from *Omphalina mutila* (= *Pleurotus mutilus*), with antibiotic properties and inhibiting the development of PR8 influenza virus. The compounds with antibiotic properties were isolated from the mycelial culture of *Cyathus helenae*. The structure of these compounds was determined as diterpenes with a cyathane skeleton, with molecules characterized by the presence of three rings: five- six- and seven-membered and defined as cyathanes, and their isomers–allocyathanes. Striatins isolated from the *Cyathus striatus* mycelium, and erinacins from the *H. erinaceus* mycelium are classified as cyathane xylosides due to the xylose molecule attached to their aglycone (Anke et al., 2002). Erinacin E, in addition to its neurotropic effect contributing to the synthesis of a nerve growth factor (NGF), also has an opioid receptor- stimulating activity (Monique, 2021). From the extract of the methanol of *Sarcodon scabrosus*, the following diterprnes were isolated: scabronins A–F, scabronin J, sarcodonines A and G, sarcodonines M and L, and neosarcodonines A and B. In the course of further work, two new cyathane diterpenoids from *S. scabrosus*, scabronins K and L and H were isolated. These compounds were assessed for NGF via the growth factor neurite using rat pheochromocytoma (PC12) cells as a model system for neuronal differentiation. The sarcodonines A and G at a concentration of 25 μM showed a significant neurite growth-promoting activity in the presence of 20 ng/mL NGF after 24 hours of treatment (Shi et al., 2011). Cyathane diterpenes were isolated from the *Sacodon glaucopus*, glaucopins A–C were isolated, and from the *S. cyrneus,* cyrneins A–D. The structure and configuration of these compounds were determined by spectral methods (^{1}HNMR, EI–MS) (Curini et al., 2005). Currently, conducted research was conducted to isolate six new cyatan diterpenoids from the liquid culture of *Cyathus hookeri*: cyjahokerine A–F and nine known analogues. Their structures were elucidated. Cyjahookerins A–F displayed differential nerve growth factor-induced neurite outgrowth-promoting activity in PC–12 cells at concentrations of 10 μM. In addition, cyahookerin B, cyathin E, cyathin B2, and cyathin Q showed significant inhibition of nitric oxide production in lipopolysaccharide (LPS)-activated BV-2 microglial cells with IC50 values of 12.0, 6.9, 10.9, and 9.1 μM, respectively (Tang et al., 2015; Tang et al., 2019).

Studies on the medicinal mushroom *Wolfiporia cocos* have led to identification of 13 diterpenes of the abietane type. All compounds were assessed for cytotoxicity (K562 and HepG2), ABA, and anti–inflammatory effects. The compounds showed cytotoxicity against K562 cells with IC_{50} and low ABA against *S. aureus*. In addition, these compounds were shown to inhibit NO release in LPS-induced RAW 264.7 cells (Chen et al., 2020).

5.4 Triterpenes

Triterpenes are isoprene hexamers. The main precursor in the synthesis of mushroom triterpenoid products is the active isoprene, isopentenyl pyrophosphate (IPP) and dimethylallyl pyrophosphate (DMAPP), which are formed from acetyl–CoA via the mevalonic acid pathway (MEV). In the 1990s, Rohmer published an alternative pathway, MEP (named after an intermediate metabolite, methylerythritol phosphate). The triterpenes known so far in mushrooms have, among others, a tetracyclic structure with the following functional groups: acid, alcohol, aldehyde, and ketone. Triterpenes are essential for the functioning of eukaryotes as part of cell membranes. Mushrooms accumulate several types of polycyclic triterpenes (lanostane, ergostane, cucurbitan) (Liu, 2014).

Numerous lanostane triterpenoids, such as lucidene acid, ganodermanodiol, and ganoderiol, which are powerful activators of the complementary system play an important role in inducing a humoral response in the human defense mechanism. They been isolated from the fruiting bodies of *G. lucidum* and *G. applanatum*. Additionally, the isolated acids: ganodermic B, ganolcidic A, and lucidumol B have the properties of an HIV protease inhibitor.

An example of an important metabolite with antiviral activity from this group of compounds is ganodermanotriol, isolated from *G. lucidum*. The minimum concentration of ganodermanotriol required for complete inhibition of HIV-1 causing a cytotoxic effect in cells is 7.8 $\mu g/mL^3$. The mechanism of antiviral action is related to the interferon formation (Zhou et al., 2006).

Polyporenic acids A–C with bacteriostatic and anti-inflammatory properties were isolated from the arboreal species *F. betulina*. The study of an ethyl acetate extract from the same species led to the isolation of a new bioactive triterpene lanostane identified as 3β–acetoxy16–hydroxy–24–oxo–5α–lanosta–8–ene–21–oic acid. In addition, ten known triterpenes were identified, including polyporenic acid A, polyporenic acid C, three derivatives of polyporenic acid A, betulinic acid, betulin, ergosterol peroxide, 9,11–dehydroergosterol peroxide, and fomefficinic acid. All isolated compounds were tested for antimicrobial activity against G(+) and G(–) bacteria, as well as against a selected strain of filamentous fungi. Novel triterpene and some other compounds have been shown to have antimicrobial activity against G(+) bacteria (Alresly et al., 2016). ATA is characterized by a triterpene called inotodiol derived from *Inonotus obliquus*. Finnish scientists demonstrated its activity in Walker sarcoma and adenoma tumor models (MCV7) in *in vitro* tests. Fasciculic acid A is a triterpene compound obtained from *Hypholoma fasciculare*, an arboreal species considered poisonous. This acid has the activity of a specific calmodulin antagonist, a regulatory protein found in the cytoplasm and cell membranes, which also participates in the biochemical transformations of ATP and cAMP. Another example is the favolon with an ergostane skeleton with AFA isolated from the genus *Favolashia* (Anke et al., 1995). A new biologically active triterpen, favolon B, was isolated from mycelial cultures of *Mycena* sp. Favolon B showed AFA against *Botrytis cinerea*, *Mucor miehei*, *Paecilomyces variotii*, and *Penicillium notatum* (Aqueveque et al., 2005). Three triterpenes were isolated from *Pleurotus eryngii*. Lupeol and other new compounds 2,3,6,23–tetrahydroxy–urs–12–en–28–oic acid and 2,3,23–trihydroxy–urs–12–en–28–oic acid have been tested on breast cancer cell lines *in vitro*. All listed compounds significantly inhibited the growth of cancer cells, and the newly isolated compounds were even more effective than lupeol (Xue et al., 2015). Another mushroom with a relatively well-known triterpenes is *I. obliquus*. The fruiting bodies contain betulin with proven anti-inflammatory effect, ABA and ATA. Another triterpene is betulinic acid which has shown significant activity against human adenocarcinoma and lung cancer cell lines. Recent studies have shown that 3β–hydroxy–8,24–diene–21–ol and inotodiol manifested potent activity in the human complement test. Inotodiol, as well as betulinic acid and betulin, had revealed antiproliferative effects against HT29–MTX colon adenocarcinoma cell line. Recently, another recently discovered triterpene, botulin–3–O–caffeate, was shown to reduce macrophage nitric oxide (NO) production. NO production signals inflammation and plays a role in autoimmune diseases (Wold et al., 2020).

5.5 Tetraterpenes

Tetraterpenes are compounds with the molecular formula $C_{40}H_{64}$. One of the groups of tetraterpenes are carotenoids. The carotenoid molecule is composed of eight isoprene units, which form two cyclohexyl rings connected by a chain in which there is a system of conjugated carbon-carbon double bonds. The precursor of synthesis is IPP (isopentenyl pyrophosphate), the isomer of which is DMAPP (dimethylallyl pyrophosphate). Although various alternative pathways have been discovered and described, IPP in most eukaryotic organisms is synthesized from acetyl-CoA by the mevalonate pathway.

The carotenoids possess antioxidant properties, act as free radical scavengers, and take part in photoprotection. In addition, they can support many beneficial processes, such as stimulation of the immune system, modulation of intercellular signaling pathways, regulation of cell cycle and apoptosis (Fiedor and Burda, 2014). Numerous studies show that the content and quality of carotenoids synthesized by fungi depend on many environmental and genetic factors. Particularly, in the efficient production of carotenoids (600–700 mg/g) was observed in the representatives of the following species: *Cystofilobasidium capitatum*, *Rhodosporidium diobovatum*, *Rhodotorula glutinis* and *R. minuta*, *Sporobolomyces roseus* (phylum Basidiomycota), and *Rhizophydium sphaerocarpum* (phylum Chytridiomycota). The proportions of carotenoids β–carotene, torulene and torularodine have also been investigated (Yurkov et al., 2008). The carotenoid biosynthesis is described, inter alia, on the example of the synthesis of astaxanthin, a dye of great importance in biotechnology, occurring in the species *Xanthophyllomyces dendrorhous* (Basidiomycota). The carotene synthesis presumably arose as a defense mechanism of cells against oxidative damage by reactive oxygen species. In the subsequent stages of 'carotenogenesis' carried out by *X. dendrorhous*, lycopene is produced by four successive phytoene desaturations, and β–carotene occurred by cyclization of lycopene. Then astaxanthin is formed by hydroxylation and introduction of a ketone group. The final reaction is catalyzed by the enzyme astaxanthin synthase. Astaxanthin has strong antioxidant activity (Loto et al., 2012). A significant amount of carotenoid, over 0.6 mg/g of dry weight, was found in the following representatives of Basidiomycota: *Cystofilobasidium capitatum*, *Rhodosporidium diobovatum* and *R. sphaerocarpum*, and *Rhodotorula glutinis*, and lower carotenoid content was reported from *Sporobolomyces roseus* and *Rhodotorula minuta* (Yurkov et al., 2008). The β–carotene and lycopene ($\leq$ 0.24 mg/100 g d.w.) were also identified in *F. betulina* and *Collybia fusipes* (Reis et al., 2011). Additionally, carotenoids were determined using spectrophotometric analysis, in the following species of mushrooms: *Agaricus arvensis*, *A. bisporus*, *A. romagnesii*, *A. silvaticus*, *A. silvicola*, *Calocybe gambosa*, *C. cibarius*, *Craterellus cornucopioides*, and *Hypholoma fasciculare*. The carotene has also been detected in *Hypsizgus marmoreus*, *Lepista nuda*, *P. ostreatus*, *Polyporus squamosus*, *Russula delica*, and *Verpa conica* using HPLC–UV method (Ferreira et al., 2009). Carotenoids were also isolated from *C. cibarius*, which mainly contain β–carotene, and in smaller amounts lycopene, α–carotene, and γ– and δ–isomers of carotene (Velišek and Cejpek, 2011). Another study showed that of the six species collected in northwest Portugal, *C. cibarius* contains the highest amounts of β–carotene (13.56 µg/g d.w.), while the content of this metabolite in other studied mushrooms ranged from 1.95 to 12.77 µg/g dry weight (Barros et al., 2007). The content of this antioxidant was analyzed in the fruiting body and mycelial cultures of *Calocera viscosa*. The content of carotenoids in fruiting bodies growing in Poland was estimated by spectrophotometry at 2.46 mg/g fresh weight (Czeczuga, 1980). The higher amount than was determined by Czeczuga was found in the biomass of *C. viscosa* obtained from mycelial cultures. The cultures were kept under various conditions to optimize the biomass growth and to establish the most favorable conditions for the accumulation of β–carotene. The content of β–carotene in biomass from solid cultures is comparable to that in fruiting bodies (7.1 and 7.5 µg/g d.w., respectively). The mycelium derived from liquid cultures contained half of the β–carotene (3.5 µg/g d.w.) (Czeczuga, 1980; Muszyńska et al., 2017).

6. Sterols

The precursor of sterols is isopentenyl pyrophosphate (IPP), which is synthesized from acetyl–CoA via the mevalonate pathway. In the subsequent stages of the pathway, IPP and its isomer, dimethylallyl pyrophosphate (DMAPP), condense to form geranyl pyrophosphate (GPP) with the use of prenyl transferase, and then attach another IPP molecule, leading to the formation of farnesyl pyrophosphate (FPP). Then squalene is formed by the condensation, which is made by the condensation of two FPP molecules with the participation of squalene synthase. In a further step, squalene is cyclized to the cyclopentanophenanthrene system. Ring closure is possible when the hydrogen cations attack C3 of the cyclic system.

The total content of sterols in mushrooms ranges from 625 to 774 mg/100 g d.w. (Mattila et al., 2002). The prevalent fungal sterols are ergosterol and its peroxide. Ergosterol (24R–methyl–cholesta–5,7,22 (E)–trienol) is present in mushrooms in both free and esterified forms, and the relative ratio of these two forms may differ between species. Ergosterol is the main component of fungal cell membranes, strongly related to the cytoplasm. It is found in most Basidiomycota species, with the highest content occurring in the saprobiotic species. It accounts for 83–89% of the total sterol content. Ergosterol undergoes photolysis and chemical rearrangement due to irradiation with UV rays, which leads to its transformation into vitamin D_2. It is also a starting product for the creation of cortisone—an anti–inflammatory and anti-rheumatic adrenal hormone (Barreira et al., 2013). It has also been shown that this compound possesses many health-enhancing effects, such as anti-inflammatory, antioxidant activity and antihyperlipidemic effects. Moreover, it may be involved in the expression of specific defense genes (Barreira et al., 2013). Since, as mentioned previously, ergosterol is an essential component of fungal cell membranes, disruption of its synthesis leads to disrupted fungal reproduction and leads to the death. In medicine, this sensitive point of fungal metabolism was used in the treatment of mycoses. A number of drugs block this process at various stages of ergosterol biosynthesis (Weete et al., 2010). When ergosterol is exposed to UV light, it is photolyzed to form various products, mainly provitamins, tachysterol, and luminasterol. Further, these compounds undergo a thermal rearrangement to vitamin D_2. This vitamin is known for regulating calcium and phosphorus levels in the human body. Clinical studies presented a relationship between vitamin D_2 levels and function of the cardiovascular system. It has been proven that low levels of vitamin D_2 metabolites may contribute to an increased risk of congestive heart failure and hence greater mortality. Additionally, there is a growing evidence of a relationship between adequate vitamin D_2 levels and a reduced risk of colorectal, prostate, breast, and ovarian cancer (Shao et al., 2010). The ergosterol content in wild mushroom species, such as *Cantharellus tubaeformis*, *C. cibarius*, and *B. edulis*, ranges from 1.4 to 4.0 mg/g d.w. On the other hand, in the species of cultivated mushrooms *A. bisporus*, *L. edode*s, and *P. ostreatus*, the ergosterol content was in the range of 3.7 to 5.1 mg/g d.w. Other sterols, such as ergosta: 7.22–dienol, ergosta–5,7–dienol, and ergosta–7–enol (fungisterol) have also been detected in the above species of fungi (Teichmann et al., 2007). Another example of a sterol compound in mushroom fruiting bodies is ergosterol peroxide. It has a wide spectrum of biological activities, such as antioxidant, antitumor, anti-inflammatory, or antimicrobial. This compound was originally isolated from the mycelium of *H. erinaceus* (Krzyczkowski et al., 2009). In other studies, the content of ergosterol peroxide in *n*–hexane extracts from the mycelium of *H. erinaceus*, *L. sulphureus*, *Morchella esculenta*, and fruiting bodies of *B. edulis*, *Suillus bovinus*, and *Boletus badius* was determined. Furthermore, ergosterol peroxide has been isolated from numerous species of mushrooms, such as *G. lucidum*, *Cordyceps sinensis*, *Volvariella volvacea, A. mellea*, microscopic fungi, e.g., *Gibberella fujikuroi* and *Aspergillus* spp., or yeasts such as *Saccharomyces cerevisiae* (Krzyczkowski et al., 2009). Other fungal sterols are fungisterol, (3β,5α, 22E)–ergosta–6,8,22–trien–3–ol, hydroxyergosterol, isoergosterol, ergosta–7,22–dienol, and ergosta–5,7–dienol (Weete et al., 2010). In fruiting bodies of *Tuber sinense*, *T. aestivum*, *T. indicum*, *T. himalayense*, and *T. borchii* var. *sphaerospermum*, 13 sterols were identified. In this species, quantitatively dominant compounds were brassicasterol and ergosterol (17–64% and 25–67%, respectively). Other compounds includes cholesterol, 5–dihydroergosterol, kampesterol, 24(28)–dihydroergosterol, stigmasterol, stigmasta–7,24(28)–dienol, fungisterol, lanosterol, β–sitosterol, and 4–α–metyloergosta–8(9),24(28)–dienol (Tang et al., 2012). In turn, in *F. betulina*, *Coriolus (=Trametes) pargamenus, T. versicolor*, *Coriolus (=Trametes) heteromorphus*, *Fomitopsis citisina*, *Microporus flabelliformis*, *Gloephyllum sepiarium*, *Crytoderma citrinum*, *G. frondosa* ergosterol, and ergosta–7,22–dien–3β–ol were determined (Yokokawa, 1980).

The examples of fungal terpenes and sterols are presented in Table 4.1.

Table 4.1. Selected examples of terpenes found in fruiting bodies and mycelial cultures of several mushrooms

Terpenes	Compound	Species	Chemical formula	References
Monoterpenes (C10)	Limonene	*Gomphidius glutinosus* *Hydnum repandum* *Lepista nuda* *Marasmius aliaceus Mycena pura* *Suillus lutetus*	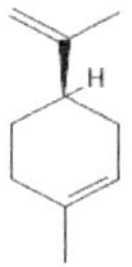 	(Breheret et al., 2019)
	α–Fenchene		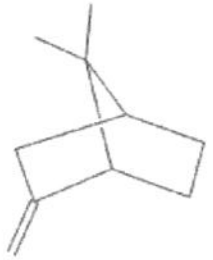	
	Camphene		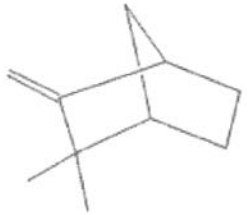	
	β–Phellandrene		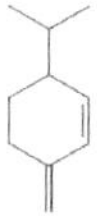	
	Terpinolene			

Table 4.1contd. ...

...*Table 4.1contd.*

Terpenes	Compound	Species	Chemical formula	References
Sesquiterpenes (C15)	*Hypnophilin*	*Lentinus crinitus* *Pleurotellus hypnophilus*	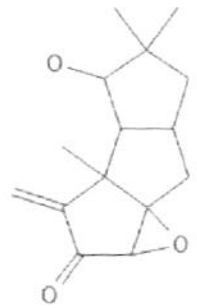	(Giannetti et al., 1986; Abate and Abraham, 1994; Qiu et al., 2018)
	Enokipodin A	*Flammulina velutipes*	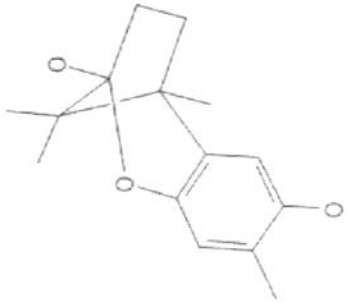	(Ishikawa et al., 2005)
	Nematolon	*Panus* sp.	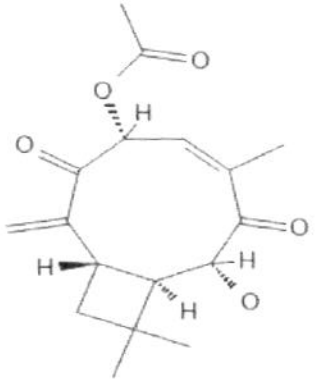	(Thakur et al., 2013)
	(−)–Complicatic acid	*Stereum complicatum*	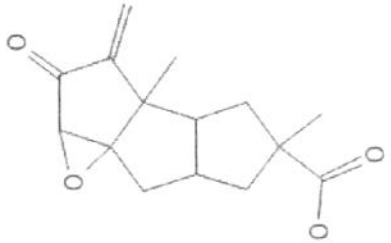	(Perera et al., 2020)
	Coriolin	*Coriolus consors*	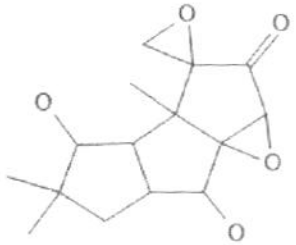	(Sushila et al., 2012)

Diterpenes (C20)	Erinacine A	*Hericium erinaceus*	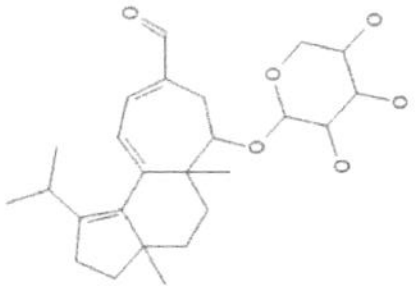	(Anke et al., 2002)
	Sarcodonin A	*Sarcodon scabrosus*	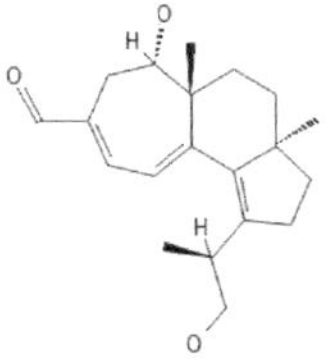	(Shi et al., 2011)
	Scabronine G	*Sarcodon scabrosus*	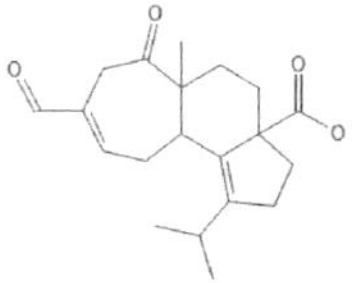	(Shi et al., 2011)
	Cyrneine A	*Sarcodon cyrneus*	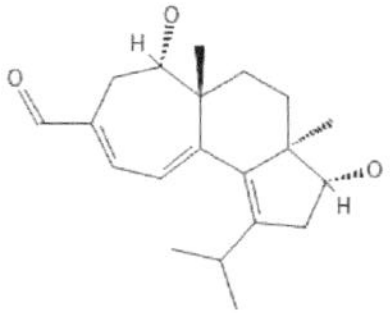	(Curini et al., 2005)
	Striatin A	*Cyathus striatus*	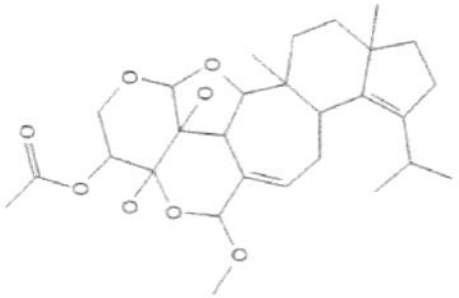	(Anke et al., 2002)

Table 4.1contd. ...

...Table 4.1contd.

Terpenes	Compound	Species	Chemical formula	References
Triterpenes (30)	Betulin	*Fomitopsis betulina*	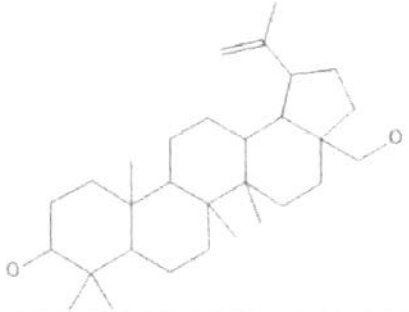	(Alresly et al., 2016)
	Betulinic acid	*Fomitopsis betulina*	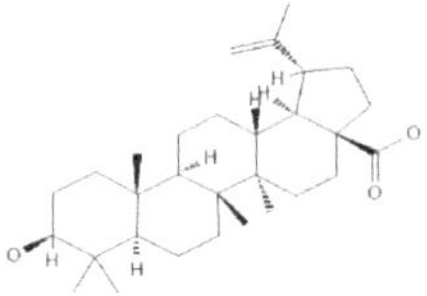	(Alresly et al., 2016)
	Lucidenic acid A	*Ganoderma lucidum* *Ganoderma applanatum*	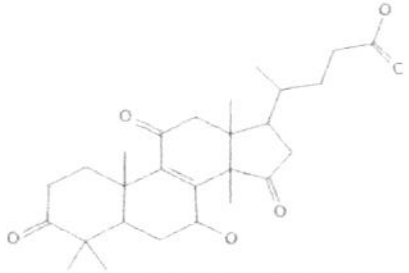	(Zhou et al., 2006)
	Ganoderic acid A	*Ganoderma lucidum*	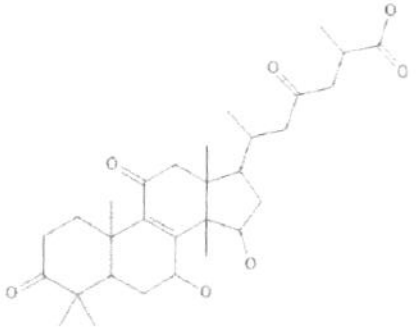	(Zhou et al., 2006)
	Polyporenic acid A	*Fomitopsis betulina*	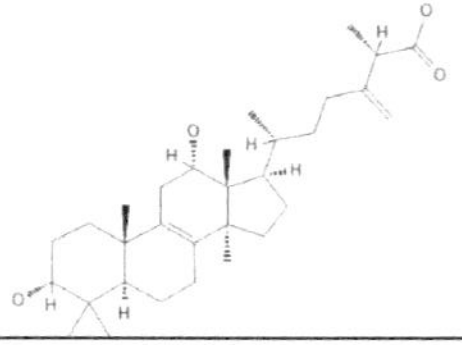	(Alresly et al., 2016)

Tetraterpenes (C40)	β–Carotene	*Cantharellus cibarius*	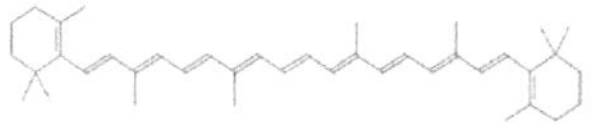	(Barros et al., 2007)
	Lycopene		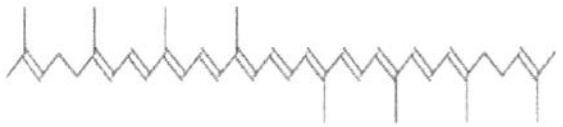	
Sterols	Ergosterol	*Agaricus bisporus* *Boletus edulis* *Cantharellus cibarius* *Cantharellus tubaeformis* *Lentinus edodes* *Pleurotus ostreatus*	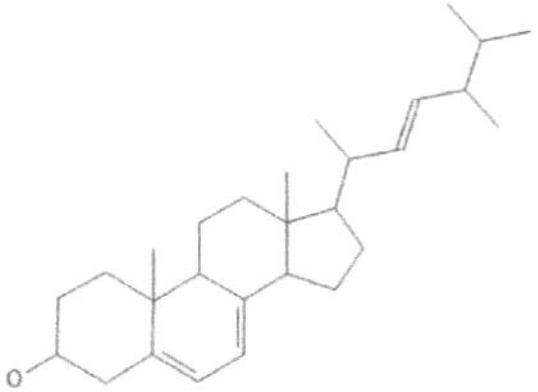	(Shao et al., 2010; Teichmann et al., 2007)
	Ergosterol peroxide	*Armillariella mellea* *Boletus badius* *Boletus edulis* *Cordyceps sinensis* *Gibberella fujikuroi* *Ganoderma lucidum* *Hericium erinaceus* *Laetiporus sulfureusi* *Morchella esculenta* *Saccharomyces cerevisiae* *Suillus bovinus* *Volvariella volvacea*	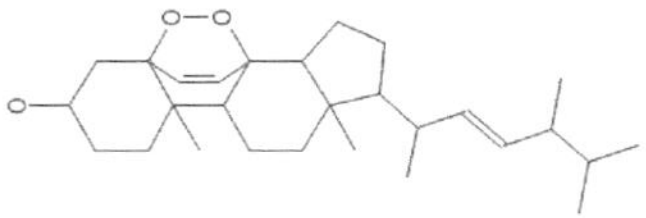	(Krzyczkowski et al., 2009)

7. Conclusion

Biopolymers are organic molecules containing monomeric units that are covalently bonded. The growing interest in biopolymers of mushroom origin comes from their widespread use in many industrial sectors. Mushrooms are a valuable source of biopolymers, including polysaccharides (glucans), peptides (lectins), isoprenoids (triterpenes, sesquiterpenes), and others.

Numerous studies have shown the high therapeutic value of these compounds. They possess antitumor, antioxidant, antibacterial, antifungal, antiviral, and immunomodulatory activities confirmed in experimental studies. In addition, they possess hepatoprotective, and neuroprotective effects, as well as support the beneficial intestinal microflora.

Mushroom biopolymers have a positive impact on the functioning of immune system, especially concerning inflammatory diseases in the human body. Recently, mycelial cultures have emerged as attractive alternatives for producing biopolymers. Mushrooms serve as a potential source of compounds especially with biopolymers for the mass production of functional foods that are recommended as an important ingredient in the human diet.

Further studies are warranted to investigate biopolymers obtained from mushrooms, which can certainly be an alternative to many pharmaceutical preparations and products.

References

Abate, D. and Abraham, W.R. (1994). Antimicrobial metabolites from *Lentinus crinitus*. *The J. Antibiot.*, 47: 1348–1350.

Abraham, W.R. (2001). Bioactive sesquiterpenes produced by fungi are they useful for humans as well. *Curr. Med. Chem.*, 8(6): 583–606.

Agger., S., Lopez-Gallego, F. and Schmidt-Dannert, C. (2009). Diversity of sesquiterpene synthases in the basidiomycete *Coprinus cinereus*. *Mol. Microbiol.*, 72(5): 1181–1195.

Alresly, Z., Lindequist, U., Lalk, M., Porzel, A., Arnold, N. and Wessjohann, L. (2016). Bioactive triterpenes from the fungus *Piptoporus betulinus*. *Rec. Nat. Prod.*, 10(1): 103–108.

Anke, T., Werle, A. and Zapf, S. (1995). Favolon, a new antifungal triterpenoid from *Favolaschia species*. *J. Antibiot.*, 48: 725–726.

Anke, T., Rabe., U., Schu, P., Eizenhoefer, T., Scharge M. and Steglich, W. (2002). Studies on the biosynthesis of striatal-type diterpenoids and the biological activity of herical. *Zeitschrift für Naturforschung*, 57: 263–271.

Aqueveque, P., Anke, T., Anke, H., Sterner, O., Becerra, J. and Silva, M. (2005). Favolon B, a new triterpenoid isolated from the Chilean *Mycena* sp. strain 96180. *J. Antibiot.*, 58(1), 61–64.

Bach, F., Helm, C.V., Bellettini, M.B., Maciel, G.M. and Haminiuk, W.I. (2017). Edible mushrooms: A potential source of essential amino acids, glucans, and minerals. *Int. J. Food Sci. Technol.*, 52(11): 2382–2392.

Barreira, J.C.M., Oliveira, M.B.P.P. and Ferreira, I.C.F.R. (2013). Development of a novel methodology for the analysis of ergosterol in mushrooms. *Food Anal. Meth.*, 7: 217–223.

Barros, L., Ferreira, M.J., Queirós, B., Ferreira, I.C.F.R. and Baptista, P. (2007). Total phenols, ascorbic acid, β–carotene end lycopene in Portuguese wild edible mushrooms and their antioxidant activities. *Food Chem.*, 103: 413–419.

Blagodatski, A., Yatsunskaya, M., Mikhailova, V., Tiasto, V., Kagansky, A. et al. (2018). Medicinal mushrooms as an attractive new source of natural compounds for future cancer therapy. *Oncotarget.*, 9(49): 29259–29274.

Breheret, S., Talou, T., Rapior, S. and Bessière, J.M. (2019). Monoterpenes in the aromas of fresh wild mushrooms (Basidiomycetes). *J. Agric. Food Chem.*, 45(3): 831–438.

Byerrum, R., Clarke, D., Lucas, E., Ringler, R., Stevens, J. and Stock, C.C. (1957). Tumor inhibitors in *Boletus edulis* and other *Holobasidiomycetes*. *Antibiot. Chemother.*, 7: 1–4.

Cai, H., Liu, X., Chen, Z., Liao, S. and Zou, Y. (2013). Isolation, purification, and identification of nine chemical compounds from *Flammulina velutipes* fruiting bodies. *Food Chem.*, 141(3): 2873–2879.

Caramelo, L., Martinez, M.J. and Martinez, A.T. (1999). A search for ligninolytic peroxidases in the fungus *Pleurotus eryngii* involving alpha-keto-gamma-thiomethylbutyric acid and lignin model dimers. *Appl. Environ. Microbiol.*, 65: 916–922.

Chen, B., Wang, S., Liu., G., Bao L., Huang, Y. et al. (2020). Anti–inflammatory diterpenes and steroids from peels of the cultivated edible mushroom *Wolfiporia cocos*. *Phytochem., Lett.*, 36: 11–16.

Cheung, P.C.K. 2013. Mini-review on edible mushrooms as source of dietary fiber: Preparation and health benefits. *Food Sci. Hum. Welln.*, 2: 162–166.

Curini, M., Maltese, F., Marcotullio, M.C., Menghini, L., Pagiotti, R. et al. (2005). Glaucopines A and B, new cyathane diterpenes from the fruiting bodies of *Sarcodon glaucopus*. *Pl. Med.*, 71(2): 194–196.

Czeczuga, B. (1980). Badania nad karotenoidami u grzybów IX Dacrymycetaceae. *Acta Mycol.*, 16(1): 115–120.

Daba, A.S. and Ezeronye, O.U. (2003). Anti-cancer effect of polysaccharides isolated from higher basidiomycetes mushrooms. *African J. of Biotechnol.*, 2(12): 672–678.

Devkota, P. and Hammerschmidt, R. (2020). The infection process of *Armillaria mellea* and *Armillaria solidipes*. *Physiol. Mol. Plant Pathol.*, 112: 101543. 10.1016/j.pmpp.2020.101543.

Donnelly, D.M., Quigley, P.F., Coveney, D.J. and Polonsky, J. (1987). Two new sesquiterpene esters from Armillaria mellea. *Phytochem.*, 26: 3075–3077.

Duru, M.E. and Çayan G.T. (2015). Biologically active terpenoids from mushroom origin: A review. *Rec. Nat. Prod.*, 9(4): 456–483.

El Enshasy, H.A. and Rajni Hatti-Kaul, R. (2013). Mushroom immunomodulators: unique molecules with unlimited applications. *Tr. Biotechnol.*, 31(12): 668–677.

Ferreira, I.F.C.R., Barros, L. and Abreu, R.M.V. (2009). Antioxidants in wild mushrooms. *Curr. Med. Chem.*, 1172: 5301–855.

Fiedor, J. and Burda, K. (2014). Potential role of carotenoids as antioxidants in human health and disease. *Nutrients*, 6(2): 466–488.

Gao, L., Li W., Zhao, Y. and Wang, J. (2009). The cultivation, bioactive components, and pharmacological effects of Armillaria mellea. *Afr. J. Biotechnol.*, 8(25): 7383–7390.

Giannetti, B., Steffan, B., Steglich, W., Kupka, J. and Anke, T. (1986). Antibiotics from basidiomycetes. Part 24.1: Antibiotics with a rearranged hirsutane skeleton from *Pleurotellus Hypnophilus* (agaricales). *Tetrahedron*, 42(13): 3587–3593.

Ho, J.C.K., Sze, S.C.W., Shen, W.Z. and Liu, W.K. (2004). Mitogenic activity of edible mushroom lectins. *Biochim. Biophys. Acta*, 1671(1–3): 9–17.

Hofrichter, M. (2002). Lignin conversion by manganese peroxidase (MnP). *Enz. Microb. Technol.*, 30: 454–466.

Howlett, J., Betteridge, V., Champ, M., Craig, S., Meheust A. and Jones, J. (2010). The definition of dietary fiber: Discussions at the ninth vahouny fiber symposium: Building scientific agreement. *Food Nutr. Res.*, 54: 5750. 10.3402/fnr.v54i0.5750.

Ishikawa, N., Yamaji, K., Ishimoto, H., Miura, K., Fukushi, Y. et al. (2005). Production of enokipodins A, B, C, and D: a new group of antimicrobial metabolites from mycelial culture of *Flammulina velutipes*. *Mycoscience*, 46: 39–45.

Ishiyama, D., Fukushi, Y., Ohnishi–Kameyama, M., Nagata, T., Mori, H., Inakuma, T. et al. (1999). Monoterpene–alcohols from a mushroom Dictyophora indusiate. *Phytochem.*, 50: 1053–1056.

Kalač, P. (2009). Chemical composition and nutritional value of European species of wild growing mushrooms: A review. *Food Chem.*, 113(1): 9–16.

Kersten, P. and Cullen, D. (2007). Extracellular oxidative systems of the lignin–degrading Basidiomycete Phanerochaete chrysosporium. *Fungal Gen. Biol.*, 44: 77–87.

Konska, G. (2006). Lectins of higher fungi (Macromycetes): Their occurrence, physiological role, and biological activity. *Int. J. Med. Mush.*, 8(1): 19–30.

Kosanić, M., Ranković, B., Rancić, A. and Stanojković, T. (2018). Evaluation of metal concentration an antioxidant, antimicrobial, and anticancer potentials of two edible mushrooms *Lactarius deliciosus* and *Macrolepiota procera*. *J. Food Drug Anal.*, 24(3): 477–484.

Krzyczkowski, W., Malinowska, E., Suchocki, P., Kleps, J., Olejnik, M. and Herold, F. (2009). Isolation and quantitative determination of ergosterol peroxide in various edible mushroom species. *Food Chem.*, 113: 351–355.

Kumari, S. and Kishor, R. (2020). Chitin and Chitosan: Origin, properties, and applications. pp. 1–33. *In*: Lambertus, A.M. and Broek, V.D. (Eds.). *Handbook of Chitin and Chitosan*, Wiley. https://www.wiley.com/en-cw/9781119450436.

Kuwahara, M., Glenn, J.K., Morgan, M.A. and Gold, M.H. (1984). Separation and characterization of two extracellular H2O2-dependent oxidases from ligninolytic cultures of Phanerochaete *chrysosporium*. *FEBS Lett.*, 169: 247–50.

Kuznetsova, O. and Vlasenko, E. (2020). Effect of natural and synthetic phytohormones on growth and development of higher Basidiomycetes. *Biotechnol. Acta.*, 13(5): 19–31.

Lee, S., Ryoo, R., Choi, J.H., Kim, J.H. Kim, S.H. and Kim, K.H. (2020). Trichothecene and tremulane sesquiterpenes from a hallucinogenic mushroom *Gymnopilus junonius* and their cytotoxicity. *Arch. Pharmacal Res.*, 43(2): 214–223.

Lemieszek, M. and Rzeski, W. (2012). Anticancer properties of polysaccharides isolated from fungi of the Basidiomycetes class. *Contemp. Oncol.*, 16(4): 285–289.

Lindequist, U., Niedermeyer, T.H.J. and Jülich W.D. (2005). The pharmacological potential of mushrooms. *Evid/Based Compl. Alt. Med.*, 2: 285–299.

Liu, D. (2014). A review of ergostane and cucurbitane triterpenoids of mushroom origin. *Nat. Prod. Res.*, 14(28): 1099–1105.

Loto, I., Gutiérrez, M.S., Barahona, S., Sepúlveda, D., Martinez-Moya, P. et al. (2012). Enhancement of carotenoid production by disrupting the C22–sterol desaturas gene (CPY61) in Xanthophyllomyces dendrorhous. *BMC Microbiol.*, 12: 235.

Lucas, E. (1957). Tumor inhibitors in *Boletus edulis* and other Holobasidiomycete. *Antibiot. Chemiother.*, 7: 1–4.

Mantovani, M.S., Bellini, M.F., Angeli, J.P.F., Oliveira, R.J., Silva, A.F. and Ribeiro, L.R. (2008). β-glucans in promoting health: Prevention against mutation and cancer. *Mut. Res.*, 658: 154–161.

Marcle, R. (2008). *Handbook of Prebiotics. Prebiotics: Concept, definition, criteria, methodologies, and products*. Boca Raton, Florida, USA: CRC Press. 10.1201/9780849381829.ch3.

Masuda, Y., Inoue, M., Miyata, A., Mizuno, S. and Nanba, H. (2009). Maitake beta-glucan enhances therapeutic effect and reduces myelosupression and nephrotoxicity of cisplatin in mice. *Int. Immuno. Pharmacol.*, 9(5): 620–626.

Matuszewska, A., Jaszek, M., Janusz, G., Osińska-Jaroszuk, M., Sulej, J. et al. (2013). Enzym lakaza wyizolowany z grzyba Cerrena unicolor do zastosowania w leczeniu raka szyjki macicy. *Nr patentu*: PL225869.

Mattila, P., Lampi, A.M., Ronkainen, R., Toivo, J. and Piironen, V. (2002). Sterol and vitamin D2 contents in some wild and cultivated mushrooms. *Food Chem.*, 76: 293–298.

Meng, X., Liang, H. and Luo, L. (2016). Antitumor polysaccharides from mushrooms: a review on the structural characteristics, antitumor mechanisms and immunomodulating activities. *Carbohydrate Research*, 424: 30–41.

Mester, T. and Field, J.A. (1998). Characterization of a novel manganese peroxidase–lignin peroxidase hybrid isozyme produced by Bjerkandera species strain BOS55 in the absence of manganese. *J. Biol. Chem.*, 273: 15412–15417.

Midland, S.L., Izac, R.R., Wing, R.M., Zaki, A.I., Munecke, D.E. and Sims, J.J. (2001). Melleolide, a new antibiotic from *Armillaria mellea*. *Tetrahed. Lett.*, 23(25): 2515–2518.

Miyazaki, T. and Nishijima, M. (1981). Studies on fungal polysaccharides. XXVII. Structural examination of a water-soluble, antitumor polysaccharide of *Ganoderma lucidum*. *Chem. Pharm. Bull.*, 29: 3611–3616.

Monique, R. 2021. Chemical and pharmacological characterization of terpenoids from *Hericium* species and other *Basidiomycota* as neuroregenerative compounds. *Monographie. Rechte vorbehalten—Freier Zugang*. 10.24355/dbbs.084-202007271046-0.

Muszyńska, B., Sułkowska-Ziaja, K., Łojewski, M., Opoka, W., Zając, M. and Rojowski, J. (2013). Edible mushrooms in prophylaxis and treatment of human diseases. *Med. Int. Rev.*, 101: 170–183.

Muszyńska, B., Kała, K. and Sułkowska-Ziaja, K. (2017). Edible mushrooms and their *in vitro* culture as a source of anticancer compounds. Biotechnology and production of anti–cancer compounds. pp. 231–251. *In*: Malik (Ed.). *Biotechnology and Production of Anti-Cancer Compounds*. Springer International Publishing AG.

Nikitina, V.E., Loshchinina, E.A. and Vetchinkina E.P. (2017). Lectins from mycelia of basidiomycetes. *Int. J. Mol. Sci.*, 18(7): 1334.

Patel, S. and Goyal, A. (2012). Recent developments in mushrooms as anticancer therapeutics: A review. *3 Biotech.*, 2: 1–15.

Pennerman, K., Guohua, Y. and Bennett, J.W. (2015). Health effects of small volatile compounds from East Asian medicinal mushrooms. *Mycobiol.*, 43(1): 9–13.

Perera, W.H., Meepagala, K.M., Wedge, D.E. and Duke, S.O. (2020). Sesquiterpenoids from culture of the fungus *Stereum complicatum* (Steraceae): Structural diversity, antifungal and phytotoxic activities. *Phytochem. Lett.*, 37: 51–58.

Perez–Boada, M., Ruiz-Duenas, F.J., Pogni, R., Basosi, R., Choinowski, T. et al. (2005). Versatile peroxidase oxidation of high redox potential aromatic compounds: Site directed mutagenesis, spectroscopic and crystallographic investigation of three long range electron transfer pathways. *J. Mol. Biol.*, 354: 385–402.

Qiu, Y., Lan, W., Li, H. and Chen, L. (2018). Linear triquinane sesquiterpenoids: Their isolation, structures, biological activities, and chemical synthesis. *Molecules*, 23(9): 2095.

Reis, F.S., Pereira, E., Barros, L., Sousa, M.J., Martins, A. and Ferreira, I.C.F.R. (2011). Biomolecule profiles in inedible wild mushrooms with antioxidant value. *Molecules*, 16: 4328–4338.

Ribeiro, B., Andrade, P.B., Silva, B.M., Baptista, P., Seabra, R.M. and Valentão, P. (2008). Comparative study on free amino acid composition of wild edible mushroom species. *J. Agric. Food Chem.*, 56(22): 10973–10979.

Riva, S. (2006). Laccases: Blue enzymes for green chemistry. *Tr. Biotechnol.*, 24: 219–226.

Roupas, P., Keogh, J., Noakes, M., Margetts, C. and Taylor, P. (2012). The role of edible mushrooms in health: Evaluation of the evidence. *J. Func. Foods*, 4: 687–709.

Ruthes, A.C., Smiderle, F.R. and Iacomini, M. (2015). D–Glucans from edible mushrooms: A review on the extraction, purification, and chemical characterization approaches. *Carbohyd. Polym.*, 117: 753–761.

Shi, X.W., Liu, L., Gao, J.M. and Zhang, A.L. (2011). Cyathane diterpenes from Chinese mushroom *Sarcodon scabrosus* and their neurite outgrowth–promoting activity. *Eur. J. Med. Chem.*, 46(7): 3112–3117.

Singh, R.S., Bhari, R. and Kaur, H.P. (2009). Mushroom lectins: Current status and future perspectives. *Crit. Rev. Biotechnol.*, 30(2): 99–126.

Singh, R.S., Bhari, R. and Kaur H.P. (2010). Mushroom lectins: Current status and future perspectives. *Crit. Rev. Biotechnol.*, 30(2): 99–126.

Singh, R.S., Kaur, H.P. and Kanwar J.R. (2016). Mushroom lectins as promising anticancer substances. *Curr. Prot. Pept. Sci.*, 17(8): 797–807.

Singh, S.S., Wang, H., Chan, Y.S., Pan, W., Dan, X. et al. (2014). Lectins from edible mushrooms. *Molecules*, 20(1): 446–469.

Singdevsachan, S.K., Auroshree, P., Mishra, J., Baliyarsingh, B., Tayung, K. and Thatoi, H. (2016). Mushroom polysaccharides as potential prebiotics with their antitumor and immunomodulating properties: A review. *Bioact. Carbohyd. Diet. Fib.*, 7: 1−14.

Slavin, J. (2013). Fiber and prebiotics: Mechanisms and health benefits. *Nutrients*, 5: 1417–1435.

Stadler, M., Anke, H. and Stener, O. (1994). Six new antimicrobial and nematicidal bisabolanes from the basidiomycete *Cheimonophyllum candidissimum*. *Tetrahedron*, 50(44): 12649–12654.

Stadler, M., Sterner, O. and Anke, H. (1995). 1,2–Dihydroxymintlactone, a new nematicidal monoterpene isolated from the basidiomycete Cheimonophyilum candidissimum (Berk & Curt.) Sing. *Zeitschrift für Naturforschung*, 50: 473–475.

Struck, C. (2015). Amino acid uptake in rust fungi. *Pl. Sci. Front.*, 6: 40. 10.3389/fpls.2015.00040.

Sułkowska-Ziaja, K., Muszyńska, B. and Końska, G. (2005). Biologically active compounds of fungal origin displaying antitumor activity. *Acta Pol. Pharma. Drug Res.*, 62(2): 153–160.

Sułkowska-Ziaja, K., Katarzyna, K., Lazur, J. and Muszyńska, B. (2018). Chemical and bioactive profiling of wild edible mushrooms. pp. 129–157. *In*: Singh, B.P., Chhakchhuak, L. and Prassari, A.K. (Eds.). *Biology of Macrofungi*. Switzerland AG: Springer Nature.

Sushila, R., Dharmender, R., Deepti, R., Vikash, K. and Permender, R. (2012). Mushrooms as therapeutic agents. *Br. J. Pharmacogn.*, 22(2): 459–474.

Tanaka, S., Ko, K., Kino, K., Tsuchiya, K., Yamashita, A. et al. (1989). Complete amino acid sequence of an immunomodulatory protein, ling zhi-8 (LZ–8). *J. Biol. Chem.*, 264(28): 16372–16377.

Tang, D., Xu, Y.Z., Wang, W.W., Yang, Z., Liu, B. et al. (2019). Cyathane diterpenes from cultures of the bird's nest fungus *Cyathus hookeri* and their neurotrophic and anti-neuroinflammatory activities. *J. Nat. Prod.*, 82(6): 1599–1608.

Tang, H.Y., Yin, X., Zhang, C.C., Jia, Q. and Gao, J.M. (2015). Structure diversity, synthesis, and biological activity of cyathane diterpenoids in higher fungi. *Curr. Med. Chem.*, 22(19): 2375–2391.

Tang, Y., Li, M.H. and Tang, Y.J. (2012). Comparison of sterol composition between Tuber fermentation mycelia and natural fruiting bodies. *Food Chem.*, 132: 1207–1213.

Teichmann, A., Dutta, P.C., Staffas, A. and Jägerstad, M. (2007). Sterol and vitamin D2 concentrations in cultivated and wild grown mushrooms: Effects of UV irradiation. *LWT–Food Sci. Technol.*, 40: 815–822.

Thakur, M.P. and Singh Harvinder, K. (2013). Mushrooms, their bioactive compounds, and medicinal uses: A review. *Med. Pl. Int. J. Phytomed. Rel. Ind.*, 5(1): 1–20.

Tian Y., Zhao, Y., Huang, J., Zeng, H. and Zheng, B. (2016). Effects of different drying methods on the product quality and volatile compounds of whole shiitake mushrooms. *Food Chem.*, 197: 714–722.

Velíšek, J. and Cejpek, K. (2011). Pigments of higher fungi: A review. *Czech J. Food Sci.*, 29(2): 87–102.

Wang, P.H., Hsu, C.I., Tang, S.C., Huang, Y.L., Lin, J.Y. and Ko, J.L. (2004). Immunomodulatory protein from Flammulina velutipes induces interferon–g production through p38 mitogen–activated protein kinase signaling pathway. *J. Agric. Food Chem.*, 52(9): 2721–2725.

Wang, Y., Bao, L., Yang, X., Dai, H., Guo, H. et al. (2012). Four new cuparene-type sesquiterpenes from *Flammulina velutipes*. *Helvetica*, 95(2): 261–267.

Wannet, W.J., Aben, E.M., van der Drift, C., van Griensven, J.L.D., Vogels, G.D. et al. (1999). Trehalose phosphorylase activity and carbohydrate levels during axenic fruiting in three *Agaricus bisporus* strains. *Curr. Microbiol.*, 39: 205–210.

Wasser, S.P. (2002). Medicinal mushrooms as a source of antitumor and immunomodulating polysaccharides. *Appl. Microbiol. Biotechnol.*, 60(3): 258–274.

Weete, J.D., Abril, M. and Blackwell, M. (2010). Phylogenetic distribution of fungal sterols. *PLOS One*, 5(5): 10898.

Wesenberg, D., Kyriakides, I. and Agathos, S.N. (2003). White rot fungi and their enzymes for the treatment of industrial dye effluents. *Biotechnol. Adv.*, 22: 161–187.

Wold, C.W., Gerwick, W.H., Wangensteen, H. and Inngjerdingen, K.T. (2020). Bioactive triterpenoids and water–soluble melanin from *Inonotus obliquus* (Chaga) with immunomodulatory activity. *J. Func. Foods*, 71: 104025.

Wong, D.W.S. (2009). Structure and action mechanism of ligninolytic enzymes. *Biotechnol. Appl. Biochem.*, 157: 174–209.

Xu, T., Beelman, R.B. and Lambert, J.D. (2012). The cancer preventive effects of edible mushrooms. *Anticancer Agents Med. Chem.*, 12(10): 1255–1263.

Xue, Z., Li, J., Cheng, A., Yu, W., Zhang, Z. et al. (2015). Structure identification of triterpene from the mushroom *Pleurotus eryngii* with inhibitory effects against breast cancer. *Pl. Foods Hum. Nutr.*, 70: 291–296.

Yan, J., Han, Z., Qu, Y., Yao, C., Shen, D. et al. (2018). Structure elucidation and immunomodulatory activity of a β-glucan derived from the fruiting bodies of *Amillariella mellea*. *Food Chem.*, 240(1): 534–543.

Yang, J.S., Su, Y.L., Wang, Y.L., Feng, X.Z., Yu, D.Q. et al. (1989). Isolation and structures of two new sesquiterpenoid aromatic esters: Armillarigin and armillarikin. *Pl. Med.*, 55: 479–481.

Ying-Chao, X., Xiao-Xia, X., Zhong-You, Z., Tao Feng, L. and Ji-Kai, L. (2018). A new monoterpene from the poisonous mushroom Trogia venenata, which has caused sudden unexpected death in Yunnan province, China. *Nat. Prod. Res.*, 32(21): 2547–2552.

Yokokawa, H. (1980). Fatty acid and sterol compositions in mushrooms of ten species of Polyporaceae. *Phytochemistry*, 19: 2615–2618.

Yurkov, A.M., Vustin, M.M., Tyaglov, B.V., Maksimova, I.A. and Sineokiy, S.P. (2008). Pigmented basidiomycetous yeasts are a promising source of carotenoids and ubiquinone Q10. *Microbiology*, 77(1): 1–6.

Zhang, M., Cui, S.W., Cheung, P.C.K. and Wang, Q. (2007). Antitumor polysaccharides from mushrooms: A review on their isolation process, structural characteristics, and antitumor activity. *Tr. Food Sci. Technol.*, 18: 4–19.

Zhao, S., Gao, Q., Rong, C., Wang, S., Zhao, Z. et al. (2020). Immunomodulatory effects of edible and medicinal mushrooms and their bioactive immunoregulatory products. *J. Fungi*, 6(4): 269.

Zhou, Y., Yang, X. and Yang, Q. (2006). Recent advances on triterpenes from *Ganoderma* mushroom. *Food Rev. Int.*, 22(3): 259–273.

Zhu, F., Du, B., Bian, Z. and Xu B. (2015). Beta-glucans from edible and medicinal mushrooms: Characteristics, physicochemical and biological activities. *J. Food Comp. Anal.*, 41: 165–173.

Zong, A., Cao, H. and Wang, F. (2012). Anticancer polysaccharides from natural resources: A review of recent research. *Carbohyd. Polym.*, 90: 1395–1410.

Metabolites and Medicine

5

Fungal Metabolites
Advances in Contemporary Industrial Scenario

J.A. Takahashi,[1,*] *C. Contigli,*[2] *B.A. Martins*[1] and *M.T.N.S. Lima*[3]

1. Introduction

The discovery of how to cure lethal diseases and the large-scale production of medicines were undoubtedly the greatest drivers of human prosperity and well-being on the planet. Contemporary society is the result of victory over major obstacles related to devastating diseases, some of them capable of taking more lives than wars themselves. The most visible example of these events is unquestionably the inspiring discovery of penicillin (**1**) (Fig. 5.1), from *Penicillium notatum*, reported by Alexander Fleming in 1929. After studies on its spectrum of action, clinical testing of penicillin (**1**) in humans started in the 1940s, when Howard Florey and Ernest Chain developed a suitable fermentation process to prepare and purify adequate amounts of this metabolite for toxicity evaluation (Aminov, 2010). The mass distribution of the antibiotic started during this period. However, most of the production was reserved for military use since, during World War II, many soldiers died due to infections caused by the low sanitary conditions in the trenches, rather than due to fatal gun shots. Countless lives were saved, from the beginning of the distribution of penicillin (**1**) to the present day.

After the penicillin (**1**) boom, cephalosporins produced by *Acremonium chrysogenum* were reported and, in 1956, Edward Abraham and Guy Newton reported the isolation and purification of cephalosporin C (**2**) (Fig. 5.1) (Abraham and Newton, 1961), an antibiotic resistant to the action of acylases, enzymes

[1] Departamento de Química, Instituto de Ciências Exatas, Universidade Federal de Minas Gerais, Belo Horizonte, MG 31270-901, Brazil.
[2] Serviço de Biologia Celular, Diretoria de Pesquisa e Desenvolvimento, Fundação Ezequiel Dias, Belo Horizonte, MG 30510-010, Brazil.
[3] Departamento de Ciência de Alimentos, Faculdade de Farmácia, Universidade Federal de Minas Gerais, Belo Horizonte, MG 31270-901, Brazil.
* Corresponding author: jat@qui.ufmg.br

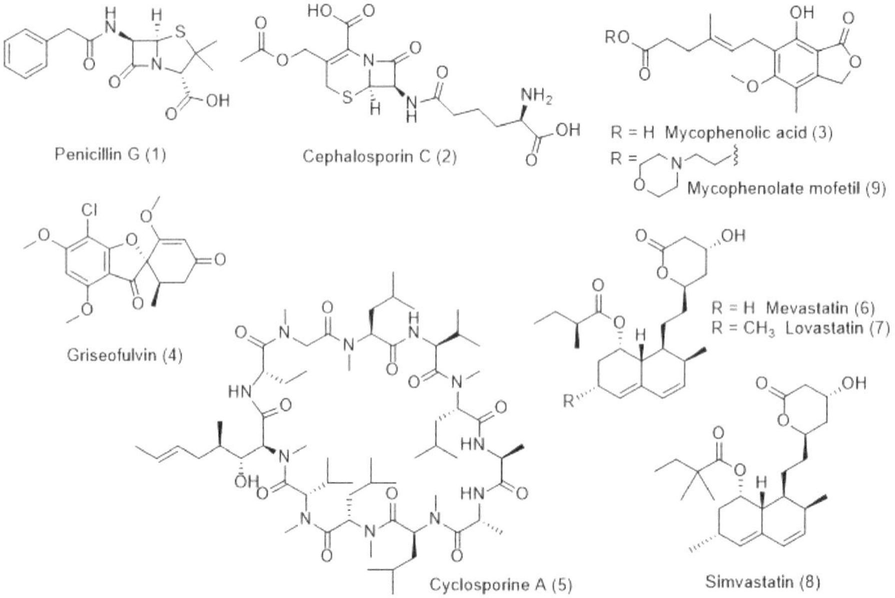

Fig. 5.1. Chemical structures of fungal metabolites **(1–9)** recognized as drugs.

produced by bacteria that break down the side chain of penicillin **(1)**, resulting in loss of antibiotic activity.

Although the importance of Fleming's milestone achievements cannot be underestimated, there are ancient accounts of the role of fungi in the production of antibiotic substances, such as in the *Papyrus of Ebers*, one of the oldest medical texts known (Hutchings et al., 2019). In this medical treatise, dated 1550 BC, there are reports on the prescription of moldy bread for the healing of wounds, probably due to the presence of antibiotic substances produced by fungi that colonize the bread. Another pre-penicillin mention found in the literature is the isolation and purification of mycophenolic acid **(3)** (Fig. 5.1) from *Penicillium brevicompactum* by Bartolomeo Gosio, an Italian researcher. Although the literature is controversial about the date of his discovery, 1913 (Page and Tait, 2015) or 1893 (Bentley, 2000), most papers on this area point mycophenolic acid **(3)** as the first purified antibiotic, i.e., before Fleming's report of penicillin **(1)** discovery, as reported in the interesting paper of Bentley (2000). Further developments in the pre-penicillin era have been reported by Aminov (2010).

Other natural fungal metabolites of historical importance that have become medicines are worthy of being cited. During the period when penicillin-related studies were on the rise (~ 1939), the isolation of griseofulvin **(4)** (Fig. 5.1), a fungistatic agent, from *Penicillium griseofulvum* (Petersen et al., 2014) was reported. Later, in 1969, Hans Peter Frey reported the production of cyclosporine A **(5)** (Fig. 5.1) by *Tolypocladium inflatum* and its immunosuppressive activity, which was of fundamental importance in medicine, enabling transplants by reducing the probability of organ rejection in recipient patients. The development of the first immunosuppressant discovered, cyclosporine A **(5)**, is very interesting and suggested as a further reading (Laupacis et al., 1982; Borel et al., 1995).

Years later, in 1973, Akira Endo isolated a metabolite from *Penicillium citrinum*, named compactin (mevastatin) **(6)** (Fig. 5.1), a competitive inhibitor of 3-hydroxy-3-methyl-glutaril-CoA reductase (HMG-CoA reductase), an enzyme responsible for the rate-control of cholesterol synthesis. In 1979, Endo reported the isolation of monacolin K, a C7 homolog of compactin **(6)**, from *Monascus ruber*. In the same period, mevinolin was isolated from *Aspergillus terreus* by Alfred Albert's group. Lovastatin **(7)** (Fig. 5.1) and mevinolin were soon assigned to the same chemical structure and the name lovastatin was adopted (Endo, 2010). Clinical trials with lovastatin **(7)** started in the early 1980s, culminating in Food and Drug Administration (FDA) approval. Natural statins and their synthetic analogs such as simvastatin **(8)** (Fig. 5.1), the result of molecular simplification, are drugs capable of mimicking the natural intermediate involved in the reduction promoted by HMG-CoA reductase. The molecular interaction between the enzyme and statins is much stronger than the interaction with the natural substrate, preventing a key step in cholesterol biosynthesis. A synthetic statin, structurally different from the natural ones, with the same HMG-CoA affinity as the molecules responsible for inhibition of cholesterol synthesis, is one of the world's best-selling drugs (Barrios-González and Miranda, 2010).

In 1982, Alisson and Fugui prepared a derivative of mycophenolic acid **(3)**, mofetil mycophenolate **(9)** (Fig. 5.1), another important immunosuppressant. A few years later, around 1986, the isolation of huperzine A **(10)** (Fig. 5.2) was reported from the fungal species *Huperzia serrata* (Ma et al., 2007). Huperzine A **(10)** is an inhibitor of acetylcholinesterase, a target enzyme of drugs intended for patients with Alzheimer's disease. Although this is still an incurable disease, acetylcholinesterase inhibitors increase the quality of life and cognitive functions of patients, being an important palliative treatment (Ding et al., 2012).

Many other fungal metabolites have become successful drugs, but the examples mentioned are already strong supporting arguments on the contribution of fungal metabolites to humanity. The development of the first industrial antibiotic was relatively rapid, considering the laboratory equipment and knowledge existing in the first half of the twentieth century. These well-succeeded enterprises have led many research groups to start broad screenings on fungal metabolites aiming at finding new drug leads. In the recent decades, many metabolites have been isolated, identified, and associated with various biological activities, such as anti-inflammatory (Xu et al., 2019), antimicrobial (Ancheeva et al., 2020), antiviral (Linnakoski et al., 2018; Deshmukh et al., 2022), antitumor (Uzma et al., 2018), and acetylcholinesterase inhibitors (Takahashi et al., 2019). Fungal metabolites effective for specific diseases such as malaria (Niu et al., 2020), Chagas disease (Do Nascimento et al., 2020), and thrombosis (Wu et al., 2009) have also been described.

Processes for industrial production and features of many fungal metabolites have been protected in patents, covering their biological potential such as antiviral [alterporriol R **(11)** and Q **(12)**], antimicrobial [beauvericin **(13)**], antitumor [camptothecin **(14)**, taxol **(15)**, and podophyllotoxin **(16)**], antifungal [cryptocandin **(17)**], and antioxidant agents [flavipin **(18)**], compounds for cardiovascular control

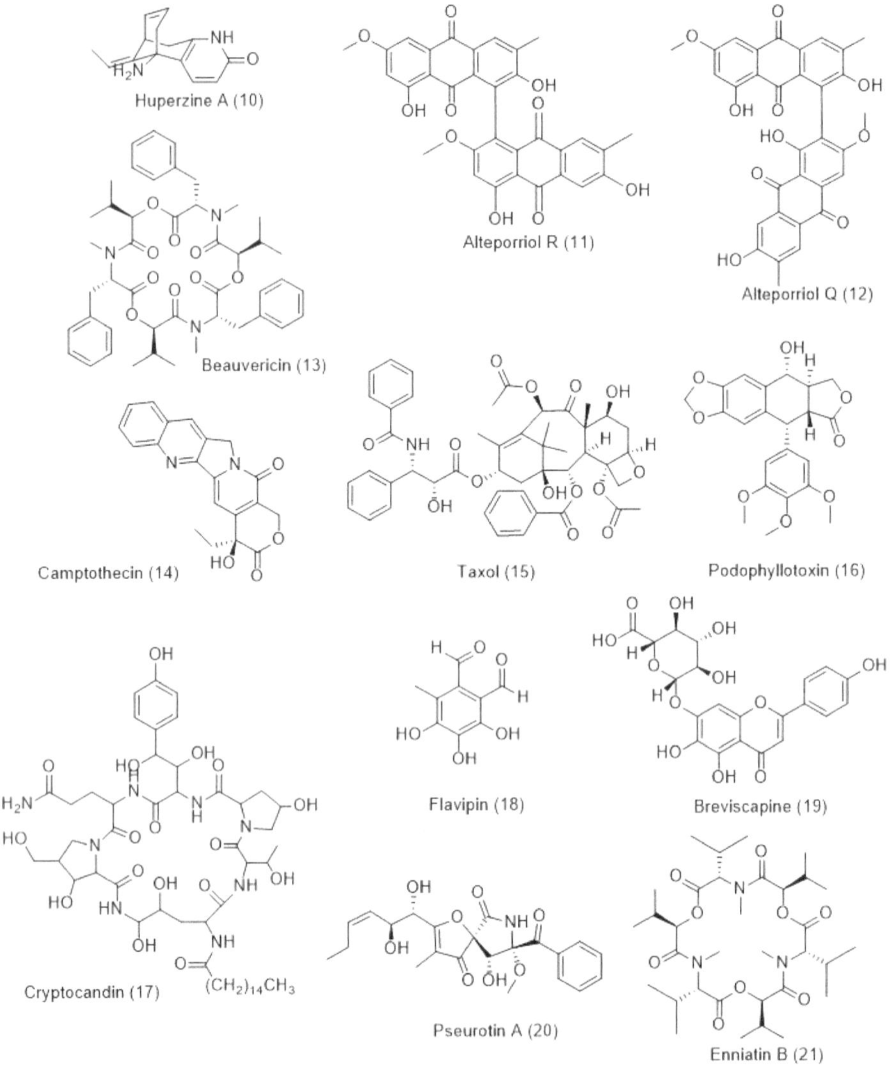

Fig. 5.2. Chemical structures of some fungal metabolites **(10–21)** protected in patents.

[breviscapine **(19)**], food preservatives [pseutorin A **(20)**], acetylcholinesterase inhibitors [huperzine A **(10)**] and intermediaries as enniatin B **(21)**, a drug used to prepare antitubercular agents (Fig. 5.2). Some representative patents are presented in Table 5.1 (Torres-Mendoza et al., 2020).

Nevertheless, the number of fungal metabolites identified and patented seems disproportional in comparison with the number of new drugs of fungal origin, despite the great technological advances achieved in recent decades in all areas related to discovery of new drugs and biotechnology. Additionally, several unprecedented environments are prolific for the isolation of new fungal species or metabolically specialized species adapted to extreme conditions, subjected to osmotic, thermal,

Table 5.1. Some fungi metabolites and derivatives subjected to patents.

Secondary Metabolite or Derivative	**Fungal Origin**	**Activity**	**Country**	**Publication Date**	**Patent Number**
Alterporriol R (**11**) and Q (**12**)	Endophytic *Alternaria* sp. (ZJ2008003)	Antiviral	China	2016-03-02	CN102643186A
Beauvericin (**13**)	Endogenetic *Fusarium* sp. (Dzf2)	Antibacterial	China	2008-08-13	CN101240249A
Breviscapine (**19**)	*Alternaria* sp.	Cardiovascular control	China	2003-06-04	CN1421522A
Camptothecin (**14**)	Endophytic fungus MTCC 5124 from *Nothapodytes foetida*	Anticancer	United States	2006-06-22	US20060134762A1
Cryptocandin (**17**)	Endophytic *Cryptosporiopsis* cf. *quercina*	Antifungal	United States	2003-09-02	US6613738B1
Enniatin B (**21**)	Endogenous *Fusarium* from *Rhizophora apiculata* Blume sp.732#	Anti-tuberculosis	China	2010-03-17	CN101669939A
Flavipin (**18**)	Endophytic *Chaetomium globosum* CDW7 from *Ginkgo biloba*	Antioxidant	China	2014-10-08	CN103087923A
Huperzine A (**10**)	Endophytic fungus from Herba *Lycopodii serrati*	Acetyl cholinesterase inhibitor	China	2008-10-01	CN101275116A
Pleuromutilin **(22)**	*Clitopilus passeckerianus*	Antibiotic	United Kingdom	1968-04-24	GB1111010A

Table 5.1 contd. ...

...Table 5.1 contd.

Secondary Metabolite or Derivative	Fungal Origin	Activity	Country	Publication Date	Patent Number
Podophyllotoxin (**16**)	Endophytic *Phialocephala fortinii* from *Podophyllum peltatum*	Anticancer	United States	2004-12-09	US20040248265A1
Pseurotin A (**20**)	Endophytic *Aspergillus fumigatus* GRP13	Antibacterial, food preservative	China	2018-01-23	CN104774774A
Rancinamycin IV (Protocatechualdehyde)	Endogenetic *A. fumigatus* HWT5 from *Schisandra chinensis*	Anticancer, antibacterial, food preservative	China	2016-04-13	CN103966109B
Taxol (**15**)	Endogenetic *Aspergillus niger* var.taxi HD86-9	Anticancer	China	2011-07-27	CN101486974A
3β,5α,9α-Trihydroxy ergot steroid-7,22-bis-alkene-6-ketone	Mushroom *Volvariella volvacea*	Anticancer	China	2017-12-15	CN107474092A

and other physicochemical challenges. Table 5.2 presents representative reports on metabolites produced by different groups, such as extremophiles (Ibrar et al., 2020), endophytes (Deshmukh et al., 2018; Ancheeva et al., 2020; Rustamova et al., 2020), endolichenic (Agrawal et al., 2020), and marine fungi species (Shabana et al., 2020; Agrawal et al., 2021a). Although some compounds may have been counted in more than one of the reviews reported in Table 5.2, but these reviews represent

Table 5.2. Representative reviews about fungi metabolites over time.

Fungi or Class of Compounds	Number of Metabolites Reported and Scope	Period	References
Endolichenic	99 novel compounds, highlighted out of 172 reported endolichenic fungi metabolites	2008–March 2019	Agrawal et al., 2020
Endophytes from terrestrial plants	221 metabolites, structurally diverse and novel secondary metabolites from 67 species of fungi, all biologically active	January 2018–June 2019	Rustamova et al., 2020
Endophytes	65 metabolites with pronounced biological activities, primarily as antimicrobial and cytotoxic agents, some of them with high selectivity or novel mechanisms of action	2010–2017	Ancheeva et al., 2020
Brazilian endophytes	303 chemical compounds from 60 Brazilian endophytic fungi strains, many of them with cytotoxic, antibacterial, antifungal, and antiparasitic activities	Up to 2019	Ribeiro et al., 2021
Extremophile	314 novel compounds from 56 extremophilic fungal strains, chemical structures, and biological potential	2005–2017	Zhang et al., 2018
	155 compounds (105 new and 50 known) isolated from 25 *Penicillium* species, 16 *Aspergillus* species, and 23 other species; bioactive compounds (77 cytotoxic, 46 anti-microbial, and 32 with nematocidal, anti-allergic, antioxidant, and anti-inflammatory potential)	2005–2020	Ibrar et al., 2020
Marine	187 novel compounds and 212 known compounds with anticancer and antibacterial activities	2011–2019	Shabana et al., 2020
	471 new terpenoids of 6 groups (mono, sesqui, di, sester, tri, and meroterpenes) from 133 marine fungal strains of 34 genera (140 research papers)	2015–2019	Jiang et al., 2020
Lasiodiplodia theobromae	134 chemically defined compounds from over 30 *L. theobromae* isolates, most of them with phytotoxic, cytotoxic, and antimicrobial activities	not mentioned	Salvatore et al., 2020
Antraquinones	166 fungal antraquinones	1966–2020	Masi and Evidente, 2020
Ganoderma lucidum	83 triterpenoids and 7 ganoderic acids with antitumor, antidiabetic, and other biological activities	1982–2018	Liang et al., 2019

the expressive number of fungal metabolites isolated, chemically characterized and assayed.

The huge gap between the number of fungal bioactive metabolites available and the compounds that effectively become drug leads and reach the pharmaceutical industry relies upon many factors that will be further discussed, including early issues related to wet-bench experiments, process optimization, and issues related to toxicity in preclinical and clinical trials.

2. Secondary Metabolites: Advances and Issues

Due to several factors, the chemistry of natural products gained new impetus from the 1990s; a scenario that was greatly influenced by the increasing attention that society, including scientists and entrepreneurs, directed to nature, sustainability, and biodiversity during that time. The diversity and structural complexity of natural products have been increasingly associated with new mechanisms of action and new applications of these molecules in various industrial sectors, predominantly in the pharmaceutical industry (Newman and Cragg, 2020).

A more effective contribution, i.e., an increase in the number of fungal metabolites used as lead molecules for drug development, depends greatly on overcoming wet-bench issues, to direct the efforts to the development stage. Without entering the scope of infrastructure and financing limitations, an initial difficulty is based on the constant re-isolation of already known substances that, in general, constitute the major metabolites produced by the fungal species. On the other hand, the isolation of novel metabolites, with new skeletons, is traditionally time-consuming and often such substances are isolated in low yield (Takahashi et al., 2013). Efforts to enhance the expression of minor metabolites have been well addressed by robust studies on the fungal nutritional requirements. The term OSMAC (one strain, many compounds) was proposed to express the susceptibility of fungal metabolism to growing conditions, since biotic and abiotic factors can modulate the expression of different biosynthetic pathways, enabling yield improvement and diversification of metabolites production (Takahashi et al., 2016; Ying et al., 2018). Several effective methods have been successfully utilized in this field, such as epigenetic induction (Guo and Zou, 2020), dereplication approaches (Abrol et al., 2021), and genome mining (El-Sayed, 2020).

Several approaches have also been proposed to overcome the initial low yields of bioactive fungal metabolites (Takahashi et al., 2020). This issue is of huge importance, since the limited availability of compounds isolated strongly slows down post-bench development, including toxicity, preclinical or clinical research. Therefore, optimization of fermentative conditions to achieve large-scale production is usually necessary (Kebede et al., 2017). Most often, by varying the carbon and/or nitrogen sources utilized in the fermentation (Tang et al., 2021), yield improvement can be accomplished in preliminary bench works using the OSMAC approach. Many efforts have been directed to the use of agricultural or food industry residues as sources of nutrients to maintain productivity while reducing process costs (Kornienko et al., 2015), as exemplified in the works of Del Valle et al. (2016) and Vasantha et al. (2018).

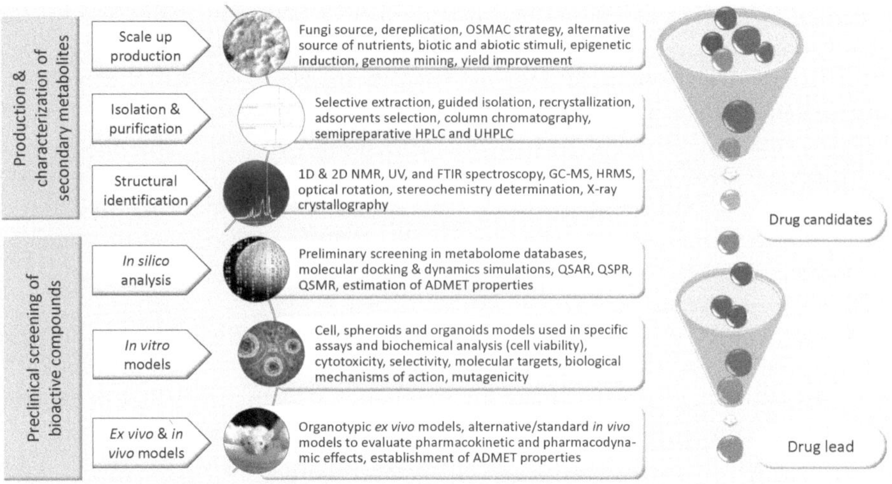

Fig. 5.3. Key steps from fungal metabolite isolation until selection of drug leads.

Going forward in the process, efficient purification techniques of organic compounds were developed since penicillin **(1)** purification, greatly advancing from the manual techniques used in the mid-twentieth century. Efficient chromatography systems utilizing High-Performance Liquid Chromatography (HPLC), Ultra-High Performance Liquid Chromatography (UHPLC), counter-current facilities, and other modern separation techniques, as well as reversed-phase and other versatile adsorbents, can be used nowadays to decrease bench time with high separation efficiency (Atanasov et al., 2021). Structural identification techniques have also made a huge breakthrough in relation to the indirect techniques used in the 1950s, although X-rays were already an option to resolve the structure of some crystalline compounds at that time. Currently, powerful mass and nuclear magnetic resonance (NMR) spectrometers speed up identification of compounds, independent of the crystalline aspect. In terms of processes, hyphenated techniques, robotic screenings, organic synthesis, metabolic profiling, dereplication, and others have also been strong allies in drug discovery and development (Emwas et al., 2020; Garcia-Perez et al., 2020; Lima et al., 2020). Figure 5.3 contains an overview of the key steps of fungal metabolites in the discovery of drug candidates.

2.1 Drug Discovery in Biotechnology of Fungi

Innovative *in silico* strategies associated with analytical methods have been improving the screening of fungal bioactive metabolites, reducing time and costs of experimentation, and even directing the selected candidates for potential applications. Metabolomics technologies consist of high throughput screening, where data obtained by mass spectrometry (MS), liquid chromatography (LC), gas chromatography (GC), and nuclear magnetic resonance (NMR) spectroscopy are processed by bioinformatics, to generate databases containing the profile of metabolites produced by an organism—the metabolome (Wishart, 2007; Atanasov et al., 2021). Computational tools such as molecular docking and molecular dynamics simulations can elucidate the chemical structures of new compounds and mimic their interactions with possible molecular targets in biological systems (Chen, 2015; Gupta et al., 2018; Hollingsworth and Dror, 2018; Oselusi et al., 2021). The structure-activity relationship (SAR), quantitative SAR (QSAR), quantitative structure-property relationship (QSPR), and quantitative structure-metabolism relationship (QSMR), among other computational techniques using data mining, machine learning, and virtual modeling, can estimate the potential pharmacokinetic and toxic effects of these molecules, the so-called ADMET (absorption, distribution, metabolism, excretion, and toxicity) properties (Rifaioglu et al., 2019; Madden et al., 2020; Wu et al., 2020).

Clearly, it is essential to confirm the computational predictions by the experimental research. In this field, several analytical methods have been developed, improved, validated, and integrated on technological platforms of chemical and biological assays, to determine a vast spectrum of cellular targets, chemical and biological interactions, toxicity, and a variety of biological mechanisms of action of selected molecules (Zhivotovsky et al., 1999; Freshney, 2005; Sachana et al., 2018; Cao et al., 2021). Metabolomics and high throughput proteomics and transcriptomics technologies allow the achievement of complete databases of secondary metabolites, proteins (proteome), or transcript RNAs (transcriptome) of biological samples in response to selected drug candidates (Weston et al., 2004; Wishart, 2007; Wang et al., 2009; Hautbergue et al., 2018).

Currently, concerns regarding preclinical experiments rely on the increasing demand of society to reduce the number of animals under experimental conditions, which imposes the establishment of alternative models to verify the effects of pharmaceuticals, cosmetics, or any other compound with industrial interest. At the same time, to overcome the limitations of *in vitro* two-dimensional cellular assays and improve simulations of *in vivo* microenvironment, several models of 3D spheroids and organoids, besides *ex vivo* organotypic models, have been developed to evaluate drug candidates on preclinical drug discovery (Gähwiler et al., 1997; Fennema et al., 2013; Lancaster and Knobich, 2014; Supadmanaba et al., 2021). Over the past years, various *in vivo* alternative models such as algae (*Volvox*), chicken chorioallantoic membrane (CAM), invertebrates as worms (*Caenorhabditis elegans*) and flies (*Drosophila melanogaster*), and simple-vertebrates such as zebrafish (*Danio rerio*), have been used to systematically assess biological effects with precision, reproducibility, and statistical confidence, while reducing the use of pain-susceptible organisms and the requirement of expensive animal facilities in early preclinical studies (Kendall et al., 2018; Lopes et al., 2021).

Nevertheless, the path of a bioactive fungal metabolite tested on *ex vivo* organotypic and the other above mentioned models has other limitations until industrial production can be undergone. The complexity of pharmacokinetic and systemic responses observed in experimental mammalian models and humans is still impossible to be replicated entirely on these alternative models. Any experimental drug formulation will present different behaviors *in vivo*, including the form of administration, time and concentration of exposure, specificity *versus* toxicity for tissues and cells, metabolic degradation, clearance, and excretion, local or systemic effects (including adverse immune response), and genotoxic and teratogenic potential

(Freshney, 2005; Doke et al., 2015). Consequently, only after these extensive preclinical studies, potential drugs will be approved for clinical trials in humans, with the aim of developing industrial production, or disapproved as toxic or ineffective.

2.2 Genetic Approaches on Fungi Secondary Metabolites Studies

The numerous bioactive natural products from fungi known to date are part of an extensive secondary metabolism repertoire (Bills and Gloer, 2016). However, even though the number of new molecules has increased in recent decades, the biosynthetic pathways of compounds with biotechnological applications are not out of print, and research is arduous. In this way, great progress has been made in the approaches in metabolomics for the purification and in-depth characterization of secondary metabolites (Hautbergue et al., 2018). So far, industrial efforts on the extensive automation of molecular bio-screening have accelerated the discovery of novel bioactive molecules (Newman, 2017). However, such techniques make use of downstream products, working on the endpoint of their biosynthesis. Hence, the adoption of upstream techniques, targeting the source of compound synthesis, narrows down key spots in the fundamental production steps.

As a point in case, molecular biology techniques allowed gene mining, gene functional characterization, and the heterologous expression of target genes coding for important secondary metabolites (Oikawa, 2020). Current knowledge on genomics, including gene manipulation, genetic editing, DNA sequencing, and molecular predictive tools are state-of-the-art technologies that have revolutionized the prospects of new industrial bioproducts (Adrio and Demain, 2010; Oikawa, 2020). In this sense, a major characteristic of the genetic setup coding for secondary metabolites in fungi is the organization of gene clusters, that are genetic units with several genes under a common expression promoter (Keller, 2019). Such clusters are considered toolkits for innovations in several biotechnological fields, and the composition of fungal genomic banks allows the characterization of secondary metabolite gene clusters (SMGC) (Theobald et al., 2018). Such recognizable genomic structures are the basis for bioinformatic predictive tools that can compare, annotate, and identify new genes with possible functional applications, and common genetic patterns distributed along the genomes. Accordingly, the recognition of similarities among genetic structures coding genes for secondary metabolites is of great interest, owing to limitations currently faced at the laboratory scale for the production and diversification of metabolites.

The expression of some genes induced by environmental triggers epigenetic control of gene expression, which is a strong guiding force for the biosynthesis of secondary metabolites (Gacek and Strauss, 2012). The discovery of gene coding for the expression of a secondary metabolite allows downstream application in genetic cloning and heterologous expression (Oikawa, 2020). Smith et al. (1990) reported a primary effort to manipulate fungal genes, and were able to clone the *Penicillium chrysogenum* gene cluster containing all the enzymes required for penicillin biosynthesis. The cluster was genetically engineered in a *cosmid*, allowing its expression in *Neurospora crassa* and *A. niger* species (Smith et al., 1990). More recently, a similar strategy was effective in the expression of pleuromutilin **(22)**

(Fig. 5.4, Table 5.1), a diterpene with human and veterinary clinical applications as an antibiotic, synthesized by the basidiomycete *C. passeckerianus* (patented by Sandoz GmbH, 1968). Based on the identification of the gene cluster responsible for antibiotic production (Bailey et al., 2016; Alberti et al., 2017) *Aspergillus oryzae* was used as a bio-platform for the production and isolation of pleuromutilin **(22)**, allowing to isolate, characterized, and easily promote derivative products.

In 2021, the genome sequence of a lovastatin **(7)** producing strain, *A. terreus* ATCC 20542, was published (Ryngajłło et al., 2021), revealing the conservation of its cluster DNA sequence, a knowledge useful for the induction of clusters. Lovastatin **(7)** production, for instance, is favored by oxidative stress, which was previously investigated in other *A. terreus* strains, and the similarities in gene structure may guide the induction methods in different species (Barrios-González et al., 2020).

A versatile tool, that is currently used in different organisms (i.e., fungi, bacteria, and mammals), is the CRISPR/Cas9 genomic editing tool. This technique can target a gene of interest with impressive specificity (Brooks and Gaj, 2018). Therefore, key applications are possible as site-specific deletions and the isolation of entire clusters. The use of CRISPR/Cas9 as a mutation tool is powerful for the proper association of phenotypes and specific modifications on the DNA sequence (Wang and Coleman, 2019). The deletion of a polyketide synthase gene (FUM1) on the plant pathogen *Fusarium proliferatum* was demonstrated to block fumonisin mycotoxin biosynthesis without further modification of its genome. In addition, encoding FUM1 has been shown to be directly implicated in the fumonisin biosynthesis pathway (Ferrara et al., 2019). Furthermore, other industrial applications are also possible. The lack of complete genome sequences of some fungal species is surpassed by interspecific genomic similarities. By way of illustration, a CRISPR/Cas9 platform developed for the *Aspergillus* species was applied to *Talaromyces atroroseus*. The successful intervention of hybrid polyketide synthase-non ribosomal peptide synthetase enzymes, involved in platelet-activator factor biosynthesis with important biomedical interest, demonstrated that CRISPR toolkits developed for other species with close gene homology were able to work in a cross-platform fashion (Nielsen et al., 2017).

Altogether, advances in the manipulation of fungal genomics have helped to elucidate cost-effective genetic tools, and faster responses in the production of molecules with industrial interest. Such discoveries must pave the way for more adequate responses to pandemics. Facing pandemic periods like the one in December 2019 with COVID-19, the set-up genetic and bioinformatic approaches helped to define the potential of fungal secondary metabolites for the treatment of severe

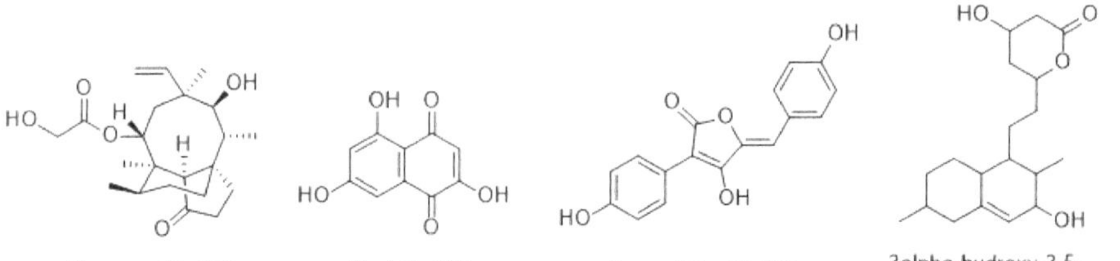

Pleuromutilin (22) Flaviolin (23) Aspergillide B1 (24) 3alpha-hydroxy-3,5-dihydromonacolin L (25)

Fig. 5.4. Chemical structures of some metabolites **(22–25)** expressed by manipulation of fungal genes.

respiratory disease associated with SARS-CoV-2 infections. Potential SARS-CoV-2 proteases ($3CL^{pro}$ and M^{pro}) inhibitors previously deposited in data banks have been readily screened since the beginning of the pandemic. Based on molecular docking of more than 100 molecules, the *Phoma wasabiae* metabolite flaviolin **(23)** (Fig. 5.4) was proposed as a ligand for SARS-CoV-2 $3CL^{pro}$ protease, preventing virus replication *in vitro* (Soga, 1982; Rao et al., 2020). *A. terreus*, previously cited for its importance in the production of lovastatin **(7)**, was shown to produce two molecules with antiviral activity. Bioinformatic analysis of aspergillide B1 **(24)** and 3αhydroxy3,5dihydromonacolin L **(25)** (Fig. 5.4) demonstrated its affinity to M^{pro} of SARS-CoV-2 (El-Hawary et al., 2021). Notably, the versatility of fungal species enables the production of secondary metabolites with diverse applications. In-depth research on the molecular mechanisms determining biosynthetic pathways is an important key for industry-aimed synthetic biology projects.

3. Fungal Metabolites to Treat Non-Communicable Diseases

Non-communicable diseases (NCDs) consist of chronic dysfunctions that affect a large part of the world population and cause approximately 71% of registered deaths. The main ones are cardiovascular diseases, cancers, diabetes, and chronic respiratory diseases, but dementia and other types of NCDs also affect the lives of many people (Fig. 5.5). Although many of them are commonly associated with aging, many premature deaths of working-age people have taken place (WHO, 2021). In addition to high mortality, due to symptoms, treatment costs and duration, patients can have

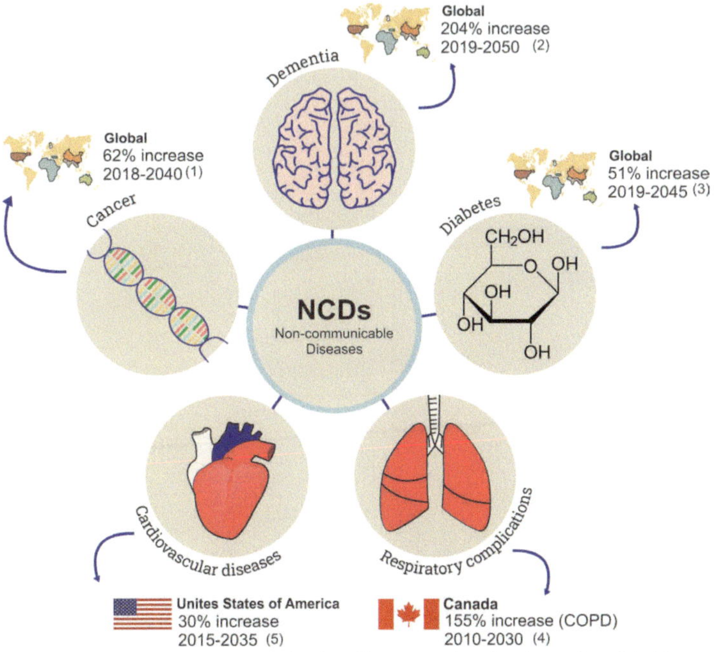

Fig. 5.5. The most common non-communicative diseases and projections of worldwide prevalence.

Notes: 1. WHO, 2020; 2. ADI, 2019; 3. IDF, 2019; 4. Khakban et al., 2017; 5. Nelson et al., 2016.

disability-adjusted life-years (DALYs), with an increased number of years of life lost and lived with disability (Roth et al., 2020). Research and industrial production of new drugs to improve the quality of life of patients with NCDs is as necessary as investing in preventive measures.

Cardiovascular diseases (CVDs) are currently the leading causes of mortality and disease burden in the world. Ischemic heart disease and stroke stand out (Roth et al., 2020), and their contribution to disability is also pronounced, as the second and third causes of global deaths in 2019 (GBD, 2019, 2020). The various modifiable causes can be a consequence of metabolism, environmental, or behavioral conditions, including high systolic blood pressure, high LDL cholesterol, and high body-mass index for the first, air pollution for the second, and dietary risks and tobacco for the third (Roth et al., 2020). In the United States, e.g., it is estimated that 45% of the population (almost 132 million people) will be affected by CVD in 2035 (Fig. 5.5). The American Heart Association foresees costs of up USD 1.1 trillion for the same year, covering medical treatment and losses of productivity associated with CVD complications (Nelson et al., 2016). China, where 30% of the world's smoking population resides, is the country with the highest number of CVD-related deaths in 2019, more than 4 million (Fig. 5.6) (Roth et al., 2020).

The incidence and mortality rates of several types of cancers are equally high. The World Health Organization (WHO) estimates that one in five people are diagnosed with cancer during their lifetime and that one in six deaths worldwide is due to cancer. An increase in cases is expected from 18.1 million to 29.4 million between 2018 and 2040 (Fig. 5.5). Lung, female breast, and colorectal cancer are the most frequently diagnosed types, and lung, colorectal, and stomach cancers are

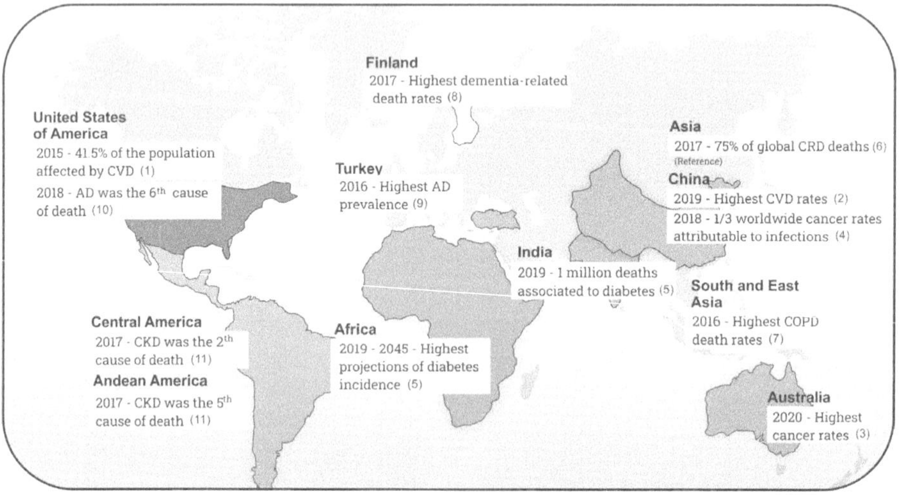

Fig. 5.6. Expressive data about non-communicative diseases in the world.

Notes: AD = Alzheimer's disease; CKD = chronic kidney disease; COPD = chronic obstructive pulmonary disease; CRD = chronic respiratory diseases; CVD = cardiovascular diseases: 1. Nelson et al., 2016; 2. Roth et al., 2020; 3. Sung et al., 2021; 4. De Martel et al., 2020; 5. IDH, 2019; 6. Baptista et al., 2021; 7. GBD 2016, 2020; 8. Eiser, 2017; 9. GBD 2016, 2019; 10. Alzheimer's Association, 2020; 11. Cockwell and Fisher, 2020.

the most common causes of death (WHO, 2020). Long and costly treatments are necessary to improve the quality of life of patients. Lung, colorectal, stomach, and prostate cancers, e.g., were among the main causes of global deaths in 2019 for the elderly population (GBD, 2019, 2020). However, it is known that these diseases are not exclusive to an age group or social class, and it is essential that risk factors are eliminated or minimized. Australia (Fig. 5.6) and New Zealand, for example, had the highest overall cancer incidence rates for men and women in 2020, attributed in part to the excessive sun exposure of a predominantly fair-skinned population (Sung et al., 2021). In the BRICS members (Brazil, Russia, India, China, and South Africa), premature cancer deaths resulted in estimated productivity losses of US$ 46.3 billion in 2012 (Pearce et al., 2018).

Diabetes mellitus is a long-term condition attributed to higher blood glucose levels, due to the lack of adequate insulin-mediated glucose cellular uptake. This disease causes numerous damages to the human body and, consequently, to the quality of life, if not controlled quickly (IDF, 2019). It is no accident that it was the eighth cause of all global deaths in 2019 (GBD, 2019, 2020). Considering the age group of 20 – 79 years, the global forecast is that from 463 million people in 2019 it could reach 700 million by 2045 (Fig. 5.5). Although the projections of an increase in the same period are more expressive for the African continent (143%) (Fig. 5.6), it is North America and the Caribbean region that holds the highest healthcare expenditure related to diabetes, around US$ 324.5 billion in 2019 (42.7% of the world) (IDF, 2019).

Chronic respiratory diseases (CRD) caused 519,100 deaths worldwide in 2016 (GBD, 2016, 2020). CRDs are related to air pollution, which can be indoor and outdoor, containing toxic particles or gases, as well as microbes or allergens, in addition to smoke from tobacco (FIRS, 2017). These diseases include chronic obstructive pulmonary disease (COPD), asthma, acute lower respiratory tract infections, tuberculosis, and lung cancer (FIRS, 2017). The first one was identified as the sixth cause of all ages global DALYs (GBD, 2019, 2020), with prominence in the number of deaths in East Asia and South Asia, 71.3% of the total of 460,080 global deaths caused by COPD in 2016 (Fig. 5.6) (GBD, 2016, 2020). In the European Union, COPD treatment comprises 6% of annual health expenditures, around € 38.6 billion (FIRS, 2017), while in the United States, direct expenditures comprise US$ 72 billion (Khakban et al., 2017), with projections of increase in the future.

Diseases that affect the central nervous system are also NCDs of great impact and are often ignored and surrounded by stigmas and stereotypes. Estimates of the number of people affected by dementia worldwide were 50 million in 2019, with the prospect of reaching 152 million by 2050 (Fig. 5.5). An increase in annual costs is also expected, which were already US$ 1 trillion in 2019 (ADI, 2019). Finland has the highest dementia mortality rate (Fig. 5.6), which can be a consequence of environmental factors and the action of specific toxins such as β-N-methyl amino-L-alanine, a neurotoxin produced by cyanobacteria present in Finnish waters (Eiser, 2017). Alzheimer's disease is the most common cause of dementia, 60–80% of all global cases. This degenerative brain disease is a considerable cause of disability, morbidity, and social dependency on other people; in 2018 it was the sixth leading cause of death in the United States (Fig. 5.6) (Alzheimer's Association, 2020).

Turkey had the highest Alzheimer's disease age-standardized prevalence per 100,000 population in 2016 (Fig. 5.6), which also had high rates of deaths (GBD, 2016, 2019). Worldwide, Alzheimer's disease is the fourth leading cause of global deaths for 75 years of age and older, after ischemic heart disease, stroke, and COPD (GBD, 2019, 2020).

3.1 Fungal Metabolites in Preclinical and Clinical Research

From the analysis of therapeutic agents approved between 1981 and 2019 for the treatment of diseases worldwide, the influence of natural products or derived compounds obtained by semi-synthetic modifications in various formulations is noteworthy (Newman and Cragg in 2020). The sum of these two sources of new chemical entities comprises, e.g., approximately 54.9% of the total of 162 compounds for antibacterial indication, 14.3% of the 63 antidiabetic compounds, and 24.7% of the 247 compounds for anticancer treatment registered in the mentioned period. Even so, for some medicine areas, drugs derived from natural products have yet not been used. Some drugs as diuretics, anxiolytics, and antihistamine originated exclusively from synthetic sources, mimic natural products (Newman and Cragg, 2020). Although there is a visible decrease in the percentage of pharmaceuticals originating from natural products every decade, because of the accelerated development of synthetic drugs, they are still considered valuable sources of lead compounds for the development of new drugs. Metabolites of fungal origin are highlighted in this search, since it is estimated that an extensive number of fungal species remain undiscovered, and the vast diversity of species and cultivation possibilities underscore this potential (Beekman and Barrow, 2014).

For some highly complex diseases, established treatments still have high failure rate, therefore, new approaches are needed (Iyengar, 2013). In addition, the discovery of new bioactive secondary metabolites that can act against diseases of a wide spectrum is desirable, as well as the definition of the appropriate form of use. For this purpose, many possibilities of fungal metabolites that remain untapped are included (Keller, 2019). Nonetheless, approval for the effective use of these metabolites and their derivatives as pharmaceuticals requires preclinical and clinical studies, which comprise full structural characterization, followed by experimentation to guarantee their safety and efficacy and, finally, the development of suitable formulations for *in vivo* drug delivery.

3.1.1 Fungal Metabolites against Cancer

Cancer is an example of a disease whose success rate of treatment is still unsatisfactory in many cases. Cancer comprises a group of genetic diseases, which have in common an abnormal proliferation of neoplastic cells, that affects the homeostasis of the primary tissue or organ, and other secondary tissues during the metastatic process (Hanahan and Weinberg, 2011; Tomasetti and Vogelstein, 2017). Since carcinogenesis is related to multiple and diverse genetic and phenotypic alterations within cell populations, evolution and severity are different among individuals, even for the same histopathological type of cancer. Moreover, this cellular heterogeneity can lead to drug resistance that, along with the stage and rate of development, in combination with preexisting health conditions, may interfere with the effectiveness of chemotherapy protocols (DeVita et al., 2018; Fares et al., 2020). The discovery of more effective and selective anticancer medicines is crucial to improve treatment strategies for this worldwide public health problem, and fungal metabolites proved to be an interesting source of chemotherapeutics, such as the successful examples of camptothecin **(14)**, taxol **(15)**, and podophyllotoxin **(16)** (Table 5.1 and Fig. 5.2).

A cytotoxic or poisonous compound cannot be considered *per se* as an anticancer candidate. Such a drug should present an intimate relationship with cancer cell targets. Selectiveness is a key feature, providing higher toxicity for cancer cells than for healthy ones. And for that guarantee, in the preclinical *in vitro* phase, this molecule must be tested in a range of different cell lines, especially multidrug or apoptosis-resistant cancer cells, searching for a lead structure that can also elicit its anticancer effects through molecular pathways not limited to apoptosis (Gomes et al., 2015).

Furthermore, after first *in vitro* screenings to determine cellular targets and selective cytotoxicity, the indicated compound must have its biological mechanism of action elucidated through *in vitro* assays and analysis, and complementary *ex vivo* and *in vivo* assays (see Section 2.2). Especially in oncology studies, it is mandatory to deeply understand the effects on cancer cells, looking at specific molecular targets, affected metabolic and signaling pathways, phenotypic and genotypic alterations, the insurgence of resistant cells, mechanisms of cell death, proliferation, migration, invasion, and angiogenesis. On the other side, undesirable side effects, including mutagenicity and toxicity to healthy cells, tissues, and organs, local and systemic inflammatory immune responses, among other systemic outcomes, should be elucidated as well. On this issue, exhaustive pharmacological features of the drug candidate must be determined in experimental *in vivo* models.

Acute myeloid leukemia (AML) is one example of cancer with high genetic complexity and whose treatment protocols are not entirely effective. Therefore, the search for new chemotherapeutic drugs from natural sources was exploited as a valid approach to improve the arsenal of treatment for this disease. The fungal polysaccharide tramesan, isolated by size exclusion chromatography of the purified liquid culture of edible mushroom *Trametes versicolor*, was evaluated in preclinical assays as a promising bioactive compound for the treatment of this type of cancer (Table 5.3). Tramesan (Table 5.3) exhibited significant viability reduction and increased production of reactive oxygen species (ROS) against human myeloid (OCI-AML3) and lymphoid (Jurkat) cancer cell lines. Although, no significant effect on cell cycle kinetics or apoptosis was observed, probably due to the previously described resistance of these cell lines to apoptosis-induced drugs, possibly indicating an alternative pathway to tramesan activity. Time- and dose-dependent cytotoxic activity of the compound was confirmed on primary cells obtained from four AML patients. This time, with a significant increase in apoptosis, demonstrating the different behaviors that may occur among tumor cells from established lines and primary origins. Equally important was the analysis of tramesan action in normal peripheral blood mononuclear cells from healthy donors, which induced non-significant pro-apoptotic activity, inferring the absence of relevant cytotoxic effects in these representative cells (Ricciardi et al., 2017). Promising results were previously obtained with the polysaccharopeptides (PSP), and the protein-bound

Table 5.3. Preclinical studies on promising bioactive fungal secondary metabolites.

Secondary Metabolite [Fungal Source]	**Experimental Parameters or Origin**	**Health Benefits Investigated [Experimental Model]**	**Compound's Test Response[1,2]**	**Reference**
1403P-3 **(33)** [*Halorosellinia* sp. 1403 (endophyte)]	Isolation with > 98% purity	Anticancer [*In vitro* bioassay with human cancer cell lines]	Cytotoxicity and apoptosis to KB: IC_{50} = 19.66 μM multi-drug resistant KBv200: IC_{50} = 19.27 μM	Zhang et al., 2007
Beauvericin **(13)** [*Beauveria bassiana*]	Fungal cultivation in PDB. Mycelium lyophilization and extraction with ethyl acetate. Residue dissolution in acetonitrile-water [80:20]. Supernatant subjected to preparative HPLC for compound isolation. Purity verified by LC-MS and NMR.	Anticancer [*In vitro* bioassay with murine (M) and human (H) malignant cell lines]	Cytotoxicity to CT-26 (M): IC_{50} = 1.8 ± 0.2 μM SW620 (H): IC_{50} = 0.7 ± 0.1 μM, with significant SI = 5.57 (*versus* control HaCaT cell line) ME-180 (H): IC_{50} = 2.2 ± 0.7 μM SW480 (H): IC_{50} = 3.3 ± 0.3 μM GH354 (H): IC_{50} = 3.6 ± 1.2 μM KB-3-1 (H): IC_{50} = 3.1 ± 0.7 μM, with G0/G1 cell cycle arrest and apoptosis at different concentrations	Heilos et al., 2017
		Anticancer [*In vitro* bioassay with murine (M) and human (H) non-malignant cells]	Cytotoxicity to NIH/3T3 (M): IC_{50} = 3.1 ± 0.2 μM HaCaT (H): IC_{50} = 3.9 ± 0.4 μM	
		Anticancer [*In vivo* allo- or xenograft mouse models, using murine (M) or human (H) carcinoma cell lines]	CT-26 (M): 52.8% reduction of mean tumor volume, and increase of apoptotic/necrotic cells, 14 d after treatment (5 mg/kg/day) KB-3-1 (H): 31.3% reduction of mean tumor volume, and increase of apoptotic/necrotic cells, after 16 d treatment	

Benzomalvin A **(38)** [*Penicillium spathulatum* B35]	Fungal cultivation in moist rice. Extraction with CH_2Cl_2-MeOH [8:2]. Isolation from extensive chromatography of organic extract.	Antidiabetic [*In vitro* α-glucosidase inhibition assay]	IC_{50} = 383.2 μM	Del Valle et al., 2016
		Antidiabetic [*In vitro* bioassay with normal cells]	Human gingival fibroblasts: IC_{50} = 54.5 ± 3.3 μM	
		Antidiabetic [*In vivo* oral sucrose tolerance test in healthy mice]	Doses of 3.1, 10.0, and 31.6 mg/kg reduced blood glucose levels 30 min after administration	
		Antidiabetic [*In vivo* oral sucrose tolerance test in hyperglycemic mice]	Doses of 10.0 mg/kg reduced blood glucose levels 30 min after administration	
		Antihyperalgesic [*In vivo* formalin assay in hyperglycemic mice]	Increment in the response to chemical stimuli	
Bostrycin[a] **(32)** [*Halorosellinia* sp. 1403 (endophyte)]	Supplied by Marine Microorganism Laboratory, Sun Yat-Sen University	Anticancer [*In vitro* bioassay with human cancer cell lines]	Cytotoxicity (10–30 μM), G0/G1 cell cycle arrest, and apoptosis on A549 cells	Chen et al., 2011
Ganoderic acid A **(35)** [*G. lucidum* (mushroom)]	Purified commercial source	Anticancer [*In vitro* bioassay with human glioblastoma cell line]	Inhibition of cell proliferation, migration and invasion, induction of apoptosis and autophagy on U251 cells	Cheng and Xie, 2019
Ganoderic acid A **(35)** /DM **(36)** [*G. lucidum* (mushroom)]	Purified commercial source	Anticancer [*In vitro* bioassay with human anaplastic meningioma cells]	Inhibition of cell proliferation, and apoptosis induction on primary cell cultures and IOMM-Lee cell line, through upregulation of NDRG2 gene	Das et al., 2020
		Anticancer [*In vivo* xenograft mouse model, using human meningioma cell line]	60% reduction of brain tumors caused by IOMM-Lee cell line, after 14-day treatment with 10 mg/kg of compounds	

Table 5.3 contd. ...

...Table 5.3 contd.

Secondary Metabolite [Fungal Source]	Experimental Parameters or Origin	Health Benefits Investigated [Experimental Model]	Compound's Test Response[1,2]	Reference
Ganoderic acid DM **(36)** [*G. lucidum* (mushroom)]	Extraction with 95% EtOH, fractionation by silica gel (SiG) column chromatography (CC), and preparative reversed phased (RP) HPLC. Identification by MS, NMR, and optical rotation data	Anticancer [*In vitro* bioassay with hormone-dependent or independent cancer cell lines]	Cytotoxic to androgen-dependent LNCaP cells androgen-independent PC-3 cells	Liu et al., 2009
		Anticancer [*In vitro* bioassay with preosteoclastic cell line]	Inhibition of RAW 264 cells differentiation, but no cytotoxic	
	Purified commercial source	Anticancer [*In vitro* bioassay with hormone-dependent or independent cancer cell lines]	Inhibition of cell proliferation and colony formation. Cell cycle arrest at G1 phase, and apoptosis induction on estrogen receptor (ER)-positive MCF-7 cells. Less cytotoxic effects on ER-negative MDA-MB-231 cells	Wu et al., 2012
	Purified commercial source	Anticancer [*In vitro* bioassay with human non-small cell lung cancer]	Autophagic apoptosis to A549 NCI-H460	Xia et al., 2020
Gliotoxin **(29)** [*Gliocladium fimbriatum*]	Purified commercial source[d]	Anticancer [*In vitro* prenyltransferase assay]	Inhibition of enzymes FTase (IC_{50} = 80 μM); GGTase I (IC_{50} = 17 μM)	Vigushin et al., 2004
		Anticancer [*In vitro* bioassay with human breast cancer cell lines]	Cytotoxicity to MCF-7: IC_{50} = 985 nM T-47D: IC_{50} = 365 nM BT-474: IC_{50} = 102 nM ZR-75-1: IC_{50} = 158 nM MDA-MB-231: IC_{50} = 38 nM	
		Anticancer [*In vitro* prenylation assay on tumor cells]	Inhibition of lamin B prenylation by Ftase and Rap1A by GGTase I on MCF-7, MDA-MB-231	
		Anticancer [*In vivo* carcinogen-induced rat mammary cancer model]	60% of animals with > 50% tumor regression	

Gliotoxin **(29)**; 5α,6-didehydroglio-toxin **(31)**, and gliotoxin G **(30)** [*Penicillium* sp. JMF034]	Fungal cultivation in broth containing yeast and malt extracts, peptone, and glucose. Broth extraction with ethyl acetate. Purification from C18 flash CC, Sephadex LH-20 column, and semipreparative ODS-HPLC	Anticancer [*In vitro* bioassay with murine cell line]	Cytotoxicity to P388 cells: IC_{50} = 0.020 to 0.056 μM	Sun et al., 2012
		Anticancer [*In vitro* histone methyltransferase assay]	Inhibition of HTM G9a enzyme (IC_{50} = 2.1 to 6.4 μM)	
Gliotoxin **(29)** [*Aspergillus* sp. YL06]	Fungal cultivation in YPG medium. Broth extraction with ethyl acetate. Purification by SiG, ODS CC, and HPLC. Purity verified by ^{1}H and ^{1,3} CNMR and HPLC	Anticancer [*In vitro* bioassay with human cancer cell lines]	Inhibition of cell proliferation, morphological changes, and apoptosis on HeLa SW1353	Nguyen et al., 2013
Leptosin C **(27)** [*Leptosphaeria* sp.]	Purified commercial source	Anticancer [*In vitro* bioassay on human lymphoblastoid cell line]	Inhibition of DNA topoisomerase I, and apoptosis induction on RPMI18402 cells	Yanagihara et al., 2005
Leptosin F **(28)** [*Leptosphaeria* sp.]			Inhibition of DNA topoisomerase I and II, and apoptosis induction on RPMI18402 cells	
Leptosin M **(26)** [*Leptosphaeria* sp. OUPS-4]	Fungal cultivation in broth containing glucose, peptone, and yeast extract. Extraction of mycelium with MeOH, fractionation by Sephadex LH-20, SiG CC, and RP HPLC	Anticancer [*In vitro* biochemical assay]	Inhibition of protein kinases PTK and CaMKIII, and DNA topoisomerase II	Yamada et al., 2002
Nigerloxin **(39)** [*A. niger*]	Cultivation in wheat bran with trisodium citrate. Extraction with ethyl acetate, re-suspension in chloroform, centrifugation, treatment with activated charcoal treatment	Antidiabetic [*In vivo* safety assessment in Wistar rats: histopathologic effects and levels of marker enzymes]	Absence of toxicity at 100 mM/kg body weight administration	Vasantha et al., 2018

Table 5.3 contd. ...

...Table 5.3 cont.

Secondary Metabolite [Fungal Source]	**Experimental Parameters or Origin**	**Health Benefits Investigated [Experimental Model]**	**Compound's Test Response**[1,2]	**Reference**
Tramesan[b] [*T. versicolor* (mushroom)]	Fungal cultivation in PDB, lyophilization, exopolysaccharide extraction, size exclusion chromatography	Anticancer [*In vitro* bioassay with leukemic cell lines]	Dose- and time-dependent reduction of cell viability (0.5–2 mg/mL, 24–72 hours), and induction of ROS at 0.5 mg/mL, but no significant cell cycle arrest or apoptosis on OCI-AML3 Jurkat cells	Ricciardi et al., 2017
		Anticancer [*In vitro* bioassay with acute myeloid leukemia primary cells]	Dose- and time-dependent reduction of cell viability, and increased apoptosis levels (77.2 ± 28.3% at 2 mg/mL *versus* 35.7 ± 13.8% in control)	
		Anticancer [*In vitro* bioassay with normal peripheral blood mononuclear cells from healthy donors]	No significant apoptosis levels (31.8 ± 15.8% at 2 mg/mL *versus* 18.3 ± 15.6% in control)	

2,4,6-Triphenylaniline **(37)** [*Alternaria longipes* VITN14G (endophyte)]	Fungal cultivation in PDB, followed by extraction with ethyl acetate. Isolation by SiGCC using n-hexane-ethyl acetate [70:30]. Compound elucidation by GC-MS, NMR, UV, and FTIR techniques	Antidiabetic [*In vitro* α-glucosidase inhibition assay]	78.88% (standard acarbose: 89.91%) IC_{50} = 39.02 µg/mL	Ranganathan and Mahalingam, 2019
		Antidiabetic [*In vitro* α-amylase inhibition assay]	51.53% (standard acarbose: 56.69%) IC_{50} = 27.05 µg/mL	
		Antidiabetic [*In vitro* bioassay with murine cell line]	L929: Cell viability of 73.96% at 1000 µg/mL	
		Antidiabetic [*In silico* molecular docking]	high affinity to mitochondrial complex 1: 7.7 kcal/mol (standard metformin: –3.6 kcal/mol)	
		Antidiabetic [*In vivo* biochemical tests of the isolated (TPA) and nano-emulsified compound (TPA load NE) on diabetic rats]	Fasting blood glucose: TPA = 175.67 ± 4.40 TPA loaded NE = 140.50 ± 2.59 Standard metformin = 118.17 ± 3.32 Diabetic = 330.00 ± 19.22 Normal = 79.50 ± 1.95	Ranganathan and Mahalingam, 2020
			Oral glucose tolerance > 3 h: TPA = 172.50 ± 3.35 TPA loaded NE = 153.30 ± 4.77 Standard metformin = 143.20 ± 6.93 Diabetic = 189.200 ± 6.64 Normal = 101.70 ± 4.60	
			Insulin: TPA = 32.83 ± 1.30 TPA loaded NE = 26.67 ± 0.95 Standard metformin = 26.67 ± 1.58 Diabetic = 54.83 ± 3.48 Normal = 16.83 ± 1.13	
		Antidiabetic [*In vivo* histopathologic effects of the isolated and nano-emulsified compound (NE) on diabetic rats]	Non-toxic effects on liver parenchyma cells, kidney cells, pancreas cells and heart	

Table 5.3 contd. ...

...Table 5.3 contd.

Secondary Metabolite [Fungal Source]	Experimental Parameters or Origin	Health Benefits Investigated [Experimental Model]	Compound's Test Response[1,2]	Reference
SZ-685C[c] **(34)** [*Halorosellinia* sp. No. 1403 (endophyte)]	Ethyl acetate extract. Identification by MS, ^{1}H NMR, ^{1,3}C NMR, 2D NMR, and X-ray crystal diffraction.	Anticancer [*In vitro* bioassays with human cancer cells]	Cytotoxicity and apoptosis on MCF-7: IC_{50} = 7.5 μM PC-3: IC_{50} = 4.1 μM LN-444: IC_{50} = 7.8 μM Hep-3B: IC_{50} = 3.2 μM Huh-7: IC_{50} = 9.6 μM NFPA: IC_{50} = 18.76 μM	Xie et al., 2010; Wang et al., 2015
		Anticancer [*In vitro* bioassays with adriamycin-sensitive and resistant human cancer cell lines]	Cytotoxicity and apoptosis on MCF-7: IC_{50} = 7.38 μM resistant MCF-7/ADR: IC_{50} = 4.17 μM resistant MCF-7/Akt: IC_{50} = 3.36 μM HL-60: IC_{50} = 1.94 μM resistant HL-60/ADR: IC_{50} = 1.76 μM K-562: IC_{50} = 1.09 μM resistant K-562/ADR: IC_{50} = 1.35 μM	Zhu et al., 2012
		Anticancer [*In vivo* xenograft mouse model, using adriamycin-resistant human breast carcinoma cell line]	64.32% inhibition of tumor growth elicited by MCF-7/ADR cancer cells, with no systemic side effects	

Notes. [1] The cytotoxicity of a compound is expressed as the minimal concentration capable of inhibiting 50% of the cells (IC_{50}) in culture condition. The Selectivity Index (SI) is the ratio between the IC_{50} of two cell lines, generally, a non-tumoral cell line and a tumoral one, that indicates the preference of a given compound for the malignant cells (SI = control cell IC_{50}/tumor cell IC_{50}). The greater the value, the more selective the substance is for a given cancer cell line. Different authors suggest an index varying from SI ≥ 2 to SI ≥ 3 as the cut-off to consider a molecule of interest to proceed to other *in vitro* and *in vivo* assays.

[2] Cancer cell lines: BT-474, MCF-7 – human breast carcinoma; MDA-MB-231, T-47D, ZR-75-1 – human breast carcinoma metastasis; MDA-MB-435 – originally identified as human breast adenocarcinoma, identified as human M14 melanoma (Korch et al., 2018); MCF-7/ADR, MCF-7/Akt – adriamycin-resistant MCF-7-derived; HeLa, KB-3-1 – human cervix carcinoma; ME-180 – human cervix metastasis; GH354 – human cervix adenocarcinoma; CNE-2 – originally identified as human nasopharyngeal carcinoma, identified as a hybrid cell line of HeLa (Strong et al., 2014); CNE-2R – drug-resistant CNE-2; PC-3, LNCaP - human prostate carcinoma metastasis; A549 – human non-small cell lung adenocarcinoma; KB – human epidermoid carcinoma; KBv200 – multi-drug-resistant KB-derived; HL-60, OCI-AML3 – human acute myeloid leukemia; HL-60/ADR – adriamycin-resistant HL-60-derived; K-562 – human chronic myeloid leukemia; K-562/ADR – adriamycin-resistant K-562-derived; Jurkat – human T-cell acute lymphoblastic leukemia; SW480 – human colorectal adenocarcinoma; SW620 – human colorectal adenocarcinoma metastasis from SW480; Hep-3B, Huh-7 – human hepatocellular carcinoma; LN-444; U251 – human glioblastoma cell line; IOMM-Lee – human meningioma; SW1353 – human chondrosarcoma; CT-26 – *Mus musculus* colon carcinoma; P388 – *Mus musculus* lymphoma. Primary cancer cells: NFPA – human nonfunctioning pituitary adenoma. Non-tumoral cell lines: HaCaT – human keratinocytes; L929 – *Mus musculus* fibroblast; NIH/3T3 – *Mus musculus* embryonic fibroblasts. Genes: NDRG2 gene – N-myc downstream-regulated gene. 2. Proteins: FTase – Farnesyltransferase; GGTase I – geranylgeranyltransferase I; Rap1A – Ras-related protein 1A; PTK – tyrosine-protein kinase; CaMKIII – calmodulin dependent protein kinase III. ROS – reactive oxygen species. [a] tautomer of SZ685C; [b] Patent number RM2012A00057; [c] tautomer of bostrycin; [d] Sigma-Aldrich # G9893, CAS Number 67-99-2

polysaccharide K (PSK) isolated from the same *T. versicolor*, justifying the interest in tramesan. For instance, PSK induced cell cycle arrest at S phase and apoptosis in HL-60 promyelomonocytic leukemia cell line and was already submitted to human clinical trials as an adjuvant to immunochemotherapy for colorectal, gastric, and breast cancer (Sakamoto et al., 2006; Ueda et al., 2006; Jiménez-Medina et al., 2008; Torkelson et al., 2012).

The class of leptosins (Table 5.3), isolated from the marine fungus strain *Leptosphaeria* sp. strain OUPS4, initially screened for the cytotoxicity to murine P388 lymphocytic cell line (Takahashi et al., 1994) and thereafter in a panel of 39 human cell lines from breast, central nervous system, colon lung melanoma, ovary, kidney, stomach, and prostate cancer (Yamada et al., 2002), offered three compounds whose biological mechanisms of action were further analyzed *in vitro* (Table 5.3). Leptosin M **(26)** (Fig. 5.7) inhibits protein kinases PTK and CaMKIII, as well as DNA topoisomerase II. Leptosin C **(27)** catalytically inhibits DNA topoisomerase I, and leptosins F **(28)** (Fig. 5.7) inhibits topoisomerase I and II. Both molecules induce apoptosis in human lymphoblastoid RPMI18402 cells by dose-dependent activation of caspase 3 and concomitant inactivation of Akt/protein kinase B pathway (Yamada et al., 2002; Yanagihara et al., 2005).

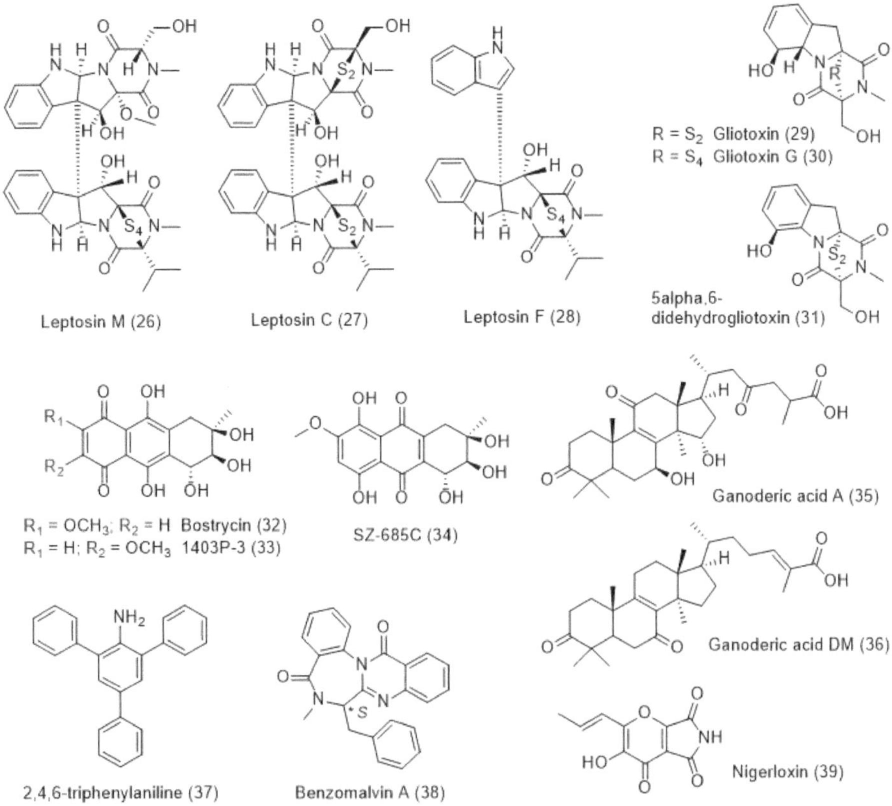

Fig. 5.7. Chemical structures of fungal metabolites **(26–39)** investigated as agents against non-communicative diseases.

Gliotoxin **(29)** is a secondary metabolite isolated from several marine fungus and, along with its derivatives gliotoxin G **(30)** and 5α,6-didehydrogliotoxin **(31)** (Fig. 5.7) (Table 5.3), are being considered as promising candidates for the development of anticancer drugs, based on their unusual biological mechanism of action. Inhibition of proliferation is observed in a panel of human breast carcinoma cell lines (MCF-7, T47D, BT-474, ZR75-1, and MDA-MB-231) treated with gliotoxin **(29)**. It was also demonstrated that gliotoxin **(29)** was able to inhibit farnesyltransferase (FTase) as well as geranylgeranyltransferase I (GGTase I), enzymes that catalyze the prenylation of diverse proteins related to tumor proliferation (Vigushin et al., 2004). Sun et al. (2012) observed a cytotoxic effect of gliotoxin **(29)** and its derivatives on P388 murine leukemia cell line and revealed a role of these molecules in the inhibition of histone methyltransferase HTM G9a, in which up-regulation seems to be involved in tumor progression. Gliotoxin **(29)** is also involved in the mechanism of death by apoptosis, elicited by caspases 3, 9, and 8 activation, leading to down-regulation of Bcl-2 and Bcl-xL proteins and up-regulation of Bax and cytochrome C, in HeLa cervical carcinoma and SW1353 human chondrosarcoma cell lines (Nguyen et al., 2013). Although, *in vitro* assays using non-tumoral cells and a more expressive panel of multidrug and apoptosis-resistant cancer cell lines, as well as *in vivo* experimental models, are needed to confirm the selectivity and low toxicity to healthy tissues by these promising molecules.

The antracenediones bostrycin **(32)**, its isomer 1403P-3 **(33)**, and its tautomer SZ-685C **(34)** (Fig. 5.7 and Table 5.3) are another group of secondary metabolites, isolated from the endophytic fungus *Halorosellinia* sp. N° 1403, with strong potential for new anticancer drug candidates. Their activities on various cancer cell lines, and more interestingly, on multidrug-resistant cell lines, are correlated to apoptosis induction, triggered *via* distinct routes of the intrinsic and extrinsic pathways. Bostrycin **(32)** was effective against human lung adenocarcinoma A549 cell line, causing dose- and time-dependent cytotoxicity, cell cycle arrest at G0/G1 phase, and apoptosis by the inhibition of phosphatidylinositol-3-kinase (PI3K)/Akt signal pathway (Chen et al., 2011). In a subsequent work, a series of synthetic derivatives exhibited cytotoxicity to a panel of cancer cell lines (MCF-7, MDA-MB-435, A549, HepG2, and HCT-116) but, unfortunately, they were also toxic to the non-tumoral MCF-10A breast cell line used as control (Chen et al., 2012). 1403P-3 **(33)** demonstrated to be cytotoxic to human epidermoid carcinoma KB and its multidrug-resistant derivative KBv200, and to induce apoptosis on both cell lines by the release of mitochondrial cytochrome *c* (the intrinsic pathway) concomitantly to caspase-8 activation (the extrinsic pathway) (Zhang et al., 2007). SZ-685C **(34)** also elicited a pro-apoptotic effect against a panel of human cancer cell lines (breast cancer cell line MCF-7, melanoma cell line MDA-MB-435, prostate adenocarcinoma cell line PC-3, glioma cell line LN-444, and hepatoma cell lines Hep-3B and Huh-7) (Xie et al., 2010) and nonfunctioning pituitary adenoma primary cells (Wang et al., 2015), and this effect was induced by the down-regulation of the Akt pathway. Moreover, in a xenograft murine cancer model using MDA-MB-435 cells (previously known as breast carcinoma, and afterward identified as melanoma, according to Korch et al., 2018), a significant reduction of tumors was observed after treatment with SZ-685C **(34)** (Xie et al., 2010). Interestingly, this compound was also able to reduce the resistance factor on adriamycin (ADR)-resistant cell lines MCF-7/ADR and MCF-7/Akt (derived from MCF-7), HL-60/ADR (from human promyelocytic leukemia HL-60), and K562/ADR (from human erythromyeloid leukemia K542), as well as to significantly reduce the growth of xenografted tumors induced by MCF-7/ADR, eliciting the intrinsic and extrinsic apoptosis pathways both *in vitro* and *in vivo* (Zhu et al., 2012). Therefore, bostrycin **(32)** and its derivatives appeared to be promising candidates against resistant tumors from different histological origins, remaining to confirm their absence of toxicity to non-tumoral cells.

The caution in the analysis of the cytotoxicity of metabolites against non-malignant cells was considered in preclinical *in vitro* and *in vivo* tests performed with beauvericin **(13)** (Table 5.1 and Fig. 5.2), a metabolite isolated from filamentous fungus *B. bassiana*, and investigated as to its therapeutic potential against several types of cancer cells (Table 5.3). *In vitro* assays showed cytotoxicity of beauvericin not only to the murine colon carcinoma CT-26 line, but also to the normal fibroblast NIH/3T3 line, with an unsatisfactory Selectivity Index (SI < 3). Concerning human cells, although greater cytotoxic effects (SI $= 5.57$) were observed for SW620 metastatic colorectal adenocarcinoma cells compared to the non-malignant HaCaT keratinocyte cells, the overall results of cytotoxicity presented no statistical difference among other malignant cells and healthy ones. In human cell lines, treatment with increased cell density (50–60% cell confluency), imitating the occurrence in tissues, generated higher compound IC_{50} values for the non-malignant cells, providing a more favorable result. To understand the mechanism of cytotoxicity, KB-3-1 cervix carcinoma cell line was further analyzed, demonstrating the occurrence of cell cycle arrest at G0/G1 phase at subtoxic beauvericin **(13)** concentrations, and apoptosis at concentrations > 2 μM, associated to up-regulation of proapoptotic cell signals and down-regulation of antiapoptotic ones. In addition to these results, *in vivo* assays using allograft and xenograft mouse carcinomas, respectively, produced through subcutaneous injection of murine CT-26 and human KB-3-1 cell lines, showed a significant reduction in tumor volume, and increased apoptotic/necrotic areas, after treatment with beauvericin **(13)**, with no indications of adverse effects in both cases. Based on previous works, it was suggested that the mechanism of action on tumor cells is related to the activation of apoptotic and necrotic pathways, increase of ROS production, and anti-angiogenic activity, among others (Heilos et al., 2017).

Several bioactive metabolites, from distinct chemical classes, have been isolated from medicinal fungi widely used in Asian natural medicine and are being investigated for their anticancer effects. These compounds can elicit death by apoptosis or autophagy, inhibit metastasis and angiogenesis, and induce immune responses against tumors from diverse histological origins. Ganoderic acid A **(35)** (Fig. 5.7, Table 5.3), a triterpenoid isolated from the edible mushroom *G. lucidum*, is correlated to inhibiting proliferation, migration, and invasion, as well as inducing apoptosis and autophagy through down-regulation of the PI3K/Akt pathway on U251 human glioblastoma cell line (Cheng and Xie, 2019). The effect of ganoderic acid DM **(36)**, also isolated from *G. lucidum*, was evaluated on hormone-dependent/independent human cancers since treatment strategy includes blocking these hormones or their cell receptors to reduce early-stage tumors, although refractory tumors are not responsive to this protocol. Ganoderic acid DM **(36)** demonstrated to be cytotoxic to the androgen-dependent LNCaP (lymph node carcinoma of the prostate) cell line, as well as to the androgen-independent prostatic carcinoma PC-3 cell line. Metastasis to bone and osteoclastogenesis is an undesirable consequence of prostate cancer migration and invasion, therefore, ganoderic acid DM **(36)** was tested on the preosteoclastic RAW 264 cell line and demonstrated to inhibit cell differentiation without being toxic to these cells, indicating a role for that bioactive compound on future chemotherapy to hormone-dependent or independent prostate tumor formation and metastasis (Liu et al., 2009). (ER)-positive MCF-7 breast cancer cells and ER-negative MDA-MB-231 were also treated with ganoderic acid DM **(36)**, and effective results were obtained only with MCF-7 cells, which showed a reduction of proliferation, cell cycle arrest at G1 phase, and apoptosis (Wu et al., 2012). More recently, autophagic apoptosis, elicited through the PI3K/Akt/mTOR pathway, was observed in human non-small cell lung cancer (NSCLC) cell lines A549 and NCI–H460 (Xia et al., 2020). Both ganoderic A **(35)** and DM **(36)** acids were effective in controlling the proliferation and inducing apoptosis on human anaplastic meningioma primary cultures and IOMM-Lee cell line, with no toxic effect on non-tumoral neurons and arachnoid cell lines. In anaplastic meningioma xenografted mice using IOMM-Lee cell line, tumors were significantly reduced by treatment with both ganoderic acids, and no hepatotoxic side effects were observed. According to this research, the primary molecular target of ganoderic A **(35)** and DM **(36)** was the tumor suppressor gene NDRG2 (N-myc downstream-regulated gene 2), usually tightly down-regulated on anaplastic meningioma, demonstrating the potential of these bioactive molecules as an alternative treatment to this rare but highly aggressive brain cancer (Das et al., 2020).

Efforts must be made so that these and other metabolites candidates for clinical trials can have their biological mechanisms of action and their biosynthetic pathways better defined. This lack delays the development and approval of formulations containing fungal metabolites or derivatives as anticancer drugs, although many of them have their activities confirmed (Yuan et al., 2020). Moreover, they must prove their selectivity to cancer cells, especially resistant ones, with no significant effects on normal cells *in vitro* and *in vivo*, and preferentially present innovative molecular mechanisms of action. For instance, although the mechanism of apoptosis through Akt inhibition plays an important role in avoiding cancer progression, according to Gomes et al. (2015), it is already explored by other drug candidates on clinical trials, so, the interest for other compounds with the same action may be at risk in the next future.

3.1.2 Fungal Metabolites against Diabetes

Many fungal metabolites have already been tested as possible inhibitors of α-glucosidase and α-amylase, target enzymes for the treatment of type II diabetes mellitus, a long-term condition (Agrawal et al., 2021b). Although many of these compounds have presented IC_{50} values lower than those presented for standard inhibitory compounds, such as acarbose and genistein (Hussain et al., 2021), in-depth investigations of this biological activity in healthy and compromised cells still seem to be restricted to a few studies. These investigations combine *in vitro* and *in vivo* assays to obtain a broader response under antidiabetic activity and cell viability of the target fungal metabolites. These parameters were considered in studies of 2,4,6-triphenylaniline **(37)** (Fig. 5.7), a metabolite isolated from the endophytic fungus *A. longipes* strain VITN14G, which in turn was obtained from leaf and stem samples of the mangrove plant *Avicennia officinalis*. The *in vitro* inhibitory activities of the two target enzymes were evaluated, as well as the cytotoxicity against L929 fibroblast cell line (Table 5.3). These results, along with molecular docking analysis demonstrating the affinity of 2,4,6-triphenylaniline to the same target of metformin standard drug, were favorable for the continuation of studies on the compound as a possible non-cytotoxic antidiabetic drug (Ranganathan and Mahalingam, 2019). *In vivo* tests were also satisfactory (Table 5.4), with the administration of the isolated compound, in its pure form and a nano-emulsion formulation, in groups of Albino Wistar rats, in addition to the administration of the metformin in a third group (all at 10 mg/kg body weight/day for 48 days). In both presentation forms of the 2,4,6-triphenylaniline **(37)** under evaluation, blood glucose levels were controlled, in addition to liver profile and kidney function, with values comparable to those obtained for the standard compound. Besides, microscopic analysis of possible histopathology changes showed no toxicity to liver and kidney cells, and pancreatic cells and myocardium remained unremarkably healthy. In general, biochemical, and histopathological studies have been more favorable for the nano-emulsion formulation than for the pure isolated compound (Ranganathan and Mahalingam, 2020).

Assays under healthy and hyperglycemic mice were also performed for the metabolite benzomalvin A **(38)** (Fig. 5.7), isolated from *P. spathulatum* B35 in a mixture of its two conformers (Table 5.3). In the *in vitro* inhibition assay against α-glucosidase, this metabolite showed moderate action. Benzomalvin A **(38)** efficiency was confirmed by *in vivo* oral sucrose tolerance tests (Table 5.3), by decreasing blood glucose levels 30 minutes after sucrose administration to healthy mice submitted to different doses of the compound. In nicotinamide-streptozotocin hyperglycemic mice submitted to 10 mg/kg of benzomalvin A **(38)**, decreased blood glucose levels were comparable to the acarbose standard. Benzomalvin A **(38)** was also considered non-toxic to human gingival fibroblasts (Del Valle et al., 2016).

Nigerloxin **(39)** (Fig. 5.7), a metabolite specially produced by the solid-state fermentation of *A. niger*, has recognized benefits against complications arising from diabetes in rat's plasma and, additionally, antioxidant characteristics. This metabolite was also evaluated for its safety (Table 5.3). After four and twelve weeks of administering a single dose of nigerloxin **(39)** in Wistar rats, no clinical signs of toxicity were noted, with no differences in body weight of the individuals on whom the drug was administered compared to the control, as well as in the weight of some vital organs analyzed. The monitoring of marker enzymes also showed no changes, pioneering good prospects for conducting clinical trials with this bioactive compound (Vasantha et al., 2018).

3.1.3 Fungal Metabolites against Other Diseases

Unfortunately, a considerable number of studies of metabolites that had advanced to clinical trials were discontinued. This occurred with the antiparasitic fumagillin **(40)** (Fig. 5.8), a mycotoxin obtained from the strains of *A. fumigatus*, approved in some

Table 5.4. Some fungal secondary metabolites with no toxicity in different experimental models.

Secondary Metabolite	Fungus Origin	Experimental Models	Toxicity	Possible applications	References
Pestalotione A **(44)**, B **(45)**	*P. theae*	Cytotoxic assay using human cancer cell lines HepG2, HeLa, MCF-7, ACHN	No effect up to 100 µM	Antitumorigenic and antiviral	Guo et al., 2020
Isosulochrin dehydrate **(46)**					
3,8-dihydroxy-6-methyl-9-oxo-9H-xanthene-1-carboxylate **(47)**					
Isosulochrin **(48)**					
Colomitide C **(49)**	*C. luteo-olivacea*	Larval caudal fin amputation assay using Zebrafish *D. rerio*	No effect over larvae viability up to 100 µM[a]	Tissue-remodel modulator	Cavanah et al., 2021
Griseaketide A **(52)**	*M. grisea*	Nematicidal assay using *C. elegans*	No effect up to 400 ppm	Antibiotics	Yang et al., 2019
Dihydropyriculol **(53)**					
Epi-dihydropyriculol **(54)**					
Pyriculin A **(55)**					
Dihydroxyisoechinulin A **(59)**	*Eurotium* sp.	Barnacle (*Balanus amphitrite*) larvae viability assay	Antifouling effect over *B. amphitrite*, no effect over barnacle larvae up to 50 ng/mL	Antifouling activity with no teratogenic effect, *B. Amphitrite* biocontrol	Chen et al., 2018
Neoechinulin A **(60)**					
Echinulin **(61)**		Zebrafish (*D. rerio*) embryo teratogenic assay	No teratogenic effect over zebrafish		
Pseurotin A_3 **(56)**, A_4 **(57)**, A_5 **(58)**	*W. carpophilus*	Nematicidal assay using *C. elegans*	No effect up to 100 µg/mL	Immunomodulation, cell differentiation	Narmani et al., 2018
		Cyototoxic assay using mouse fibroblast (L929) and human cancer cells (HeLa)			
Akanthol **(50)**	*A. novoguineensis*	Nematicidal assay using *C. elegans*	No effect up to 100 µg/mL	Design of antimicrobial or antitumor agents	Helaly et al., 2017
Akanthozine **(51)**		Cytotoxic assay using human cancer cell KB-3-1 and mouse fibroblast L929			
Hydroxamic acid derivatives					
Phenazine **(62)**	Pseudomonas and *Burkholderia* spp.	Cell viability using rat INS-1 pancreatic β-cells	No effect up to 24 h[b]	Treatment of cystic fibrosis-related diabetes	Nisr et al., 2011
Pyrrolnitrin **(63)**					

Notes. [a]Fin regeneration inhibition; [b]Diabetogenic potentiated insulin production and Ca^{2+} accumulation.

Cell lines: HepG2 - human hepatoblastoma; HeLa - human endocervical adenocarcinoma; MCF-7 - human breast carcinoma; ACHN - human renal carcinoma; L929 - murine (*Mus musculus*) immortalized fibroblasts; KB-3-1 - human endocervical adenocarcinoma derived from HeLa; INS-1 - rat (*Rattus norvegicus*) insulinoma.

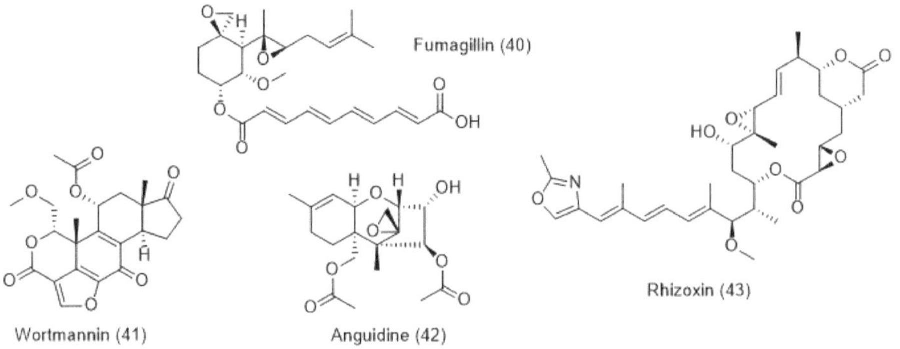

Fig. 5.8. Chemical structures of fungal metabolites **(40–43)** that had clinical trials discontinued due to toxicity and/or inactivity detected.

countries as a chemical treatment against the bee parasite *Nosema* spp. Fumagillin **(40)** presented inhibitory effects on angiogenesis and cell proliferation, interesting antitumor properties, but was also considered toxic in clinical trial studies, with a side effect of weight loss. Thus, fumagillin analogs were developed to obtain greater safety, maintaining the properties of interest (Guruceaga et al., 2020), but some derivatives, such as TNP-470, PPI-2458, and CKD-732, were already discarded in clinical trials due to neurotoxic adverse effects (Yuan et al., 2020). In addition, to assessing the toxicity of compounds under healthy cells and the possible side effects for organisms, other parameters must be considered as they progress to clinical phases. Wortmannin **(41)** (Fig. 5.8) is another metabolite isolated from *Penicillium wortmannii*, and from *Fusarium torulosum* and *Trichoderma* sp., that did not reach clinical approval because of its high toxicity, besides problems in stability and water solubility. Currently, its synthetic analog PX-866 is under Phase II clinical trial as an oncological agent candidate (Kornienko et al., 2015; Yuan et al., 2020). Other compounds, although relatively safe, did not show the expected response in *in vitro* tests, such as anguidine **(42)**, isolated from *Fusarium* spp., and rhizoxin **(43)** (Fig. 5.8), isolated from *Rhizopus chinensis* (Kornienko et al., 2015).

Alongside, there is a massive effort to discover and characterize fungal secondary metabolites, once they are considered as primary sources for novel commercial molecules with applications as pesticides, drugs, food additives, pigments, and cosmetics (Madariaga-Mazón et al., 2019). Such interest is exemplified by the successful drugs currently available, originated and/or derived from fungi, such as antibiotics (e.g., penicillin **(1)** and its derivative amoxicillin), cholesterol reducers (e.g., lovastatin **(7)** and its synthetic derivative simvastatin **(8)**), and immunomodulators (e.g., cyclosporine **(5)**), among many others (Handsfield et al., 1973; Barrios-González and Miranda, 2010; Anjum et al., 2012). Therefore, the diversity of secondary metabolites under investigation for biotechnological applications requires exhaustive testing of their *in vitro* and *in vivo* toxicological effects, once they are directly related to animal and human health (Madariaga-Mazón et al., 2019).

4. Non-Toxic Fungal Metabolites and Potential Uses

Besides the well-known mycotoxins from different chemical classes (non-ribosomal peptides, terpenes polyketides) (Cano et al., 2016), several other secondary metabolites of industrial interest are being investigated and do not present toxicity in several experimental models. Despite the relentless search for molecules with cytotoxic action, especially with antitumoral activity, low toxicity, or the lack of it opens perspectives for applications as distinct as the diversity of existing fungal species on the possibility of producing new metabolites. The scope of the search for non-toxic compounds is to find molecules with wide applications for human or animal uses.

Table 5.4 and Fig. 5.9 provide examples of non-toxic compounds produced by fungi against different models, including invertebrates, murine, and human

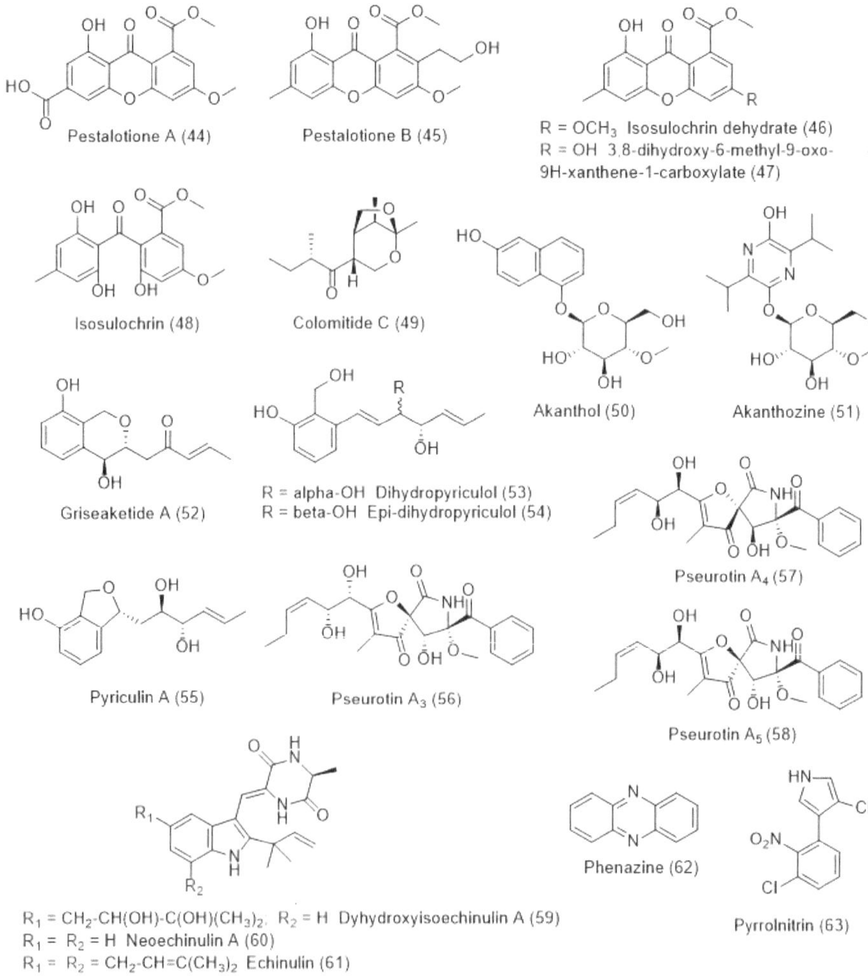

Fig. 5.9. Chemical structures of promising **(44–63)** non-toxic fungal metabolites.

cancer cell cultures. The plant pathogen *Pestalotiopsis theae*, a widely distributed endophytic species associated with *Camelia sinensis*, for instance, had nine novel structurally diverse metabolites isolated, including xanthone derivatives and diphenyl ether, characterized according to its cytotoxic activity. In an antitumor screening including four human cancer cell lines from endocervical adenocarcinoma (HeLa), breast carcinoma (MCF-7), hepatoblastoma (HepG2), and renal carcinoma (ACHN), these newly identified compounds, pestalotione A **(44)** and B **(45)**, isosulochrin dehydrate **(46)**, 3,8-dihydroxy-6-methyl-9-oxo-9H-xanthene-1-carboxylate **(47)**, and isosulochrin **(48)** (Fig. 5.9) had no detectable inhibitory effect over any cell type up to 100 μM. On the other hand, pestalotione A **(44)** and B **(45)** presented moderate 2,2-diphenyl1-picrylhydrazyl (DPPH) radical scavenge IC_{50} values of 54.2 μg/mL and 59.2 μg/mL (control: IC_{50} = 6 μg/mL), respectively (Guo et al., 2020). Notably, xanthones can have diverse applications, like antitumorigenic and antiviral (Miladiyah et al., 2018), and antimalarial activities (Hay et al., 2004). The production of non-toxic xanthones by *P. theae* is an interesting commercial trait, considering the relative effortless manipulation of fungi on industrial conditions to produce high-yield target molecules with larvicidal or antibiotic activity.

Under a zebrafish *D. rerio* larvae fin regeneration protocol, colomitide C **(49)** (Fig. 5.9), a diastereoisomeric bicyclic ketal from the Antarctic fungus *Cadophora luteo-olivacea*, was unable to inhibit the vascularization of fish fin after amputation and larvae viability up to 100 μM. However, fin regeneration was compromised due to fibroblast growth factor receptors (FGFR) pathway block. Such receptors are also present in human breast cancer cells, but a later discovery demonstrated that the translational FGFR blocking effect was not observed in any of the three human breast cancer cell lines (CAMA-1, HCC38, and MDA-MB-361) used, having no cytotoxic effect over the purposing cell cultures. Colomitide C **(49)** turns up, therefore, as a possible tissue-remodel modulator with positive implications on vascularization (Cavanah et al., 2021).

A non-explored spider-parasite fungal species *Akanthomyces novoguineensis* was demonstrated to produce novel glycosylated β-naphthol and pyrazine, akanthol **(50)** and akanthozine **(51)** (Fig. 5.9), respectively, and hydroxamic acid derivatives (non-identified). For decades, hydroxamic acid derivatives have been used as anticancer compounds, as much as β-naphthol and pyrazine-conjugates have been used as antimicrobials. However, in the nematocidal effect over *C. elegans* (up to 100 μg/mL) or human cervix adenocarcinoma cell line HeLa KB-3-1 and murine immortalized fibroblast line L929 (up to 37 μg/mL), no differential phenotype was observed. Therefore, the authors suggest that the chemical backbones 2-naphthol can be used for the design of antimicrobial or antitumor agents with no toxic effect (Helaly et al., 2017).

Aromatic polyketides griseaketide A **(52)**, dihydropyriculol **(53)**, epi-dihydropyriculol **(54)**, and pyriculin A **(55)** (Fig. 5.9) from the rice pathogen *Magnaporthe grisea* did not affect *C. elegans* up to 400 ppm, which reinforces the diversity of fungi with potential production of non-toxic compounds of industrial interest (Yang et al., 2019). Polyketides are so far recognized as an important class of molecules with pharmaceutical importance, as lovastatin **(7)** for the treatment of hypercholesterolemia (Wang et al., 2021), or antibiotic effect, as dihydropyriculol (Furuyama et al., 2021). Concerning structure diversity, encouraging data was also observed from the recently described pseurotin A_3 **(56)**, besides the two new 1-oxa-7-azaspiro [4.4] non-2-ene-4,6-dione core, pseurotin A_4 **(57)** and A_5 **(58)** (Fig. 5.9) isolated from *Wilsonomyces carpophilus* brown rot-causing species. *Wilsonomyces* is an under-explored genus and the uncommon metabolites isolated presented no cytotoxic effect in mouse fibroblast line L929 and human endocervical adenocarcinoma cell line HeLa KB-3-1, nor nematocidal activity on *C. elegans* (Narmani et al., 2018). Some biological activities have been addressed to the family of pseurotins, including cell differentiation and immunomodulation mediated by lymphocyte activation suppression (Rubanova et al., 2021).

5. Concluding Remarks

The contribution of secondary metabolites for the development of new drug leads has been very important to the history of mankind. This contribution is strongly linked to the creative biosynthesis of fungi, organisms capable of producing structurally complex molecules, often with unprecedented skeletons and great pharmacological potential. However, despite the large number of research institutes that prospect bioactive fungal metabolites, most studies do not progress after structural identification and preliminary evaluation of metabolites in *in vitro* assays. In this way, there is a large amount of information, including toxicity studies about fungi bioactive compounds still unexplored that deserve attention, mainly in the contemporary scenario where the advance of chronic non-communicable diseases has become a world concern.

In this context, it is essential to accelerate screening larger numbers of fungi metabolites that could be applied in the development of new, more effective, and specific medicines, among other biotechnological uses. The later efforts in this area are directed to improve preclinical studies, using *in silico* and high throughput *in vitro* methods. Security regarding toxic or ecotoxicological effects, defined biological mechanisms of action, and *in vivo* pharmacokinetic properties, as well as scalable production, are essential aspects to indicate potential drug candidates for translational clinical trials. On the other side, approaches for yield improvement and metabolites diversification, appliable to the fermentation process, have been advanced. Innovation is well represented by the recent evolution on the manipulation of genetic aspects in model organisms (e.g., CRISPR technology), allowing advances in the exploration of mutant organisms as well as industrial production of target molecules.

Acknowledgements

The authors gratefully acknowledge financial support and grants from Fundação de Amparo à Pesquisa do Estado de Minas Gerais, Conselho Nacional de Desenvolvimento Científico e Tecnológico and Coordenação de Aperfeiçoamento de Pessoal de Nível Superior (Brazil).

References

Abraham, E.P. and Newton, G.G. (1961). The structure of cephalosporin C. *Biochem. J.*, 79(2): 377–393.

Abrol, V., Kushwaha, M., Arora, D., Mallubhotla, S. and Jaglan, S. (2021). Mutation, chemoprofiling, dereplication, and isolation of natural products from *Penicillium oxalicum*. *ACS Omega*, 6(25): 16266–16272.

[ADI] Alzheimer's Disease International. (2019). *World Alzheimer Report 2019: Attitudes to dementia*. London: Alzheimer's Disease International. Available on https://www.alz.co.uk/research/world-report-2019.

Adrio, J.L. and Demain, A.L. (2010). Recombinant organisms for production of industrial products. *Bioeng. Bugs*, 1(2): 116–131.

Agrawal, S., Deshmukh, S.K., Reddy, M.S., Prasad, R. and Goel, M. (2020). Endolichenic fungi: A hidden source of bioactive metabolites. *S. Afr. J. Bot.*, 134: 163–186.

Agrawal, S., Barrow, C.J., Adholeya, A. and Deshmukh, S.K. (2021a). Unveiling the dermatological potential of marine fungal species components: Antioxidant and inhibitory capacities over tyrosinase. *Biotechnol. Appl. Biochem.*, 10.1002/bab.2201.

Agrawal, S., Samanta, S. and Deshmukh, S.K. (2021b). The anti-diabetic potential of endophytic fungi: Future prospects as therapeutic agents. *Biotechnol. Appl. Biochem.*, 10.1002/bab.2192.

Alberti, F., Khairudin, K., Venegas, E.R., Davies, J.A., Hayes, P.M., Willis, C.L., Bailey, A.M. and Foster, G.D. (2017). Heterologous expression reveals the biosynthesis of the antibiotic pleuromutilin and generates bioactive semi-synthetic derivatives. *Nat. Commun.*, 8(1): 1831.

Alzheimer's Association. (2020). 2020 Alzheimer's disease facts and figures. *Alzheimers Dement.*, 16(3): 391–460.

Aminov, R.I. (2010). A brief history of the antibiotic era: lessons learned and challenges for the future. *Front. Microbiol.*, 1: 134.

Ancheeva, E., Daletos, G. and Proksch, P. (2020). Bioactive secondary metabolites from endophytic fungi. *Curr. Med. Chem.*, 27(11): 1836–1854.

Anjum, T., Azam, A. and Irum, W. (2012). Production of cyclosporine a by submerged fermentation from a local isolate of *Penicillium fellutanum*. *Indian J. Pharm. Sci.*, 74(4): 372–374.

Atanasov, A.G., Zotchev, S.B., Dirsch, V.M. and Supuran, C.T. (2021). Natural products in drug discovery: Advances and opportunities. *Nat. Rev. Drug Discov.*, 20(3): 200–216.

Bailey, A.M., Alberti, F., Kilaru, S., Collins, C.M., De Mattos-Shipley, K., Hartley, A.J., Hayes, P., Griffin, A., Lazarus, C.M., Cox, R.J., Willis, C.L., O'Dwyer, K., Spence, D.W. and Foster, G.D. (2016). Identification and manipulation of the pleuromutilin gene cluster from *Clitopilus passeckerianus* for increased rapid antibiotic production. *Sci. Rep.*, 6(1): 25202.

Baptista, E.A., Dey, S. and Pal, S. (2021). Chronic respiratory disease mortality and its associated factors in selected Asian countries: Evidence from panel error correction model. *BMC Public Health*, 21(1): 53.

Barrios-González, J. and Miranda, R.U. (2010). Biotechnological production and applications of statins. *Appl. Microbiol. Biotechnol.*, 85(4): 869–883.

Barrios-González, J., Pérez-Sánchez, A. and Bibián, M.E. 2020. New knowledge about the biosynthesis of lovastatin and its production by fermentation of *Aspergillus terreus*. *Appl. Microbiol. Biotechnol.*, 104(21): 8979–8998.

Beekman, A.M. and Barrow, R.A. (2014). Fungal metabolites as pharmaceuticals. *Aust. J. Chem.*, 67(6): 827–843.

Bentley, R. (2000). Mycophenolic Acid: A one-hundred-year odyssey from antibiotic to immunosuppressant. *Chem. Rev.*, 100(10): 3801–3826.

Bills, G.F. and Gloer, J.B. (2016). Biologically active secondary metabolites from the fungi. *Microbiol. Spectr.*, 4(6): 10.1128.

Borel, J.F., Kis, Z.L. and Beveridge, T. (1995). The history of the discovery and development of Cyclosporine (Sandimmune®). pp. 27–63. *In*: Merluzzi, V.J. and Adams, J. (Eds.). *The Search for Anti-Inflammatory Drugs*, Boston: Birkhäuser.

Brooks, A.K. and Gaj, T. (2018). Innovations in CRISPR technology. *Curr. Opin. Biotechnol.*, 52: 95–101.

Cano, P.M., Puel, O. and Oswald, I.P. (2016). Mycotoxins: Fungal secondary metabolites with toxic properties. pp. 318–371. *In*: Deshmukh, S.K., Misra, J.K., Tewari, J.P. and Papp, T. (Eds.). *Fungi: Applications and Management Strategies* (1st Edn.). CRC Press.

Cao, Y., Li, S. and Chen, J. (2021). Modeling better *in vitro* models for the prediction of nanoparticle toxicity: A review. *Toxicol. Mech. Methods*, 31(1): 1–17.

Cavanah, P., Itou, J., Rusman, Y., Tahara, N., Williams, J.M., Salomon, C.E. and Kawakami, Y. (2021). A nontoxic fungal natural product modulates fin regeneration in zebrafish larvae upstream of FGFWNT developmental signaling. *Dev. Dyn.*, 250(2): 160–174.

Chen, H., Zhong, L., Long, Y., Li, J., Wu, J., Liu, L., Chen, S., Lin, Y., Li, M., Zhu, X. and She, Z. (2012). Studies on the synthesis of derivatives of marine-derived bostrycin and their structure-activity relationship against tumor cells. *Mar. Drugs*, 10(4): 932–952.

Chen, M., Wang, K.-L. and Wang, C.-Y. (2018). Antifouling indole alkaloids of a marine-derived fungus *Eurotium* sp. *Chem. Nat. Compd.*, 54(1): 207–209.

Chen, W.-S., Hou, J.-N., Guo, Y.-B., Yang, H.-L., Xie, C.-M., Lin, Y.-C. and She, Z.-G. (2011). Bostrycin inhibits proliferation of human lung carcinoma A549 cells *via* downregulation of the PI3K/Akt pathway. *J. Exp. Clin. Canc. Res.*, 30(1): 17.

Chen,Y.-C. (2015). Beware of docking! *Trends Pharmacol. Sci.*, 36(2): 78–95.

Cheng, Y. and Xie, P. (2019). Ganoderic acid A holds promising cytotoxicity on human glioblastoma mediated by incurring apoptosis and autophagy and inactivating PI3K/AKT signaling pathway. *J. Biochem. Mol. Toxicol.*, 33(11): e22392.

Cockwell, P. and Fisher, L.-A. (2020). The global burden of chronic kidney disease. *Lancet*, 395(10225): 662–664.

Das, A., Alshareef, M., Henderson, F., Jr., Martinez Santos, J.L., Vandergrift, W.A., 3rd, Lindhorst, S.M., Varma, A.K., Infinger, L., Patel, S.J. and Cachia, D. (2020). Ganoderic acid A/DM-induced NDRG2 over-expression suppresses high-grade meningioma growth. *Clin. Transl. Oncol.*, 22(7): 1138–1145.

Del Valle, P., Martínez, A.-L., Figueroa, M., Raja, H.A. and Mata, R. (2016). Alkaloids from the fungus *Penicillium spathulatum* as α-glucosidase inhibitors. *Planta Med.*, 82(14): 1286–1294.

De Martel, C., Georges, D., Bray, F., Ferlay, J. and Clifford, G.M. (2020). Global burden of cancer attributable to infections in 2018: A worldwide incidence analysis. *Lancet Glob. Health*, 8(2): e180–e190.

Deshmukh, S.K., Gupta, M.K., Prakash, V. and Saxena, S. (2018). Endophytic fungi: A source of potential antifungal compounds. *J. Fungi*, 4(3): 77.

Deshmukh, S.K., Agrawal, S., Gupta, M.K., Patidar, R.K. and Ranjan, N. (2022). Recent advances in the discovery of antiviral metabolites from fungi. *Curr. Pharm. Biotechnol.*, 23(4): 495–537.

DeVita, V.T., Rosenberg, S.A. and Lawrence, T.S. (2018). *DeVita, Hellman, and Rosenberg's Cancer: Principles & practice of Oncology*, (11th Edn.).

Ding, R., Sun, B.-F. and Lin, G.-Q. (2012). An efficient total synthesis of (-)-huperzine A. *Org. Lett.*, 14(17): 4446–4449.

Do Nascimento, J.S., Silva, F.M., Magallanes-Noguera, C.A., Kurina-Sanz, M., Dos Santos, E.G., Caldas, I.S., Luiz, J.H.H. and Silva, E.O. (2020). Natural trypanocidal product produced by endophytic fungi through co-culturing. *Folia Microbiol.*, 65(2): 323–328.

Doke, S.K. and Dhawale, S.C. (2015). Alternatives to animal testing: A review. *Saudi Pharm. J.*, 23(3): 223–229.

Eiser, A.R. (2017). Why does Finland have the highest dementia mortality rate? Environmental factors may be generalizable. *Brain Res.*, 1671: 14–17.

El-Hawary, S.S., Mohammed, R., Bahr, H.S., Attia, E.Z., El-Katatny, M.H., Abelyan, N., Al-Sanea, M.M., Moawad, A.S. and Abdelmohsen, U.R. (2021). Soybean-associated endophytic fungi as potential source for anti-COVID-19 metabolites supported by docking analysis. *J. Appl. Microbiol.*, 10.1111/jam.15031.

El-Sayed, A.S., El-Sayed, M.T., Rady, A.M., Zein, N., Enan, G., Shindia, A., El-Hefnawy, S., Sitohy, M. and Sitohy, B. (2020). Exploiting the biosynthetic potency of taxol from fungal endophytes of conifers plants; genome mining and metabolic manipulation. *Molecules*, 25(13): 3000.

Emwas, A.-H., Szczepski, K., Poulson, B.G., Chandra, K., McKay, R.T., Dhahri, M., Alahmari, F., Jaremko, L., Lachowicz, J.I. and Jaremko, M. (2020). NMR as a "gold standard" method in drug design and discovery. *Molecules*, 25(20): 4597.

Endo, A. (2010). A historical perspective on the discovery of statins. *Proc. Jpn. Acad., Ser B, Phys. Biol. Sci.*, 86(5): 484–493.

Fares, J., Fares, M.Y., Khachfe, H.H., Salhab, H.A. and Fares, Y. (2020). Molecular principles of metastasis: A hallmark of cancer revisited. *Signal Transduct. Target. Ther.*, 5(1): 28.

Fennema, E., Rivron, N., Rouwkema, J., Van Blitterswijk, C. and De Boer, J. (2013). Spheroid culture as a tool for creating 3D complex tissues. *Trends Biotechnol.*, 31(2): 108–15.

Ferrara, M., Haidukowski, M., Logrieco, A.F., Leslie, J.F. and Mulè, G. (2019). A CRISPR-Cas9 system for genome editing of *Fusarium proliferatum*. *Sci. Rep.*, 9(1): 19836.

[FIRS] Forum of International Respiratory Societies. (2017). *The Global Impact of Respiratory Disease*, 2nd Edn. Sheffield, European Respiratory Society.

Freshney, R.I. (2005). *Culture of Animal Cells: A manual of basic techniques* (5th Edn.), John Wiley & Sons.

Furuyama, Y., Motoyama, T., Nogawa, T., Kamakura, T. and Osada, H. (2021). Dihydropyriculol produced by *Pyricularia oryzae* inhibits the growth of *Streptomyces griseus*. *Biosci. Biotechnol. Biochem.*, 85(5): 1290–1293.

Gacek, A. and Strauss, J. (2012). The chromatin code of fungal secondary metabolite gene clusters. *Appl. Microbiol. Biotechnol.*, 95(6): 1389–1404.

Gähwiler, B.H., Capogna, M., Debanne, D., McKinney, R.A. and Thompson, S.M. (1997). Organotypic slice cultures: A technique has come of age. *Trends Neurosci.*, 20(10): 471–477.

Garcia-Perez, I., Posma, J.M., Serrano-Contreras, J.I., Boulangé, C.L., Chan, Q., Frost, G., Stamler, J., Elliott, P., Lindon, J.C., Holmes, E. and Nicholson, J.K. (2020). Identifying unknown metabolites using NMR-based metabolic profiling techniques. *Nat. Protoc.*, 15(8): 2538–2567.

[GBD 2016] Global Burden Disease (Institute). (2016). Dementia Collaborators. (2019). Global, regional, and national burden of Alzheimer's disease and other dementias, 1990–2016: A systematic analysis for the Global Burden of Disease Study 2016. *Lancet Neurol.*, 18(1): 88–106.

[GBD 2016] GBD 2016 Occupational Chronic Respiratory Risk Factors Collaborators. (2020). Global and regional burden of chronic respiratory disease in 2016 arising from non-infectious airborne occupational exposures: A systematic analysis for the Global Burden of Disease Study 2016. *Occup. Environ. Med.*, 77(3): 142–150.

[GBD 2019] GBD 2019 Diseases and Injuries Collaborators. (2020). Global burden of 369 diseases and injuries in 204 countries and territories, 1990–2019: A systematic analysis for the Global Burden of Disease Study 2019. *Lancet*, 396(10258): 1204–1222.

Gomes, N.G.M., Lefranc, F., Kijjoa, A. and Kiss, R. (2015). Can some marine-derived fungal metabolites become actual anticancer agents? *Mar. Drugs*, 13(6): 3950–3991.

Guo, L., Lin, J., Niu, S., Liu, S. and Liu, L. (2020). Pestalotiones A–D: Four new secondary metabolites from the plant endophytic fungus *Pestalotiopsis theae*. *Molecules*, 25(3): 470.

Guo, Z. and Zou, Z.M. (2020). Discovery of new secondary metabolites by epigenetic regulation and NMR comparison from the plant endophytic fungus *Monosporascus eutypoides*. *Molecules*, 25(18): 4192.

Gupta, M., Sharma, R. and Kumar, A. (2018). Docking techniques in pharmacology: How much promising? *Comput. Biol. Chem.*, 76: 210–217.

Guruceaga, X., Perez-Cuesta, U., Abad-Diaz de Cerio, A., Gonzalez, O., Alonso, R.M., Hernando, F.L., Ramirez-Garcia, A. and Rementeria, A. (2019). Fumagillin, a mycotoxin of *Aspergillus fumigatus*: Biosynthesis, biological activities, detection, and applications. *Toxins*, 12(1): 7.

Hanahan, D. and Weinberg, R.A. (2011). Hallmarks of cancer: The next generation. *Cell*, 144(5): 646–674.

Handsfield, H.H., Clark, H., Wallace, J.F., Holmes, K.K. and Turck, M. (1973). Amoxicillin, a new penicillin antibiotic. *Antimicrob. Agents Ch.*, 3(2): 262–265.

Hautbergue, T., Jamin, E.L., Debrauwer, L., Puel, O. and Oswald, I.P. (2018). From genomics to metabolomics, moving toward an integrated strategy for the discovery of fungal secondary metabolites. *Nat. Prod. Rep.*, 35(2): 147–173.

Hay, A.E., Hélesbeux, J.J., Duval, O., Labaïed, M., Grellier, P. and Richomme, P. (2004). Antimalarial xanthones from *Calophyllum caledonicum* and *Garcinia vieillardii*. *Life Sci.*, 75(25): 3077–3085.

Heilos, D., Rodríguez-Carrasco, Y., Englinger, B., Timelthaler, G., Van Schoonhoven, S., Sulyok, M., Boecker, S., Süssmuth, R.D., Heffeter, P., Lemmens-Gruber, R., Dornetshuber-Fleiss, R. and

Berger, W. (2017). The natural fungal metabolite beauvericin exerts anticancer activity *in vivo*: A pre-clinical pilot study. *Toxins*, 9(9): 258.

Helaly, S.E., Kuephadungphan, W., Phongpaichit, S., Luangsa-Ard, J.J., Rukachaisirikul, V. and Stadler, M. (2017). Five unprecedented secondary metabolites from the spider parasitic fungus *Akanthomyces novoguineensis*. *Molecules*, 22(6): 991.

Hollingsworth, S.A. and Dror, R.O. (2018). Molecular dynamics simulation for all. *Neuron*, 99: 1129–1143.

Hussain, H., Nazir, M., Saleem, M., Al-Harrasi, A., Elizbit and Green, I.R. (2021). Fruitful decade of fungal metabolites as anti-diabetic agents from 2010 to 2019: Emphasis on α-glucosidase inhibitors. *Phytochem. Rev.*, 20(1): 145–179.

Hutchings, M.I., Truman, A.W. and Wilkinson, B. (2019). Antibiotics: Past, present, and future. *Curr. Opin. Microbiol.*, 51: 72–80.

Ibrar, M., Ullah, M.W., Manan, S., Farooq, U., Rafiq, M. and Hasan, F. (2020). Fungi from the extremes of life: An untapped treasure for bioactive compounds. *Appl. Microbiol. Biotechnol.*, 104(7): 2777–2801.

[IDF] International Diabetes Federation. (2019). Diabetes Atlas (9th Edn.).

Iyengar, R. (2013). Complex diseases require complex therapies. *EMBO Rep.*, 14(12): 1039–1042.

Jiang, M., Wu, Z., Guo, H., Liu, L. and Chen, S. (2020). A review of terpenes from marine-derived fungi: 2015–2019. *Mar. Drugs*, 18(6): 321.

Jiménez-Medina, E., Berruguilla, E., Romero, I., Algarra, I., Collado, A., Garrido, F. and Garcia-Lora, A. (2008). The immunomodulator PSK induces *in vitro* cytotoxic activity in tumour cell lines *via* arrest of cell cycle and induction of apoptosis. *BMC Cancer*, 8: 78.

Kebede, B., Wrigley, S.K., Prashar, A., Rahlff, J., Wolf, M., Reinshagen, J., Gribbon, P., Imhoff, J.F., Silber, J., Labes, A. and Ellinger, B. (2017). Establishing the secondary metabolite profile of the marine fungus: *Tolypocladium geodes* sp. MF458 and subsequent optimisation of bioactive secondary metabolite production. *Mar. Drugs*, 15(4): 84.

Keller, N.P. (2019). Fungal secondary metabolism: Regulation, function, and drug discovery. *Nat. Rev. Microbiol.*, 17(3): 167–180.

Kendall, L.V., Owiny, J.R., Dohm, E.D., Knapek, K.J., Lee, E.S., Kopanke, J.H., Fink, M., Hansen, S.A. and Ayers, J.D. (2018). Replacement, refinement, and reduction in animal studies with biohazardous agents. *ILAR J.*, 59(2): 177–194.

Khakban, A., Sin, D.D., FitzGerald, J.M., McManus, B.M., Ng, R., Hollander, Z. and Sadatsafavi, M. (2017). The projected epidemic of chronic obstructive pulmonary disease hospitalizations over the next 15 years: A population-based perspective. *Am. J. Respir. Crit. Care Med.*, 195(3): 287–291.

Korch, C., Hall, E.M., Dirks, W.G., Ewing, M., Faries, M., Varella-Garcia, M., Robinson, S., Storts, D., Turner, J.A., Wang, Y., Burnett, E.C., Healy, L., Kniss, D., Neve, R.M., Nims, R.W., Reid, Y.A., Robinson, W.A. and Capes-Davis, A. (2018). Authentication of M14 melanoma cell line proves misidentification of MDA-MB-435 breast cancer cell line. *Int. J. Cancer*, 142(3): 561–572.

Kornienko, A., Evidente, A., Vurro, M., Mathieu, V., Cimmino, A., Evidente, M., Van Otterlo, W.A.L., Dasari, R., Lefranc, F. and Kiss, R. (2015). Toward a cancer drug of fungal origin. *Med. Res. Rev.*, 35(5): 937–967.

Lancaster, M.A. and Knoblich, J.A. (2014). Organogenesis in a dish: Modeling development and disease using organoid technologies. *Science*, 345(6194): 1247125.

Laupacis, A., Keown, P.A., Ulan, R.A., McKenzie, N. and Stiller, C.R. (1982). Cyclosporin A: A powerful immunosuppressant. *Can. Med. Assoc. J.*, 126(9): 1041–1046.

Liang, C., Tian, D., Liu, Y., Li, H., Zhu, J., Li, M., Xin, M. and Xia, J. (2019). Review of the molecular mechanisms of *Ganoderma lucidum* triterpenoids: Ganoderic acids A, C2, D, F, DM, X, and Y. *Eur. J. Med. Chem.*, 174: 130–141.

Lima, N.M., Andrade, T.J.A.S. and Silva, D.H.S. (2020). Dereplication of terpenes and phenolic compounds from *Inga edulis* extracts using HPLC-SPE-TT, RP-HPLC-PDA and NMR spectroscopy. *Nat. Prod. Res.*, 1–5.

Linnakoski, R., Reshamwala, D., Veteli, P., Cortina-Escribano, M., Vanhanen, H. and Marjomäki, V. (2018). Antiviral agents from fungi: Diversity, mechanisms, and potential applications. *Front. Microbiol.*, 9: 2325.

Liu, J., Shiono, J., Shimizu, K., Kukita, A., Kukita, T. and Kondo, R. (2009). Ganoderic acid DM: Anti-androgenic osteoclastogenesis inhibitor. *Bioorg. Med. Chem. Lett.*, 19(8): 2154–2157.

Lopes, A.M., Dahms, H.U., Converti, A. and Mariottini, G.L. (2021). Role of model organisms and nanocompounds in human health risk assessment. *Environ. Monit. Assess.*, 193(5): 285.

Ma, X., Tan, C., Zhu, D., Gang, D.R. and Xiao, P. (2007). Huperzine A from *Huperzia* species: An ethnopharmacological review. *J. Ethnopharmacol.*, 113(1): 15–34.

Madariaga-Mazón, A., Hernández-Alvarado, R.B., Noriega-Colima, K.O., Osnaya-Hernández, A. and Martinez-Mayorga, K. (2019). Toxicity of secondary metabolites. *Phys. Sci. Rev.*, 4(12): 20180116.

Madden, J.C., Enoch, S.J., Paini, A. and Cronin, M. (2020). A review of *in silico* tools as alternatives to animal testing: Principles, resources, and applications. *Altern. Lab. Anim.*, 48(4): 146–172.

Masi, M. and Evidente, A. (2020). Fungal bioactive anthraquinones and analogues. *Toxins*, 12(11): 714.

Miladiyah, I., Jumina, J., Haryana, S.M. and Mustofa, M. (2018). Biological activity, quantitative structure-activity relationship analysis, and molecular docking of xanthone derivatives as anticancer drugs. *Drug Des. Devel. Ther.*, 12: 149–158.

Narmani, A., Teponno, R.B., Arzanlou, M., Babai-Ahari, A. and Stadler, M. (2018). New secondary metabolites produced by the phytopathogenic fungus *Wilsonomyces carpophilus*. *Phytochem. Lett.*, 26: 212–217.

Nelson, S., Whitsel, L., Khavjou, O., Phelps, D. and Leib, A. (2016). *Projections of cardiovascular disease prevalence and costs: 2015–2035, technical report*. Research Triangle Park, NC: RTI International.

Newman, D. (2017). Screening and identification of novel biologically active natural compounds. *F1000Research*, 6: 783.

Newman, D.J. and Cragg, G.M. (2020). Natural products as sources of new drugs over the nearly four decades from 01/1981 to 09/2019. *J. Nat. Prod.*, 83(3): 770–803.

Nguyen, V.-T., Lee, J.S., Qian, Z.-J., Li, Y.-X., Kim, K.-N., Heo, S.-J., Jeon, Y.-J., Park, W.S., Choi, I.W., Je, J.-Y. and Jung, W.-K. (2013). Gliotoxin isolated from marine fungus *Aspergillus* sp. induces apoptosis of human cervical cancer and chondrosarcoma cells. *Mar. Drugs*, 12(1): 69–87.

Nielsen, M.L., Isbrandt, T., Rasmussen, K.B., Thrane, U., Hoof, J.B., Larsen, T.O. and Mortensen, U.H. (2017). Genes linked to production of secondary metabolites in *Talaromyces atroroseus* revealed using CRISPR-Cas9. *PLOS One*, 12(1): e0169712.

Nisr, R.B., Russell, M.A., Chrachri, A., Moody, A.J. and Gilpin, M.L. (2011). Effects of the microbial secondary metabolites pyrrolnitrin, phenazine and patulin on INS-1 rat pancreatic β-cells. *FEMS Immunol. Med. Microbiol.*, 63(2): 217–227.

Niu, G., Hao, Y., Wang, X., Gao, J.-M. and Li, J. (2020). Fungal metabolite asperaculane B inhibits malaria infection and transmission. *Molecules*, 25(13): 3018.

Oikawa, H. (2020). Reconstitution of biosynthetic machinery of fungal natural products in heterologous hosts. *Biosci. Biotechnol. Biochem.*, 84(3): 433–444.

Oselusi, S.O., Christoffels, A. and Egieyeh, S.A. (2021). Cheminformatic characterization of natural antimicrobial products for the development of new lead compounds. *Molecules*, 26(13): 3970.

Page, S.J.S. and Tait, C.P. (2015). Mycophenolic acid in dermatology a century after its discovery. *Aust. J. Dermatol.*, 56(1): 77–83.

Pearce, A., Sharp, L., Hanly, P., Barchuk, A., Bray, F., De Camargo Cancela, M., Gupta, P., Meheus, F., Qiao, Y.-L., Sitas, F., Wang, S.-M. and Soerjomataram, I. (2018). Productivity losses due to premature mortality from cancer in Brazil, Russia, India, China, and South Africa (BRICS): A population-based comparison. *Cancer Epidemiol.*, 53: 27–34.

Petersen, A.B., Rønnest, M.H., Larsen, T.O. and Clausen, M.H. (2014). The chemistry of griseofulvin. *Chem. Rev.*, 114(24): 12088–12107.

Ranganathan, N. and Mahalingam, G. (2019). Secondary metabolite as therapeutic agent from endophytic fungi *Alternaria longipes* strain VITN14G of mangrove plant *Avicennia officinalis*. *J. Cell. Biochem.*, 120(3): 4021–4031.

Ranganathan, N. and Mahalingam, G. (2020). Validation of the antidiabetic potential of isolated and nanoemulsified endophytic fungal metabolite 2,4,6-triphenylaniline through AMPK activation pathway. *J. Mol. Liq.*, 316: 113836.

Rao, P., Shukla, A., Parmar, P., Rawal, R.M., Patel, B.V., Saraf, M. and Goswami, D. (2020). Proposing a fungal metabolite-flaviolin as a potential inhibitor of $3CL^{pro}$ of novel coronavirus SARS-CoV-2 identified using docking and molecular dynamics. *J. Biomol. Struct. Dyn.*, 1–13.

Ribeiro, B.A., Sa Mata, T.B., Canuto, G.A.B. and Silva, E.O. (2021). Chemical diversity of secondary metabolites produced by Brazilian endophytic fungi. *Curr. Microbiol.*, 78(1): 33–54.

Ricciardi, M.R., Licchetta, R., Mirabilii, S., Scarpari, M., Parroni, A., Fabbri, A.A., Cescutti, P., Reverberi, M., Fanelli, C. and Tafuri, A. (2017). Preclinical antileukemia activity of tramesan: A newly identified bioactive fungal metabolite. *Oxid. Med. Cell. Longev.*, 2017: 5061639.

Rifaioglu, A.S., Atas, H., Martin, M.J., Cetin-Atalay, R., Atalay, V. and Doğan, T. (2019). Recent applications of deep learning and machine intelligence on *in silico* drug discovery: Methods, tools, and databases. *Brief. Bioinform.*, 20(5): 1878–1912.

Roth, G.A., Mensah, G.A., Johnson, C.O., Addolorato, G., Ammirati, E., Baddour, L.M., Barengo, N.C., Beaton, A.Z., Benjamin, E.J., Benziger, C.P., Bonny, A., Brauer, M., Brodmann, M., Cahill, T.J., Carapetis, J., Catapano, A.L., Chugh, S.S., Cooper, L.T., Coresh, J., Criqui, M., DeCleene, N., Eagle, K.A., Emmons-Bell, S., Feigin, V.L., Fernández-Solà, J., Fowkes, G., Gakidou, E. Grundy, S.M., He, F.J., Howard, G., Hu, F., Inker, L., Karthikeyan, G., Kassebaum, N., Koroshetz, W., Lavie, C., Lloyd-Jones, D., Lu, H.S., Mirijello, A., Temesgen, A.M., Mokdad, A., Moran, A.E., Muntner, P., Narula, J., Neal, B., Ntsekhe, M., de Oliveira, G.M., Otto, C., Owolabi, M., Pratt, M., Rajagopalan, S., Reitsma, M., Ribeiro, A.L.P., Rigotti, N., Rodgers, A., Sable, C., Shakil, S., Sliwa-Hahnle, K., Stark, B., Sundström, J., Timpel, P., Tleyjeh, I.M., Valgimigli, M., Vos, T., Whelton, P.K., Yacoub, M., Zuhlke, L., Murray, C. and Fuster, V. (2020). Global burden of cardiovascular diseases and risk factors, 1990–2019: Update from the GBD 2019 Study. *J. Am. Coll. Cardiol.*, 76(25): 2982–3021.

Rubanova, D., Dadova, P., Vasicek, O. and Kubala, L. (2021). Pseurotin D inhibits the activation of human lymphocytes. *Int. J. Mol. Sci.*, 22(4): 1938.

Rustamova, N., Bozorov, K., Efferth, T. and Yili, A. (2020). Novel secondary metabolites from endophytic fungi: Synthesis and biological properties. *Phytochem. Rev.*, 19(2): 425–448.

Ryngajłło, M., Boruta, T. and Bizukojć, M. (2021). Complete genome sequence of lovastatin producer *Aspergillus terreus* ATCC 20542 and evaluation of genomic diversity among *A. terreus* strains. *Appl. Microbiol. Biotechnol.*, 105(4): 1615–1627.

Sachana, M. and Hargreaves, A.J. (2018). Toxicological testing: *in vivo* and *in vitro* models. pp. 145–161. *In*: Gupta, R.C. (Ed.). *Veterinary Toxicology: Basic and Clinical Principles* (3rd Edn.), Academic Press.

Sakamoto, J., Morita, S., Oba, K., Matsui, T., Kobayashi, M., Nakazato, H. and Ohashi, Y. (2006). Efficacy of adjuvant immunochemotherapy with polysaccharide K for patients with curatively resected colorectal cancer: A meta-analysis of centrally randomized controlled clinical trials. Cancer Immunol. *Immunother.*, 55(4): 404–411.

Salvatore, M.M., Alves, A. and Andolfi, A. (2020). Secondary metabolites of *Lasiodiplodia theobromae*: Distribution, chemical diversity, bioactivity, and implications of their occurrence. *Toxins*, 12(7): 457.

Shabana, S., Lakshmi, K.R. and Satya, A.K. (2021). An updated review of secondary metabolites from marine fungi. *Mini Rev. Med. Chem.*, 21(5): 602–642.

Smith, D.J., Burnham, M.K., Edwards, J., Earl, A.J. and Turner, G. (1990). Cloning and heterologous expression of the penicillin biosynthetic gene cluster from *Penicillium chrysogenum*. *Nat. Biotechnol.*, 8(1): 39–41.

Soga, O. (1982). Stimulative production of flaviolin by *Phoma wasabiae*. *Agric. Biol. Chem.*, 46(4): 1061–1063.

Sun, Y., Takada, K., Takemoto, Y., Yoshida, M., Nogi, Y., Okada, S. and Matsunaga, S. (2012). Gliotoxin analogues from a marine-derived fungus, *Penicillium* sp., and their cytotoxic and histone methyltransferase inhibitory activities. *J. Nat. Prod.*, 75(1): 111–114.

Sung, H., Ferlay, J., Siegel, R.L., Laversanne, M., Soerjomataram, I., Jemal, A. and Bray, F. (2021). Global Cancer Statistics 2020: GLOBOCAN estimates of incidence and mortality worldwide for 36 cancers in 185 countries. *Cancer J. Clin.*, 71(3): 209–249.

Supadmanaba, I., Comandatore, A., Morelli, L., Giovannetti, E. and Lagerweij, T. (2021). Organotypic-liver slide culture systems to explore the role of extracellular vesicles in pancreatic cancer metastatic behavior and guide new therapeutic approaches. *Expert Opin. Drug Metab. Toxicol.*, 1–10.

Takahashi, C., Numata, A., Ito, Y., Matsumura, E., Araki, H., Iwaki, H. and Kushida, K. (1994). Leptosins, antitumour metabolites of a fungus isolated from a marine alga. *J. Chem. Soc., Perkin Trans.*, 1, 13: 1859–1864.

Takahashi, J.A., Teles, A.P.C., Bracarense, A.D.A.P. and Gomes, D.C. (2013). Classical and epigenetic approaches to metabolite diversification in filamentous fungi. *Phytochem. Rev.*, 12(4): 773–789.

Takahashi, J.A., Gomes, D.C., Lyra, F.H. and Dos Santos, G.F. (2016). Modulation of fungal secondary metabolites biosynthesis by chemical epigenetics. pp. 117–133. *In*: Deshmukh, S.K., Misra, J.K., Tewari, J.P. and Papp, T. (Eds.). *Fungi: Applications and Management Strategies* (1st Edn.). CRC Press.

Takahashi, J.A., Sande, D., Da Silva Lima, G., Moura, M.A.F.E. and Lima, M.T.N.S. (2019). Fungal metabolites as promising new drug leads for the treatment of Alzheimer's Disease. *Stud. Nat. Prod. Chem.*, 62: 1–39.

Takahashi, J.A., Barbosa, B.V., Martins, B.D.A., Guirlanda, C.P. and Moura, A.F.M. (2020). Use of the versatility of fungal metabolism to meet modern demands for healthy aging, functional foods, and sustainability. *J. Fungi*, 6(4): 223.

Tang, P.-J., Zhang, Z.-H., Niu, L.-L., Gu, C.-B., Zheng, W.-Y., Cui, H.-C. and Yuan, X.-H. (2021). *Fusarium solani* G6, a novel vitexin-producing endophytic fungus: Characterization, yield improvement and osteoblastic proliferation activity. *Biotechnol. Lett.*, 43: 1–13.

Theobald, S., Vesth, T.C., Rendsvig, J.K., Nielsen, K.F., Riley, R., De Abreu, L.M., Salamov, A., Frisvad, J.C., Larsen, T.O., Andersen, M.R. and Hoof, J.B. (2018). Uncovering secondary metabolite evolution and biosynthesis using gene cluster networks and genetic dereplication. *Sci. Rep.*, 8(1): 17957.

Tomasetti, C., Li, L. and Vogelstein, B. (2017). Stem cell divisions, somatic mutations, cancer etiology, and cancer prevention. *Science*, 355(6331): 1330–1334.

Torkelson, C.J., Sweet, E., Martzen, M.R., Sasagawa, M., Wenner, C.A., Gay, J., Putiri, A. and Standish, L.J. (2012). Phase 1 clinical trial of *Trametes versicolor* in women with breast cancer. *Int. Sch. Res. Notices*, 2012: 251632.

Torres-Mendoza, D., Ortega, H.E. and Cubilla-Rios, L. (2020). Patents on endophytic fungi related to secondary metabolites and biotransformation applications. *J. Fungi*, 6(2): 58.

Ueda, Y., Fujimura, T., Kinami, S., Hirono, Y., Yamaguchi, A., Naitoh, H., Tani, T., Kaji, M., Yamagishi, H. and Miwa, K. (2006). A randomized phase III trial of postoperative adjuvant therapy with S-1 alone *versus* S-1 plus PSK for stage II/IIIA gastric cancer: Hokuriku-Kinki Immunochemo-Therapy Study Group-Gastric Cancer (HKIT-GC). *Jpn. J. Clin. Oncol.*, 36(8): 519–522.

Uzma, F., Mohan, C.D., Hashem, A., Konappa, N.M., Rangappa, S., Kamath, P.V, Singh, B.P., Mudili, V., Gupta, V.K., Siddaiah, C.N., Chowdappa, S., Alqarawi, A.A. and Abd_Allah, E.F. (2018). Endophytic fungi: Alternative sources of cytotoxic compounds: A review. *Front. Pharmacol.*, 9: 309.

Vasantha, K.Y., Singh, R.P. and Sattur, A.P. (2018). A preliminary pharmacokinetic and toxicity study of nigerloxin. *Indian J. Biochem. Biophys.*, 55(1): 44–51.

Vigushin, D.M., Mirsaidi, N., Brooke, G., Sun, C., Pace, P., Inman, L., Moody, C.J. and Coombes, R.C. (2004). Gliotoxin is a dual inhibitor of farnesyltransferase and geranylgeranyltransferase I with antitumor activity against breast cancer *in vivo*. *Med. Oncol.*, 21(1): 21–30.

Wang, J., Liang, J., Chen, L., Zhang, W., Kong, L., Peng, C., Su, C., Tang, Y., Deng, Z. and Wang, Z. (2021). Structural basis for the biosynthesis of lovastatin. *Nat. Commun.*, 12(1): 867.

Wang, Q. and Coleman, J.J. (2019). Progress and challenges: development and implementation of CRISPR/Cas9 technology in filamentous fungi. *Comput. Struct. Biotechnol. J.*, 17: 761–769.

Wang, X., Tan, T., Mao, Z.-G., Lei, N., Wang, Z.-M., Hu, B., Chen, Z.-Y., She, Z.-G., Zhu, Y.-H. and Wang, H.-J. (2015). The marine metabolite SZ-685C induces apoptosis in primary human nonfunctioning pituitary adenoma cells by inhibition of the Akt pathway *in vitro*. *Mar. Drugs*, 13(3):

Wang, Z., Gerstein, M. and Snyder, M. (2009). RNA-Seq: A revolutionary tool for transcriptomics. *Nat. Rev. Genet.*, 10(1): 57–63.

Weston, A.D. and Hood, L. (2004). Systems biology, proteomics, and the future of health care: Toward predictive, preventative, and personalized medicine. *J. Proteome Res.*, 3(2): 179–96.

[WHO] World Health Organization. (2020). WHO Report on Cancer: Setting priorities, investing wisely and providing care for all. Available on https://apps.who.int/iris/handle/10665/330745. 149p.

[WHO] World Health Organization. (2021). Noncommunicable Diseases. Available on https://www.who.int/news-room/fact-sheets/detail/noncommunicable-diseases. Accessed 27 April 2021.

Wishart, D.S. (2007). Current progress in computational metabolomics. Brief. *Bioinform.*, 8 (5): 279–93.

Wu, B., Wu, L., Ruan, L., Ge, M. and Chen, D. (2009). Screening of endophytic fungi with antithrombotic activity and identification of a bioactive metabolite from the endophytic fungal strain CPCC 480097. *Curr. Microbiol.*, 58(5): 522–527.

Wu, F., Zhou, Y., Li, L., Shen, X., Chen, G., Wang, X., Liang, X., Tan, M. and Huang, Z. (2020). Computational approaches in preclinical studies on drug discovery and development. *Front. Chem.*, 8: 726.

Wu, G.S., Lu, J.J., Guo, J.J., Li, Y.B., Tan, W., Dang, Y.Y., Zhong, Z.F., Xu, Z.T., Chen, X.P. and Wang, Y.T. (2012). Ganoderic acid DM, a natural triterpenoid, induces DNA damage, G1 cell cycle arrest, and apoptosis in human breast cancer cells. *Fitoterapia*, 83(2): 408–414.

Xia, J., Dai, L., Wang, L. and Zhu, J. (2020). Ganoderic acid DM induces autophagic apoptosis in non-small cell lung cancer cells by inhibiting the PI3K/Akt/mTOR activity. *Chem.-Biol. Interact.*, 316: 108932.

Xie, G., Zhu, X., Li, Q., Gu, M., He, Z., Wu, J., Li, J., Lin, Y., Li, M., She, Z. and Yuan, J. (2010). SZ685C, a marine anthraquinone, is a potent inducer of apoptosis with anticancer activity by suppression of the Akt/FOXO pathway. *Br. J. Pharmacol.*, 159(3): 689–697.

Xu, J., Yi, M., Ding, L. and He, S. (2019). A review of anti-inflammatory compounds from marine fungi, 2000–2018. *Mar. Drugs*, 17(11): 636.

Yamada, T., Iwamoto, C., Yamagaki, N., Yamanouchi, T., Minoura, K., Yamori, T., Uehara, Y., Andoh, T., Umemura, K. and Numata, A. (2002). Leptosins M–N1, cytotoxic metabolites from a *Leptosphaeria* species separated from a marine alga: Structure determination and biological activities. *Tetrahedron*, 58(3): 479–487.

Yanagihara, M., Sasaki-Takahashi, N., Sugahara, T., Yamamoto, S., Shinomi, M., Yamashita, I., Hayashida, M., Yamanoha, B., Numata, A., Yamori, T. and Andoh, T. (2005). Leptosins isolated from marine fungus *Leptoshaeria* species inhibit DNA topoisomerases I and/or II and induce apoptosis by inactivation of Akt/protein kinase B. *Cancer Sci.*, 96(11): 816–824.

Yang, Y.-H., Yang, D.-S., Lei, H.-M., Li, C.-Y., Li, G.-H. and Zhao, P.-J. (2019). Griseaketides A–D, new aromatic polyketides from the pathogenic fungus *Magnaporthe grisea*. *Molecules*, 25(1): 72.

Ying, Y.M., Huang, L., Tian, T., Li, C.Y., Wang, S.L., Ma, L.F., Shan, W.G., Wang, J.W. and Zhan, Z.J. (2018). Studies on the chemical diversities of secondary metabolites produced by *Neosartorya fischeri via* the OSMAC Method. *Molecules*, 23(11): 2772.

Yuan, S., Gopal, J.V., Ren, S., Chen, L., Liu, L. and Gao, Z. (2020). Anticancer fungal natural products: Mechanisms of action and biosynthesis. *Eur. J. Med. Chem.*, 202: 112502.

Zhang, J., Wu, H., Xia, X., Liang, Y., Yan, Y., She, Z., Lin, Y. and Fu, L. (2007). Anthracenedione derivative 1403P-3 induces apoptosis in KB and KBv200 cells *via* reactive oxygen species-independent mitochondrial pathway and death receptor pathway. *Cancer Biol. Ther.*, 6(9): 1413–1421.

Zhang, X., Li, S.-J., Li, J.-J., Liang, Z.-Z. and Zhao, C.-Q. (2018). Novel natural products from extremophilic fungi. *Mar. Drugs*, 16(6): 194.

Zhivotovsky, B., Samali, A. and Orrenius, S. (1999). Determination of apoptosis and necrosis. pp. 2.2.1–2.2.34. *In*: Maines, M.D., Costa, L.G., Reed, D.J., Sassa, S. and Sipes, I.G. (Eds.). *Current Protocols in Toxicology*. New York: Wiley.

Zhu, X., He, Z., Wu, J., Yuan, J., Wen, W., Hu, Y., Jiang, Y., Lin, C., Zhang, Q., Lin, M., Zhang, H., Yang, W., Chen, H., Zhong, L., She, Z., Chen, S., Lin, Y. and Li, M. (2012). A marine anthraquinone SZ685C overrides adriamycin-resistance in breast cancer cells through suppressing Akt signaling. *Mar. Drugs*, 10(4): 694–711.

Patents

CN101240249A. 2008. *Dioscorea zingiberensis* endogenesis fusarium capable of producing beauvericin and antibacterial activity thereof. China Patent, 13 Aug 2008.

CN101275116A. 2008. Mixed endophyte and method for preparing huperzine A by using the endophyte. China Patent, 01 Oct 2008.

CN101486974A. 2011. Endogenetic fungus for producing paclitaxel. China Patent, 27 Jul 2011.

CN101669939A. 2010. Application of enniatine compound for preparing anti-drug-resistant tubercle bacillus drugs. China Patent, 17 Mar 2010.

CN102643186A. 2016. Anthraquinone dimer derivative and preparation method and application thereof. Ocean University of China. China Patent, 02 Mar 2016.

CN103087923A. 2014. Preparation and application of *Chaetomium globosum* and metabolite flavipin thereof. China Patent, 08 Oct 2014.

CN103966109B. 2016. The shizandra berry endogenetic fungus of rancinamycin IV is produced in one strain. Heilongjiang University. China Patent, 13 Apr 2016.

CN104774774A. 2018. Licorice endophytic fungus capable of producing pseurotin A. Heilongjiang University. China Patent, 23 Jan 2018.

CN107474092A. 2017. A kind of straw mushroom fructification active component and its application. Jilin Agricultural University. China Patent, 15 Dec 2017.

CN1421522A. 2003. Fleabane endogenous fungus. China Patent, 04 Jun 2003.

GB1111010A. Sandoz. 1968. Process for the production of the antibiotic pleuromutilin. United Kingdom Patent, 24 Apr 1968 (expired).

US20040248265A.1Porter J., A. Eyberger. 2004. Endophytes for production of podophyllotoxin. US Patent, 09 Dec 2004.

US20060134762A1. Puri S., V. Vijeshwar, T. Amna, G. Handa, V. Gupta, N. Verma, R. Khajuria, A. Saxena, G. Qazi, M. Spiteller. 2006. Novel endophytic camptothecin and camptothecinoid producing fungi and process of producing the same. US Patent, 22 Jun 2006.

US6613738B1. Strobel, G.A. 2003. Cyclic lipopeptide from *Cryptosporiopsis quercina* possessing antifungal activity. US Patent, 02 Sep 2003.

6

Global Manufacturers of Echinocandins, Echinocandin Intermediates, Market and Future Perspectives

A Review

*Pradipta Tokdar** and *Saji George*

1. Introduction

Echinocandins are antifungal drugs that inhibit the synthesis of β-1, 3-D-glucan in the cell wall of fungi. These drugs show low toxicity, high potency with a rapid fungicidal activity especially against *Candida* species. Echinocandins are derived from fungi by fermentation. The fermented product is synthetically modified to produce different classes of echinocandin molecules. Amphotericin B and the azole drugs, such as voriconazole, fluconazole, and itraconazole were the mainstays of the treatment of fungal infections. However, they are associated with limitations, such as toxicity to red blood cells (RBCs); nephrotoxicity and arrhythmias associated with amphotericin B; hepatic toxicity associated with fluconazole; gastrointestinal disturbances and cardiac toxicity associated with itraconazole; hepatic toxicity and neurological toxicity associated with voriconazole; and elevation in the serum aminotransferase level leading to hepatic toxicity associated with posaconazole. Amphotericin B acts by binding to ergosterol on the cell membrane of fungi that causes leakage in the cell fungal wall. Azoles act by inhibiting α-demethylase, which is essential for producing ergosterol.

Echinocandin B was the first echinocandin class of molecule discovered and isolated to show antifungal activity. Echinocandin B can be biosynthetically produced from the *Aspergillus nidulans* strain (NRRL 8112) and other sources, such as pneumocandin B_0, aculeacin A, and papulacandins (Huttel et al., 2016).

KRIBS-BIONEST (RGCB Campus-3), Kerala Technology Innovation Zone, BTIC Building, KINFRA Hi-Tech Park, Kalamessary, Kochi, India 683503.

* Corresponding author: pradiptatokdar@rgcb.res.in

Its haemolytic activity negated the therapeutic potential. Thus, the central skeletal structure was semi-synthetically modified to make cilofungin that showed lower haemolytic activity than the parent molecule. Thereafter, pneumocandins, including pneumocandin A_0, B_0, C_0, D_0, D_1, and E_5, belonging to echinocandin family were synthesized. Although pneumocandins show antifungal effects, they have low physicochemical properties. Pneumocandin B_0 is the most preferred drug due to its high potency and antifungal activity (Connors and Pollard, 2004). Pneumocandin B_0 was used for the synthesis of caspofungin acetate. Furthermore, FR901379 belongs to the echinocandin family and is structurally like pneumocandin A_0. It is the crucial starting material for synthesis of micafungin. FR901379 has a sulphate moiety that makes it aqueous soluble. FR131535 and FR179642 are the two intermediates synthesized by modifying the acyl side chains of FR901379. These intermediates showed a reduction in lytic activity and improved antifungal activity (Hashimoto, 2009). FK463 was synthesized by further optimization of these intermediates which showed no lysis but a potent activity towards *Candida* and *Aspergillus* (Fujie, 2007). The nucleus of echinocandin B was modified structurally by deacylation to introduce a reactive amino group, which was then treated with terphenyl acid. This was done to introduce the alkoxy triphenyl group at the amino group, which is reactive on the nucleus of echinocandin B. This modification led to the invention of a drug compound known as anidulafungin, with increased antifungal activity and reduced lysis of RBCs. Currently, anidulafungin has been approved as an antifungal therapy (Sunsan and Jose, 2008). All the echinocandins can penetrate well into all the tissues but marginally into the central nervous system (CNS) and the eye. It has shown lower bioavailability when administered via the oral route (Patil and Majumdar, 2017). Echinocandins show high antifungal activity, fewer drug–drug interactions and are less susceptible to resistance compared to other antifungals. The three echinocandins, namely, caspofungin, micafungin, and anidulafungin are antifungal drugs that target the cell wall of the fungi. As mammalian cells lack cell wall, echinocandins do not have any action in the mammals. Echinocandins were obtained as a by-product during the fermentation process of pneumocandins. Although pneumocandin shows antifungal activity, it has very poor physicochemical properties, which led to the discovery of newer analogues of echinocandins (Biernasiuk et al., 2015). These new analogues have improved physicochemical properties, safety profiles, a broad-spectrum nature, and good pharmacokinetic profiles. Due to resistance of the fungi to the present drugs, the development of echinocandins was of great importance (Patil and Majumdar, 2017).

Caspofungin was the first drug to be licensed among echinocandins followed by micafungin and anidulafungin. Caspofungin is used for the treatment of invasive aspergillosis. Due to their high molecular weight, echinocandins cannot be absorbed orally and hence are given intravenously (IV). Interaction and cross reactivity with other drugs are lesser in echinocandins compared to azoles (Denning, 2002).

Beta-1, 3-D-glucan, a major fungal cell wall component, is the target of echinocandins activity. The cell wall of fungi is rigid and consists of a several polysaccharides, mannan, chitin, and galactomannan. The gene that codes for glucan synthase is *FKS1*. The deletion of this gene results in a phenotype that is highly sensitive to tacrolimus. It also produces many mutant phenotypes that can lead to the cloning of many genes identical to *FKS1*. The cloning of a closely related gene *FKS2*, also known as *GSC2*, was done, and was found to be nearly 80% identical to the *FKS1* gene. *FKS2* or *GSC2* gene, upon deletion, is lethal and the mutated *FKS1* is resistant to caspofungin. The products obtained from *FKS1* and *FKS2* genes are found to be the alternate subunits of the enzyme β-1, 3-D-glucan synthase (Denning, 2003). This review describes the current global therapeutic and market potential of echinocandin classes of molecules and its future scope for the manufacture of the drug.

2. Types of Echinocandins

Echinocandin B has a cyclic lipopeptide in the nucleus with the most important structural feature of N-linked acyl fatty acid chain in the echinocandin class. Amino acid residues including 3, 4 dihydroxy-ornithine; 3, 4-dihydroxy homotyrosine; 3, 4-dihydroxy proline; 3-methyl-4-hydroxy proline and two threonine residues are present in the nucleus core of the hexacyclic lipopeptide. The amino acid residues show antifungal activity along with the physicochemical properties of the nucleus of echinocandin B as well as the analogues. Homotyrosine shows antifungal properties and hence inhibits glucan synthase. Proline enhances the antifungal potency and the core of the amino acid residue, which has the hydroxyl groups, increases the stability and water solubility but does not have any antifungal activity. Therefore, during the modification of the drug, the core of the nucleus remains unchanged. The echinocandin B nucleus is linked with the linoleoyl fatty acid chain that is important mainly for not only its antifungal activity but also for a haemolytic activity. To overcome this disadvantage, the structure of the echinocandin B was modified. Pneumocandin A_0 and pneumocandin B_0 are products of the structurally modified echinocandin B where the nucleus is altered by the incorporation of 1-3-hydroxy-glutamine residue instead of a threonine residue and the linoleoyl side chain is replaced by the dimethylmyristoyl side chain to reduce the haemolytic susceptibility of RBCs. Commercially available echinocandin drugs do not have echinocandin B as their core ingredient. Instead, they have either pneumocandin B_0 or pneumocandin A_0 because of their antifungal potency (Yao, 2012). Several types of echinocandins are reviewed below.

2.1 Caspofungin

It belongs to the echinocandin family that is derived from pneumocandin B_0 and it is a semi-synthetic water-soluble drug. It has a pneumocandin B_0 nucleus and does not show haemolytic activity (Hashimoto, 2009). It is a drug that binds to high-protein regions distributed in the tissues. It undergoes degradation by hepatic metabolism and Nacetylation (Aguilar et al., 2015). The chemical structure of caspofungin is shown in Fig. 6.1.

2.2 Micafungin

It has a lipopeptide core like caspofungin and consists of pneumocandin A_0 with a difference in a fatty acid side chain. Micafungin has a fatty acid side chain of

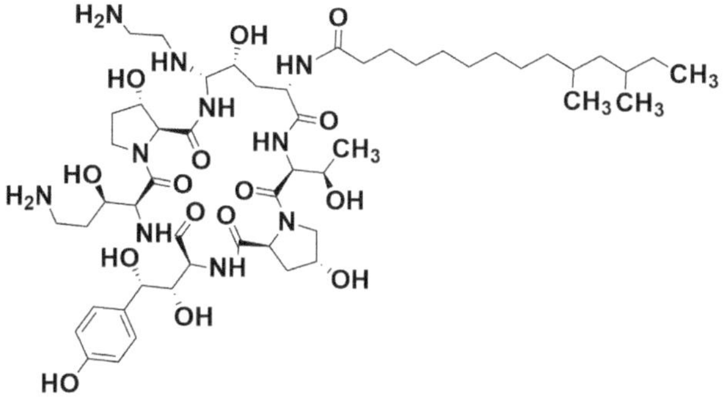

Fig. 6.1. Chemical structure of caspofungin.

3-diphenyl-substituted isoxazole ring that reduces the RBC haemolytic activity and maintains the antifungal property against the *Aspergillus* and *Candida* species. It also has an amino acid residue sulphated tyrosine that shows high water solubility (Hashimoto, 2009). It showed linear pharmacokinetics with a dose of over 50–150 mg and consists of three metabolites, namely, M1, M2, and M5. M1 is a product of the metabolic reaction of arylsulfatase and the breakdown of M1 by catechol-o-methyl-transferase forms M2. M5 is formed by hydrolysis of the micafungin side chain by an isoenzyme (Aguilar et al., 2015). The chemical structure of micafungin is shown in Fig. 6.2.

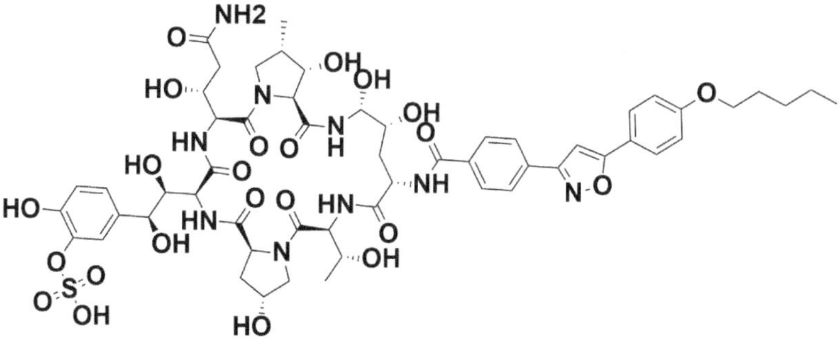

Fig. 6.2. Chemical structure of micafungin.

2.3 *Anidulafungin*

It contains a lipopeptide nucleus with amino acid residues like that present in echinocandin B. The antifungal property is maintained by the core nucleus; the water-soluble property is maintained by the hydroxyl group and amino linkers. Present in anidulafungin is a lipophilic side chain, Alkoxy triphenyl, with lower water-soluble properties compared to other echinocandins, that disrupts the fungal cell wall (Vazquez and Sobel, 2006). A single dose of about 35–100 mg of anidulafungin showed linear pharmacokinetics with maximum concentration, a long half-life, and a large distribution volume (Estes et al., 2009). The chemical structure of anidulafungin is shown in Fig. 6.3.

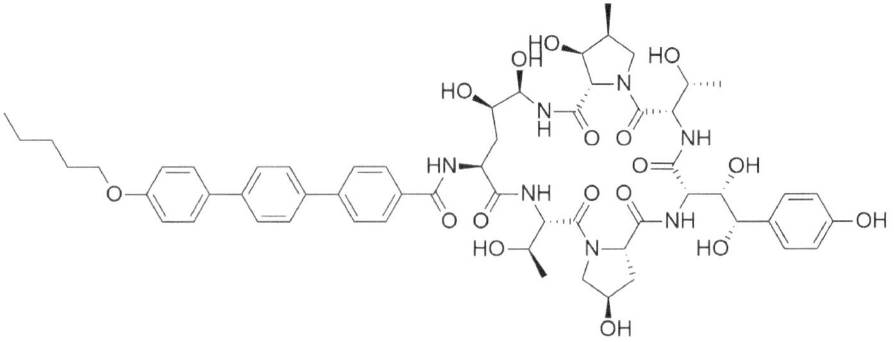

Fig. 6.3. Chemical structure of anidulafungin.

The concentration of echinocandins used for the inhibition of the *Candida* species is much lower than that of amphotericin B and fluconazole. All three types of echinocandins show antifungal activity against the *Candida* species. *C. famata* and *C. rugosay* are more susceptible to echinocandins, whereas *C. fermentati* and *C. guilliermondii* are less vulnerable. To inhibit the growth of *Candida* species by 50%, the concentration of echinocandins should be less than or equal to 0.5 μg/mL and to inhibit it by 90%, the echinocandins concentration should be less than 2 μg/mL. Echinocandins show antifungal activity against *Aspergillus* species by damaging the branches and hyphae of growing cells, thereby reducing their invasion (Ioannou et al., 2016). Echinocandins do not show their antifungal activity against the *Fusarium, Mucorales*, and *Scedosporium* species due to diminished β-1, 3-D-glucan synthase activity and against *Cryptococcus neoformans* and *Trichosporon* species due to the β-1, 6-D glucan linkage (Aguilar et al., 2015).

3. Structural Diversity and Biosynthesis

The pharmacological and chemical properties of the echinocandins are shown by the combination of elements in their structure. An amino-acid sequence and a fatty acid are commonly shared by all echinocandins, but frequent changes are present with respect to the side chain and most of the products are bioactive. The increase in the resistance of the echinocandins is due to the changes made in the structure of the parent molecule.

The acylation of echinocandins is done by attaching the fatty acid to the α-amino group of dihydroxy ornithine that is required for anchorage in the cell membrane to ensure bioactivity. The fatty acids such as myristic acid, palmitic acid, and linoleic acid are produced by the strains during primary metabolism. The branched chain fatty acid namely 10R, 12S-10, 12-dimethylmyristate present in the *Leotiomycetes* species is mainly produced by a methyl transferase domain.

Pneumocandins that have a myristate branched side chain show no haemolytic effect on RBCs; hence, the fatty acid side chain present in the echinocandins is replaced in some agents such as anidulafungin, cilofungin, micafungin, but it is not replaced in caspofungin. This modification was done by Eli Lilly and Company. It is an *in vivo* process where the deacylation was done in echinocandin B by the *Actinoplanes* species (Boeck et al., 1989; Debono et al., 1989).

The nucleus of echinocandin is further re-acylated chemically to cilofungin and other products of echinocandin B by using an active ester of their respective products. Micafungin is produced from FR901379 by the same process used by cilofungin. In the last few years, the process of acylation has been investigated by utilizing genetic methods to delete the fatty acid ligase gene in *Glarea lozoyensis* (Boeck et al., 1989). The nucleus of pneumocandin was not found; hence, it stated that the enzyme is essential for the biosynthesis of echinocandins. If a mutated strain lacking the *PKS* gene responsible for the biosynthesis of dimethyl myristate can be compensated by C_{14} to C_{16} fatty acids, then it proves that the enzyme ligase is not specific to a particular substrate (Chen et al., 2016).

4R, 5R-4, 5-dihydroxy ornithine is the default amino acid. In addition, there are many other minor products where the ornithine is not hydroxylated. The N-acyl-hemiaminal bridge of the terminal part of ornithine at position 6 of the hydroxy-proline is sensitive to hydrolysis. Due to the reduction in the amide nitrogen basicity, the hemiaminal group is stable at a weakly acidic and neutral state and a pH of more than 7 can lead to hydrolysis. The resulting compound on hydrolysis is 5-cyclic hemiaminal which has the α-amino group of ornithine that is acylated; it is a much more stable compound. Dihydroxy ornithine that has hydroxyl groups is not essential for bioactivity.

Echinocandins that are derived from ornithine, such as echinocandin D, pneumocandins, and other synthetically derived echinocandins, show high bioactivity. The amino acid in the echinocandins present at position 2 is threonine, except in sporiofungin A, sporiofungin C, and FR209602, where serine is present. Serine is also present in pneumocandins in small amounts. Different hydroxy-L-prolines are present at positions 3 and 6, trans-4-hydroxy-L-proline is present at position 3, and 3-hydroxy-4-methyl-L-proline is present at position 6. Like other amino acids in the echinocandins, specific substitution on proline is not needed to maintain bioactivity. Pneumocandin B_0 has the best bioactivity compared to the other pneumocandins. The genomes of pneumocandin A_0 and pneumocandin B_0 are independently sequenced, which allowed the identification of point mutations. The production strain of pneumocandin A_0 and pneumocandin B_0 is *Glarea lozoyensis* ATCC 74030 (wild type). The only change in *G. lozoyensis* ATCC 74030 is the lack of leucine dioxygenase that causes changes in the pneumocandin biosynthesis but not the hydrolysis of proline. The deletion of the gene in the wild type of *G. lozoyensis* created a mutant that was not able to produce pneumocandin A_0. The 3S, 4S-3, 4-dihydroxy-L-homotyrosine is the preferred amino acid at position 4 in echinocandins (Masurekar et al., 1992). There are several variants without the hydroxylation of sidechain at positions 3 and 4, and a sulphated aromatic hydroxyl group or the sulphated group may be present beside it, the side-chain hydroxylation is not necessary for the bioactivity, but a longer homotyrosine side chain is essential. In the homotyrosine chain, the ether unit ($-CH_2-O-CH_2-$) can replace the ethylene unit, unlike the tyrosine side chain, where a methylene group replaces an ethylene unit. The phenolic hydroxyl group is also important as it can be replaced by phosphate or amino groups for the bioactivity. The 3S-3-hydroxy glutamine is present at position 5 in echinocandins derived from *Leotiomycetes* and by threonine in echinocandins derived from *Eurotiomycetes*. 3-hydroxy glutamine is a non-proteogenic gene that is biosynthesized by the hydrolysis of glutamine by α-KG dependent glutamine hydroxylase (Obermaier and Muller, 2020). The biosynthesis and structural diversity of each antifungal drug is schematically elaborated in Fig. 6.4.

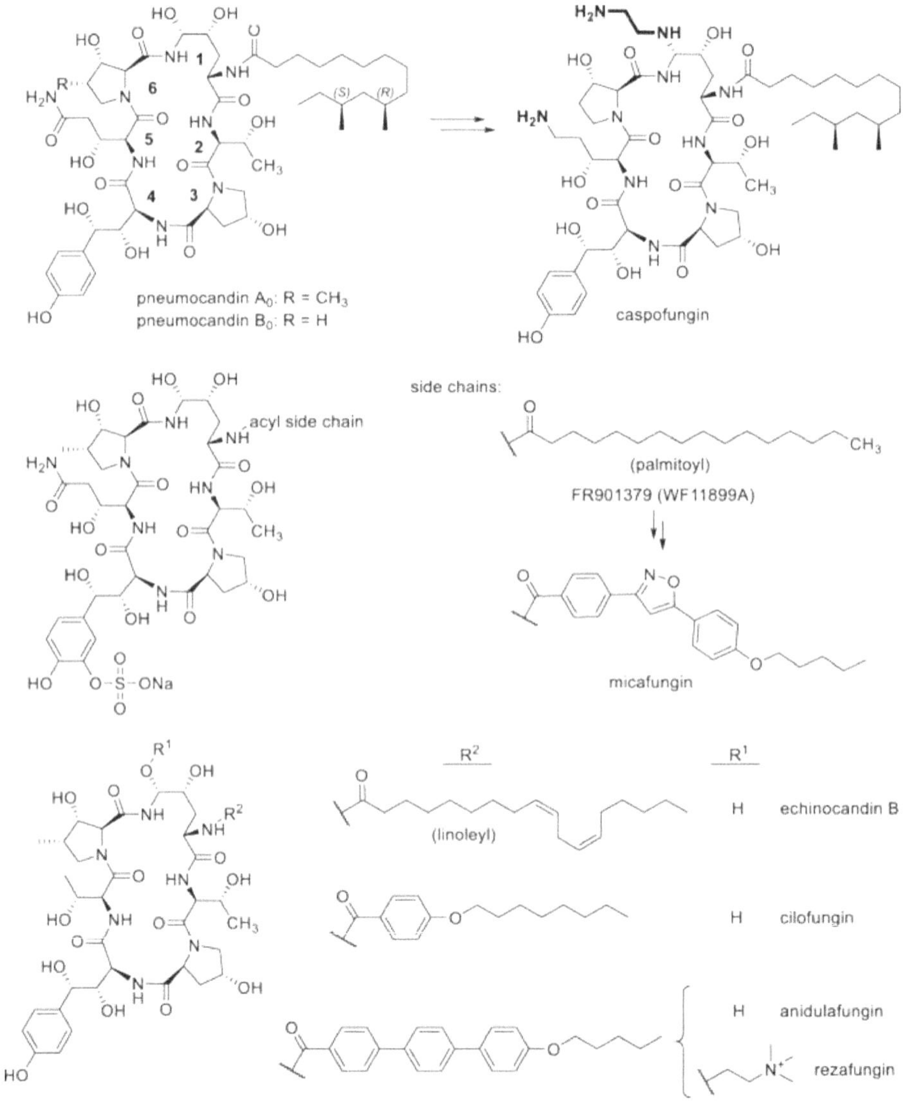

Fig. 6.4. Biosynthesis and structural diversity of echinocandins.

3.1 Biosynthesis

The biosynthesis of echinocandin was first elaborated by Merck and Co. while experimenting with one of the pneumocandin producer, *Glarea lozoyensis*. *Aspergillus nidulans* strain (NRRL 8112), isolated in India, is a weak producer of echinocandins. A comparison with the current echinocandins revealed a core set of genes encoding the enzymes required for biosynthesis, although there is no particular evidence the genes present in the periphery were used for the biosynthesis of echinocandins (Huttel et al., 2016). Tang and Walsh discovered and characterized the presence of a biosynthetic gene of echinocandin B in *A. pachycristatus* NRRL 11440, which allowed the deletion of the target gene for the first time and resulted in mutated genes that produce the parent echinocandin derivative and biosynthetic heterologous enzymes (Cacho et al., 2012). The cluster of the gene (from *A. nidulans* NRRL 8112) was divided into two gene sequences, namely, *Ecd* and *Hty*, that are located at a particular site of the genome. Further, the polymerase chain reaction (PCR) experiments indicated it to be a biosynthetic coherent gene cluster. At present, only ten biosynthetic gene clusters found in *Leotiomycetes* have shown five, and the other five gene clusters are found in *Eurotiomycetes*. In recent years, the biosynthesis of acrophiarin in *Penicillium arenicola* is a structural hybrid of *Leotiomycetes* and *Eurotiomycetes*. This is caused by the phenomena of horizontal gene transfer between million years' old ancestor *P. arenicola* and the currently available *Leotiomycetes* strain (Lan et al., 2020). Due to the adverse effect of haemolysis, the therapeutic use of echinocandin B as an antifungal decreased and cilofungin was produced as a semi-synthetic analogue. Cilofungin had an antifungal property, and the adverse effects were fewer for the drug. However, clinical trials with cilofungin could not be conducted for a long period due to increasing events of adverse effects. Consequently, pneumocandin A_0 and Pneumocandin B_0 were synthesized (Meunler et al., 1989; Schmatz et al., 1992). The structures of pneumocandin A_0 and echinocandin B are presented in Fig. 6.5. The, the differences in structures of each echinocandin are presented in Table 6.1. The details of products and their biosynthetic pathways are presented in Table 6.2.

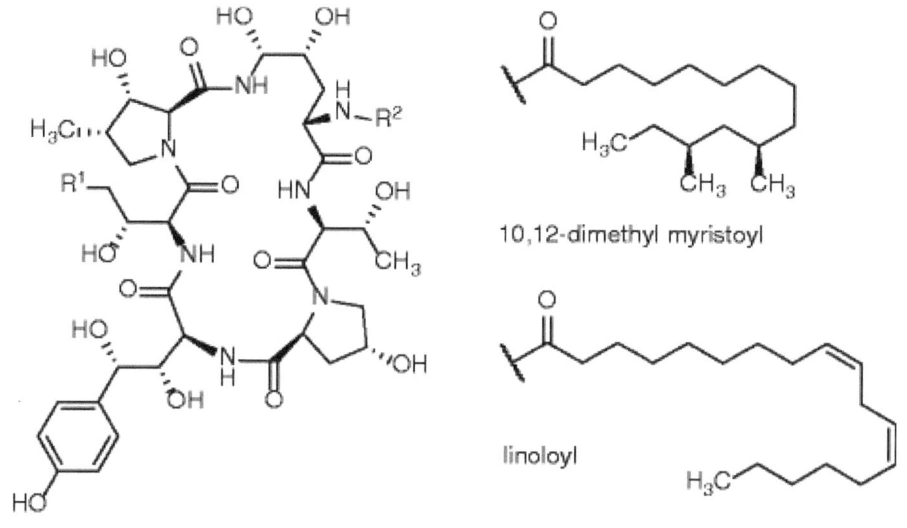

Pneumocandin A_0: R^1 = $CONH_2$, R^2 = 10,12-dimethyl myristoyl
Echinocandin B: R^1 = H, R^2 = linoleoyl

Fig. 6.5. Structures of pneumocandin A_0 and echinocandin B (Lan et al., 2020).

Table 6.1. Differences in structures of echinocandins.

Echinocandins	R1	R2	R
Echinocandin B	$-CH_3$	$-CH_3$	
Pneumocandin A_0	$-CH_3$	$-CH_2CONH_2$	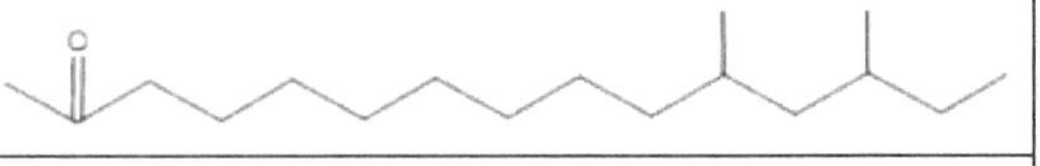
Pneumocandin B_0	-H	$-CH_2CONH_2$	

4. Medical Conditions Associated with Echinocandins

The three main echinocandins, i.e., caspofungin, micafungin, and anidulafungin have been approved by the Food and Drug Administration (FDA) for use in several diseases mentioned in Table 6.3. Caspofungin is the first drug from the echinocandins group approved by FDA in January 2001 to be used as an antifungal drug. It is also approved it for use as an empirical treatment in neutropenia, fever, candidemia, and oesophageal candidiasis. Micafungin and anidulafungin were approved in March 2005 and February 2006, respectively, for the treatment of invasive candidiasis, oesophageal candidiasis, and candidaemia. In addition, micafungin is used for the *Candida* infection prophylaxis during stem cell transplantation (Pappas et al., 2009; Brueggeman et al., 2009; Krishnan and Chandrasekar, 2008).

4.1 Oesophageal Candidiasis

Caspofungin, micafungin, and anidulafungin are effective for the treatment of both oesophageal and oropharyngeal candidiasis. The efficacy of the three echinocandins was assessed through a randomized control trial compared with fluconazole for the treatment of oesophageal candidiasis. The study showed similar efficacies with echinocandins and fluconazole. In another randomized double-blind study, caspofungin and amphotericin B were compared. Caspofungin, given at three different doses in mostly human immunodeficiency virus (HIV)-infected patients, showed a similar response. In an open-label study with anidulafungin for the treatment of azole-refractory mucosal candidiasis, about 19 HIV-infected patients were first given a loading dose of 200 mg/day followed by 100 mg/day for about 14–21 days. The study showed a clinical success rate of 95% and an endoscopic success of 92% (Fig. 6.6.) (Arathoon et al., 2002).

4.2 Invasive Candidiasis

Five double-blind randomized control trials were conducted to assess the use of echinocandins in the treatment of candidaemia and other invasive candidiasis. Three randomized clinical trials were done to compare the efficiency of micafungin, caspofungin, and anidulafungin with fluconazole, amphotericin B, or liposomal amphotericin B. The studies showed an equal response towards treatment of microbial infection. Based on the results of these trials, it was concluded that

Table 6.2. Summary of products and details of biosynthetic pathway of echinocandins (Yue et al., 2015).

Primary Product of Echinocandin Pathways	Strain Analyzed	Family Classification	Acyl Side Chain	Amino Acid in Position					
				1	2	3	4	5	6
Echinocandin B	*Aspergillus pachycristatusa* NRRL 11440 (ATCC58397)	Aspergillaceae	Linoleic acid	4R,5R-Dihydroxy-l-Orn	l-Thr	4R-Hydroxy-l-Pro	3S,4S-Dihydroxy-l-homo-Tyr	l-Thr	3S-Hydroxy-4S-methyl-l-Pro
Echinocandin B	*Aspergillus nidulans* var. echinulatus NRRL3860	Aspergillaceae	Linoleic acid	4R,5R-Dihydroxy-l-Orn	l-Thr	4R-Hydroxy-l-Pro	3S,4S-Dihydrocy-l-homo-Tyr	l-Thr	3S-Hydroxy-4S-methyl-l-Pro
Mulundocandin	*Aspergillus mulundensisb* DSMZ 5745	Aspergillaceae	12-Methylmyristic acid	4R,5R-Dihydroxy-l-Orn	l-Thr	4R-Hydroxy-l-Pro	3S,4S-Dihydrocy-l-homo-Tyr	l-Ser	3S-Hydroxy-4S-methyl-l-Pro
Aculeacin A	*Aspergillus aculeatus* ATCC 16872 (NRRL 5094)	Aspergillaceae	Palmitic acid	4R,5R-Dihydroxy-l-Orn	l-Thr	4R-Hydroxy-l-Pro	3S,4S-Dihydrocy-l-homo-Tyr	l-Thr	3S-Hydroxy-4S-methyl-l-Pro
Pneumocandin A0	*Glarea lozoyensis* ATCC 20868	Helotiaceae	10,12-Di methylmyristic acid	4R,5R-Dihydroxy-l-Orn	l-Thr	4R-Hydroxy-l-Pro	3S,4S-Dihydroxy-l-homo-Tyr	3R-Hydroxy-l-Gln	3S-Hydroxy-4S-methyl-l-Pro
FR901379 (WF11899 A)	*Coleophoma empetri* FERM BP 6252	Dermataceae	Palmitic acid	4R,5R-Dihydroxy-l-Orn	l-Thr	4R-Hydroxy-l-Pro	3S,7-Dihydroxy-l-homo-Tyr-7-O-sulfate	3R-Hydroxy-l-Gln	3S-Hydroxy-4S-methyl-l-Pro
FR209602	*Coleophoma crateriformis* FERM BP 5796	Dermataceae	Palmitic acid	4R,5R-Dihydroxy-l-Orn	l-Ser	4R-Hydroxy-l-Pro	3S,4S,7-Trihydroxy-l-homo-Tyr-7-O-sulfate	3R-Hydroxy-l-Gln	3S-Hydroxy-4S-methyl-l-Pro
FR190293	*Phialophora cf. hyalinac* FERM BP 5553	Heloltiales, family unknown	10,12-Di methylmyristic acid	4R,5R-Dihydroxy-l-Orn	l-Thr	4R-Hydroxy-l-Pro	3S,7-Dihydroxy-l-homo-Tyr-7-O-sulfate	3R-Hydroxy-l-Gln	4R,5R-Dihydroxy-l-Orn
FR227673d	*Chalara* sp.	Heloltiales, family unknown	12,14-Di methylmyristic acid	4R,5R-Dihydroxy-l-Orn	l-Thr	4R-Hydroxy-l-Pro	3S,7-Dihydroxy-l-homo-Tyr-7-O-sulfate	3R-Hydroxy-l-Gln	4R,5R-Dihydroxy-l-Orn

Table 6.3. Use of echinocandins as per FDA.

Caspofungin	**Micafungin**	**Anidulafungin**
• Intra-abdominal abscesses • Peritonitis • Pleural space infections	• Acute disseminated candidiasis • Abscesses • Peritonitis	• Intra-abdominal abscesses • Peritonitis

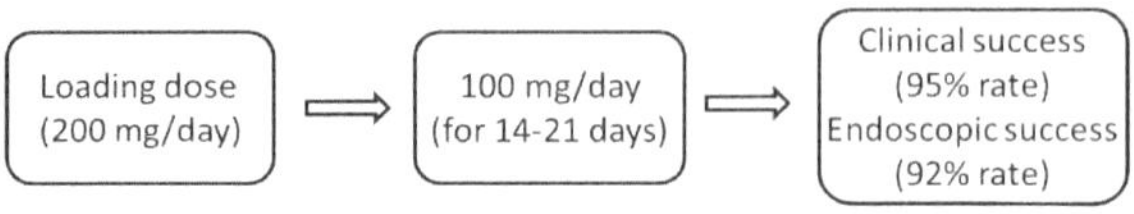

Fig. 6.6. A study involving anidulafungin for the treatment of azole-refractory mucosal candidiasis on 19 HIV-infected patients.

micafungin, caspofungin, and anidulafungin drugs are a favourable choice for treatment of fungal infection. In the studies with anidulafungin and caspofungin compared with either fluconazole or amphotericin B, patient outcomes were better in the echinocandin group using anidulafungin showing up to 75.6% response at the end of the treatment compared to fluconazole that showed only 60.2% positivity. In addition, echinocandins showed lower toxicity compared to amphotericin B. The other two studies compared echinocandin at higher doses of the same agent. Caspofungin showed a similar efficacy when given at two different doses without a loading dose. Several studies stated that the use of caspofungin was effective without the use of other antifungals in the treatment of endocarditis caused by *Candida* or by other species. This indicates the use of caspofungin as a monotherapy for fungal endocarditis. Echinocandins cannot penetrate the cerebrospinal fluid; hence, they cannot be used in the treatment of meningitis or brain abscess. However, caspofungin is used in the treatment of endophthalmitis caused by *Candida*, but it cannot be used in the treatment of *Candida*-associated eye infections due to poor penetration into the vitreous fluid (Pappas et al., 2009; Pappas et al., 2016; Eschenauer et al., 2007).

4.3 Invasive Aspergillosis

A study was conducted in patients on high-risk stem cell therapy and those who are likely to be infected with invasive aspergillosis. Out of the 17 patients who were treated with caspofungin, 1 patient had fatal *Aspergillus* infection. The efficiency of caspofungin was also studied in patients with confirmed invasive aspergillosis having haematological malignancies. In this case, 85% of the patients of the study group, were reported to have neutropenia, 75% of the patients were reported with cancer that was not in the remission stage. Overall, at the end of the treatment, caspofungin showed a response rate of 33%. In another study, both caspofungin and micafungin efficiency was studied (Viscoli et al., 2009). The efficacy of caspofungin at 33% was shown to be achieved at 12 weeks if used as the first-line therapy in invasive aspergillosis (Herbrecht et al., 2010). Maertens evaluated the efficiency of caspofungin in patients who did not respond to other therapies. A favourable response to treatment was seen in 45% of the patients. A positive response was seen in about 50% of the patients with invasive pulmonary aspergillosis. Using caspofungin in 11 European countries, data collected on the treatment of invasive aspergillosis, showed a favourable response in 56.4% of the patients. To understand the efficiency of micafungin when used as monotherapy or in combination with other drugs, a single non-comparative open-label study was conducted for the treatment of invasive aspergillosis. A 50% response was seen in the treatment using micafungin as a monotherapy and 40% response in a combination therapy. In patients with invasive aspergillosis who received micafungin as primary therapy in combination with another drug, a response of 34% was seen. The low loading dose of micafungin was administered in these studies (Higashiyama and Kohno, 2004; Ikeda et al., 2000; Denning et al., 2006).

4.4 Febrile Neutropenia

A large-scale double-blind randomized control trial was conducted where only caspofungin was studied as the treatment of febrile neutropenia. Caspofungin was shown to be well tolerated and was as effective as liposomal amphotericin B in patients with neutropenic fever. The efficiency and safety of caspofungin and micafungin were studied in several cohort studies, and it was concluded that there were no adverse effects that required the discontinuation of the drug. The efficacy was acceptable (Cohen et al., 2009).

4.5 Antifungal Prophylaxis

The use of echinocandins in antifungal prophylaxis was studied in patients who underwent stem cell therapy and liver transplantation. Additional studies are required to study the role of echinocandins in invasive fungal infection (IFI) prophylaxis. A double-blind randomized control trial was done to compare the prophylaxis of micafungin and fluconazole in neutropenic patients. The prophylaxis was followed and continued until the neutropenia resolved. Micafungin was proven to be the better drug compared to fluconazole. However, no difference was found in the mortality rate after using the two mentioned drugs. A small, randomized control trial was conducted for the comparison of prophylaxis of micafungin and fluconazole in the treatment of neutropenia, but no difference was seen in the prevention of IFIs. Again, in a retrospective analysis of stem cell therapy, where the subjects were on caspofungin therapy, there was a breakthrough in 7.3% of IFI patients (caused mainly due to the molds). The studies concluded that both caspofungin and micafungin are effective for use as antifungal prophylaxis. However, micafungin was recommended for the treatment of candidiasis (Doring et al., 2012; Hiramatsu et al., 2008). A comparison between itraconazole and caspofungin was studied for prophylaxis in patients undergoing chemotherapy for acute myeloid leukaemia. Itraconazole and caspofungin showed similar results. Caspofungin was well tolerated in patients with liver transplantation. A non-comparative study using caspofungin for prophylaxis was conducted in patients with intra-abdominal invasive candidiasis. Caspofungin showed an outbreak of infection in some patients (Barchiesi et al., 2006). A summary of indications and dosage of each echinocandin is given below in Table 6.4.

Table 6.4. Summary of indications and dosage of echinocandins.

Indications	Caspofungin	Micafungin	Anidulafungin
Febrile neutropenia (empirical treatment)	Loading dose: 70 mg daily Subsequent dose: 50 mg daily	Not approved for use	Not approved for use
Candidemia	Loading dose: 70 mg daily Subsequent dose: 50 mg daily	100 mg daily	Loading dose: 200 mg daily Subsequent dose: 100 mg daily
Other Candida-associated infections	Loading dose: 70 mg daily Subsequent dose: 50 mg daily	100 mg daily	Loading dose: 200 mg daily Subsequent dose: 100 mg daily
Invasive aspergillosis	Loading dose: 70 mg daily Subsequent dose: 50 mg daily	Not approved for use	Not approved for use
Oesophageal candidiasis	50 mg daily	150 mg daily	Loading dose: 100 mg daily Subsequent dose: 50 mg daily
Prophylaxis against Candida in stem cell transplant	Not approved for use	50 mg daily	Not approved for use
Paediatric patients aged more than 2 months of age	Loading dose: 70 mg daily Subsequent dose: 50 mg daily Maximum daily dose 70 mg	Not approved for use	Not approved for use

5. Mechanism of Action: Pharmacokinetics and Pharmacodynamics

The cell wall of fungi consists of β-1, 3-D-glucan synthase which is a heteromeric glycosyltransferase enzyme composed of a subunit that belongs to Rho GTPase and a catalytic *Fks* p-subunit. β-1, 3-D-glucan synthase and Rho GTPase are bound to GTP and UDP-glucose, respectively. The UDP-glucose is polymerized to β-1, 3-D-glucan. It constitutes about 30–60% of the cell wall of fungi and provides the strength of fungal cell wall. Echinocandins bind to the *Fks* p-subunit non-competitively and block the synthesis of β-1, 3-D-glucan. This leads to cell wall leakage making it highly permeable, thereby affecting the intracellular osmotic pressure balance and subsequent cell lysis. All the echinocandin derivatives show similar fungicidal effects on *Candida* and *Saccharomyces* species and fungistatic effects on the *Aspergillus* species; this difference in the action of echinocandins on different species is due to the variable content of glucan in the different species of fungi. The fungistatic effect is by the change in morphology commonly seen at the cell wall and the fungal hyphae (Bowman et al., 2002; Chandrasekar and Sobel, 2006; Miron et al., 2014).

5.1 Pharmacokinetics

5.1.1 Caspofungin

It is highly bound to the protein and undergoes degradation spontaneously along with hepatic metabolism through hydrolysis of the peptide and N-acetylation. It shows a beta half-life of 10–15 hours which allows for a daily dosage. Caspofungin need not be adjusted for renal insufficiency. Caspofungin is infused intravenously and subsequently, there is a decrease in the fall in the plasma level of caspofungin due to its distribution to the tissues. Slowly, the drug is released again from extravascular sites. It shows dose-proportional plasma pharmacokinetics and a half-life of 10–15 hours. It can be given once every day. The clearance of saspofungin is found to be lower in both adults and neonates but remain higher in children. This occurs due to the rate of distribution into hepatic tissues from the plasma (Walsh et al., 2005).

5.1.2 Micafungin

At the therapeutic dosage of 50–150 mg, micafungin shows linear pharmacokinetics. Micafungin also manifested penetration into the CNS at a dose of more than 2 mg/kg. It could not be detected in the cerebrospinal fluid. It has also been stated that the maximal antifungal property is shown by micafungin at a dose of 12–15 mg/kg in neonates (Hope et al., 2008). Micafungin gets metabolized into three metabolites: M1, M2, and M5. The metabolic reaction of micafungin with arylsulfatase results in the formation of M1 metabolite, which is again catabolized to M2 by catechol-O-methyltransferase. The enzyme cytochrome P450 isoenzyme hydrolyses the side chain of M2 which results in the formation of M5 (Cappelletty and Eiselstein, 2007). Recent research has shown that linear pharmacokinetic followed with micafungin, shows an inverse relationship between age and clearance of the drug. It has been found that the dosage of 3–4 mg/kg once daily (QD) and 2–3 mg/kg QD yield the same exposure among 2–8-year-old children and 9–17-year-old children, respectively. Penetration into the tissues of the CNS depends on the concentration gradient of the drug. Generally, neonates can achieve the therapeutic level at the dose of 8–5 mg/kg for CNS infections (Seibel et al., 2005).

5.1.3 Anidulafungin

A single dose of 35–100 mg shows linear pharmacokinetics with a large volume of distribution, a half-life of approximately 40 hours, maximum concentration (Sandhu et al., 2004).

5.1.4 Adult Populations

Due to the higher molecular weights, caspofungin, micafungin, and anidulafungin have poor oral bioavailability and hence they are given intravenously. Enfumafungin (a triterpene glycoside isolated from the Hormonema species by fermentation that inhibits glucan synthesis in the fungal cell wall) is a derivative of echinocandin. It is shown to be the first orally active drug among echinocandins in clinical trials. Enfumafungins can penetrate various tissues but cannot penetrate the prostrate, cerebrospinal fluid, brain, and eyes. However, micafungin can penetrate the eye. The poor penetration of these drugs into the tissues is due to high molecular weight and subsequently high protein binding (Eschenauer et al., 2007).

Upon administration, echinocandins are routed to the liver to undergo metabolism by N-acetylation and hydrolysis and form inactive metabolites eliminated by bile in the form of faeces. Echinocandins cannot undergo renal metabolism to be eliminated through urine as they cannot be dialyzed by kidneys. Caspofungin is metabolized in the liver through the hydrolysis of peptide bonds and N-acetylation. The resultant metabolites are eliminated through the urine and faeces. Micafungin undergoes metabolism in the liver by catechol-O-methyltransferase and arylsulfatase, and the resultant metabolites are eliminated through the urine and faeces. Anidulafungin does not undergo hepatic metabolism; it degrades over time and forms a ring-opened chemical moiety which also undergoes faecal elimination. As the echinocandins do not undergo renal metabolism, the dose of the drug need not be changed in renal insufficiency (Sandhu et al., 2004). Each of the three drugs and their pharmacokinetic properties are summarized in Table 6.5.

Table 6.5. Summary of pharmacology of echinocandins (Cappelletty and Eiselstein, 2007).

Pharmacology	**Caspofungin**	**Micafungin**	**Anidulafungin**
Administration	Intravenous (IV) route	IV route	IV route
Pharmacokinetics	Linear	Linear	Linear
Drug distribution	High distribution: liver and kidney; moderate distribution: spleen and lung; low distribution: eye and brain	High distribution: liver, kidney, and lung; low distribution: eye	High distribution: liver and lung; moderate distribution: kidney and spleen; low distribution: CSF, eye, and brain
Half life	Average 10–15 hours	Average 36–52 hours	Average 11–17 hours
Bioavailability	Less than 10%	Less than 10%	2%–7%
Metabolism	Spontaneous hepatic degradation; hydrolysis and N-acetylation	Spontaneous hepatic degradation; hydrolysis and N-acetylation	No hepatic degradation; degradation in blood
Adult dosage	Loading dose: 70 mg (IV) Subsequent dose: 50 mg	100 mg (IV)	Loading dose: 200 mg (IV) Subsequent dose: 100 mg
CYP 3A4 inhibition	No	Weak inhibition	No

5.1.5 Special Populations

Hepatic Insufficiency

When caspofungin was administered to patients with mild hepatic insufficiency patients, there was about a 55% increase in the area under the curve, which was found to be similar in the control group. Thus, no change in the dosage is needed in patients with hepatic insufficiency. In a study, a dosage of 100 mg of micafungin was given to patients with moderate and severe hepatic insufficiency via IV route, the area under the curve was found to be 22% and 30% in both groups. Thus, no change in dosage is recommended. Anidulafungin is not a substrate of hepatic metabolism and hence there is no need for dose adjustment (Stone et al., 2002).

Renal Insufficiency

Echinocandins do not undergo renal metabolism and hence no dose adjustment is needed in mild, moderate, and severe renal insufficiency. Echinocandins may be administered without considering the haemodialysis regimen, as all the echinocandins are non-dialysable (Stone et al., 2002).

5.1.6 Paediatric Population

Micafungin is approved for use in infants and neonates as antifungal therapy. The clearance value, half-life, and volume of distribution are the same as seen in the adults. A similar dose response is seen in children and adults administered about 1 mg/kg per and 50 or 70 mg/kg day dose of caspofungin, respectively. Anidulafungin is not used in the paediatric population (Walsh et al., 2005).

5.1.7 Geriatric Population

No adjustments in the dose of echinocandins are required in elderly patients. Anidulafungin showed differences in the clearance with other parameters as in the younger population (Walsh et al., 2005).

5.1.8 Lactating and Pregnant Women

Echinocandins have been categorized as class C drugs of pregnancy; they pass through the placental barrier and hence should be used cautiously in pregnant women. The administration should be avoided in lactating women as it was found in the milk of lactating rats (Jimenez et al., 2013).

5.2 Pharmacodynamics

All the three drugs of echinocandins show fungicidal activity depending on the concentration of the drug for the *Candida* species. This is best compared with maximum concentration/minimum inhibitory concentration (C_{max}/MIC) or area under the curve as mentioned in Table 6.6. The post antifungal effect by echinocandins is greater than 12 hours if the concentration exceeds MIC. To exhibit a post antifungal

Table 6.6. MIC and MIC 90% for each echinocandin in various *Candida* species.

Different *Candida* species	**Caspofungin**		**Micafungin**		**Anidulafungin**	
	MIC (range)	MIC_{90}	MIC (range)	MIC_{90}	MIC (range)	MIC_{90}
Candida krusei	0.125–2	1	0.125–0.25	0.125–0.25	0.0078–0.015	0.015
Candida tropicalis	0.03–2	0.125–1	0.0156–0.125	0.0313–0.125	0.002–0.015	0.0015
Candida lusitaniae	0.12–2	1–2	--	--	0.0015	0.0015
Candida albicans	0.007–>0.8	0.125–0.5	<0.0039–0.0625	0.0156–0.0625	< 0.001–0.015	0.0078
Candida glabrata	0.007–>0.8	0.25–1	0.0078–0.0625	0.0156–0.0039	0.0039–0.03	0.03

Notes: MIC = Minimum inhibitory concentration (μg/mL); MIC_{90}, = MIC required to inhibit 90% of organisms compared to growth control.

effect over 0.9 hours to greater than 20.1 hours, the micafungin concentration should be four times greater than the MIC. A post antifungal effect that stays for more than 9 hours has been observed in anidulafungin. The therapeutic concentration of caspofungin is present at the site of infection in the kidneys after there is a fall of serum concentration below the MIC. The pharmacodynamics for inhibiting the *Aspergillus* species is not defined clearly. There is a slight increase in the growth of *Aspergillus* species when caspofungin followed by micafungin is used. This was studied by Lewis and colleagues in mice. Both micafungin and caspofungin showed a dose-dependent effect by reducing the fungal growth, but a dose greater than 4 mg/kg day of caspofungin showed a steeper dose-response cure; this exhibits an increase in fungal load. Echinocandins showed the pharmacodynamic properties mainly against *Candida* and *Aspergillus* species; their effect on other pathogens is low. Caspofungin is found to reduce the *Rhizopus oryzae* growth in the brain (Lewis et al., 2008; Jeong et al., 2015; Ibrahim et al., 2005).

6. Side Effects and Therapeutic Use

6.1 Side Effects

The side effects that occurred on treatment with echinocandins are compared with those of fluconazole and are found to be fewer in amphotericin B. The side effects that lead to the decision of discontinuing the drugs are fewer when compared with other antifungal therapy. Facial flushing, rash, thrombophlebitis, pruritis, fever, and hypotension are the common side effects that occur immediately after injecting the drug. These are observed similarly in all three echinocandins but differ in frequency in each patient. Fever frequently occurs in 30% of the patients treated with echinocandin, but only in 1% of patients treated with micafungin. To reduce the side effects, the speed of application of the drug should be reduced. Gastrointestinal problems, such as vomiting, diarrhoea, and nausea are the most common side effects found in less than 7% of the patients; 3–25% of patients treated with caspofungin showed phlebitis, but when treated with micafungin and anidulafungin, phlebitis is seen in less than 2% of the patients. Leukopenia, anaemia, thrombocytopenia, and neutropenia are seen in less than 10% of the patients treated with echinocandins. Laboratory test abnormalities in alkaline transferase and aminotransferase levels are observed but these are more commonly seen in patients treated with azole drugs. Histamine levels are elevated when polypeptide-like compounds are used due to histaminic reactions when anidulafungin is injected at a rate exceeding 1.1 mg/min. Embryotoxicity is reported by using echinocandins and hence it should not be used in pregnant women (Aguilar et al., 2015; Grover, 2010; Denning, 2003).

6.1.1 Caspofungin

The administration of caspofungin has revealed several cases of allergy to anaphylactic reactions. Infusion of more than 10% of the drug may lead to rash, fever, shivering, hypotension, septic shock, increase in alkaline phosphatase, and respiratory failure. With 1–10% of the drug infusion, there have been reports of vomiting, anaemia, chills, neutropenia, hypokalaemia, nausea, pain in the abdomen, pleural effusion, erythema, haematuria, dizziness, myalgia, paraesthesia, pruritis, facial oedema and flushing, hyperbilirubinemia, myalgia, tachycardia, and sepsis. If less than 1% of the drug is infused, hepatic necrosis, nephrotoxicity, renal impairment, pancreatitis, liver failure, erythema multiforme, anaphylaxis, and Stevens-Johnson syndrome are seen. Disorders of the blood and lymphatic system, such as coagulopathy, febrile neutropenia, anaemia, neutropenia, and thrombocytopenia have been reported. Disorders of the gastrointestinal system, such as constipation, abdominal pain, dyspepsia, abdominal distension, and heart disorders including atrial fibrillation, tachycardia, cardiac arrest, arrhythmia, bradycardia, myocardial infarction, cardiac arrest are seen. Abnormalities in liver functions and bile formations have also been reported including hepatomegaly, jaundice, hepatotoxicity, hepatic failure, and hyperbilirubinemia (Senn et al., 2009). Infections such as urinary tract infection (UTI), sepsis and bacteraemia; nutritional and metabolic disorders, such as decrease in appetite, hypomagnesemia, anorexia, hypercalcemia, hypokalaemia, fluid overload, and hyperglycaemia are seen. Urinary and renal disorders, including renal failure and haematuria and subcutaneous tissue and skin disorders, such as petechiae, exfoliation, erythema, urticaria and skin lesions are seen. Some cases of psychiatric disorders such as depression, insomnia, confusion, and anxiety are seen. Abnormalities such as hypoxia, dyspnoea, tachypnoea, and epistaxis, have also been reported. Thoracic, respiratory, and mediastinal disorders have been observed. Vascular disorders, such as hypertension, phlebitis, and flushing appeared in some cases. Musculoskeletal, bone, and connective tissue disorders including pain in the extremities, back pain, arthralgia also occurred in many cases (Cleary and Stover, 2015). Disorders of the nervous system such as tremors, convulsions, dizziness were also reported. General manifestations, such as fatigue, peripheral oedema, pain, swelling, pruritis and asthenia were also noticed (Sun and Tong, 2014).

6.1.2 Micafungin

There may be redness, irritation, and swelling at the site of injection. Other manifestations which may occur and were reported are vomiting, nausea, headache, trouble sleeping, and diarrhoea. Micafungin use has led to serious side effects like signs of liver abnormalities such as dark urine, yellow skin or eyes, nausea, vomiting and loss of appetite. Micafungin use has also shown signs of kidney diseases, such as change urine output, tiredness, and breathlessness; signs of infections, such as fever and chills; and other manifestations such as bruising, mental disturbance, mood changes, irregular heartbeat, and bleeding. Serious allergic reactions are rare with the use of this drug (Yeoh et al., 2018).

6.1.3 Anidulafungin

Some of the common side effects reported after using this drug are dizziness, constipation, headache, vomiting, flushing, nausea, pain and swelling, and irritation at the site of injection. Some serious side effects are also reported. These include breathlessness, severe pain in the abdomen, dark-coloured urine, nausea, vomiting, yellow-coloured skin or eyes, spasm of muscle, swelling of legs or arms, bleeding, bruising, seizures, irregular heartbeat, increase in urine output, and mood swings (Hinske et al., 2012).

6.2 Therapeutic Uses

6.2.1 Caspofungin

It is used in the treatment of *Candida* infections such as candidemia, peritonitis, intra-abdominal abscesses, and pleural space infection. A dose of over 70 mg is infused for 1 hour intravenously and a maintenance dose of 50 mg is infused for 1 hour/day, continued for 14 days. In the treatment of oesophageal candidiasis, a dose of 50 mg is infused for 1 hour/day intravenously; this is to be continued for 7–14 days until the resolution of the symptoms (Kuhn et al., 2002). In HIV patients, oral therapy is given because of the relapse of oropharyngeal candidiasis. Caspofungin is used for the treatment of invasive aspergillosis in patients who are intolerant to other antifungal therapies; a loading dose of 70 mg is infused for 1 hour as a single dose and 50 mg of maintenance dose is given for 1 hour intravenously. The duration of treatment mainly depends on the clinical response of the patient, recovery from immunosuppression, and the underlying disease severity. Caspofungin is an empirical therapy for the treatment of febrile neutropenia; a loading dose of 70 mg is infused for 1 hour intravenously and a maintenance dose of 50 mg is given over 1 hour intravenously. The treatment is continued for 14 days minimum after the confirmation of fungal infection until 7 days after the clinical symptoms and neutropenia are resolved. Caspofungin is used as initial therapy in the treatment of *C. auris* infections; on the first day, a loading dose of 70 mg/m^2 is infused intravenously, and thereafter a maintenance dose of 50 mg/m^2 is given intravenously, based on body surface area. Most strains of *C. auris* found in the United States have been susceptible to echinocandins although reports of echinocandin- or pan-resistant cases are increasing. This organism appears to develop resistance quickly. Patients on antifungal treatment should be carefully monitored for clinical improvement. Follow-up cultures and repeat susceptibility testing should be conducted (Aguilar et al., 2015).

6.2.2 Micafungin

It is used in the treatment of acute disseminated candidiasis, candidemia, candida abscesses, and peritonitis in paediatric patients less than 4 months old and in adults. It is also used for treatment in oesophageal candidiasis and as prophylaxis of candida infections in paediatric patients less than 4 months of age and in adults. In adults, micafungin is used in the treatment of acute disseminated candidiasis, candidemia, candida abscesses and peritonitis at a dosage of 100 mg/day intravenously, for 10–47 days (mean 15 days). In the treatment of oesophageal candidiasis, a dosage of 150 mg/day is given intravenously for 10–30 days (mean 15 days). It is also used as prophylaxis of candida infections in patients who have undergone haematopoietic stem cell transplantation (HSCT) at a dosage of 50 mg/day intravenously for 6–51 days (mean 19 days). In patients below the age of 4 months, 2 mg/kg intravenous QD of micafungin is given for the treatment of candidemia, acute disseminated candidiasis, candida peritonitis, and abscesses. In case of oesophageal candidiasis, for a body weight ≤ 30 kg, a dose of 3 mg/kg IV QD, and for body weight > 30 kg, 2.5 mg/kg IV QD is given; the dose should not exceed 150 mg/day. For prophylaxis of candida infections in HSCT recipients, a dose is 1 mg/kg IV QD is needed, which should not exceed 50 mg/day (Cornely et al., 2008).

6.2.3 Anidulafungin

In adults, for the treatment of candidemia, intra-abdominal abscess, and peritonitis caused by *Candida* infections, the usual dose pattern is 200 mg IV infusion on day 1, 100 mg/day IV from day 2 onwards until 14 days after the last negative culture. In case of oesophageal candidiasis, the usual dose pattern is 100 mg IV infusion on day 1, then 50 mg/day IV from day 2 onwards. The treatments last for a minimum 14 days or at least 7 days following resolution of symptoms. Owing to the risk of relapse of oesophageal candidiasis in patients with HIV infection, consider suppressive antifungal therapy after the course treatment. In paediatric patients aged ≥ 1 month, for the treatment of candidemia, intra-abdominal abscess, and peritonitis caused by *Candida* infections, the usual dose pattern is 3 mg/kg (not to exceed 200 mg/dose) IV infusion, then 1.5 mg/kg (not to exceed 100 mg/dose) IV from day 2 onwards until 14 days after the last positive culture (Cancidas®. Merck and Co. Inc., 2009).

7. Resistance to Echinocandins and the Reasons

The studies conducted by Myra Kurtz and Cameron Douglas on *C. albicans* and *S. cerevisiae* with caspofungin have shown that the main target of echinocandins was *Fks1*, the major subunit of glucan synthase. The studies stated that the cause of reduction in susceptibility is due to the modification of the target site. The mutants of *C. albicans* resistant to N2-PnB0, an analogue of caspofungin, showed an elevation in the values of MIC and half-maximal inhibitory concentration (IC_{50}). To reduce 99% of the kidney burden, the mutants require 20-fold increase in the drug dosage. Mutants that reduced the susceptibility of echinocandins in *C. albicans* and *S. cerevisiae* are due to *Fks1*. The elevated levels of MIC with the use of caspofungin in the treatment of *C. albicans* sowed mutations in *Fks1*. These strains of *Candida* species showed two to three log shifts in the values of effective dose (ED_{99}) and a similar shift in the values of IC_{50}, which inhibits the synthesis of glucan synthase (Coste et al., 2020; Kurtz and Marrinan, 1989). The substitutions in amino acids defined a region named hotspot 1 (*HS1*). The mutations of *Fks1* were confirmed to obtain resistance when they were introduced into the strains. Genetic studies of the other mutant *S. cerevisiae* contain a substitution R1357S that helped in identifying a separate region with reducing susceptibility termed hotspot 2 (*HS2*) (Perlin, 2007). Isolates of *C. albicans* from patients who failed to respond to the treatment showed the substitution of amino acid at ser645 in *HS1* of *Fks*. Mutation of the hotspot is observed only in resistant strains. In some patients, multiple strains of *C. albicans* were isolated that showed substitution in *Fks1* conferred a reduction in the susceptibility. Some studies state that the mutant strains are identical to each other and to the susceptible strains of the same patient, which suggests that the resistance is due to different exposure to drugs. The mutations that are observed in the *Fks1* hotspot region from the *C. albicans* isolates showed an elevation in the values of MIC; these mutations show the substitution of a non-conservative amino acid in *HS1* and *HS2* (Perlin, 2007). The strains of *Candida* are cross-resistant to anidulafungin and micafungin, which indicates that the modification in *Fks1* enhances the echinocandin drugs. The evidence that the modification of the *Fks1* is the result of the resistance is the biochemical observation stating that the hotspot mutations decrease the sensitivity of the drug to glucan synthase. The enzyme that undergoes mutation is 1000-fold less sensitive to the drug, making it resistant to the drug. The wild type of enzyme is inhibited in low concentration of the drug but not the mutant enzyme, this enzyme requires high drug concentrations for inhibition. *Fks1* encodes an integral membrane protein of molecular weight of 215-kDa, which is the main catalytic subunit of the enzyme glucan synthase. In *HS1*, the resistant mutant is within 89 amino acids that are present on the cytoplasmic membrane; this region may indirectly block the action of the drug (Mio et al., 1997; Douglas et al., 1994).

7.1 *Reasons*

One of the reasons for resistance is modification that results in reducing antifungal effects. In the specific regions of *Fks1* genes that encode a catalytic subunit, the formation of point mutations is the reason for the reduction in the sensitivity or resistance to echinocandins. The catalytic subunit has three genes encoding them *Fks1*, *Fks2*, and *Fks3*. The mutations are present in the *Fks1* region in *Candida* species and the *Fks2* region in *C. glabrata*. Mutations that affect the susceptibility are present in the major regions named hotspot. In the case of mutations in *Candida*, species mutations are seen in Ser645 and Phe64. Mutations are seen in Ser629, Ser663, and Phe659 in the *Fks1* region in the *C. glabrata* which is the result of mutations in *Fks2*. Mutations that occur in the hotspot region result in 10–100 times increase in MIC and a reduction in glucan synthase sensitivity for echinocandin drugs. As a result of mutations, there is a reduction in the catalytic efficiency of glucan synthase that results in a change in cell wall morphology and composition. *C. albicans* having homozygous *Fks1* and with a mutation in the hotspot possess a thicker chitinous cell wall, showed a reduction in growth (Ben-Ami et al., 2011; Berkow and Lockhart, 2018). The mapping mutations showed that the substitution of amino acid occurs at the surface of the extracellular membrane, this is where echinocandins interact with the enzyme that would not enter the cell. Another reason for the resistance to echinocandins is the stress response of the cell. The fungi cannot survive without a cell wall, so it is essential to maintain it. The decrease in the synthesis of β-1, 3-D-glucan is responsible for the stress due to lack of cell wall continuity. To decrease the stress, adaptive mechanisms are activated which transmit the signals to a protein. Protein kinase C is responsible for the integrity of the cell wall by synthesizing chitin, an increase of which results in the destruction of the cell wall (Forastiero et al., 2015; Grossman et al., 2014).

8. Caspofungin

8.1 Introduction and Development

Due to the chemical synthesis and complex structure of echinocandins, the developments of the non-hydroxylated side chains are not economically feasible. In such a case, fermentation is done along with chemical modification which is the only economically feasible option. The fungal strains that produce echinocandins are cultivated to give reasonable amounts of end products. Developmental efforts are also considered in producing the echinocandins on an industrial scale. The process of cultivating echinocandins includes fermentation that is optimized, mutagenesis to improve the parent producer strain, and chemical modification to increase the efficacy and water solubility of the drug. In micafungin and anidulafungin, the linear fatty acid chain was replaced and modified structurally to reduce the haemolytic effects on RBCs. In caspofungin, amino acid groups, which help increase the antifungal activity and water solubility of the drug, were introduced. Cilofungin was the first echinocandin drug developed by Eli Lilly and Co. It is a derivative of echinocandin B that contains a 4-(n-octyloxy) benzoyl side acyl chain essential for reducing the haemolytic effect of RBCs. Since it is not water soluble, it is administered along with polyethylene glycol used as a co-solvent to increase solubility, but it proved to be toxic. Despite the failure, Merck and Co. did massive research on pneumocandins which finally led to the development of caspofungin that was approved in 1990. Caspofungin is the result of structural modification in pneumocandin B_0. It is produced by the interconversion of N-ethyl aminal and hemiaminal and by reducing hydroxy glutamine into an amine. These structural changes increased the solubility, antifungal activity, and stability mainly against *Aspergillus* (Agarwal et al., 2006; Keating and Figgitt, 2003; Saravolatz et al., 2003; Morrison, 2005).

8.2 Structural Diversity

The development of inhibitors for the synthesis of glucan synthase of fungal cell wall is a new advent in the chemotherapy of antifungal agents. Terpenoids, echinocandins, papulocandins, and cyclic hexapeptides are some of the known inhibitors of β-1, 3-D-glucan synthase (Fesel and Zuccaro, 2016). Caspofungin, a water-soluble drug, is produced by the fermentation of *Glarea lozoyensis*. It is a semi-synthetic compound and a derivative of pneumocandin B_0. Caspofungin contains the nucleus of pneumocandin B_0 that reduces the haemolytic effect of RBCs. Caspofungin is substituted with ethylenediamine that reduces the potency of the drug but increases the water solubility (Chen et al., 2013; Letscher and Herbrecht, 2003). Caspofungin, being an antimycotic drug, can alter the function of cardiomyocytes and affect the ciliated epithelium of the trachea. It has been found that caspofungin can release calcium ions from the storage of the endoplasmic reticulum due to the activation of the ryanodine receptor (RyR) pathway (Muller et al., 2020).

Caspofungin comes under a new class of antifungals known as echinocandins. For many years, the mainstay of treatment for fungal infection remained as

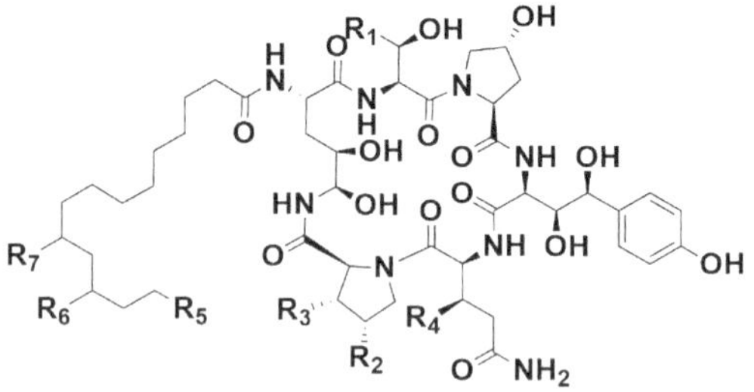

Pneumocandin A_0 $R_1=R_2=R_6=R_7=Me$, $R_3=R_4=OH$, $R_5=H$
Pneumocandin B_0 $R_1=R_6=R_7=Me$, $R_2=R_5=H$, $R_3=R_4=OH$
Pneumocandin C_0 $R_1=R_6=R_7=Me$, $R_2=R_4=OH$, $R_3=R_5H$

Fig. 6.7. The differences in structures of pneumocandin A0, B0, and C0.

voriconazole and amphotericin B. Caspofungin has the advantage of having an antifungal effect on the genus *Candida*, which is even resistant to accepted and known antifungal medications like amphotericin B and other azole drugs. It has been noted that caspofungin is as effective as the existing medications for aspergillosis and invasive candidiasis. The tolerability of caspofungin has been proved to be better than liposomal amphotericin B. More studies have revealed that caspofungin has been effective in 45% of the cases of invasive aspergillosis which are not responding to the existing azole drugs. Caspofungin is also useful in combination with amphotericin B and triazoles as it shows synergistic activity. Finally, caspofungin has shown a high safety profile, good tolerance, and other advantages over the existing drugs against systemic fungal infections (Agarwal et al., 2006). The structures of different pneumocandins are summarized in Fig. 6.7.

8.3 Biosynthesis

Caspofungin is produced as a fermentation product of *Glarea lozoyensis*. It is produced by the modification of pneumocandin B_0. It is a drug that binds to high-protein regions distributed in the tissues. It undergoes spontaneous degradation, hepatic metabolism by hydrolysis of peptide bonds and N-acetylation (Pianalto et al., 2019). Caspofungin can inhibit cell wall biosynthesis and especially prevent the upregulation of β-1, 3-D-glucan synthase, and some cell wall-modifying enzymes, which leads to the altered glucan content. As these types of specific inhibitors can interfere with the *Ras* protein and chito-oligomer synthesis, they have a new scope in the future as a treatment for fungal diseases, whether prescribed alone or in combination (Pianalto et al., 2019).

8.4 Pneumocandins and Pneumocandin B_0

Pneumocandin A_0 and pneumocandin B_0 are the products of the structurally modified echinocandin B where the nucleus is modified by the incorporation of 1-3-hydroxy-glutamine residue instead of a threonine residue and the linoleoyl side chain is replaced by the dimethylmyristoyl side chain to reduce the haemolytic susceptibility of the RBCs. The commercially available echinocandin drugs do not have echinocandin B as their cores; instead, they have either pneumocandin B_0 or pneumocandin A_0 because of their antifungal potency. The cell wall of the fungi contains a β-1, 3-D-glucan synthase that is composed of the *Fks* p-subunit and a regulatory subunit. Echinocandins bind to the *Fks* p-subunit non-competitively and block the synthesis of the β-1, 3-D-glucan synthase resulting in the disruption of the cell wall (Patil and Majumdar, 2017). It is used in the treatment of *Candida* infections such as candidemia, peritonitis, intra-abdominal abscesses, and pleural space infection (Adefarati et al., 1992; Chen et al., 2014).

8.5 Global Status of Caspofungin Acetate

Caspofungin acetate was approved as New Drug Application (NDA; 021227) product under the brand name Cancidas® in the USA on 26 January 2001 (https://www.accessdata.fda.gov/drugsatfda_docs/nda/2001/21227_cancidas.cfm). After US approval, the first marketing authorization in Europe and European Economic Area (EEA) was granted on 24 October 2001. It is available as an intravenous injection at a dose strength of 50 mg/vial and 70 mg/vial. Both strengths are white to off-white lyophilized cake or powder for reconstitution in a single-dose glass vial. The NDA product was developed and distributed by Merck and Co., Inc. USA, Jiangsu Hengrui Med, Xellia Pharmaceuticals APS, Gland Pharmaceuticals and Mylan Lab Ltd. The worldwide sale of the caspofungin acetate by 31 March 2017 and 2018 was found to be USD 541.5 million and 492.3 million, respectively, while during the same period the active pharmaceutical ingredient (API) consumption was found to be 116.9 kg and 127.5 kg, respectively. The details of leading caspofungin acetate manufacturers along with DMF status are presented in Table 6.7 and the patent status is presented in Table 6.8. The market is estimated to value over USD 487 million by the end of 2027 and register a CAGR of over 0.16% during the forecast period between 2020 and 2027. The expansion of the caspofungin market in the sub-continent over the forecast timeline can be credited to the huge patient population base and presence of key pharmaceutical firms in the countries such as the US and Canada. In addition, a prominent surge in the ageing population in these countries will increase the growth of the caspofungin business in the North American region over the forecasted timespan. The market for caspofungin has been fragmented based on indication (thrush and candidiasis), the distribution channel (online, retails, and hospital pharmacies), and even region (North America, Europe, Asia Pacific, Latin America, Middle East, and Africa). Based on the indication, the market has been bifurcated into use for thrush and candidiasis. With this segment, candidiasis-related used is estimated to hold some of the largest market shares in the worldwide market during the forecasted timespan. Some of the key players in the caspofungin market include Gland Pharma Limited, Xellia Pharmaceuticals, Merck & Co., Inc., Fresenius SE & Co., Mylan N.V., and Teva Pharmaceutical Industries Ltd. According to the trade data released by the Govt. of India, India has exported 21,153.50 kg, 31.39 kg, 428.35 kg, 11.17 kg, 29.26 kg, and 2,370.0 kg of caspofungin to the Netherlands, Spain, Turkey, Greece, US, Brazil, and China, respectively, during the year 2016, whereas nearly 37.0 kg, 0.24 kg, 18.40 kg, and 2.0 kg of API was imported from China, Hungary, US, and the Netherlands during the same year, respectively (https://www.pharmacompass.com).

Table 6.7. Leading caspofungin manufacturers and DMF status.

Manufacturer	**Country**	**Manufacturing Status**	**Active US DMF**	**Japanese DMF**	**Korean DMF**	**Available for Ref US DMF**	**Active COS**
Argus Pharmaceuticals Ltd.	China	Unconfirmed API activity	No	No	No	No	No
Assia Chemical Industries Ltd -Teva Tech Ltd.	Israel	Not manufacturing	No	No	No	No	No
Avanthera SA	Switzerland	Not manufacturing	No	No	No	No	No
Biocon Ltd.	India	Commercially available	No	No	No	No	No
Bright Gene Bio-Medical Technology Co. Ltd.	China	Commercially available	No	No	No	No	No
CKD Bio Corp	South Korea	Under development	No	No	No	No	No
Chunghwa Chemical Synthesis and Biotech Co. Ltd.	Taiwan	Commercially available	Yes	No	No	No	No
Concord Biotech Ltd.	India	Commercially Available	No	No	No	No	No
DSM Sinochem Pharmaceuticals	Netherlands	Commercially available	No	No	No	No	No
Esteve Quimica SA	Spain	Not manufacturing	No	No	No	No	No
Euticals S.p.A	Italy	Unconfirmed API activity	No	No	No	No	No
Gland Pharma Ltd.	India	Unconfirmed API activity	Yes	No	No	No	No
Gufic Biosciences Ltd.	India	Unconfirmed API activity	No	No	No	No	No
Jiangsu Hengrui Medicine Co. Ltd.	China	Commercially available	Yes	No	No	Yes	No
Lek Pharmaceuticals	Slovenia	Commercially available	No	No	No	No	No
MSN Laboratories Ltd.	India	Unconfirmed API activity	No	No	No	No	No
Merck and Co. Inc.	USA	Innovator or marketer	No	No	No	No	No
QR Pharmaceuticals Ltd.	China	Not manufacturing	No	No	No	No	No
RPG Life Sciences Limited	India	Not manufacturing	No	No	No	No	No
Ranbaxy Laboratories Ltd.	India	Commercially available	No	No	No	No	No

Table 6.7 contd. ...

...Table 6.7 contd.

Manufacturer	Country	Manufacturing Status	Active US DMF	Japanese DMF	Korean DMF	Available for Ref US DMF	Active COS
Sai Life Sciences Ltd.	India	Commercially available	No	No	No	No	No
Shanghai SIPI Pharmaceutical Co. Ltd.	China	Not manufacturing	No	No	No	No	No
Shanghai Techwell Biopharmaceutical Co. Ltd.	China	Commercially available	Yes	No	No	No	No
Suzhou No 4 Pharmaceutical Factory	China	Unconfirmed API activity	No	No	No	No	No
Teva API India Ltd.	India	Commercially available	No	No	No	No	No
Teva Pharmaceutical Works Private Ltd. Company	Hungary	Commercially available	Yes	No	No	Yes	No
Xellia Pharmaceuticals ApS	Denmark	Commercially available	Yes	Yes	No	No	No
Yacht Bio-Tech	China	Unconfirmed API activity	No	No	No	No	No
Zhejiang HISOAR Pharmaceutical Company Ltd.	China	Not manufacturing	No	No	No	No	No
Zhejiang Hisun Pharmaceutical Factory	China	Not manufacturing	No	No	No	No	No
Zhejiang Medicine Co. Ltd. Xinchang Pharmaceutical Factory	China	Not manufacturing	No	No	No	No	No

Table 6.8. Patent status of caspofungin.

Sr. No	Patent No	Country	Drug Substance Claim	Drug Product Claim	Patent Expiry Date
1	9636407	US	No	Yes	2032/12/21
2	2251928	Canada	No	Yes	2017/04/15
3	2118757	Canada	No	Yes	2014/10/03
4	2014O142032A1	US	No	Yes	2032/12/20
5	WO 2017/185030 Al	PCT	No	Yes	2037/04/20
6	201001 68415A1	US	Yes	No	2029/12/21
7	CN2012105812232A	China	Yes	No	2032/12/26

Table 6.9. Import details of caspofungin.

Indian Port	Description	Quantity	Units	Unit USD	Origin Port
Chennai Air Cargo	Caspofungin acetate for injection 70 mg/vial	20	VLS	797.90	France
Bengaluru Air Cargo	sml0425–5 mg caspofungin diacetate (organic chemical)	3	NOS	76.27	United States
Chennai Air Cargo	Caspofungin acetate for injection 50 mg/vial	20	VLS	747.40	France
Chennai Air Cargo	Caspofungin cs 32 ww s30 laboratory reagent	1	NOS	65.29	France
Delhi Air Cargo	Caspofungin cs 32 ww s30 (diagnostic kits)	1	NOS	64.60	France
Delhi Air Cargo	Caspofungin cs 32 ww s30 (diagnostic kits)	1	NOS	63.53	France
Delhi Air Cargo	Caspofungin cs 32 ww s30 (diagnostic kits)	3	NOS	63.53	France
Delhi Air Cargo	Caspofungin cs 32 ww s30 (diagnostic kits)	1	NOS	62.85	France
Delhi Air Cargo	Caspofungin cs 32 ww s30 (diagnostic kits)	3	NOS	62.48	France
Hyderabad Air Cargo	Caspofungin (Cancidas 50 mg/mL vial) (quantity: 3; price: 150 USD each)	0.1	KGS	5505.13	United States
Chennai Air Cargo	Caspofungin acetate for injection 70 mg/vial	20	VLS	502.73	United States
Delhi Air Cargo	921540 50 Caspofungin cas 0.002-32 [lab reagent for ruo]	2	PAC	230.71	Italy
Delhi Air Cargo	Caspofungin cs 32 ww f100 (diagnostic kits)	1	NOS	171.12	France
Hyderabad Air Cargo	Caspofungin acetate for injection 70 mg (Cancidas)	2	PAC	1284.20	United States
Chennai Air Cargo	(kb1-intermediate of caspofungin acetate) [(10r,12s)-n-((2r,6s,9s,11r,12r,14as,15s,20s,23s,25as)-20-((r)-3-amino-1-hydro]	4000	GMS	126.25	China
Chennai Air Cargo	Caspofungin cs 32 ww s30 (diagnostic kits) 1 pc	0.5	KGS	126.09	France
Chennai Air Cargo	Caspofungin cs 32 ww s30 (diagnostic kits) 2 pcs	1.36	KGS	105.50	France
Hyderabad Air Cargo	Caspofungin acetate impurity b1+b2 impurity standard (lab reagent)	20	MGS	0.51	Hungary
Hyderabad Air Cargo	Caspofungin impurity b1+b2 impurity standard (lot# st520051062-64)	20	MGS	0.31	Hungary
Hyderabad Air Cargo	Caspofungin contaminated marker for caspofungin c0 isomer (lot 1210-71)	50	MGS	0.12	Hungary

Notes: VLS: Vials; NOS: Numbers; KGS: kg; GMS: gm; MGS: mg; PAC: Packet.

The import and export details of caspofungin are also summarized here. The global market of caspofungin is projected to reach USD 560.3 million by 2026 from USD 417.7 million in 2019. Mostly, caspofungin has been imported from France and United States along with other countries such as Italy and China. India imports caspofungin in the form of caspofungin acetate for injection 70 mg/vial, caspofungin diacetate [organic chemical], caspofungin cs 32 ww s30 (diagnostic kits), caspofungin (Cancidas 50 mg/mL vial). Regarding the export of caspofungin, India exports caspofungin to Turkey, Russia, Taiwan, and Kazakhstan. Caspofungin is exported as caspofungin acetate under different brand names. The details of caspofungin import and export are presented in Table 6.9 and Table 6.10, respectively (https://www.seair.co.in/).

Table 6.10. Export details of caspofungin.

Indian Port	Description	Quantity	Unit USD	Total Value USD	Destination Port	Country
Hyderabad Air Cargo	Caspofungin acetate	500	105.75	52875.00	Istanbul	Turkey
Hyderabad Air Cargo	Caspofungin acetate	500	105.75	52875.00	Istanbul	Turkey
Delhi Air Cargo	Caspofungin acetate	50	98.80	4939.82	Amsterdam, Schiphol	Netherlands
Delhi Air Cargo	Pharmaceutical medicinal products - Cagin 70 (caspofungin acetate for injection 70 mg)	40	1.19	47.40	Yangon	Myanmar
Delhi Air Cargo	Caspofungin acetate	26	123.65	3214.85	Taipei	Taiwan
Delhi Air Cargo	Pharmaceutical medicinal products - Cagin 50 (caspofungin acetate for injection 50 mg)	70	4.26	298.00	Almaty	Kazakhstan
Mumbai Air Cargo	Caspofungin acetate for injection 50 mg	750	29.50	22127.50	Sheremetyevo	Russia
Mumbai Air Cargo	Caspofungin acetate	2	1.00	2.00	Algiers	United States
Delhi Air Cargo	Caspofungin acetate-0.500 g each	4	4.62	18.47	Amsterdam, Schiphol	Netherlands
Delhi Air Cargo	Caspofungin acetate-0.500 g each	3	5.00	15.00	Amsterdam, Schiphol	Netherlands
Bengaluru Air Cargo	Caspofungin injection 50 mg casporan	1	144.97	144.97	Georgetown	United States

9. Micafungin

9.1 Introduction and Development

Micafungin is the second drug to get approval as antifungal in the echinocandin class. Micafungin is a semi-synthesised and water-soluble compound produced from the acylated cyclic hexapeptide FR901379. FR901379 is a naturally occurring substance derived from *Coleophoma empetri* F-11899, in a sample from the soil in the city of Iwaki (Hashimoto, 2009). The enzymatic deacylation reaction is used to derive micafungin from FR901379, which is followed by re-acylation with the optimized N-acyl side chain. Micafungin is a stronger inhibitor of β-1, 3-D-glucan synthase which is required for the synthesis of the cell wall of several pathogenic fungi (Hashimoto, 2009).

Micafungin and FR901379 were discovered in 1989 by Fujisawa Pharmaceutical Co., Ltd. by a screening of nearly 6,000 samples. FR901379 is the parent compound of micafungin. These two new compounds were recognized as the echinocandin class members. Pneumocandin B_0, echinocandin B, and other groups of echinocandin lipopeptides are structurally similar. These molecules have a cyclic hexapeptide that is acylated with a long chain and show a good antifungal (against *Candida* sp.) activity by inhibiting the synthesis of the enzyme β-1, 3-D-glucan synthase. Their main drawback is that they are insoluble in water. FR901379 and other compounds that are related to it show high water solubility and high antifungal activity against *Candida* species. FR901379 differs from other echinocandins by the presence of a sulphate moiety in its molecule (Hashimoto, 2009). The presence of the sulphate moiety is the main reason for its water solubility. FR901379 has a higher IC_{50} value compared to echinocandin B on the β-1, 3-D-glucan synthase. FR901379 and other compounds that are related to it show antifungal activity against both *Aspergillus fumigatus* and *C. albicans*. It shows the least antifungal activity against *A. fumigatus*. None of the compounds of FR901379 show antifungal activity against *C. neoformans*. FR901379 was infused continuously for 4 days in a model for the treatment against murine C. *albicans*. It prolonged the life of the infected mice. Although FR901379 has a good antifungal effect and high water solubility, it is not developed because of the haemolytic activity.

9.2 Biosynthesis

FR901379 has a good antifungal activity and water solubility, but it shows haemolytic activity of RBCs. To overcome the haemolytic activity of the RBCs, the acyl side chain is substituted. As other compounds are not water soluble in comparison to FR901379, the sulphate moiety is kept intact, and the side chain is replaced. FR901379 was treated with acylase derived from *Actinoplanes utahensis* to yield FR179642 by removing the palmitoyl group and adding a new acyl side chain (Iwamoto, 1996). The new compound FR131535 is yielded by re-acylation of FR179642. FR131535 is water soluble even after the replacement of the acyl side chain. FR131535 inhibits the synthesis of the β-1,3-D-glucan synthase in a non-competitive manner. FR131535 showed anti-aspergillus activity which made it the first echinocandin to do so. FR131535 showed less haemolytic activity than FR901379 (Fujie et al., 2001).

FR901379, the parent compound of micafungin, was manufactured at Fujisawa Pharmaceuticals in Japan. It was structurally similar to pneumocandin A_0. To increase the water solubility, the Echinocandin B nucleus is modified by incorporation of hydroxy glutamine residue. FR901379 was optimized structurally for the incorporation of a sulphate moiety that helps to overcome the drawback of echinocandins by increasing the solubility in water, but it has a haemolytic effect on the RBCs that is overcome by using a lead compound. FR131535 and FR179642 are the two important intermediates that are synthesized by modifying the acyl side chains through an optimization process; these intermediates showed a decrease in the lytic activity and increased antifungal activity against the *Candida* and *Aspergillus* species. These intermediates undergo further optimization to obtain a final product FK463 that has an isoxazole ring in the side chain. It has no lytic activity and has potent antifungal activity against the *Candida* and *Aspergillus* species (Hashimoto, 2009).

The clinical efficiency of FK463 is proven in *Aspergillus* and *Candida*-associated infections. The clinical improvement was seen in HIV-positive patients, with oesophageal candidiasis, who were given with micafungin. The safety profile is also appreciated. The symptoms of the patients either improved or resolved within 3–5 days, irrespective of the CD4+ lymphocyte count. The average rate of resolution with micafungin (83.5%) can be compared to the existing drugs like fluconazole (86.7%), and hence the efficiency of micafungin is also considerable in HIV-positive patients (Pettengell et al., 2004; De Wet et al., 2004; De Wet et al., 2005). Micafungin is approved for use in *Candida* infection as a prophylactic treatment as well as for oesophageal infection (Micafungin—an overview, ScienceDirect Topics, 2008).

9.3 Structural Diversity

Micafungin is a cyclic lipopeptide; it consists of a lipopeptide core of pneumocandin A_0 with a difference in the side chain fatty acid that consists of a 3,5-diphenyl-substituted isoxazole ring and helps in decreasing the haemolytic activity. It has good antifungal activity against both *Candida* and *Aspergillus*. To increase the solubility, a sulphated tyrosine residue is essential (Hashimoto, 2009). This side chain is composed mainly of methyl 4-(5-(4-pentyl-oxyphenyl) isoxazol-3-yl) benzoate which was produced by regioselective 1, 3-dipolar cycloaddition of 4-methyl carbonyl benzhydroxamic acid chloride to 4-pentyl oxyphenyl acetylene (Micafungin - an overview, ScienceDirect Topics, 2008).

9.4 FR901379

FR901379 is an antibiotic with a lipopeptide chain. It is produced from the *Coleophomo empetri* F-11899. It is the parent compound of micafungin. Micafungin is the second antifungal agent that is approved for the treatment of *Candida*-associated infections. It is water soluble and has good antifungal activity. It acts by inhibiting the synthesis of the β-1, 3-D-glucan synthase in a non-competitive manner. The cell wall of fungi consists of β-1, 3-D-glucan synthase consisting of the *Fks*p-subunit. FR901379 binds to the *Fks*p-subunit in a non-competitive manner and blocks the synthesis of β-1, 3-D-glucan synthase which in turn disrupts the cell wall of fungi. FR901379 is a cyclic hexapeptide and consists of a sulphate moiety, which is the main reason for its water soluble property. The acyl side chain is substituted to reduce the haemolytic effect on RBCs. The re-acylation of FR901379 forms FR131535, which shows less haemolytic activity. Upon optimization, FR901379 yields micafungin. It is used in the treatment of acute disseminated candidiasis, candidemia, candida abscesses, and peritonitis in paediatric patients of less than 4 months of age and in adults. In adults, micafungin is used at a dosage of 100 mg/day intravenously in the treatment of acute disseminated candidiasis, candidemia, candida abscesses, and peritonitis. In the treatment of oesophageal candidiasis, a dosage of 150 mg/day is given intravenously (Hashimoto, 2009).

9.5 Global Status of Micafungin

The global market for micafungin (used as micafungin sodium) is expected to grow from 2020 to 2025 from US$ 335.9 million US$ to 368 million. The market segmentation of micafungin is done by dosage (50 mg single-use vial or 100 mg single-use vial) or application (respiratory mycosis, gastrointestinal mycosis, or candidemia). Mostly the market demand comes from North America including United States, Mexico, and Canada. The other regions where micafungin is in demand are Russia, Italy, United Kingdom, Japan, Korea, Australia, Brazil, Colombia, Saudi Arabia, Nigeria, and South Africa (Global Micafungin sodium for injection market 2021 by Manufacturers, Regions, Type, and Application, Forecast to 2026 - Marketsandresearch.biz, 2021).

Micafungin was approved as an NDA (021506) product under the brand name Mycamine in the US on 16 March 2005. After US approval, the first marketing authorization in Europe and EEA was granted on 25 April 2008. It is available as an intravenous freeze-dried powder injection at doses of 50 mg/vial and 100 mg/vial. The NDA product was developed and distributed by Astellas Pharma Tech Co., Ltd., Japan. The worldwide sale of the micafungin by September 2016 and 2017 was found to be US$ 347 million and 366.4 million, respectively, while during the same period the API consumption was found to be 303.3 kg and 315.8 kg, respectively. The leading generic players of micafungin are Fresenius Kabi US, Apotex Pharmachem Inc, Xellia Pharmaceuticals, Teva Pharmaceuticals, Hikma Pharmaceuticals, and Jiangsu Hansoh Pharmaceuticals. The details of leading micafungin manufacturers along with DMF status are presented in Table 6.11 and the patent status is presented in Table 6.12. According to the trade data released by the Govt. of India, India has exported 0.02 kg, 1.17 kg, 14.25 kg, and 1.01 kg of micafungin to Turkey, US, Norway, and Germany respectively, during the year 2016, while nearly 1,00,227.0 kg, 17.0 kg, and 0.15 kg of API were imported from China, Malta, and Hongkong during the same year (https://www.pharmacompass.com).

Table 6.11. Leading micafungin manufacturers and DMF status.

Manufacturer	USDMF	Japanese DMF	EU DMF	Korean DMF
Shanghai Techwell Biopharmaceutical Co. Ltd.	Yes	Yes	No	No
Medichem Manufacturing MALTA Ltd.	Yes	No	No	No
Biocon Ltd.	Yes	No	Yes	No
Teva Pharmaceutical Industries Ltd.	Yes	No	No	No
Jiangsu Hansoh Pharmaceutical Group Co., Ltd.	Yes	No	No	No
MSN Life Sciences Private Ltd.	Yes	No	No	No
Yung Shin Pharmaceutical Industrial Co. Ltd.	Yes	No	No	No
Brightgene Bio-Medical Technology Co. Ltd.	Yes	Yes	No	Yes
Sichuan Novales Pharmaceuticals Co.	No	Yes	No	No
Gufic Biosciences Ltd	No	No	Yes	No
Astellas Pharma Inc.	No	No	No	Yes

Table 6.12. Patent status of micafungin.

Sr. No	Patent No	Country	Drug Substance Claim	Drug Product Claim	Patent Expiry Date
1	6774104	US	No	Yes	08/01/2021
2	6774104*PED	US	No	No	08/07/2021
3	2202058	Canada	No	Yes	29/09/2015
4	2044746	Canada	No	Yes	17/06/2011
5	2341568	Canada	No	Yes	17/06/2020
6	WO2012143293A1	Patent Cooperation Treaty	Yes	No	20/04/2031
7	US9115177B2	United States	Yes	No	12/05/2031

Micafungin is expected to reach US$ 290 million by 2023. Micafungin is mostly imported in the form of micafungin sodium injection from Japan and from China, Italy, and United States, while the export of micafungin is mostly to the United States followed by Croatia, Japan Brazil, Taiwan, and United Kingdom (https://www.seair.co.in/). The details are given below in Table 6.13 and Table 6.14, respectively.

10. Anidulafungin

10.1 Introduction and Development

Anidulafungin is a newly introduced echinocandin agent used as an antifungal with broad-spectrum antifungal properties against *Candida* and *Aspergillus* species. This

Table 6.13. Import details of micafungin.

Indian Port	Description	Quantity	Unit USD	Total Value USD	Origin Port
Hyderabad Air Cargo	Micafungin sodium	500	8.48	4240	China
Chennai Air Cargo	fr179642 (intermediate of micafungin sodium)(5-((1s,2s)-2-((2r,6r,9s,11r,12r,14ar,15r,16s, 20r,23s,25ar)-9-amino-20-((r)-	250	151.50	37875	China
Chennai Air Cargo	fr179642 (intermediate of micafungin sodium)(5-((1s,2s)-2-((2r,6r,9s,11r,12r,14ar,15r,16s, 20r,23s,25ar)-9-amino-20-((r)-	250	151.39	37847.5	China
Mumbai Air Cargo	Micafungin sodium (test lic. no. tl/az/16/000142) for research and development purposes	0.2	272700.00	54540	Malta
Ahmedabad	Mycamine micafungin sodium injection powder lyophilized for solution dose. 20 mg in 1 mL, count/bottle 5 mL in 1 vial	10	211.67	2116.7	United States
Ahmedabad	Mycamine micafungin sodium injection powder lyophilized for solution dose; 10 mg in 1 mL count/bottle 5 mL in 1 vial	10	104.22	1042.2	United States
Chennai Air Cargo	FR179642 (intermediate of micafungin sodium)(5-((1s,2s)-2-((2r,6r,9s,11r,12r,14ar,15r, 16s,20r,23s,25ar)-9-amino-20-((r)-	250	151.50	37875	China
Mumbai Air Cargo	Mycamine 50 mg vials micafungin sodium for injection (each vial contains micafungin sodium) (equivalent to micafungin 50 mg)	300	10.01	3003	Japan
Mumbai Air Cargo	Micafungin impurity VII	22	229.55	5050.1	Italy
Mumbai Air Cargo	Micafungin impurity VI	22	229.55	5050.1	Italy
Mumbai Air Cargo	Micafungin impurity V	22	229.55	5050.1	Italy
Mumbai Air Cargo	Micafungin impurity IV	22	229.55	5050.1	Italy
Mumbai Air Cargo	Micafungin impurity III	22	229.55	5050.1	Italy
Mumbai Air Cargo	Micafungin impurity II	22	229.55	5050.1	Italy
Mumbai Air Cargo	Micafungin impurity I	22	229.55	5050.1	Italy
Delhi Air Cargo	Micafungin myc 32 ww b30 (diagnostic kits)	1	67.48	67.48	France
Mumbai Air Cargo	Mycamine 50 mg vials micafungin sodium for injection (each vial contains micafungin sodium) (equivalent to micafungin 50 mg)	42130	10.03	422563.9	Japan
Chennai Air Cargo	FR179642 (intermediate of micafungin sodium)(5-((1s,2s)-2-((2r,6r,9s,11r,12r,14ar,15r, 16s,20r,23s,25ar)-9-amino-20-((r)-	250	152.39	38097.5	China
Hyderabad Air Cargo	Micafungin sodium (145 grams) (chemicals)	0.145	353500.00	51257.5	Malta
Bengaluru Air Cargo	Mycamine micafungin sodium injection 100 mg for research and development purposes) dcgi (perishable cargo)	1	1065.69	1065.69	United Kingdom

Table 6.14. Export details of micafungin.

Indian Port	Description	Quantity	Unit USD	Total Value USD	Destination Port	Country
Bengaluru Air cargo	Micafungin sodium	0.42	237.19	99.62	Tokyo	Japan
Hyderabad Air cargo	Micafungin sodium	1000	7.30	7300.00	Oslo	United States
Bengaluru Air cargo	Micafungin sodium	0.05	57.15	2.86	Zagreb	Croatia
Hyderabad Air cargo	Micafungin sodium	60	7.29	437.67	Oslo	United States
Hyderabad Air cargo	Micafungin sodium	35	7.30	255.50	Zagreb	Croatia
Hyderabad Air cargo	Micafungin placebo (200 mL)	2	1.00	2.00	Rio de Janeiro-gal	Brazil
Hyderabad Air cargo	Micafungin sodium for injection (100 mg/vial)	300	0.10	30.00	Rio de Janeiro-gal	Brazil
Ahmedabad	Micafungin sodium-free sample for research and development use only no commercial value	0.25	40.00	10.00	New York	United States
Ahmedabad	Micafungin sodium	0.25	1205.77	301.44	New York	United States

drug exhibits lower toxicity, and its fungicidal effect depends upon the concentration of the drug. Anidulafungin exhibits a concentration-dependent killing effect on the residual fungal burden in visceral organs such as the spleen, kidney, lung, and liver. The clinical profile of the anidulafungin has shown to be safer and effective against the treatment of oesophageal infections due to *Candida* and candidemia (Pfaller, 2004).

Anidulafungin was produced by optimizing the chemical structure of cilofungin. The nucleus of echinocandin B was modified structurally by deacylation at the acyl side chain using *Actinoplane* species. An alkoxy terphenyl group is introduced on the nucleus of echinocandin B at the amino group which gets reactive by treating the amino acid with terphenyl acid. By introducing this alkoxy terphenyl, the antifungal activity was increased and the haemolytic activity of the RBCs was reduced. This result is called anidulafungin (Boeck et al., 1989).

It was found that anidulafungin clearance was not affected by the treatment of inhibitors and inducers of the cytochrome P450 metabolic pathway. The interaction of anidulafungin and cyclosporine was also documented in healthy adults. However, a higher incidence of hepatic transaminase levels is reported (Dowell et al., 2004).

The administration of anidulafungin with amphotericin B is also proved to be safe. In a recent trial, anidulafungin and liposomal amphotericin B administration led to successful resolution of invasive aspergillosis after 90 days of therapy. No adverse effect reported (Herbrecht et al., 2010).

10.2 Structural Diversity

Like the other two echinocandins, anidulafungin contains a lipopeptide nucleus that is composed of amino acid residues present in the echinocandin B nucleus. The core nucleus containing the amino acid residues is responsible for the antifungal activity. The core of the nucleus also contains amino linkers and the hydroxyl groups that are mainly essential for the water solubility of the drug. Anidulafungin consists of an alkoxy triphenyl side chain that is mainly important for the drug to get in contact with the fungal cell wall. The alkoxy triphenyl side chain is lipophilic compared to other side chains in the echinocandins, which is mainly responsible for water solubility of the drug. Anidulafungin consists of an octyloxy triphenyl side chain which is the main reason for the attachment of the drug to the fungal cell wall (Vazquez and Sobel, 2006).

10.3 Biosynthesis

Anidulafungin is a semi-synthetic drug that is produced by the fermentation of *Aspergillus nidulans*. It possesses antifungal activity unlike itraconazole and amphotericin B against the species of *Aspergillus*. Anidulafungin can penetrate the oesophagus and salivary glands and shows dose-dependent antifungal activity against the species that are resistant to itraconazole and fluconazole in oesophageal and oropharyngeal candidiasis (Denning, 2003).

10.4 Echinocandin B

Echinocandin B was first developed in the year 1974 as a metabolite of *Aspergillus delacroxii* by Ciba Geigy and group. Researchers from Sandoz AG reported that echinocandin B was developed from the *A. rugulose* strain NRRL 8113. Echinocandin B contains an N-linked acyl fatty acid chain which is the main important structural feature and a cyclic lipopeptide core is present. The hexacyclic lipopeptide nucleus contains different amino acid residues such as 3,4-dihydroxy ornithine, 3,4-dihydroxy homotyrosine, 3-hydroxy proline, 3,4-dihydroxy proline, and two residues of threonine; these show antifungal activity and the physicochemical properties of the echinocandin B and its congeners are determined. The homotyrosine amino acid residue is responsible for inhibition of glucan synthase and antifungal activity. Proline residue shows antifungal potency. The hydroxyl groups present at the three amino acid residues are the core of the nucleus; it does not have antifungal property but increases the stability and water solubility. The linoleic fatty acid present in the nucleus of echinocandin B is essential for antifungal activity as it acts as an anchorage for the cell wall but has a haemolytic effect on RBCs. The N-linked acyl fatty acid chain is modified structurally to further improve the safety of the echinocandin class drugs. The cell wall of the fungi contains a β-1,3-D-glucan synthase that is composed of the *Fks* p-subunit and a regulatory subunit. Echinocandins bind to the *Fks* p-subunit non-competitively and block the synthesis of β-1,3-D-glucan synthase that results in the disruption of the cell wall. Echinocandin B is used in the treatment of invasive candidiasis, invasive aspergillus, and neutropenic fever. Fever, rash, phlebitis, and nausea are the common side effects at the injection site. Histamine reaction is also caused when the drug is infused rapidly. Elevated levels of aminotransferase and alkaline phosphatase are associated with the administration of echinocandin B (Patil and Majumdar, 2017). Therefore, anidulafungin proves to have several advantages over the others. It has a broad-spectrum fungicidal activity against a wide variety of *Candida* species including those which are azole- and polyene-resistant. Anidulafungin dosage is not adjusted while administering to the patients with liver and renal abnormalities receiving other drugs. Anidulafungin does not stimulate the reaction of cross-resistance with other antifungals. It can be given comfortably with other antifungals such as voriconazole and amphotericin B. Anidulafungin is considered the first-line drug for use in case of severe fungal infections, candidemia, mucosal candidiasis resistant to azole drugs, etc. (Vazquez and Sobel, 2006).

10.5 Global Status of Anidulafungin

Anidulafungin was approved as NDA (021632 and 021948) product under the brand name Eraxis® in the US on 17 February 2006. After US approval, the first marketing authorization in Europe and EEA was granted on 20 September 2007. It is available as an intravenous powder injection at dose strength of 50 mg/vial and 70 mg/vial. Also, it is available in an injectable form at a dose of 50 mg. Eraxis® injection is a sterile, lyophilized product for intravenous infusion that contains anidulafungin. The NDA product was developed and distributed by Roerig, Division of Pfizer Inc., US. The worldwide sale of the anidulafungin by September 2016 and 2017 was found to be US$ 161.2 million and 175.7 million, respectively, while during the same period the API consumption was found to be 87.9 kg and 105.0 kg, respectively. The details of leading anidulafungin manufacturers along with DMF status are presented in Table 6.15 and the patent status is presented in Table 6.16. According to the trade data released by the Govt. of India, India has exported 3.01 kg, 0.01 kg, 0.02 kg, and 5.0 kg of anidulafungin to Turkey, Bangladesh, Pakistan, and the UK, respectively during the year 2016, while nearly 10.0 kg of API was imported from China during the same year (https://www.pharmacompass.com).

Anidulafungin is imported to India mostly from the United States and France. It is mainly imported in the form of injection and few shares are imported in the form of laboratory reagent. It is exported to a limited number of countries. Egypt and Turkey are the two countries where anidulafungin is being exported from India as presented in Table 6.17 and Table 6.18, respectively (https://www.seair.co.in/).

Table 6.15. Leading anidulafungin manufacturers and DMF status.

Manufacturer	**Country**	**Manufacturing Status**	**Active US DMF**	**Japanese DMF**	**Korean DMF**	**Available for Ref US DMF**	**Active COS**
Apicore Pharmaceuticals Pvt. Ltd.	India	Commercially available	No	No	No	No	No
Aptuit Inc.	Unites States	Not manufacturing	No	No	Yes	No	No
Bright Gene Bio-Medical Technology Co. Ltd.	China	Unconfirmed API activity	Yes	No	No	No	No
Chunghwa Chemical Synthesis and Biotech Co. Ltd.	Taiwan	Under development	No	No	No	No	No
Concord Biotech Ltd.	India	Commercially available	No	No	No	No	No
DSM Sinochem Pharmaceuticals	Netherlands	Commercially available	No	No	No	No	No
Euticals S.p.A	Italy	Not manufacturing	No	No	No	No	No
Gufic Biosciences Ltd.	India	Early API activity	No	No	No	No	No
Pharmacia & Upjohn Company	Unites States	Innovator or marketer	No	No	Yes	No	No
Shandong New Time Pharmaceutical Company Ltd.	China	Early API activity	No	No	No	No	No
Shanghai SIPI Pharmaceutical Co. Ltd.	China	Unconfirmed API activity	No	No	No	No	No
Shanghai Techwell Biopharmaceutical Co. Ltd.	China	Not manufacturing	No	No	No	No	No
Suzhou No 4 Pharmaceutical Factory	China	Early API activity	No	No	No	No	No
Teva Pharmaceutical Works Private Limited Company	Hungary	Commercially available	No	No	No	No	No
Xingcheng Chempharm Co. Ltd.	China	Under development	No	No	No	No	No
Zhejiang Medicine Co. Ltd. Xinchang Pharmaceutical Factory	China	Not manufacturing	No	No	No	No	No

Table 6.16. Patent status of anidulafungin.

Sr. No	Patent No	Country	Drug Substance Claim	Drug Product Claim	Patent expiry Date
1	6960564	Unites States	No	Yes	04/12/2021
2	7709444	Unites States	No	Yes	04/12/2021
3	2091663	Canada	No	Yes	15/03/2013
4	2362481	Canada	No	Yes	03/02/2020
5	EP3464319A2	Europe	Yes	No	07/10/2035
6	ES2694561T3	Spain	No	Yes	23/02/2035
7	US20090238867A1	Unites States	No	Yes	13/12/2007

Table 6.17. Import details of anidulafungin.

Indian Port	Description	Quantity	Unit USD	Total Value USD	Origin Port
Mumbai Air Cargo	Eraxis 100 mg spo 1 × 30 mL gvl in (anidulafungin) injection li. rsp item list 3 sr. no. 16	7000	56.65	396550	United States
Mumbai Air Cargo	Eraxis 100 mg spo 1 × 30 mL gvl in (anidulafungin) injection li. rsp item list 3 sr. no. 16	5000	56.78	283900	United States
Mumbai Air Cargo	Eraxis 100 mg spo 1 × 30 mL gvl in (anidulafungin) injection li. rsp item list 3 sr. no. 16	3500	57.37	200795	United States
Mumbai Air Cargo	Eraxis 100 mg spo 1 × 30 mL gvl in (anidulafungin) injection li. rsp item list 3 sr. no. 16	3500	57.16	200060	United States
Mumbai Air Cargo	Eraxis 100 mg spo 1 × 30 mL gvl in (anidulafungin) injection list 3 sr. no.161	3000	56.65	169950	United States
Mumbai Air Cargo	Eraxis 100 mg spo 1 × 30 mL gvl in (anidulafungin) injection li. rsp item list 3 sr. no. 161	3000	56.45	169350	United States
Mumbai Air Cargo	Eraxis 100 mg spo 1 × 30 mL gvl in (anidulafungin) injection li. rsp item list 3 sr. no.161	3000	56.45	169350	United States
Mumbai Air Cargo	Eraxis 100 mg spo 1 × 30 mL gvl in (anidulafungin) injection li. rsp item list 3 sr. no.161	2808	56.91	159803.28	United States
Mumbai Air Cargo	Eraxis 100 mg spo 1 × 30 mL gvl in (anidulafungin) injection li. rsp item list 3 sr. no. 16	2654	56.65	150349.1	United States
Mumbai Air Cargo	Eraxis 100 mg spo 1 × 30 mL gvl in (anidulafungin) injection li. rsp item list 3 sr. no. 16	1815	56.65	102819.75	United States
Delhi Air Cargo	Anidulafungin and 32 ww b100 (diagnostic kits)	10	2640.84	26408.4	France

Table 6.17 contd. ...

...*Table 6.17 contd.*

Indian Port	Description	Quantity	Unit USD	Total Value USD	Origin Port
Chennai Air Cargo	Anidulafungin and 32 ww b100 laboratory reagent	10	2618.17	26181.7	France
Chennai Air Cargo	Anidulafungin and 32 ww b100 (diagnostic kits) 10 pcs	4.69	5570.83	26127.1927	France
Mumbai Air Cargo	Eraxis 100 mg spo 1 × 30 mL gvl in (anidulafungin) injection li. rsp item list 3 sr. no. 16	192	56.40	10828.8	United States
Delhi Air Cargo	Anidulafungin and 32 ww b100 (diagnostic kits)	1	2716.22	2716.22	France
Delhi Air Cargo	Anidulafungin and 32 ww b100 (diagnostic kits)	1	2675.95	2675.95	France
Chennai Air Cargo	(foc) Anidulafungin - eraxis, ecalta (qty: 0.9785 mg) (research materials for laboratory analysis purpose)	1	2040.93	2040.93	United States
Delhi Air Cargo	Anidulafungin and 32 ww b100 (diagnostic kits)	1	179.86	179.86	France
Chennai Air Cargo	Anidulafungin and 32 ww b30 laboratory reagent	1	59.65	59.65	France

Table 6.18. Export details of anidulafungin.

Indian Port	Quantity	Unit USD	Total Value USD	Destination Port	Country
Mumbai Air Cargo	8	889.76	7118.06	Istanbul	Turkey
Mumbai Air Cargo	10	580.05	5800.45	Alexandria	Turkey
Ahmedabad	7	800.00	5600.00	Istanbul	Turkey

11. Conclusion and Future Prospects

Being a novel class of antifungals, echinocandins can override the disadvantages of the existing antifungals. Echinocandins are the antifungal drugs that inhibit the synthesis of β-1,3-D-glucan in the cell walls of fungi. These drugs show low toxicity, favourable effects at low dose and rapid fungicidal activity mainly against the *Candida* species. They are successful in the treatment of invasive aspergillosis and candidiasis, oesophageal candidiasis in both adults and paediatric patients. Echinocandin B that belongs to a class of echinocandins was the first compound to be discovered. Despite its antifungal activity, the haemolytic activity on the RBCs acts a limitation in its use as an antifungal. Thus, a semisynthetic compound called cilofungin was developed, which had reduced haemolytic activity. Pneumocandin A_0 and pneumocandin B_0 belong to the pneumocandin class having antifungal effects with low physicochemical properties. Caspofungin belongs to the family of echinocandins that is derived from pneumocandin B_0; it is a semi-synthetic water soluble drug, as it has a pneumocandin B_0 nucleus, and does not show haemolytic activity on RBCs. Micafungin has a lipopeptide core similar to caspofungin. It consists of a pneumocandin A_0 with a difference in a fatty acid side chain. Anidulafungin contains a lipopeptide nucleus with amino acid residues same as other echinocandins present in echinocandin B. The antifungal property is maintained by the core nucleus; the water soluble property is maintained by the hydroxyl group and amino linkers. The pharmacological and chemical properties of the echinocandins are attributed to a combination of elements in the structure. This review article also summarizes the global market status including import and export data of caspofungin acetate, micafungin, and anidulafungin. This review also includes the details of the leading manufacturers and DMF as well as the patent status of the three newly approved drugs in various countries. The treatment outcomes of tolerability as well as safety profiles are significant and encouraging over the existing antifungals. There are unexplored aspects of this class of drugs despite available proven data. Nevertheless, these drugs are not widely used in practice, as they should be. The acceptability of the clinicians would be important to make use of this class of drugs at a wider range. Also, there is a need of conducting more clinical trials against different fungal infections. Attempts should be made in the future to develop more drugs of this class. Particularly, research should be conducted in drug designing to develop drugs of this class keeping in mind the side effects already documented. If the minimal side effects can be managed, then, this antifungal class will be accepted more easily in the scientific community as well as among clinicians. This review provides explanations and documentation of echinocandins with a detailed explanation of each of the members, pharmacotherapeutic value, pharmacodynamics, and pharmacokinetics. There is a prospective market for echinocandins already and its growth is expected to rise in the coming years. Given the safety, tolerability, and efficacy, echinocandins can be well considered the next generation of antifungals.

Acknowledgements

We thank Ranjani Bharadwaj for editorial and proofreading services.

References

Adefarati, A.A., Hensens, O.D., Jones, E.T.T. and Tkacz, J.S. (1992). Pneumocandins from *Zalerion arboricola* V Glutamic acid and leucine derived amino acids in pneumocandin A0 (L-671,329) and distinct origins of the substituted proline residues in pneumocandins A0 and B0. *J. Antibiot.*, 45(12): 1953–1957. https://doi.org/107164/antibiotics451953.

Agarwal, M.B., Rathi, S.A., Ratho, N. and Subramanian, R. (2006). Caspofungin: A major breakthrough in treatment of systemic fungal infections. *J. Assoc. Physicians. India.*, 54: 943–948. https://pubmed.ncbi.nlm.nih.gov/17334012/.

Aguilar-Zapata. D., Petraitiene, R. and Petraitis, V. (2015). Echinocandins: The expanding antifungal armamentarium. *Clin Infect Dis.*, 61 (Suppl_6): S604–S611. https://doi.org/101093/cid/civ814.

Application, Forecast to 2026 – Markets and research biz (2021). wwwmarketsandresearch.biz. Retrieved 31 July 2021, from https://www.marketsandresearch.biz/report/161951/global.-micafungin-sodium-for-injection-market-2021-by-manufacturers-regions-type-and-application-forecast-to-2026.

Arathoon, E.G., Gotuzzo, E., Noriega, L.M., Berman, R.S., DiNubile, M.J. and Sable, C.A. (2002). Randomized, double-blind, multi-center study of caspofungin versus amphotericin B for treatment of oropharyngeal and esophageal candidiases. *Antimicrob. Agents. Chemother.*, 46(2): 451–457. https://doi.org/101128/aac462451-4572002.

Barchiesi, F., Spreghini, E., Tomassetti, S., Della, V.A., Arzeni, D., Manso, E. and Scalise, G. (2006). Effects of Caspofungin against *Candida guilliermondii* and *Candida parapsilosis*. *Antimicrob. Agents Chemother.*, 50(8): 2719–2727. https://doi.org/101128/aac00111-06.

Ben-Ami, R., Garcia-Effron, G., Lewi, R.E., Gamarra, S., Leventakos, K., Perlin, D.S. and Kontoyiannis, D.P. (2011). Fitness and virulence costs of *Candida albicans* FKS1 hot spot mutations associated with echinocandin resistance. *J. Infect. Dis.*, 204(4): 626–635. https://doi.org/101093/infdis/jir351.

Berkow, E.L. and Lockhart, S.R. (2018). Activity of CD101, a long acting echinocandin, against clinical isolates of *Candida auris*. *Diagn. Microbiol. Infect. Dis.*, 90(3): 196–197. https://doi.org/101016/jdiagmicrobio201710021.

Biernasiuk, A., Dobiecka, E., Zdzienicka, G. and Malm, A. (2015). The activity of micafungin against clinical isolates of non-albicans *Candida* spp. *Curr. Issues Pharm. Med. Sci.*, 28(1): 13–16. https://doi.org/101515/cipms-2015-0033.

Boeck, V.D., Fukuda, D.S., Abbott, B.J. and Debono, M. (1989). Deacylation of echinocandin B by *Actinoplanes utahensis*. *J. Antibiot.*, 42(3): 382–388. https://doi.org/107164/antibiotics42382.

Bowman, J.C., Hicks, P.S., Kurtz, M.B., Rosen, H., Schmatz, D.M., Liberator, P.A. and Douglas, C.M. (2002). The antifungal echinocandin caspofungin acetate kills growing cells of *Aspergillus fumigatus in vitro*. *AAC.*, 46(9): 3001–3012. https://doi.org/101128/aac4693001-30122002.

Brueggeman, R.J.M., Alffenaar, J.W. and Bliijlevens, N.M.A. (2009). Clinical relevance of the pharmacokinetic interactions of azole antifungal drugs with other co-administered agents. *Clin. Infect. Dis.*, 48: 1441–1458.

Cacho, R.A., Jiang, W., Chooi, Y.H., Walsh, C.T. and Tang, Y. (2012). Identification and characterization of the echinocandin B biosynthetic gene cluster from *Emericella rugulosa* NRRL 11440. *J. Am. Chem. Soc.*, 134(40): 16781–16790. https://doi.org/101021/ja307220z.

Cancidas (caspofungin) [package insert] 2009 Jul Whitehouse Station (NJ): Merck & Co Inc.

Cappelletty, D. and Eiselstein-McKitrick, K. (2007). The echinocandins: Pharmacotherapy. *J. Human Pharma. Drug Thera.*, 27(3): 369–388.

Capsofungin Market: Industry Trends, Share, Size and Forecast Report (2021). www.futurewiseresearch.com Retrieved 30 July 2021, from https://www.futurewiseresearch.com/heal.thcare-market-research/Caspofungin-Market-by/3480.

Chandrasekar, P.H. and Sobel, J.D. (2006). Micafungin: A new echinocandin. 42(8): 1171–1178. https://doi.org/101086/501020.

Chen, L., Yue, Q., Zhang, X., Xiang, M., Wang, C., Li, S., Che, Y., Ortiz-López, F., Bills, G.F., Liu, X. and An, Z. (2013). Genomics-driven discovery of the pneumocandin biosynthetic gene cluster in the fungus *Glarea lozoyensis*. *BMC. Genomics.*, 14(1): 339. https://doi.org/101186/1471-2164-14-339.

Chen, L., Yue, Q., Li, Y., Niu, X., Xiang, M., Wang, W., Bills, G.F., Liu, X. and An, Z. (2014). Engineering of *Glarea lozoyensis* for exclusive production of the pneumocandin B0 precursor of the antifungal drug caspofungin acetate. *Appl. Environ. Microbiol.*, 81(5): 1550–1558. https://doi.org/101128/aem03256-14.

Chen, L., Li, Y., Yue, Q., Loksztejn, A. Yokoyama, K., Felix, E.A., Liu, X., Zhang, N., An, Z. and Bills, G.F. (2016). Engineering of new pneumocandin side-chain analogues from *Glarea lozoyensis* by mutasynthesis and evaluation of their antifungal activity. *ACS. Chem. Biol.*, 11: 2724–2733. https://doi.org/101021/acschembio6b00604.

Cleary, J.D. and Stover, K.R. (2015). Antifungal-associated drug-induced cardiac disease. *Clin. Infect. Dis.*, 61 (suppl._6): S662–S668. https://doi.org/101093/cid/civ739.

Cohen-Wolkowiez, M., Moran, C., Benjamin, D.K. and Smith, P.B. (2009). Pediatric antifungal agents. *Curr. Opin. Infect. Dis.*, 22(6): 553–558. https://doi.org/101097/qco0b013e3283321ccc.

Connors, N. and Pollard, D. (2004). Pneumocandin B0 production by fermentation of the fungus *Glarea lozoyensis*: Physiological and engineering factors affecting titer and structural analogue formation. pp. 515–538. *In*: Zhiqiang An (ed.). Handbook of Industrial Mycology, CRC press, Boca Raton.

Cornely, O.A., Sidhu, M., Odeyemil, van Engen, A.K., van der Waal, J.M. and Schoeman, O. (2008). Economic analysis of micafungin versus liposomal amphotericin B for treatment of candidaemia and invasive candidiasis in Germany. *Curr. Med. Res. Opin.*, 24: 1743–1753. doi:101185/03007990802124889.

Coste, AT., Kritikos, A., Khanna, N., Goldenberger, D., Garzoni, C., Zehnder, C., Boggian, K., Neofytos, D., Riat, A., Bachmann, D., Sanglard, D., Lamoth, F., Lamoth, F., Khanna, N., Boggian, K. and Sanglard, D. (2020). Emerging echinocandin-resistant *Candida albicans* and *glabrata* in Switzerland. *Infection*, 48(5): 761–766. https://doi.org/101007/s15010-020-01475-8.

De Wet, N.T.E., Cuentas, A.L., Suleiman, J., Baraldi, E., Krantz, E.F., Della Negra, M. and Diekmann-Berndt, H. (2004). A randomized, double-blind, parallel-group, dose-response study of micafungin compared with fluconazole for the treatment of esophageal candidiasis in HIV-positive patients. *Clin. Infect. Dis.*, 39: 842–849.

De Wet, N.T.E., Bester, A.J., Viljoen, J.J., Filho, F., Suleiman, J.M., Ticona, E. and Buell, D. (2005). A randomized, double blind, comparative trial of micafungin (FK463) vs. fluconazole for the treatment of oesophageal candidiasis. *Pharmacol. Ther.*, 21(7): 899–907.

Debono, M., Abbott, B.J., Fukuda, D.S., Barnhart, M., Willard, K.E., Molloy, R.M., Michel, K.H., Turner, J.R., Butler, T.F. and Hunt, A.H. (1989). Synthesis of new analogs of echinocandin B by enzymatic deacylation and chemical reacylation of the echinocandin B peptide: synthesis of the antifungal agent cilofungin (LY121019). *J. Antibiot.*, 42(3): 389–397. https://doi.org/107164/antibiotics42389.

Denning, D.W. (2002). Echinocandins: A new class of antifungal. *J. Antimicrob. Chemother.*, 49(6): 889–891. https://doi.org/101093/jac/dkf045.

Denning, D.W. (2003). Echinocandin antifungal drugs. *The Lancet*, 362(9390): 1142–1151. https://doi.org/101016/s0140-6736(03)14472-8.

Denning, D.W., Marr, K.A., Lau, W.M., Facklam, D.P., Ratanatharathorn, V., Becker, C., Ullmann, A.J., Seibel, N.L., Flynn, P.M., van Burik, J.A.H., Buell, D.N. and Patterson, T.F. (2006). Micafungin (FK463), alone or in combination with other systemic antifungal agents for the treatment of acute invasive aspergillosis. *J. Infect.*, 53(5): 337–349. https://doi.org/101016/jjinf200603003.

Doring, M., Hartmann, U., Erbacher, A., Lang, P., Handgretinger, R. and Muller, I. (2012). Caspofungin as antifungal prophylaxis in paediatric patients undergoing allogeneic hematopoietic stem cell transplantation: a retrospective analysis. *BMC. Infect. Dis.*, 12(1): 1–9. https://doi.org/101186/1471-2334-12-151.

Douglas, C.M., Foor, F., Marrinan, J.A., Morin, N. and Nielsen, J.B. (1994). The *Saccharomyces cerevisiae* FKS1 (ETG1) gene encodes an integral membrane protein which is a subunit of 1, 3-beta-D-glucan synthase. *Proc. Natl. Acad. Sci.*, 91(26): 12907–12911.

Dowell, J.A., Knebel, W. and Ludden, T. (2004). Population pharmacokinetic analysis of anidulafungin, an echinocandin antifungal. *J. Clin. Pharmacol.*, 44: 590–598.

Eschenauer, G., DePestel, D.D. and Carver, P.L. (2007). Comparison of echinocandin antifungals. *Ther. Clin. Risk. Manag.*, 3(1): 71–97. https://doi.org/102147/tcrm20073171.

Estes, K.E., Penzak, S.R., Calis, K.A. and Walsh, T.J. (2009). Pharmacology and antifungal properties of anidulafungin, a new Echinocandin. *Pharmacotherapy*, 29: 17–30.

Fesel, P.H. and Zuccaro, A. (2016). β-glucan: Crucial component of the fungal cell wall and elusive MAMP in plants. *Fungal. Genet. Biol.*, 90: 53–60. https://doi.org/101016/jfgb201512004.

Forastiero, A., Garcia, G.V., Rivero, M.O., Garcia, R.R., Monteiro, M.C., Alastruey, I.A., Jordan, R., Agorio, I. and Mellado, E. (2015). Rapid development of *Candida krusei* echinocandin resistance during caspofungin therapy. *Antimicrob. Agents Chemother.*, 59(11): 6975–6982. https://doi.org/101128/aac01005-15.

Fujie, A., Iwamoto, T., Sato, B., Muramatsu, H., Kasahara, C., Furuta, T. and Hashimoto, S. (2001). FR131535, a novel water-soluble echinocandin-like lipopeptide: Synthesis and biological properties. *Bioorg. Med. Chem. Lett.*, 11(3): 399–402.

Fujie, A. (2007). Discovery of Micafungin (FK463): A novel antifungal drug derived from a natural product lead. *Pure. Appl. Chem.*, 79(4): 603–614. doi: https://doi.org/10.1351/pac200779040603.

Global Micafungin Sodium for Injection Market 2021 by Manufacturers, Regions, Type and Grossman, N.T., Chiller, T.M. and Lockhart, S.R. (2014). Epidemiology of echinocandin resistance in *Candida. Curr. Fungal. Infect. Rep.*, 8(4): 243–248. https://doi.org/101007/s12281-014-0209-7.

Grover, N. (2010). Echinocandins: A ray of hope in antifungal drug therapy. *Indian J. Pharmacol.*, 42(1): 9. https://doi.org/104103/0253-761362396.

Hashimoto, S. (2009). Micafungin: A sulfated echinocandin. *J. Antibiot.*, 62(1): 27–35. https://doi.org/101038/ja20083.

Herbrecht, R., Maertens, J., Baila, L., Aoun, M., Heinz, W., Martino, R., Schwartz, S., Ullmann, A.J., Meert, L., Paesmans, M., Marchetti, O., Akan, H., Ameye, L, Shivaprakash, M. and Viscoli, C. (2010). Caspofungin first-line therapy for invasive aspergillosis in allogeneic hematopoietic stem cell transplant patients: A European organisation for research and treatment of cancer study. *Bone Marrow Transplant.*, 45(7): 1227–1233. https://doi.org/101038/bmt2009334.

Higashiyama, Y. and Kohno, S. (2004). Micafungin: A therapeutic review. *Expert. Rev. Anti. Infect. Ther.*, 2(3): 345–355.

Hinske, L.C., Weis, F., Heyn, J., Hinske, P. and Beiras, F.A. (2012). The role of micafungin and anidulafungin in the treatment of systemic fungal infections: Applications and patents for two novel echinocandins. *Recent. Pat. Antiinfect. Drug. Discov.*, 7(1): 1–7. doi: 102174/157489112799829747 PMID: 22044354.

Hiramatsu, Y., Maeda, Y., Fuji, N., Saito, T., Nawa, Y., Hara, M., Yano, T., Asakura, S., Sunam, K., Tabayashi, T., Miyata, A., Matsuoka, K., Shinagawa, K., Ikeda, K., Matsuo, K. and Tanimoto, M. (2008). Use of micafungin versus fluconazole for antifungal prophylaxis in neutropenic patients receiving hematopoietic stem cell transplantation. *Int. J. Hematol.*, 88(5): 588–595. https://doi.org/101007/s12185-008-0196-y.

Hope, W.W., Mickiene, D., Petraitis, V., Petraitiene, R., Kelaher, A.M., Hughes, J.E., Cotton, M.P., Bacher, J., Keirns, J.J., Buell, D., Heresi, G., Benjamin, Jr., D.K., Groll, A.H., Drusano, G.L. and Walsh, T.J. (2008). The pharmacokinetics and pharmacodynamics of micafungin in experimental Hematogenous *Candida* Meningoencephalitis: Implications for Echinocandin therapy in neonates. *J. Infect. Dis.*, 197(1): 163–171. https://doi.org/101086/524063.

https://www.pharmacompass.com.

https://wwwquickcompanyin/patents/mutant-strain-of-glarea-lozoyensis-and-use-thereof-for-preperation-of-pneumocandin-b0.

https://www.seair.co.in

Huttel, W., Youssar, L., Gruning, B.A., Gunther, S. and Hugentobler, K.G. (2016). Echinocandin B biosynthesis: A biosynthetic cluster from *Aspergillus nidulans* NRRL 8112 and reassembly of the sub-clusters *Ecd* and *Hty* from *Aspergillus pachycristatus* NRRL 11440 reveals a single coherent gene cluster. *BMC. Genom.*, 17(1): 1–8. https://doi.org/101186/s12864-016-2885-x.

Ibrahim, A.S., Bowman, J.C., Avanessian, V., Brown, K., Spellberg, B., Edwards, J.E. and Douglas, C.M. (2005). Caspofungin inhibits *Rhizopus oryzae* 1,3-β-d-Glucan synthase, lowers burden in brain measured by quantitative PCR, and improves survival at a low but not a high dose during murine disseminated zygomycosis. *Antimicrob. Agents Chemother.*, 49(2): 721–727. https://doi.org/101128/aac492721-7272005.

Ikeda, F., Wakai, Y., Matsumato, S., Maki, K., Watabe, E., Tawara, S., Goto. T., Watanabe, Y., Matsumoto, F. and Kuwahara, S. (2000). Efficacy of FK463, a new lipopeptide antifungal agent, in mouse models of disseminated candidiasis and aspergillosis. *Antimicrob. Agents Chemother.*, 44(3): 614–618. doi: https://doi.org/10.1128/aac.44.3.614-618.2000.

Ioannou, P., Andrianaki, A., Akoumianaki, T., Kyrmizi, I., Albert, N., Perlin, D., Samonis, G., Kontoyiannis, D.P. and Chamilos, G. (2016). Albumin enhances caspofungin activity against *Aspergillus* species by facilitating drug delivery to germinating hyphae. *Antimicrob. Agents Chemother.*, 60(3): 1226–1233. https://doi.org/101128/aac02026-15.

Iwamoto, T. (1996). *Studies on Antifungal Antibiotics*. PhD Thesis, University of Tokyo.

Jeong, S.H., Kim, D.Y., Jang, J.H., Mun, Y.C., Choi, C.W., Kim, S.H., Kim, J.S. and Park, J.S. (2015). Efficacy and safety of micafungin versus intravenous itraconazole as empirical antifungal therapy for febrile neutropenic patients with haematological malignancies: a randomized, controlled, prospective, multicenter study. *Ann. Hematol.*, 95(2): 337–344. https://doi.org/101007/s00277-015-2545-2.

Jimenez, O.C., Paderu, P., Motyl, M.R. and Perlin, D.S. (2013). Enfumafungin derivative MK-3118 shows increased *in vitro* potency against clinical echinocandin-resistant *Candida* species and *Aspergillus* species isolates. *Antimicrob. Agents. Chemother.*, 58(2): 1248–1251. https://doi.org/101128/aac02145-13.

Keating, G.M. and Figgitt, D.P. (2003). Caspofungin. *Drugs*, 63(20): 2235–2263. https://doi.org/102165/00003495-200363200-00008.

Krishnan, N.S. and Chandrasekar, P.H. (2008). Current and future therapeutic options in the management of invasive aspergillosis. *Drugs*, 68(3): 265–282. https://doi.org/102165/00003495-200868030-00002.

Kuhn, D.M., George, T., Chandra, J., Mukherjee, P.K. and Ghannoum, M.A. (2002). Antifungal susceptibility of *Candida* biofilms: Unique efficacy of amphotericin B lipid formulations and echinocandins. *Antimicrob. Agents Chemother.*, 46: 1773–1780.

Kurtz, M.B. and Marrinan, J. (1989). Isolation of Hem3 mutants from *Candida albicans* by sequential gene disruption. *Mol. Gen. Genet.*, 217(1): 47–52. https://doi.org/101007/bf00330941.

Lan, N., Perlatti, B., Kvitek, D.J., Wiemann, P., Harvey, C.J.B., Frisvad, J., An, Z. and Bills, G.F. (2020). Acrophiarin (antibiotic S31794 /F-1) from *Penicillium arenicola* shares biosynthetic features with both *Aspergillus* and Leotiomycete type echinocandins. *Environ. Microbiol.*, 22(6): 2292–2311. https://doi.org/101111/1462-292015004.

Letscher, B.V. and Herbrecht, R. (2003). Caspofungin: The first representative of a new antifungal class. *J. Antimicrob. Chemother.*, 51(3): 513–521. https://doi.org/101093/jac/dkg117.

Lewis, R.E., Albert, N.D. and Kontoyiannis, D.P. (2008). Comparison of the dose-dependent activity and paradoxical effect of caspofungin and micafungin in a neutropenic murine model of invasive pulmonary aspergillosis. *J. Antimicrob. Chemother.*, 61(5): 1140–1144. https://doi.org/101093/jac/dkn069.

Masurekar, P.S., Fountoulakis, J.M., Hallada, T.C., Sosa, M.S. and Kaplan, L. (1992). Pneumocandins from *Zalerion arboricola* II modification of product spectrum by mutation and medium manipulation. *J. Antibiot.*, 45(12): 1867–1874.

Meunler, F., Lambert, C. and Van Der, A.P. (1989). *In-vitro* activity of cilofungin (LY121019) in comparison with amphotericin B. *J. Antimicrob. Chemother.*, 24(3): 325–331.

Mio, T., Adachi, S.M., Tachibana, Y., Tabuchi, H. and Inoue, S.B. (1997). Cloning of the *Candida albicans* homolog of *Saccharomyces cerevisiae* GSC1/FKS1 and its involvement in beta-1,3-glucan synthesis. *J. Bacteriol.*, 179: 4096–4105.

Miron, D., Battisti, F., Silva, F.K., Lana, A.D., Pippi, B., Casanova, B., Gnoatto, S., Fuentefria, A., Mayorga, P. and Schapoval, E.E.S. (2014). Antifungal activity and mechanism of action of monoterpenes against dermatophytes and yeasts. *Brazilian Journal of Population Studies*, 24(6): 660–667. https://doi.org/101016/jbjp201410014.

Morrison, V.A. (2005). Caspofungin: An overview. *Expert Rev. Anti. Infect. Ther.*, 3(5): 697–705. https://doi.org/101586/1478721035697.

Muller, S., Koch, C., Weiterer, S., Weigand, M.A., Sander, M. and Henrich, M. (2020). Caspofungin induces the release of Ca2+ ions from internal stores by activating ryanodine receptor-dependent pathways in human tracheal epithelial cells. *Sci. Rep.*, 10(1): 1–15. https://doi.org/101038/s41598-020-68626-7.

Obermaier, S. and Muller, M. (2020). Ibotenic acid biosynthesis in the fly agaric is initiated by glutamate hydroxylation. *Angewandte. Chemie. International. Edition.*, 59(30): 12432–12435. https://doi.org/101002/anie202001870.

Pappas, P.G., Kauffman, C.A. and Andes, D. (2009). Clinical practice guidelines for the management of candidiasis: 2009 update by the infectious diseases society of America. *Clin. Infect. Dis.*, 48: 503–535.

Pappas, P.G., Kauffman, C.A., Andes, D.R., Clancy, C.J., Marr, K.A., Ostrosky, Z.L., Reboli, A.C., Schuster, M.G., Vazquez, J.A., Walsh, T.J., Zaoutis, T.E. and Sobel, J.D. (2016). Executive summary: clinical. practice guideline for the management of candidiasis: 2016 Update by the infectious diseases society of America. *Clin. Infect. Dis. S.*, 62(4): 409–417. https://doi.org/101093/cid/civ1194.

Patil, A. and Majumdar, S. (2017). Echinocandins in antifungal pharmacotherapy. *J. Pharm. Pharmacol.*, 69(12): 1635–1660. https://doi.org/101111/jphp12780.

Perlin, D.S. (2007). Resistance to echinocandin-class antifungal drugs. *Drug. Resist. Update*, 10(3): 121–130. https://doi.org/101016/jdrup200704002.

Pettengell, K., Mynhard, J., Kluyts, T., Lau, W., Facklam, D. and Buell, D. (2004). Successful treatment of oesophageal candidiasis by micafungin: A novel systemic antifungal agent Aliment. *Pharmacol. Ther.*, 20: 475–481.

Pfaller, M.A. (2004). Anidulafungin: An echinocandin antifungal. *Expert Opin. Investig. Drugs*, 13(9): 1183–1197. https://doi.org/101517/135437841391183.

Pianalto, K.M., Billmyre, R.B., Telzrow, C.L. and Alspaugh, J.A. (2019). Roles for stress response and cell wall biosynthesis pathways in caspofungin tolerance in *Cryptococcus neoformans*. *Genetics*, 213(1): 213–227. https://doi.org/101534/genetics119302290.

Sandhu, P., Xu, X., Bondiskey, P.J., Balani, S.K., Morris, M.L., Tang, Y.S., Miller, A.R. and Pearson, P.G. (2004). Disposition of caspofungin, a novel antifungal agent, in mice, rats, rabbits, and monkeys. *Antimicrob. Agents. Chemother.*, 48(4): 1272–1280. https://doi.org/101128/aac4841272-12802004.

Saravolatz, L.D., Deresinski, S.C. and Stevens, D.A. (2003). Caspofungin. *Clin. Infect. Dis.*, 36(11): 1445–1457. https://doi.org/101086/375080.

Schmatz, D.M., Abruzzo, G., Powles, M.A., McFadden, D.C., Balkovec, J.M., Black, R.M. and Bartizal, K.D. (1992). Pneumocandins from *Zalerion arboricola* IV. Biological evaluation of natural and semisynthetic pneumocandins for activity against *Pneumocystis carinii* and *Candida* species. *J. Antibiot.*, 45(12): 1886–1891.

Science Direct Topics (2008). Micafungin: An overview. https://wwwsciencedirectcom/topics/chemistry/micafungin.

Seibel, N.L., Schwartz, C., Arrieta, A., Flynn, P., Shad, A., Albano, E. and Walsh, T.J. (2005). Safety, tolerability, and pharmacokinetics of micafungin (FK463) in febrile neutropenic paediatric patients. *Antimicrob. Agents Chemother.*, 49(8): 3317–3324.

Senn, L., Eggimann, P., Ksontini, R., Pascual, A., Demartines, N., Bille, J., Calandra, T. and Marchetti, O. (2009). Caspofungin for prevention of intra-abdominal candidiasis in high-risk surgical patients. *Intensive Care Med.*, 35(5): 903–908. https://doi.org/101007/s00134-009-1405-8.

Stone, E.A., Fung, H.B. and Kirschenbaum, H.L. (2002). Caspofungin: An echinocandin antifungal agent. *Clin. Ther.*, 24(3): 351–377. https://doi.org/101016/s0149-2918(02)85039-1.

Sun, P. and Tong, Z. (2014). Efficacy of caspofungin, a 1,3-β-D-glucan synthase inhibitor, on *Pneumocystis carinii* pneumonia in rats. *Med. Mycol. J.*, 52(8): 798–803. https://doi.org/101093/mmy/myu060.

Sunsan, L.A. and Jose, A.V. (2008). Anidulafungin: An evidence-based review of its use in invasive fungal infections. *Core. Evidence*, 2(4): 241–249.

Vazquez, J.A. and Sobel, J.D. (2006). Reviews of anti-infective agents: Anidulafungin: A novel echinocandin. *Clin. Infect. Dis.*, 43(2): 215–222. https://doi.org/10.1086/505204.

Viscoli, C., Herbrecht, R., Akan, H., Baila, L., Sonet, A., Gallamini, A., Giagounidis, A., Marchetti, O., Martino, R., Meert, L., Paesmans, M., Ameye, L., Shivaprakash, M., Ullmann, A.J. and Maertens, J. (2009). An EORTC Phase II study of caspofungin as first-line therapy of invasive aspergillosis in haematological patients. *J. Antimicrob. Chemother.*, 64(6): 1274–1281. https://doi.org/101093/jac/dkp355.

Walsh, T.J., Adamson, P.C., Seibel, N.L., Flynn, P.M., Neely, M.N., Schwartz, C., Shad, A., Kaplan, S., Roden, M.M., Stone, J.A., Miller, A., Bradshaw, S.K., Li, S.X., Sable, C.A. and Kartsonis, N.A. (2005). Pharmacokinetics, safety, and tolerability of caspofungin in children and adolescents. *Antimicrob. Agents Chemother.*, 49(11): 4536–4545. https://doi.org/10.1128/aac.49.11.4536-4545.2005.

Yao, J. (2012). Total synthesis and structure–activity relationships of caspofungin-like macrocyclic antifungal lipopeptides. *Tetrahedron*, 68(14): 3074–3085.

Yeoh, S.F., Lee, T.J., Chew, K.L., Lin, S., Yeo, D. and Setia, S. (2018). Echinocandins for management of invasive candidiasis in patients with liver disease and liver transplantation. *Infect. Drug. Resist.*, 11: 805–819. https://doi.org/10.2147/idr.s165676.

Yue, Q., Chen, L., Zhang, X., Li, K., Sun, J., Liu, X., An, Z. and Bills, G.F. (2015). Evolution of chemical diversity in echinocandin lipopeptide antifungal metabolites. *Eukaryot. Cell.*, 14(7): 698–718. https://doi.org/10.1128/ec.00076-15.

7

Metabolites of *Ganoderma* and their Applications in Medicine

Revanth Babu Pallam and *Vemuri V. Sarma**

1. Introduction

The genus *Ganoderma* are usually plant pathogenic with large, bracket-shaped fruiting bodies that mainly degrade wood logs and live as saprophytes throughout the world (Richter et al., 2015). They are known to cause root and stem rots on a wide range of monocots, dicots, and gymnosperms, which may lead to the death of affected trees. *Ganoderma* spp. have been used in traditional medicine for treatment of cancer, bronchitis, bronchial asthma, arterial hypertension, and chronic kidney diseases. *Ganoderma* powder is an important ingredient in medicinal formulations, tonics, and sedatives (Lee et al., 2005). Medicinal formulations are usually prepared from different parts of mycelia, fruiting bodies, and spores of *Ganoderma*. *G. lucidum* has drawn attention in the market as a nutraceutical and dietary supplement due to its anticancer, immunomodulatory, and antioxidant effects (Wasser et al., 2002; Wachtel- Galor et al., 2004). In addition, *G. applanatum*, *G. colossum*, *G. concinna*, *G. neo-japonicum*, *G. pfeifferi*, and *G. tsugae* are known for their medicinal potential (Kleinwächter et al., 2001; Gonzalez et al., 2002).

Ganoderma lucidum (Lingzhi or Reishi) has been known in China for more than 2,000 years due to its medicinal properties. Lingzhi has been described as *G. lucidum* in Europe (Cao et al., 2012). The bioactive compounds present in specific *Ganoderma* spp. are responsible for their medicinal properties. Terpenoids, alkaloids, polysaccharides, peptides, and sterols are present in *Ganoderma* spp. The synthesis of these biochemicals depends on temperature, substrate, humidity, incubation period. *Ganoderma lucidum* is usually cultivated on wooden logs, saw dust, bottles, or bags with different substrates (Chen, 1999). In this chapter, we emphasize cultivation aspects, types of fermentation, and medicinal properties of *Ganoderma* spp.

Department of Biotechnology, Pondicherry University, Kalapet, Pondicherry 605014, India.
* Corresponding author: sarmavv@yahoo.com

2. Cultivation

2.1 Cultivation in the Natural Environment

Ganoderma usually grows on trunks of trees, stumps, wood logs in the natural environment and requires humid conditions for growth. Researchers tried to cultivate *Ganoderma* in laboratory conditions, but their attempts have failed as it could not form a pileus. In nature, *Ganoderma* spp. form fruiting bodies but in *in vitro* conditions they could not form a cap. The Fungal Research Laboratory of the Institute of Microbiology, Chinese Academy of Sciences discovered in 1991 that a higher relative humidity (85–95%) was an important factor for cultivation of *Ganoderma* in a laboratory. Higher humidity is needed for formation of basidiocarps and release of spores from *Ganoderma*. Researchers performed the artificial large-scale cultivation of *Ganoderma* in 1992. Furthermore, the research institute concentrated on popularizing the artificial cultivation method free of cost which is the most important contribution to the rapid research on *Ganoderma* and its products (Galor et al., 2011). Traditional cultivation of *Ganoderma* involves inoculation of *Ganoderma* on a meter-long wood logs without any sterilization, followed by the burial of logs in a shallow trough. Fruiting bodies emerge from logs after 6–24 months, but their harvesting can be performed up to five years (Pegler, 2002). In the late 1980s, people started using wood pieces of short length, about 15 cm, for cultivating *Ganoderma* in China, Japan, and Korea (Chen, 2002).

There are several other methods that have been tried to grow *Ganoderma*, such as lime tree cultivation (Zhang et al., 2004), and sawdust cultivation. The artificial cultivation of *G. lucidum* was carried out with substrates, including cereals, sawdust, logs (Chang and Buswell, 1999; Wasser, 2005; Boh et al., 2007), tea waste (Peksen and Yakupoglu, 2009), cottonseed pods or crop residues (Zhang and Wang, 2010), cork residues (Riu et al., 1997), sunflower seed husks Matute et al., 2002), corn cobs (Ueitele et al., 2014), olive oil cake (Gregori and Pohleven, 2015) and wheat straw (Priya and Robinka, 2014). The cultivation of *Ganoderma* on natural logs results in the production of the highest quality fruit bodies and fetches the best prices in Southeast Asian markets. However, the sawdust culture in the laboratory needs a longer duration for fruit body production. Thus, the preservation of forest logs supports *Ganoderma* production, while the harvest of logs curtails its production (Tan et al., 2015).

Ganoderma is frequently found in subtropical zones compared to temperate zones. Growth parameters are very crucial for optimal growth, such as temperature, humidity, air, and light. The growth of mycelia at 15–35°C is suitable and optimum temperature is 25–30°C. The initial growth of *Ganoderma* requires 18–25°C and 24–28°C is required for the development of the fruiting body, which turns yellow and stops growing below 20°C. The most important factor in the cultivation of *Ganoderma* is humidity; 60–65% of moisture should be present in the substrate. Relative humidity for mycelial growth is about 60–67% and 85–90% is required for primordial growth. Different development stages of *Ganoderma*, based on the growth stage, require various concentrations of oxygen and carbon dioxide. Oxygen is not required for the mycelial growth of *Ganoderma*, but high concentrations of

Fig. 7.1. *Ganoderma* sp. on a tree bark.

oxygen are necessary for the growth of fruiting bodies. If the concentration of CO_2 is below 0.1%, the development of fruiting bodies will be abnormal. The normal development of fruiting bodies is observed when concentrations of CO_2 are above 0.1% (Lin, 1999; Chang and Miles, 2004). The pictures of *Ganoderma*, from the Pondicherry University campus, India, are presented in Fig. 7.1.

2.2 Types of Cultivation

The methods of *Ganoderma* cultivation are divided into two types: liquid-state cultivation (LSC) and solid-state cultivation (SSC). The SSC is further divided into two types based on the use of raw materials: log (basswood) and substituted cultivation. Initially, the cultivation was carried out with *G. capense*, *G. japonicum*, and *G. lucidum* (Zhou and Lin, 1999). The growth of *Ganoderma* requires carbon sources, nitrogen sources, inorganic salts, and growing factors. *Ganoderma* uses various organic sources, such as sugar, starch, cellulose, hemicelluloses, and lignin. Mycelial culture is usually obtained by glucose or sucrose-based media. The fruiting bodies are cultivated on agricultural byproducts, including cotton seed husks, straw, and corn cobs (Yan, 2000; Wei et al., 2007). *Ganoderma* mycelia also use small molecular weight compounds, such as amino acids, urea, and nitrogen. In addition, yeast powder and peptone are also added in the medium. Meanwhile, wheat bran, corn powder, coarse powders of rice bran, ammonium sulfate, and urea are added for cultivation of fruiting bodies. The ratio of carbon to nitrogen (C/N) 15–45:1 in the substrate is favorable for the cultivation of mycelia, and 30–40:1 is preferred for fruiting bodies (Lin and Zhou, 1999; Han et al., 2003; Wu et al., 2008).

Inorganic salts also play a crucial role in *Ganoderma* cultivation, including potassium, nitrogen, calcium, magnesium, phosphorus, sulfur, and zinc. Among them, nitrogen, phosphorus, and magnesium are the key elements. Even though these elements are present in some substrates, the concentration of 100–150 mg/l of inorganic salts, $CaSO_4$, KH_2PO_4, and $MgSO_4$ are added to the medium. Most importantly, the concentration of $CaSO_4$ is up to 1% of the total substrate. $CaSO_4$ plays an important role in adjusting the pH value of medium, changing the substrate porosity, increasing air flow, fixing nitrogen, and increasing the amounts of calcium and sulfur elements. The growth factors, Vitamins B1 and B6, which are required for the metabolism of *Ganoderma*, may not be added to the medium since most of the substrates possess them. Sterile water should be provided for cultivation, 65–70% of water content is required (Juan, 1999; Han et al., 2003). There are two popular methods for cultivation of *Ganoderma*: liquid state fermentation (LSF), and solid-state fermentation (SSF).

2.2.1 Liquid State Fermentation

Liquid-state fermentation (LSF) is also known as submerged fermentation, which is suitable for mycelial growth in liquid medium but not on the surface. The submerged fermentation for mycelial growth was developed in the 1970s, initially for lower fungi which do not have basidiocarps. This technique was adopted for economical production of various products (Yang and Liau, 1998). The mycelial growth in LSF has a higher stability. The physiochemical parameters, such as temperature, dissolved oxygen, and pH are easier to regulate. Maintaining these parameters plays a key role in producing the desired products and is beneficial for producing mushroom-based preparations. Those harvested by LSF are considered to have high quality standards and safety (Wasser and Weis, 1999). The production of fungal metabolites through this method has received attention due to its short cultivation time, high yield, less contamination, and recovery methods of metabolites (Kim et al., 2007; Huang and Liu, 2008).

2.2.2 Solid State Fermentation

Solid-state fermentation (SSF) has also been used for the cultivation of *Ganoderma*. SSF is cultivation of microorganisms under controlled conditions in the absence of free water. It facilitates the production of industrial enzymes (Wu et al., 2000), such as biofuels, biopesticides, and nutrient-enriched animal feeds (Habijanic and Berovic, 2000). *Ganoderma* in SSF could be cultivated in three types of containers: test tubes, 500 ml flasks, and polypropylene bags. In test tubes, C/N ratio was maintained at 80, whereas in test tubes, the mycelium growth was observed as 6 mm/day. In a 500 ml flask, C/N ratio was maintained at 70–80, and mycelial growth was observed as 7.5 mm/day (Hsieh and Yang, 2004). The tea waste, along with sawdust in different proportions (75S:25TW, 80S:20TW, 85S:15TW, and 90S:10TW), was used as a substrate for SSF of *G. lucidum*. The effect of substrate on yield, biological efficiency, and chemical composition of fruiting bodies was studied. The results of this study revealed significant differences in yield and biological efficacy among different proportions of substrate (Peksen and Yakupoglu, 2009).

In bidirectional SSF, the medicinal substrate not only provides nutrients for fungal growth but also affects the production of enzymes. Hence, novel compounds are expected when the herbs are used as substrates for fermentation (Zhuang et al., 2004). Since the invention of bidirectional SSF, *Ganoderma* has been the most preferred species for study by Chinese scientists. SSF was carried out with *G. lucidum* with a medium-containing *Radix astragali*, selenium-rich medium, and normal medium, respectively. The polysaccharide content was observed after the experiment. The results showed that 4.5%, 3.76%, and 4.65% of polysaccharides were recovered from the total fermented products, respectively. The secondary metabolism of *G. lucidum* was the most active on the 28th day which was considered a stop point for fermentation. Many combinations of *Ganoderma* spp. and other medicinal herbs, such as Lingzhi with *Astragalus membranaceus* (Zhu et al., 2010), *R. glycyrrhizae* (Zhu et al., 2009), and *R. astragali* have also been tested.

3. The Extraction and Purification of Bioactive Compounds

3.1 Polysaccharides

Fruiting bodies of *G. lucidum* were dissected into small pieces, then extracted with 95% ethanol for 24 hours to remove ethanol-soluble impurities. The crude extract was partially purified through a weak anion exchanger column (DEAE-cellulose 52 column). The gradient elution was carried out by 0.1 M NaCl to 1.0 M NaCl. The eluted fractions were further collected, concentrated and eluted through size exclusion chromatography column Sephadex G-100. The elution was carried out by 0.05 M NaCl solution. The eluted fractions were concentrated by lyophilization. UV-visible spectroscopy was carried out for fractions. No absorbance was observed at 260 and 280 nm indicating that there was no contamination of nucleic acids and proteins, respectively. HPLC was performed to analyze the molecular weights of fractions: the average molecular weights of two fractions were confirmed to be 1.926 and 1086 kDa. Infrared spectroscopy was carried out to find the types of bonds present in purified polysaccharides (Zhao et al., 2010).

Previous study reported the gradient ethanol extraction to derive crude polysaccharides from *G. lucidum*. The crude extracts were purified as GLP-1 and GLP-2 polysaccharides by Q-sepharose fast-flow column, a strong anion exchange column. Later, high performance gel permeation chromatography–multi-angle laser light scattering–refractive index (HPGPC-MALLS-RI), fourier transform–infrared spectroscopy (FT-IR), atomic force microscopy (AFM), methylation analysis and nuclear magnetic resonance (NMR) techniques were used to characterize GLP-1 and GLP-2 polysaccharides. The results have shown that GLP-1 was a unique hetero galacoglucan, whereas GLP-2 was a β-glucan (Li et al., 2020). In another study, *G. capense* was obtained from submerged fermentation and its mycelial powder was extracted with hot water to obtain a water-soluble polysaccharide (GCP50-1). It was purified by a weak anion exchanger diethylaminoethyl (DEAE) Sepharose CL-6B and a size exclusion chromatography column sephadex G-75. The average molecular weight of the polysaccharide confirmed by HPGPC was 1.5×10^4 Da. It was characterized by FT-IRGC-MS, NMR. The polysaccharide was identified as α-D-glucan (Li et al., 2013).

In another study, a polysaccharide crude extract was obtained by a hot water extraction from spores of *G. lucidum*. The crude polysaccharide was purified by anion exchange and size exclusion chromatography. The polysaccharide was characterized by methylation analysis, periodate oxidation Smith degradation and NMR spectroscopy. The polysaccharide was identified as β-D-(1→3)-glucan which contains branches at C-6 carbon and branching proportion was observed as 20% approximately (Bao et al., 2002). Polysaccharide crude extract was obtained by hot water extraction from fruiting bodies of *G. lucidum* in the study of Liu et al. (2010). The polysaccharide preparation (GLPP) was purified by DEAE-32 anion exchange chromatography and sepharcyl S-200 HR gel permeation chromatography columns. Two peaks were observed in gel permeation chromatography, GLP-1 and GLP-2. The molecular weights of GLP-1 and GLP-2 estimated by a size-exclusion HPLC chromatography instrument were 5.2 kDa and 15.4 kDa, respectively. GLP-1 was found to be a glucan, whereas GLP2 was composed of glucose, galactose, and mannose (Liu et al., 2010).

3.2 Sterols

Mendoza et al. (2015) characterized the compounds of *G. oerstedii* fruiting bodies and mycelium by spectroscopy techniques mainly based on NMR spectroscopy. Initially, they lyophilized and extracted the compounds with hexane, chloroform, and methanol. Each of the extract was purified by column chromatography. Five sterol compounds were purified from fruiting bodies: ergosta-7, 22-dien-3β-ol, ergosterol peroxide, ergosterol, cerevisterol, and ergosta-7, 22-dien-3-one; and sterols were purified from mycelium, ergosterol, and cerevisterol. The authors of another study have extracted *G. lucidum* sterols by an alcohol/salt aqueous two-phase system (ATPS) (Xu et al., 2021). The characterization of sterols was carried out by a high performance liquid chromatography–evaporative light scattering detector (HPLC-ELSD) analysis. The sterols were further identified as ergosterol and ergosterol peroxide.

The mycelia of *G. lucidum* and *G. sinense* were extracted by pressurized liquid extractions for crude sterols (Lv et al., 2012). The GC-MS analysis revealed that both fungal species contained ergosterol, a common component of the fungal cell membrane. The ergosterol content was higher in *G. lucidum* than in *G. sinense*. *G. australe* basidiomata were extracted with hexane, chloroform, and ethyl acetate (Zhao et al., 2005). These extracts (hexane, chloroform and ethyl acetate) were further fractionated by silica gel column chromatography. The fractions were pooled and analyzed by thin layer chromatography (TLC). A total of five sterols, 5α-ergost-7-en-3β-ol;5α-ergost-7, 22-dien-3β-ol;5, 8-epidioxi-5α, 8α-ergost-6, 22-dien-3β-ol, have been identified.

Ganoderma applanatum mycelium was collected from the fermentation medium and the mycelium extracted with MeOH extract (2.25 kg) was suspended in water and successively partitioned with *n*-hexane, dichloromethane, EtOAc, and BuOH (Lee et al., 2011). A novel sterol, a polyoxygenated ergostane-type sterol, 3β,5α,6β,8β,14α-pentahydroxy-(22*E*,24*R*)-ergost-22-en-7-one has been purified from the mycelium of *G. applanatum*. The four known sterols, 3β,5α,9α-trihydroxy-(22*E*,24*R*)-ergosta-7, 22-dien-6-one ergosterol peroxide, 6-dehydrocerevisterol, and cerevisterol were also purified. 3β, 5α, 9α-trihydroxy-(22*E*, 24R)-ergosta-7, 22-dien-6-one, and 6-dehydrocerevisterol were originally purified from *G. applanatum*. The structure of these sterols was confirmed by one-dimensional (1D) and two-dimensional (2D) NMR spectroscopy.

3.3 Meroterpenoids

The fruiting bodies of *G. lucidum* were extracted with ethanol and suspended in water followed by extraction with ethyl acetate to get the crude extract. The compounds were purified by silica gel column chromatography followed by Sephadex LH-20 size exclusion chromatography. The chiral HPLC was used to separate the individual (−)- and (+)-antipodes. The compounds were characterized by UV-VIS spectrophotometer, CD spectrum, and the structure was confirmed by H-NMR.

A total of six new terepenoid chizhines, A–F, were identified through this study (Luo et al., 2015). In another study, the fruiting bodies of *G. leucocontextum* were initially extracted with ethanol, and later the extract was partitioned between ethyl acetate and water (Wang et al., 2017). The ethyl acetate soluble fraction was eluted in a silica gel column followed by Sephadex LH-20 column chromatography. Preparative HPLC was performed to purify the compounds. The characterization of compounds was performed by UV-VIS, IR, FT-IR, and NMR. Three novel meroterpenoids, Ganoleucin A–C, were identified together with five known meroterpenoids: ganomycin I, ganomycin B, fornicin C, fornicin B, and ganomycin C (Wang et al., 2019).

The dried fruiting bodies *G. sinensis* were extracted using 70% ethanol followed by the suspension of extract in water. The extraction was further carried out by ethyl acetate. MCI gel CHP 20P column chromatography was used for partial purification of the crude extract. Then column chromatography was carried out through Sephadex LH-20. Further purification was carried out by semi-preparative HPLC. Previous study identified six pairs of novel meroterpenoid enantiomers, Zizhines A–F (Cao et al., 2016).

3.4 Alkaloids

Ganderma sinense fruiting bodies were extracted with 95% ethanol followed by filtration (Liu et al., 2011). The extract was further suspended in petroleum, ethyl acetate, and n-butanol, respectively. The extracted ethyl acetate was initially eluted through a silica gel, RP-18 column, followed by the Sephadex LH-20 column. The samples were further analyzed by a high-resolution electron spray ionization mass spectroscopy (HRESIMS) and one and two-dimensional nuclear magnetic resonance spectroscopy, X-ray crystallography. Four new alkaloids, sinensines B–E, together with known alkaloid sinensine were isolated from the fruiting bodies of *G. sinense* (Liu et al., 2011). *G. luteomarginatum* fruiting bodies were dried and extracted with 95% of ethanol, suspended in water and partitioned with ethyl acetate to obtain an ethyl acetate soluble extract (Li et al., 2019). The extract was eluted through silica gel column, and fractions were further eluted by Sephadex LH-20. The thin layer chromatography was carried out followed by a reverse phase semi-preparative HPLC. The planar structure of alkaloids was confirmed by extensive spectroscopic analysis. The absolute configuration was obtained by calculation and experimental electronic circular dichorism (ECD). These alkaloids possess a phenyl-substituted 6, 7-dihydro-5H-cyclopenta [c]-pyridine structure which has only been described in the genus *Ganoderma*. (+) - and (–) - 1 were new alkaloids obtained from *Ganoderma*, while (+) - and (–) - 2 were originally isolated from *G. luteomarginatum* (Li et al., 2019).

The fruiting bodies of *G. lucidum* were dried, powdered, and macerated with methanol. The crude methanol extract was separated by a solvent partition between water and ethyl acetate. The extract was eluted through a medium pressure liquid chromatography (MPLC) column. Further purification was carried out in Sephadex-LH20. The preparative HPLC (Prep-HPLC) was further performed. Four new polycyclic alkaloids, lucidimine A–D, were identified from the fruiting bodies of

G. lucidum. The structure of alkaloids was confirmed by 1D, 2D NMR, and high-resolution electrospray ionization mass spectroscopy (HRESIMS) (Zhao et al., 2015). The fruiting bodies of *G. australe* were soaked in 95% ethanol to obtain a crude extract which was further suspended in water followed by extraction with ethyl acetate (Zhang et al., 2019). The extract was eluted through MCI gel CHP 20P column, and purification was carried out with the Sephadex LH-20 column followed by silica gel column chromatography. Further purification was carried out by semi-preparative HPLC column. A new alkaloid, australine, was identified. The absolute configuration of compounds and structures was revealed by spectroscopic techniques and electronic circular dichorism (ECD) calculations (Zhang et al., 2019).

The fruiting bodies of *G. calidophilum* were dried, powdered, and extracted with ethyl acetate (Huang et al., 2016). The crude extract was eluted in silica gel column chromatography, followed by Sephadex-LH20 column. The semi-preparative HPLC was carried out for further purification. Two new alkaloids, ganocalicine A and B were purified. The structural characterization was obtained from spectroscopic studies, as well as one and two-dimensional (2D) NMR and mass spectroscopy (MS). (Huang et al., 2016).

3.5 Terpenoids

The fruiting bodies of *G. lingzhi* were extracted with ethanol and suspended in water followed by extraction with chloroform (Amen et al., 2016). Initially, the crude extract was eluted through a silica gel column followed by elution through the RP-18 silica gel column. Further purification was carried out in Sephadex LH-20 size-exclusion chromatography. The structure of purified compounds was confirmed by NMR, HR-ESI-MS, and liquid chromatography coupled to ion trap time-of-flight *mass spectrometry* (LC-MS-IT-TOF). The new alkaloid was identified as lucidumol C (Amen et al., 2016). *Ganoderma sinense* fruiting bodies (FB) were dried, powdered, and extracted with methanol (Wang et al., 2010). The methanol extract was later partitioned between water and ethyl acetate. The extract was purified through silica gel and RP-gel columns followed by the Sephadex LH-20 column. Further purification was carried out by HPLC. The structure of the three new tritepenoids: methyl ganosinensate A (1), ganosinensic acid A (1a), and ganosinensic acid B (2) was studied by 1D, 2D NMR, and X-ray crystallography. The isolated triterpenoids were found to have an unusual four-membered ring skeleton which was by a bond formation between C-1 and C-11 carbons (Wang et al., 2010).

The air-dried fruiting bodies of *G. theacolum* were extracted with ethanol (Liu et al., 2017). The crude extract was eluted through icroporous resin chromatography followed by a silica gel column chromatography. The triterpenoid structure was confirmed by a high-resolution electron spray ionization mass spectroscopy (HRESIMS), nuclear magnetic resonance spectroscopy (NMR) and X-ray crystallography. Three new triterepenoids, ganoderic acid XL_3, ganoderic acid XL_4, and ganoderic acid XL_5, were isolated from fruiting bodies of *G. theacolum* (Liu et al., 2017). *G. leucocontextum* fruiting bodies were air-dried, powdered, and extracted with ethanol (Zhang et al., 2018). The ethanol extract was further partitioned between ethyl acetate and water to obtain crude extract. The crude extract was eluted through silica gel column followed by Sephadex LH-20 column. Further purification was performed by HPLC. A total of eight new triterepnoids, ganoleucoins T–Z, were isolated from fruiting bodies of *G. leucocontextum*. The structure of these triterpenoids was revealed by CD Spectra, NMR, and HRESIMS.

The fruiting bodies of *G. sinense* were dried and cut into small pieces extracted with chloroform (Sato et al., 2009). The crude extract was eluted through silica gel column. Further purification was carried out by HPLC. Five new oxygenated lanostane type triterpenoids, ganoderic acid GS-1, ganoderic acid GS-2, ganoderic acid GS-3, 20-dehydrolucidenic acid N, and 20-hydroxylucidenic acid A were isolated from fruiting bodies of *G. sinense*. The structure of triterpenoids was confirmed by HRESIMS and NMR.

4. Medicinal Properties

Ganoderma spp. possess a variety of medicinal properties, such as immunomodulatory, hepatoprotective, antitumor, anti-diabetic, anti-obesity, neuroprotective, antiviral, and antimicrobial properties (Table 7.1).

4.1 Immunomodulatory Activity

The polysaccharide water extract obtained from *G. sinense* (GS) was evaluated for immunomodulatory activity in human peripheral mononuclear cells (PBMC). The water extract of GS could enhance the proliferation of PBMC. Immunomodulation was augmented by increasing the production of tumor necrosis factor-α (TNF-α), interleukin (IL)-10, transforming the tumour growth factor–β (TGF-β), and the generation of $CD14^+$ monocyte subpopulation within the PBMC was consecutively increased. The production of IL-10 and IL-12 in monocyte-derived dendritic cells were increased. The study originally demonstrated the immunostimulatory effect of the fraction enriched in GS-Stipe polysaccharide on PBMCs and dendritic cells (Yue et al., 2013).

A polysaccharide derived from *G. lucidum*, D-galactoglucan has shown immunomodulatory effects by providing better protection to the spleen and thymus and has also increased the levels of immunoglobulin A (IgA) in the serum. The immunomodulatory effects highly corresponded to the structural features of D-galactoglucan, such as molecular weight, composition, and advanced conformation (Li et al., 2020). A water-soluble polysaccharide GLP-3 was isolated from *G. leucocontextum*. It was found that GLP-3 binds to toll-like receptor 2 (TLR-2) and thereby exerts immunomodulatory activity via mitogen-activated protein kinases (MAPKs), phosphatidylinositol-3-kinase (PI3K)/Akt, and nuclear factor-κB (NF-κB) signaling pathways (Gao et al., 2020). A proteoglycan namely Fudan Yueyang *Ganoderma lucidum* (FYGL) was isolated from fruiting bodies of *G. lucidum*. The immunoregulative mechanism of FYGL was studied in macrophages. The study has shown that by activating NF-κB and MAPK signaling pathways, FYGL exerts its immunomodulatory effect (Teng et al., 2020). A polysaccharide of *G. atrum*

Table 7.1. Medicinal properties of *Ganoderma* spp.

	Compounds	Medicinal Effects	References
G. ahmadii	Meroterepenoids	Antidiabetic	Guo et al., 2020
G. annulare	Applanoxidic acids A, C, and F	Antifungal	Smania et al., 2003
G. applanatum	Terepenes	Hepatoprotective	Ma et al., 2011
G. applanatum	Polysaccharide	Antitumor	Sun et al., 2015
G. applanatum	Polysaccharide	Antitumor	Hanyu et al., 2020
G. applanatum	Crude extract	Antitumor	Elkhateeb et al., 2018
G. applanatum	(±)-Applandimeric acid D	Anti-obesity	Peng et al., 2021
G. applanatum	Ethanol and water extracts	Hypouricemic	Yong et al., 2018
G. applanatum	Ganoapplanin, Sphaeropsidin D, and Cytosporone C	Neuroprotective	Hossen et al., 2021
G. applanatum	Ethanol, methanol, and water extracts	Antibacterial and antifungal	Jonathan and Awotona, 2010
G. applanatum	Triterpenes	Antidiabetic	Chen et al., 2019
G. atrum	Ethanol extract	Antibacterial	Li et al., 2012
G. atrum	Polysaccharide	Immunomodulatory	Zhang et al., 2013
G. australe	Astralic acid and methyl austrulate	Antibacterial	Smania et al., 2007
G. australe	Ethanol, methanol, and water extracts	Antibacterial and antifungal	Jonathan and Awotona, 2010
G. cochlear	Fornactins (A, D, F) and Fredelin	Hepatoprotective	Peng et al., 2014
G. leucocontextum	Triterpenes	Antidiabetic	Chen et al., 2019
G. leucocontextum	Ganomycin I	Antidiabetic	Wang et al., 2017
G. leucocontextum	Polysaccharide	Immunomodulatory	Gao et al., 2020
G. linghzi	Ganoderic acid TQ and TR	Antiviral	Zhu et al., 2015
G. lucidium	Ethanol, methanol, and water extracts	Antibacterial and antifungal	Jonathan and Awotona, 2010
G. lucidum	Triterepenoids	Hepatoprotective	Li et al., 2019
G. lucidum	Triterepenoids	Hepatoprotective	Wu et al., 2016
G. lucidum	Terepenes	Antitumor	Liu et al., 2020
G. lucidum	Ganoderic acid H	Antitumor	Qu et al., 2017
G. lucidum	Flavonoids, terepenoids, alkaloids, phenolics	Antitumor	Lai et al., 2010
G. lucidum	Ganodermanontriol, Lucidumol A, Ganoderic acid C2 and Ganosporeric acid A	Antiviral	Bharadwaj et al., 2019
G. lucidum	Lanosta-7,9(11),24-trien-3-one,15;26-dihydroxy and Ganoderic acid Y	Antiviral	Zhang et al., 2014

Table 7.1 contd. ...

...Table 7.1 contd.

	Compounds	Medicinal Effects	References
G. lucidum	Ganoderat acid-B*	Antiviral	Kang et al., 2015
G. lucidum	Plasmepsin I*	Antiviral	Kang et al., 2015
G. lucidum	Hesperetin and Ganocin B*	Antiviral	Lim et al., 2020
G. lucidum	Polysaccharides	Antidiabetic	Agius, 2007
G. lucidum	Polysaccharides + proteins	Antidiabetic	Xiao et al., 2012
G. lucidum	Low molecular weight polysaccharide	Antidiabetic	Oliver-Krasinski et al., 2009
G. lucidum	Polysaccharide	Antidiabetic	Zheng et al., 2012
G. lucidum	Polysaccharides	Antidiabetic	Liu et al., 2019
G. lucidum	Triterpenes	Antidiabetic	Chen et al., 2019
G. lucidum	Ganoderlactone B, Ganoderlactone D, Ganoderlactone E, Ganodernoid A and 11 β-hydroxy-3,7-dioxo-5 α-lanosta8,24(E)-dien-26-oic acid	Antidiabetic	Zhao et al., 2015
G. lucidum	Ganoderic acid C, Methyl ganoderate K, lucidenic acid H, 23-Dihydroganoderic acid N, Methyl lucidenate F, Ganoderic acid B, Lucidenic acid E	Antidiabetic	Chen et al., 2017
G. lucidum	Proteoglycans	Antidiabetic	Teng et al., 2011
G. lucidum	High molecular weight polysaccharide	Anti-obesity	Chang et al., 2015
G. lucidum	β-D-glucan	Anti-obesity	Sang et al., 2021
G. lucidum	Crude powder of *Ganoerma*	Anti-obesity	Lee et al., 2020
G. lucidum	*Ergosterol peroxide*	Anti-obesity	Jeong and Park, 2020
G. lucidum	Hexane, dichloromethane, ethyl acetate and methanol extracts	Antibacterial	Kumara and Bhatt, 2012
G. lucidum	Ganodermin	Antifungal	Wang and Ng, 2006
G. lucidum	Aqueous and methanol extracts	Antibacterial and antifungal	Sridhar et al., 2011
G. lucidum	Chloroform extract	Antibacterial	Keypour et al., 2008
G. lucidum	Triterpenoids	Anti-Alzheimer's	Yu et al., 2020
G. lucidum	Ganoderic acid and Lucidone A	Anti-Alzheimer's	Lai et al., 2019
G. lucidum	Triterepenoids	Cytoprotective	Wang et al., 2019
G. lucidum	Ganodermanondiol	Anti-melanogenetic	Kim et al., 2016.
G. lucidum	Terepenoids	Anti-protozoal	Oluba, 2019
G. lucidum	Ganoderol B	Anti-androgenic	Liu et al., 2007
G. lucidum	Ethyl ganoderate C2	Nephroprotective	Su et al., 2017
G. lucidum	D-galactoglucan	Immunomodulatory	Li et al., 2020

Table 7.1 contd. ...

...*Table 7.1 contd.*

	Compounds	Medicinal Effects	References
G. lucidum	Proteoglycon	Immunomodulatory	Teng et al., 2020
G. mediosinense	Methanol extract	Anti-amnesic	Kaur et al., 2017
G. neo-japonicum imazeki	Polysaccharide	Antiviral	Ang et al., 2021
G. pfeifferi	Ganomycin A and B	Antibacterial	Mothana et al., 2000
G. ramosissimum	Methanol extract	Anti-amnesic	Kaur et al., 2017
G. resinaceum	Ganodrol B	Hepatoprotective	Peng et al., 2013
G. resinaceum	Phytosterol α-Spinasterol	Antitumour	Sedky et al., 2018
G. resinaceum	Resinacein S	Anti-obesity	Huang et al., 2020
G. sinense	Triterpenes	Antidiabetic	Chen et al., 2019
G. sinense	Cyathisterol	Antitumour	Zheng et al., 2018
G. sinense	Polysaccharide	Antitumour	Wu et al., 2018
G. sinense	Polysaccharide	Immunomodulatory	Yue et al., 2013
G. tsugae	Triterpenes	Antidiabetic	Chen et al., 2019

Note: (* Computational studies).

(PSG-1) was found to activate macrophages by toll-like receptor 4 (TLR4) dependent signaling pathway, leading to the strengthening of immunity (Zhang et al., 2013).

4.2 *Hepatoprotective Effect*

Triterpenoids are major constituents of these fruiting bodies. These compounds possess cytotoxic, antibacterial, antiviral, antitumor, and anti-osteoplastic effects. Among 14 isolated compounds, Ganoderesin B, Ganodrol B, and Lucidone A have hepatoprotective effects. It has been hypothesized that this effect was achieved by lowering the alanine amino transferase (ALT) and aspertate amino transferase (AST) levels. CYP3A4 is a predominant enzyme in Cyt P450 involved in detoxification of xenobiotics and lipophilic compounds which enhance the expression of CYP3A4 by binding to its transcription factor PXR. Ganodrol B has been shown to increase the levels of CYP3A4 by binding to PXR which resulted in a decrease in ALT and AST levels (Peng et al., 2013).

The protective effect of triterpenoids of fruiting bodies of *G. lucidum* on cadmium-induced liver damage has been reported (Li et al., 2019). The study revealed that the levels of TNF-α, IL-1β, and IL-6 were higher in chickens fed with Cadmium compared to those fed with Cd (Cadmium) + GT (Ganoderma triterepenoids). These results have shown that *Ganoderma* triterepenoids protected the liver from cadmium-induced damage. Malondidealdehyde (MDA) content was significantly higher in Cd group of chickens and lower in Cd+GT chickens. The levels of free radical-scavenging enzymes, superoxide dismutase (SOD), and glutathione peroxidase (GSH-Px) were higher in Cd+GT chickens. The results have shown that triterpenoids were responsible for the expression of free radical-scavenging enzymes in Cd+GT chickens (Li et al., 2019).

The cytoprotective effects of GTs isolated from fruiting bodies of *G. lucidum* were tested in HepG2 cell lines (Wu et al., 2016). The cell lines were incubated with t-BHP, an organic hydroperoxidant which is metabolized into free radical intermediates, causing lipid peroxidation and oxidative damage. It leads to the leakage of aminotransferase and lactate dehydrogenase enzymes, resulting in the formation of malondialdehyde (MDA). The efficacy of GTs was checked by incubating HepG2 cells with and without GTs for 4 hours. The cells were further incubated with t-BHP (60 μMol/L) for four hours. The cells pre-incubated with 50, 100, and 200 μg/μl of GTs concentration have shown increased viability. This experiment revealed that GTs could protect cells from oxidative damage. The treatment of cells with t-BHP resulted in a decrease in the levels of SOD, GSH and increase in the levels of MDA (Wu et al., 2016).

Terpenes of *G. applanatum* (GAT) have been used in several medical formulations. Previous study evaluated the hepatoprotective effect of GAT against benzo(a) pyrene (BaP)-induced oxidative stress and inflammation in a mouse liver, as well as its mechanism of action. The results have shown that GAT decreased the levels of ALT and AST enzymes in serum and liver in BaP-treated mice. GAT could inhibit the inflammation by suppressing the expression of IL-1β and COX-2, thereby inhibiting the translocation of NF-κB in the liver of BaP-treated mice. These results suggest that GAT may protect the mouse liver from BaP-induced damage by improving liver function, mitigating histopathological changes, lowering ROS and MDA levels, renewing activities of antioxidant enzymes, and suppressing the inflammatory response (Ma et al., 2011). Two novel trinorlanstanes, cohaltes A and B, were found in *G. cochlear*. These compounds possess 3, 4–seco-9, 10, seco-9-19 cyclo skeleton. Six new triterpenoids, fornicatins D–F and ganodercochlearins A–C, have also been identified. The fornactins (A, D, F) and fredelin lowered ALT and AST levels in HepG2- and H202-treated cells. This study proved that triterepenoids of *G. cochlear* had hepatoprotective effects by decreasing levels of ALT and AST enzymes (Peng et al., 2014).

4.3 Antitumour Activity

The tumor-bearing BALB/c nude mouse model was used to evaluate the *G. lucidum* triterpenoids (GLTs) capacity as an enhancer of tumor-suppressive quality of Gefitinib (GEF) in A549 cell lines (Liu et al., 2020). The study revealed that GLTs suppressed tumor angiogenesis. The microvessel density (MVD) of the tumour was decreased when the cells were separately treated with GLTs and GEF. The decrease in MVD was particularly significant when both GLTs and GEFs were used. The expression level of VEGFR2 protein was decreased when the cells were treated with both. In the control saline group, the level of endostatin and angiostatin expression were higher. GLTs inhibited the growth of lung cancer in nude mice (Liu et al., 2020). *G. lucidum* triterpenes have been reported to inhibit the metastasis of several cancer cells. DU-145 prostate cancer cell lines were selected to check GLT ability to inhibit metastasis. The study has shown that GLT inhibited prostate cancer cells in a dose-dependent manner; a high concentration of 2 mg/ml is effective against

DU-145 cell lines. The GLT suppressed the invasion of DU-145 cancer cells via inhibiting matrix metalloproteinases (Qu et al., 2017).

The steroids (22E, 24R)-Ergosta-7, 9(11), 22-triene-3β, 5α, 6β-triol (compound 1), ergosta-4, 6, 8(14), 22-tetraen- 3-one (compound 2), cyathisterol (compound 3), were purified from fruiting bodies of *G. sinense* and evaluated for their inhibiting activity of Isocitrate dehyrogenase (IDH) (Zheng et al., 2018). The cyathisterol had the highest binding affinity with IDH1 and significant binding-free energy. Enzyme kinetics revealed that cyathisterol inhibits IDH1 in a non-competitive inhibitory manner. Further knockdown studies of IDH1 in HT1080 have shown a decreased anti-proliferative sensitivity to cyathisterol (Zheng et al., 2018).

The *G. lucidum* polysaccharides (GLPs) have been known to elicit anti-oncogenic and pro-apoptotic activities. Non-steroidal anti-inflammatory drug-activated gene-1 (NAG-1) plays a key role in pro-apoptotic mechanism (Wu et al., 2018). Flow cytometry studies revealed that GLP mediated late apoptosis in PCa cells accompanied by poly ADP-ribose polymerase (PRAP) cleavage and inhibition of pro-capsase 3, 6, and 9 expressions. Luciferase assay revealed that GLP induces the promoter of NAG-1. Inhibition of NAG-1 with a small interfering RNA significantly and partially inhibited GLP-induced apoptosis. The expression of PARP and pro-capsases reversed the effects of GLP. Furthermore, GLP inhibited the phosphorylation of protein-kinase B and mitogen-activated protein kinase/ extracellular signal-regulated kinase signaling in PCa cells (Wu et al., 2018).

The polysaccharides of *G. applanatum* were isolated from fruiting bodies (FGAP) and submerged fermentation (SGAP) (Sun et al., 2015). *In vivo* studies were performed to check anti-tumor activities of FGAP and SGAP on sarcoma 180 cell lines. FGAP showed a more significant anti-tumor activity compared to SGAP since it contained more carboxylating groups. Further studies were carried out to introduce carboxyl groups to SGAP by carboxymethylation to study the effect of carboxyl groups on antitumor activity. The carboxylated SGAP (CSGAP) have shown a more prominent anti-tumor activity compared to SGAP. The study has shown that carboxyl groups had a key role in antitumor activity of *G. applanatum* polysaccharides (Sun et al., 2015). Polysaccharides derived from *G. applanatum* (GAP) were purified and evaluated for their antitumor activity against human breast cancer cell lines MCF-7, and the molecular mechanism has also been elucidated (Hanyu et al., 2020). The results have shown that GAP inhibited the proliferation and migration of MCF-7 cancer cell lines in a dose and time-dependent manner. The maximal inhibition rate was observed as 50.2% at 500 μg/mL in 48 hours. These results suggest that GAP may regulate MAPK/ERK pathway, thereby inducing the early autophagy. Thus, it was concluded that GAP possessed an anti-tumor effect by regulating apoptosis and autophagy in MCF-7 cell lines through the MAPK/ERk signaling pathway (Hanyu et al., 2020).

Cervical cancer is caused by human papilloma virus (HPV). The oncoprotein expressed by E6 region of HPV genome is responsible for cervical cancer pathogenesis. The crude extracts of *G. lucidum* were assessed for their inhibitor activity on the expression of E6 HPV16 oncoprotein (Lai et al., 2010). Cervical cancer cell lines, CaSki, were used in this study which revealed that crude extracts suppressed the expression of E6 oncoprotein. The flavonoids, terpenoids, phenolics, and alkaloids were present in a crude extract (Lai et al., 2010). The ethanol and methanol extracts of *G. applanatum* were obtained and compounds from them were assessed for their cytotoxic activity against human colon cancer cell line (Caco-2). The ethanol extract showed cytotoxic activity with IC_{50} value of 160 ± 4.08 μg/ml, whereas the methanol extract revealed different effects. The levels of glutathione increased in Caco-2 cells. *G. applanatum* extracts significantly increased the Bax/Bcl-2 ratio on apoptotic Caco-2 cells, through the p53 independent pathway, and upregulation of Cas-3 expression was observed. An *in vivo* study was conducted on Solid Ehrlich tumor (SEC) Cells. The tumor mass developed after 5 days of treatment with extracts. The p53-dependent apoptotic pathway was confirmed by an increased Bax/Bcl-2 ratio and upregulation of p53 and Cas-3. Thus, *G. applanatum* extracts may exert antitumor effects by p53-independent pathway in Caco-2 cells, and p53-dependent pathway in SEC cells (Elkhateeb et al., 2018).

Previous study was conducted with a phytosterol α-spinasterol compound isolated from mycelium of Egyptian *G. resinaceum*. The antitumor activity of the compound was evaluated on human breast cancer cell lines, MCF-7, MDA-MB-231, and ovarian cancer cell lines, SKOV-3. The findings of this study have shown that α-spinasterol possessed the highest antitumor activity against MCF-7 compared to SKOV-3 cell lines and showed the lowest antitumor effect on MDA-MB-231 cell lines. The upregulation of p53 and Bax was observed in α-spinasterol treated cell lines, whereas cdk 4/6 were downregulated. The analysis revealed that α-spinasterol inhibited the cell cycle at G0-G1. Thus, α-spinasterol was suggested as a potential therapeutic agent for the treatment of breast and ovarian cancer (Sedky et al., 2018).

4.4 Antidiabetic Activity

Polysaccharides derived from *G. lucidum* have shown to enhance the activity of hepatic enzymes glucokinase, glucose-6-phosphate dehydrogenase, and phosphofructokinase. These polysaccharides decrease the levels of hepatic glucose by inhibiting glycogen synthetase activity, thereby preventing the development of hyperglycemia (Agius, 2007). *G. lucidum* polysaccharides also play a role in decreasing the levels of mRNA of glycogenphosphorylase, fructose-1,6-bisphosphatase, phosphoenolpyruvate carboxykinase, and glucose-6-phosphatase involved in glycogenolysis and/or gluconeogenesis (Xiao et al., 2012). The cardiovascular complications in diabetic patients develop due to apoptosis of endothelial cells (He et al., 2005) which is induced by hyperglycemia in diabetic mice (Recchioni et al., 2002). Low molecular weight polysaccharides of *G. lucidum* may exert hypoglycemic activity, inhibit pancreatic cell death, upregulate Bcl-2 (antiapoptotic protein), and PDX1 (pancreatic and duodenal homeobox 1), thereby promoting B-cell regeneration (Oliver-Krasinski et al., 2009). Furthermore, the polysaccharides were also involved in downregulation of nitric oxide synthases and capsae-3 in STZ-induced diabetic rats (Streptozotocin) (Zheng et al., 2012). The polysaccharides of *G. lucidum* were also involved in the inhibition of NF-kB, thereby protecting the alloxan-induced damage in pancreatic cells (Zhang et al., 2003). The wound-healing ability of *G. lucidum* polysaccharides has been investigated in STZ-induced diabetic mice (Tie et al., 2012). They were involved in the healing of 21% of wounds in diabetic mice. Increased angiogenesis was achieved by inhibition and nitration of manganese superoxide dismutase (MnSOD), suppression of glutathione peroxidase activity, decrease in redox enzyme p66Shc expression and phosphorylation. The mechanism of antidiabetic effect of *G. lucidum* polysaccharides is illustrated in Fig. 7.2.

The polysaccharides (GLPs) extracted from fruiting bodies of *G. lucidum* were evaluated for their hypoglycemic activity. They were administered into mice for 7 days and the results were analyzed. Both low and high-dose of GLPs reduced fasting serum glucose, and insulin levels, as well as body weight and epididymal white adipose tissue. The hepatic mRNA levels of glycogen phosphorylase (GP), fructose-1, 6-bisphosphatase (FBPase), phosphoenolpyruvate carboxykinase (PEPCK), and glucose-6-phosphatase (G6Pase) genes were significantly lowered

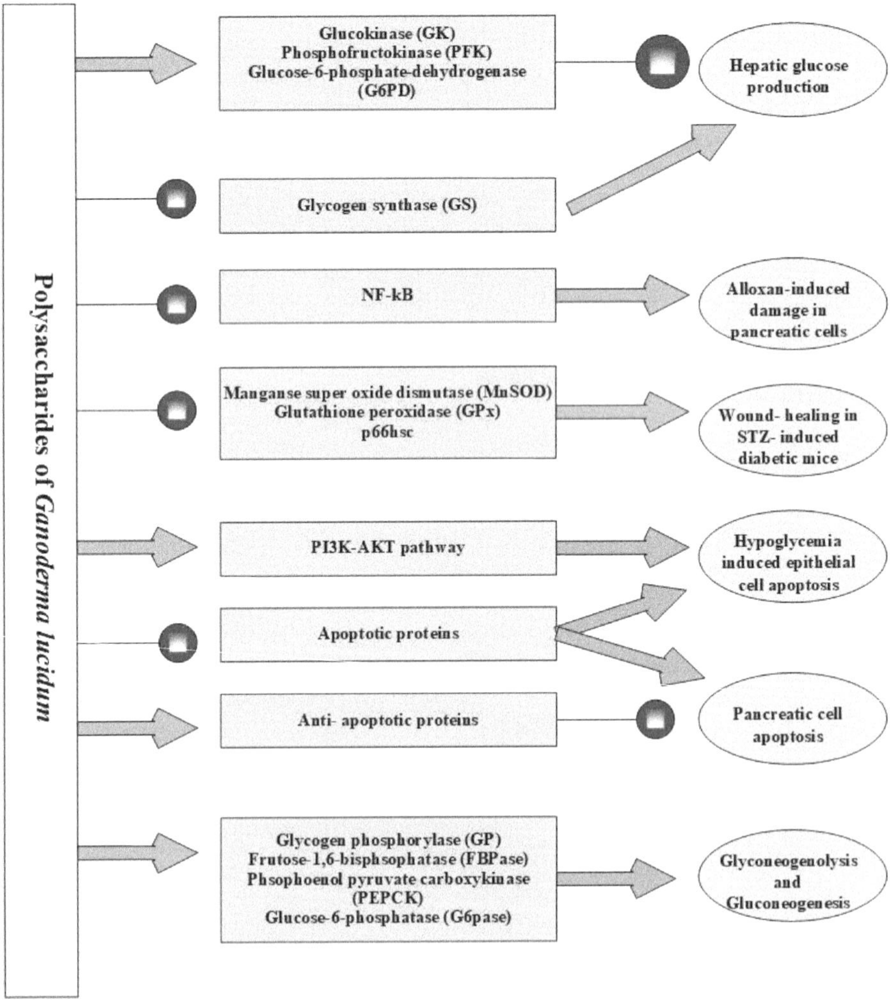

Fig. 7.2. The mechanism of antidiabetic effect of *G. lucidum* polysaccharides.

Note: (⇒: upregulation or activation; —●: downregulation or inactivation) (adopted from Ma et al., 2015).

when compared to the control group by both GLPs which was confirmed by RT-PCR. The cause of low serum glucose levels may be the decreased expression of several enzymes involved in gluconeogenesis and/or glycogenolysis (Xiao et al., 2012). The combined hypoglycemic activities of commercially available polysaccharides (70%) of *G. lucidum* and inulin were studied in type 2 diabetes mellitus (DM) rats. This disease was experimentally induced by high fat diet and injecting Streptozotocin for 5 weeks. The hypoglycemic effect was achieved by regulating proteins involved in the PI3K/AKT signaling pathway. Inulin and polysaccharides improved the glucose and lipid metabolism in diabetic rats. The hypoglycemic effect was due to an increase in insulin sensitivity, glycogen synthesis, and glucose transport through the PI3K/AKT signaling pathway (Liu et al., 2019).

The 7 Ganomycin I of *G. leucocontextum* was studied for its inhibition of HMG-COA reductase and α-glucosidase. The ganomycin I was the most potent HMG-COA reductase and α-glucosidase inhibitor. The obtained results indicated that ganomycin I (4) exerted potent hypoglycemic, hypolipidemic, and insulin-sensitizing effects in KK-A^y mice (Wang et al., 2017). A comparative study was carried out between triterpenes and polysaccharides of five different species of Ganoderma (*G. applanatum*, *G. leucocontextum*, *G. lucidum*, *G. sinense*, and *G. tsugae*) to investigate their anti-α-glucosidase and anti-α-amylase activities. The triterpenes of *G. lucidum* have shown maximum anti-α-glucosidase and anti-α-amylase effects with IC_{50} values of 10.02 ± 0.95 µg/mL and 31.82 ± 4.30 µg/mL, respectively. The quantity of triterpenes positively correlated with anti-α-glucosidase and anti-α-amylase activities. The study suggested that triterpenes of *G. lucidum* were potential antidiabetic agents (Chen et al., 2019). The highly oxygenated Lanostane derivatives obtained from *G. lucidum* were evaluated for their α-glucosidase inhibitory activity. Overall, five compounds, ganoderlactone B, ganoderlactone D, ganoderlactone E, ganodernoid A, and 11 β-hydroxy-3,7-dioxo-5 α-lanosta8,24(E)-dien-26-oic acid showed significant α-glucosidase inhibitory activity. The structural analysis revealed that these compounds consist of a pentatomic 20(24)-γ-lactone ring. The study of new pentatomic 20(24)-γ-lactone based drugs for the treatment of diabetes may be promising (Zhao et al., 2015).

Protein tyrosine phosphatase 1B (PTP1B) is an enzyme belonging to the protein tyrosinase family. Recent studies have shown that type 2 diabetes is regulated by PTP1B. Four meroterepenoids have shown a significant inhibitory activity against PTP1B. The structural analysis revealed their molecular formulas, $C_{30}H_{32}O_9$ (Compound 1), $C_{30}H_{34}O_{10}$ (Compound 2), $C_{32}H_{36}O_{10}$ (Compound 3), and $C_{32}H_{36}O_{10}$ (Compound 4). These compounds exert their inhibitory activity on PTP1B with IC50 values of 17, 20, 19, and 23 µM, respectively. The compound 1 formed H-bonds with ALA-217 and GLN-266 and 7″-OH bond was formed with ARG-24 which was present in the secondary binding site of PTP1B (Guo et al., 2020). Three new meroterpenoids were isolated from the fruiting bodies of *G. ahmadii*. The structural analysis was performed by NMR and HRESIMS. These compounds were identified as ganoduriporols C-E. They have shown an anti-PTP1B activity with IC_{50} values of 19.1, 17.81, and 29.6 $\mu mol{\cdot}L^{-1}$, respectively (Guo et al., 2019). Chen et al. 2017 studied meroterpenoids and triterpenoids derived from *G. lucidum* fruiting bodies. Among these compounds, ganoderic acid C (16), methyl ganoderate

K, lucidenic acid H, 23-dihydroganoderic acid N, methyl lucidenate F, ganoderic acid B, and lucidenic acid E displayed inhibitory effect on PTP1B within the range of 7.6–41.9 μM (Chen et al., 2017). A novel water-soluble macromolecular proteoglycan was isolated from the fruiting bodies of *G. lucidum*. It was named FYGL (Fudan–Yueyang–G. lucidum) which showed significant PTP1B activity with a IC50 value of 5.12 ± 0.05 μg/mL. The level of plasma glucose was reduced in FYGL-treated mice compared to diabetic control mice. The toxicity level of FYGL was low which may be considered as a potential antidiabetic drug (Teng et al., 2011).

4.5 Anti-Obesity Effect

Overall, four pairs of meroterpenoids, (±)-applandimeric acids A–D (**1**–**4**) were isolated from *G. applanatum*. These compounds possess an unprecedented Spiro[furo[3,2–*b*] benzofuran-3,2′-indene] core. Previous studies have shown that (±)-applandimeric acid D binds with FRP2 with hydrogen bonds and inhibits FRP2 with IC_{50} at 7.93 μm when compared to control LiCl 20 mM. It has also shown a comparable anti-lipogenetic effect at 20 mM concentration. The compound (±)-applandimeric acid D activates the MP-activated protein kinase (AMPK) signaling pathway and thereby inhibits protein levels of peroxisome proliferators-activated receptor-γ (PPAR-γ), CCAAT/enhancer-binding protein-β (C/EBP-β), and adipocyte fatty acid-binding protein 4 (FABP4). These results suggest that (±) applandimeric acid D could be a leading compound in the treatment of obesity and obesity-related diseases by inhibiting the accumulation of lipids in adipocytes and reducing inflammatory response (Peng et al., 2021).

Recent studies have shown that *Ganoderma* triterpenoids (GTs) possess anti-obesity effects. GTs isolated from *G. resinaceum* growing in China were evaluated for this effect on brown/beige adipocytes *in vitro*. Four new Ganoderenses H–K (**1**–**4**) and known compounds (**5**–**8**) derived from *G. resinaceum* were analyzed by spectroscopic techniques. The results have shown that Resinacein S, without effecting the differentiation of C3H10T1/2 adipocytes, reduced the size of lipid droplets. Resinacein S was responsible for mitochondrial biogenesis and increased oxygen consumption rate (OCR) in differentiated C3H10T1/2 adipocytes. This compound exerted its effects by regulating the AMPK/PGC1α signaling pathway, thereby preventing obesity-related diseases (Hung et al., 2020). A high fat diet (HFD) was selected for evaluation of anti-obese, and anti-inflammatory effects of water extracts of *G. lucidum* mycelium (WEGL) in mice. The results have shown that WEGL improved the beneficial gut microbiota which was confirmed by decreased Firmicutes-to-Bacteroidetes ratios and the levels of endotoxin-bearing Proteobacteria. The effects of WEGL, anti-obesity, and gut microbiota-modulating effects were transmissible through feces of WEGL-treated mice to HFD-treated mice. Further purification studies revealed that WEGL contained a polysaccharide with a molecular weight > 300 kDa and was responsible for the anti-obesity effect and maintenance of beneficial microbiota in the gut (Chang et al., 2015).

The polysaccharide purified from Sporoderm broken spores of *G. lucidum* (BSGLP) has shown an anti-obesity effect. Particularly, C57BL/6 J mice were selected for studying anti-obesity effects fed with HFD. The results have shown that

BSGLP was beneficial for the treatment of obesity, hyperlipidemia, inflammation, and fat accumulation in C57BL/6 J mice. BSGLP was also responsible for increasing the beneficial gut microflora, maintaining intestinal barrier function, increasing the production of short chain fatty acids, and expression of G-protein coupled receptor 43 (GPR43). The fecal microbiota study revealed that BSGLP-induced increase in beneficial bacteria was responsible, at least in part, for its anti-obesity effect. BSGLP showed its activity by regulating the TLR4/Myd88/NF-κB signaling pathway in adipose tissue. It also exerted its anti-obesity effect by modulating inflammation, gut microflora, and barrier function (Sang et al., 2021).

The anti-obesity effects of 3% *G. lucidum* extract powder (GEP) were studied in C57BL/6 mice. GEP improved glucose metabolism, reduced the lipid levels in liver, and adipocytes. These effects were possible by an activated AMPK which was responsible for transcription and translation of fatty acid synthase (FAS), stearoyl-CoA desaturase 1 (SCD1), and sterol regulatory element-binding protein-1c (SREBP1c) genes. The activation of AMPK leads to increased acetyl-CoA carboxylase (ACC), insulin receptor (IR), IR substrate 1 (IRS1), and Akt protein expression and glucose transporter 1/4 (GLUT ¼) expression levels, thereby affecting obesity-induced insulin resistance in C57BL/6 mice (Lee et al., 2020).

The erogosterol peroxide purified from *G. lucidum* was evaluated for its anti-obesity activity. The effect of ergosterol on triglyceride synthesis at transcription and translation levels and differentiation of 3T3-L1 adipocytes was studied. The expression of gamma receptor activated by peroxisome proliferators (PPARγ) and the enhancer binding protein CCAT/alpha (C/EBPα), and the expression of the protein binding to the regulatory element of sterols-1c (SREBP-1c) which promotes PPARγ activity and leads to inhibition of differentiation have also been studied. It also inhibits the expression of fatty acid synthase (FAS), fatty acid transloases (FAT), and acetyl coenzyme A carboxylase (ACC) which are lipogenic factors. The ergosterol peroxidase phosphorylates mitogen-activated protein kinases (MAPKs) which are responsible for cell proliferation and differentiation, as well as proliferation of transcription factors involved in the mitotic clonal expansion. Thus, it was concluded that ergosterol peroxidase inhibited the synthesis of triacylglycerides and 3T3-L1 adipocytes (Jeong and Park, 2020).

4.6 Neuroprotective Activity

The triterpenoids of *G. lucidum* were examined for anti-Alzheimer's activity. APP/PSI mouse and Aβ-induced hippocampal neuron cell model of AD were used *in vivo* and *in vitro* models, respectively. *In vivo* experiments in APP/PSI revealed that GLTs reduced the cognitive impairment in AD mice. GLTs exert their effects by inhibiting apoptosis, reducing oxidative damage, and inhibiting the ROCK-signaling pathway. The hippocampus region of APP/PSI mouse was severely damaged which was reversed by treatment with GLTs. *In vitro* cell experiments were carried out by amyloid *β-protein*, to induce hippocampasl neuron cells into the AD model. GLTs promoted cell proliferation, facilitated the expression of superoxide dismutase (SOD), inhibited the expression of malondialdehyde (MDA) and lactic acid dehydrogenase (LDH) in neurons. The study has shown that

GLTs improve cognitive function, attenuate neuronal damage, and inhibit apoptosis in hippocampal tissues and cells in AD by inhibiting the ROCK signaling pathway (Yu et al., 2020).

The petroleum ether extract obtained from *G. lucidum* was evaluated for its anti-aging property and effects on DNA methylation (Lai et al., 2019). The aging process in rats was induced by 100 mg/kg/day of D-galactose, and alcohol extracts of *G. lucidum* have shown the highest anti-aging effect in mice. The regulators of methylation, including Histone H3, DNMT3A, and DNMT3B in brain tissues were upregulated with treatment of ethanol extract. This study has shown that the alcohol extract of *G. lucidum* contained ganoderic acid and lucidone A which were responsible for delaying AD progression (Lai et al., 2019).

The aqueous and ethanol extracts of *G. applanatum* were evaluated for their neurodepressant, analgesic, and anxiolytic activities. The water and ethanol extracts were named AEGA (Aqueous extract of *G. applanatum*) and EEGA (Ethanol extract of *G. applanatum*), respectively. They showed a significant depressant activity in all employed methods. Molecular docking revealed that ganoapplanin, sphaeropsidin D, and cytosporone C showed the best binding affinity to the selected receptors. The study suggested that AEGA and EEGA have potential neurodepressant, anxiolytic, and analgesic activities (Hossen et al., 2021).

4.7 *Antiviral Activity*

DENV-NS2B-N53 is an enzyme which participates in the formation of viral particles by the cleavage of DENV polyprotein. *In silico* analysis revealed that four compounds, viz. Ganodermanontriol, Lucidumol A, Ganoderic acid C2, and Ganosporeric acid A were effective against viral protease DENV-NS2B-N53. *In vitro* studies suggested that ganodermanontriol was a potential anti-DENV-NS2B-N53 inhibiting agent (Bharadwaj et al., 2019). Enterovirus 71 (EV71) caused hand, mouth and foot disease (HFMD). Two triterepenoids of *G. lucidum*, Lanosta-7,9(11),24-trien-3-one,15; 26-dihydroxy (GLTA) and Ganoderic acid Y (GLTB) were evaluated as antiviral agents against EV71 infection. GLTA and GLTB blocked the viral particle adsorption into the cell, thereby preventing viral replication. Further molecular docking studies revealed that GLTA and GLTB bind to the viral capsid protein at hydrophobic site on F pocket. In this way, both compounds prevented the RNA replication of EV 71 virus (Zhang et al., 2014).

HFMD is a highly contagious disease which leads to neurological complications. This disease is caused by Enterovirus A71 (EVA71) and Coxsackievirus A16 (CV-A16). The antiviral effects of *G. neojaponicum* Imazeki (GNJI) crude extracts (S1-S4) against EV-A71, CV-A16, CV-A10, and CV-A6 have been evaluated *in vitro*. The aqueous extract of compound S2 has shown the maximal antiviral activity against all viruses in human primary oral fibroblast cells. Except CV-A6, the replication of all viruses was inhibited from 2 to 71 hours after infection, and CV-A6 was inhibited 2 hours post-infection. The aqueous crude S2 fraction was purified in later studies. The polysaccharides purified from S2 showed similar antiviral effects to S2 crude fraction. Hence, the polysaccharides present in S2 fraction were responsible for inhibition of viruses (Ang et al., 2021).

The triterepenoids of *G. linghzi* were analyzed for neuraminidase (NA) inhibitors, which led to the discovery of Ganoderic acid TQ and TR, inhibitors of H5N1 and H1N1 of neuraminadase. The studies of structure-activity relation revealed that the triterpenoid scaffold is a key for designing neuraminidase-inhibitor drugs. Further studies showed that ARG292 and/or Glu119 were crucial in inhibition of H5N1 H1N1 (Zhu et al., 2015). The computational studies were carried out for *G. lucidum* triterpenoids against HIV-I1protease and Plasmepsin I by using Autodock 4.2. Nelfinavir and KNI-1006 were used as control drugs for HIV-1 and Plasmepsin I, respectively. Ganoderat acid-B showed the maximal affinity for HIV-1 protease and found to be better compared to control drug Nelfinavir. Although ganodermanondiol has shown a maximal affinity with Plasmepsin I, it was found to be better than control drug KNI-1006. Therefore, the triperepdnoids of *G. lucidum* may be used in potential drugs against HIV-1 and *Plasmodium falciparum* parasite (Kang et al., 2015). The extraction of water was carried out to derive compounds from normal *G. lucidum* and its antler form. The water extract of *G. lucidum* and its antler form displayed a potent inhibitory activity on DENV2 NS2B-NS3 protease compared to the normal *G. lucidum* water extract. The results of LC-MS, molecular docking confirmed that hesperetin and ganocin B were the strongest inhibitor of DENV2 NS2B-NS3 protease (Lim et al., 2020).

4.8 Antimicrobial Activity

Previous study investigated the antimicrobial activities of australic acid and its methyl ester methyl australate (Smania et al., 2007). These compounds were isolated from Brazilian *G. australe.* Both compounds, australic acid and methyl australate, have shown MIC (minimal inhibitory concentration) values against fungal species. The antibacterial activity was investigated against four bacterial species, *Bacillus cereus* (ATCC 11778), *Escherichia coli* (ATCC 25922), *Pseudomonas aeruginosa* (ATCC 27853), and *Staphylococcus aureus* (ATCC 25923). Both compounds have shown similar MIC against bacterial spp., except *S. aureus*, australic acid has shown less MIC compared to methyl australate (Smania et al., 2007).

G. pfeifferi was collected from Germany. Two compounds, Ganomycin A and Ganomycin B, were isolated from dried fruiting bodies of *G. pfeifferi.* Both exerted a moderate inhibitory effect on bacterial gram-positive strains, such as *Bacillus subtilis*, *Staphylococcus aureus*, and *Micrococcus flavus*. The zone of inhibition was found between 15–25 mm. The MIC of both compounds against *M. flavus* was estimated 2.5 μg/mL and 25 μg/mL for both against *S. aureus* (Mothana et al., 2000). *G. atrum* was collected from China, from which ethanol-soluble acidic components (ESAC) were extracted. ESAC was further investigated for its antibacterial activity. Four bacterial spp., *Staphylococus aureus*, *Escherichia coli*, *Bacillus subtilis*, and *Proteus vulgaris* were chosen for this study. At 5 mg/ml of ESAC concentration, the inhibition zones were observed at 0.87 ± 0.36, 7.53 ± 0.30, 8.09 ± 0.11, and 6.15 ± 0.45 for *S. aureus*, *E. coli*, *B. subtilis*, and *P. vulgaris*, respectively. The diameter of inhibition was dependent on the concentration of ESAC. *P. vulgaris* was less sensitive to ESAC compared to *S. aureus. E. coli* and *B. subtilis* (Li et al., 2012).

The methanol and aqueous extracts of *G. lucidum* fruiting bodies have shown significant antibacterial activity at 0.5 mg/ml of concentration, whereas the remaining three extracts showed insignificant effects. These results have shown that *P. vulgaris* (23.66 ± 0.57 mm zone of inhibition in methanol extract) was the most susceptible followed by *Pseudomonas aeruginosa* (23.13 ± 0.49 mm inhibitory zone in aqueous extract). The MIC value for *Streptococcus mutans* was 62.50 μg/m and 31.25 μg/ml for the remaining bacterial spp. (Kumara and Bhatt 2012).

A 15 KD protein, ganodermin, was purified from the fruiting bodies of *G. lucidum*. The ganodermin was evaluated for antifungal activity. It significantly inhibited the growth of *Botrytis cinerea, Fusarium oxysporum*, and *Physalospora piricola* at IC50 values of 15.2 mM, 12.4 mM, and 18.1 mM, respectively. Further studies revealed that ganodermin was devoid of inhibitory activities of hemagglutination, deoxyribonuclease, ribonuclease, and protease (Wang and Ng, 2006). *G. annulare* was collected from the decaying wood in Southern Brazil. Three sterols, 5a-ergost-7-en-3b-ol, 5a-ergosta-7,22-dien-3b-ol and 5,8-epidioxy-5a,8aergosta-6,22-dien-3b-ol and five triterpenes, applanoxidic acids A, C, F, G, and H were isolated from *G. annulare*. All these compounds were investigated for antifungal activity against *Microsporum cannis* and *Trichophyton mentagrophytes* fungal spp. These sterols and triterpenes exerted antifungal activities. Among them, triterpenes applanoxidic acids A, C, and F significantly inhibited fungal growth at concentrations of 500–100 mg/ml (Smania et al., 2003).

The aqueous and methanol extracts of *G. lucidum* were evaluated for their antimicrobial activity. The aqueous extract has shown the highest inhibitory activity against *S. aureus* and *Salmonella typhi* with the minimum inhibitory zone of 31 mm at 200 mg of concentration. The minimum inhibitory activity against *E. coli* was estimated 10 mm against 50 mg of aqueous extract. The methanol extract showed the highest antifungal activity at 200 mg of concentration with 30 mm zone against *Mucor indicus*, and the lowest antifungal activity was observed at 50 mg of concentration with 3 mm inhibition zone against *Aspergillus flavus* (Sridhar et al., 2011). Different *Ganoderma* spp., *G. lucidium, G. applanatum*, and *G. australe* were collected from decaying wood logs in the University of Ibadan Botantical Gardens, Nigeria. The highest inhibitory zone (20.3 mm) on Proteus mirabilis was shown by methanolic extract of *G. lucidum*. The crude ethanolic extract manifested the highest antifungal activity against *Aspergillus niger*, inhibition zone was observed at 24.33 mm. The lowest antibacterial activity was exerted by aqueous extract of *G. australe* against *Escherichia coli* (2.3 mm), and the lowest antifungal activity revealed a purified extract of *G. australe* against *Penicillum oxalium*. The MIC for ethanol extract was in the range between 1.7 –5.0 m/ml for bacteria, whereas it was 2.0–6.0 mg/ml in case of fungi (Jonathan and Awotona, 2010). The chloroform extract of fruiting bodies of *G. lucidum* was investigated for antibacterial activity. The MIC and minimal bactericidal concentration (MBC) for *Staphylococcus aureus*, were 8 mg/ml and for *Bacillus subtillis* were 8 and 16 mg/ml, respectively. Further studies have shown that the chloroform extract was composed of various sterols and triterepnoids (Keypour et al., 2008).

5. Further Applications in Medicine

5.1 Cytoprotective Effect

The protective effects of GTs on cadmium-induced testicular injury has been investigated. In this study, Hyline egg-lying chickens were selected; GTs were effective in reducing cadmium, IL-1, IL-6, and TNF-α levels and increasing the levels of antioxidant enzymes, super oxide dismutase (SOD), and glutathione peroxidase (GSG-Px). These enzymes are responsible for reduction of malondialdehyde (MDA). The GTs also regulated the expression levels of apoptotic enzymes: Bax, Bcl2, and Capsase-3. The regulation of these enzymes plays an important role in protecting testis from cell damage (Wang et al., 2019).

The medicinal properties of *Ganoderma* compounds are presented in Table 7.1.

5.2 Anti-melanogenic Activity

The compound ganodermanondiol was purified from dried *G. lucidum* by ethanol extraction. Its ability of inhibiting melanogenesis was identified in B16F10 melanoma cells. Ganodermanondiol possesses the capacity of inhibiting tyrosine-related proteins (TRP) TRP1 and TRP2 as well as microphthalmic-associated transcription factor (MITF). Thereby, ganodermanondiol could inhibit the process of melanogenesis and affect mitogen-activated protein kinase (MAPK) pathway and cyclic AMP-dependent signaling pathways. These pathways are responsible for melanogenesis in B16F10 melanoma cells. Thus, ganodermanondiol may be used in the manufacturing of skin-care cosmetic products and formulations (Kim et al., 2016).

5.3 Hypouricemic Effect

The hypouricemic effects of ethanol (GAE) and water (GAW) extracts of *G. applanatum* were studied on chemically-induced hyperuricemia mice. The GAE and GAW were administered orally in different doses to hyperuricemic mice. They decreased the serum uric acid (SUA) levels in hyperuricemia control mice, and urine uric acid (UUA) levels increased. These compounds have revealed hupouricemic effects by regulating the levels of OAT1 GLUT9, URAT1 and gastrointestinal CNT2 that may increase the uric acid secretions and decrease the absorption of purine in the gastrointestinal tract (Yong et al., 2018).

5.4 Anti-protozoal Activity

The terpenoids were extracted from fruiting bodies of *G. lucidum* (GT) by Oluba (2019) and assessed for their antiparasitic activity against *Plasmodium berghei* in mice. The study suggests that anti-plasmodial effects of GT may involve mechanisms associated with its hypolipidemic activity. The chloroquine, when given in combination with GT, has also been shown to improve its healing properties in mice infected with *P. berghei* (Oluba, 2019).

5.5 Antiandrogenic Effect

Ganoderol B was purified from the ethanol extract of fruiting bodies of *G. lucidum* and its anti-androgenic effect was evaluated. This compound binds with the androgen receptor (AR) thereby inhibiting androgen-induced lymph node carcinoma of prostate (LNCaP) cell growth and is responsible for suppression of ventral prostate regrowth induced by testosterone in rats. Ganoderol B also possesses 5α-reductase inhibitory activity. The antiandrogenic activity was achieved by downregulation of AR signaling by Ganoderol B. The prostate specific antigen (PSA), a biomarker of prostate cancer, was downregulated. This study suggests that Ganoderol B may be a key drug in treating prostate cancer and benign prostatic hyperplasia (BPH) by suppressing androgens and androgen receptors (Liu et al., 2007).

5.6 Anti-Amnesic Activity

The methanol extracts derived from fruiting bodies of *G. mediosinense* and *G. ramosissimum* were evaluated for their anti-amnesic activity. Methanol extracts of *G. mediosinense* showed strong antioxidant and AChE activities and were selected for *in vivo* studies. The methanol extracts reversed the scopolamine-induced amnesia in mice which was confirmed by a decrease ($p < 0.05$) in the transfer latency period and increase in object recognition index in PSA and NOR methods, respectively. The methanol extracts significantly reduced the brain AChE activity and oxidative stress. Histopathological examination of brain tissue showed a decrease in vacuolated cytoplasm and an increase in pyramidal cells in the cerebral hippocampus and cortical regions. GME revealed an anti-amnesic effect by inhibiting AChE and its antioxidant effect (Kaur et al., 2017).

5.7 Nephroprotective Effect

G. lucidum triterpenes (GT) were assessed for their activity on inhibition of cysts in kidneys. Renal cysts were commonly observed in autosomal dominant polycystic kidney disease (ADPKD). GT reduced cyst formation in kidneys in two ADPKD mouse models and induced epithelial cell differentiation by enhancing the formation of epithelial tubules in MDCK cells. The study revealed that *in vivo* and *in vitro* GT showed similar results: GT without interfering in the mTOR signaling pathway downregulated the Ras/MAPK pathway. The downregulation of Ras/MAPK may be achieved by decreasing the levels of intracellular cAMP. Among 15 screened triterpenes, ethyl ganoderate C2 (CBLZ-7) had the highest inhibitory activity of cyst formation *in vitro*. In addition, CBLZ-7 was able to downregulate forskolin-induced activation of the Ras/MAPK signaling pathway (Su et al., 2017).

6. Conclusion

Ganoderma spp. are known for their medicinal properties in traditional Chinese medicine (TCM). Among *Ganoderma* species, the medicinal applications of *G. lucidum* have been extensively studied. Inventions on artificial cultivation techniques of *Ganoderma* spp. facilitated the study of novel medicinal applications of different species. Recent advances in medicinal applications of 15 *Ganoderma* species: *G. ahmadii, G. annulare, G. applanatum, G. atrum, G. australe, G. cochlear, G. lucidum, G. leucocontextum, G. mediosinense, G. neo-japonicum, G. pfeifferi, G. ramosissimum, G. resinaceum, G. sinense,* and *G. tsugae* have been discussed in this chapter. Cultivation methods, purification of medicinally valued compounds and medicinal values of *Ganoderma* spp. have been elucidated. Further research is warranted to perform purification of metabolites and study their medicinal properties. Although we have considerable amount of information on the metabolites derived from fruiting bodies of *Ganoderma* spp. in the wild, the same cannot be applied to artificial cultures. The novelty of artificial cultures needs to be further explored.

References

Agius, L. (2007). New hepatic targets for glycaemic control in diabetes. *Best Pract. Res. Clin Endocrinol.*, 21(4): 587–605. 10.1016/j.beem.2007.09.001.

Amen, Y.M., Zhu, Q., Tran, H.B., Afifi, M.S., Halim, A.F., Ashour, A., Mira, A. and Shimizu, K. (2016). Lucidumol C, a new cytotoxic lanostanoid triterpene from *Ganoderma lingzhi* against human cancer cells. *J. Nat. Med.*, 70(3): 661–666. 10.1007/s11418-016-0976-2.

Ang, W.X., Sarasvathy, S., Kuppusamy, U.R., Sabaratnam, V., Tan, S.H., Wong, K.T., Perera, D. and Ong, K.C. (2021). Research article: *In vitro* antiviral activity of medicinal mushroom *Ganoderma neo-japonicum* Imazeki against enteroviruses that caused hand, foot, and mouth disease. *Trop. Biomed.*, 38(3): 239–247.10.47665/tb.38.3.063.

Bao, X.F., Wang, X.S., Dong, Q., Fang, J.N. and Li, X.Y. (2002). Structural features of immunologically active polysaccharides from *Ganoderma lucidum*. *Phytochemistry*, 59(2): 175–181. 10.1016/S0031-9422(01)00450-2.

Bharadwaj, S., Lee, K.E., Dwivedi, V.D., Yadava, U., Panwar, A., Lucas, S.J., Pandey, A. and Kang, S.G. (2019). Discovery of *Ganoderma lucidum* triterpenoids as potential inhibitors against Dengue virus NS2B-NS3 protease. *Sci. Rep.*, 9(1): 1–12. 10.1038/s41598-019-55723-5.

Boh, B., Berovic, M., Zhang, J. and Zhi-Bin, L. (2007). *Ganoderma lucidum* and its pharmaceutically active compounds. *Biotechnol. Annu. Rev.*, 13: 265–301. 10.1016/S1387-2656(07)13010-6.

Cao, W.W., Luo, Q., Cheng, Y.X. and Wang, S.M. (2016). Meroterpenoid enantiomers from *Ganoderma sinensis*. *Fitoterapia*, 110: 110–115. 10.1016/j.fitote.2016.03.003.

Cao, Y., Wu, S.H. and Dai, Y.C. (2012). Species clarification of the prize medicinal *Ganoderma* mushroom "Lingzhi". *Fungal Divers.*, 56(1): 49–62.10.1007/s13225-012-0178-5.

Chang, C.J., Lin, C.S., Lu, C.C., Martel, J., Ko, Y.F., Ojcius, D.M., Tseng, S.F., Wu, T.R., Chen, Y.Y.M., Young, J.D. and Lai, H.C. (2015). *Ganoderma lucidum* reduces obesity in mice by modulating the composition of the gut microbiota. *Nat. Commun.*, 6(1): 1–19. 10.1038/ncomms8489.

Chang, S.T. and Buswell, J.A. (1999). *Ganoderma lucidum* (Curt.: Fr.) P. karst. (Aphyllophoromycetideae): A mushrooming medicinal mushroom. *Int. J. Med. Mushrooms*, 1(2): 139–146. 10.1615/IntJMedMushrooms.v1.i2.30.

Chang, S.T. and Miles, P.G. (2004). *Mushrooms: Cultivation, Nutritional Value, Medicinal Effect, and Environmental Impact.* (2nd Ed.). Boca Raton (FL): CRC press, pp. 125–140. 10.1201/9780203492086.

Chen, A.W. (1999). Cultivation of the medicinal mushroom *Ganoderma lucidum* (Curt.: Fr.) P. karst. (Reishi) in North America. *Int. J. Med. Mushrooms*, 1(3): 263–282.10.1615/IntJMedMushrooms.v1.i3.90.

Chen, A.W. (2002). Natural log cultivation of the medicinal mushroom, *Ganoderma lucidum* (Reishi). *Mushroom Growers' Newsl.*, 3(9): 2–6.

Chen, B., Tian, J., Zhang, J., Wang, K., Liu, L., Yang, B., Bao, L. and Liu, H. (2017). Triterpenes and meroterpenes from *Ganoderma lucidum* with inhibitory activity against HMGs reductase, aldose reductase and α-glucosidase. *Fitoterapia*, 120: 6–16. 10.1016/j.fitote.2017.05.005.

Chen, S.D., Yong, T.Q., Zhang, Y.F., Hu, H.P. and Xie, Y.Z. (2019). Inhibitory effect of five *Ganoderma* species (Agaricomycetes) against key digestive enzymes related to Type 2 diabetes mellitus. *Int. J. Med. Mushrooms*, 21(7): 703–711. 10.1615/IntJMedMushrooms.v21.i7.70.

Elkhateeb, W.A., Zaghlol, G.M., El-Garawani, I.M., Ahmed, E.F., Rateb, M.E. and Moneim, A.E.A. (2018). *Ganoderma applanatum* secondary metabolites induced apoptosis through different pathways: *In vivo* and *in vitro* anticancer studies. *Biomed. Pharmacother.*, 101: 264–277. 10.1016/j.biopha.2018.02.058.

Galor, S.W., Yuen, J., Buswell, J.A. and Benzie, I.F.F. (2011). *Ganoderma lucidum* (Lingzhi or Reishi), a medicinal mushroom. *In*: *Herbal Medicine: Biomolecular and clinical aspects* (2nd Edn.) Boca Raton (FL): CRC Press/Taylor & Francis, Chapter 9. https://www.ncbi.nlm.nih.gov/books/NBK92757/.

Gao, X., Qi, J., Ho, C.T., Li, B., Mu, J., Zhang, Y., Hu, H., Mo, W., Chen, Z. and Xie, Y. (2020). Structural characterization and immunomodulatory activity of a water-soluble polysaccharide from *Ganoderma leucocontextum* fruiting bodies. *Carbohydr. Polym.*, 249: 116874. doi.org/10.1016/j.carbpol.2020.116874.

González, A.G., León, F., Rivera, A., Padrón, J.I., González-Plata, J., Zuluaga, J.C., Quintana, J., Estévez, F. and Bermejo, J. (2002). New lanostanoids from the fungus *Ganoderma c oncinna*. *J. Nat. Prod.*, 65(3): 417–421. 10.1021/np010143e.

Gregori, A. and Pohleven, F. (2015). Cultivation of three medicinal mushroom species on olive oil press cakes containing substrates. *Acta Agric. Slov.*, 103(1): 49–54. 10.14720/aas.2014.103.1.05.

Guo, J., Ma, Q., Kong, F., Xie, Q., Zhou, L., Ding, Q., Wu, Y. and Zhao, Y. (2019). Meroterpenoids from the fruiting bodies of *Ganoderma ahmadii* steyaret and their protein tyrosine phosphatase 1B inhibitory activities. *Chinese J. Org. Chem.*, 39(11): 3264–3268. 10.6023/cjoc201905010.

Guo, J., Kong, F., Ma, Q., Xie, Q., Zhang, R., Dai, H., Wu, Y. and Zhao, Y. (2020). Meroterpenoids with protein tyrosine phosphatase 1B inhibitory activities from the fruiting bodies of *Ganoderma ahmadii*. *Front. Chem.*, 8: 279. 10.3389/fchem.2020.00279.

Habijanič, J. and Berovič, M. (2000). The relevance of solid-state substrate moisturizing on *Ganoderma lucidum* biomass cultivation. *Food Technol. Biotechnol.*, 38(3): 225–228.

Han, X.H., Wang, M.C., Wang, H.Z. and He, B. (2003). A preliminary study on nutritional conditions for the strain mycelium growth of several cultivars of *Ganoderma*. *J. Hainan Nor. Univ. (Nat Sci).*, 16: 88–92.

Hanyu, X., Lanyue, L., Miao, D., Wentao, F., Cangran, C. and Hui, S. (2020). Effect of *Ganoderma applanatum* polysaccharides on MAPK/ERK pathway affecting autophagy in breast cancer MCF-7 cells. *Int. J. Biol. Macromol.*, 146: 353–362. 10.1016/j.ijbiomac.2020.01.010.

He, Z., Naruse, K. and King, G.L. (2005). Effects of diabetes and insulin resistance on endothelial functions. *In*: *Diabetes and Cardiovascular Disease*. Totawa (NJ); Humana Press, 25: 46.

Hossen, S.M., Islam, M.J., Hossain, M.R., Barua, A., Uddin, M.G. and Emon, N.U. (2021). CNS anti-depressant, anxiolytic and analgesic effects of *Ganoderma applanatum* (mushroom) along with ligand-receptor binding screening provide new insights. *Biochem. Biophys. Rep.*, 27: 101062 10.1016/j.bbrep.2021.101062.

Hsieh, C. and Yang, F.C. (2004). Reusing soy residue for the solid-state fermentation of *Ganoderma lucidum*. *Bioresour. Technol.*, 91(1): 105–109. 10.1016/S0960-8524 (03)00157-3.

Huang, H.C. and Liu, Y.C. (2008). Enhancement of polysaccharide production by optimization of culture conditions in shake flask submerged cultivation of *Grifola umbellata*. *Chin. J. Chem. Eng.*, 39(4): 307–311. 10.1016/j.jcice.2008.01.003.

Huang, S.Z., Cheng, B.H., Ma, Q.Y., Wang, Q., Kong, F.D., Dai, H.F., Qiu, S.Q., Zheng, P.Y., Liu, Z.Q. and Zhao, Y.X. (2016). Anti-allergic prenylated hydroquinones and alkaloids from the fruiting body of *Ganoderma calidophilum*. *RSC Adv.*, 6(25): 21139–21147. 10.1039/C6RA01466F.

Huang, Y., Wei, G., Peng, X., Hu, G., Su, H., Liu, J., Chen, X. and Qiu, M. (2020). Triterpenoids from functional mushroom *Ganoderma resinaceum* and the novel role of Resinacein S in enhancing the activity of brown/beige adipocytes. *Food Res. Int.*, 136: 109303. 10.1016/j.foodres.2020.109303.

Jeong, Y.U. and Park, Y.J. (2020). Ergosterol peroxide from the medicinal mushroom *Ganoderma lucidum* inhibits differentiation and lipid accumulation of 3T3-L1 adipocytes. *Int. J. Mol. Sci.*, 21(2): 460. 10.3390/ijms21020460.

Jonathan, S.G. and Awotona, F.E. (2010). Studies on antimicrobial potentials of three *Ganoderma* species. *Afr J. Biomed Res.*, 13(2): 131–139.

Juan, Z.X.L. 1999. Resources of Wild *Ganoderma* spp. in China and their development and utilization. *Acta Edulis Fungi.*, 6(1): 60.

Kamra, A. and Bhatt, A.B. (2012). Evaluation of antimicrobial and antioxidant activity of *Ganoderma lucidum* extracts against human pathogenic bacteria. *Int. J. Pharm. Pharm.*, 4(2): 359–362.

Kang, D., Mutakin, M. and Levita, J. 2015. Computational study of triterpenoids of *Ganoderma lucidum* with aspartic protease enzymes for discovering HIV-1 and plasmepsin inhibitors. *Int. J. Chem.*, 7(1): 62. 10.5539/ijc.v7n1p62.

Kaur, R., Singh, V. and Shri, R. (2017). Anti-amnesic effects of *Ganoderma* species: a possible cholinergic and antioxidant mechanism. *Biomed. Pharmacother.*, 92: 1055–1061. 10.1016/j.biopha.2017.06.029.

Keypour, S., Riahi, H., Moradali, M.F. and Rafati, H. (2008). Investigation of the antibacterial activity of a chloroform extract of Ling Zhi or Reishi medicinal mushroom, *Ganoderma lucidum* (W. Curt.: Fr.) P. Karst. (Aphyllophoromycetideae), from Iran. *Int. J. Med. Mushrooms*, 10(4): 345–349. 10.1615/IntJMedMushr.v10.i4.70.

Kim, J.W., Kim, H.I., Kim, J.H., Kwon, O., Son, E.S., Lee, C.S. and Park, Y.J. (2016). Effects of ganodermanondiol, a new melanogenesis inhibitor from the medicinal mushroom *Ganoderma lucidum*. *Int. J. Mol. Sci.*, 17(11): 1798. 10.3390/ijms17111798.

Kim, S.W., Hwang, H.J., Lee, B.C. and Yun, J.W. (2007). Submerged production and characterization of *Grifola frondosa* polysaccharides: A new application to cosmeceuticals. *Food Technol. Biotechnol.*, 45(3): 295–305.

Kleinwächter, P., Anh, N., Kiet, T.T., Schlegel, B., Dahse, H.M., Härtl, A. and Gräfe, U. (2001). Colossolactones, new triterpenoid metabolites from a Vietnamese mushroom *Ganoderma colossum*. *J. Nat. Prod.*, 64(2): 236–239. 10.1021/np000437k.

Lai, G., Guo, Y., Chen, D., Tang, X., Shuai, O., Yong, T., Wang, D., Xiao, C., Zhou, G., Xie, Y. and Yang, B.B. (2019). Alcohol extracts from *Ganoderma lucidum* delay the progress of Alzheimer's disease by regulating DNA methylation in rodents. *Front. Pharmacol.*, 10: 272. 10.3389/fphar.2019.00272.

Lai, L.K., Abidin, N.Z., Abdullah, N. and Sabaratnam, V. (2010). Anti-human papillomavirus (HPV) 16 E6 activity of Ling Zhi or Reishi medicinal mushroom, *Ganoderma lucidum* (W. Curt.: Fr.) P. Karst. (Aphyllophoromycetideae) extracts. *Int. J. Med. Mushrooms*, 12(3): 279–286. 10.1615/IntJMedMushr.v12.i3.70.

Lee, H.A., Cho, J.H., Afinanisa, Q., An, G.H., Han, J.G., Kang, H.J., Choi, S.H. and Seong, H.A. (2020). *Ganoderma lucidum* extract reduces insulin resistance by enhancing AMPK activation in high-fat diet-induced obese mice. *Nutrients*, 12(11): 3338. 10.3390/nu12113338.

Lee, H.H., Itokawa, H. and Kozuka, M. (2005). Asian herbal products: The basis for development of high-quality dietary supplements and new medicines. *Asian Functional Foods*, 21–72.

Lee, S.Y., Kim, J.S., Lee, S. and Kang, S.S. (2011). Polyoxygenated ergostane-type sterols from the liquid culture of *Ganoderma applanatum*. *Nat. Prod. Res.*, 25(14): 1304–1311. 10.1080/14786419.2010.503190.

Li, J., Gu, F., Cai, C., Hu, M., Fan, L., Hao, J. and Yu, G. (2020). Purification, structural characterization, and immunomodulatory activity of the polysaccharides from *Ganoderma lucidum*. *Int. J. Biol. Macromol.*, 143: 806–813. 10.1016/j.ijbiomac.2019.09.141.

Li, N., Yan, C., Hua, D. and Zhang, D. (2013). Isolation, purification, and structural characterization of a novel polysaccharide from *Ganoderma capense*. *Int. J. Biol. Macromol.*, 57: 285–290. 10.1016/j.ijbiomac.2013.03.030.

Li, T., Yu, H., Song, Y., Zhang, R. and Ge, M. (2019). Protective effects of *Ganoderma* triterpenoids on cadmium-induced oxidative stress and inflammatory injury in chicken livers. *J. Trace Elem. Med. Biol.*, 52: 118–125. 10.1016/j.jtemb.2018.12.010.

Li, W.J., Nie, S.P., Liu, X.Z., Zhang, H., Yang, Y., Yu, Q. and Xie, M.Y. (2012). Antimicrobial properties, antioxidant activity and cytotoxicity of ethanol-soluble acidic components from *Ganoderma atrum*. *Food Chem. Toxicol.*, 50 (3-4): 689–694. 10.1016/j.fct.2011.12.011.

Li, X.C., Liu, F., Su, H.G., Guo, L., Zhou, Q.M., Huang, Y.J., Peng, C. and Xiong, L. (2019). Two pairs of alkaloid enantiomers from *Ganoderma luteomarginatum*. *Biochem. Syst. Ecol.*, 86: 103930. 10.1016/j.bse.2019.103930.

Lim, W.Z., Cheng, P.G., Abdulrahman, A.Y. and Teoh, T.C. (2020). The identification of active compounds in *Ganoderma lucidum* var. antler extract inhibiting dengue virus serine protease and its computational studies. *J. Bimol. Struct. Dyn.*, 38(14): 4273–4288. 10.1080/07391102.2019.1678523.

Lin, J. and Zhou, X.W. (1999). Artificial cultivation of organizational separation of *Ganoderma lucidum*. *Edible Fungi*, 2: 10–11.

Liu, J., Shimizu, K., Konishi, F., Kumamoto, S. and Kondo, R. (2007). The anti-androgen effect of ganoderol B isolated from the fruiting body of *Ganoderma lucidum*. *Bioorg. Med. Chem.*, 15(14): 4966–4972. 10.1016/j.bmc.2007.04.036.

Liu, J.Q., Wang, C.F., Peng, X.R. and Qiu, M.H. (2011). New alkaloids from the fruiting bodies of *Ganoderma sinense*. *Nat. Prod. Bioprospect.*, 1(2): 93–96. 10.1007/s13659-011-0026-4.

Liu, L.Y., Chen, H., Liu, C., Wang, H.Q., Kang, J., Li, Y. and Chen, R.Y. (2014). Triterpenoids of *Ganoderma theaecolum* and their hepatoprotective activities. *Fitoterapia*, 98: 254–259. 10.1016/j.fitote.2014.08.004.

Liu, L.Y., Yan, Z., Kang, J., Chen, R.Y. and Yu, D.Q. (2017). Three new triterpenoids from *Ganoderma theaecolum*. *J. Asian Nat. Prod. Res.*, 19(9): 847–853. 10.1080/10286020.2016.1271793.

Liu, W., Wang, H., Pang, X., Yao, W. and Gao, X. (2010). Characterization and antioxidant activity of two low-molecular-weight polysaccharides purified from the fruiting bodies of *Ganodermalucidum*. *Int. J. Biol. Macromol.*, 46(4): 451–457. 10.1016/j.ijbiomac.2010.02.006.

Liu, W., Yuan, R., Hou, A., Tan, S., Liu, X., Tan, P., Huang, X. and Wang, J. (2020). *Ganoderma* triterpenoids attenuate tumour angiogenesis in lung cancer tumour-bearing nude mice. *Pharm. Biol.*, 58(1): 1070–1077. 10.1080/13880209.2020.1839111.

Liu, Y., Li, Y., Zhang, W., Sun, M. and Zhang, Z. (2019). Hypoglycemic effect of inulin combined with *Ganoderma lucidum* polysaccharides in T2DM rats. *J. Funct. Foods*, 55: 381–390. 10.1016/j.jff.2019.02.036.

Luo, Q., Wang, X.L., Di, L., Yan, Y.M., Lu, Q., Yang, X.H., Hu, D.B. and Cheng, Y.X. (2015). Isolation and identification of renoprotective substances from the mushroom *Ganoderma lucidum*. *Tetrahedron*, 71(5): 840–845. 10.1016/j.tet.2014.12.052.

Lv, G.P., Zhao, J., Duan, J.A., Tang, Y.P. and Li, S.P. (2012). Comparison of sterols and fatty acids in two species of *Ganoderma*. *Chem. Cent. J.*, 6(1): 1–8. 10.1186/1752-153X-6-10.

Ma, H.T., Hsieh, J.F. and Chen, S.T. (2015). Anti-diabetic effects of *Ganoderma lucidum*. *Phytochemistry*, 114: 109–113. 10.1016/j.phytochem.2015.02.017.

Ma, J.Q., Liu, C.M., Qin, Z.H., Jiang, J.H. and Sun, Y.Z. (2011). *Ganoderma applanatum* terpenes protect mouse liver against benzo (α) pyren-induced oxidative stress and inflammation. *Environ. Toxicol. Pharmacol.*, 31(3): 460–468. 10.1016/j.etap.2011.02.007.

Matute, R.G., Figlas, D., Devalis, R., Delmastro, S. and Curvetto, N. (2002). Sunflower seed hulls as a main nutrient source for cultivating *Ganoderma lucidum*. *Micol Aplicada Int.*, 14(2): 19–24.

Mendoza, G., Suárez-Medellín, J., Espinoza, C., Ramos-Ligonio, A., Fernández, J.J., Norte, M. and Trigos, Á. (2015). Isolation and characterization of bioactive metabolites from fruiting bodies and mycelial culture of *Ganoderma oerstedii* (Higher Basidiomycetes) from Mexico. *Int. J. Med. Mushrooms*, 17: (6). 10.1615/IntJMedMushrooms.v17.i6.10.

Mothana, R.A., Jansen, R., Jülich, W.D. and Lindequist, U. (2000). Ganomycins A and B, new antimicrobial farnesyl hydroquinones from the basidiomycete *Ganoderma pfeifferi*. *J. Nat. Prod.*, 63(3): 416–418.

Oliver-Krasinski, J.M., Kasner, M.T., Yang, J., Crutchlow, M.F., Rustgi, A.K., Kaestner, K.H. and Stoffers, D.A. (2009). The diabetes gene Pdx1 regulates the transcriptional network of pancreatic endocrine progenitor cells in mice. *J. Clin. Investig.*, 119(7): 1888–1898. 10.1172/JCI37028.

Oluba, O.M. (2019). *Ganoderma* terpenoid extract exhibited anti-plasmodial activity by a mechanism involving reduction in erythrocyte and hepatic lipids in *Plasmodium berghei* infected mice. *Lipids Health Dis.*, 18(1): 1–9. 10.1186/s12944-018-0951-x.

Pegler, D.N. (2002). Useful Fungi of the World: The Ling-zhi-The mushroom of immortality. *Mycologist*, 16(3): 100–101. 10.1017/s0269915x0200304X.

Peksen, A. and Yakupoglu, G. (2009). Tea waste as a supplement for the cultivation of *Ganoderma lucidum*. *World J. Microbiol. Biotechnol.*, 25(4): 611–618. 10.1007/s11274-008-9931-z.

Peng, X.R., Liu, J.Q., Han, Z.H., Yuan, X.X., Luo, H.R. and Qiu, M.H. (2013). Protective effects of triterpenoids from *Ganoderma resinaceum* on H2O2-induced toxicity in HepG2 cells. *Food Chem.*, 141(2): 920–926. 10.1016/j.foodchem.2013.03.071.

Peng, X.R., Liu, J.Q., Wang, C.F., Li, X.Y., Shu, Y., Zhou, L. and Qiu, M.H. (2014). Hepatoprotective effects of triterpenoids from *Ganoderma cochlear*. *J. Nat. Prod.*, 77(4): 737–743. 10.1021/np400323u.

Peng, X.R., Wang, Q., Wang, H.R., Hu, K., Xiong, W.Y. and Qiu, M.H. (2021). FPR2-based anti-inflammatory and anti-lipogenesis activities of novel meroterpenoid dimers from *Ganoderma*. *Bioorg. Chem.*, 105338. 10.1016/j.bioorg.2021.105338.

Priya, B. and Robinka, K. (2014). Supplementation of nitrogen source in wheat straw for improving cellulolytic potential of *Ganoderma lucidum*. *Int. J. Pharma Bio Sci.*, 5(2): 90–99.

Qu, L., Li, S., Zhuo, Y., Chen, J., Qin, X. and Guo, G. (2017). Anticancer effect of triterpenes from *Ganoderma lucidum* in human prostate cancer cells. *Oncol. Lett.*, 14(6): 7467–7472. 10.3892/ol.2017.7153.

Recchioni, R., Marcheselli, F., Moroni, F. and Pieri, C. (2002). Apoptosis in human aortic endothelial cells induced by hyperglycemic condition involves mitochondrial depolarization and is prevented by N-acetyl-L-cysteine. *Metab. Clin. Exp.*, 51(11): 1384–1388. 10.1053/meta.2002.35579.

Richter, C., Wittstein, K., Kirk, P.M. and Stadler, M. (2015). An assessment of the taxonomy and chemotaxonomy of *Ganoderma*. *Fungal Divers*, 71(1): 1–15. 10.1007/s13225-014-0313-6.

Riu, H., Roig, G. and Sancho, J. (1997). Production of carpophores of *Lentinus edodes* and *Ganoderma lucidum* grown on cork residues. *Microbiol (Madrid, Spain).*, 13(2): 185–192.

Sang, T., Guo, C., Guo, D., Wu, J., Wang, Y., Wang, Y., Chen, J., Chen, C., Wu, K., Na, K. and Li, K. (2021). Suppression of obesity and inflammation by polysaccharide from sporoderm-broken spore of *Ganoderma lucidum* via gut microbiota regulation. *Carbohydr. Polym.*, 256: 117594. 10.1016/j.carbpol.2020.117594.

Sato, N., Zhang, Q., Ma, C.M. and Hattori, M. (2009). Anti-human immunodeficiency virus-rotease activity of new lanostane-type triterpenoids from *Ganoderma sinense*. *Chem. Pharm. Bull.*, 57(10): 1076–1080. 10.1248/cpb.57.1076.

Sedky, N.K., El Gammal, Z.H., Wahba, A.E., Mosad, E., Waly, Z.Y., El-Fallal, A.A., Arafa, R.K. and El-Badri, N. (2018). The molecular basis of cytotoxicity of α-spinasterol from *Ganoderma resinaceum*: Induction of apoptosis and overexpression of p53 in breast and ovarian cancer cell lines. *J. Cell. Biochem.*, 119(5): 3892–3902.10.1002/jcb.26515.

Smania, E., Delle Monache, F., Smania Jr., A., Yunes, R.A. and Cuneo, R.S. (2003). Antifungal activity of sterols and triterpenes isolated from *Ganoderma annulare*. *Fitoterapia*, 74(4): 375–377. 10.1016/S0367-326X (03)00064-9.

Smania, E.D.F.A., Delle Monache, F., Yunes, R.A., Paulert, R. and Smania Junior, A. (2007). Antimicrobial activity of methyl australate from *Ganoderma australe*. *Rev. bras. Farmacogn.*, 17: 14–16. 10.1590/S0102-695X2007000100004.

Sridhar, S., Sivaprakasam, E., Balakumar, R. and Kavitha, D. (2011). Evaluation of antibacterial and antifungal activity of *Ganoderma lucidum* (Curtis) P. Karst fruit bodies extracts. *World J. Sci. Technol.*, 1(6): 8–11.

Su, L., Liu, L., Jia, Y., Lei, L., Liu, J., Zhu, S., Zhou, H., Chen, R., Lu, H.A.J. and Yang, B. (2017). *Ganoderma* triterpenes retard renal cyst development by downregulating Ras/MAPK signaling and promoting cell differentiation. *Kidney Int.*, 92(6): 1404–1418. 10.1016/j.kint.2017.04.013.

Sun, X., Zhao, C., Pan, W., Wang, J. and Wang, W. (2015). Carboxylate groups play a major role in antitumor activity of *Ganoderma applanatum* polysaccharide. *Carbohydr. Polym.*, 123: 283–287. 10.1016/j.carbpol.2015.01.062.

Tan, W.C., Kuppusamy, U.R., Phan, C.W., Tan, Y.S., Raman, J., Anuar, A.M. and Sabaratnam, V. (2015). *Ganoderma neo-japonicum* Imazeki revisited: Domestication study and antioxidant properties of its basidiocarps and mycelia. *Sci. Rep.*, 5(1): 1–10. 10.1038/srep12515 (2015).

Teng, B.S., Wang, C.D., Yang, H.J., Wu, J.S., Zhang, D., Zheng, M., Fan, Z.H., Pan, D. and Zhou, P. (2011). A protein tyrosine phosphatase 1B activity inhibitor from the fruiting bodies of *Ganoderma lucidum* (Fr.) Karst and its hypoglycemic potency on streptozotocin-induced type 2 diabetic mice. *J. Agric. Food Chem.*, 59(12): 6492–6500. 10.1021/jf200527y.

Teng, Y., Liang, H., Zhang, Z., He, Y., Pan, Y., Yuan, S., Wu, X., Zhao, Q., Yang, H. and Zhou, P. (2020). Biodistribution and immunomodulatory activities of a proteoglycan isolated from *Ganoderma lucidum*. *J. Funct. Foods*, 74: 104193. 10.1016/j.jff.2020.104193.

Tie, L., Yang, H.Q., An, Y., Liu, S.Q., Han, J., Xu, Y., Hu, M., Li, W.D., Chen, A.F., Lin, Z.B. and Li, X.J. (2012). *Ganoderma lucidum* polysaccharide accelerates refractory wound healing by inhibition of mitochondrial oxidative stress in type 1 diabetes. *Cell. Physiol. Biochem.*, 29(3-4): 583–594. 10.1159/000338512.

Ueitele, I.S.E., Kadhila-Muanding, N.P. and Matundu, N. (2014). Evaluating the production of *Ganoderma* mushroom on corn cobs. *Afr. J. Biotechnol.*, 13(22): 2215–2219. 10.5897/AJB2014.13650.

Wachtel-Galor, S., Szeto, Y.T., Tomlinson, B. and Benzie, I.F. (2004). *Ganoderma lucidum* ('Lingzhi'); acute and short-term biomarker response to supplementation. *Int J. Food Sci. Nutr.*, 55(1): 75–83. 10.1080/09637480310001642510.

Wang, C.F., Liu, J.Q., Yan, Y.X., Chen, J.C., Lu, Y., Guo, Y.H. and Qiu, M.H. (2010). Three new triterpenoids containing four-membered ring from the fruiting body of *Ganoderma sinense*. *Org. Lett.*, 12(8): 1656–1659. 10.1021/ol100062b.

Wang, H. and Ng, T.B. (2006). Ganodermin, an antifungal protein from fruiting bodies of the medicinal mushroom *Ganoderma lucidum*. *Peptides*, 27(1): 27–30. 10.1016/j.peptides.2005.06.009.

Wang, H., Zhang, R., Song, Y., Li, T. and Ge, M. (2019). Protective effect of ganoderma triterpenoids on cadmium-induced testicular toxicity in chickens. *Biol. Trace Elem. Res.*, 187(1): 281–290. 10.1007/s12011-018-1364-4.

Wang, K., Bao, L., Ma, K., Zhang, J., Chen, B., Han, J., Ren, J., Luo, H. and Liu, H. (2017). A novel class of α-glucosidase and HMG-CoA reductase inhibitors from *Ganoderma leucocontextum* and the anti-diabetic properties of ganomycin I in KK-Ay mice. *Eur. J. Med. Chem.*, 127: 1035–1046. 10.1016/j.ejmech.2016.11.015.

Wang, X.L., Wu, Z.H., Di, L., Zhou, F.J., Yan, Y.M. and Cheng, Y.X. (2019). Renoprotective phenolic meroterpenoids from the mushroom *Ganoderma cochlear*. *Phytochemistry*, 162: 199–206. 10.1016/j.phytochem.2019.03.019.

Wasser, S.P. and Weis, A.L. (1999). Medicinal properties of substances occurring in higher basidiomycetes mushrooms: current perspectives. *Int. J. Med.*, 1(1): 31–62. 10.1615/IntJMedMushrooms.v1.i1.30.

Wasser, S.P. (2002). Medicinal mushrooms as a source of antitumor and immunomodulating polysaccharides. *Appl. Microbiol. Biotechnol.*, 60(3): 258–274. doi.org/10.1007/s00253-002-1076-7.

Wasser, S.P. (2005). Reishi or ling zhi (*Ganoderma lucidum*). *Encyclopedia of Dietary Supplements*, 1: 603–622.

Wei, S.J., Xu, X.R., Huang, L. and Tu, G.Q. (2007). Study on the production of polysaccharide from *Ganoderma lucidum* mycelia in deep-liquid fermentation. *Jiangxi Sci.*, 25(292294): 301.

Wu, B.F., Liu, L.L., Fang, Z.H. and Liu, X.Y. (2008). Effects of nutrition factors on mycelium growth of 51427 in *Ganoderma lucidum*. *Anhui. Agr. Sci. Bull.*, 14: 57–58.

Wu, D.Z., Zhang, L.X., Xu, R. and Zhang, K.C. (2000). Production of bacterial alpha-amylase by solid-state fermentation. *J. Wuxi Univ. Light Ind.*, 19: 54–57.

Wu, J.G., Kan, Y.J., Wu, Y.B., Yi, J., Chen, T.Q. and Wu, J.Z. (2016). Hepatoprotective effect of ganoderma triterpenoids against oxidative damage induced by tert-butyl hydroperoxide in human hepatic HepG2 cells. *Pharm. Biol.*, 54(5): 919–929. 10.3109/13880209.2015.1091481.

Wu, K., Na, K., Chen, D., Wang, Y., Pan, H. and Wang, X. 2018. Effects of non-steroidal anti-inflammatory drug-activated gene-1 on *Ganoderma lucidum* polysaccharides-induced apoptosis of human prostate cancer PC-3 cells. *Int. J. Oncol.*, 53(6): 2356–2368. 10.3892/ijo.2018.4578.

Xiao, C., Wu, Q.P., Cai, W., Tan, J.B., Yang, X.B. and Zhang, J.M. (2012). Hypoglycemic effects of *Ganoderma lucidum* polysaccharides in type 2 diabetic mice. *Arch. Pharm. Res.*, 35(10): 1793–1801. 10.1007/s12272-012-1012-z.

Xu, J., Xiao, C., Xu, H., Yang, S., Chen, Z., Wang, H., Zheng, B., Mao, B. and Wu, X. (2021). Anti-inflammatory effects of *Ganoderma lucidum* sterols via attenuation of the p38 MAPK and NF-κB pathways in LPS-induced RAW 264.7 macrophages. *Food Chem. Toxicol.*, 150: 112073. 10.1016/j.fct.2021.112073.

Yan, M.H. (2000). Cultivation *Ganoderma lucidum* using oak leaf and waste tea. *Edible Fungi*, 2: 22–23.

Yang, F.C. and Liau, C.B. (1998). Effects of cultivating conditions on the mycelial growth of *Ganoderma lucidum* in submerged flask cultures. *Bioprocess Eng.*, 19(3): 233–236. 10.1007/PL00009014.

Yong, T., Chen, S., Xie, Y., Chen, D., Su, J., Shuai, O., Jiao, C. and Zuo, D. (2018). Hypouricemic effects of *Ganoderma applanatum* in hyperuricemia mice through OAT1 and GLUT9. *Front. Pharmacol.*, 8: 996. 10.3389/fphar.2017.00996. 10.1155/2020/9894037.

Yu, N., Huang, Y., Jiang, Y., Zou, L., Liu, X., Liu, S., Chen, F., Luo, J. and Zhu, Y. (2020). *Ganoderma lucidum* Triterpenoids (GLTs) reduce neuronal apoptosis via inhibition of ROCK signal pathway in APP/PS1 transgenic Alzheimer's disease mice. *Oxid. Med. Cell. Longev.*, Article ID 9894037: 11. 10.1155/2020/9894037.

Yue, G.G., Chan, B.C., Han, X.Q., Cheng, L., Wong, E.C., Leung, P.C., Fung, K.P., Ng, M.C., Fan, K., Sze, D.M. and Lau, C.B. (2013). Immunomodulatory activities of *Ganoderma sinense* polysaccharides in human immune cells. *Nutr. Cancer*, 65(5): 765–774. 10.1080/01635581.2013.788725.

Zhang, H.J., Cao, L.S. and Ye, S.Q. (2004). Measures of log-cultivated *Ganoderma lucidum* for high-yielding and quality. *J. Agr Sci.*, 3: 136–138.

Zhang, H.N., He, J.H., Yuan, L. and Lin, Z.B. (2003). *In vitro* and *in vivo* protective effect of *Ganoderma lucidum* polysaccharides on alloxan-induced pancreatic islets damage. Life Sci., 73(18): 2307–2319. 10.1016/S0024-3205(03)00594-0.

Zhang, J., Ma, K., Han, J., Wang, K., Chen, H., Bao, L., Liu, L., Xiong, W., Zhang, Y., Huang, Y. and Liu, H. (2018). Eight new triterpenoids with inhibitory activity against HMG-CoA reductase from the medical mushroom *Ganoderma leucocontextum* collected in Tibetan plateau. *Fitoterapia*, 130: 79–88. 10.1016/j.fitote.2018.08.009.

Zhang, J.J., Dong, Y., Qin, F.Y., Yan, Y.M. and Cheng, Y.X. (2019). Meroterpenoids and alkaloids from *Ganoderma australe*. *Nat. Prod. Res.*, 35: 3226–3232. 10.1080/14786419.2019.1693565.

Zhang, L.H. and Wang, S.X. (2010). Study on the binding and packing cultivation technology of the *G. lucidum*'s artificial alternative compost. *Agr. Tech. Serv.*, 27: 516–517.

Zhang, S., Nie, S., Huang, D., Li, W. and Xie, M. (2013). Immunomodulatory effect of *Ganoderma atrum* polysaccharide on CT26 tumor-bearing mice. *Food Chem.*, 136(3-4): 1213–1219. 10.1016/j.foodchem.2012.08.090.

Zhang, W., Tao, J., Yang, X., Yang, Z., Zhang, L., Liu, H., Wu, K. and Wu, J. (2014). Antiviral effects of two *Ganoderma lucidum* triterpenoids against enterovirus 71 infection. *Biochem. Biophys. Res. Commun.*, 449(3): 307–312. 10.1016/j.bbrc.2014.05.019.

Zhao, H.B., Wang, S.Z., He, Q.H., Yuan, L., Chen, A.F. and Lin, Z.B. (2005). *Ganoderma* total sterol (GS) and GS1 protect rat cerebral cortical neurons from hypoxia/reoxygenation injury. *Life Sci.*, 76(9): 1027–1037. 10.1016/j.lfs.2004.08.013.

Zhao, L., Dong, Y., Chen, G. and Hu, Q. (2010). Extraction, purification, characterization, and antitumor activity of polysaccharides from *Ganoderma lucidum*. *Carbohydr. Polym.*, 80(3): 783–789. 10.1016/j.carbpol.2009.12.029.

Zhao, X.R., Huo, X.K., Dong, P.P., Wang, C., Huang, S.S., Zhang, B.J., Zhang, H.L., Deng, S., Liu, K.X. and Ma, X.C. (2015). Inhibitory effects of highly oxygenated lanostane derivatives from the fungus *Ganoderma lucidum* on P-glycoprotein and α-glucosidase. *J. Nat. Prod.*, 78(8): 1868–1876. 10.1021/acs.jnatprod.5b00132.

Zhao, Z.Z., Chen, H.P., Feng, T., Li, Z.H., Dong, Z.J. and Liu, J.K. (2015). Lucidimine AD, four new alkaloids from the fruiting bodies of *Ganoderma lucidum*. *J. Asian Nat. Prod. Res.*, 17(12): 1160–1165. 10.1080/10286020.2015.1119128.

Zheng, J., Yang, B., Yu, Y., Chen, Q., Huang, T. and Li, D. (2012). *Ganoderma lucidum* polysaccharides exert anti-hyperglycemic effect on streptozotocin-induced diabetic rats through affecting β-cells. *Comb. Chem. High Throughput Screen*, 15(7): 542–550. 10.2174/138620708784534815.

Zheng, M., Tang, R., Deng, Y., Yang, K., Chen, L. and Li, H. (2018). Steroids from *Ganoderma sinense* as new natural inhibitors of cancer-associated mutant IDH1. *Bioorg. Chem.*, 79: 89–97. 10.1016/j.bioorg.2018.04.016.

Zhu, Q., Xiong, X.H. and Wang, F. (2009). Effects of radix glycyrrhizae on production of *Ganoderma lucidum* polysaccharide. *Chin. Brewing.*, 2: 86–88.

Zhu, Q., Xia, Y.Q., Dong, K.K., Yang, C.F. and Wang, Z.J. (2010). Optimization of *Ganoderma lucidum* medicinal solid fermentation medium by response surface test. *Chin. Biotechnol.*, 30: 75–79.

Zhu, Q., Bang, T.H., Ohnuki, K., Sawai, T., Sawai, K. and Shimizu, K. (2015). Inhibition of neuraminidase by *Ganoderma* triterpenoids and implications for neuraminidase inhibitor design. *Sci. Rep.*, 5(1): 1–9. 10.1038/srep13194 (2015).

Zhuang, Y., Chi, Y.M. and Chen, S.B. (2004). Preparation of medicinal fungal new type bi-directional solid fermentation engineering and Huai Qi fungal substance. *Chin. Pharm. J.*, 39(3): 175–178.

8

Bioactive Properties of Malaysian Medicinal Mushrooms *Lignosus* spp.

Hui-Yeng Yeannie Yap,[1] *Boon Hong Kong*[2] and *Shin Yee Fung*[2,3,4*]

1. Introduction

The tiger milk mushroom *L. rhinocerus* is traditionally used by aborigines and locals to treat various human diseases and improve general health. It has been originally documented more than 400 years ago in The Diary of John Evelyn (31 October 1620–27 February 1706), a gentleman Royalist and virtuoso of the seventeenth century. Three species of the genus *Lignosus* found in Malaysia include *L. cameronensis*, *L. rhinocerus*, and *L. tigris*. The tiger milk mushroom was originally described by Cubitt who recorded the taxonomy of a sample which he discovered in Penang Island of Malaysia and named it *Polyporus rhinocerus* (Mycobank MB534962) (Cooke, 1879). This mushroom is known by a variety of local names, e.g., Cendawan Susu Harimau, Betes Kismas, Hurulingzhi, Hijiritake (Burkill and Haniff, 1930; Burkill et al., 1966; Haji Taha, 2006; Huang, 1999a,b; Yokota, 2011). The synonyms of tiger milk mushrooms include *Fomes rhinocerus*, *Microporus rhinoceros*, *Polystictus rhinocerus*, *Polyporus sacer* var. *rhinocerus*, and *Scindalma rhinocerus* (Sacchardo, 1888; Kuntze, 1898; Lloyd, 1920; Imazeki, 1952). Its accepted name is *L. rhinocerus* (Cooke) (Ryvarden, 1972). Taxonomically, *L. rhinocerus* is distinct from the rest of the *Lignosus* species.

[1] Department of Oral Biology and Biomedical Sciences, Faculty of Dentistry, MAHSA University, Bandar Saujana Putra, Selangor, Malaysia
[2] Medicinal Mushroom Research Group (MMRG), Department of Molecular Medicine, Faculty of Medicine, University of Malaya, Kuala Lumpur, Malaysia
[3] Center for Natural Products Research and Drug Discovery (CENAR), University of Malaya, Kuala Lumpur, Malaysia
[4] University of Malaya Centre for Proteomics Research (UMCPR), University of Malaya, Kuala Lumpur, Malaysia
* Corresponding author: syfung@ummc.edu.my

Fig. 8.1. Basidiomes of *Lignosus rhinocerus* collected from Peninsular Malaysia. Picture courtesy of LiGNO Biotech Sdn.Bhd.

Prior to the discovery and identification of *L. cameronensis* and *L. tigris* as two distinct species derived from *L. rhinocerus*, these were often described and interchangeably used as *L. rhinocerus*. In 2013, *L. cameronensis* and *L. tigris* were described based on collections from the tropical forests of Pahang, Malaysia (Tan et al., 2013) (Fig. 8.1). Morphologically, these species are alike, but upon detailed inspection, differ in their pore and basidiospore dimensions. *L. cameronensis* possesses 2–4 pores/mm with basidiospore 2.4–4.8 × 1.9–3.2 μm. On the other hand, *L. tigris* possesses 1–2 pores/mm, with basidiospore 2.5–5.5 × 1.8–3.6 μm. The comparison of these species with its sister species revealed *L. rhinocerus*, 7–8 pores/mm with basidiopore 3–3.5 × 2.5–3 μm. To date, wild types of these species are still referred to as *L. rhinocerus*, since the sclerotia are often collected without an intact stipe and cap, which makes the differentiation of these species difficult among these species. The taxonomic tree described by Fung and Tan (2019) have shown that *L. rhinocerus* is distinctly set apart from the rest of *Lignosus* sp. leading to the assumption that its bioactive properties may possibly be different in these species. This chapter addresses the cultivation, pharmacological significance, and biomedical applications of tiger milk mushrooms with its usage as a functional food.

2. Cultivation of *Lignosus* species

The deforestation to support modern development and environmental pollution have placed many wild fungi under the threat of existence (Vikineswary and Chang, 2013). However, efforts to cultivate *Lignosus* spp., particularly the well-known species *L. rhinocerus*, have been initiated since 1999. Huang (1999b) cultivated *P. rhinocerus* (= *L. rhinocerus*) using substrate bags inoculated with spawn (young *P. rhinocerus* mycelia grown on sawdust-wheat bran medium) which formed sclerotia after 18 months. Most of other cultivation efforts are focused on mycelium. Rahman and colleagues tried to cultivate *L. rhinocerus* mycelium in mushroom complete medium (MCM) using the submerged culture technique (Rahman et al., 2012). Two years later, Lai and colleagues (Lai et al., 2014) further optimized the submerged culture conditions for production of mycelial biomass and exopolysaccharides derived from *L. rhinocerus* by controlled culture conditions and modified medium composition. Other techniques of mycelial cultivation of tiger milk mushroom use sawdust, paddy straw, and spent yeast (Abdullah et al., 2013) in small quantities' using flasks (Lau et al., 2013; Lau et al., 2011) and bioreactors in the process of submerged fermentation (Chen et al., 2013)

A quick cultivation method leading to mass production of *L. rhinocerus* was used in 2009 by LiGNO™ Biotech Sdn. Bhd., a Malaysian Small Medium Enterprise. They utilized a formulated culture medium consisting of rice, water and other nutrient-based materials and incubated the inoculated medium in an environmentally controlled culture room for up to six months to allow the formation of sclerotia before harvesting. This in-house proprietary method has enabled the commercial production of *L. rhinocerus* (trade name, TM02®) and promoted scientific studies ranging from safety studies (preclinical and pre-commercialization), leading to the registration of *L. rhinocerus* as a dietary supplement. The same method was used to successfully cultivate S4 strain of *L. cameronensis* and K strain of *L. tigris* in small scales. Similar to *L. rhinocerus* TM02®, the advent of cultivation has also enabled researchers to perform a study related to their sub-acute toxicity (Kong et al., 2016a; Lee et al., 2017). Nutritional properties of *L. cameronensis* (Fung et al., 2019) and *L. tigris* (Kong et al., 2016b) have been found to be different compared to *L. rhinocerus*. Both species have been shown to be safe for consumption with no adverse effects on blood biochemical parameters and human organs. The study on the bioactivity of *L. tigris* showed an interesting comparison with *L. rhinocerus*, especially in the treatment of breast cancer (Kong et al., 2020).

3. *Lignosus rhinocerus*—A Medicinal Fungi

The scientific studies to validate the ethno-mycological uses of *L. rhinocerus* and subsequent commercialisation have been made possible since its mass cultivation. The revelation of *L. rhinocerus* genome has provided valuable insights leading to the discovery of biomolecules. It is particularly enriched with sesquiterpenoid biosynthesis genes and may encode for 1,3-β- and 1,6-β-glucans, as well as bioactive proteins, including lectins and fungal immunomodulatory proteins (Yap et al., 2014). Using the *L. rhinocerus* genome database, genome-wide expression profiling *via* RNA-seq, or whole transcriptome, and shotgun sequencing were conducted to identify the genes expressed in the sclerotium of *L. rhinocerus* (Yap et al., 2015a). A few highly expressed genes encoding cysteine-rich ceratoplatanin, hydrophobins, and sugar-binding lectins were identified and promoted research in this area. The research regarding this species was initiated from the beginning of the millenia (*circa* 2000), including studies related to the immunomodulatory, neuroprotective, anti-asthmatic, anti-inflammatory, anti-tumour, and antimicrobial effects. A review of its research status has been conducted twice to consolidate the scientifically validated data for prospecting practical applications of the mushroom (Lau et al., 2015; Nallathamby et al., 2018). The following subsections summarise the latest research activities conducted on *L. rhinocerus* mushroom. These data form a strong basis for pharmacological and industrial applications of *L. rhinocerus*.

3.1 Immunomodulatory Effect

Mushrooms are often associated with immune modulation due to their ability to stimulate the production of hematopoietic stem cells, lymphocytes, macrophages,

T cells, dendritic cells, and natural killer (NK) cells (Moradali et al., 2007). In 2016, a double-blind, placebo-controlled study investigated the combined effect of *L. rhinocerus* supplementation (LRS; 500 mg/day for 8 weeks) and resistance training (RT) on isokinetic muscular strength and power, anaerobic and aerobic fitness level, as well as immunological status in young males. Combined RT with LRS and RT alone showed beneficial effects on all described parameters, except for the strength of immune response. A significant increase in the total lymphocyte, B and T lymphocyte levels was only observed because of RT. Thus, LRS alone and in combination with RT may not provide benefits on these immune parameters (Chen et al., 2016).

However, in vitroly, several polysaccharide fractions with immunomodulatory properties were isolated from the sclerotium of *L. rhinocerus* and characterised by their β-glucan content, glycosidic linkages and/or enzymatic reaction. Two specific water-soluble polysaccharide fractions, namely fractions D and E, inhibited the secretion of tumour necrosis factor-α (TNF-α) in lipopolysaccharide-stimulated RAW 264.7 macrophages in a dose-dependent manner suggesting an anti-inflammatory effect of these fractions (Keong et al., 2016). Previous study reported four polysaccharide fractions, LRP-1, LRP-2, LRP-3, and LRP-4, which prevented cyclophosphamide-induced immunosuppression in mice, as well as stimulated the synthesis of TNF-α and INF-γ cytokines. LRP-1 and LRP-2 were polysaccharide-protein complexes mostly composed of β-d-glucose (Hu et al., 2017). Previous studies isolated polysaccharides *via* hot water extraction. Meanwhile, Sum et al. (2020) performed cold water extraction to obtain CWE from mushroom sclerotia. In addition to proteins, CWE contains linear polysaccharides with 1,4-linkages. Its rhinoprolycan Sephadex G-50® fractions, HMW and MMW, possess 1,4-Glcp and 1,6-Glcp backbone and a branched chain (1,3,6-Glcp, 1,4,6-Glcp, 1,3,6-Glcp, 1,2,4,6-Glcp). CWE, HMW, and MMW have shown to regulate the release of cytokines (i.e., interleukins, TIMP-1) in RAW 264.7 macrophages.

In addition to polysaccharides, fungal immunomodulatory protein (FIP) also plays a significant role in enhancing the immune response. A genome-based proteomic analysis of *L. rhinocerus* TM02® sclerotial proteins revealed the expression of a protein with amino acid sequence homolog to lingzhi-8, an immunomodulatory protein isolated from *Ganoderma lucidum* with a Fve domain, a major fruiting body protein derived from *Flammulina velutipes*, which possesses immunomodulatory activity (Yap et al., 2015c). The corresponding gene was subsequently cloned and characterized. The product, 6xHisFIP-Lrh, possesses a haemagglutinative effect which has shown to exhibit cytotoxicity on MCF-7, HeLa, and A549 cancer cell lines (Pushparajah et al., 2016).

3.2 Anti-cancer and Anti-proliferative Effects

Previous studies have demonstrated that (medicinal) mushrooms possess potent anti-cancer properties by functioning as inducers of reactive oxygen species, mitotic kinase inhibitors, anti-mitotic agents, angiogenesis inhibitors, topoisomerase inhibitors, leading to apoptosis and decrease in cancer proliferation (Patel and Goyal, 2012). This property is also found in tiger milk mushroom. Several groups have reported the cytotoxic effects of hot and cold aqueous extracts of mushroom's sclerotium against a panel of human cancer cell lines. The cytotoxic component(s) in cold extract were further determined to be protein/peptide(s) and/or protein-carbohydrate complex of high molecular weight (Lee et al., 2012; Lau et al., 2013; Yap et al., 2013). It has also been reported that the cultivated strain (i.e., TM02®) was more cytotoxic than the wild type (90 μg/ml vs. 206 μg/ml) against human breast carcinoma (MCF-7) cells (Yap et al., 2013). Furthermore, purification of the sclerotial cold water extract *via* sequential chromatography methods resulted in the production of cytotoxic fungal serine protease fraction, termed F5, which exhibited potent selective cytotoxicity against MCF7 cells with IC_{50} value of 3.00 ± 1.01 μg/ml (Yap et al., 2015b). It was further suggested that F5 induced apoptosis in MCF7 cells by upregulating caspase-8 and -9 activities with a significant decrease of pro-survival Bcl-2. The levels of pro-apoptotic Bax, BID and cleaved BID were reportedly increased accompanied by an observable actin cleavage (Yap et al., 2018b). Due to the genome database availability of *L. rhinocerus*, a lectin gene from *L. rhinocerus* TM02® was cloned and a recombinant lectin, Rhinocelectin, was obtained with a selective anti-proliferative activity against triple negative breast cancer MDA-MB-231 and MCF7 cells in a concentration-dependent manner (Cheong et al., 2019). Lectins are major proteins in the sclerotium of *L. rhinocerus* with 23.1 to 39.1% of the total extracted protein reported in earlier studies (Yap et al., 2015b,c).

Most studies have described the anti-proliferative and/or anti-cancer effects of *L. rhinocerus* sclerotial proteins from cold water extraction, others have reported different extraction methods. Fauzi et al. (2015) used pressurized liquid extraction to obtain LRME (methanol extract) and LRAE (hot aqueous extract) which demonstrated a selective mild toxic effect on human colorectal cancer cells (HCT 116), specifically the former (IC_{50}: 600 μg/ml) with induction of apoptosis and G0/G1-phase arrest. Both extracts are rich in alkaloids. The mycelium and culture broth of the mushroom have also been suggested as substitutes for the naturally occurring sclerotium regarding the cytotoxic effect (Lau et al., 2014). The group adapted methanol extraction for sample preparation. It is not surprising that overall cytotoxicity was weaker as an earlier study showed inactivity of *L. rhinocerus* methanol extract against several human cancer cell lines (Fung et al., unpublished).

3.3 Respiratory Health-Enhancing Effect

Tiger milk mushroom has historically been used to treat pulmonary diseases, such as bronchial asthma. Since its commercialisation, various studies have investigated its effects on airway relaxation and inflammation. Using airway segments isolated from Sprague Dawley rat in an organ bath setup, Lee et al. (2018c) revealed that the polysaccharide-protein complex or proteins found in HMW and MMW fractions (refer to Section 3.1) contributed to broncho-relaxative effect of CWE mediated by calcium signalling pathway downstream of $G_{\alpha q}$-coupled protein receptors (Lee et al., 2018a; Lee et al., 2018b). Meanwhile, Johnathan et al. (2016) focused on hot water extract (HWE) of *L. rhinocerus* sclerotium and investigated its effect on ovalbumin (OVA)-sensitized asthmatic Sprague Dawley rats, an airway inflammation model. The group outlined the presence of volatile constituents, including alkanes and linoleic acid, as major components upon sequential liquid-liquid extraction in HWE. Interestingly, another group described HWE as carbohydrates, despite of differences in the method of (hot water) extraction used (data unpublished). HWE was reported to increase the total serum IgE levels and suppressed eosinophil infiltrations in the lungs. Low levels of interleukins (IL-4, IL-5, IL-13) and eosinophils were observed in the isolated bronchoalveolar lavage fluid. The presence of polyunsaturated fatty acids (PUFA), i.e., linoleic acid, may have contributed to these events (Johnathan et al., 2016). Intranasal administration of HWE decreases airway inflammation in a murine model of allergic bronchial asthma (Muhamad et al., 2019) which is congruent with earlier findings reporting similar effects of HWE in an airway hyper-responsiveness study using house dust mite-induced asthma in Balb/c mice (Johnathan et al., 2021).

Recent clinical study suggested that certain tiger milk mushroom supplementation (formulation and source to be verified) effectively improves respiratory health, immunity, and antioxidant status in a group of 50 healthy volunteers (aged 30–50 years). The results of this study have shown that the levels of pro-inflammatory cytokines, IL-1β and IL-8, malondialdehyde and respiratory symptoms were attenuated, whereas the level of immunogenic IgA, total antioxidant capacity and respiratory function improved (Tan et al., 2021a). These findings provide evidence to support the traditional use of tiger milk mushroom for treatment of bronchial asthma and cough.

3.4 Anti-Inflammatory Effect

Inflammation is a normal response to infection and/or physical trauma. However, an over-inflammatory state may lead to an increased risk of various diseases and death (Sherwood and Toliver-Kinsky, 2004). Non-steroidal anti-inflammatory drugs (NSAIDS) are widely used for treatment of inflammation. However, due to their severe side effects (Raskin, 1999), herbs and dietary supplements may be a solution to a safer and possibly more effective alternative (continuous) treatment (Maroon et al., 2010).

The anti-inflammatory activity of *L. rhinocerus* has been attributed to the protein component (containing 8 parts of carbohydrate to 1 part of protein, mostly α-glucan) of its HMW (refer to Section 3.1) in a carrageenan-induced paw edema test where Lee et al. (2014) described the inhibition of TNF-α production in lipopolysaccharide-induced RAW 264.7 macrophage cells by HMW (IC_{50} = 0.76 μg/ml). Further studies were mainly focused on their anti-inflammatory activities. Among the tested preparations, hot aqueous extract and its n-butanol and ethyl acetate fractions showed the maximal inhibition of nitric oxide (NO) production in lipopolysaccharide-stimulated BV2 microglia (Seow et al., 2017). Meanwhile, the presence of linoleic acid in ethyl acetate fraction of *L. rhinocerus* sclerotium has been shown to suppress inflammation by reducing NO production and down-regulated the expression of neuroinflammatory iNOS and COX2 genes in brain microglial (BV2) cells *via* either NF-κB pathway, STAT3 pathway, or both pathways (Nallathamby et al., 2016). In view of these findings, the anti-inflammatory activity of *L. rhinocerus* may be an outcome of synergistic effect of various components present in mushroom sclerotium, including proteins, carbohydrates (polysaccharides), fatty acids, and secondary metabolites.

3.5 Neuroprotective Effect

The neuritogenic activity of *L. rhinocerus* has been described by several research groups across Asia. This raises interest in *L. rhinocerus* as a potential preventive and/or therapeutic agent for neurodegenerative diseases. Seow et al. (2015) suggested that the mushroom sclerotium contains neuroactive compound(s) that mimic the neuritogenic activity of nerve growth factor (NGF). In particular, hot aqueous extract of *L. rhinocerus* at 25 μg/ml exhibited a stimulatory effect comparable to a nerve growth factor (NGF; 50 ng/ml) in pheochromocytoma (PC-12) cells which could be mediated through the phosphorylation of TrkA receptor and ERK1/2 signalling pathway. Furthermore, the group reported expression of neuritin, a key protein in neuritogenesis, neurite arborization and extension in *L. rhinocerotis* (syn. *L. rhinocerus*) aqueous co-treatment with NGF-treated PC-12 cells which suggests that neuritin modulation is involved in neurite outgrowth (Tan et al., 2021b). In addition to PC-12 cells, stimulation of neurite outgrowth by *L. rhinocerus* has been described in dissociated brain cells, spinal cord, and retina (Samberkar et al., 2015). The oral administration of a low-dose of hot aqueous sclerotial extract enhanced functional recovery in *in vivo* model of a sciatic nerve crush injury and showed no adverse effects on nervous tissue (Farha et al., 2019).

Kittimongkolsuk et al. (2021a) investigated the neuroprotective effect of *L. rhinocerus*. The results of this study have shown that sclerotial ethanol extract exhibited neuroprotective effect in glutamate-induced oxidative stress in a mouse hippocampal (HT22) cells (*in vitro* model) and prevented neurotoxicity in *C. elegans in vivo*. The presence of phenolics and phospholipids may have instigated these beneficial effects. This data is in line with the findings of a previous study which showed that methanol extract, among other extracts of *L. rhinocerus* sclerotium, exhibited the highest neuroprotective effect (with the possible involvement of Akt signalling) in dexamethasone-induced toxicity in human embryonic stem cell-derived neural stem cells (Yeo et al., 2019). In addition, the mycelial aqueous extract of *L. rhinocerus* at 20 μg/ml has been shown to stimulate neurite outgrowth by 21.1% in PC-12 cells. Further supplementation with 1 μg/ml of curcumin resulted in 27.2% neurite extension (John et al., 2013). Thus, the findings of previous studies have shown that neuritogenetic potential of *L. rhinocerus* is determined by carbohydrates (e.g., polysaccharide-including glucans) and secondary metabolites, including phenolics, triterpenes, terpenoids, and alkaloids present in hot aqueous and methanol extracts. Further exploration and elucidation of these potential neuroactive compounds are warranted.

3.6 Antioxidant and Anti-Ageing Effect

Mushrooms are rich in antioxidants which enable them to neutralize free radicals. The content of these antioxidant components varies within species and can be found in fruiting bodies of both mycelium and culture, which include polysaccharides, tocopherols, phenolics, carotenoids, ergosterol, ergothioneine, and ascorbic acid among others (Sánchez, 2016). The sclerotial extracts of *L. rhinocerus* have been reported to contain phenolics and significant superoxide anion radical-scavenging activity comparable to rutin in both wild and cultivated strains (Yap et al., 2013). Lau et al. (2014) have observed that aqueous methanol extracts of its mycelium, culture broth, and sclerotium demonstrated antioxidant effects based on their radical-scavenging properties, metal-chelating and inhibitory effects on lipid peroxidation. On the other hand, Nallathamby et al. (2016) reported that ethyl acetate fraction derived from *L. rhinocerus* sclerotial hydroethanol extract possessed ferrum-reducing capacity and radical (i.e., $ABTS^{\bullet+}$, DPPH) scavenging activity.

A genome-based proteomic analysis of *L. rhinocerus* sclerotium reported the presence of manganese-superoxide dismutase (SOD) and catalases (CAT) that function as antioxidants to reduce cytotoxic reactive oxygen species (Yap et al., 2015c). The feeding of streptozotocin-induced diabetic rats with powder tiger milk mushroom sclerotial has been shown to significantly increase the activities of glutathione, CAT, and SOD and decrease lipid peroxidation (Nyam et al., 2017). The group also reported a significant decrease in the levels of elevated blood glucose concentrations to a normal range in the rat model, indicating an antidiabetic potential of the product. This is further supported by its ability to suppress protein glycation which may be partly associated with its strong superoxide anion radical-scavenging activity. The MMW fraction (see Section 3.1) obtained from a combination of cold-water extract and Sephadex® G-50 (fine) gel filtration chromatography of *L. rhinocerus* sclerotia powder has been shown to inhibit the formation of Nε-(carboxymethyl) lysine, pentosidine, and other advanced glycation end-product structures in a human serum albumin-glucose system (IC_{50} value = 1 μg/ml) (Yap et al., 2018a). In an open-label prospective study, tiger milk mushroom supplementation further induced total antioxidant capacity of participants with regard to their gender and body mass index (BMI) (Tan et al., 2021a). The antioxidant profile and anti-glycating activity of *L. rhinocerus* demonstrated its anti-ageing potential. Its ethanol, cold and hot water sclerotial extracts have been shown to enhance stress resistance, decrease intracellular ROS and extend lifespan in *Caenorhabditis elegans via* DAF-16/FoxO signalling pathway (Kittimongkolsuk et al., 2021b).

3.7 A Wound-Healing Effect

The tiger milk mushroom has been traditionally used by local Malay and Chinese communities to treat wounds. Ahmad et al. (2014) reported considerable protease and fibrinolytic activities in LR-1 of wild type *L. rhinocerus* sclerotium in which a clear zone was observed on skim milk agar and fibrin plates. This finding signifies the potential role of this mushroom in the wound-healing process during the breakdown of damaged ECM proteins and foreign material for new tissue formation and wound closure. Besides, β-glucans obtained from the mushroom sclerotium by an enzyme-assisted extraction method, facilitated the healing of intestinal mucosal wounds by hastening intestinal epithelial cell proliferation and migration *via* the activation of Rho-dependent pathway (Veeraperumal et al., 2021). The oral administration of *L. rhinocerus* sclerotial powder exhibited dose-dependent anti-ulcer effect in the gastric mucosa and demonstrated a significant protection against gastric ulceration in ethanol and aspirin-induced, as well as water immersion-restraint stress-induced ulcer Sprague-Dawley rat models (Nyam et al., 2016).

3.8 Antimicrobial and Antiviral Activities

Mushrooms are regarded as sources of natural antibiotics due to the presence of secondary metabolites including terpenes and sesquiterpenes, steroids, anthraquinone, benzoic acid derivatives, quinolines, oxalic acid, as well as peptides and proteins (Lima et al., 2016). Shopana et al. (2012) have previously reported a mild to moderate antimicrobial activity of *L. rhinocerus* sclerotium against 15 bacteria and four fungi. The tested samples were petroleum ether, chloroform, methanol, and water extracts. The ethanol and water extracts of *L. rhinocerus* contain fatty acids, peptides, and terpenoids and have been reported to show human immunodeficiency virus type-1 (HIV-1) protease (PR) and reverse transcriptase (RT) inhibitions at 1 mg/ml (Sillapachaiyaporn and Chuchawankul, 2020). The authors of another study reported the antiviral activity of aqueous and hexane fractions from an ethanol-extracted liquid fermented *L. rhinocerus* mycelium against H1N1 Influenza virus infections suggesting the presence of antiviral components in *L. rhinocerus* with different polarities (Lin et al., 2020). Khazali et al. (2021) also reported the inhibitory activity of CWE against dengue serotype 2 infection in Vero cells in a dose-dependent manner on virus replication, particularly during the early stages of infection. In addition, the extract exhibited significant antiviral and mild prophylactic effects suggesting its potential as an alternative medicine in decreasing the risk and treatment of dengue infection. Table 8.1 shows an overview of pharmacological studies related to *L. rhinocerus* sclerotia.

4. The Biomedical Application of *Lignosus tigris*

Previous study on the chemical composition of wild strain *L. tigris* sclerotium revealed a higher content of proteins, carbohydrates, and minerals compared to the wild strain of *L. rhinocerus* (Yap et al., 2013, 2014b). The findings by Yap and colleagues (Yap et al., 2013, 2014b) suggest that the medicinal value of *L. tigris* is equivalent to *L. rhinocerus*. An upscale production of *L. tigris* sclerotium using the cultivation technology by Tan (2009) enables further studies of pharmacological properties of this mushroom. Preliminary studies on the cultivated *L. tigris* sclerotium (Ligno TG-K) revealed that it contains biologically active proteins, carbohydrates, and secondary metabolites, such as phenolics and terpenoids (Kong et al., 2016b).

The amount of protein in Ligno TG-K was three times higher compared to the wild strain (Yap et al., 2014b; Kong et al., 2016b), sclerotium of *L. rhinocerus* TM02 (13.8 g/100 g DW) (Yap et al., 2013), as well as *Pleurotus tuber-regium* (12.4 g/100g DW) (Oranusi et al., 2014). A preliminary study showed that cold water extract of Ligno TG-K inhibited the proliferation of breast cancer MCF7 cells and its activity was superior to *L. rhinocerus* TM02 (Lee et al., 2012; Kong et al., 2020; Yap et al., 2013). Further studies demonstrated that the cytotoxicity of Ligno TG-K was attributed to a high molecular weight protein fraction which contains several cytotoxic proteins, including lectins, serine proteases, and nucleases (Kong et al., 2020). A breast cancer MCF7 cell line exposed to Ligno TG-K protein

Table 8.1. An overview of pharmacological studies of *L. rhinocerus* sclerotia by country

MALAYSIA				
Strain	**Name of Specific Constituent**	**Bioactive Compounds**	**Pharmacological Effects**	**References**
Wild type	Fraction D, Fraction E	Water-soluble polysaccharides	Immunomodulatory, anti-inflammatory	Keong et al., 2016
Wild type	LR1 (homogenized water extract)	Not investigated	Fibrinolytic	Ahmad et al., 2014
Wild type	Petroleum ether, chloroform, methanol, and water extracts	Not investigated	Antimicrobial	Shopana et al., 2012
TM02®	CWE, HMW, MMW	Protein-polysaccharides	Immunomodulatory	Sum et al., 2020
TM02®	LR-CW (high-molecular-weight fraction)	Proteins or protein-carbohydrate complex	Anti-proliferative	Lee et al., 2012
TM02®	MMW	Protein-polysaccharides	Anti-glycation	Yap et al., 2018a
TM02®	F5	Serine protease	Anti-cancer	Yap et al., 2018b
TM02®	CWE	Proteins or protein-carbohydrate complex	Airway relaxation	Lee et al., 2018c
TM02®	HWE	Probable volatile components like alkanes and fatty acids	Airway inflammation inhibition, hyperresponsiveness	Johnathan et al., 2016; Johnathan et al., 2021
TM02®	CWE, HMW	Protein-polysaccharides (protein component)	Anti-(acute) inflammation	Lee et al., 2014
TM02®	Ethyl acetate fraction	Linoleic acid	Anti-neuroinflammatory, antioxidant	Nallathamby et al., 2016
TM02®	Hot aqueous extract	Not investigated	Neuritogenic activity in PC-12 cells	Seow et al., 2015
TM02®	Methanol extract	Probable phenolic compounds and phospholipids	Neuroprotective activities against DEX-induced toxicity in hESC-derived NSCs	Yeo et al., 2019
TM02®	CWE	Proteins or protein-carbohydrate complex	Dengue virus replication and infection inhibition	Khazali et al., 2021
KUM61075	LR-CA	Thermo-labile, water-soluble protein/peptide(s)	Antiproliferative	Lau et al., 2013
ND	Hot aqueous extract	n-Butanol, ethyl acetate fractions	Antiinflammatory	Seow et al., 2017

Table 8.1 contd. ...

...Table 8.1 contd.

THAILAND				
Strain	**Name of Specific Constituent**	**Bioactive Compounds**	**Pharmacological Effects**	**References**
TM02®	Ethane, hot, cold extracts	Not investigated	Antioxidant and anti-ageing	Kittimongkolsuk et al., 2021b
TM02®	Ethane extract	Probable phenolic compounds and phospholipids	Neuroprotective *in vivo* and *in vitro*	Kittimongkolsuk et al., 2021a
TM02®	Crude hexane, ethanol, water extracts	Fatty acids, peptides, and terpenoids	HIV-1 protease and reverse transcriptase inhibition	Sillapachaiyaporn and Chuchawankul, 2020

CHINA				
Strain	**Name of specific constituent**	**Bioactive Compounds**	**Pharmacological Effects**	**References**
TM02®	Polysaccharides	β-glucans	Intestinal mucosal wound-healing	Veeraperumal et al., 2021
ND	LRP-1, LRP-2, LRP-3, LRP-4	Polysaccharides	Immunomodulatory	Hu et al., 2017

Note: ND – not disclosed.

fraction manifested morphological changes of cells, an increase in pro-apoptotic caspases and Bax protein indicative of apoptotic cell death (Kong et al., 2020). An experimental study using MCF7 breast cancer xenograft mouse model demonstrated the inhibitory effect of Ligno TG-K protein on tumour growth with a significant number of apoptotic cells detected in tumour biopsies (Kong et al., 2020). However, the anti-tumour mechanism has not been sufficiently investigated yet. The cytotoxic nucleases have not been reported in *L. rhinocerus*. Interestingly, these enzymes are found abundantly in Ligno TG-K at approximately 27.6% of the total high molecular weight proteins (Kong et al., 2020). Previous research has shown that mushroom-derived nucleases, including *Hohenbuehelia geogenius*, *Ganoderma lucidum*, *Pholiota nameko*, and *Ramaria botrytis* may induce cell cycle arrest, apoptosis, and suppress an autophagy in human cancer cells (Zhang et al., 2014a,b; Dan et al., 2016; Zhou et al., 2017). An anti-tumour effect of a mixture of deoxyribonuclease and proteases has also been tested. Almost half of the mice-bearing human colorectal tumours showed complete tumour regression after receiving treatment with DNase I and proteases cocktail over two months (Trejo-Becerril et al., 2016). Furthermore, previous studies have reported that degradation of extracellular DNA or destruction of neutrophil extracellular traps by DNase could prevent tumour metastasis (Alexeeva et al., 2016; Najmeh et al., 2017). Serine proteases and DNase-like protein obtained from Ligno TG-K may be successfully developed into new drugs for cancer treatment, however more research is required to disclose an insight into the underlying mechanisms of their anti-tumour effects.

The aqueous extracts of Ligno TG-K showed antioxidant capacity, similar to *L. rhinocerus* TM02 (Yap et al., 2013; Kong et al., 2016b). *L. rhinocerus* TM02 possesses antioxidant proteins, including catalase, manganese-superoxide dismutase, and gluthathione transferase. Ligno TG-K possibly contains the same proteins (Yap et al., 2015b,c). Mushroom-derived bioactive compounds, including phenolics, terpenoids, polysaccharides (alpha and beta glucans) and polysaccharide complexes, also found in Ligno TG-K, have been reported to function as antioxidants to mitigate oxidative stress (Kozarski et al., 2012; Ahmad et al., 2014; Duru and Cayan, 2015; Wang et al., 2016).

Thus, studies have shown that Ligno TG-K contains numerous biologically active components (Yap et al., 2014; Kong et al., 2016b; Kong et al., 2020). However, research on bio-pharmacological properties of Ligno TG-K is lacking. The existing data on anti-carcinoma activity of Ligno TG-K suggests that it may be used as an adjuvant therapy for breast cancer treatment. Further studies are warranted to validate the bioactivity of Ligno TG-K to support its development into a functional food in the future.

5. Functional Food Potential of Tiger Milk Mushrooms in Malaysia

Presently, a rapid lifestyle and an advent of massive growth of fast-food industry have contributed to emergence of chronic diseases. Simultaneously, the realization of living a healthy lifestyle and maintaining balanced diet have also led consumers to expect functional food products which are foods and beverages that are advantageous (Arshad, 2002). Functional foods possess a variety of nutrients not found in ordinary food (Siro et al., 2008).

The tropics boost a variety of natural effective foods that proposes a novel remedy for a variety of diseases. Proponents of simple and basic food support the idea that they promote optimal health and help reduce the risk of disease. The functional food should not be seen as an alternative to balanced diet and healthy lifestyle but may be viewed as a holistic approach to diseases since they have multifaceted effects with medicinal benefits. The advent of scientific research with analysis of genomic, transcriptomic, and proteomic data has opened a perspective for downstream work in developing new useful nutraceuticals. The Malaysian medicinal gem, tiger milk mushroom, has been investigated *in vitro*, *ex vivo*, and *in vivo*, leading to interesting applicable results.

The tiger milk mushroom may be categorized as a functional food by Rincón-León (2003) according to the following criteria

1. It is derived from natural ingredients.
2. It can and should be consumed as part of the daily diet.
3. It has a particular function in the regulation of processes, such as enhancement of biological defence mechanisms, prevention and recovery from specific diseases, control of physical and mental disorders, and slowing of the aging process.

These criteria describe the scientific evidence behind the biomedical properties of tiger milk mushroom, starting from well-known species, *L. rhinocerus* to other two species found in Malaysia. The discovery of bioactive compounds, subsequent investigations into their mechanistic roles and the structure-function relationship of *Lignosus* sp. (*L. rhinocerus*, *L. tigris*, and *L. cameronensis*) may lead to the development of novel functional foods, nutraceuticals, and pharmacological products in the future.

6. Conclusion and Future Prospects

The studies conducted to date for *Lignosus* sp. have established a solid foundation for future research to the study of *L. rhinocerus*, *L. tigris*, and *L. cameronensis* in pharmacological and industrial applications. Together with clinical trials to support pre-clinical data, the perspectives for a future therapeutic usage of *Lignosus* sp. are promising. Future trials should be directed to the modernization of herbal medicine(s)-derived components, as alternative therapeutic agents. It is still uncertain whether these medicinal principles need to be exploited further as once these active bio-compounds are mined, they become "too powerful with significant side effects" (Lam, 2001). Previous reports validating the biopharmaceutical potential of *L. rhinocerus* directed us to the possibility of synergistic effect played by various components in the mushroom. Single compound purification has often resulted in a higher potency accompanied by increased toxicity. It is prudent that natural product scientists continue to understand the function of these medicinal principles from nature, to discover optimal ways for their application as functional foods and/or supplements to contribute to the improvement of human health.

References

Abdullah, N., Haimi, M.Z.D., Lau, B.F. and Annuar, M.S.M. (2013). Domestication of a wild medicinal sclerotial mushroom, *Lignosus rhinocerotis* (Cooke) Ryvarden. *Ind. Crops. Prod.*, 47: 256–261.

Ahmad, M.S., Noor, Z.M. and Ariffin, Z.Z. (2014). New thrombolytic agent from endophytic fungi and *Lignosus rhinocerus*. *Open. Conf. Proc. J.* 4: 95–98.

Ahmad, R., Muniandy, S., Shukri, N.I.A., Alias, S.M.U., Hamid, A.A. et al. (2014). Antioxidant properties and glucan compositions of various crude extract from *Lentinus squarrosulus* mycelial culture. *Adv. Biosci Biotechnol.*, 5(10): 805.

Alexeeva, L.A., Patutina, O.A., Sen'kova, A.V., Zenkova, M.A. and Mironova, N.L. (2017). Inhibition of invasive properties of murine melanoma by bovine pancreatic DNase I *in vitro* and *in vivo*. *Mol. Biol.*, 51(4): 562–570.

Arshad, F. (2002). Functional foods from the dietetic perspective. *Malaysia Journal of Community Health*, 8(S): 8–13.

Burkill, I.H. and Haniff, M. (1930). Malay village medicine. *Gard. Bull. Straits Settlements*, 6: 165–321.

Burkill, I.H., Birtwistle, W., Foxworthy, F.W., Scrivenor, J.B. and Watson, J.G. (1966). *A Dictionary of the Economic Products of the Malay Peninsula* (Vol. 1). Governments of Malaysia and Singapore by the Ministry of Agriculture and Co-operatives, Kuala Lumpur.

Chen, C.K., Hamdan, N.F., Ooi, F.K. and Wan Abd Hamid, W.Z. (2016). Combined effects of *Lignosus rhinocerotis* supplementation and resistance training on isokinetic muscular strength and power, anaerobic and aerobic fitness level, and immune parameters in young males. *Int. J. Prev. Med.*, 7: 107.

Chen, T.I., Zhuang, H.W., Chiao, Y.C. and Chen, C.C. (2013). Mutagenicity and genotoxicity effects of *Lignosus rhinocerotis* mushroom mycelium. *J. Ethnopharmacol.*, 149(1): 70–74.

Cheong, P.C.H., Yong, Y.S., Fatima, A., Ng, S.T., Tan, C.S. et al. (2019). Cloning, overexpression, purification, and modeling of a lectin (Rhinocelectin) with antiproliferative activity from tiger milk mushroom, *Lignosus rhinocerus*. *IUBMB Life*, 71(10): 1579–1594.

Cooke, M.C. (1879). January. XV. Enumeration of polyporus. *In*: *Transactions of the Botanical Society of Edinburgh* (Vol. 13, Nos. 1–4, pp. 131–159). UK: Taylor & Francis Group.

Dan, X., Liu, W., Wong, J.H. and Ng, T.B. (2016). A ribonuclease isolated from wild *Ganoderma lucidum* suppressed autophagy and triggered apoptosis in colorectal cancer cells. *Front. Pharmacol.*, 7: 217.

Duru, M.E. and Çayan, G.T. (2015). Biologically active terpenoids from mushroom origin: A review. *Rec. Nat. Prod.*, 9(4): 456.

Farha, M., Parkianathan, L., Abdul Amir, N.A.I., Sabaratnam, V. and Wong, K.H. (2019). Functional recovery enhancement by tiger milk mushroom, *Lignosus rhinocerotis* in a sciatic nerve crush injury model and morphological study of its neurotoxicity. *J. Anim. Plant. Sci.*, 4(29): 930–942.

Fauzi, S.Z.C., Rajab, N.F., Leong, L.M., Pang, K.L., Nawi, N.M. et al. (2015). Anti proliferative effect of *Lignosus rhinocerus* extract on colorectal cancer cells via apoptosis and cell cycle arrest. *Int. J. Pharm. Sci. Rev. Res.*, 33(1): 13–17.

Fung, S.Y., Cheong, P.C.H., Tan, N.H., Ng, S.T. and Tan, C.S. (2019). Nutritional evaluation on *Lignosus cameronensis* CS Tan, a medicinal polyporaceae. *IUBMB Life*, 71(7): 821–826.

Fung, S.Y. and Tan, C.S. (2019). Tiger milk mushroom (The *Lignosus* Trinity) in Malaysia: A medicinal treasure trove. pp. 349–370. *In*: Agrawal, D.C. and Dhanasekaran, M. (Eds.). *Medicinal Mushrooms: Recent Progress in Research and Development*, Springer Nature.

Haji Taha, A. (2006). *Orang Asli: The hidden treasure*. Department of Museums, Malaysia.

Hu, T., Huang, Q., Wong, K. and Yang, H. (2017). Structure, molecular conformation, and immunomodulatory activity of four polysaccharide fractions from *Lignosus rhinocerotis* sclerotia. *Int. J. Biol. Macromol.*, 94: 423–430.

Huang, N.L. (1999a). Identification of the scientific name of Hurulingzhi. *Acta Edulis Fungi*, 6(1): 32–34.

Huang, N.L. (1999b). A new species of medicinal mushroom: Cultivation of Hurulingzhi. *Edible Fungi China*, 18: 8–9.

Imazeki, R. (1952). A contribution to the fungus flora of Dutch New Guinea. *Bull. Gov. For. Exp. Stn. (Jpn.)*, 57: 87–128.

John, P.A., Wong, K.H., Naidu, M., Sabaratnam, V. and David, P. (2013). Combination effects of curcumin and aqueous extract of *Lignosus rhinocerotis* mycelium on neurite outgrowth stimulation activity in PC-12 Cells. *Nat Prod Commun.*, 8(6): 711–714.

Johnathan, M., Gan, S.H., Ezumi, M.F.W., Faezahtul, A.H. and Nurul, A.A. (2016). Phytochemical profiles and inhibitory effects of tiger milk mushroom (*Lignosus rhinocerus*) extract on ovalbumin-induced airway inflammation in a rodent model of asthma. *BMC Complement Altern. Med.*, 16(1): 167.

Johnathan, M., Muhamad, S.A., Gan, S.H., Stanslas, J., Mohd Fuad, W.E. et al. (2021). *Lignosus rhinocerotis* Cooke Ryvarden ameliorates airway inflammation, mucus hypersecretion and airway hyperresponsiveness in a murine model of asthma. *PLOS ONE*, 16(3): e0249091.

Keong, C.Y., B.V., Daker, M., Hamzah, M.Y., Mohamad, S.A., Lan, J. et al. (2016). Fractionation and biological activities of water-soluble polysaccharides from sclerotium of tiger milk medicinal mushroom, *Lignosus rhinocerotis* (Agaricomycetes). *Int. J. Med. Mushrooms*, 18(2): 141–154.

Khazali, A.S., Nor Rashid, N., Fung, S.Y. and Yusof, R. (2021). *Lignosus rhinocerus* TM02® sclerotia extract inhibits dengue virus replication and infection. *J. Herb. Med.*, In press, 100505.

Kittimongkolsuk, P., Pattarachotanant, N., Chuchawankul, S., Wink, M. and Tencomnao, T. (2021a). Neuroprotective effects of extracts from tiger milk mushroom *Lignosus rhinocerus* against glutamate-induced toxicity in HT22 hippocampal neuronal cells and neurodegenerative diseases in *Caenorhabditis elegans*. *Biology*, 10(1): 30.

Kittimongkolsuk, P., Roxo, M., Li, H., Chuchawankul, S., Wink, M. et al. (2021b). Extracts of the tiger milk mushroom (*Lignosus rhinocerus*) enhance stress resistance and extend lifespan in caenorhabditis elegans via the DAF-16/FoxO signaling pathway. *Pharmaceuticals*, 14(2): 93.

Kong, B.H., Tan, N.H., Fung, S.Y. and Pailoor, J. (2016a). Sub-acute toxicity study of tiger milk mushroom *Lignosus tigris* chon S. Tan cultivar E sclerotium in Sprague Dawley rats. *Front. Pharmacol.*, 7: 246.

Kong, B.H., Tan, N.H., Fung, S.Y., Pailoor, J., Tan, C.S. et al. (2016b). Nutritional composition, antioxidant properties, and toxicology evaluation of the sclerotium of tiger milk mushroom *Lignosus tigris* cultivar E. *Nutr. Res.*, 36(2): 174–183.

Kong, B.H., Teoh, K.H., Tan, N.H., Tan, C.S., Ng, S.T. et al. (2020). Proteins from *Lignosus tigris* with selective apoptotic cytotoxicity towards MCF7 cell line and suppresses MCF7-xenograft tumor growth. *PeerJ.* 8: e9650.

Kozarski, M., Klaus, A., Nikšić, M., Vrvić, M.M., Todorović, N. et al. (2012). Antioxidative activities and chemical characterization of polysaccharide extracts from the widely used mushrooms *Ganoderma applanatum, Ganoderma lucidum, Lentinus edodes* and *Trametes versicolor. J. Food Compos. Anal.*, 26(1-2): 144–153.

Kuntze, C.E.O. (1898). *In*: *Revisio generum plantarum* (Leipzig). 3(2): 519.

Lai, W.H., Salleh, S.M., Daud, F., Zainal, Z., Othman, A.M. et al. (2014). Optimization of submerged culture conditions for the production of mycelial biomass and exopolysaccharides from *Lignosus rhinocerus. Sains Malaysiana*, 43(1): 73–80.

Lam, T.P. (2001). Strengths and weaknesses of traditional Chinese medicine and Western medicine in the eyes of some Hong Kong Chinese. *J. Epidemiology Community Health*, 55(10): 762.

Lau, B.F., Aminudin, N. and Abdullah, N. (2011). Comparative SELDI-TOF-MS profiling of low-molecular-mass proteins from *Lignosus rhinocerus* (Cooke) Ryvarden grown under stirred and static conditions of liquid fermentation. *J. Microbiol. Methods*, 87(1): 56–63.

Lau, B.F., Abdullah, N., Aminudin, N. and Lee, H.B. (2013). Chemical composition and cellular toxicity of ethnobotanical-based hot and cold aqueous preparations of the tiger's milk mushroom (*Lignosus rhinocerotis*). *J. Ethnopharmacol.*, 150(1): 252–262.

Lau, B.F., Abdullah, N. and Aminudin, N. (2013). Chemical composition of the tiger's milk mushroom, *Lignosus rhinocerotis* (Cooke) Ryvarden, from different developmental stages. *J. Agric. Food Chem.*, 61(20): 4890–4897.

Lau, B.F., Abdullah, N., Aminudin, N., Lee, H.B., Yap, K.C. et al. (2014). The potential of mycelium and culture broth of Lignosus rhinocerotis as substitutes for the naturally occurring sclerotium with regard to antioxidant capacity, cytotoxic effect, and low-molecular-weight chemical constituents. *PLOS ONE*, 9(7): e102509.

Lau, B.F., Abdullah, N., Aminudin, N., Lee, H.B. and Tan, P.J. (2015). Ethnomedicinal uses, pharmacological activities, and cultivation of *Lignosus* spp. (tiger milk mushrooms) in Malaysia: A review. *J. Ethnopharmacol.*, 169: 441–458.

Lee, M.K., Li, X., Yap, A.C.S., Cheung, P.C.K., Tan, C.S. et al. (2018c). Airway relaxation effects of water-soluble sclerotial extract from *Lignosus rhinocerotis. Front. Pharmacol.*, 9: 461.

Lee, M.K., Lim, K.H., Millns, P., Mohankumar, S.K., Ng, S.T. et al. (2018a). Bronchodilator effects of *Lignosus rhinocerotis* extract on rat isolated airways is linked to the blockage of calcium entry. *Phytomedicine*, 42: 172–179.

Lee, M.K., Millns, P., Mbaki, Y., Ng, S.T., Tan, C.S. et al. (2018b). Data on the *Lignosus rhinocerotis* water soluble sclerotial extract affecting intracellular calcium level in rat dorsal root ganglion cells. *Data in Brief*, 18: 1322–1326.

Lee, M.L., Tan, N.H., Fung, S.Y., Tan, C.S. and Ng, S.T. (2012). The antiproliferative activity of sclerotia of *Lignosus rhinocerus* (tiger milk mushroom). *Evid. Based Complement. Alternat. Med.*, 2012, 697603.

Lee, S.S., Tan, N.H., Fung, S.Y., Sim, S.M., Tan, C.S. et al. (2014). Anti-inflammatory effect of the sclerotium of *Lignosus rhinocerotis* (Cooke) Ryvarden, the tiger milk mushroom. *BMC Complement. Altern. Med.*, 14: 359.

Lee, S.S., Tan, N.H., Pailoor, J. and Fung, S.Y. (2017). Safety evaluation of sclerotium from a medicinal mushroom, *Lignosus cameronensis* (cultivar): Preclinical toxicology studies. *Front. Pharmacol.*, 8: 594.

Lima, C.U.J.O., Gris, E.F. and Karnikowski, M.G.O. (2016). Antimicrobial properties of the mushroom *Agaricus blazei*: Integrative review. *Rev Bras Farmacogn.*, 26(6): 780–786.

Lin, J., Zhao, C., Wu, C.J. and Chen, C. (2020). Evaluation of effectiveness of liquid fermented *Lignosus rhinocerus* mycelium against H1N1 Influenza virus infections. *Hans J. Food Nutr. Sci.*, 9(2): 201–210.

Lloyd, C.G. 1920. *In*: *Mycological Writers*, 6 (Letter 65), p. 1037.
Maroon, J.C., Bost, J.W. and Maroon, A. (2010). Natural anti-inflammatory agents for pain relief. *Surgical Neurology International*, 1: 80.
Moradali, M.F., Mostafavi, H., Ghods, S. and Hedjaroude, G.A. (2007). Immunomodulating and anticancer agents in the realm of macromycetes fungi (macrofungi). *Int. Immunopharmacol.*, 7(6): 701–724.
Muhamad, S.A., Muhammad, N.S., Ismail, N.D.A., Mohamud, R., Safuan, S. et al. (2019). Intranasal administration of *Lignosus rhinocerotis* (Cooke) Ryvarden (tiger milk mushroom) extract attenuates airway inflammation in murine model of allergic asthma. *Exp. Ther. Med.*, 17(5): 3867–3876.
Najmeh, S., Cools-Lartigue, J., Rayes, R.F., Gowing, S., Vourtzoumis, P. et al. (2017). Neutrophil extracellular traps sequester circulating tumor cells via β1-integrin mediated interactions. *Int. J. Cancer Res.*, 140(10): 2321–2330.
Nallathamby, N., Serm, L.G., Raman, J., Malek, S.N.A., Vidyadaran, S. et al. 2016. Identification and *in vitro* evaluation of lipids from sclerotia of *Lignosus rhinocerotis* for antioxidant and anti-neuroinflammatory activities. *Nat. Prod. Commun.*, 11(10): 1934578X1601101016.
Nallathamby, N., Phan, C.W., Seow, S.L.S., Baskaran, A., Lakshmanan, H. et al. (2018). A status review of the bioactive ctivities of tiger milk mushroom *Lignosus rhinocerotis* (Cooke) Ryvarden. *Front. Pharmacol.*, 8: 988.
Nyam, K.L., Chang, C.Y., Tan, C.S. and Ng, S.T. (2016). Investigation of the tiger milk medicinal mushroom, *Lignosus rhinocerotis* (Agaricomycetes), as an Antiulcer Agent. *Int. J. Med. Mushrooms*, 18(12): 1093–1104.
Nyam, K.L., Chow, C.F., Tan, C.S. and Ng, S.T. (2017). Antidiabetic properties of the tiger milk medicinal mushroom, *Lignosus rhinocerotis* (Agaricomycetes), in streptozotocin-induced diabetic rats. *Int. J. Med. Mushrooms*, 19(7): 607–617.
Oranusi, S.U., Ndukwe, C.U. and Braide, W. (2014). Production of *Pleurotus tuber-regium* (Fr.) Sing Agar, chemical composition and microflora associated with sclerotium. *Int. J. Curr. Microbiol. Appl. Sci.*, 3(8): 115–126.
Patel, S. and Goyal, A. (2012). Recent developments in mushrooms as anti-cancer therapeutics: A review. *3 Biotech*, 2(1): 1–15.
Pushparajah, V., Fatima, A., Chong, C.H., Gambule, T.Z., Chan, C.J. et al. (2016). Characterisation of a new fungal immunomodulatory protein from tiger milk mushroom, *Lignosus rhinocerotis*. *Scientific Reports*, 6(1): 30010.
Rahman, N.A., Daud, F., Kalil, M.S. and Ahmad, S. (2012). Tiger milk mushroom cultivation by using submerged culture technique. *WSEAS Trans. Biol. and Biomed.*, 9(3): 83–92.
Raskin, J.B. (1999). Gastrointestinal effects of nonsteroidal anti-inflammatory therapy. *Am. J. Med.*, 106(5): 3S–12S.
Rincón-León, F. (2003). Functional Foods. pp. 2827–2832. *In*: Benjamin Caballero (Ed.). *Encyclopedia of Food Sciences and Nutrition* (2nd Edn.), UK: Elsevier.
Ryvarden, L. (1972). *Lignosus* goetzei (Henn.) Ryvarden. *Norwegian Journal of Botany*, 19: 232.
Saccardo, P.A. (1888). *Sylloge Hymenomycetum*, Vol. II. *Polyporeae, Hydneae, Thelephoreae, Clavarieae, Tremellineae. Sylloge Fungorum* Vol. 6, pp. 1–928.
Samberkar, S., Gandhi, S., Naidu, M., Wong, K.H., Raman, J. and Sabaratnam, V. (2015). Lion's Mane, *Hericium erinaceus* and tiger milk, *Lignosus rhinocerotis* (higher basidiomycetes) medicinal mushrooms stimulate neurite outgrowth in dissociated cells of brain, spinal cord, and retina: An *in vitro* study. *Int. J. Med. Mushrooms*, 17(11): 1047–1054.
Sánchez, C. (2016). Reactive oxygen species and antioxidant properties from mushrooms. *Synth. Syst. Biotechnol.*, 2(1): 13–22.
Seow, S.L.S., Eik, L.F., Naidu, M., David, P., Wong, K.H. et al. (2015). *Lignosus rhinocerotis* (Cooke) Ryvarden mimics the neuritogenic activity of nerve growth factor via MEK/ERK1/2 signaling pathway in PC-12 cells. *Sci. Rep.*, 5(1): 16349.
Seow, S. L.S., Naidu, M., Sabaratnam, V., Vidyadaran, S. and Wong, K.H. (2017). Tiger's milk medicinal mushroom, *Lignosus rhinocerotis* (agaricomycetes) sclerotium inhibits nitric oxide production in LPS-stimulated BV2 microglia. *Int. J. Med. Mushrooms*, 19(5): 405–418.
Sherwood, E.R. and Toliver-Kinsky, T. (2004). Mechanisms of the inflammatory response. *Best Pract. Res. Clin. Anaesthesiol.*, 18(3): 385–405.

Shopana, M., Sudhahar, D. and Anandarajagopal, K. (2012). Screening of *Lignosus rhinocerus* extracts as antimicrobial agents against selected human pathogens. *J. Pharm. Biomed. Sci.*, 18(11): 1–4.

Sillapachaiyaporn, C. and Chuchawankul, S. (2020). HIV-1 protease and reverse transcriptase inhibition by tiger milk mushroom (*Lignosus rhinocerus*) sclerotium extracts: *In vitro* and *in silico* studies. *J. Tradit. Complement. Med.*, 10(4): 396–404.

Siró, I., Kápolna, E., Kápolna, B. and Lugasi, A. (2008). Functional food. Product development, marketing, and consumer acceptance: A review. *Appetite*, 51(3): 456–467.

Sum, A.Y.C., Li, X., Yeng, Y.Y.H., Razif, M.F.M., Jamil, A.H.A. et al. (2020). The immunomodulating properties of tiger milk medicinal mushroom, *Lignosus rhinocerus* TM02® Cultivar (agaricomycetes) and its associated carbohydrate composition. *Int. J. Med. Mushrooms*, 22(8): 803–814.

Tan C.S. (2009). *Setting-up pilot-plant for up-scaling production of 'Tiger-Milk'-mushroom as dietary functional food.* MOA TF0109M004, Government of Malaysia. 2009.

Tan, C.S., Ng, S.T. and Tan, J. (2013). Two new species of *Lignosus* (Polyporaceae) from Malaysia-*L. tigris* and *L. cameronensis*. *Mycotaxon*, 123(1): 193–204.

Tan, E.S.S., Leo, T.K. and Tan, C.K. (2021a). Effect of tiger milk mushroom (Lignosus rhinocerus) supplementation on respiratory health, immunity, and antioxidant status: An open-label prospective study. *Sci. Rep.*, 11(1): 11781.

Tan, Y.H., Lim, C.S.Y., Wong, K.H. and Sabaratnam, V. (2021b). Neuritin protein expression is positively correlated with neurite outgrowth induced by the tiger milk mushroom, *Lignosus rhinocerotis* (agaricomycetes), in PC12 cells. *Int. J. Med. Mushrooms*, 23(6): 1–11.

Trejo-Becerril, C., Pérez-Cardenas, E., Gutiérrez-Díaz, B., De La Cruz-Sigüenza, D., Taja-Chayeb, L. et al. (2016). Antitumor effects of systemic DNAse I and proteases in an *in vivo* model. *Integr. Cancer Ther.*, 15(4): NP35-NP43.

Veeraperumal, S., Qiu, H.M., Tan, C.-S., Ng, S.T., Zhang, W. et al. (2021). Restitution of epithelial cells during intestinal mucosal wound healing: The effect of a polysaccharide from the sclerotium of *Lignosus rhinocerotis* (Cooke) Ryvarden. *J. Ethnopharmacol.*, 274(114024).

Vikineswary, S. and Chang, S.T. (2013). Edible and medicinal mushrooms for sub-health intervention and prevention of lifestyle diseases. *Tech Monitor*, 3: 33–43.

Wang, J., Hu, S., Nie, S., Yu, Q. and Xie, M. (2016). Reviews on mechanisms of *in vitro* antioxidant activity of polysaccharides. *Oxid. Med. Cell. Longev.*, 2016.

Yap, H.Y.Y., Tan, N.H, Fung, S.Y., Aziz, A.A., Tan, C.S. et al. (2013). Nutrient composition, antioxidant properties, and anti-proliferative activity of *Lignosus rhinocerus* Cooke sclerotium. *J. Sci. Food Agric.*, 93(12): 2945–52.

Yap, H.Y.Y, Chooi, Y.H., Firdaus-Raih, M., Fung, S.Y., Ng, S.T. et al. (2014a). The genome of the tiger milk mushroom, *Lignosus rhinocerotis*, provides insights into the genetic basis of its medicinal properties. *BMC Genomics*, 15: 635.

Yap, H.Y.Y., Aziz, A.A., Fung, S.Y., Ng, S.T., Tan, C.S. et al. (2014b). Energy and nutritional composition of tiger milk mushroom (*Lignosus tigris* Chon S. Tan) sclerotia and the antioxidant activity of its extracts. *Int. J. Med. Sci.*, 11(6): 602.

Yap, H.Y.Y., Chooi, Y.H., Fung, S.Y., Ng, S.T., Tan, C.S. et al. (2015a). Transcriptome analysis revealed highly expressed genes encoding secondary metabolite pathways and small cysteine-rich proteins in the sclerotium of *Lignosus rhinocerotis*. PLOS ONE, 10(11): e0143549.

Yap, H.Y.Y., Fung, S.Y., Ng, S.T., Tan, C.S. and Tan, N.H. (2015b). Shotgun proteomic analysis of tiger milk mushroom (*Lignosus rhinocerotis*) and the isolation of a cytotoxic fungal serine protease from its sclerotium. *J. Ethnopharmacol.*, 174: 437–451.

Yap, H.Y.Y., Fung, S.Y., Ng, S.T., Tan, C.S. and Tan, N.H. (2015c). Genome-based proteomic analysis of *Lignosus rhinocerotis* (Cooke) Ryvarden sclerotium. *Int. J. Med. Sci.*, 12(1): 23–31.

Yap, H.Y.Y., Tan, N.H., Ng, S.T., Tan, C.S. and Fung, S.Y. (2018a). Inhibition of protein glycation by tiger milk mushroom [*Lignosus rhinocerus* (Cooke) Ryvarden] and search for potential anti-diabetic activity-related metabolic pathways by genomic and transcriptomic data mining. *Front. Pharmacol.*, 9: 103–103.

Yap, H.Y.Y., Tan, N.H., Ng, S.T., Tan, C.S. and Fung, S.Y. (2018b). Molecular attributes and apoptosis-inducing activities of a putative serine protease isolated from tiger milk mushroom (*Lignosus rhinocerus*) sclerotium against breast cancer cells *in vitro*. *PeerJ*, 6: e4940.

Yeo, Y., Tan, J.B.L., Lim, L.W., Tan, K.O., Heng, B.C. et al. (2019). Human embryonic stem cell-derived neural lineages as *in vitro* models for screening the neuroprotective properties of *Lignosus rhinocerus* (Cooke) Ryvarden. *BioMed Res. Int.* : 3126376.

Yokota, A. (2011). Tropical forests and people's livelihood: Stalls of traditional medicine vendors in Kota Kinabalu. *JIRCAS Newsl.*, 62: 9–10.

Zhang, R., Zhao, L., Wang, H. and Ng, T.B. (2014). A novel ribonuclease with antiproliferative activity toward leukemia and lymphoma cells and HIV-1 reverse transcriptase inhibitory activity from the mushroom, *Hohenbuehelia serotina. Int. J. Mol. Med.*, 33(1): 209–214.

Zhang, Y., Liu, Z., Ng, T.B., Chen, Z., Qiao, W. et al. (2014). Purification and characterization of a novel antitumor protein with antioxidant and deoxyribonuclease activity from edible mushroom *Pholiota nameko. Biochimie*, 99: 28–37.

Zhou, R., Han, Y.J., Zhang, M.H., Zhang, K.R., Ng, T.B. et al. (2017). Purification and characterization of a novel ubiquitin-like antitumour protein with hemagglutinating and deoxyribonuclease activities from the edible mushroom *Ramaria botrytis. Amb Express*, 7(1): 1–11.

Food, Feed and Nutrition

9

Filamentous Fungi and Yeasts as Sources of Coenzyme Q_{10} and Its Applications

Pradipta Tokdar,[1,*] *Prafull Ranadive*[2] and *Saji George*[1]

1. Introduction

Coenzyme Q (CoQ) is an important biomolecule usually found in most of the animal, plant, and microbial cells. It plays a crucial role in generation of cellular energy and free radical scavenging. This molecule comprises of a benzoquinone ring with a hydrophobic isoprenoid side chain attached to it and is mainly confined to a mitochondrial inner membrane in animal and yeast cells. CoQ is species specific and is differentiated by the number of isoprenyl units present on the isoprenoid side chain. The maximum, ten isoprenyl units Coenzyme Q_{10} (CoQ_{10}) are found in human and in few microbes like fission yeast *Schizosaccharomyces pombe*, but fewer units are found in other species, CoQ_8 in *Escherichia coli*, CoQ_9 in *Arabidopsis thaliana*, and CoQ_6 in *Saccharomyces cerevisiae* (Lee et al., 2017). Due to physiological importance and myriad functions of CoQ_{10} in the human body, its deficiency can result in numerous diseases pertaining to neurological, cardiological, or metabolic disorders. The concentration of this molecule in the body gets depleted significantly in the people who are taking prolong treatment with cholesterol lowering statin drugs. Also with age, the CoQ_{10} levels in the body declines and results in age-related disorders, hence it is an indicator molecule of our health. Owing to its therapeutic relevance, exogenous CoQ_{10} emerged as an important biomolecule in

[1] KRIBS-BIONEST (RGCB Campus-3), Kerala Technology Innovation Zone, BTIC Building, KINFRA Hi-Tech Park, Kalamessary, Kochi, India, 683503.
[2] Organica Biotech Pvt. Ltd., 36, Ujagar Industrial Estate, W.T. Patil Marg, Govandi, Mumbai, India, 400 088.
* Corresponding author: pradiptatokdar@rgcb.res.in

health supplements for pharmaceutical and cosmeceutical applications. Over the last 30 years, several studies have been conducted and published on CoQ_{10} benefits for maintaining optimum health. It has been proved that CoQ_{10} supplements strengthens the immune system and fosters resistance to disease (https://www.kanekanutrients.com/kaneka-q10).

The industrial importance of CoQ_{10} attracted the attention of the scientific community to focus their research on synthesis, formulation, and clinical studies on this molecule. Many decades of research resulted in commercial production and marketing of this molecule and its various formulations. Out of the different routes of CoQ_{10} synthesis explored, like extraction from biological tissues (Laplante et al., 2009), chemical synthesis (Iehud and Doron, 1988), and microbial fermentation (Yoshida et al., 1998), the microbial route is most preferred due to its simplicity, feasibility, high levels of purity, efficacy, and cost effectiveness (Tian et al., 2010).

The microbial synthesis involves the use of a microbial strain that naturally produces CoQ containing ten isoprene units, as CoQ_{10}. Several approaches have been used to improve the fermentative production of CoQ_{10}, which have relied predominantly on bacterial and yeast mutants selected for their high CoQ_{10} content combined with genetic and metabolic engineering (Ndikubwimana and Lee, 2014). These strains are fermented in large quantities under optimized process conditions followed by extraction and purification of CoQ_{10} till the desired purity levels. Although most work has been done on bacterial CoQ_{10} synthesis at the academic level, the commercial production using yeast has been preferred and reported (Cluis et al., 2007). The CoQ_{10} production pathway in bacteria and yeast differs significantly, but the yeast CoQ_{10} pathway closely resembles that in humans (Moriyama et al., 2015). Therefore, few fungi and yeast strains become a promising and preferred source of natural CoQ_{10} derived from the eukaryote kingdom, having a high degree of conservation with the mammalian system and its biosynthetic pathways (Choi et al., 2005). This review explains the properties and functions of CoQ_{10}, its biosynthesis from the wild, as well as its recombinant strains and their physiological significance and industrial applications.

2. CoQ_{10}: General Information

CoQ_{10} is the other name of ubiquinone which is an ATP (adenosine phosphate), a generating component of the respiratory chain (Kawamukai, 2002; Kazunori et al., 2004; Takahashi et al., 2006). CoQ_{10} is soluble in lipid and is present in almost all animal cells including cells of the human body (Siemieniuk and Skrzydlewska, 2005). It is a phenolic component and has a potent activity against oxidative cellular injury. This compound can inhibit the accretion of anions like superoxide. Therefore, it reduces the cellular stress during aerobic respiration (Poon et al., 1999). It is also known as ubiquinone because it is present ubiquitously in nature and it has a quinone structure. CoQ_{10} has a similar chemical structure to vitamin K but because it is the only antioxidant that is lipid-soluble and synthesized in the body. CoQ_{10} is supplied to the cells through biosynthesis (Ernster and Dallner, 1995; Tran and Clarke, 2007).

2.1 Occurrence of CoQ_{10}

CoQ_{10} is mainly found in mitochondria (Fig. 9.1) (Doring et al., 2007). However, it is also found inside the cell membrane, specifically inside the hydrophobic part of the phospholipid bilayer (Yen and Shih, 2009). It is an essential compound found in the cell membrane of every cell of the human body. CoQ_{10} can also be found inside lysosomes and Golgi bodies. In lysosomes, CoQ_{10} acts by transferring protons. The functioning of CoQ_{10} in the Golgi body is almost the same (Gille and Nohl, 2000). Inside the human body, CoQ_{10} is encoded by a specific gene, named COQ_4. This component has been found to have three different oxidation states (Fig. 9.2). The completely oxidized form of CoQ_{10} is known as ubiquinone whereas the partially oxidized form of COQ_{10} is named semiubiquinone. Ubiquinol is the reduced form

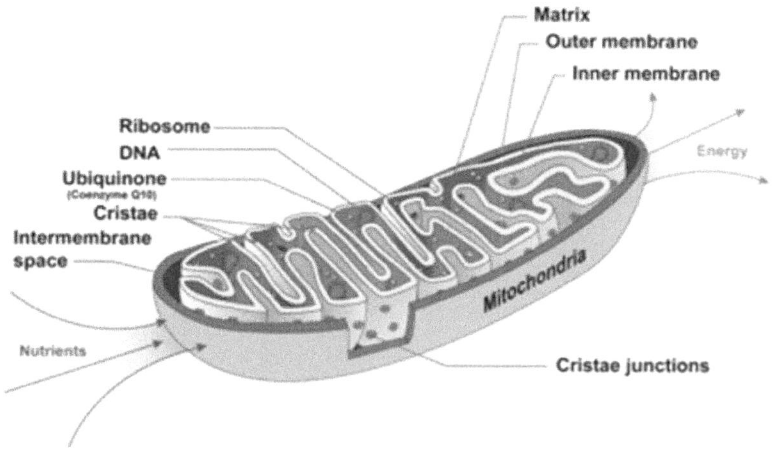

Fig. 9.1. Anatomic Structure of a mitochondrion illustrating CoQ_{10} (Hiebert et al., 2012).

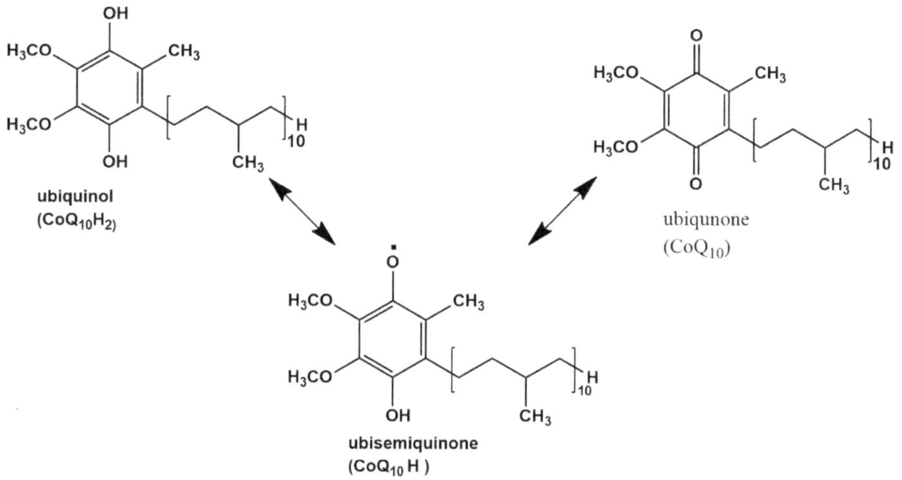

Fig. 9.2. CoQ_{10} exists in three oxidation states: the fully reduced ubiquinol form ($CoQ_{10}H_2$), the radical semiquinone intermediate ($CoQ_{10}H$) and the fully oxidized ubiquinone form (CoQ_{10}).

of CoQ_{10} (Siemieniuk and Skrzydlewska, 2005). Recent researchers revealed that seventeen genes can stimulate CoQ_{10}. These genes are functionally interrelated by G-protein receptors and their signaling pathways (Schmelzer et al., 2007). In this way, ubiquinol gives away an electron to the iron-sulphur protein. Subsequently, the unstable ubi semiquinone transports an electron to cytochrome b (Wang and Hekimi, 2016). Therefore, the ubiquinol gets oxidized to ubiquinone. Ubiquinol then gets restored by accepting electrons either from complex I and complex II of the electron transport mechanism or from acyl coenzyme A dehydrogenase. CoQ_{10} acts as a pro-oxidant because the unstable form ubi semiquinone reacts with the molecular form of oxygen and forms superoxide radicals. They cannot cross the membranes but are released at the inter membranous space of mitochondria (Linnane et al., 2007).

2.2 Chemistry of CoQ_{10}

The chemical name of CoQ_{10} is 2, 3-dimethoxy, 5-methyl, 6-decaprenyl benzoquinone (Fig. 9.3a) (Vaghari et al., 2016). CoQ_{10} is yellowish orange in color and the texture is like a crystalline, fine powder. Its molecular weight is 863.34 g/mol and has a melting point of 48°C. It is heat-labile and is also affected by exposure to oxygen and light (Terao et al., 2006; Balakrishnan et al., 2009; Nepal et al., 2010). CoQ_{10} can be synthesized by various methods such as microbial biosynthesis, chemical synthesis, and semi-synthesis. Microbial synthesis of CoQ_{10}, involving fermentation, is the most feasible among all these methods (Clarke, 2000; Tian et al., 2010; Yuting et al., 2010; Qiu et al. 2012; Lu et al. 2013). In this method, CoQ_{10} can be synthesized more efficiently at low expenditure. Also, the formed CoQ_{10} is optical-isomer-free in this way (Tian et al., 2010). Coenzyme Q is an essential cofactor that takes part in oxidative phosphorylation reactions in mitochondria. This cofactor is species-specific. Coenzyme Q consists of isoprenoid side chains linked to the number of isoprenyl groups which are in turn attached to a quinone head group. The number of isoprenyl groups varies with the species. CoQ in humans is found with ten isoprenyl groups. The structure of CoQ_{10} with a typical quinone head and isoprenoid chain is shown Fig. 9.3b. The lipid solubility to the coenzyme Q is imparted by the isoprenoid side chain (Lipshutz et al., 2002; Lankin et al., 2007; Bentinger et al., 2010). Chemical synthesis of CoQ_{10} is carried out mostly in commercial production. It involves coupling of chemicals which possess a quinone structure attached to an isoprenoid tail. The tail is derived from tobacco or potato by isolating as solanesol. Then it is converted into CoQ_{10}. However, these synthetic methods of CoQ_{10} production are very expensive and attributable to cause environmental degradation due to production of waste material (Lipshutz et al., 2005). The chemical synthesis is schematically summarized in Fig. 9.4.

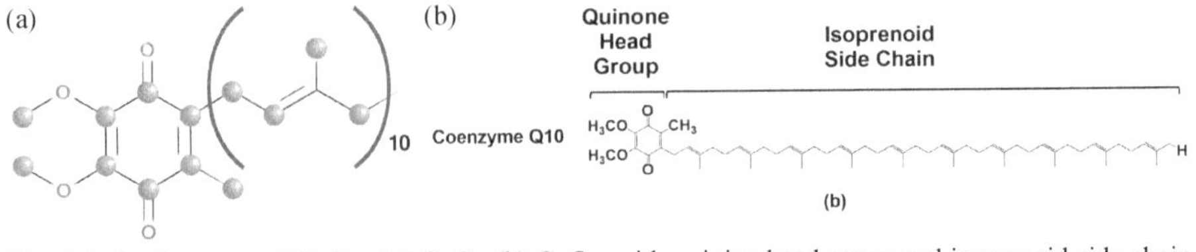

Fig. 9.3a,b. Structure of CoQ_{10}: (a) CoQ_{10} (b) CoQ_{10} with quinine head group and isoprenoid side chain (Bentinger et al., 2010).

The stages of CoQ_{10} biosynthesis are aromatic group synthesis, isoprene tail synthesis, forming attachment between the two structures (aromatic group and isoprene tail), and finally the formation of CoQ_{10}. In the human body, this synthesis occurs in both the mitochondria and Golgi apparatus. The aromatic group mentioned above is regarded as the quinone head which is derived from the chorismate precursor found in prokaryotes shikimate pathway of CoQ_{10} synthesis and case of eukaryotes, from tyrosine. Both the aromatic head and isoprene tail are species specific. Methyl erythritol 4-phosphate or 2-C-methyl-D-erythritol-4-phosphate (MEP) gives the isoprene tail in prokaryotes and from acetyl CoA in the mevalonate pathway in eukaryotes. All these are the entry or starting points of biosynthesis of CoQ_{10} (Choi et al., 2005; Jeya et al., 2010).

CoQ_{10} is formed by the combination of the quinonoid nucleus and an isoprenoid side chain (Cluis et al., 2011; Lu et al., 2013). It is synthesized in the cytosol where the isoprene tail is produced by the conversion of an important intermediate at the cellular level during the synthesis of cholesterol. The quinone ring present in CoQ_{10} is derived from tyrosine (Acosta et al., 2016). The final step of the biosynthesis process is catalyzed by a domain containing a protein known as *UbiA* prenyl transferase. *UbiA* is present in several sites including the endoplasmic reticulum, mitochondria, and Golgi apparatus. *UbiA* has the function to protect cardiovascular tissues against oxidative stress via the synthesis of CoQ_{10}. Generally, the plasma level of ubiquinone ranges between 0.40 and 2.0 μmol/L. CoQ_{10} gets absorbed from the intestine by entering the lymphatic system and then to the systemic circulation (Palamakula et al., 2005).

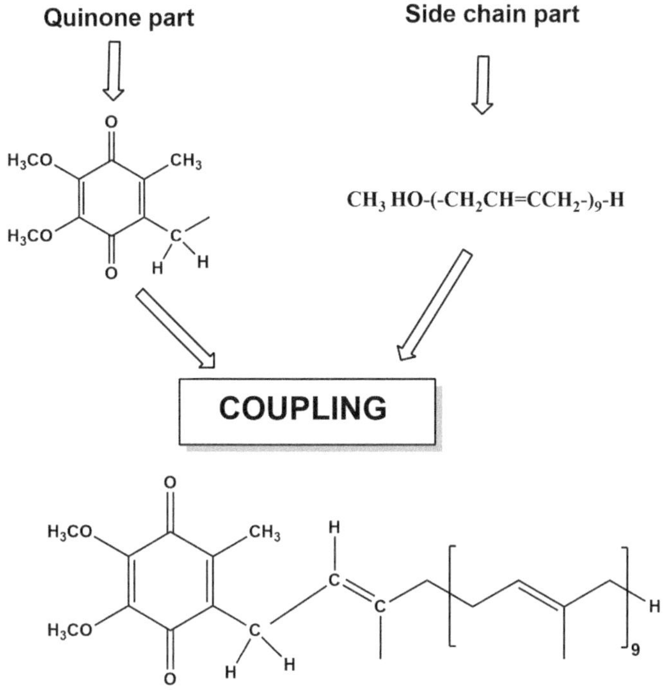

Fig. 9.4. Chemical synthetic pathway of CoQ_{10} (Taylor and Fraser, 2011).

Vitamin E may slow the absorption of CoQ_{10}, and the bioavailability depends upon the carrier lipids present (Lopez et al., 2019). The synthesis of CoQ_{10} decreases in the body after the age of 20 years. The concentration in the myocardium reduces to half during the 80s. This was evidenced by a clinical trial involving a group of elderly who were given a combination of selenium and ubiquinone over 4 years and a much-improved physical outcome was achieved. CoQ_{10} supplementary proved to be associated with benefits among the elderly population, especially in terms of prevention of oxidative stress and cardiovascular manifestations (Gonzalez et al., 2015).

2.3 Functions of CoQ in Humans

CoQ or Ubiquinone plays a crucial role as an electron transporter in mitochondria between complexes I, II, and III of the respiratory chain (Fig. 9.5) (Mitchell, 1975; Acosta et al., 2016). To produce ATP and cellular respiration, adequate amounts of CoQ_{10} is necessary. It also functions as an antioxidant. It affects gene expression that affects the overall metabolism of tissue. CoQ_{10} has been found to reduce the free radicals and stabilizes them in the extracellular space (Gomez et al., 1997; Formigli et al., 2003). It takes part in the cellular signaling pathways and the genetic expressions (Ernster and Dallner, 1995; Crane, 2001). The antioxidant activity of CoQ_{10} is better observed in the form of Ubiquinol after it undergoes reduction. Ubiquinol, which is the reduced form of CoQ_{10}, has been found to have a shielding effect towards the serum low density lipoprotein (LDL) and phospholipids from the lipid peroxidation. It prevents the oxidative damage of the mitochondrial DNA and other proteins.

Studies have revealed that it can enhance the action of vitamin E by recycling it from its oxidized state. A specific biochemical process known as the mevalonate pathway (MVP) is associated with the regulation of ubiquinone at the tissue level inside the human body. The amount of CoQ_{10} is directly proportional to the degree of oxidative stress and it reduces as age progresses (Ernster and Dallner, 1995). CoQ_{10} can be produced inside the cell and can function as a fat-soluble antioxidant (Mellors and Tappel, 1966; Ernster and Dallner, 1995). CoQ_{10} has been

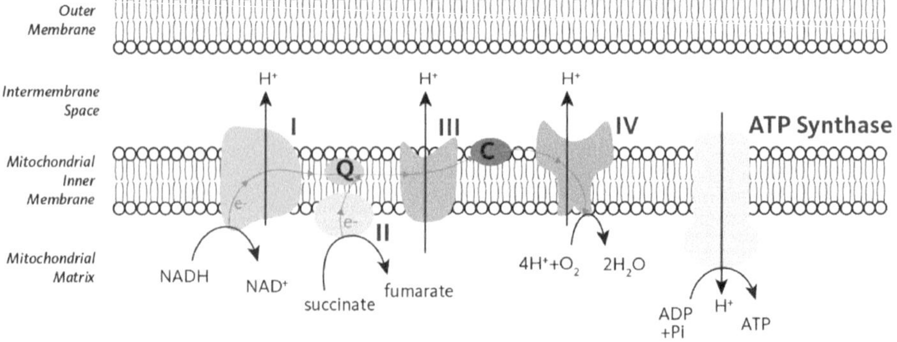

Fig. 9.5. CoQ_{10} is critical to electron transport in the mitochondrial respirator chain. It carries electron from Complex I and II to complex III, thus participating in ATP production.

Note. C: Cytochrome C; e^-: Electron; H^+: Proton; Q: CoQ_{10}.

proven to have a preventive action against cell death by the apoptotic process. This particular property of CoQ_{10} is associated with the inhibition of the mitochondrial channel, named permeability transition pore (PTP) (Papucci et al., 2003). CoQ_{10} is necessary for potentiating the uncoupling process of some of the proteins inside the mitochondria. CoQ_{10} can control inflammation by modulating the CoQ_{10} inducible genes (Schmelzer et al., 2007).

As the amount of CoQ_{10} reduces in certain conditions including ageing, the normal metabolic activities of the cell get disrupted and the degree of cellular dysfunction increases (Turunen et al., 2004). Since high density lipoprotein (HDL) can reduce the chance of atherosclerosis, CoQ_{10} can prevent atherosclerosis by protecting the HDL from getting oxidized (Thomas et al., 1999). Apart from the antioxidant property, CoQ_{10} can modify the β-2 integrins, lipids and this aids in the prevention of atherosclerosis of the blood vessels (Turunen et al., 2004). CoQ_{10} is associated with enhanced endothelial function for its ability to reduce endothelial injury due to oxidative stress (Hamilton et al., 2007). It is a very important cofactor for dihydroorotate dehydrogenase and therefore it is needed for the endogenous synthesis of the nucleotide known as pyrimidine (Jones, 1980). CoQ_{10} is beneficial for the treatment of disorders related to the mitochondria (Sacconi et al., 2010). The hydrophobic nature of CoQ_{10} is responsible for its delayed and partial absorption in the gastrointestinal tract. This is the reason for decreased bioavailability of CoQ_{10} via the oral route (Miles, 2007). The absorption of CoQ_{10} is governed by zero-order kinetics and is completed through a series of active and passive transport processes inside the small intestinal part of the gastrointestinal tract (Miles, 2007). Entero-hepatic recirculation is thought to be responsible for the second spike of the plasma concentration of CoQ_{10} after 24 hours of its oral consumption (Kalenikova et al., 2008). CoQ_{10} can protect the lipids such as LDL, HDL, and very low-density lipoprotein (VLDL) from the oxidative stress of the cells (Tomasetti et al., 1999).

Different studies have revealed an association between the plasma-CoQ_{10} and LDL or total serum cholesterol (Lopez et al., 2019). After reaching the liver, CoQ_{10} is transported to the lipoproteins and is distributed back to the systemic circulation from the liver. This typical phenomenon is believed to be responsible for the longer elimination period of CoQ_{10} which has a half-life ($t_{1/2}$) of 33 hours (Tomono et al., 1986). The enzyme cytochrome P450 can metabolize CoQ_{10} (Zaki, 2017). CoQ_{10} remains in a reduced state in organs other than the brain and liver, having high oxidative stress (Aberg et al., 1992). A reduced state of CoQ_{10} accounts for 95% of its total amount (Bhagavan and Chopra, 2006; 2007). The insufficiency of CoQ_{10} leads to cellular dysfunction since it is a major component of respiration at the cellular level. Even triggering of an oncogenic signal is associated with the deficiency of CoQ_{10} (Meganathan, 2001; Kawamukai, 2002; Potgieter et al., 2013). Controlling the ratio of NAD^+/NADH and thus maintaining the oxidative status of the cells is a key function of CoQ_{10}. It has an antioxidant property and can effectively reduce the oxidative stress of the cells. The proper genetic expression is controlled by CoQ_{10}. It is therapeutically used to protect people from various diseases and age-related difficulties (Prieme et al., 1997; Oytun et al., 2000; Zahiri et al., 2006; Cocheme et al., 2007; Rosenfeldt et al., 2007; Gao et al., 2012). Insufficiency of CoQ_{10} has been proven to have an association with the causation of diseases and conditions which involve increased oxidative stress including hypertension, Parkinson's disease, Huntington's disease, and type-2 diabetes (Bentinger et al., 2010). Transporting electrons in the respiratory chain is the most important function of CoQ_{10} (Ernster and Dallner, 1995; Choi et al., 2009; Kawamukai, 2009). CoQ_{10} has been found to have an association with the formation of a disulfide bond, pyrimidine metabolism, and oxidation of sulfide (Lee et al., 2004; Kawamukai, 2009). CoQ_{10} with its anti-oxidative property can slow down the production of free radicals and inhibit the oxidation-related alterations of certain components like DNA, lipids, and proteins (Siemieniuk and Skrzydlewska, 2005). The structure of CoQ_{10} is quite like that of vitamins (Mukama et al., 2017). The mammals can produce CoQ_{10} by the metabolism of tyrosine which is an essential amino acid and therefore it is not regarded as a vitamin in the case of mammals (Meganathan, 2001). CoQ_{10} taken from an outside source is absorbed exclusively in the liver tissue. It has been found that all human tissues have the potential to produce CoQ_{10}. This ability remains slow and less efficient during the embryonic period. Certain conditions like decreased catabolism and increased production can lead to the CoQ_{10} build-up in the liver. A negative feedback loop controls the synthesis of CoQ_{10} when there is a surplus of this component in the liver (Ramasarma, 2012). Various studies revealed that CoQ_{10} has multiple functions in mmuno modulation and lowering cholesterol levels. It also has well-known nutrigenomic and nutraceutical properties (Gopi et al., 2015). CoQ_{10} has been found to have an association with muscle regeneration and anti-ageing properties. The endogenous synthesis of CoQ_{10} is inhibited by substances like antimycin, stigmatellin, and lovastatin (Shukla and Dubey, 2018).

2.4 Plant and Animal Source

CoQ_{10} is present in plants, animals, and microorganisms, and in the human body. The chief sources of CoQ_{10} are fish, meat, different nuts, and oils. The net amount of CoQ_{10} in these plant and animal products varies from 10 to 50 mg/kg. The amount of CoQ_{10} is quite high (ranging from 30 to 200 mg/kg) in the animal organs like the heart and liver (Eskin and Tamir, 2005; Nabavi and Silva, 2018). The utilization of microbes, plants, and animals is in large use for producing CoQ_{10}. The microbial production of CoQ_{10} imparts the advantage of processing an optically pure form of CoQ_{10} with high yield and above all its an inexpensive process. It is known that a variety of plants and animals contain CoQ_{10} but specifically its concentration is higher in major organs of animals such as kidneys, liver, and heart. Beef contains a higher amount of CoQ_{10}. Reig et al. (2015) suggested that fish tissues and viscera contain the highest amount of CoQ_{10}. Among animals, the chicken heart contains CoQ_{10} at the concentration of 116 mg/kg to 132 mg/kg while the beef heart contains 113 mg/kg. The concentration of CoQ_{10} in the pork heart ranges from 11 mg/kg to 128 mg/kg. The full details are mentioned in Table 9.1.

Among plants, soybean oil may contain the highest concentration of CoQ_{10} which is as high as 280 mg/kg. This is followed by olive oil (160 mg/kg) and grape seed oil (73 mg/kg). Among nuts, peanuts contain 27 mg/kg of CoQ_{10} followed by sesame seeds (23 mg/kg). Among vegetables, parsley can have the highest concentration of CoQ_{10} (26 mg/kg) followed by spinach (10 mg/kg). Among fruits, avocado contains the highest concentration of CoQ_{10} (10 mg/kg). However, all these levels of concentration are the upper limits of their range (Pravst et al., 2010). The different sources of CoQ_{10} derived from are shown below in Table 9.2.

Table 9.1. Concentration of CoQ_{10} in different animal foods (Pravst et al., 2010).

Organ in Animal	CoQ_{10} Concentration (mg/kg)
Beef Heart	113
Beef Liver	39–50
Beef Muscle	26–40
Pork Liver	22.7–54
Chicken Heart	116.2–132.2
Sardine fish	5–64
Mackerel Red flesh	43–67

Table 9.2. Concentration of CoQ_{10} in different plant-based foods (Cluis et al., 2007).

Plant	CoQ_{10} Concentration (mg/kg)
Soybean oil	54–280
Olive oil	4–160
Sesame seeds	18–23
Peanuts	27
Walnuts	19
Almond nuts	5–14
Avocado	10
Grape seed oil	64–73
Sunflower oil	4–15

2.5 Microbial Sources

Among all microorganisms, CoQ_{10} is primarily produced by fungi and bacteria. Fungi such as *Aspergillus*, *Rhodotorula*, *Candida*, *Neurospora*, *Sporidobolus*, and others and bacteria such as *Rhodobacter*, *Tricosporon*, *Agrobacterium*, *Paracoccus*, *Cryptococcus*, and others are mostly known for the producer strain of CoQ_{10}. The biosynthesis of CoQ is largely controlled by the oxidation-reduction potential (ORP) of the fermentation medium. ORP can be lowered by reducing the supply of oxygen to the microbial culture and hence raising the production of CoQ_{10}. Natural selection processes and chemical mutagenesis are employed in commercial practice for developing the required strain (Shukla and Dubey, 2018). Chemical mutagenesis of these strains can lead to an increase in CoQ_{10} production as high as 770 mg/L. The physical and environmental factors that can improve the generation of CoQ_{10} are the temperature, the ratio between carbon and nitrogen, aeration supplied to the culture, and viscosity. These factors are very useful in commercial manufacture of CoQ_{10}

(Jeya et al., 2010; Cluis et al., 2012). Metabolic engineering approaches have been used in microbial biosynthesis to increase the titer of CoQ_{10} and overexpress some of the rate-limiting steps along the biosynthetic pathway. These approaches initially employ chemical mutagenesis-based selection and chemical engineering procedures that center on manipulating the substrate flux (Moriyama et al., 2015).

3. Prokaryotes as a Source of CoQ_{10}

3.1 Wild Type Strains

Several native producers of CoQ_{10} have been exploited for studies on production of CoQ_{10}. It includes mainly few bacterial and yeast strains including *S. pombe*, *S. johnsonii*, *Rhodobacter sphaeroides*, and *Agrobacterium tumefaciens*. Several other organisms, like *Pseudomonas*, *Paracoccus denitrificans*, *Candida*, and *Saitoella* yeasts also produce CoQ_{10} natively (Lee et al., 2017).

In prokaryotes, methyl erythritol 4-phosphate (MEP) pathway initiates using pyruvate and glyceraldehyde-3-phosphate as substrates. Both substrates get converted to 1-deoxy-D-xylulose-5-phosphate (DXP) using the catalyzing enzyme as DXP synthase (DXS). Subsequently, DXP converts into 2-C-methyl-D-erythritol -4-phosphate (MEP), followed by 4-(cytidine 5'-diphosphate)-2-C-methyl-D-erythritol (DCME), 2-phospho-4-(cytidine 5'-diphosphate)-2-C-methyl-D-erythritol (DCME-2-P). The MECP synthase catalyzes the conversion of DCME-2-P into 2-C-methyl-D-erythritol-2, 4-cyclodiphosphate (MECP) and then 1-hydroxy- 2-methyl-2-butenyl-4-diphosphate (HMBPP). Then HMBPP converts into isopentenyl-5-diphosphate (IPP). In eukaryotes, the mevalonate pathway is found which starts with acetyl CoA and acetoacetyl-CoA as substrates. They convert into 3-hydroxy-3-methylglutaryl-CoA (HMG-CoA) followed by mevalonate, later mevalonate-5-phosphate to mevalonate-5-pyrophosphate. Finally, it converts into IPP catalyzed by mevalonate pyrophosphate decarboxylase. In a few studies it was concluded that yeasts, fungi, and some other eukaryotic microorganisms depend upon the mevalonate pathway (MVA pathway) during the synthesis of CoQ_{10} (Meganathan, 2001; Opitz et al., 2014). In both the eukaryotes and prokaryotes, IPP is the common step from where geranyl diphosphate (GPP) is formed and followed by farnesyl diphosphate, decaprenyl diphosphate, decaprenyl-4-hydroxybenzoic acid, and finally to CoQ_{10}. The whole process is schematically represented in Fig. 9.6.

In the synthesis of coenzyme Q, the formation of 4-hydroxybenzoate is the primary step. Wild type yeast produces 4-hydroxybenzoate mainly in two different ways, from chorismate catalyzed by chorismate pyruvate-lyase and from tyrosine. Wild type yeast obtains 4-hydroxybenzoate from chorismate as the precursor substance for coenzyme Q (Meganathan, 2001). The two pathways namely, mevalonate (MVA) and non-mevalonate (non-MVA) are present by which CoQ_{10} is produced. In both, the prokaryotes and eukaryotes, the isoprene subunits can be made by IPP and its isomer dimethyl allyl pyrophosphate (DMAPP). IPP and DMAPP are the two precursors which are essential participants in the MVA pathway (HMG-CoA reductase pathway). Alternative to the MVA pathway, other bacteria like *E. coli* and *Mycobacterium tuberculosis* can produce isoprenoid chains by using the

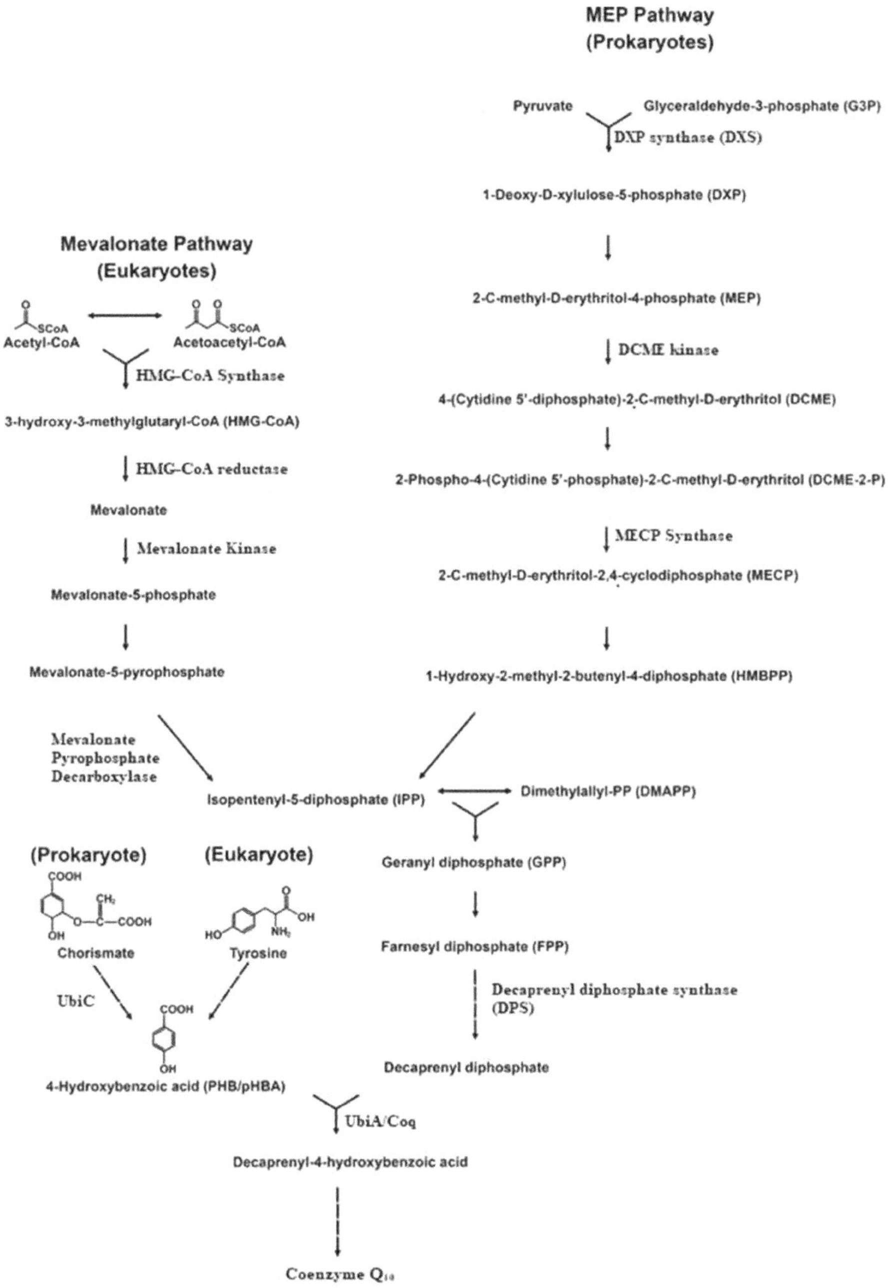

Fig. 9.6. Biosynthesis of CoQ_{10} in wild type strains.

MEP pathway. This pathway produces high amounts of IPP and DMAPP. Strains like *A. tumefaciens* and *R. sphaeroides* are found to be efficient natural producers of CoQ_{10} (Meganathan, 2001; Bentinger and Dallner, 2010; Yen et al., 2010). Basically, DXS is the initiator of the MEP pathway. In the reduction reaction of this pathway, a cyclic diphosphate ring opens for HMBPP, the step which is followed by the second reduction reaction in which there is the removal of hydroxyl group, leading to the formation of a mixture of IPP and DMAPP. Both IPP and DMAPP are the precursors of CoQ_{10} and other similar compounds such as hemes and lipids (Testa et al., 2006; Lee et al., 2007). Several aspects of the CoQ_{10} synthesis pathway are explained above section.

3.2 Recombinant Strains

In recombinant strains, CoQ_{10} production is increased by bringing changes in the naturally existed biosynthetic pathways. These changes may include the addition and alteration of some genes externally by bioengineering. In the *Agrobacterium tumefaciens* A603-35 strain, a certain modification is brought about by increasing NADH generating enzymes. As a result, the strain shows a higher $NADH/NAD^+$ ratio as compared to the original strain C58. $NADH/NAD^+$ ratio is directly proportional to the CoQ_{10} content in the A603-35 strain (Koo et al., 2010). Originally, it is known that the CoQ_{10} is produced through the MEP and mevalonate pathway. The overexpression of CoQ_{10} biosynthetic enzymes involved in the above pathway can lead to the increased production of CoQ_{10}. There has been the utilization of the cloning process to increase the production of CoQ_{10}. One of the examples of this is the cloning of genes producing polyprenyl diphosphate synthase that makes long-chain isoprenoid like *ddsA* (Okada et al., 1996; Kawamukai, 2009). Generally, *E. coli* is more capable of producing CoQ_8 instead of CoQ_{10}. Bacteria like *E. coli* is suitable for genetic modification to make it produce CoQ_{10} by the cloning process of E-decaprenyl diphosphate synthase (*DdsA*) and deleting *ispB* (responsible of coding of octaprenyl diphosphate synthase) to eliminate the process of synthesizing CoQ_8. Overexpression of 1-deoxy-D-xylulose-5-phosphate synthase (*DXS*) that takes part in the MEP pathway of CoQ_{10} production can make *E. coli* to generate CoQ_{10} in a significant amount (Cluis et al., 2007). However, genetically engineered *E. coli* still produces less CoQ_{10} (2.5 mg/g of DCW) as compared to the mutant strain of *R. sphaeroides* (8.7 mg/g of DCW) and *A. tumefaciens* (11.8 mg/g of DCW) (Cluis et al., 2007; 2011). The biosynthesis of CoQ in *E. coli* is preferable over the biosynthesis of menaquinone (MK) and dimethyl menaquinone (DMK) under aerobic conditions. But the same quinone pool is shared by CoQ along with MK and DMK. Hence, blocking the biosynthesis of MK and DMK will decrease the competition of synthesizing CoQ and there will be more available quinone for the biosynthesis of CoQ. It would be the metabolic strategy to block or down-regulate the biosynthesis of MK and DMK to favor the biosynthesis of CoQ in *E. coli* (Cluis et al., 2007). Another method is developed in which metabolic pathways are genetically modified into mevalonate-based isopentenyl pyrophosphate biosynthesis pathway in *E. coli* inducing the production of a large amount of isoprenoid. In this method, there is the possibility of accumulation of intermediates which can limit the growth of the cell. This accumulation occurs due to the insufficient availability of the HMG-CoA reductase. Therefore, modulation of HMG-CoA reductase may eliminate the problem and mevalonate production can be restored significantly. Hence, there can be increased isoprenoid synthesis and finally CoQ_{10} production (Pitera et al.,

2007). A potential organism for industrial production of CoQ_{10} *Agrobacterium tumefaciens* (KCCM 10413) was used to isolate gene (*dxs11*) from it. This gene was cloned in *E. coli* and subsequently, its nucleotide sequence was analyzed. The analysis of *E. coli* had shown to produce a heterodimer enzyme (a polypeptide of 640 amino acids) which is overexpressed in *E. coli* containing the cloned gene. This cloned gene, then transformed into *A. tumefaciens* (KCCM 10413) and the recombinant (*A. tumefaciens* pGX11) that resulted, started producing a higher amount of CoQ_{10} as compared to *A. tumefaciens* (KCCM 10413) (Lee et al., 2007).

4. Fungi and Yeast as a Source of CoQ_{10}

Several native producers of CoQ_{10} like yeast and fungi have also been exploited for studies on the production of CoQ_{10}. It includes mainly the yeasts *S. pombe*, *S. johnsonii*, *Candida*, and *Saitoella* that generate CoQ_{10} natively. *S. cerevisiae*, common fodder yeast produces CoQ_6, hence several genetic expressions of pathway genes are needed to produce CoQ_{10} using this yeast. The yeast CoQ_{10} biosynthesis pathway is very close to that of humans; therefore few yeast strains have been exploited as a model organism to study the biosynthesis and carry out manipulations in the biosynthetic pathway for overexpression and overproduction of CoQ_{10} (Lee et al., 2017).

COQ_1-COQ_{11} genes are needed for the effective production of CoQ_{10} by using the yeast-model. Yeast-model is an extremely useful and simple model for studying the mutations and polymorphism of the COQ gene. The yield of CoQ_{10} is boosted to a large extent if some naturally occurring producers are used as the sources of CoQ_{10} production. The factors influencing the production of CoQ_{10} in yeast and human cells are CoQ, intermediates of CoQ, and high molecular weight complexes that contain the CoQ gene. The CoQ reaction cascade in yeast is crucial for the biological production of CoQ. Yeast genes COQ_1 to CoQ_{10} can be used to restore the synthesis of CoQ (Awad et al., 2018). Biosynthesis of CoQ_{10} can be divided into three parts, namely, making of the quinonoid ring, production of decaprenyl diphosphate, and modifying quinonoid ring (Jeya et al., 2010). Yamada et al. (1983) examined yeast strains that have a CoQ_{10} system. They found that *P. lactucaedebilis*, *P. pachydermus*, *Protomyces inouyei*, and *Taphrina wiesneri* can produce CoQ_{10}. Other species which were found to have the CoQ_{10} system are *T. minor var. flava*, *T. lilacina*, *Cryptococcus hungaricus*, *Bullera alba*, *Tilletiopsis crimea*, and *T. minor var. minor* (Yamada et al., 1983). There are several other producers of CoQ_{10}, like *Rhodotorula*, *Saitoella*, and *Candida* are the ones known to have the capacity to produce on their own, but they cannot be considered as the generating hosts because they cannot create significantly without the introduction of other expensive constituents (Lee et al., 2017). *Schizosaccharomyces pombe* has shown to increase two-fold higher than the wild-type cells under normal physical environmental parameters (Nishida et al., 2019). Many features made *S. pombe* a good choice for eukaryotic gene study. It can be readily grown and modified in the laboratory setting as required (Zhao and Lieberman, 1995). The *S. pombe* is a unicellular rod-shaped eukaryote. It varies from 2 to 3 microns in diameter and 7 to 14 microns in length. This yeast is mainly found in fermented alcohol that contains sugar. This yeast shares similar features with that of many complicated yeasts (Crichton et al., 2007). Moriyama et al., 2015, suggested that *S. pombe* is the most efficient organism for genetic engineering to produce CoQ_{10} in higher amounts. CoQ_{10} productivity was improved in the fission yeast *S. pombe* by cloning and overexpressing ten CoQ biosynthetic genes. Simultaneous overexpression of precursor genes *Eco_ubiC* (encoding chorismate lyase), *Eco_aroFFBR* (encoding 3-deoxy-D-arabino-heptulosonate 7-phosphate synthase), or *Sce_thmgr1* (encoding truncated HMG-CoA reductase) resulted in two-fold increases in CoQ_{10} production. Another study showed that the cloned *ppt1* gene encodes p-hydroxy benzoate and the subsequent modification of *ERppt1* gene on endoplasmic reticulum in yeasts. This has increased CoQ_{10} production. The study manifested that even the recombinant yeasts with the capacity to produce CoQ_{10} in a higher amount than the wild type, are also resistant to copper ions, sodium chloride, and hydrogen peroxide. At lower temperatures, the recombinant strain grows faster than the wild type (Zhang et al., 2007). The native biosynthetic pathway of CoQ_{10} depends highly on the host organism. This is evidenced by higher growth rates and are shown with genetic modifications of the host organism (Ng, 2021). The *S. pombe* has been developed in a way to produce CoQ_{10} more efficiently by overexpressing certain genes like *HMGR* which are directly involved in the biosynthesis of CoQ_{10}. Originally, the mechanism of biosynthesis of CoQ is carried out by feeding cholesterol to the yeasts. A variety of genetic engineering approaches have been tried to improve the metabolic efficiency in the host organism. These approaches may include utilizing high-performance liquid chromatography (HPLC) to the whole-cell metabolomics. This typical process can enhance the expression of the enzyme and hence increase the amount of coenzyme produced. Another method to significantly increase production of CoQ_{10} from *S. pombe* is by modulation of the coenzyme expression using RNA interference (Ng, 2021). Researchers described the genus *Sporidiobolus* in detail as ballistoconidial yeast with clamp connections and the presence of thick-walled chlamydospores. *Sporidiobolus* is the sexual state of the species of *Sporobolomyces* (Kurtzman, 2014). One of the recently discovered species in this genus is *S. johnsonii*. It was regarded as one of the natural producers of CoQ_{10}. It is found that *S. johnsonii* can produce CoQ_{10} at 0.8 to 3.3 mg/g of DCW. The production of CoQ_{10} was observed at 10 mg/g of DCW in media supplemented with 4-hydroxy benzoic acid (HBA). This shows *S. johnsonii* has the potential to create CoQ_{10} at a higher amount naturally (Dixson et al., 2011). *Saccharomyces cerevisiae* is another unicellular fungus and is also known as budding yeast. The genome is well sequenced, and its genes are mostly a result of lateral gene transfer. The strains of *S. cerevisiae* are often selected for the fermentation process in industries to impart aroma and flavor to the final products like bread (Parapouli et al., 2020). Originally, *S. cerevisiae* could not produce CoQ_{10} but metabolic engineering has enabled us to modify the length determining the function of the DPS enzyme. This modification can allow *S. cerevisiae* to create CoQ_{10} given that the step of the DPS reaction would become the rate-limiting one. Therefore, there is a probability that *S. cerevisiae* can produce CoQ_{10} effectively without generating unnecessary products (Cluis et al., 2012; Brown et al., 2015). Some of the microbial sources of CoQ_{10} are presented in Table 9.3.

Table 9.3. Some of the microbial sources of CoQ_{10}.

Source	**Type of Strain**
Pseudomonas N84	Wild
Protaminobacter ruber	Wild
Rhodospirillum rubrum ATCC 25852	Wild
E. coli BL21	Recombinant
Paracoccus denitrificans ATCC 19367	Wild
Pseudomonas diminuta NCIM 2865	Wild
Schizosaccharomyces pombe	Wild
Sporidiobolus johnsonii	Wild

5. Biotechnological Production of CoQ_{10}

5.1 Scientific and Academic Reports

In general, the wild strains have been found as good sources of CoQ_{10} (Ha et al., 2007a; Bule and Singhal, 2009). The natural producers of CoQ_{10} are *Agrobacterium radiobacter* (or *A. tumefaciens*), *Rhodopseudomonas sphaeroides*, *Paracoccus denitrificans*, *Sporobolomyces salmonicolor*, and *Cryptococcus laurentii*. However, the ability to produce CoQ_{10} by these organisms cannot be commercially viable as their production ranges much below 500 mg/L. A study concluded that there are two ways of carrying out optimized production of CoQ_{10} through the process of microbial fermentation. The first way is to depend on the natural production of CoQ_{10} which occurs spontaneously but is not suitable for industrial production. The second way is to genetically modify them to make the microbes more favorable to produce CoQ_{10}. There are many attempts to genetically modify the microbes but few of them could increase the production titers of CoQ_{10} successfully (Cluis et al., 2012). Therefore, the adoption of several techniques like cellular-regulatory mechanisms, optimization conditions, and mutagenesis in the wild type may make them commercially viable to produce CoQ_{10} (Cluis et al., 2007). The manufacture of CoQ_{10} in industries is mostly done using mutants of wild-type strains due to their capacity to produce a higher amount of CoQ_{10}. Fed-batch fermentation is quite a popular technique of obtaining CoQ_{10} at low expenditure (Qiu et al., 2012). The extensive use of CoQ_{10} in the cosmetic and pharmaceutical sectors has raised its demand at its peak (Parmar et al., 2015). Hence, the synthesis of CoQ_{10} on a large scale is needed to fulfill its surging demand (Cluis et al., 2011; Lu et al., 2013; deDieu et al., 2014). *A. tumefaciens* was subjected to mutagenesis using high hydrostatic pressure (HHP) treatment, ultraviolet (UV) irradiation, and diethyl sulfate (DES) treatment to obtain mutant with higher CoQ_{10} content than wild-type strains. A mutant containing four genetic markers showed 52.83% increase in CoQ_{10} titer as compared to the original strain. The sucrose at concentration of 30 g/L, yeast extract at concentration of 30 g/L and 10 g/L of ammonium sulfate was identified as most favorable media ingredients for CoQ_{10} production using fed-batch culture strategy (Ha et al., 2007b; Yuan et al., 2012). Production of CoQ_{10} by *Candida glabrata* using an optimized medium, a compound of agro-industrial wastes (whey and corn steep liquor), was evaluated.

C. glabrata was found to be an excellent producer of CoQ_{10} by a researcher (Lima et al., 2015).

There have been efforts to produce CoQ_{10} using mutant strains. For this, attempts have been made by mutagenesis followed by optimization of the fermentation strategies. The development of the mutant strain was made with the property of drug resistance. These strains show a higher concentration of CoQ_{10} production. According to Choi et al. (2005) and Jeya et al. (2010), *Trichosporon* sp., *Sporobolomyces salmonicolor*, *R. Sphaeroides*, *Cryptococcus laurentii*, and other species of yeast are the microorganisms that can be used for producing CoQ_{10} using fermentation.

The main enzymes that take part in CoQ_{10} production have been cloned and made to express in *E. coli*, which could produce CoQ_{10} in higher amounts successfully (Choi et al., 2005). Another study confirmed that most of the long-chain part in CoQ_{10} does not participate in the electron transport chain (ETC). Overall CoQ functions in the respiratory chain in mitochondria where it serves as a lipophilic antioxidant. It was found that the function of CoQ_{10} from yeasts (like *Saccharomyces cerevisiae*) to human is quite significant in the respiratory chain (Tran and Clarke, 2007). In the synthesis of CoQ_{10}, the para-hydroxy benzoate (PHB) ring which is derived from tyrosine is condensed with a side chain. This condensation reaction is catalyzed by the PHB polyprenyl transferase. The genes that encode the PHB polyprenyl transferase were cloned in yeasts and bacteria. This has resulted in increased production of CoQ_{10} from these yeasts and bacteria (Szkopinska, 2000). Another study suggested that CoQ_{10} can be synthesized directly from solanesol and PHB by utilizing a two-phase conversion system (using water-organic solvent) in the *Sphingomonas* species. It was found that CoQ_{10} concentration was much higher in the organic solvent phase than that in the cell. The study also found that soybean oil improves CoQ_{10} release from the gel entrapped cells of *Sphingomonas* species. These gel entrapped cells can be re-used for CoQ_{10} production by using repeated batch culture. The two-phase conversion system is highly capable of enhancing the production of CoQ_{10} in the *Sphingomonas* species (Zhong et al., 2011).

In *Rhizobium radiobacter* T6102, the production of CoQ_{10} was improved by controlling the agitation speed, rate of aeration, and the dissolved oxygen supply. Due to the reduced oxygen supply, the production of CoQ_{10} was increased. It was also found that increasing the level of sodium azide and hydrogen peroxide, up to the level of 0.4 mM and 10 µm concentration, respectively (Seo and Kim, 2010). Tian et al. (2010) concluded that using *Rhodospirillum* as the source, a maximum amount of specific CoQ_{10} can be generated. Adding hydroxy butyrate to the culture medium increases the CoQ_{10} content even more by 5.27 mg/g of DCW. The precursor of coenzyme Qs in yeasts comes from the mevalonate pathway described above. According to Zahiri et al. (2006) mevalonate is found in an increased amount during the overexpression of CoQ_{10}. According to Hoffman et al. (2015), fission yeast or *Schizosaccharomyces pombe* is used as a model organism for the synthesis of CoQ_{10} as it has a human-like genetic mechanism and exhibits similar molecular properties, in terms of chemical reactions. Ioana et al. (2009) and Lambrechts and Siebrecht (2013) concluded that the synthesis of CoQ_{10} by the fermentation of yeast is the dominant method for the industrial production of CoQ_{10} and this involves Vitamin B during the culture process.

Sporidiobolus johnsonii ATCC 20490, a heterobasidiomycetes yeast strain, was explored for generating mutant phenotypes with high CoQ_{10} content (Ranadive et al., 2011). An atorvastatin resistant mutant strain-UF16 (phenotype) was generated from *S. johnsonii* by sequential induced mutagenesis, rational selection and screening process, resulting in 2.3-fold improvement in CoQ_{10} content (Ranadive et al., 2014). The glycerol- and tryptone-based medium was designed for optimum expression of CoQ_{10} (13.35 mg/L, 0.834 mg/g of DCW) by UF16. UF16. It showed improved flux of isoprene precursors on the mevalonate pathway leading to overproduction of CoQ_{10} (2.3-fold), ergosterol (2.6-fold), and total carotenoids (1.4-fold), whereas down-regulation of the fatty acid biosynthetic pathway was observed in this study. The HMG-CoA reductase gene (HMG1), which is a gene coding for the rate-limiting enzyme in the isoprenoid pathway, was found to be up-regulated in UF16 during submerged fermentation, as seen from the gene expression studies using RT-qPCR. It was found by this group that acquired atorvastatin resistance in UF16 due to induced mutagenesis facilitated the overexpression of HMG-CoA reductase gene, resulting in an improved flux of isoprene units to key downstream metabolites like CoQ_{10}, leading to overproduction. The fermentation and downstream purification process which was optimized for production of CoQ_{10} by UF16 in a laboratory fermentor (10L), resulted in 20.90 mg/L of CoQ_{10} having specific CoQ_{10} content of 0.86 mg/g of DCW. The outcome of this academic research may lead to an industrial process for commercial production of CoQ_{10} derived from eukaryote kingdom (8th Conference, International CoQ_{10} Association, Bologna, Italy).

Another CoQ_{10} known producer wild-type strain *Paracoccus denitrificans* ATCC 19367 was used to generate mutant strains using sequential rounds of mutagenesis followed by genome shuffling through protoplast fusion technology. The generated fusant strain PF-P1 showed the capability of CoQ_{10} production, specifically in a much higher amount as compared to the wild strain (Tokdar et al., 2014a,b). Further, the process to produce CoQ_{10} using the fusant strain PF-P1 has been scaled up to 2 L laboratory fermenter level. The strain PF-P1 produced 113.68 mg/L of CoQ_{10} having a specific content of 2.44 mg/g of DCW (Tokdar et al., 2017). Generally, in microbial fermentation generation of exoplysaccharides is a common phenomenon which hinders production of desired metabolites. In another study, *Agrobacterium tumefaciens* ATCC 4452 was found to produce excessive amounts of exopolysaccharides, which reduces the yield of CoQ_{10}. So, this problem was addressed by modifying the media and follow a classical mutagenesis approach to develop a mutant strain which can produce a higher amount of CoQ_{10} with less amount of accumulation of exoplolysaccharides (Tokdar et al., 2013).

5.2 Commercial Production of CoQ_{10} at the Industrial Scale

Presently a company in Japan is producing an active form of CoQ_{10} using their proprietary fermentation-based technology utilizing natural yeast strain, which is being used in supplements in over 450 products sold in Japan, the US, Europe, and other Asian countries. The detailed process is not disclosed by the company, but they claim to produce genetically modified organism (GMO)-free CoQ_{10} which is the only yeast-fermented CoQ_{10} and thus does not contain the impurities that the synthetically processed CoQ_{10} has. They also claim that their product is bio-identical to the CoQ_{10} produced within the body as it is derived from yeast and not from bacteria or tobacco derivatives used by other companies (https://www.kanekanutrients.com/kaneka-q10). Additionally, commercial production in Asia using a hyper-producing mutant strain of bacteria is known from an unpublished source.

5.3 CoQ_{10} Research Community

International CoQ_{10} Association having their headquarters in Europe, is a global organization that promotes academic and industrial research on CoQ_{10}. Every alternate year they hold a meeting to discuss different activities of CoQ_{10}. They give a platform to those institutes conducting clinical trials globally (usually by clinicians) to present their data and they generally give a recommendation. The research work carried out globally on CoQ_{10} biosynthesis aspects using yeast and recombinant strains as well as work on its basic biological functions is also discussed and published by the association (https://icqaproject.org/wp-content/uploads/2016/04/book3_all.pdf).

6. Applications of CoQ_{10}

The diseases that are associated with the deficiency of CoQ_{10} like fibromyalgia, cardiovascular diseases, primary and secondary CoQ_{10} deficiencies, cancer, neurodegenerative diseases, male infertility, diabetes mellitus, periodontal diseases are treated with the supplementation of CoQ_{10}. The decline in serum levels of CoQ_{10} and its deficiency in the tissues is due to medical conditions like primary inadequacy of CoQ_{10} and in mitochondrial diseases of secondary insufficiency of CoQ_{10} (Emmanuele et al., 2012). The levels of CoQ_{10} decrease with an increase in age. The CoQ_{10} deficiency is due to the nutritional deficiency that affects the synthesis of CoQ_{10}, a genetically acquired defect in the synthesis of CoQ_{10}, and increased tissue demands due to some disease. Encephalopathy, cerebral ataxia, isolated myopathy, leigh syndrome associated with retardation of growth, and severe infantile multisystem diseases occur with severe shortage of CoQ_{10}. The CoQ_{10} levels can be increased in the tissues by oral supplementation of the nutrient, hence it can be possible to correct the diseases that occur by the deficiency of CoQ_{10} (Quinzii et al., 2007). As CoQ_{10} has a role in the ATP synthesis process, it can affect the functions of all the cells in the body, mainly those cells which have a higher energy demand. It is the only antioxidant that is synthesized in the body. CoQ_{10} prevents the oxidation of DNA, lipids, and proteins (Littarru and Tiano, 2010). CoQ_{10} is used to treat different cardiovascular, neurodegenerative, mitochondrial, and neuromuscular diseases. Statin-associated myopathy can be treated with CoQ_{10} (Garrido et al., 2014). An overview of applications of CoQ_{10} is presented in the Fig. 9.7 and subsequent sections contain the applications of CoQ_{10} in details.

6.1 Treatment for the Deficiencies of CoQ_{10}

The deficiency of CoQ_{10} can be well managed and hence an early diagnosis is important in infants by the neurologists and pediatricians. The diagnosis can be made by measuring the amount of CoQ_{10} in the muscles and by the reduction in

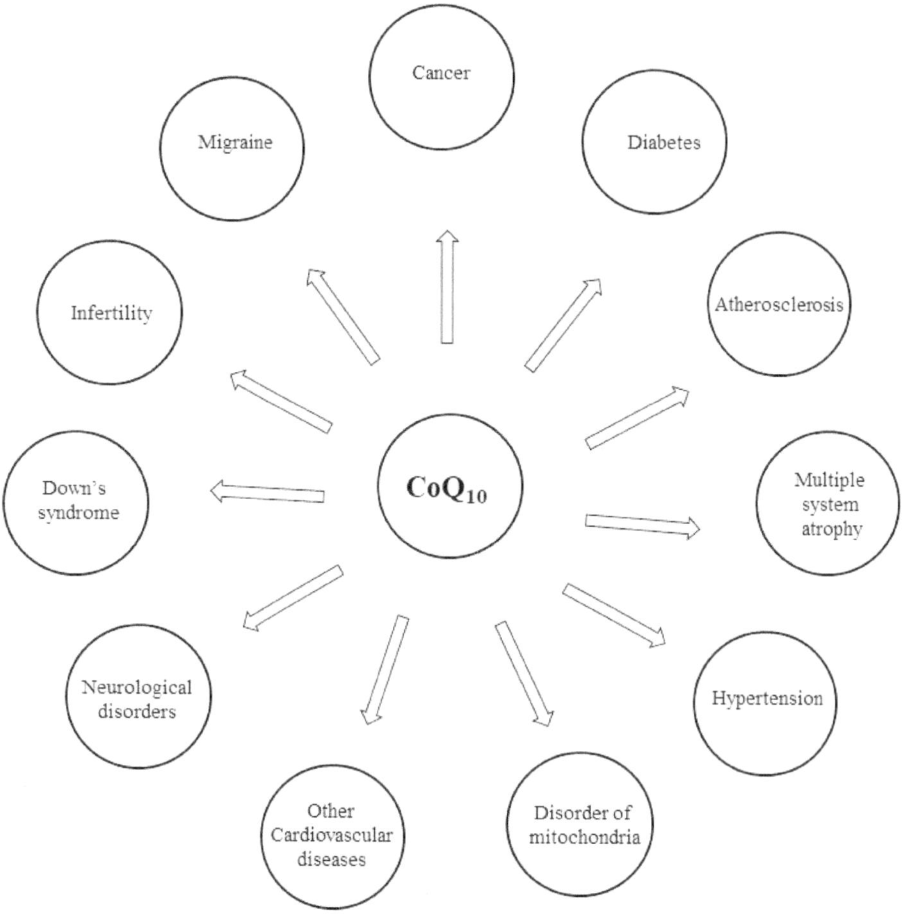

Fig. 9.7. An overview of the applications of CoQ_{10}.

the biochemical activities. The patients showed improvements with all forms of coenzyme Qs, especially a marked improvement was seen in CoQ_{10} deficiency by oral supplementation, except in the cerebral symptoms. This may occur because of the irreversible damage of the brain cells before the treatment or poor penetration across the blood-brain barrier by CoQ_{10} (Rotig et al., 2007). In hypercholesterolemia, the known drug presently is statin. However, there is no proven efficacy in statin therapy. Moreover, there are cases of skeletal complaints, creatine kinase elevation, and sometimes rhabdomyolysis. Some research papers are showing that statin therapy has inhibited CoQ_{10} synthesis by affecting mitochondrial biosynthesis leading to cardiomyopathy. This condition can only be reversed by providing the supplementation therapy of CoQ_{10}. Therefore, the patient who already has symptoms of CoQ_{10} deficiency, statin therapy should be avoided (Littarru and Langsjoen, 2007). Cardiomyopathies, neuronal and degenerative muscle diseases are also involved with the deficiency of CoQ_{10}. Encephalomyopathy, cerebral ataxia, isolated myopathy, leigh syndrome associated with abnormal features like retardation of growth, and severe infantile multisystem diseases are the phenotypes that are caused by the deficiency of CoQ_{10} (Quinzii et al., 2011).

The cerebellum is the first tissue to be affected by the pathological storage of CoQ_{10} as it has the least safety. The most severe deficiencies of CoQ_{10} in humans are due to the mutations in the autosomal recessive strains that are classified as primary and secondary deficiencies. When the CoQ_{10} biosynthetic genes are affected, it is called the primary deficiency. If the cause of the deficiency is due to other defects in the gene it is called the secondary deficiency. The response to the treatment of CoQ_{10} varies from patient to patient. The normal schedule for recommended doses for oral supplementation of CoQ_{10} includes a dosage up to 2400 mg/day in adults and up to 30 mg/kg/day in children three times a day (Naini et al., 2003).

6.2 Disorders of Mitochondria

In patients with mitochondrial myopathy, there is a reduction in CoQ_{10} in the muscles. The supplementation of CoQ_{10} is important for the treatment of primary mitochondrial disorders. The studies conducted on the doses of CoQ_{10} stated that larger doses are essential for the treatment of disorders of mitochondria (Kerr, 2010; Spinazzi et al., 2019; Tiano and Busciglio, 2011).

6.3 Fibromyalgia

It is a chronic pain syndrome. It does not have any specific etiology. Fibromyalgia can have various symptoms which may include migraine, allodynia, joint stiffness, and debilitating fatigue. Cellular abnormal features like oxidative stress are also one of the symptoms of fibromyalgia. The recent studies showed that there is a reduction in CoQ_{10} levels, a decrease in membrane potential of mitochondria with an increase in the superoxide levels of mitochondria in patients with fibromyalgia. The patients with fibromyalgia showed a reduction in the symptoms with the treatment of oral supplementation of CoQ_{10} (Cordero et al., 2011; 2012).

6.4 Cardiovascular Diseases

Oxidative stress plays an important role in the pathogenesis of cardiovascular diseases like hypertension and heart failure. Heart failure is caused by loss of contractile function due to the depletion of energy in the mitochondria which can be caused due to the decreased levels of CoQ_{10}. In patients with the cardiomyopathic condition, the results of endomyocardial biopsy samples state that there is a deficiency of CoQ_{10} in the myocardium of the heart. This can be improved with oral supplementation of CoQ_{10} that can increase the contractility of the myocardium (Cordero et al., 2012). Many studies also suggested that the supplementation of CoQ_{10} improves cardiac functions by increasing the production of energy, by its antioxidant property, improving the cardiac muscle contractility, and preventing the oxidation of LDL which is predominantly responsible for heart diseases. In chronic heart failure patients, there is a reduction in CoQ_{10} levels in both serum and myocardial tissue. Dilated cardiomyopathy is a form of disease of the cardiac muscle with symptoms like ventricular dilation, congestive heart failure, and contractile dysfunction (Folkers et al., 1970; 1985). In patients with stable but moderate congestive heart failure, the oral supplementation of CoQ_{10} can help in improving cardiac contractility and endothelial dysfunction (Littarru and Langsjoen, 2007).

6.5 Diabetes

It causes an increase in oxygen-free radical production which leads to oxidative stress that plays an important role in the pathogenesis of diabetes and its complications. An increase in the oxygen-free radical levels and depletion of antioxidant defense mechanisms may lead to damage to the cells. This may increase in peroxidation of the lipids and the development of resistance to insulin and hence may result in the complications of diabetes. As CoQ_{10} is an antioxidant and scavenger of free radicals, it may act as an indicator for oxidative stress and is used as a treatment modality for diabetes. CoQ_{10} acts as a self-protection mechanism during oxidative stress (Kucharska et al., 2000; Wei et al., 2009). The major complication of type 2 diabetes is cardiovascular disease caused due to endothelial dysfunction. Oxidative stress is the main reason for the development of endothelial dysfunction. Many studies state that the supplementation of CoQ_{10} improves endothelial function by activating mitochondrial oxidative phosphorylation reaction and endothelial nitric acid synthase enzyme (Watts et al., 2002; Chew and Watts, 2004).

6.6 Multiple Systems Atrophy

It is a neurodegenerative condition that is not curable and is progressive. Recent studies showed that patients with multiple system diseases *(MSA)* have low levels of CoQ_{10} because of mutations in the CoQ_2 gene and low concentrations of CoQ_{10} in the cerebellum. Recent findings show that a low level of CoQ_{10} present in cerebrospinal fluid serves as a biomarker for MSA. The patients with multiple system diseases had significantly low levels of CoQ_{10} in the cerebrospinal fluid when compared to patients with Parkinsonism or other healthy individuals. Therefore, ubiquinol is used in the treatment of patients with multiple system diseases (Brownlee, 2001).

6.7 Atherosclerosis

Ubiquinol is the reduced form of CoQ_{10} that inhibits the protein and oxidation of DNA. Ubiquinol inhibits the peroxidation of lipoprotein lipids and cell membrane lipids that are present in the circulation. CoQ_{10} supplementation results in the increased resistance of LDL that initiates the peroxidation of lipids. A dose of 150 mg/day of CoQ_{10} can increase the antioxidant activity of the enzyme, lower oxidative stress, and decrease the inflammatory cytokines like IL-6 marker in patients suffering from atherosclerosis (Li et al., 2017; Lee et al., 2012).

6.8 Hypertension

Depending on the type of various anti-hypertensive drugs, various side effects like cough, cardiac and renal dysfunction, and depression can be found. Many patients need to take more than one drug to control their hypertension that has many side effects. Many researchers believe the use of CoQ_{10} reduces the need to take many anti-hypertensive drugs (Langsjoen et al., 1994). CoQ_{10} is found to have a lowering effect on blood pressure (BP). Although the mechanism is not well studied, there are several theories and one of them is that CoQ_{10} has the effect of reducing the peripheral resistance of vessels by increasing the level of nitric oxide. The presence of increased nitric oxide levels can lead to the relaxation of peripheral arteries and hence leading to lowering the (BP). As CoQ_{10} has anti-oxidation properties, it can also minimize the effect of superoxide radicals and help in keeping the nitric oxide levels higher and active. In a randomized placebo-controlled double-blind trial, it was found that CoQ_{10} has a positive effect on BP control. This infers that adding CoQ_{10} as adjunctive therapy along with anti-hypertensive agents is a better way to control hypertension (Young et al., 2012). CoQ_{10} may help in increasing the production of prostaglandin, prostacyclin (prostaglandin I2) which helps in vasodilation and inhibition of aggregation of platelets, or it may increase the sensitivity of the arterial muscle to prostaglandin prostacyclin (Rosenfeldt et al., 2007).

6.9 Parkinsonism

Researchers found that the CoQ_{10} plays an important role in the cellular dysfunction of Parkinsonism. In patients with Parkinsonism, reduced levels of CoQ_{10} are found in the plasma and mitochondria of blood and platelet (Shults et al., 1997; Sohmiya et al., 2004). Several preclinical studies involving *in vitro* and *in vivo* experiments showed that CoQ_{10} has protective effects on the dopaminergic system. The clinical trials also showed that CoQ_{10} has a defensive mechanism on the neurological system in patients of Parkinsonism at the early and mid-stage. Recent advances in the research of CoQ_{10} in neurology found that the CoQ_{10} can mitigate the effect of cellular pathophysiological alterations which lead to abnormal mitochondrial functions among the patients of the familial type of Parkinsonism (Cooper et al., 2012; Mischley et al., 2012).

6.10 Cancer

CoQ_{10} has an antioxidant property that can help in preventing the peroxidation of lipids and oxidative damage to proteins and DNA. In the early stages of cancer, there is oxidative damage to DNA along with cytogenic and mutagenic lesions. CoQ_{10} can be useful in reducing the susceptibility to cells in developing cancer. In the tissues of breast tumor, a low level of CoQ_{10} is observed. The patients with melanoma and those with metastasis also have low levels of CoQ_{10}. There is a relation between the thickness of the tumor and the levels of CoQ_{10} with the highest levels of CoQ_{10} being observed in thinner tumors (Portakal et al., 2000; Valko et al., 2004). CoQ_{10} can have beneficial effects on cancer in several ways. It has positive effects on immunity and possesses antioxidant activity. CoQ_{10} supplements at the dosage of 100–200 mg/day may prevent cardiac abnormalities and can cause diarrhea and stomatitis. However, it has been found that CoQ_{10} concentration can be lowered by doxorubicin which is one of the useful chemotherapeutic agents in cancer. The normal chemotherapy regimen has the probability to cause liver and cardio toxicity whose effect can be mitigated by using CoQ_{10} supplements (Domae et al., 1981; Roffe et al., 2004; Brea et al., 2006; Schmelzer et al., 2008).

6.11 Migraine

It is the result of the impairment of mitochondria. As CoQ_{10} increases the mitochondrial oxidative phosphorylation, it can be used successfully in the treatment of migraines to reduce the frequency, severity, and length of the headaches. Oxidative stress is present in patients with migraines that can deplete the levels of antioxidants. Patients with migraines have low levels of CoQ_{10}. There is an increase in levels of inflammation in patients with migraines, so CoQ_{10} supplements are given that help in decreasing the levels of endogenous inflammatory mediators like tumor necrosis factor. The studies finally concluded that the supplementation of CoQ_{10} can reduce the duration of migraines but do not reduce the frequency and the severity of attacks (Dahri et al., 2018). The combination of CoQ_{10} and riboflavin has shown to be effective as a prophylactic treatment for migraines. There is some evidence that migraine condition is triggered due to abnormalities in energy impairment in the brain which explains the reason that CoQ_{10} (being energy supplements) can effectively improve the symptoms of migraine. A dosage of 150 mg of CoQ_{10} given once a daily basis for three months along with anti-migraine drugs can reduce the features of a migraine almost up to 50% in patients experiencing aura and without aura. It has been found that both in pediatric and adolescent patients of migraine, the deficiency of CoQ_{10} is usually found, and providing supplementation of CoQ_{10} to these patients has shown improvement in the symptoms of migraine (Rozen et al., 2002; Hershey et al., 2007).

6.12 Infertility in Males

The sperms produce reactive oxygen species in small quantities that play an important role in the homeostasis and signaling of the cells. If there is an increase in the production of reactive oxygen species, it causes an increase in damage to DNA and peroxidation of lipid of sperm membrane resulting in infertility. As CoQ_{10} is an antioxidant it helps in balancing the production of reactive oxygen species. Antioxidants are present in semen and sperms that help in the production of energy. Due to the presence of antioxidants, it is pro-motile and an antioxidant molecule (Wright et al., 2014; Hosen et al., 2015). CoQ_{10} concentration can be used to determine the sperm count and sperm motility by quantification in the seminal fluid. A higher level of oxidative stress is found in the patients of varicocele whose CoQ_{10} distribution is abnormal. There were studies conducted that showed that there is an inverse correlation between CoQ_{10} concentration and abnormal sperm formation (Mancini et al., 2005; Safarinejad, 2012).

6.13 Down's Syndrome

It is a chromosomal abnormality of trisomy 21. Oxidative stress plays an important role in its pathology. In patients with Down's syndrome, the main pathology is contributed by genetic and epigenetic factors. In this syndrome, there are structural changes of mitochondria. The use of CoQ_{10} in the treatment of Down's syndrome shows its antioxidant effect and it also repairs the DNA (Tiano et al., 2011).

6.14 Ageing

The decrease in the levels of CoQ_{10} may be one of the reasons for the development of chronic diseases in old people. As CoQ_{10} not only has antioxidant property but is also involved in the cellular process, hence supplementation of CoQ_{10} is important for the activity of the cells in the old people (Lopez et al., 2019). The ageing process in humans involves several aspects which result from the interaction of environmental, physiological, and genetic components. One of the hypotheses to explain ageing is the occurrence of imbalanced oxidative status due to the production of reactive oxygen species (ROS) and anti-oxidation mechanism. Many studies have shown that the overall progression of the ageing process can be arrested by CoQ_{10}. The formation of anion radical superoxide ions is inversely proportional to the presence of CoQ_{10}. It has been suggested that a lower level of CoQ_{10} might contribute to a reduction in the electron transport function and hence acceleration of the ageing process. Moreover, oxidative stress reduction occurs due to CoQ_{10} which is largely thought of as a reason behind considering that CoQ_{10} has a protective role to play against ageing (Takahashi et al., 2016; Barcelos and Haas, 2019).

6.15 Alzheimer's Disease

Alzheimer's disease is caused by mitochondrial dysfunction associated with oxidative damage. CoQ_{10} helps in the prevention of mitochondrial dysfunction and oxidative damage in patients with Alzheimer's disease. Advanced studies conducted by Galasko et al. (2012) showed that the disease can be associated with oxidative damage due to dysfunction in the mitochondrial system Alzheimers is thought to be enhanced due to the increased oxidative stress which contributes significantly to its pathogenesis. Although amyloid deposition is one of the main pathologies, the mitochondrial dysfunction owing to the increased ROS occurs much before. CoQ_{10} is actively involved in reducing the oxidative stress. Recent studies have shown that there is improvement in the behavioral performance of mice models with Alzheimers after giving CoQ_{10}. Moreover, the plaque areas got reduced while the mice were on CoQ_{10} therapy. The supplementation of CoQ_{10} also decreased the amyloid-β protein precursor (AβPP) level. Hence the cognitive performance increased significantly. There are events of human clinical trials of CoQ_{10} in patients with Parkinson's disease and Huntington's disease. Since CoQ_{10} is tolerable in humans, it may be seen as a prospective treatment for Alzheimer's disease (Dumont et al., 2011).

7. Conclusion and Future Perspectives

This chapter reviewed various aspects of CoQ_{10}, its functions and industrial applications. CoQ_{10} plays an important role in the production of cellular energy (in the form of ATP) and therefore is an indicator of health. Given the wide use of CoQ_{10} in many diseases as supplements, the supply of CoQ_{10} from diets and other natural sources are not sufficient. Production of CoQ_{10} through fermentation using microbes that synthesize CoQ_{10} is a favorable option. Few bacterial and fungal strains have been exploited academically and commercially for its synthesis and these details have been elaborated. Production of CoQ_{10} in large amounts could be achieved by bringing modifications at the genetic level in the wild strains leading to changes in biosynthetic pathways in mutated strains. These mutated strains of yeasts are created so that they can produce substantial amount of CoQ_{10} than the parent strain. The physical properties of CoQ_{10} were discussed. CoQ_{10} endogenous levels deplete with age and are responsible for age-related metabolic disorders. The pathophysiology regarding various diseases wherein CoQ_{10} has protective effects has been discussed. The diseases that are associated with increased oxidative stress can be prevented by maintaining the concentration of CoQ_{10}. This study described the significance of CoQ_{10} supplementation therapy. Overall, the antioxidant property of CoQ_{10} has been successful in managing various conditions including ageing and neurological diseases. However, CoQ_{10} is a valuable supplement and a significant product from a commercial point of view. A concise review on the recent developments in the biosynthesis of this molecule using genetically improved microbial strains clubbed with their process optimization will facilitate mass and feasible production of this molecule commercially. The clinical trials on this molecule have been successful and this approved molecule is being produced commercially by a few companies. Additionally, the mindset of clinicians for prescribing CoQ_{10} is also an important factor to successfully facilitate the therapeutic use of CoQ_{10}. The information covered herein could benefit the academic as well as the industrial research community in furthering technological advancement in this field, especially generating hyper-producing fungal or GMO strain for commercial applications.

Acknowledgements

We thank Ranjani Bharadwaj for editorial and proofreading services.

References

Aberg, F., Appelkvist, E.L., Dallner, G. and Ernster, L. (1992). Distribution and redox state of ubiquinones in rat and human tissues. *Arch. Biochem. Biophys.*, 295(2): 230–234.

Acosta, M.J., Fonseca, L.V., Desbats, M.A., Cerqua, C., Zordan, R., Trevisson, E. and Salviati, L. (2016). Coenzyme Q biosynthesis in health and disease. *Biochim. Biophys. Acta. Bioenerg.*, 1857(8): 1079–1085.

Awad, A.M., Bradley, M.C., Fernández-del-Río, L., Nag, A., Tsui, H.S. and Clarke, C.F. (2018). Coenzyme Q10 deficiencies: Pathways in yeast and humans. *Essays Biochem.*, 62(3): 361–376. https://doi.org/10.1042/ebc20170106.

Balakrishnan, P., Lee, B.J., Oh, D.H., Kim, J.O., Lee, Y.I., Kim, D.D. and Choi, H.G. (2009). Enhanced oral bioavailability of coenzyme Q10 by self-emulsifying drug delivery systems. *Int. J. Pharm.*, 374(1-2): 66–72.

Barcelos, I.P. and Haas, R.H. (2019). CoQ10 and aging. *Biology*, 8(2): 28.

Bentinger, M., Tekle, M. and Dallner, G. (2010). Coenzyme Q–Biosynthesis and functions. *Biochem. Biophys. Res. Commun.*, 396(1): 74–79. https://doi.org/10.1016/j.bbrc.2010.02.147.

Bhagavan, H.N. and Chopra, R.K. (2006). Coenzyme Q10: Absorption, tissue uptake, metabolism and pharmacokinetics. *Free Radic. Res.*, 40(5): 445–453.

Bhagavan, H.N. and Chopra, R.K. (2007). Plasma coenzyme Q10 response to oral ingestion of coenzyme Q10 formulations. *Mitochondrion*, 7: S78–S88.

Brea, C.G., Rodríguez, H.A., Fernandez, A.D.J., Navas, P. and Sanchez, A.J.A. (2006). Chemotherapy induces an increase in coenzyme Q10 levels in cancer cell lines. *Free. Radic. Biol. Med.*, 40(8): 1293–1302.

Brown, S., Clastre, M., Courdavault, V. and O'Connor, S.E. (2015). *De-novo* production of the plant derived alkaloid strictosidine in yeast. *PNAS*, 112(11): 3205–3210.

Brownlee, M. (2001). Biochemistry and molecular cell biology of diabetic complications. *Nature*, 414(6865): 813–820.

Bule, M.V. and Singhal, R.S. (2009). Use of carrot juice and tomato juice as natural precursors for enhanced production of ubiquinone-10 by *Pseudomonasdiminuta* NCIM 2865. *Food. Chem.*, 116(1): 302–305.

Chew, G.T. and Watts, G.F. (2004). Coenzyme Q10 and diabetic endotheliopathy: Oxidative stress and the 'recoupling hypothesis'. *Qjm.*, 97(8): 537–548.

Choi, J.H., Ryu, Y.W. and Seo, J.H. (2005). Biotechnological production and applications of coenzyme Q10. *Appl. Microbiol. Biotechnol.*, 68(1): 9–15. https://doi.org/10.1007/s00253-005-1946-x.

Choi, H.K., Pokharel, Y.R., Lim, S.C., Han, H.K., Ryu, C.S., Kim, S.K. and Kang, K.W. (2009). Inhibition of liver fibrosis by solubilized coenzyme Q10: Role of Nrf2 activation in inhibiting transforming growth factor-β1 expression. *Toxicol. Appl. Pharmacol.*, 240(3): 377–384.

Clarke, C.E. (2000). New advances in coenzyme Q biosynthesis. *Protoplasma*, 213: 134–147.

Cluis, C.P., Burja, A.M. and Martin, V.J. (2007). Current prospects for the production of coenzyme Q10 in microbes. *Trends Biotechnol.*, 25: 514–521.

Cluis, C.P., Ekins, A., Narcross, L., Jiang, H., Gold, N.D., Burja, A.M. and Martin, V.J. (2011). Identification of bottlenecks in *Escherichia coli* engineered for the production of CoQ10. *Metab. Eng.*, 13: 733–744.

Cluis, C.P., Pinel, D. and Martin V.J. (2012). The production of coenzyme Q10 in microorganisms. pp. 303–326. *In*: Wang, X., Chen, J. and Quinn, P. (Eds.). *Reprogramming Microbial Mtabolic Pathways*.

Cocheme, H.M., Kelso, G.F., James, A.M., Ross, M.F., Trnka, J., Mahendiran, T., Asin, C.J., Blaikie, F.H., Manas, A.R., Porteous, C.M., Adlam, V.J., Smith, R.A. and Murphy, M.P. (2007). Mitochondrial targeting of quinones: Therapeutic implications. *Mitochondrion.*, 7(Suppl): S94–S102.

Cooper, O., Seo, H., Andrabi, S, Guardia, L.C., Graziotto, J., Sundberg, M. and Isacson, O. (2012). Pharmacological rescue of mitochondrial deficits in iPSC derived neural cells from patients with familial Parkinson's disease. *Sci. Transl. Med.*, 4(141): 141ra90–141ra90.

Cordero, M.D., Alcocer, G.E., de Miguel, M., Cano, G.F.J., Luque, C.M., Fernandez, R.P. and Sanchez, A.J.A. (2011). Coenzyme Q10: A novel therapeutic approach for Fibromyalgia? Case series with 5 patients. *Mitochondrion*, 11(4): 623–625.

Cordero, M.D, Cano, G.F.J, Alcocer, G.E., De, M.M. and Sanchez A.J.A. (2012). Oxidative stress correlates with headache symptoms in fibromyalgia: Coenzyme Q10 effect on clinical improvement. PlOS ONE, 7(4): e35677.

Crane, F.L. (2001). Biochemical functions of coenzyme Q10. *J. Am. Coll. Nutr.*, 20(6): 591–598.

Crichton, P.G., Affourtit, C. and Moore, A.L. (2007). Identification of a mitochondrial alcohol dehydrogenase in *Schizosaccharomyces pombe*: New insights into energy metabolism. *Biochem. J.*, 401(Pt 2): 459–464. https://doi.org/10.1042/BJ20061181.

Dahri, M., Tarighat, E.A., Asghari, J.M. and Hashemilar, M. (2018). Oral coenzyme Q10 supplementation in patients with migraine: Effects on clinical features and inflammatory markers. *Nutr. Neurosci.*, 1–9. https://doi.org/10.1080/1028415X.2017.1421039.

de Dieu, N.J. and Lee, B.H. (2014). Enhanced production techniques, properties, and uses of coenzyme Q10. *Biotechnol. Lett.*, 36(10): 1917–1926.

Dixson, D.D., Boddy, C.N. and Doyle, R.P. (2011). Reinvestigation of coenzyme Q10 isolation from *Sporidiobolus johnsonii*. *Chem. Biodivers.*, 8: 1033–1051.

Domae, N., Sawada, H., Matsuyama, E., Konishi, T. and Uchino, H. (1981). Cardiomyopathy and other chronic toxic effects induced in rabbits by doxorubicin and possible prevention by coenzyme Q10. *Cancer. Treat. Rep.*, 65(1-2): 79–91.

Doring, F., Schmelzer, C., Lindner, I., Vock, C. and Fujii, K. (2007). Functional connections and pathways of coenzyme Q10-inducible genes: An *in silico* study. *IUBMB. Life*, 59(10): 628–633.

Dumont, M., Kipiani, K., Yu, F, Wille, E., Katz, M., Calingasan, N.Y. and Beal, M.F. (2011). Coenzyme Q10 decreases amyloid pathology and improves behavior in a transgenic mouse model of Alzheimer's disease. *J. Alzheimer's Dis.*, 27(1): 211–223.

Emmanuele, V., Lopez, L.C., Berardo, A., Naini, A., Tadesse, S., Wen, B. and Hirano, M. (2012). Heterogeneity of coenzyme Q10 deficiency: Patient study and literature review. *Arch. Neurol.*, 69(8): 978–983.

Ernster, L. and Dallner, G. (1995). Biochemical, physiological, and medical aspects of ubiquinone function. *Biochim. Biophys. Acta.*, 1271: 195–204.

Eskin, M. and Tamir, S. (2005). *Dictionary of Nutraceuticals and Functional Foods*. Boca Raton, CRC Press.

Folkers, K., Littarru, G.P., Ho, L., Runge, T.M., Havanonda, S. and Cooley, D. (1970). Evidence for a deficiency of coenzyme Q10 in human heart disease. *Int. Z. Vitaminforsch.*, 40: 380–390.

Folkers, K., Vadhanavikit, S. and Mortensen, S.A. (1985). Biochemical rationale and myocardial tissue data on the effective therapy of cardiomyopathy with coenzyme Q10. *PNAS*, 82(3): 901–904.

Formigli, L., Zecchi, O.S., Orlandini, G., Carella, G., Brancato, R., Papucci, L. and Capaccioli, S. (2003). Coenzyme Q10 prevents apoptosis by inhibiting mitochondrial depolarization independently of its free radical scavenging property. *J. Biol. Chem.*, 278(30): 28220–28228.

Galasko, D.R., Peskind, E., Clark, C.M., Quinn, J.F., Ringman, J.M., Jicha, G.A. and Aisen, P. (2012). Antioxidants for Alzheimer disease: A randomized clinical trial with cerebrospinal fluid biomarker measures. *Arch. Neurol.*, 69(7): 836–841.

Gao, L., Mao, Q., Cao, J., Wang, Y., Zhou, X. and Fan, L. (2012). Effects of coenzyme Q10 on vascular endothelial function in humans: A meta-analysis of randomized controlled trials. *Atherosclerosis*, 221(2): 311–316.

Garrido, M.J., Cordero, M.D., Oropesa, A.M., Vega, A.F., De, L.M.M., Pavon, A.D. and Sanchez, A.J.A. (2014). Coenzyme Q10 therapy. *Mol. Syndromol.*, 5(3-4): 187–197.

Gille, L. and Nohl, H. (2000). The existence of a lysosomal redox chain and the role of ubiquinone. *Arch. Biochem. Biophys.*, 375(2): 347–354.

Gomez, D.C., Rodriguez-Aguilera, J.C., Barroso, M.P., Villalba, J.M., Navarr, F., Crane, F.L. and Navas, P. (1997). Antioxidant ascorbate is stabilized by NADH-coenzyme Q10 reductase in the plasma membrane. *J. Bioenerg. Biomembr.*, 29(3): 251–257.

Gonzalez, G.L., Yubero, S.E.M., Delgado, L.J., Perez, M.P., Garcia, R.A., Marin, C. and Lopez, M.J. (2015). Effects of the Mediterranean diet supplemented with coenzyme Q10 on metabolomic profiles in elderly men and women. *J. Gerontol. A. Biol. Sci. Med. Sci.*, 70(1): 78–84.

Gopi, M., Purushothaman, M.R. and Chandrasekaran, D. (2015). Influence of coenzyme Q10 supplementation in high energy broiler diets on production performance, hematological and slaughter parameters under higher environmental temperature. *Asian. J. Anim. Vet. Adv.*, 10(7): 311–322.

Ha, S.J., Sang, Y.K., Jin, H.S., Deok, K.O. and Jung, K.L. (2007a). Optimization of culture conditions and scale-up to pilot and plant scales for coenzyme Q10 production by *Agrobacterium tumefaciens*. *Appl. Microbiol. Biotechnol.*, 74: 974–980.

Ha, S.J., Kim, S.Y., Seo, J.H., Moon, H.J., Lee, K.M. and Lee, J.K. (2007b). Controlling the sucrose concentration increases coenzyme Q10 production in fed-batch culture of *Agrobacterium tumefaciens*. *Appl. Microbiol. Biotechnol.*, 76(1): 109–116.

Hamilton, S.J., Chew, G.T. and Watts, G.F. (2007). Therapeutic regulation of endothelial dysfunction in type 2 diabetes mellitus. *Diabetes. Vasc. Dis. Res.*, 4(2): 89–102.

Hershey, A.D., Powers, S.W., Vockell, A.L.B., Le, C.S.L., Ellinor, P.L., Segers, A. and Kabbouche, M.A. (2007). Coenzyme Q10 deficiency and response to supplementation in pediatric and adolescent migraine. *Headache*, 47(1): 73–80.

Hiebert, J., Shen, Q. and Pierce, J. (2012). Application of coenzyme Q10 in clinical practice. *The Internet Journal of Internal Medicine*, 9(2): 1–10.

Hoffman, C.S., Wood, V. and Fantes, P.A. (2015). An ancient yeast for young geneticists: A primer on the *Schizosaccharomyces pombe* model system. *Genetics*, 201(2): 403–423.

Hosen, M.B., Islam, M.R., Begum, F., Kabir, Y., Howlader, M.Z.H. (2015). Oxidative stress induced sperm DNA damage, a possible reason for male infertility. *Iran. J. Reprod. Med.*, 13(9): 525.

Iehud, I. and Doron, E. (1988). Total synthesis of polyprenoid natural products via Pd(O)-catalyzed oligomerizations. *Pure and Applied Chemistry*, 60: 89–98.

International CoQ_{10} Association. (2015). 8th Conference, Bologna, Italy (8–11 October 2015). https://icqaproject.org/wp-content/uploads/2016/04/book3_all.pdf.

Ioana, V.S., Lasio, V. and Uivarosan, D. (2009). Stimulation of biosynthesis of coenzyme Q10 by *Sacharomyces cerevisiae* under the influence of vitamin B1. *Analele. Univ. din Oradea. Fasc. Biol.*, 2: 693–700.

Jeya, M., Moon, H.J., Lee, J.L., Kim, I.W. and Lee, J.K. (2009). Current state of coenzyme Q10 production and its applications. *Appl. Microbiol. Biotechnol.*, 85(6): 1653–1663. https://doi.org/10.1007/s00253-009-2380-2.

Jones, M.E. (1980). Pyrimidine nucleotide biosynthesis in animals: Genes, enzymes, and regulation of UMP biosynthesis. *Annu. Rev. Biochem.*, 49(1): 253–279.

Kalenikova, E.I., Gorodetskaya, E.A. and Medvedev, O.S. (2008). Pharmacokinetics of coenzyme Q10. *Bull. Exp. Biol. Med.*, 146(3): 313. https://www.kanekanutrients.com/kaneka-q10.

Kawamukai, M. (2002). Biosynthesis, bioproduction, and novel roles of ubiquinone. *J. Biosci. Bioeng.*, 94(6): 511–517.

Kawamukai, M. (2009). Biosynthesis and bioproduction of coenzyme Q10 by yeasts and other organisms. *Biotechnol. Appl. Biochem.*, 53: 217–226.

Kazunori, O., Kazuaki, O., Kazufumi, Y., Kouhei, N., Naonori, U., Makoto, K. and Hisakazu, Y. (2004). The AtPPT1 gene encoding 4-hydroxybenzoate polyprenyl diphosphate transferase in ubiquinone biosynthesis is required for embryo development in *Arabidopsis thaliana*. *Plant. Mol. Biol.*, 57: 567–577.

Kerr, D.S. (2010). Treatment of mitochondrial electron transport chain disorders: A review of clinical trials over the past decade. *Mol. Genet. Metab.*, 99(3): 246–255.

Koo, B.S., Gong, Y.J., Kim, S.Y., Kim, C.W. and Lee, H.C. (2010). Improvement of coenzyme Q10 production by increasing the NADH/NAD (+) ratio in *Agrobacterium tumefaciens*. *Biosci. Biotechnol. Biochem.*, 74(4): 895–898.

Kucharska, J., Braunova, Z., Ulicna, O., Zlatos, L. and Gvozdjakova, A. (2000). Deficit of coenzyme Q in heart and liver mitochondria of rats with streptozotocin induced diabetes. *Physiol. Res.*, 49(4): 411–418.

Kurtzman, C.P. (2014). Use of gene sequence analyses and genome comparisons for yeast systematics. *International Journal of Systematic and Evolutionary Microbiology*, 64: 325–332.

Lambrechts, P. and Siebrecht, S. (2013). Coenzyme Q10 and ubiquinol as adjunctive therapy for heart failure. *Agro. Food. Ind. Hi Tech.*, 24: 60–62.

Langsjoen, P., Willis, R. and Folkers, K. (1994). Treatment of essential hypertension with coenzyme Q10. *Mol. Asp. Med.*, 15: s265–s272.

Lankin, V.Z., Tikhaze, A.K., Kapel'ko, V.I., Shepel'kova, G.S., Shumaev, K.B., Panasenko, O.M., Konovalova, G.G. and Belenkov, Y.N. (2007). Mechanisms of oxidative modification of low-density lipoproteins under conditions of oxidative and carbonyl stress. *Biochemistry (Mosc).*, 72(10): 1081–1090. https://doi.org/10.1134/s0006297907100069.

Laplante, S., Souchet, N. and Bryl, P. (2009). Comparison of low-temperature processes for oil and coenzyme Q10 extraction from mackerel and herring. *European Journal of Lipid Science and Technology*, 111(2): 135–141.

Lee, J.K., Oh, D.K. and Kim, S.Y. (2004). Cloning and functional expression of the dps gene encoding decaprenyl diphosphate synthase from *Agrobacterium tumefaciens*. *Biotechnol. Prog.*, 20: 51–56.

Lee, J.K., Oh, D.K. and Kim, S.Y. (2007). Cloning and characterization of the dxs gene, encoding 1-deoxy-d-xylulose 5-phosphate synthase from *Agrobacterium tumefaciens*, and its overexpression in *Agrobacterium tumefaciens*. *J. Biotechnol.*, 128(3): 555–566. https://doi.org/10.1016/j.jbiotec.2006.11.009.

Lee, B.J., Huang, Y.C., Chen, S.J. and Lin, P.T. (2012). Effects of coenzyme Q10 supplementation on inflammatory markers (high-sensitivity C-reactive protein, interleukin-6, and homocysteine) in patients with coronary artery disease. *Nutrition*, 7-8: 767–772.

Lee, S.Q.N., Tan, T.S., Kawamukai, M. and Chen, E.S. (2017). Cellular factories for coenzyme Q10 production. *Microb. Cell. Fact.*, 16: 39. DOI 10.1186/s12934-017-0646-4.

Li, X., Guo, Y., Huang, S., He, M., Liu, Q., Chen, W. and He, P. (2017). Coenzyme Q10 prevents the interleukin-1 beta induced inflammatory response via inhibition of MAPK signaling pathways in rat articular chondrocytes. *Drug. Dev. Res.*, 78(8): 403–410.

Lima, R.A., Saconi, A., Campos-Takaki, G.M., Andrade, R.F.S. and Montero, D.R. (2015). Optimized submerged batch fermentation for co-enzyme Q Production by *Candida glabrata* using renewable substrates. *International Journal of Innovative Research in Engineering and Management (IJIREM)*, 2(5).

Linnane, A.W., Kios, M. and Vitetta, L. (2007). Coenzyme Q10: Its role as a prooxidant in the formation of superoxide anion/hydrogen peroxide and the regulation of the metabolome. *Mitochondrion*, 7: S51–S61.

Lipshutz, B.H., Mollard, P., Pfeiffer, S.S. and Chrisman, W. (2002). A short highly efficient synthesis of coenzyme Q10. *J. Am. Chem. Soc.*, 124: 14282–14283.

Lipshutz, B.H., Lower, A., Berl, V., Schein, K. and Wetterich, F. (2005). An improved synthesis of the "miracle nutrient" coenzyme Q10. *Org. Lett.*, 7(19): 4095–4097.

Littarru, G.P. and Langsjoen, P. (2007). Coenzyme Q10 and statins: Biochemical and clinical implications. *Mitochondrion*, 7: S168–S174.

Littarru, G.P. and Tiano, L. (2010). Clinical aspects of coenzyme Q10: An update. *Nutrition*, 26(3): 250–254.

Lopez, L.G., del Pozo, C.J., Sanchez, C.A., Cortes, R.A.B. and Navas, P. (2019). Bioavailability of coenzyme Q10 supplements depends on carrier lipids and solubilization. *Nutrition*, 57: 133–140. https://doi.org/10.1016/J.NUT.2018.05.020.

Lu, W., Shi, Y., He, S., Fei, Y., Yu, K. and Yu, H. (2013). Enhanced production of CoQ10 by constitutive overexpression of 3-demethyl ubiquinone-9 3-methyltransferase under tac promoter in *Rhodobacter sphaeroides*. *Biochem. Eng. J.*, 72: 42–47.

Mancini, A., De Marinis, L., Littarru, G.P. and Balercia, G. (2005). An update of coenzyme Q10 implications in male infertility: Biochemical and therapeutic aspects. *Biofactors*, 25(1–4): 165–174.

Meganathan, R. (2001). Ubiquinone biosynthesis in microorganisms. *FEMS. Microbiol. Lett.*, 203(2): 131–139.

Mellors, A.A. and Tappel, A.L. (1966). The inhibition of mitochondrial peroxidation by ubiquinone and ubiquinol. *J. Biol. Chem.*, 241(19): 4353–4356.

Miles, M.V. (2007). The uptake and distribution of coenzyme Q10. *Mitochondrion*, 7: S72–S77.

Mischley, L.K., Allen, J. and Bradley, R. (2012). Coenzyme Q10 deficiency in patients with Parkinson's disease. *J. Neurol. Sci.*, 318(1-2): 72–75.

Mitchell, P. (1975). The protonmotive Q cycle: A general formulation. FEBS. *Letters*, 59(2): 137–139.

Moriyama, D., Hosono, K., Fujii, M., Washida, M., Nanba, H., Kaino, T. and Kawamukai, M. (2015). Production of CoQ10 in fission yeast by expression of genes responsible for CoQ10 biosynthesis. *Biosci. Biotechnol. Biochem.*, 79(6): 1026–1033.

Mukama, O., Sinumvayo, J.P., Shamoon, M., Shoaib, M., Mushimiyimana, H., Safdar, W. and Wang, Z. (2017). An update on aptamer-based multiplex system approaches for the detection of common foodborne pathogens. *Food Analytical Methods*, 10(7): 2549–2565.

Nabavi, S.M. and Silva, A.S. (Eds.) (2018). *Nonvitamin and Nonmineral Nutritional Supplements*. London, United Kingdom, Academic Press.

Naini, A., Lewis, V.J., Hirano, M. and DiMauro, S. (2003). Primary coenzyme Q10 deficiency and the brain. *Biofactors*, 18(1–4): 145–152.

Ndikubwimana, J. de D. and Lee, B.H. (2014). Enhanced production techniques, properties, and uses of coenzyme Q10. *Biotechnol. Lett.*, 36(10): 1917–26. DOI 10.1007/s10529-014-1587-1.

Nepal, P.R., Han, H.K. and Choi, H.K. (2010). Preparation and *in vitro–in vivo* evaluation of Witepsol® H35 based self-nanoemulsifying drug delivery systems (SNEDDS) of coenzyme Q10. Eur. *J. Pharm. Sci.*, 39(4): 224–232.

Ng, W. (2021). Production of Coenzyme Q10 in recombinant *schizosaccharomyces pombe* through metabolic engineering approaches. Papers.ssrn.com. https://papers.ssrn.com/sol3/papers.cfm?abstract_id=3814228.

Nishida, I., Yokomi, K., Hosono, K., Hayashi, K., Matsuo, Y., Kaino, T. and Kawamukai, M. (2019). CoQ10 production in *Schizosaccharomyces pombe* is increased by reduction of glucose levels or deletion of pka1. *Appl. Microbiol. Biotechnol.*, 103(12): 4899–4915. https://doi.org/10.1007/s00253-019-09843-7.

Okada, K., Suzuki, K., Kamiya, Y., Zhu, X., Fujisaki, S., Nishimura, Y. and Matsuda, H. (1996). Polyprenyl diphosphate synthase essentially defines the length of the side chain of ubiquinone. *Biochim. Biophys. Acta. Mol. Cell. Biol. Lipids*, 1302(3): 217–223.

Opitz, S., Nes, W.D. and Gershenzon, J. (2014). Both methylerythritol phosphate and mevalonate pathways contribute to biosynthesis of each of the major isoprenoid classes in young cotton seedlings. *Phytochemistry*, 98: 110–119.

Oytun, P.O.O., Mine, E.I., Berrin, B., Muberra, K. and Iskender, S. (2000). Coenzyme Q10 concentrations and antioxidant status in tissues of breast cancer patients. *Clin. Biochem.*, 33: 279–284.

Palamakula, A., Soliman, M. and Khan, M.M.A. (2005). Regional permeability of coenzyme Q10 in isolated rat gastrointestinal tracts. *Die Pharmazie-An Int. J. Pharm. Sci. Res.*, 60(3): 212–214.

Papucci, L., Schiavone, N., Witort, E., Donnini, M., Lapucci, A., Tempestini, A., Formigli, L., Zecchi, S., Orlandini, G., Giuseppe Carella, G., Rosario Brancato, R. and Capaccioli, S. (2003). Coenzyme Q10 prevents apoptosis by inhibiting mitochondrial depolarization independently of its free radical scavenging property. *The Journal of Biological Chemistry*, 278(30): 25: 28220–28228.

Parapouli, M., Vasileiadi, A., Afendra, A.S. and Hatziloukas, E. (2020). *Saccharomyces cerevisiae* and its industrial applications. *AIMS. Microbiology*, 6(1): 1–32. https://doi.org/10.3934/microbiol.2020001.

Parmar, S.S., Jaiwal, A., Dhankher, O.P. and Jaiwal, P.K. (2015). Coenzyme Q10 production in plants: Current status and future prospects. *Crit. Rev. Biotechnol.*, 35(2): 152–164.

Pitera, D.J., Paddon, C.J., Newman, J.D. and Keasling, J.D. (2007). Balancing a heterologous mevalonate pathway for improvedisoprenoid production in *Escherichia coli. Metab. Eng.*, 9: 193–207.

Poon, W.W., Barkovich, R.J., Hsu, A.Y., Frankel, A., Lee, P.T., Shepherd, J.N. and Clarke, C.F. (1999). Yeast and rat Coq3 and *Escherichia coliUbiG* polypeptides catalyze both O-methyltransferase steps in coenzyme Q biosynthesis. *J. Biol. Chem.*, 274(31): 21665–21672.

Portakal, O., Ozkaya, O., Bozan, B., Koşan, M. and Sayek, I. (2000). Coenzyme Q10 concentrations and anti-oxidant status in tissues of breast cancer patients. *Clin. Biochem.*, 33(4): 279–284.

Potgieter, M., Pretorius, E. and Pepper, M.S. (2013). Primary and secondary coenzyme Q10 deficiency: The role of therapeutic supplementation. *Nutr. Rev.*, 71: 180–188.

Pravst, I., Zmitek, K. and Zmitek, J. (2010). Coenzyme Q10 contents in foods and fortification strategies. *Crit. Rev. Food. Sci. Nutr.*, 50(4): 269–280.

Prieme, H., Loft, S., Nyyssonen, K., Salonen, J.T. and Poulsen, H.E. (1997). No effect of supplementation with vitamin E, ascorbic acid, or coenzyme Q10 on oxidative DNA damage estimated by 8-oxo-7, 8-dihydro-2'-deoxyguanosine excretion in smokers. *Am. J. Clin. Nutr.*, 65(2): 503–507.

Qiu, L., Ding, H., Wang, W., Kong, Z., Li, X., Shi, Y. and Zhong, W. (2012). Coenzyme Q10 production by immobilized *Sphingomonas* sp. ZUTE03 via a conversion–extraction coupled process in a three-phase fluidized bed reactor. *Enzyme Microb. Technol.*, 50(2): 137–142.

Quinzii, C.M., Di, M.S. and Hirano, M. (2007). Human coenzyme Q10 deficiency. *Neurochem. Res.*, 32(4): 723–727.

Quinzii, C.M. and Hirano, M. (2011). Primary and secondary CoQ10 deficiencies in humans. *Biofactors*, 37(5): 361–365.

Ramasarma, T. (2012). A touch of history and a peep into the future of the lipid-quinone known as coenzyme Q and ubiquinone. *Curr. Sci.*, 102(10): 1459–1471.

Ranadive, P., Mehta, A. and George, S. (2011). Strain improvement of *Sporidiobolus johnsonii* ATCC 20490 for biotechnological production of coenzyme Q10. *International Journal of Chemical Engineering and Applications*, 2(3): 216–220.

Ranadive, P., Mehta, A., Chavan, Y., Marx, A. and George, S. (2014). Morphological and molecular differentiation of *Sporidiobolus johnsonii* ATCC 20490 and its coenzyme Q10 overproducing mutant strain UF16. *Indian Journal of Microbiology*, 54: 343–357. DOI: 10.1007/s12088-014-0466-8.

Reig, M., Aristoy, M.C. and Toldra, F. (2015). Sources of variability in the analysis of meat nutrient coenzyme Q10 for food composition databases. *Food Control*, 48: 151–154.

Roffe, L., Schmidt, K. and Ernst, E. (2004). Efficacy of coenzyme Q10 for improved tolerability of cancer treatments: A systematic review. *J. Clin. Oncol.*, 22(21): 4418–4424.

Rosenfeldt, F.L., Haas, S.J., Krum, H., Hadj, A., Ng, K., Leong, J.Y. and Watts, G.F. (2007). Coenzyme Q10 in the treatment of hypertension: A meta-analysis of the clinical trials. *J. Hum. Hypertens.*, 21(4): 297–306.

Rotig, A., Mollet, J., Rio, M. and Munnich, A. (2007). Infantile and pediatric quinone deficiency diseases. *Mitochondrion*, 7: S112–S121.

Rozen, T.D., Oshinsky, M.L., Gebeline, C.A., Bradley, K.C., Young, W.B., Shechter, A.L. and Silberstein, S.D. (2002). Open label trial of coenzyme Q10 as a migraine preventive. *Cephalalgia*, 22(2): 137–141.

Sacconi, S., Trevisson, E., Salviati, L., Ayme, S., Rigal, O., Redondo, A.G. and Desnuelle, C. (2010). Coenzyme Q10 is frequently reduced in muscle of patients with mitochondrial myopathy. *NMD*, 20(1): 44–48.

Safarinejad, M.R. (2012). The effect of coenzyme Q10 supplementation on partner pregnancy rate in infertile men with idiopathic oligo astheno teratozoospermia: An open label prospective study. *Int. Urol. Nephrol.*, 44(3): 689–700.

Schmelzer, C., Lindner, I., Vock, C., Fujii, K. and Doring, F. (2007). Functional connections and pathways of coenzyme Q10 inducible genes: An *in silico* study. *IUBMB. Life*, 59: 628–633. 10.1080/15216540701545991.

Schmelzer, C., Lindner I., Rimbach, G., Niklowitz, P., Menke, T. and Doring, F. (2008). Functions of coenzyme Q10 in inflammation and gene expression. *Biofactors*, 32(1–4): 179–183.

Seo, M.J. and Kim, S.O. (2010). Effect of limited oxygen supply on coenzyme Q10 production and its relation to limited electron transfer and oxidative stress in *Rhizobium radiobacter* T6102. *J. Microbiol. Biotechnol.*, 20(2): 346–349.

Shukla, S. and Dubey, K.K. (2018). CoQ10 a super-vitamin: Review on application and biosynthesis. *Biotech.*, 8(5): 1–11.

Shults, C.W., Haas, R.H., Passov, D. and Beal, M.F. (1997). Coenzyme Q10 levels correlate with the activities of complexes I and II/III in mitochondria from parkinsonian and non-parkinsonian subjects. *Ann. Neurol.*, 42(2): 261–264.

Siemieniuk, E. and Skrzydlewska, E. (2005). Coenzyme Q10: Its biosynthesis and biological significance in animal organisms and in humans. *Postepy. Hig. Med. Dosw.*, 59: 150–159.

Sohmiya, M., Tanaka, M., Tak, N.W., Yanagisawa, M., Tanino, Y., Suzuki, Y. and Yamamoto, Y. (2004). Redox status of plasma coenzyme Q10 indicates elevated systemic oxidative stress in Parkinson's disease. *J. Neurol. Sci.*, 223(2): 161–166.

Spinazzi, M., Radaelli, E., Horre, K., Arranz, A.M., Gounko, N.V., Agostinis, P. and De, S.B. (2019). PARL deficiency in mouse causes Complex III defects, coenzyme Q depletion, and Leigh-like syndrome. *PNAS*, 116(1): 277–286.

Szkopinska, A. (2000). Ubiquinone: Biosynthesis of quinone ring and its isoprenoid side chain: Intracellular localization. *Acta. Biochim. Pol.*, 47(2).

Takahashi, S., Ogiyama, Y., Kusano, H., Shimada, H., Kawamukai, M. and Kadowaki, K.I. (2006). Metabolic engineering of coenzyme Q by modification of isoprenoid side chain in plant. *FEBS Letters*, 580(3): 955–959.

Takahashi, K., Ohsawa, I., Shirasawa, T. and Takahashi, M. (2016). Early-onset motor impairment and increased accumulation of phosphorylated α-synuclein in the motor cortex of normal aging mice are ameliorated by coenzyme Q10. *Exp. Gerontol.*, 81: 65–75.

Taylor, M.A. and Fraser, P.D. (2011). Solanesol: Added value from solanaceous waste. *Phytochemistry*, 72(11-12): 1323–1327.

Terao, K., Nakata, D., Fukumi, H., Schmid, G., Arima, H., Hirayama, F. and Uekama, K. (2006). Enhancement of oral bioavailability of coenzyme Q10 by complexation with γ-cyclodextrin in healthy adults. *Nutr. Res.*, 26(10): 503–508.

Testa, C.A., Lherbet, C., Pojer, F., Noel, J.P. and Poulter, C.D. (2006). Cloning and expression of IspDF from *Mesorhizobium loti*: Characterization of a bifunctional protein that catalyzes non-consecutive steps in the methylerythritol phosphate pathway. *Biochim. Biophys. Acta. Proteins. Proteom.*, 1764(1): 85–96.

Thomas, S.R., Witting, P.K. and Stocker, R. (1999). A role for reduced coenzyme Q in atherosclerosis. *Biofactors*, 9(2–4): 207–224.

Tian, Y., Yue, T., Yuan, Y., Soma, P.K., Williams, P.D., Machado, P.A., Fu, H., Kratochvil, R.J., Wei, C. and Lo, Y.M. (2010). Tobacco biomass hydrolysate enhances coenzyme Q10 production using photosynthetic *Rhodospirillum rubrum*. *Bioresour. Technol.*, 101(20): 7877–7881. https://doi.org/10.1016/j.biortech.2010.05.020.

Tiano, L. and Busciglio, J. (2011). Mitochondrial dysfunction and Down's syndrome: Is there a role for coenzyme Q10? *Biofactors*, 37(5): 386–392.

Tiano, L., Carnevali, P., Padella, L., Santoro, L., Principi, F., Bruge, F. and Littarru, G.P. (2011). Effect of coenzyme Q10 in mitigating oxidative DNA damage in Down syndrome patients, a double blind randomized controlled trial. *Neurobiol. Aging*, 32(11): 2103–2105.

Tokdar, P., Wani, A., Kumar, P., Ranadive, P. and George, S. (2013). Process and strain development for reduction of broth viscosity with improved yield in coenzyme Q10 fermentation by *Agrobacterium tumefaciens* ATCC 4452. *Ferment. Technol.*, 2(1). https://doi.org/10.4172/2167-7972.1000110.

Tokdar, P., Sanakal, A., Ranadive, P., Khora, S.S., George, S. and Deshmukh, S.K. (2014a). Molecular, physiological, and phenotypic characterization of *Paracoccus denitrificans* ATCC 19367 mutant strain P-87 having improved coenzyme Q_{10} production. *Indian Journal of Microbiology*, DOI: 10.1007/s12088-014-0506-4.

Tokdar, P., Vanka, R., Ranadive, P., George, S., Khora, S.S. and Deshmukh, S.K. (2014b). Protoplast fusion technology for improved production of coenzyme Q10 using *Paracoccus denitrificans* ATCC 19367 mutant strains. *J. Biochem. Technol.*, 5: 685–692.

Tokdar, P., Ranadive, P., Deshmukh, S.K. and Khora, S.S. (2017). Optimization of fermentation process conditions for the production of CoQ_{10} using *Paracoccus denitrificans* ATCC 19367 fusant strain PF-P1. *International Journal of Engineering Research and Technology (IJERT)*, 6(7): 135–143.

Tomasetti, M., Alleva, R., Solenghi, M.D. and Littarru, G.P. (1999). Distribution of antioxidants among blood components and lipoproteins: Significance of lipids/CoQ10 ratio as a possible marker of increased risk for atherosclerosis. *Biofactors*, 9(2–4): 231–240.

Tomono, Y., Hasegawa, J., Seki, T., Motegi, K. and Morishita, N. (1986). Pharmacokinetic study of deuterium-labelled coenzyme Q10 in man. *Int. J. Clin. Pharmacol. Ther. Toxicol.*, 24(10): 536–541.

Tran, U.C. and Clarke, C.F. (2007). Endogenous synthesis of coenzyme Q in eukaryotes. *Mitochondrion*, 7(Suppl): S62–S71. https://doi.org/10.1016/j.mito.2007.03.007.

Turunen, M., Olsson, J. and Dallner, G. (2004). Metabolism and function of coenzyme Q10. *Biochem. Biophys. Acta. Biomembr.*, 1660(1-2): 171–199.

Vaghari, H., Vaghari, R., Jafarizadeh, M.H. and Berenjian, A. (2016). Coenzyme Q10 and its effective sources. *Am. J. Biochem. Biotechnol.*, 12(4): 214–219. https://doi.org/10.3844/ajbbsp.2016.214.219.

Valko, M., Izakovic, M., Mazur, M., Rhodes, C.J. and Telser, J. (2004). Role of oxygen radicals in DNA damage and cancer incidence. *Mol. Cell. Biochem.*, 266(1): 37–56.

Wang, Y. and Hekimi, S. (2016). Understanding ubiquinone. *Trends Cell. Biol.*, 26(5): 367–378.

Watts, G.F., Playford, D.A., Croft, K.D., Ward, N.C., Mori, T.A. and Burke, V. (2002). Coenzyme Q10 improves endothelial dysfunction of the brachial artery in Type II diabetes mellitus. *Diabetologia*, 45(3): 420–426. https://doi.org/10.1007/s00125-001-0760-y.

Wei, W., Liu, Q., Tan, Y., Liu, L., Li, X. and Cai, L. (2009). Oxidative stress, diabetes, and diabetic complications. *Hemoglobin*, 33(5): 370–377.

Wright, C., Milne, S. and Leeson H. (2014). Sperm DNA damage caused by oxidative stress: Modifiable clinical lifestyle and nutritional factors in male infertility. *Reprod. Biomed. Online*, 28: 684–703. https://doi.org/10.1016/j.rbmo.2014.02.004.

Yamada, Y., Ohishi, T. and Kondo, K. (1983). The coenzyme Q system in strains of some yeasts and yeast-like fungi. *J. Gen. Appl. Microbiol.*, 29(1): 51–57. https://doi.org/10.2323/jgam.29.51.

Yen, H.W. and Shih, T.Y. (2009). Coenzyme Q10 production by *Rhodobacter sphaeroides* in stirred tank and in airlift bioreactor. *Bioproc. Biosyst. Eng.*, 32: 711–716.

Yen, H.W., Feng, C.Y. and Kang J.L. (2010). Cultivation of *Rhodobacter sphaeroides* in the stirred bioreactor with different feeding strategies for CoQ10 production. *Appl. Biochem. Biotechnol.*, 160(5): 1441–1449.

Yoshida, H., Kotani, Y., Ochiai, K. and Araki, K. (1998). Production of ubiquinone-10 using bacteria. *Journal of General and Applied Microbiology*, 44(1): 19–26.

Young, J.M., Florkowski, C.M., Molyneux, S.L., McEwan, R.G., Frampton, C.M., Nicholls, M.G. and George, P.M. (2012). A randomized, double-blind, placebo-controlled crossover study of coenzyme Q10 therapy in hypertensive patients with the metabolic syndrome. *Am. J. Hypertens.*, 25(2): 261–270.

Yuan, Y., Tian, Y. and Yue, T. (2012). Improvement of coenzyme Q10 Production: Mutagenesis induced by high hydrostatic pressure treatment and optimization of fermentation conditions. *Journal of Biomedicine and Biotechnology*, 607329. 10.1155/2012/607329.

Yuting, T., Tianli, Y., Jinjin, P., Yahong, Y., Juhai, L. and Martin, L.Y. (2010). Effects of cell lysis treatments on the yield of coenzyme Q10 following *Agrobacterium tumefaciens* fermentation. *Food. Sci. Technol. Int.*, 16: 195–203.

Zahiri, H.S., Yoon, S.H., Keasling, J.D., Lee, S.H., Kim, S.W., Yoon, S.C. and Shin, Y.C. (2006). *Metab. Eng.*, 8: 406–416.

Zaki, M.E., El-Bassyouni, H.T., Tosson, A., Youness, E. and Hussein, J. (2017). Coenzyme Q10 and pro-inflammatory markers in children with Down syndrome: clinical and biochemical aspects. *J. Pediatr.*, 93: 100–104.

Zhang, D., Shrestha, B., Niu, W., Tian, P. and Tan, T. (2007). Phenotypes and fed-batch fermentation of ubiquinone-overproducing fission yeast using ppt1 gene. *J. Biotechnol.*, 128(1): 120–131. https://doi.org/10.1016/j.jbiotec.2006.09.012.

Zhao, Y. and Lieberman, H.B. (1995). *Schizosaccharomyces pombe*: A Model for molecular studies of eukaryotic genes. *DNA. Cell. Biol.*, 14(5): 359–371. https://doi.org/10.1089/dna.1995.14.359.

Zhong, W., Wang, W., Kong, Z., Wu, B., Zhong, L., Li, X. and Zhang, F. (2011). Coenzyme Q10 production directly from precursors by free and gel-entrapped *Sphingomonas* sp. ZUTE03 in a water-organic solvent, two-phase conversion system. *Appl. Microbiol. Biotechnol.*, 89(2): 293–302.

10

Fungal Probiotics and Prebiotics

Kandikere R. Sridhar[1,*] and *Shivannegowda Mahadevakumar*[2]

1. Introduction

Microbial communities have an intimate association, and interaction with different niches of the human body, including integumentary, digestive, upper respiratory, and urinogenital systems (Ward et al., 2017). Normal microbiotas carry out metabolic functions and prevent the entry or colonization of exogenous microbes. The gastrointestinal tract (GIT) is one of the active immune systems which depends on the normal endogenous microbiota to prevent colonization of pathogens and facilitate the development of mucosal systemic immunity. Disturbance of normal microbiota of GIT leads to dysbiosis which requires biotherapy for restoration of the normal intestinal ecosystem. The concept of probiotics had been proposed by the Russian immunologist Metchnikoff (1907), who suggested that lactobacilli, present in yogurt, show a positive influence on the favorable microbiota of GIT, and improve human health. Probiotics represent selective viable microbes (e.g., bifidobacteria, lactobacilli, and yeasts) which assist in the treatment of pathological conditions (Havenaar et al., 1992). Nowadays, many functional foods are known to possess probiotic microbiota in view of their positive role in human nutrition and health, including beverages and dairy products.

Prebiotics are non-viable, non-digestible, and non-toxic components, usually oligosaccharides derived from polysaccharides obtained from plants, animals, and microbes, which benefit the host by stimulating proliferation, as well as function of probiotics in GIT without harming the equilibrium of normal microbiota (Gibson and Roberfroid, 1995; Gibson et al., 2004). Unlike probiotics, the term prebiotic had been coined about two decades ago based on the significance of inulin and oligosaccharides (Aida et al., 2009). Although many prebiotics are derived from

[1] Department of Biosciences, Mangalore University, Mangalore, 574199, Karnataka, India.
[2] Department of Studies in Botany, University of Mysore, Mysore, 570006, Karnataka, India.
* Corresponding author: kandikere@gmail.com

plants or animals, interest has been drawn towards fungal prebiotics as an alternative source. Fungal prebiotics are used in food and pharmaceutical industries due to their immunomodulatory and antitumor effects and induction of lymphocyte proliferation and antibody production (Boa et al., 2001; Wasser, 2002; Tao et al., 2006). Prebiotics are also used solely or with probiotics in many functional foods for improvement of human nutritional health (e.g., bakery/meat products and livestock feeds).

Several terms have been proposed to understand the significance of non-pathogenic beneficial microbes in the intestine, like the Greek word 'probiotics' (= for life) followed by other related terms, such as 'prebiotics', 'synbiotics', 'parabiotics', and 'postbiotics' (Box 10.1). The concept of probiotics extended to microbiota having a positive effect on gut immunity to stimulate health is also designated as 'immunobiotics' (Clancy, 2003). Prebiotics need to fulfill several criteria: low effective dose, stability in upper GIT, functionality in colon, stimulation of probiotics, regulation of gut microbes, and stability during food processing (Aida et al., 2009; Roberfroid, 2008; Wang, 2009). The term synbiotics (= acting together) is referred to a product which consists of a mixture of probiotics and prebiotics (Markowiak and Slizewska, 2017). A simple example of a synbiotic is a combination of probiotics, like *Bifidobacterium* or *Lactobacillus* with prebiotic fructo-oligosaccharides (FOS).

Box 10.1. Terminology.

Probiotics: Probiotics are precise viable, non-pathogenic stable microorganisms proliferate in the gastrointestinal tract (GIT) and have positive effects like immunostimulation, antiviral activity, reduction of cancer-inducing enzymes and normal bowel/intestinal functionality of the host.

Prebiotics: Prebiotics are non-digestible carbohydrate components (oligosaccharides) capable to selectively stimulate growth/functions of probiotic microorganisms in the GIT, those not digested/absorbed in the upper intestinal tract, resistant to the acids/bile/enzymes in the stomach and fermentable by the normal microbiota of the host's intestine.

Synbiotics: Synbiotics are desired dietary supplements composed of a perfect combination of probiotics (living microorganism/s) and prebiotics (an ingredient of oligosaccharide/s) function synergistically in the GIT. The chosen prebiotic component supports the proliferation of probiotics that in turn enhances the target oriented benefits to the host.

Parabiotics: Parabiotics (or paraprobiotics or ghost probiotics) are the non-viable prebiotic microbe/s which have the ability to regulate innate immunity, decrease inflammation possess antioxidant potential and inhibit the growth of pathogens in the GIT of the host.

Postbiotics: Postbiotics are a mixture of metabolic products (and components of cell wall) produced by probiotic organism/s like amino acids, biosurfactants, enzymes, fatty acids, organic acids, peptides, proteins and vitamins in the GIT as supplements to support the host's health.

Nutritional and health-enhancing effects of fungi have been known before the discovery of the precise role of fungal prebiotics (e.g., mushrooms and ethnic foods) (Hamajima et al., 2016; Swangwan et al., 2018; Kurahashi, 2021). The consumption of edible mushrooms leads to prebiotic functions of their polysaccharides/oligosaccharides in the intestines; hence, they have been generally regarded safe (GRAS) as a potential functional food. Although previous studies have shown the importance of bacterial probiotics in human health, the information on the significance of fungi as probiotics, prebiotics, and synbiotics are meagre (Pandey et al., 2015). This review presents a compilation of available information on the subject, highlighting the significance and the likelihood of prospects of fungal probiotics, prebiotics and synbiotics on human nutrition and health.

2. Probiotics

Probiotics are viable microorganisms which serve as adjuncts on digestion and are used to treat pathological conditions (Havenaar et al., 1992). The Russian immunologist Metchnikoff (1907) has shown that lactobacilli present in yogurt have a positive influence on normal microbiota of GIT leading to improvement of human health. Subsequently, the concept of probiotics has been expanded to those microorganisms which have a positive impact on the gut's immune system with a health-enhancing effect. These were also designated as 'immunobiotics' by Clancy (2003).

Most studies have focused on the colonization of GIT by the bacterial community rather than another group of microbes like fungi (Ward et al., 2017). About 0.1% of microbiota in adult GIT consists of fungi represented by about 60 unique species (Hoffmann et al., 2013; Rajilić-Stojnović and De Vos, 2014). Several yeast populations in GIT have possess health-enhancing effects. The yeast *Saccharomyces boulardii* is an extensively studied fungus with proven probiotic qualities which may be beneficial in the treatment of GIT diseases (Berg, 1996). It has been originally isolated in 1923 from the edible fruits of litchi (*Litchi chinensis* Sonn.) (*Sapindaceae*) and mangosteen (*Garcinia mangostana* L.) (*Clusiaceae*). Chewing the outer coat of these fruits has shown to control cholera amongst the native people of Southeast Asia. *S. boulardii* is considered a potential probiotic by effective colonization of GIT, stimulation of bacterial growth, and maintenance of gut health (supplement digestion and stimulate immunity) (Kumar et al., 2017; Yadav et al., 2017).

Although *S. boulardii* colonizes < 1% of normal microbiota of GIT, it has several new properties which differ from endogenous bacteria: the presence of chitin, mannose (phosphopeptidomannan and phospolipomannan) and glucan (instead of peptidoglycan, lipopolysaccharide and lipoteichoic acid in bacteria), optimum pH = 4.5–6.5 (optimal pH in bacteria, 6.5–7.5), resistance to antibiotics, and no genetic transmittance (e.g., genes resistant to antibiotics will be transferred to bacteria) (Czerucka et al., 2007). Sixteen species of yeasts belonging to eight genera have been reported to possess a probiotic potential (*Candida*, *Cryptococcus*, *Debaryomyces*, *Issatchenkia*, *Kluyveromyces*, *Meyerozyma*, *Sacchaomyces*, and *Torulaspora*) (Banik et al., 2019). The authors of this study categorized probiotic yeasts into five groups depending on their nutritional and health-promoting effects. They have

Table 10.1. An overview of clinical trials on the therapeutic effects of probiotic yeast *Saccharomyces boulardii.*

Clinical Syndromes/Diseases	References
Acquired immunodeficiency syndrome (AIDS)	Saint-Marc et al., 1991
Acute paediatric diarrhea	Kurugol and Koturoglu, 2005; Billoo et al., 2006; Villarruel et al., 2007
Antibiotic-associated diarrhea (AAD)	Adam, 1976; Surawicz et al., 1989; McFarland et al., 1994, 1995; D'Souza et al., 2002; Kotowska et al., 2005
Travellers' diarrhea	Sanders and Tribble, 2001
Tube-feeding-associated diarrhea	Bleichner et al., 1997
Inflammatory bowel diseases	Guslandi et al., 2003
Irritable bowel syndrome	Maupas et al., 1983

particularly emphasized the significance of yeasts in the treatment of gastroenteritis, giardiasis, and candidiasis, as well as their anti-proliferative, anti-inflammatory, and anti-diarrhea effects. Previous trials have also demonstrated the efficacy of a probiotic yeast *S. boulardii* in the treatment of different diseases (Table 10.1).

3. Prebiotics

Prebiotics are oligosaccharides (chain of 3–10 sugar molecules) derived from polysaccharides consisting of monosaccharide units with specific linkages possessing a diverse degree of polymerization, which serve as food ingredients without increasing the caloric value. Such prebiotics are produced by hydrolysis of complex polysaccharides into oligosaccharides. Natural sources of prebiotics include vegetables, fruits, milk, honey, seeds, and sugarcane (Mussatto and Mahcilha, 2007). Although many prebiotics have been derived from plants and animals, research has been directed towards fungal prebiotics as an alternative source. The significance of fungi in human nutrition and health has been known before identifying the role of fungal prebiotics (e.g., mushrooms and ethnic foods) (Hamajima et al., 2016; Swangwan et al., 2018; Kurahashi, 2021). Fungal prebiotics are valued in food and pharmaceutical industries due to their biologically active features (immunomodulation, lymphocyte proliferation, antibody production, and antitumor effects) (Boa et al., 2001; Wasser, 2002; Tao et al., 2006). As a source of prebiotics, fungi can be cultivated or used in fermentation to produce large amounts of prebiotics with uniform quality.

The criteria necessary for designation of a prebiotic include: (1) efficacy at a low dose; (2) tolerance in upper gut conditions; (3) persistence in the colon; (4) selective stimulation of prebiotics; (5) regulation of gut microflora; (6) stability during food procession (Gibson et al., 2004; Aida et al., 2009; Roberfroid, 2008; Wang, 2009) (Fig. 10.1). The mechanism of function of fungal prebiotics in GIT mainly include: (1) increased production of short chain fatty acids; (2) stimulation of growth of probiotics; (3) strengthening intestinal immunity of the host (e.g., cytokine modulation and production of immunoglobulin A) (Nagpal et al., 2012). These

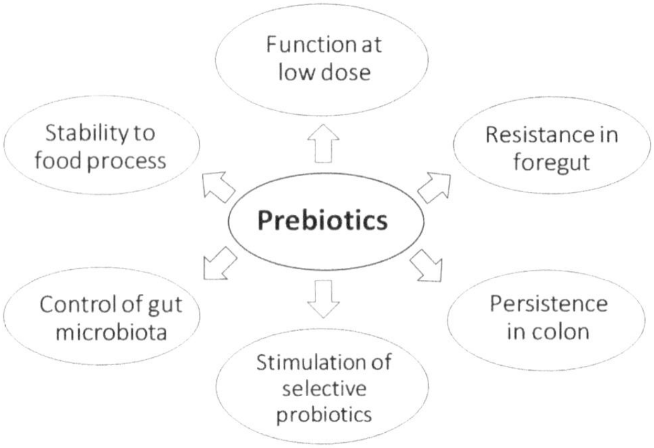

Fig. 10.1. The quality criteria of prebiotics.

prebiotics may be incorporated into many food products (e.g., beverages, dairy and bakery products, meat, and livestock feeds) (Gibson and Roberfroid, 1995). Prebiotics also possess several therapeutic effects (hypolipidemic, anticarcinogenic, and antimicrobial) and may be used in the nutrition of infants deprived of breast milk. Assessment of the nutritional and health benefits of prebiotics needs to be evaluated using the prebiotic score. This quantitative approach is also referred to as prebiotic index (PI). This index is used to compare the prebiotic efficiency of dietary prebiotics (Palframan et al., 2003). It considers the quantities of bifidobacteria (Bi), bacteroids (Ba), lactobacilli (La), and clostridia (Cl):

$$PI = (Bi/Total) - (Ba/Total) + (La/Total) - (Cl/Total)$$

(Bi: number of bifidobacteria at sampling time/their total numbers at inoculation time; Ba: number of bacteroides at sample time/their total numbers at inoculation time; La: number of lactobacilli at sampling time/their total numbers at inoculation time; Cl: number of clostridia at sample time/their total numbers at inoculation time).

3.1 Fungal Prebiotics

Recently, the demand for prebiotics or non-digestible oligosaccharides has increased to US $27 billion to nutrition and health awareness (Belorkar, 2020). Compared to plants and animal sources, fungal prebiotics are lucrative due to easy handling and bulk production. Mushrooms have served the humanity as dietary and pharmaceutical sources for 2,000 years and received high attention (Aida et al., 2009). The cultivated, and wild mushrooms possess a variety of health-enhancing bioactive constituents, including those having the potential to combat lifestyle diseases. The consumption of wild or cultivated edible mushrooms is beneficial, as their polysaccharides and their products (oligosaccharides) serve as prebiotics in the intestine. The well-known mushroom prebiotics currently include α-glucans, β-glucans, β-d-glucan, chitin, galactans, hemicellulose, mannans, and xylans. Interestingly, the non-edible mushrooms (e.g., *Coriolus versicolor*) are also regarded as potential sources of prebiotics used in the functional foods or medicine (Cui and Chisti, 2003; Matijašević et al., 2016; Sknepnek et al., 2021).

The mushroom polysaccharides or oligosaccharides which serve as prebiotics are not hydrolyzed by enzymes of GIT and possess a selective stimuatory effect to improve the gut health (Yin et al., 2020). Yang et al. (2020) have described the prebiotic potential of polysaccharides of the polypore mushroom, *Ganoderma lucidum*. Nowak et al. (2018) have demonstrated the migration of mushroom polysaccharides to the colon and stimulation of proliferation of the beneficial bacteria. Jayachandran et al. (2017) reviewed GIT health significance of selected mushrooms as prebiotics (*Coriolus*, *Ganoderma*, *Grifola*, and *Inonotus*). The supplementation of the edible mushroom *Agaricus bisporus* with probiotic bacteria, has partially or fully treated dyslipidemia in a rat model (Asad et al., 2020). The growth of this mushroom positively influenced the host, as well as the endogenous bacterial metabolism in a mice model (Tian et al., 2018). Swangwan et al. (2018) recorded growth stimulation of probiotics by the aqueous-ethanol extracts of edible mushrooms (*Lactobacillus acidophilus* and *L. plantarum*). These mushroom extracts showed growth stimulation of prebiotic bacteria which resulted in a higher inhibition of *Salmonella paratyphi*. An extract of *Auricularia auricula-judae* has stimulated *L. acidophilus*, which inhibited the pathogen *Bacillus cereus*. Thus, edible mushrooms also stimulate probiotic bacteria in GIT to prevent the growth of pathogenic bacteria.

In addition to mushrooms, several filamentous fungi serve as potential sources of prebiotics (e.g., *Aspergillus*, *Aureobasidium*, *Paecilomyces*, *Penicillium*, and *Saccharomyces*) (Belorkar, 2020). The prebiotic potential of exopolysaccharides of three entomopathogenic fungi proved its efficiency in GIT and usefulness in food industry (Prathumpaí et al., 2012).

3.2 Fungal Oligosaccharides

Roberfroid (2008) listed the oligosaccharide prebiotics fully or partially fulfilling the prebiotic criteria. The most frequently used prebiotics in food industries are inulin (or oligofructose or oligofructan) and trans-galacto-oligosaccharides (TOS). The commercially used prebiotics also include fructo-oligosaccharides (FOS), lactulose, mannan-oligosaccharides (MOS) and xylo-oligosaccharides (XOS) (Rawat et al., 2017; Palai et al., 2020). The selected microfungi and macrofungi-producing prebiotics (fructo-oligosaccharide, β-glucan, and β-d-glucan) are listed in Table 10.2.

Mushroom-derived β-glucans possess several health-enhancing effects (activation of leukocytes and immune cells as well as hypocholesterolaemic effect) (Khan et al., 2018). Mushroom derived D-glucans as dietary fiber provide new opportunities to develop novel prebiotics with nutritional and health benefits (Ruthes et al., 2021). Amer et al. (2021) have proposed an extraction method to acquire β-glucan in high amounts from *Saccharomyces cerevisiae* using a strong base and a weak acid extraction method. The isolated β-glucan has antitoxic, antimicrobial, anticancer, and immunomodulatory properties. *Aspergillus quadrilineatus* endo-β-mannanase-derived mannan-oligosaccharides (MOS) possess prebiotic activity in GIT (Suryawanshi and Kango, 2021). The species of *Aspergillus* possess the capacity

Table 10.2. A list of micro- and macro-fungal prebiotics.

Fungus	Prebiotic	References
Microfungi		
Aspergillus japonicas	Fructo-oligosaccharide	Mussatto and Teixeira, 2010
Aspergillus oryzae	Fructo-oligosaccharide	Mable et al., 2007
Aspergillus phoenicus	Fructo-oligosaccharide	Van Balken et al., 1991
Aureobasidium pullulans	Fructo-oligosaccharide	Lateef et al., 2007; Yoshikawa et al., 2007
Schwanniomyces occidentalis	Fructo-oligosaccharide	Álvaro-Benito et al., 2007
Macrofungi		
Agaricus bisporus	β-glucan	Navegantes et al., 2013
Agaricus blazei	β-glucan	Ishii et al., 2011
Coriolous (= Trametes) versicolor	β-glucan	Oba et al., 2007
Entoloma lividoalbum	β-d-glucan	Maity et al., 2014
Flammulina velutipes	β-d-glucan	Yang et al., 2012
Ganoderma lucidum	β-d-glucan	Wang et al., 2014
Gastrodia elata	β-d-glucan	Ming et al., 2012
Geastrum saccatum	β-glucan	Dore et al., 2007
Hypsizigus marmoreus	β-glucan	Zhang et al., 2012
Lentinus squarrosulus	β-glucan	Ahmad et al., 2014
Phellinus ribis	β-glucan	Liu et al., 2014
Pleurotus ostreatus	β-glucan	Kanagasabapathy et al., 2014
Pleurotus tuber-regium	β-glucan	Huang et al., 2012
Pleurotus sajor-caju	β-glucan	Kanagasabapathy et al., 2014
Poria cocos	β-glucan	Du and Xu, 2014
Russula albonigra	β-glucan	Nandi et al., 2014
Sparassis crispa	β-glucan	Sun et al., 2013
Tricholoma matsutake	β-glucan	Lu et al., 2008

to produce large amounts of stable MOS (Jana et al., 2021). Hong et al. (2019) studied the impact of phthalyl pullulan nanoparticle treated *Lactobacillus plantarum* against pathogenic *Escherichia coli*, as well as *Listeria monocytogenes*, and found a higher inhibitory activity than untreated or pullulan-treated *L. plantarum*.

3.3 Fungal Fructans

The results of *in vitro* and *in vivo* experiments have shown that fungal inulin-type fructans and transgalacto-oligosaccharides (TOS) possess the following properties: (1) resistance to gastric acid, mammalian enzyme hydrolysis and gastrointestinal absorption; (2) fermentation only by intestinal probiotics; (3) stimulation of growth and activity of selected probiotics (Roberfroid, 2008). Therefore, the inulin-type fructans are considered model prebiotics.

Aspergillus japonicas, *A. niger*, *A. oryzae*, *A. tamarii*, and *Aureobasidium pullulans* can generate large-scale fructo-oligosaccharides (FOS) by fructosyl transferase (FTase) (Hidaka et al., 1988; Kurakake et al., 1996; Yun, 1996; Madlova et al., 1999; Chien et al., 2001; Sangeetha et al., 2004a,b; Zambelli et al., 2014; Choukade and Kango, 2019). The microbiological and biotechnological techniques of enzyme production necessary for generation of FOS have been described by Mairorano et al. (2008). The β-fructosidase of *A. niger* can produce a prebiotic FOS by enzymatic transfructosylation (Roberfroid, 2008). The production of FOS using bioreactors has been optimized by Katapodis et al. (2004) with a temperature range from 280–300°C and pH range 4–6. Yun (1996) studied the immobilized cells of *A. pullulans* for FOS production in stirred tank bioreactors.

The FTase is one of the necessary enzymes for production of FOS, and fungi are the main source of FTase production. Its production is more feasible due to its thermostability, as well as pH tolerance (Yun, 1996). Belorkar (2020) has listed up to 12 fungi for production and purification of FTase, while Maiorano et al. (2008) have reported nine FTase-producing fungi. Most interestingly, *Aspergillus* spp. possess the highest capacity to produce fungal FTase which has shown a wide tolerance to temperature, and pH conditions. These are important criteria for its application in industries (Yun, 1996). In addition to the production of FTase, many fungi and yeasts produce the inulinolytic enzyme inulinase which is widly used in food, feed, and biofuel industries (Rawat et al., 2017).

3.4 Fungal Chitin

Chitin constitutes a structural polymer in filamentous fungi and yeasts. As an insoluble fiber, it serves as a prebiotic for microbiota in GIT and performs several functions (e.g., integrity of intestinal mucosa, immune protection, and energy metabolism). Similarly, mushroom-derived chitosans restore the beneficial microbiota in GIT (Neyrinck et al., 2009). The chitin-glucan complex has a positive impact on gut microbiota and reduces inflammation. This complex may be produced as a raw material for a broad range of applications using mycelial biomass of *A. niger* and subcritical water (without harsh alkali as an environmental friendly approach) (Ordoñez et al., 2013). Such complexes have served as prebiotics and increased the growth of *Bifidobacterium* in infant gut and a rat model (Alessandri et al., 2019). Previous study has shown the inhibition of growth of *Listeria monocytogenes* by three commercial chitosan-glucan complexes derived from fungi (Zimoch-Korzycka et al., 2016).

4. Synbiotics

Synbiotics are products which represent a combination of probiotics and prebiotics (= acting together) (Markowiak and Slizewska, 2017). One of the common examples of synbiotics is a blend probiotic *Bifidobacterium* or *Lactobacillus* with a prebiotic fructo-oligosaccharide (FOS). Human mycobiome represents a complex group of various fungi in different parts of the human body (Ward et al., 2017; Wu et al., 2021). Apart from GIT, fungi also colonize skin, breast, genitalia, and oral cavity. The oral cavity and GIT consist of a maximum number of fungal genera (Ward et

Table 10.3. An overview of clinical studies on the therapeutic effects of synbiotics.

Infections/Diseases	Synbiotics	References
Antibiotic-associated diarrhea (AAD)	*S. boulardii* + lactobacilli	D'Souza et al., 2002
Candidiasis	*S. boulardii* + lactobacilli, streptococci + bifidobacteria	Manzoni et al., 2015; Kumar and Singhi, 2016
Clostridium difficile infection	*S. boulardii* + *L. rhamnosus*	Beaugerie et al., 2003
Helicobacter pylori	*S. boulardii* + *Lactobacillus acidophilus* and *L. rhamnosus*	Palai et al., 2000
Travellers' diarrhea	*S. boulardii* + *L. acidophilus* and *Bifidobacterium bifidum*	McFarland, 2007
Acute gastroenteritis	*S. boulardii*, *Lactobacillus reuteri* and *L. rhamnosus*	McFarland, 2007

al., 2017). The mycobiome is influenced by probiotics, prebiotics, and antibiotics. Human bacteriobiome physically (for space) and metabolically (for nutrition) competes with mycobiome, which in turn shapes the microbiome of the human body.

Synbiotics provide more significant health benefits than either probiotics or prebiotics alone. There are several examples of a blend of single or multiple probiotics (either bacteria or fungi) with prebiotics to confer better health benefits. The administration of bacterial probiotics (e.g., lactobacilli, streptococci, and bifidobacteria) along with *S. boulardii* prevent candidiasis in GIT (Kumar and Singhi, 2016; Manzoni et al., 2015). Similarly, administration of *S. boulardii* along with a prebiotic (β-glucan) to human and/or animal subjects resulted in a reduction in obesity-related diseases (Nicolosi et al., 1999; Neyrinck et al, 2012; Everad et al., 2014; De Araújo et al., 2016). Furthermore, *S. boulardii* in association with probiotic lactobacilli (*L. acidophilus* and *L. rhamnosus*) has shown to completely exterminate *Helicobacter pylori* (Palai et al., 2020). Likewise, a combination of *S. boulardii* and *L. rhamnosus* may be used to prevent *Clostridium difficile* infections, which are responsible for AAD (Beaugerie et al., 2003). A combination of *S. boulardii* with lactobacilli (*L. reuteri* and *L. rhamnosus*) is a probiotic choice to treat acute gastroenteritis. The combination of *S. boulardii* with *L. acdophilus* and *Bifidobacterium bifidum* is efficient in preventing travellers' diarrhea (McFarland, 2007). Previous clinical trials have demonstrated the efficiency of synbiotics, consisting of *S. boulardii*, in the treatment of various diseases (Table 10.3).

5. The Role of Fungi in Human Health

The yeast *S. boulardii* is one of the most studied probiotics among fungi. It has the capacity to combat diseases like diarrhea, AAD (e.g., cycline and β-lactam) and bowel syndrome in children and adults (D'Souza et al., 2002; Czerucka et al., 2007; Zanello et al., 2009; Moré and Swdsinski, 2015). Previous studies have shown the importance of *S. boulardii* to fight against enteric intracellular pathogens in humans and animal models (e.g., *Clostridium difficile*, *Escherichia coli*, *Salmonella*, *Shigella*, and

Vibrio cholerae) (Czerucka and Rampal, 2002). The bacterial toxins, toxin A (*C. difficile*) and cholera toxin (*V. cholerae*), have shown to be neutralized by *S. boulardii* (Pothoulakiset al., 1993; Brandão et al., 1998). Furthermore, *S. boulardii* produces phosphatase, which detoxifies endotoxins of *E. coli* (Buts et al., 2006).

The host cell signaling is modified by *S. boulardii* which results in pro-inflammatory cytokine synthesis (Czerucka et al., 2000; Dahan et al., 2003). *S. boulardii* has the potential to increase the production of IgA in the small intestine (Buts et al., 1990), stimulate the production of enzymes by the brush-border membrane (Buts et al., 1986), and possesses an anti-inflammatory effect due to inhibition of pro-inflammatory cytokine production in colon cells (Powrie et al., 1993; Neurath et al., 1995; Lee et al., 2005). The suppression of growth of pathogenic *Candida albicans* which decreases colonization and inflammation by *S. boulardii* has been reported in mice models (Berg et al., 1993; Algin et al., 2005; Jawahara and Poulain, 2007). *S. boulardii* has shown a beneficial effect in the treatment of pediatric diarrhea and AAD (Szajewska and Mrukowicz, 2005; Kelesidis and Pothoulakis, 2012; Feizizadeh et al., 2014). Oral administration of *S. boulardii* in infants resulted in a decrease in fungal infection rate (Demirel et al., 2013). Kumar et al. (2013) have demonstrated a reduction in *C. albicans* in infants (colonization of GIT and candiduria) receiving broad-spectrum antibiotics by oral administration with a mixture of *S. thermophiles* and *S. boulardii*, each with bifidobacteria and lactobacilli. Up to 16 species of yeasts serve as probiotics in the human system (Banik et al., 2019). The beneficial yeasts have been categorized into five groups based on their probiotic potential. Many yeast species also possess anti-proliferative and anti-inflammatory effects and may be successfully used in the treatment of diarrhea, gastroenteritis, giardiasis, and candidiasis.

Similar to *S. boulardii*, other endogenous or food-borne fungal strains and varieties of *Saccharomyces* possess probiotic potential. Three species of yeasts (*Issatchenkia orientalis*, *Pichia kudriavzevii*, and *S. cerevisiae*) have probiotic properties, which may be subjects of future probiotic research (Lohith and Anu-Appaiah, 2014). *I. occidentalis* inhibits *C. albicans* and non-*albicans Candida* spp. (Kunyeit et al., 2019). Several probiotic yeasts are capable of combating diabetes, biofortification of folate, and absorption or destruction of mycotoxins (Banik et al., 2019).

Prebiotics are important for human health due to their anti-carcinogenic, anti-inflammatory, anti-neoplastic, and hypolipidemic properties (Mano et al., 2017). They also have clinical significance due to their enrichment of probiotic microbiota in GIT, as well as antibacterial, anti-osteoporotic, and antiviral properties (Kango and Jain, 2011). Prebiotics are well-known for improving immune response and colon integrity, enhancement of beneficial bacteria (bifidobacteria and lactobacilli), and decrease in the risk of diseases (Douglas and Sanders, 2008; Macfarlane et al., 2008). The increase in bifidobacteria and lactobacilli in GIT due to the impact of prebiotics will result in a battery of functions: production of B complex vitamins and pathogen growth inhibitors; reduction of ammonia and cholesterol in the blood; decreased constipation; check pediatric diarrhea treatment of irritable bowel syndrome and prophylaxis of *Salmonella* infections (Gibson et al., 1995; Manning and Gibson, 2004).

The treatment of bowel irritable syndromes, inflammatory bowel disorder, lactose intolerance, diabetes, cancer, enterocolitis, arterial hypertension, and cardiovascular diseases has been mainly connected to bacterial probiotics (Pandey et al., 2015; Palai et al., 2020). The use of probiotics decreases the rate of respiratory infections in infants. However, the effect of prebiotics on the upper respiratory tract has not been sufficiently investigated (Markowiak and Slizewska, 2017). Furthermore, chitin and β-glucan derived from fungi possess immunomodulatory effects to suppress airway inflammation (Catalli and Kulka, 2010). The effect of prebiotics on cardiovascular, urinary, and respiratory systems was mainly associated with bacterial probiotics (Pandey et al., 2015; Palai et al., 2020). Aida et al. (2009) reported the clinical significance of fungal probiotics and prebiotics in different health conditions. Particularly, *Helicobacter pylori* has been totally eradicated by a combination of *S. boulardii* with lactobacilli (*L. acidophilus* and *L. rhamnosus*) (Palai et al., 2020). Furthermore, different types of diarrheas (infants, travellers, and AAD) have been combated by a combination of probiotic yeast *S. boulardii* and bacteria (Szajewska and Mrukowicz, 2005; Szajewska et al., 2005; McFarland, 2007). The infection with *Clostridium difficile* which causes AAD may be prevented by *S. boulardii* and *L. rhamnosus* (Beaugerie et al., 2003). Similarly, *S. boulardii* has been shown to fight enteric pathogens in the cell and animal models (Czerucka and Rampal, 2002). Possessing a prebiotic β-glucan, *S. boulardii* may be used for treatment of obesity-related diseases (Neyrinck et al, 2012; Everad et al., 2014; De Araújo et al., 2016).

6. Functional Foods

The challenge of 21st century is to develop novel functional foods to cater to the needs of nutrition and health of human population (Miranda, 2020). A prebiotic component present in food products (e.g., beverages, confectioneries, fermented foods, milk, and sauces), may stimulate probiotics in GIT of healthy individuals. However, immune-compromised patients need probiotic microbes with prebiotics to restore the normal function of GIT.

Yeasts are very common in fermented foods, but their probiotic potential has not been sufficiently explored in this regard. Some of the yeast strains occurring in dairy products, apple cider, fermented coconut water and wine, have been reported to survive and attach to the epithelium of GIT (Kumura et al., 2004; Lohit and Anu-Appaiah, 2014; Gopal et al., 2021). The yeast β-glucan and glucan complexes of mushrooms possess growth-stimulating effects towards *Bifidobacterium bifidus* and *L. acidophilus* which leads to the formation of fermented clots in milk to develop functional foods (Nikitina et al., 2017). Fungal fructans serve as an important component in functional food due to their low calories and prebiotic potential (Choukade and Kango, 2020). The Amazake, a traditional Japanese beverage prepared from *Aspergillus oryzae*, has a prebiotic effect in humans (Kurahashi, 2021). A traditional Japanese cuisine, Koji, has a prebiotic potential due to the presence of glycosylceramide which supports the growth of *Blautia coccoides* in the intestine (Hamajima et al., 2016). The traditional fermented food of Goa (India), Bollo, possesses *Saccharomyces cerevisiae*, which serves as a probiotic-like *S. boulardii* (Pereira et al., 2021). Kombucha represents a cocktail of metabolites derived from a blend of bacteria (*Acetobacter intermedius*, *A. nitrogenifigens*, *Gluconacetobacter kombuchae*, and *G. xylinus*) and yeast (*Brettanomyces*, *Pichia*, *Schizosaccharomyces*, *Torulaspora*, and *Zygosaccharomyces*). It has been used over 2,000 years due to its health-enhancing effect (Kozyrovska et al., 2012).

Mushrooms are currently at the forefront of production of several functional foods possessing nutraceutical significance. A wide spectrum of properties of mushroom-derived β-glucans (*Lentinus edodes*) make them useful as functional foods (Sobierlski et al., 2012). The herbal tea from edible mushrooms has prebiotic effects in the modulation of mycobiome of human GIT which has been shown to have beneficial effects on human health (Vamanu et al., 2021). The non-alcoholic fermented beverage Kombucha, obtained from *Coriolus* (= *Trametes*) *versicolor* and *L. edodes* mushrooms, has shown an immunomodulatory effect due to its polysaccharide content (Sknepnek et al., 2021). This beverage prepared from hot water extract of *Ganoderma lucidum* has shown to confer several health benefits (Sknepnek et al., 2018).

7. Conclusion and Future Perspectives

The probiotics serve as alternatives to antibiotics and prevent the growth of drug-resistant pathogenic bacteria. The probiotic potential of bacteria has been extensively studied, compared only to a few fungi (mainly yeasts) which have been investigated in this regard. Despite the immense probiotic potential of *S. boulardii*, other *Saccharomyces* spp. and non-*Saccharomyces* spp., these properties have not been investigated so far. There is a perspective to develop a combination of appropriate fungal probiotics with prebiotics to combat diseases of skin, oral cavity, respiratory, and urinogenital system. There is also a perspective to use parabiotics (dead probiotics) of desired fungi or bacteria (or both) to improve human health. There are plenty of avenues to study fungal cell wall parabiotics in human and livestock health. Probiotics or parabiotics may be also used to transmit desired viral antigens (or epitopes) to the gut to increase immunity against pathogenic bacteria and viruses. With the development of new food products with preferred probiotics and prebiotics, the availability of these products in the market will become economical for consumers due to the abundance of inexpensive fungal mycelia (e.g., mushrooms and other filamentous fungi).

The nutritive value of existing fermented foods may be enhanced using probiotics and prebiotics, such as bakery and dairy products, confectioneries, sauces, and beverages. The fermented foods containing probiotics and prebiotics are easy to prescribe to infants and to immune-compromised patients for nutraceutical benefits. There is already a precedence of production of fermented indigenous foods (e.g., Bollo, Koji, Swiss cheese, and Tempeh) and beverages (e.g., Amazake and Kombucha) with probiotic fungi or a combination of probiotic fungi and bacteria. *Penicillium camemberti* and *Rhizopus oligosporus* serve as potential fungal candidates used in food fermentation. Fructans derived from fungi serve as low-calorie sweeteners in the production of desired foods. KetoZyme, a novel biopolymer extracted from cell walls of fungi (https://www.kitozyme.com/en/), is a chitin-glucan blend, which is regarded as safe for production of novel nutraceutical food formulations. The coating of edible fruits with desired probiotics and prebiotics is another option for gut inoculation. The association of probiotic fungi with livestock (e.g., ruminants and monogastrics), insects (e.g., termites, ants, and others), and wild animals is important for adopting different approaches to the future probiotic research.

Acknowledgements

The first author (KRS) is grateful to Mangalore University for academic support. The second author (SM) is grateful to the Council of Scientific and Industrial Research, New Delhi for the award of Research Associateship and support provided by the Department of Studies in Botany, University of Mysore. The authors are thankful to referees for chapter revision.

References

Adam, P. (1976). Essais cliniques contrôlés en double insu de l'ultra-levure lyophilisée (étude multicentrique par 25 médecins de 388 cas). *Méd. Chir. Dig.*, 5: 401–406.

Ahmad, R., Muniandy, S., Shukri, N.I.A., Alias, S.M.U., Hamid, A.A. et al. (2014). Antioxidant properties and glucan compositions of various crude extract from *Lentinus squarrosulus* mycelial culture. *Adv. Biosc. Biotechnol.*, 5: 805–814.

Aida, F.M.N.A., Shuhaimi, M., Yazid, M. and Maaruf, A.G. (2009). Mushroom as a potential source of prebiotics: A review. *Tr. Food Sci. Technol.*, 20: 567–575.

Alessandri, G., Milani, C., Duranti, S., Mancabelli, L., Ranjanoro, T. et al. (2019). Ability of bifidobacteria to metabolize chitin-glucan and its impact on the gut microbiota. *Sci. Rep.*, 9: 5755. doi: 10.1038/s41598-019-42257-z.

Algin, C., Sahin, A., Kiraz, N., Sahinturk, V. and Ihtiyar, E. (2005). Effectiveness of bombesin and *Saccharomyces boulardii* against the translocation of *Candida albicans* in the digestive tract in immunosuppressed rats. *Surg. Today*, 35: 869–873.

Álvaro-Benito, M., de Abreu, M., Fernández-Arrojo, L., Plou, F.J., Jiménez-Barbero, J. et al. (2007). Characterization of a β-fructosidase from *Schwanniomyces occidentalis* with transfructosylating activity yielding the prebiotic 6-kestose. *J. Biotechnol.*, 132: 75–81.

Amer, E.M., Saber, S.H., Markeb, A.A., Elkhawage, A.A., Makhemer, I.M.A. et al. (2021). Enhancement of β-glucan biological activity using a modified acid-base extraction method from *Saccharomyces cerevisiae*. *Molecules*, 26: 2113. doi: 10.3390/molecules26082113.

Asad, F., Anwar, H., Yassine, H.M., Ullah, M.I., Azizpul-Rahman et al. (2020). White button mushroom, *Agaricus bisporus* (Agaricomycetes), and a probiotics mixture supplementation correct dyslipidemia without influencing the colon microbiome profile in hypercholesterolemic rats. *Int. J. Med. Mush.*, 22: 235–244.

Banik, A., Halder, S.K., Ghosh, C. and Mondal, K.C. (2019). pp. 101–117. *In*: Yadav, A.N., Mishra, S.S., Singh, S. and Gupta, A. (Eds.). *Fungal Probiotics: Opportunity, challenge, and prospects*. Switzerland AG: Springer Nature.

Bao, X., Duan, J., Fang, X. and Fang, J. (2001). Chemical modification of the (1/3)-a-D-glucan from spore of *Ganoderma lucidum* and an investigation of their physicochemical properties and immunological activity. *Carbohydr. Res.*, 336: 127–140.

Beaugerie, L., Flahault, A., Barbut, F., Altan, P., Lalande, V. et al. (2003). Antibiotic-associated diarrhea and *Clostridium difficile* in the community. *Aliment. Pharmacol. Ther.*, 17: 905–912.

Belorkar, S.A. (2020). Fungal production of prebiotics. pp. 239–254. *In*: Hesham, A.E.-L., Upadhyay, R.S., Sharma, G.D., Manoharachary, C. and Gupta, V.K. (Eds.). *Fungal Biotechnology and Bioengineering, Fungal Biology*. Switzerland AG: Springer Nature.

Berg, R., Bernasconi, P., Fowler, D. and Gautreaux, M. (1993). Inhibition of *Candida albicans* translocation from the gastrointestinal tract of mice by oral administration of *Saccharomyces boulardii*. *J. Infect. Dis.*, 168: 1314–1318.

Berg, R.D. (1996). The indigenous gastrointestinal microflora. *Tr. Microbiol.*, 4: 430–435.

Billoo, A.G., Memon, M.A., Khaskheli, S.A., Murtaza, G., Iqbal, K. et al. (2006). Role of a probiotic (*Saccharomyces boulardii*) in management and prevention of diarrhea. *World J. Gastroenterol.*, 28: 4557–4560.

Bleichner, G., Blehaut, H. and Mentec, H. (1997). *Saccharomyces boulardii* prevents diarrhea in critically ill tube-fed patients. *Inten. Care Med.*, 23: 517–523.

Brandão, R.L., Castro, I.M., Bambirra, E.A., Amaral, S.C., Fietto, L.G. et al. (1998). Intracellular signal triggered by cholera toxin in *Saccharomyces boulardii* and *Saccharomyces cerevisiae*. *Appl. Environ. Microbiol.*, 64: 564–568.

Buts, J.P., Bernasconi, P., Van Craynest, M.P., Maldague, P. and De Meyer, R. (1986). Response of human and rat small intestinal mucosa to oral administration of *Saccharomyces boulardii*. *Pediatr. Res.*, 20: 192–196.

Buts, J.P., Bernasconi, P., Vaerman, J.P. and Díve, C. (1990). Stimulation of secretory IgA and secretory component of immunoglobulins in the small intestine of rats treated with *Saccharomyces boulardii*. *Dig. Dis. Sci.*, 35: 251–256.

Buts, J.P., Dekeyser, N., Stilmant, C., Delem, E., Smets, F. and Sokal, E. (2006). *Saccharomyces boulardii* produces in rat small intestine a novel protein phosphatase that inhibits *Escherichia coli* endotoxin by dephosphorylation. *Pediatric. Res.*, 60: 24–29.

Catalli, A. and Kulka, M. (2010). Chitin and β-glucan polysaccharides as immunomodulators of airway inflammation and atopic disease. *Rec. Pat. Endocr. Met. Immune Drug Disc.*, 4: 175–189.

Chien, C.S., Lee, W.C. and Lin, T.J. (2001). Immobilization of *Aspergillus japonicus* by entrapping cells in gluten for production of fructooligosaccharides. *Enz. Microb. Technol.*, 29: 252–257.

Choukade, R. and Kango, N. (2019). Characterization of a mycelial fructosyltransferase from *Aspergillus tamari* NKRC 1229 for efficient synthesis of fructooligosaccharides. *Food Chem.*, 286: 434–440.

Choukade, R. and Kango, N. (2020). Applications of fungal inulinases. *Encyclop. Mycol.*, doi: 10.1016/B978-0-12-819990-9.00016-0.

Clancy, R. (2003). Immunobiotics and the probiotic evolution. *FEMS Immunol. Med. Microbiol.*, 38: 9–12.

Cui, J. and Chisti, Y. (2003). Polysaccharopeptides of *Coriolus versicolor*: Physiological activity, uses, and production. *Biotechnol. Adv.*, 21: 109–122.

Czerucka, D., Dahan, S., Mograbi, B., Rossi, B. and Rampal, P. (2000). *Saccharomyces boulardii* preserves the barrier function and modulates the transduction pathway induced in enteropathogenic *Escherichia coli* infected T84 cells. *Inf. Immun.*, 68: 5998–6004.

Czerucka, D. and Rampal, P. (2002). Experimental effects of *Saccharomyces boulardii* on diarrheal pathogens. *Microbes Infect.*, 4: 733–739.

Czerucka, D., Piche, T. and Rampal, P. (2007). Yeast as probiotics: *Saccharomyces boulardii*. *Alim. Pharm. Therp.*, 26: 676–778.

D'Souza, A.L., Rajkumar, C., Cooke, J. and Bulpitt, C.J. (2002). Probiotics in prevention of antibiotic associated diarrhea: meta-analysis. *BMJ*, 324: 1–6.

Dahan, S., Dalmasso, G., Imbert, V., Peyron, J.-F., Rampal, P. and Czerucka, D. (2003). Saccharomyces *boulardii* interferes with enterohemorrhagic *Escherichia coli* induced signalling pathways in T84 cells. *Inf. Immun.*, 71: 766–773.

De Araújo, T.V., Andrade, E.F., Lobato, R.V., Orlando, D.R., Gomes, N.F. et al. (2016). Effects of beta-glucans ingestion (*Saccharomyces cerevisiae*) on metabolism of rats receiving high-fat diet. *J. Anim. Physiol. Anim. Nutr.*, 101: 349–358.

Demirel, G., Celik, I.H., Erdeve, O., Saygan, S., Dilmen, U. and Canpolat, F.E. (2013). Prophylactic *Saccharomyces boulardii* versus nystatin for the prevention of fungal colonization and invasive fungal infection in premature infants. *Eur. J. Pediatr.*, 72: 1321–1326.

Dore, C.M.P.G., Azevedo, T.C.G., Souza, M.C.R., Rego, L.A., Dantas, J.C.M. et al. (2007). Antiinflammatory, antioxidant and cytotoxic actions of β-glucan-rich extract from *Geastrumsaccatum* mushroom. *Int. Immunopharmacol.*, 7: 1160–1169.

Douglas, L.C. and Sanders, M.E. (2008). Probiotics and prebiotics in dietetics practice. *J. Am. Diet. Assoc.*, 108: 510–521.

Du, B. and Xu, B. (2014). Oxygen radical absorbance capacity (ORAC) and ferric reducing antioxidant power (FRAP) of β-glucans from different sources with various molecular weight. *Bioact. Carbohydr. Diet. Fib.*, 3: 11–16.

Everard, A., Matamoros, S., Geurts, L., Delzenne, N.M. and Cani, P.D. (2014). *Saccharomyces boulardii* administration changes gut microbiota and reduces hepatic steatosis, low-grade inflammation, and fat mass in obese and Type 2 diabetic db/db mice. *mBio.*, 5: 1011–1014.

Feizizadeh, S., Salehi-Abargouei, A. and Akbari, V. (2014). Efficacy and safety of *Saccharomyces boulardii* for acute diarrhea. *Pediatrics*, 134: 176–191.

Gibson, G.R. and Roberfroid, M.B. (1995). Dietary modulation of the colonic microbiota: Introducing the concept of prebiotics. *J. Nutr.*, 125: 1401–1412.

Gibson, G.R., Scott, K.P., Rastall, R.A., Tuohy, K.M., Hotchkiss, A. et al. (1995). Dietary prebiotics: Current status, and new definition. *Food Sci. Technol. Bull. Func. Foods*, 7: 1–19.

Gibson, G.R., Probert, H.M., Rastall, R.A. and Roberfroid, M.B. (2004). Dietary modulation of the human colonic microbiota: updating the concept of prebiotics. *Nutr. Res. Rev.*, 17: 259–275.

Gopal, M., Shil, S., Gupta, A., Hebbar, K.B. and Arivalagan, M. (2021). Metagenomic investigation uncovers presence of probiotic-type microbiome in kalparasa® (fresh unfermented coconut inflorescence sap). *Front. Microbiol.*, 12: 662783. doi: 10.3389/fmicb.2021.662783.

Guslandi, M., Giollo, P. and Testoni, P.A. (2003). A pilot trial of *Saccharomyces boulardii* in ulcerative colitis. *Eur. J. Gastroenterol. Hepatol.*, 15: 697–698.

Hamajima, H., Matsunaga, H., Fujikawa, A., Sato, T., Mitsutake, S. et al. (2016). Japanese traditional dietary fungus koji *Aspergillus oryzae* functions as a prebiotic for *Blautia coccoides* through glycosylceramide: Japanese dietary fungus koji is a new prebiotic. *SpringerPlus*, 5: 1321. doi: 10.1186/s40064-016-2950-6.

Havenaar, R. and Huis in't Veld, J.H.J. (1992). Probiotics: A general view. pp. 209–224. *In*: Wood, B. (Ed.). *The Lactic Acid Bacteria in Health and Disease*. London: Elsevier.

Hidaka, H., Hirayama, M. and Sumi, S.A. (1988). Fructooligosaccharide-producing enzyme from *Aspergillus niger* ATCC 20611. *Agric. Biol. Chem.*, 52: 1187–1988.

Hoffmann, C., Dollive, S., Grunberg, S., Chen, J., Li, H. et al. (2013). Archaea and fungi of the human gut microbiome: Correlations with diet and bacterial residents. *PLOS One*, 8: e66019. doi: 10.1371/journal.pone.0066019.

Hong, L., Kim, W.-S., Lee, S.-M., Kang, S.-K., Choi, Y.-J. and Cho, C.S. (2019). Pullulan nanoparticles as prebiotics enhance the antibacterial properties of *Lactobacillus plantarum* through the induction of mild stress in probiotics. *Front. Microbiol.*, 10: 142. doi: 10.3389/fmicb.2019.00142.

Huang, H.Y., Korivi, M., Chaing, Y.Y., Chien, T.Y. and Tsai, Y.C. (2012). *Pleurotus tuber-regium* polysaccharides attenuate hyperglycemia and oxidative stress in experimental diabetic rats. *Evid. Based Compl. Alt. Med.*, 856381. doi: 10.1155/2012/856381.

Ishii, P.L., Prado, C.K., Mauro, M.O., Carreira, C.M., Mantovani, M.S. et al. (2011). Evaluation of *Agaricus blazei in vivo* for antigenotoxic, anticarcinogenic, phagocytic and immunomodulatory activities. *Reg. Toxicol., Pharmacol.*, 59: 412–422.

Jana, U.K., Suryawanshi, R.K., Prajapati, B.P. and Kango, N. (2021). Prebiotic mannooligosaccharides: Synthesis, characterization, and bioactive properties. *Food Chem.*, 342. 128328. doi: 10.1016/j.foodchem.2020.128328.

Jawhara, S. and Poulain, D. (2007). *Saccharomyces boulardii* decreases inflammation and intestinal colonization by *Candida albicans* in a mouse model of chemically induced colitis. *Med. Mycol.*, 45: 691–700.

Jayachandran, M., Xiao, J. and Xu, B. (2017). A critical review on health promoting benefits of edible mushrooms through gut microbiota. *Int. J. Mol. Sci.*, 18: 1934. doi: 10.3390/ijms18091934.

Kanagasabapathy, G., Chua, K.H., Malek, S.N.A., Vikineswary, S. and Kuppusamy, U.R. (2014). AMP-activated protein kinase mediates insulin-like and lipo-mobilising effects of β-glucan-rich polysaccharides isolated from *Pleurotus sajor-caju* (Fr.), Singer mushroom, in 3T3-L1 cells. *Food Chem.*, 145: 198–204.

Kango, N. and Jain, S.C. (2011). Production and properties of microbial inulinases: Recent advances. *Food Biotechnol.*, 25: 165–212.

Katapodis, P., Kalogeris, E., Kekos, D. and Macris, B.J. (2004). Biosynthesis of fructo-oligosaccharides by *Sporotrichum thermophile* during submerged batch cultivation in high sucrose media. *Appl. Microbiol. Biotechnol.*, 63: 378–382.

Kelesidis, T. and Pothoulakis, C. (2012). Efficacy and safety of the probiotic *Saccharomyces boulardii* for the prevention and therapy of gastrointestinal disorders. *Ther. Adv. Gastroenterol.*, 5: 111–125.

Khan, A.A., Gani, A., Khanday, F.A. and Masoodi, F.A. (2018). Biological and pharmaceutical activities of mushroom β-glucan discussed as a potential functional food ingredient. *Bioactive Carbohydr. Diet. Fib.*, 16: 1–13.

Kotowska, M., Albrecht, P. and Szajewska, H. (2005). *Saccharomyces boulardii* in the prevention of antibiotic-associated diarrhea in children: Randomized double-blind placebo-controlled trial. *Alim. Pharmacol. Ther.*, 21: 583–590.

Kozyrovska, N.O., Reva, O.M., Goginyan, V.B. and de Vera, J.-P. (2012). Kombucha microbiome as a probiotic: A view from the perspective of post-genomics and synthetic ecology. *Biopolym. Cell*, 28: 103–113.

Kumar, S. and Singhi, S. (2016). Role of probiotics in prevention of *Candida* colonization and invasive candidiasis. *J. Matern. Fetal Neon. Med.*, 29: 818–819.

Kumar, S.B.A., Chakrabarti, A. and Singhi, S. (2013). Evaluation of efficacy of probiotics in prevention of *Candida* colonization in a PICU: A randomized controlled trial. *Crit. Care Med.*, 41: 565–572.

Kumar, V., Yadav, A.N., Verema, P., Sangwan, P., Abhishake, S. and Singh, B. (2017). β-Propeller phytases: Diversity, catalytic attributes, current developments, and potential biotechnological applications. *Int. J. Biol. Macromol.*, 98: 595–609.

Kumura, H., Tanoue, Y., Tsukahara, M., Tanaka, T. and Shimazaki, K. (2004). Screening of dairy yeast strains for probiotic applications. *J. Dairy Sci.*, 87: 4050–4056.

Kunyeit, L., Kurrey, N.K., Anu-Appaiah, K.A. and Rao, R.P. (2019). Probiotic yeasts inhibit virulence of non-*albicans Candida* species. *mBio*, 10: 2307–2719.

Kurahashi, A. (2021). Ingredients, functionality, and safety of the Japanese traditional sweet drink Amazake. *J. Fungi*, 7: 469. doi: 10.3390/jof7060469.

Kurakake, M., Onoue, T. and Komaki, T. (1996). Effect of pH on transfructosylation and hydrolysis by b-fructofuranosidase from *Aspergillus oryzae*. *Appl. Microbiol. Biotechnol.*, 45: 236–239.

Kurugol, Z. and Koturoglu, G. (2005). Effects of *Saccharomyces boulardii* in children with acute diarrhea. *Acta Pediatr.*, 94: 44–47.

Lateef, A., Oloke, J.K. and Prapulla, S.G. (2007). The effect of ultrasonication on the release of fructosyl transferase from *Aureobasidium pullulans* CFR 77. *Enz. Microb. Technol.*, 40: 1067–1070.

Lee, S.K., Kim, H.J., Chi, S.G., Jang, J.Y., Nam, K.D. et al. (2005). *Saccharomyces boulardii* activates expression of peroxisome proliferators-activated receptor-gamma in HT29 cells. *Kor. J. Gastroenterol.*, 45: 328–324.

Liu, Y., Du, Y.Q., Wang, J.H., Zha, X.Q. and Zhang, J.B. (2014). Structural analysis and antioxidant activities of polysaccharide isolated from Jinqian mushroom. *Int. J. Biol. Macromol.*, 64: 63–68.

Lohith, K.A. and Anu-Appaiah, K.A. (2014). *In vitro* probiotic characterization of yeasts of food and environmental origin. *Int. J. Probiot. Prebiot.*, 9: 1–6.

Lu, J.H., Meng, Q.F., Ren, X.D., Chen, Y.G., Li, T.T. and Teng, L.R. (2008). Extraction and antitumor activity of polysaccharides from *Tricholoma matsutake*. *J. Biotechnol.*, 136: 5469. doi: 10.1016/j.jbiotec.2008.07.1091.

Mabel, M.J., Sangeetha, P.T., Kalpana, P., Srinivasan, K. and Prapulla, S.G. (2007). Physicochemical characterization of fructo-oligosaccharides and evaluation of their suitability as a potential sweetener for diabetic. *Carbohydr. Res.*, 343: 56–66.

Macfarlane, G.T., Steed, H. and Macfarlane, S. (2008). Bacterial metabolism and health-related effects of galacto-oligosaccharides and other prebiotics. *J. Appl. Microbiol.*, 104: 305–344.

Madlova, A., Antosova, M., Barathova, M., Polakovic, M., Stefuca, V. and Bales, V. (1999). Screening of microorganisms for transfructosylating activity and optimization of biotransformation of sucrose to fructooligosaccharides. *Chem. Pap.*, 53: 366–369.

Maiorano, A.E., Piccoli, R.M., Da Silva, E.S. and De Andrade Rodrigues, M.F. (2008). Microbial production of fructosyltransferases for synthesis of prebiotics. *Biotechnol. Lett.*, 1867–1877.

Maity, P., Samanta, S., Nandi, A.K., Sen, I.K., Paloi, S. et al. (2014). Structure elucidation and antioxidant properties of a soluble β-d-glucan from mushroom *Entoloma lividoalbum*. *Int. J. Biol. Macromol.*, 63: 140–149.

Manning, T.S. and Gibson, G.R. (2004). Microbial-gut interactions in health and disease - Prebiotics. *Best Pract. Res. Clin. Gastroenterol.*, 18: 287–298.

Mano, M.C.R., Neri-Numa, I.A., da Silva, J.B., Paulino, B.N., Pessoa, M.G. and Pastore, G.M. (2017). Oligosaccharide biotechnology: An approach of prebiotic revolution on the industry. *Appl. Microbiol. Biotechnol.*, 102: 17–37.

Manzoni, P., Mostert, M. and Castagnola, E. (2015). Update on the management of *Candida* infections in preterm neonates. *Arch. Dis. Child Fetal Neonatal Ed.*, 100: 454–459.

Markowia, k.P. and Slizewska, K. (2017). Effects of probiotics, prebiotics, and synbiotics on human health. *Nutrients*, 9: 1021. doi: 10.3390/nu9091021.

Matijašević, D., Pantić, M., Rašković, B., Pavlović, V., Duvnjak, D. et al. (2016). The antibacterial activity of *Coriolus versicolor* methanol extract and its effect on ultrastructural changes of *Staphylococcus aureus* and *Salmonella enteritidis*. *Front. Microbiol.*, 7: 1226. doi: 10.3389/fmicb.2016.01226.

Maupas, J.L., Champemont, P. and Delforge, M. (1983). Treatment of irritable bowel syndrome with *Saccharomyces boulardii*: A double-blind, placebo-controlled study. *Med. Chir. Dig.*, 12: 77–79.

McFarland, L.V. (2007). Meta-analysis of probiotics for the prevention of traveler's diarrhea. *Travel Med. Infect. Dis.*, 5: 97–105.

McFarland, L.V., Surawicz, C.M., Greenberg, R.N., Elmer, G.W., Moyer, K.A. et al. (1994). Prevention of b-lactam-associated diarrhea by *Saccharomyces boulardii* compared with placebo. *Am. J. Gastroenterol.*, 90: 439–448.

Metchnikoff, E. (1907). *The Prolongation of Life*. London: Heinemann.

Ming, J., Liu, J., Wu, S., Guo, X., Chen, Z. and Zhao, G. (2012). Structural characterization and hypolipidemic activity of a polysaccharide PGEB-3H from the fruiting bodies of *Gastrodia elata* Blume. *Proc. Eng.*, 37: 169–173.

Miranda, J.M. (2020). Analytical technology in nutrition analysis. *Molecules*, 25: 1362. doi: 10.3390/molecules25061362.

Moré, M.I. and Swidsinski, A. (2015). *Saccharomyces boulardii* CNCM I-745 supports regeneration of the intestinal microbiota after diarrheic dysbiosis: A review. *Clin. Exp. Gastroenterol.*, 8: 237–255.

Mussatto, S.I. and Mancilha, I.M. (2007). Non-digestible oligosaccharides: A review. *Carbohydr. Polym.*, 68: 587–597.

Mussatto, S.I. and Teixeira, J.A. (2010). Increase in the fructo-oligosaccharides yield and productivity by solid-state fermentation with *Aspergillus japonicus* using agro-industrial residues as support and nutrient source. *Biochem. Eng. J.*, 53: 154–157.

Nagpal, R., Kumar, A., Kumar, M., Behare, P.V., Jain, S. and Yadav, H. (2012). Probiotics, their health benefits, and applications for developing healthier foods: A review. *FEMS Microbiol. Lett.*, 334: 1–15.

Nandi, F.A.K., Samanta, S., Maity, S., Sen, I.K., Khatua, S. et al. (2014). Antioxidant and immunostimulant β-glucan from edible mushroom *Russula albonigra* (Krombh.) *Fr. Carbohydr. Polym.*, 99: 774–782.

Navegantes, K.C., Albuquerque, R.F.V., Dalla-Santa, H.S., Soccol, C.R. and Monteiro, M.C. (2013). *Agaricus brasiliensis* mycelium and its polysaccharide modulate the parameters of innate and adaptive immunity. *Food Agric. Immunol.*, 24: 393–408.

Neurath, M.F., Fuss, I., Kelsall, B.L., Stüber, E. and Strober, W. (1995). Antibodies to interleukin 12 abrogate established experimental colitis in mice. *J. Exp. Med.*, 182: 1281–1290.

Neyrinck, A.M., Bindels, L.B., De Backer, F., Pachikian, B.D., Cani, PD. and Delzenne, N.M. (2009). Dietary supplementation with chitosan derived from mushrooms changes adipocytokine profile in diet induced obese mice, a phenomenon linked to its lipid lowering action. *Int. Immunopharmacol.*, 9: 767–773.

Neyrinck, A.M., Possemiers, S., Verstraete, W., De Backer, F., Cani, P.D. and Delzenne, N.M. (2012). Dietary modulation of clostridial cluster XIVa gut bacteria (*Roseburia* spp.) by chitin-glucan fiber improves host metabolic alterations induced by high-fat diet in mice. *J. Nutr. Biochem.*, 23: 51–59.

Nicolosi, R., Bell, S.J., Bistrian, B.R., Greenberg, I., Forse, R.A. and Blackburn, G.L. (1999). Plasma lipid changes after supplementation with beta-glucan fiber from yeast. *Am. J. Clin. Nutr.*, 70: 208–212.

Nikitina, O., Cherno, N., Osolina, S. and Naumenko, K. (2017). Yeast glucan and glucan-containing mushroom biopolymer complexes–stimulators of microflora growth. *Int. Food Res. J.*, 24: 2220–2227.

Nowak, R., Nowacka-Jechalke, N., Juda, M. and Malm, A. (2018). The preliminary study of prebiotic potential of Polish wild mushroom polysaccharides: the stimulation effect on *Lactobacillus* strains growth. *Eur. J. Nutr.*, 57: 1511–1521.

Oba, K., Teramukai, S., Kobayashi, M., Matsui, T., Kodera, Y. and Sakamoto, J. (2007). Efficacy of adjuvant immunochemotherapy with polysaccharide K for patients with curative resections of gastric cancer. Canc. *Immunol. Immunother.*, 56: 905–911.

Ordoñez, L., García, J. and Bolaños, G. (2013). *Producing chitin and chitin-glucan complexes from Aspergillus niger biomass using subcritical water*. III Ibeoamerican Conference on Supercritical Fluids, Colombia, pp. 1–7.

Palai, S., Derecho, C.M.P., Kesh, S.S., Egbuna, C. and Onyeike, P.C. (2020). Prebiotics, probiotics, synbiotics and its importance in the management of diseases. pp. 173–196. *In*: Egbuna, C. and Dable-Tapas, G. (Eds.). *Functional Foods and Nutraceuticals*, Switzerland AG, Springer Nature.

Palframan, R., Gibson, G.R. and Rastall, R.A. (2003). Development of a quantitative tool for comparison of the prebiotic effect of dietary oligosaccharides. *Lett. Appl. Microbiol.*, 37: 281–284.

Pandey, K.R., Naik, S.R. and Vakil, B.B. (2015). Probiotics, prebiotics and synbiotics: A review. *J. Food Sci. Technol.*, 52: 7577–7587.

Pereira, R.P., Jadhav, R., Baghela, A. and Barretto, D.A. (2021). *In vitro* assessment of probiotic potential of *Saccharomyces cerevisiae* DABRP5 isolated from *Bollo* batter, a traditional Goan fermented food. *Prob. Antimicrob. Prot.*, 13: 796–808.

Pothoulakis, C., Kelly, C.P., Joshi, M.A. et al. (1993). *Saccharomyces boulardii* inhibits *Clostridium difficile* toxin A binding and enterotoxicity in rat ileum. *Gastroenterol.*, 104: 1108–1115.

Powrie, F., Leach, M.W., Mauze, S., Caddle, L.B. and Coffman, R.I. (1993). Phenotypically distinct subsets of CD4+T cells induce or protect from chronic intestinal inflammation in C.B-17 scid mice. *Int. Immunol.*, 5: 1461–1471.

Prathumpaí, W., Rachathewee, P., Khajeerani, S., Tanjak, P. and Methacanon. (2012). Exobiopolymer application of three enthomopathogenic fungal strains as prebiotic use. *KKU Res. J.*, 17: 743–753.

Rajilić-Stojanović, M. and De Vos, W.M. (2014). The first 1000 cultured species of the human gastrointestinal microbiota. *FEMS Microbiol. Rev.*, 38: 996–1047.

Rawat, H.K., Soni, H. and Kango, N. (2017). Fungal inulinolytic enzymes: A current appraisal. pp. 279–293. *In*: Satyanarayana, T., Deshmukh, S.K. and Johri, B.N. (Eds.). *Developments in Fungal Biology and Applied Mycology*. Singapore: Springer Nature Pte Ltd.

Roberfroid, M.B. (2008). Prebiotics: Concept, definition, criteria, methodologies, and products. pp. 40–68. *In*: Gibson, G.R. and Roberfroid, M.B. (Eds.). *Handbook of Prebiotics*, CRC Press.

Ruthes, A.C., Cantu-Jungles, T.M., Cordeiro, L.M.C. and Iacomini, M. (2021). Prebiotic potential of mushroom D-glucans: Implications of physicochemical properties and structural features. *Carbohydr. Polym.*, 262: 117940. doi: 10.1016/j.carbpol.2021.117940.

Saint-Marc, T., Rosello-Prats, L. and Touraine, J.L. (1991). Efficacité de *Saccharomyces boulardii* dans le traitement des dirrhées du SIDA. *Ann. Med. Int.*, 142: 64–65.

Sanders, J.W. and Tribble, D.R. (2001). Diarrhea in the returned traveller. *Curr. Gastroenterol. Rep.*, 3: 304–314.

Sangeetha, P.T., Ramesh, M.N. and Prapulla, S.G. (2004a). Production of fructo-oligosaccharides by fructosyl transferase from *Aspergillus oryzae* CFR 202 and *Aureobasidium pullulans* CFR 77. *Process Biochem.*, 39: 753–758.

Sangeetha, P.T., Ramesh, M.N. and Prapulla, S.G. (2004b). Production of fructosyl transferase by *Aspergillus oryzae* CFR 202 in solid-state fermentation using agricultural by-products. *Appl. Microbiol. Biotechnol.*, 65: 530–537.

Sawangwan, T., Wansanit, W., Pattani, L. and Noysang, C. (2018). Study of prebiotic properties from edible mushroom extraction. *Agric. Nat. Resour.*, 52: 519–524.

Sknepnek, A., Pantić, M., Matijšević, D., Miletić, D., Lević, S. et al. (2018). Novel Kombucha beverage from Lingzhi or Reishi medicinal mushroom, *Ganoderma lucidum*, with antibacterial and antioxidant effects. *Int. J. Med. Mush.*, 20: 243–258.

Sknepnek, A., Tomić, S., Miletić, D., Lević, D., Čolić, M. et al. (2021). Fermentation characteristics of novel *Coriolus versicolor* and *Lentinus edodes* kombucha beverages and immunomodulatory potential of their polysaccharide extracts. *Food Chem.*, 342: 128344. doi: 10.1016/j.foodchem.2020.128344.

Sobierlski, K., Siwulski, M., Lisiecka, J., Jędryczka, M., Sas-Golak, I. and Frużyńska-Jóźwiak. D. (2012). Fungi-derived β-glucans as a component of functional food. *Acta Sci. Pol.*, 11: 111–128.

Sun, Y., Sun, T., Wang, F., Zhang, J., Li, C. et al. (2013). A polysaccharide from the fungi of Huaier exhibits anti-tumor potential and immunomodulatory effects. *Carbohydr. Polym.*, 92: 577–582.

Surawicz, C.M., Elmerm G.W., Speelman, P., McFarland, L.V., Chinn, J. and Van Belle, G. (1989). Prevention of antibiotic-associated diarrhea by *Saccharomyces boulardii*: A prospective study. *Gastroenterol.*, 96: 981–988.

Suryawanshi, R.K. and Kango, N. (2021). Production of mannooligosaccharides from various mannans and evaluation of their prebiotic potential. *Food Chem.*, 334: 127428. doi: 10.1016/j.foodchem.2020.127428.

Szajewska, H. and Mrukowicz, J. (2005). Meta-analysis: non-pathogenic yeast *Saccharomyces boulardii* in the prevention of antibiotic-associated diarrhea. *Aliment. Pharmacol. Ther.*, 22: 365–372.

Tao, Y., Zhang, L. and Cheung, P.C.K. (2006). Physicochemical properties and antitumor activities of water-soluble native and sulfated hyperbranched mushroom polysaccharides. *Carbohydr. Res.*, 341: 2261–2269.

Tian, Y., Nichols, R.G., Roy, P., Gui, W., Smith, P.B. et al. (2018). Prebiotic effects of white button mushroom (*Agaricus bisporus*) feeding on succinate and intestinal gluconeogenesis in C57BL/6 mice. *J. Func. Foods*, 45: 223–232.

Vamanu, E., Dinu, L.D., Pelinescu, D.R. and Gatea, F. (2021). Therapeutic properties of edible mushrooms and herbal teas in gut microbiota modulation. *Microorganisms*, 9: 1262. doi: 10.3390/microorganisms9061262.

Van Balken, J., Van Dooren, T., Van Den, T.W., Kamphuis, J. and Meijer, E.M. (1991). Production of 1-kestose with intact mycelium of *Aspergillus phoenicis* containing sucrose-1 fructosyltransferase. *Appl. Microbiol. Biotechnol.*, 35: 216–221.

Villarruel, G., Rubio, D.M., Lopez, F. et al. (2007). *Saccharomyces boulardii* in acute childhood diarrhea: A randomized, placebo-controlled study. *Acta Pediatr.*, 96: 538–541.

Wang, J., Yuan, Y. and Yue, T. (2014). Immunostimulatory activities of β-d-glucan from *Ganoderma lucidum*. *Carbohydr. Polym.*, 102: 47–54.

Wang, Y. (2009). Prebiotics: Present and future in food science and technology. *Food Res. Int.*, 42: 8–12.

Ward, T.L., Knights, D. and Gale, C.A. (2017). Infant fungal communities: Current knowledge and research opportunities. *BMC Medicine*, 15: 30. doi: 10.1186/s12916-017-0802-z.

Wasser, S.P. (2002). Medicinal mushrooms as a source of antitumor and immonomudulating polysaccharides. *Appl. Microbiol. Biotechnol.*, 60: 258–274.

Wu, X., Xia, Y., He, F., Zhu, C. and Ren, W. (2021). Intestinal mycobiota in health and diseases: From a disrupted equilibrium to clinical opportunities. *Microbiome*, 9: 60. doi: 10.1186/s40168-021-01024-x.

Yadav, A.N., Kumar, R., Kumar, S., Kumar, V., Sugitha, T. et al. (2017). Beneficial microbiomes: Biodiversity and potential biotechnological applications for sustainable agriculture and human health. *J. Appl. Biol. Biotechnol.*, 5: 1–13.

Yang, K., Zhang, Y., Cai, M., Guan, R., Neng, J. et al. (2020). *In vitro* prebiotic activities of oligosaccharides from the by-products in *Ganoderma lucidum* spore polysaccharide extraction. *Royal Soc. Chem. Adv.*, 10: 14794–14802.

Yang, W.J., Pei, F., Shi, Y., Zhao, L.Y., Fang, Y. and Hu, Q.H. (2012). Purification, characterization, and anti-proliferation activity of polysaccharides from *Flammulina velutipes*. *Carbohydr. Polym.*, 88: 474–480.

Yin, C., Noratto, G.D., Fan, X., Chen, Z., Yao, F. et al. (2020). The impact of mushroom polysaccharides on gut microbiota and its beneficial effects to host: A review. *Carobhydr. Polym.*, 250: 116942. doi: 10.1016/j.carbpol.2020.116942.

Yoshikawa, J., Amachi, S., Shinoyama, H. and Fujii, T. (2007). Purification and some properties of β-fructofuranosidase I formed by *Aureobasidium pullulans* DSM 2404. J. *Biosci. Bioeng.*, 103: 491–493.

Yun, J.W. (1996). Fructooligosaccharides: Occurrence, preparation, and application. *Enz. Microb. Technol.*, 19:107–117.

Zambelli, P., Fernandez-Arrojo, L., Romano, D., Santos-Moriano, P., Gimeno-Perez, M. et al. (2014). Production of fructooligosaccharides by mycelium-bound transfructosylation activity present in *Cladosporium cladosporioides* and *Penicillium sizovae*. *Proc. Biochem.*, 49: 2174–2180.

Zanello, G., Meurens, F., Berri, M. and Salmon, H. (2009). *Saccharomyces boulardii* effects on gastrointestinal diseases. *Curr. Issues Mol. Biol.*, 11: 47–58.

Zhang, B., Yan, P., Chen, H. and He, J. (2012). Optimization of production conditions for mushroom polysaccharides with high yield and antitumor activity. *Carbohydr. Polym.*, 87: 2569–2575.

Zimoch-Korzycka, A., Gardrat, C., Kharboutly, M.A., Castellan, A., Pianet, I. et al. (2016). Chemical characterization, antioxidant and anti-listerial activity of non-animal chitosan-glucan complexes. *Food Hydrocoll.*, 61: 338–343.

11

Application of Mushrooms in Beverages

*Aleksandra Sknepnek** and *Dunja Miletić*

1. Introduction

Macrofungi represent a group of fungal species, within the Kingdom Fungi, that belong to the phylum Basidiomycota and Ascomycota. They produce visible, morphologically diverse epigeous or hypogenous fruit bodies, carpophores (Govorushko et al., 2019). It is estimated that fungi represent between 2.2 and 3.8 million species (Hawksworth and Lücking, 2017), while the macrofungi account up to 15,000 species (Mešić et al., 2020). Among the 2,000 known macrofungi, many are edible and/or medicinal. However, they still have not been examined enough, since only 35 species of them have reached production on a commercial scale (Meenu and Xu, 2019). According to the Food and Drug Administration (FDA), China is the leader in macrofungi production with up to 80% of the total global manufacture (FAO, 2016). The FAO data revealed progressive increase in global macrofungal cultivation (FAO, 2018). In the past 30 years, the global mushroom production increased nearly 10% (El Sheikha and Hu, 2018). About 90% of edible mushrooms distributed all over the world, belong to six main genera: *Agaricus*, *Auricularia*, *Flammulina*, *Lentinula*, *Pleurotus*, and *Volvariella* (Royse et al., 2017). On the other hand, fruit bodies of *Ganoderma lucidum* and *Trametes versicolor* possess a woody texture or bitter taste and cannot be classified as edible. They are known only for their medical properties, and useful in treatment or prevention of many diseases. Up to 700 species of medicinal mushrooms are considered to be safe (Smith et al., 2002; Morris et al., 2017; Hapuarachchi, 2018). Due to the presence of biologically active compounds in their fruit bodies, several edible mushrooms are also classified as medicinal (e.g., *Lentinula*, *Pleurotus*, *Agaricus*, *Auricularia*) (Fig. 11.1). Calculating

Institute of Food Technology and Biochemistry, Faculty of Agriculture, University of Belgrade, 11000 Belgrade, Serbia.
* Corresponding author: aleksandras@agrif.bg.ac.rs

Fig. 11.1. Edible and medicinal mushrooms.

Note. Medicinal and edible mushrooms: (A). *Agaricus bisporus* (Button mushroom), (B). *Lentinus edodes* (Shiitake mushroom), (C). *Pleurotus ostreatus* (Oyster mushroom), Medicinal and non-edible mushrooms: (D). *Coriolus* (*Trametes*) *versicolor* (Turkey tail mushroom), (E). *Ganoderma lucidum* (Lingzhi or Reishi mushroom).

the total mushroom production, cultivated edible mushrooms reached 54% of global mushroom industry, while the medicinal mushrooms comprised up to 38% (Royse et al., 2017).

The *Agaricus* genus is one of the major cultivated macrofungi. Within this genus, *A. bisporus* (known as button mushroom) is the most cultivated edible mushroom all over the world (Atila et al., 2017). On the other hand, *A. brasiliensis* (almond mushroom: also known as *A. blazei* or *A. subrufescens*) is an edible mushroom that originated from Brazil, and is cultivated all over the world, due to its medicinal properties (Lisiecka et al., 2013; Llarena-Hernández et al., 2013). It has been used traditionally for cancer treatment (Wang et al., 2013; Woraharn et al., 2015). The second most cultivated mushroom that is edible as well as medicinal is *Lentinus edodes* (shiitake), with significant biological properties (Valverde et al., 2015). The low-cost production technology and very high biological value placed the oyster mushroom (*Pleurotus* spp.) on the third place of the largest commercially produced mushrooms in the world market, and the second most consumed edible mushroom worldwide (Obodai et al., 2003; Mane et al., 2007; Knop et al., 2015). The most important *Pleurotus* species cultivated on a large scale include *P. ostreatus* and *P. pulmonarius*. Besides, within *Pleurotus* genera, *P. citrinopieatus*, *P. cystidiosus*, *P. djamor*, *P. eryngii*, *P. sajor-caju*, and *P. sapidus* are also commercially available (dos Santos Bazanella et al., 2013; Knop et al., 2015). According to the data reported by Koutrotsios et al. (2017), production of *P. ostreatus* mushroom corresponds up to 30% of the total world's mushroom manufacture. Royse et al. (2017) revealed that mushrooms of the genus *Auricularia* is on the third place of the most cultivated mushroom genera accounting about 17% of world mushroom production, after

Lentinula (22%) and *Pleurotus* (19%). *Flammulina* produced up to 15% of world's edible mushroom production (Royse et al., 2017). Among this genus, *F. velutipes* (golden needle mushroom or enokitake) is a widely cultivated edible mushroom, and it has numerous medicinal properties (Zimmermanovà et al., 2001; Kalač, 2012; Zhang et al., 2018). *Volvariella volvacea* (paddy straw mushroom), is considered as one of the easiest mushrooms to be cultivated. However, its growth is restricted to the tropical and subtropical regions, due to requirements for high temperature and rainy climate, and thus, it takes the sixth place among cultivated mushrooms (5%) (Rajapakse, 2011; Royse et al., 2017).

Macrofungi, as saprobes, grow on lignocellulosic material and are cultivated on most of the agricultural wastes or byproducts. Industrial solid-state production of mushrooms represents sustainable production, since it uses recycled agricultural waste products (Rajapakse, 2011; Kertesz and Thai, 2018). *Agaricus* species naturally grow on partially degraded organic matter (El Sebaaly et al., 2019). Its cultivation procedure is the most complex process of mushroom industrial cultivation (Kertesz and Thai, 2018). The substrate for *Agaricus* spp. cultivation is a compost, that consists of a horse and broiler chicken manure, water, and gypsum. But this mushroom can also be cultivated on a compost made up of a mixture of wheat straw and broiler chicken manure (Sánchez, 2010). Cultivation technique on non-composted wheat straw is not commercially applied due to low yields (Mamiro et al., 2007). The *L. edodes* is traditionally grown on wood logs, however it is also cultivated on artificial compact logs made from sterilized sawdust. *Auricularia* sp. and *Flammulina* sp. are cultivated on sterile, partly composted mixture of sawdust, bran, straw, and corncobs (Chang and Miles, 2004; Sánchez, 2010). *Pleurotus* and *Volvariella* species are grown on non-composted or partially composted substrates not rigorously sterilized. These genera are adaptable and fast-growing, and thus are capable for bioconversion of different substrates such as rice straw, bagasse, cornstalks, waste cotton, stalks, and banana leaves (Chang and Miles, 2004; Thongklang and Luangharn, 2016).

Some medicinal mushrooms are host-specific, slow growing, or rare in nature (Duvnjak et al., 2016). For example, using traditional cultivation on wood logs, it takes several months to obtain *G. lucidum* fruit bodies. Additionally, a long growing period, variations in the quality of the growing material and in combination with different stages of development of fruit bodies, the quality of the final products gets affected (Chang and Miles, 1989). In such circumstances, submerged mushroom cultivation is a process that is suitable option for the production of macrofungal biomass and culture broth-based products, that guarantees continuous production throughout the year with standard quality (Sanodiya et al., 2009). On the world's market, a number *G. lucidum* products obtained from fruit bodies, spores, and mycelium could be found as commercial dietary supplements or other different food stuffs (Lai et al., 2004). Although the growing period in the submerged process is shorter, and cultivation is easier to control, but since the entire process is on a large-scale in a bioreactor, mycelium growth in liquid culture is influenced by numerous factors (Cui and Chisti, 2003; Turło, 2014). Obtaining a significant amount of mushroom mycelium biomass or isolation of specific bioactive compounds, from mycelium or fermentation broth, always requires optimization of the process to maximize benefits as well as to reduce the costs (Lübbert, 2003). Additionally, as the bioactive properties between mushroom fruit bodies and mycelium significantly vary, mushroom mycelia from the submerged culture represents a source that could be used in foods and beverages (Tešanović et al., 2017).

The current chapter gives an overview on the some of the most valuable edible and medicinal mushrooms, their chemical composition, impact on nutrition, contribution to the bioactive properties and specific aroma and flavor on the mushroom beverages. The different technologies for production of mushroom beverages is given and discussed herewith. The influence of the mushroom chemical composition and enzymatic activity on the fermented alcoholic and non-alcoholic beverages and the mushroom functional beverages are also reviewed.

2. Usefulness of Macrofungi in Beverages

Mushrooms are a valuable source of chemical compounds which can be divided into two groups: high-molecular weight compounds (polysaccharides, polysaccharopeptides, proteins, and lipids) and low-molecular compounds (polyphenols, terpenoids, and alkaloids) (Vetter, 2019; Mešić et al., 2020). Due to their high nutritional value and favorable impact on the human health, mushrooms have been used in human diet for centuries (Vetter, 2019). Fresh mushroom fruit bodies contain a high quantity of moisture (up to 90%) (Lu et al., 2020a). The exceptions to this are *Ganoderma* species, which have characteristic hard and dry fruit bodies (Vetter, 2019). They have less moisture (varies within the species) compared to fleshy mushrooms (Győri, 2007; Reis et al., 2012), however it is dependent on several factors such as maturity stage and storage conditions (Tsai et al., 2007; El Sebaaly et al., 2019). Usually, macrofungal fruit bodies possess the highest content of carbohydrates followed by proteins on a dry mass basis (Rathore et al., 2017).

2.1 Carbohydrates and Bioactive Polysaccharides

A large number of structurally-diverse bioactive compounds were isolated and characterized from a number of cultivated or wild growing macrofungi (Mešić et al., 2020). Based on a dry matter, carbohydrates are the most dominant mushrooms fractions present in fruit bodies or mycelia. Mushroom carbohydrate fractions include monosaccharides and disaccharides (glucose, fructose, mannose, and arabinose) their derivatives (sugar alcohols, mainly mannitol), oligosaccharides, and polysaccharides. Their content can vary depending on the mushroom species. Based on the dry matter, mushrooms contain between 50 and 65% of carbohydrates (Lu et al., 2009). In *Pleurotus* spp., the content of carbohydrates ranges between 11–42%, and mainly consist of polysaccharides (glycogen, and indigestible fibers: chitin, α- and β-glucans), hemicelluloses (mannans, xylans, and galactans), and glycoproteins (Khan and Tania, 2012; Maftoun et al., 2015). Chang and Hayes (2013) revealed a very high carbohydrate content (81%) in *Auricularia auricula-judae*. Carbohydrate content during fruit body growth and formation is subject to change. By analyzing a proximate composition of *A. bisporus* mushroom harvested at different stages of maturity, it was determined that the total carbohydrate content increased from 38.3% to 48.9%. Additionally, content of soluble sugars and polyols also increase with the maturation process. This is an important fact to highlight, since soluble sugars and polyols (mannitol) are responsible for sweet taste (Tsai et al., 2007). High content of soluble sugars and polyols were also found in *Volvariella volvacea* (349–458 mg/kg) (Mau et al., 1997) and in *Auricularia* spp. (98.7–316 mg/g) (Mau et al., 1998), while *G. lucidum* contained a low amount (16.8–83.7 mg/kg) (Mau et al., 2001). However, *G. lucidum* polysaccharides (GLPs) have been extensively studied due to various biological activities (antioxidant, antitumor, anti-inflammatory, antiviral, anti-diabetes, and immunomodulation) (Hu et al., 2012; Sohretoglu and Huang, 2018; Lu et al., 2020b).

The cell wall of macrofungi consists of three layers: external layer made of glucans and polysaccharides bounded to glycoproteins, middle layer consisting of glucans and the inner layer built up with glucans and chitin (Chen and Cheung, 2014). The β-glucans being the main structural polysaccharides of the fungal cell wall (Smiderle et al., 2013), consist of D-glucose chains linked with β-(1→ 3) glucoside bonds and β-(1→ 6) glycosidic bonds in side chains (Friedman, 2016). Structurally, macrofungal polysaccharides are non-digestible high molecular biopolymers, which exhibit biological activities including immunopotentiation, anticancer, and tumor inhibition. They can be obtained as polysaccharides, polysaccharo peptides, and proteoglucans (Chen and Cheung, 2014). When mushrooms are used as substrate for the production of fermented beverages, it cannot be fermented by yeasts without previous enzymatic treatment (Lao et al., 2020). However, their presence in the beverages could have a significant positive impact on the health of their consumers, due to the biological activities (e.g., serve as prebiotics). Various polysaccharides, polysaccharo peptides, and β-glucans are commercially available in the market. The most known and widely used include those obtained from *Pleurotus* spp. (pleuran), *L. edodes* (lentinan, LEM), *Schizophyllum commune* (schizophylan), *G. lucidum* (Gl – 1), *T. versicolor* (krestin), *Grifola fondosa* (grifolan), and *Flammulina velutipes* (flammulin) (Rop et al., 2009). Lentinan is a structural polysaccharide present in a cell wall of *L. edodes* fruit bodies. It is highly purified, free of nitrogen and thus, free of proteins, has a high molecular weight polysaccharide that contains only d-glucose bonded with β-d-(1→3) glycosidic bonds in the regularly branched backbone, and β-d-(1→ 6) glycosidic bonds in side chains (Hobbs, 2000). Lentinan has been approved as an adjuvant therapy for cancer, and it is mainly used against lung, gastric, colorectal, ovarian, cervical, pancreatic, cardiac, and nasopharyngeal cancers, and also against non-Hodgkin's lymphoma (Zhang et al., 2019). The LEM is also obtained from the *L. edodes* mushroom by hot water extraction of mycelium. Since it is not a pure polysaccharide compound such as lentinan, but a major bioactive compound which is protein-bound polysaccharide. It consists of around 24.6% proteins and 44% sugars. It also contains B vitamins (especially B1-thiamine and B2-riboflavine) and ergosterol (Hobbs, 2000). Pleuran is an insoluble polymer isolated from the *P. ostreatus*, with valuable immunomodulatory properties. It was shown that pleuran significantly reduced the incidence of upper respiratory tract infections (Bergendiova et al., 2011). Polysaccharide peptides krestin (PSK) and polysaccharide peptide (PSP) are commercial products of *C. versicolor* obtained by mycelial extraction from submerged batch fermentation. Many researchers reported a wide range of different biological activities of PSP such as antitumor, hepatoprotective, and analgesic activities (Ohwada et al., 2006).

To obtain these macromolecules, different extraction procedures are used followed by the separation and purification process (Lin et al., 2008). Their biological activities depend on several structural characteristics: chain conformation, presence of double or triple helixes or random coils, branching degree, presence of non-sugar compounds (i.e., protein complexes or sulfate groups), and methods of extraction (Giavasis, 2014; Meng et al., 2016; Shi, 2016; Miletić et al., 2020a; Miletić et al., 2021). For example, aqueous extracts of *Agaricus blazei* and *G. lucidum* activate immune cells, while their ethanolic extracts inhibit them (Lu et al., 2016). The significance of studying the structure of isolated polysaccharides lies in the fact that an inhibitory effect of lentinan was not observed when the single helical conformation of polysaccharide was tested, while its triple-helical conformation inhibited the growth of solid tumors, i.e., Sarcoma-180 (Zhang et al., 2005). Further, presence of sulfate radicals in polysaccharides could contribute to its higher antiviral activity. For instance, lentinan expresses remarkable anti-tumor effect, and possesses low antiviral activity against HIV, but when sulphated, the antiviral effect significantly increases (Uryu et al., 1992).

2.2 Proteins

In some countries mushrooms are called "forest meat" or "meat of poverty" and they are considered as a valuable, non-meat source of proteins (Dimitrijevic et al., 2018). Mushrooms consist of a significant amount of protein with all essential amino acids included (valine, leucine, isoleucine, threonine, methionine, lysine, phenylalanine, tryptophan, and histidine), thus enabling their use as a meat substitute (Kakon et al., 2012). In addition, mushrooms are rich in lysine and leucine, which are not present in most cereal foods (Chang and Buswell, 1996). Although the chemical composition of *Pleurotus* varies, the composition of cultivating substrates are considered as a source of nutritionally valuable proteins, since the content in *Pleurotus* ranges from 25to 40% (Sarangi et al., 2006) with a fair composition of essential and non-essential amino acids (Deepalakshm and Sankaran, 2014; Maftoun et al., 2015). Further, the digestibility of *Pleurotus* proteins is similar to those of plant proteins and can be assimilated up to 90% which is comparable to meat with 99% digestibility (Wani et al., 2010). Among the essential amino acids, *Pleurotus* contains the highest amount of lysine, leucine, phenylalanine, and threonine, while it is lowest in methionine and tryptophan (Mukhopadhyay and Guha, 2015). According to Liu et al. (2008) *Agaricus brasiliensis* contains a high quantity of proteins (38.5%) on dry mass basis, which is higher than in *A. bisporus*. *L. edodes* contains 20–23% of proteins and up to 80–87% are digestible (Wasser, 2005). Some mushrooms of *Ganoderma* and *Auricularia* contain less than 10% protein, and are a poor protein source (Ulziijargal and Mau, 2011). Due to a high protein content, mushrooms could be used as an alternative nitrogen source for yeasts during alcohol fermentations (Lin et al., 2010), or for other microorganisms involved in the fermentation process. Protein content, or more importantly the amino acids of mushrooms affect the flavor, which contributes to the taste of the mushroom-derived product. During mushroom consumption, a umami or palatable taste evokes a sense of satisfaction, and is a food seasoning sensation generated or enhanced by the presence of monosodium glutamate (MSG)

(Yamaguchi et al., 1971). Free amino acids of mushrooms (glutamic acid and aspartic acid) and the 5'-nucleotides (inosine 5'-monophosphate and guanosine 5'-monophosphate) are identified as the main umami flavoring substances. Phat et al. (2016) showed that among the 17 commercial and edible mushrooms, the highest content of MSG-like components was found in *A. bisporus* (42.4 ± 6.90 mg/g), while *P. ostreatus* consists of the highest quantity of the flavor 5'-nucleotides (14.8 ± 0.05 mg/g). The highest score in human sensory evaluation was obtained for *P. ostreatus*, while *F. velutipes* was the best in the electronic tongue measurement (Phat et al., 2016). The *Pleurotus* species are also rich in glutamic and aspartic acids, the amino acids responsible for the palatable taste and flavor (Mukhopadhyay and Guha, 2015).

Studies revealed that the most abundant amino acid of many mushrooms is glutamic acid: *L. sajor-caju* (24.22% of total amino acids), *P. ostreatus* (24.04%), and *A. auricula-judae* (18.55%) (Afiukwa et al., 2015). Glutamic acid is also a precursor of γ-aminobutyric acid (GABA), a non-protein bioactive amino acid that serves as the inhibitory neurotransmitter in the mammalian nervous system (Lu et al., 2009). Although some mushrooms contain a significant quantity of this amino acid depending on the maturation stage (e.g., *A. bisporus*: range, 2.16–5.79 mg/g) (Tsai et al., 2007). Some studies showed the presence of significant amount of GABA in beverages obtained by mushroom fermentations by tea leaf waste (Bai et al., 2013). The GABA-rich mushroom-based beverages are mainly produced by glutamic acid-rich mushrooms as a base for the lactic acid fermentation (Woraharn et al., 2015; Park et al., 2017).

Mushrooms, their mixtures and extracts (water and/or alcohol) are also recognized to have a good potential to be used for getting specific flavor and aroma in beverages, thus they were used in production of mushroom beers, wines and spirits, prophylactic drinks as a mushroom-based flavor enhancer or additive (Zivanovic, 2006; Diamantopoulou and Philippoussis, 2015). To ensure flavor compounds to be recovered by extracting optimal temperature and extraction, time is important. Aqueous extraction of *L. edodes* showed high extraction temperatures to get good amount of umami 5'-ribonucleotides, contributing to the greater umami taste (Dermiki et al., 2013). By analyzing six different mushrooms including *A. bisporus*, *L. edodes*, and *P. ostreatus*, it was found that the temperature optima for extraction of MSG-like amino acids and 5′-mononucleotides are not compatible. Thus, the extraction at room temperature provided a higher yield of MSG-like amino acids, while the highest nucleotides yields were provided at 70°C (Poojary et al., 2017).

2.3 Minerals

The mineral composition of mushroom fruit bodies depends on several factors including genetics, growing area, chemical composition of the substrate for cultivated mushrooms, and perhaps the distance from the pollution source for wild mushrooms (Bellettini et al., 2019). Some of contaminants of a wild macrofungi growing area and raw materials that are used for cultivation may possess heavy metals such as cadmium (Cd), lead (Pb), silver (Ag), and arsenic (As). Although these heavy metals do not have a biological role, they are very highly toxic (Zhao et al., 2016).

Studies showed that mushrooms grown in industrial areas, contain these hazardous elements (Zimmermanovà et al., 2001). On the other hand, the great significance of mushrooms is their richness in microelements, which are deficient in human diet such as potassium (K), calcium (Ca), phosphorus (P), and magnesium (Mg). *Agaricus* spp. and *Pleurotus* spp. are cultivated macrofungi with the highest K levels. The content of Mg and Ca varies in the range between 500–1500 mg/kg of dry matter. The Mg content is usually higher than the Ca content, with a few exceptions (Vetter, 2019). The low content of sodium (Na) and high content of K (or low Na/K ratio) makes them an excellent choice for consumption by people who suffer from hypertension (Miletić et al., 2020b). Studies showed that increased intake of K and Mg reduces blood pressure in patients with hypertension (Houston and Harper, 2008). Due to the significant fiber content (8–10%) and high K content, *P. ostreatus* represents a good diet choice for the reduction of body weight and lowering the blood pressure (Papaspyridi et al., 2012). Macrofungi, as the natural decomposers of lignocellulosic material, naturally contain some heavy metals in a trace amount, which are essential for fungal metabolism. For example, white-rot fungi degrade lignin by the action of enzymes containing copper (Cu) (i.e., multicopper oxidases) and manganese (Mn) (i.e., manganese-dependent peroxidase), as ions prosthetic group (Zhao et al., 2016). The *P. ostreatus* is a good source of iron (Fe), and thus, it can be used in the treatment of anemia. Mushrooms are also a good source of selenium (Se), one of the essential micronutrients for humans (Rayman, 2000). The Se concentration in soil is often insufficient to enable its proper intake through a regular diet (Miletić et al., 2019). Since mushrooms are rich in proteins, they can accumulate Se and successfully biotransform it into a bioactive Se-methionine. There are numerous researches performed in order to enrich the mushrooms with organically-bonded Se to enhance its bioavailability (Turło et al., 2007; Miletić et al., 2019; Miletić et al., 2020b). Significantly a higher content of L-aspartic (31.41 mg/g when selenourea was used; 27.73 mg/g when sodium selenite was used) and L-glutamic acids (41.39 mg/g when selenourea was used; 35.94 mg/g when sodium selenite was used) was established in *C. versicolor* mycelium after the cultivation in selenium fortified liquid medium, compared to Se non-enriched mycelium (L-aspartic acid 8.34 mg/g; L-glutamic acid 8.64 mg/g). Based on the results, *C. versicolor* was recommended as a food additive and as a potential flavor enhancer in food (Miletić et al., 2019; Miletić et al., 2020b). Additionally, a high content of L-Se-methionine which is an organic Se source with high bioavailability was established in samples (Miletić et al., 2019). Due to the ability of yeasts to bio-transform inorganic Se, this fact was used to obtain Se-enriched beer, wine, or kvass from Se-enriched barley, wort, and must (Gibson et al., 2006; Pérez-Corona et al., 2011; Sánchez-Martínez et al., 2012). However, the data on using Se-enriched mushrooms in the beverage industry are scarce. Based on the above-stated data, using Se-enriched mushrooms as a substrate in the beverage industry is an unexplored field with many possibilities for obtaining different drinks with pleasant flavors and the possibility to improve their functional characteristics.

2.4 Fatty Acids

Mushrooms are rich in fatty acids, especially polyunsaturated with 75% of the total fatty acids content. Oleic and linoleic acids are highly desirable fatty acids in human diet because they are involved in human health promotion by reducing the high-density lipoprotein (HDL) cholesterol in the blood leading to prevention of atherosclerosis (Saiqa et al., 2008). *Pleurotus* spp. are rich in essential unsaturated fatty acids (e.g., oleic, linoleic, and α-linolenic) (Farghaly and Mostafa, 2015), with the linoleic acid being the major polyunsaturated fatty acid (especially in *P. ostreatus*) (Maftoun et al., 2015). Similarly, *A. bisporus* has a low-fat content, but contains essential fatty acids like linoleic acid (44.19% of total fatty acids), caprylic, palmitic, stearic, oleic, eicosanoid, and erucic acids (Barros et al., 2008; Santos et al., 2003). Interestingly, in *A. bisporus* it was 20- and 5-folds higher in concentrations of linoleic acid compared to *G. lucidum* and *P. ostreatus*, respectively (Hossain et al., 2007). Monounsaturated fatty acids are present in a higher proportion (37.17–68.29%) than the saturated ones (26.07–47.77%) in the *Pleurotus* spp. (Atri et al., 2013), and the species *P. ostreatus* and *P. eryngii* contain higher amounts of monounsaturated fatty acids than other mushrooms (*A. bisporus*, *L. edodes*, and *F. velutipes*) (Reis et al., 2012).

It has been reported that linoleic acid exhibits anti-carcinogenic effects on breast, prostate, and colon cancers (Kim et al., 2005). In addition, linoleic acid in mushrooms contributes to the flavor as it serves as a precursor of 1-octen-3-ol, which is the aromatic mushroom compound (Maga, 1981). Apart from mushrooms, 1-octen-3-ol was detected in beverages such as black tea, brandy, coffee, red wines, orange essence oil, rum, whiskeys, and white wines (Maggi et al., 2010). The 1-octen-3-ol is an aromatic compound that is included in the FDA food additive database (US FDAs Center for Food Safety and Applied Nutrition, 2008) and can be used as a flavoring agent in food industries. According to the Takahashi et al. (2007), in alcoholic beverages (e.g., sake) obtained by *A. luchuensis* activity, 1-octen-3-ol is an important flavoring component. There is also literature related to the detection and characterization of flavoring compounds in beverages produced by macrofungi, which were determined as fatty acid derivates, aldehydes, carboxylic acid, ketones, or terpen alcohols (Zhang et al., 2014).

2.5 Vitamins

Mushrooms are a good source of vitamins. They contain high amounts of B complex vitamins such as B1 (thiamin), B2 (riboflavin), B3 (niacin), and B12 (cobalamin), folates, and have trace amounts of vitamin C, D, and E. Studies showed that mushrooms are very rich in ergosterol, a precursor that converts to ergocalciferol (vitamin D2) under UV light exposure. This is the reason why wild mushrooms represent a good source of vitamin D2, compared to the cultivated ones grown in the dark. The ergosterol content in cultivated mushrooms is negligible and amounts to less than 0.01 mg/100 g of fresh weigh (Urbain et al., 2011; Valverde et al., 2015). Studies showed that *L. edodes* cultivated under natural climatic conditions had high amounts of D2 vitamin (22 μg/100 g to 110 μg/100 g dry weight), most probably as a result of sunlight or UV irradiation (Takamura et al., 1991).

Besides the vitamin D2, mushrooms are the only the non-meat source of vitamin B12. The synthesis of vitamin B12 is restricted to some bacteria, which are part of the food chain. Although some blue-green algae can synthetize vitamin B12, it is actually a pseudo-vitamin B12 that is an inactive compound, poorly absorbed in the human intestine (Koyyalamudi et al., 2009). By analyzing B12 vitamin in *A. bisporus*, it was concluded that it was derived from the bacteria present in the compost (Koyyalamudi et al., 2009). Calculated on dry matter, *L. edodes* contains 1.8 mg/100 g of vitamin B2, and 31 mg/100 g of vitamin B3; *P. ostreatus* 2.5 mg/100 g of vitamin B2, and 65 mg/100 g of vitamin B3; *A. bisporus* 5.1 mg/100 g of vitamin B2, and 53 mg/100 g of vitamin B3, while B12 and D2 vitamins are in trace amounts (Mattila et al., 2001). As mushrooms are a good source of vitamins, they are used for the production of soy mushroom health drink powder that contains high quality proteins as well as vitamins of B complex (Farzana et al., 2017).

Vitamin E is a lipid soluble vitamin and is considered as a major natural antioxidant located in the membranes; thus it plays an important role in the prevention of lipid peroxidation. Among the eight forms of vitamin E, α-tocopherol is the most active form in humans (Ferreira et al., 2009). Bouzgarrou et al. (2018) used *G. lucidum*, *P. ostreatus*, and *P. eryngii* as alternative sources of tocopherols to fortify yogurt with vitamin E. They established that *P. eryngii* and *P. ostreatus* mainly contained β-tocopherol, while the mycelium of *G. lucidum* was rich in δ-tocopherol as well as β-tocopherol. Not only the prepared yogurt maintained the nutritional properties, but also the antioxidant capacity of such a product was improved, especially when *G. lucidum* extracts were used, most probably due to its tocopherol content.

Mushrooms also contain a small quantity of vitamin C. For example, a total content of vitamin C in *L. edodes* is 14.68 mg/100 g of wet weight, while in *Pleurotus* sp. its content varies between 5.38 and 16.1 mg/100 g of wet weight (Çaglarlrmak et al., 2002).

3. Mushroom Beverages

Beverages represent various compositions of ingredients in water and they are being recognized as a good choice to supply nutrients and/or bioactive compounds to the human body (Ignat et al., 2020). It is estimated that people daily consume tens of billions of various beverages' servings worldwide. Bearing this in mind, beverages might stand out as the cost-effective way for people to reach health improvement status through good nutrition (Wilson and Temple, 2016). A consumer's choice of food products is constantly changing with food science and technology development, globalization, economic status, and personal lifestyle, but it is also influenced by cultural backgrounds and previous sensory experience with concrete food (Ignat et al., 2020). Production of a macrofungi on a large scale enables their economic use, not only as food, but also to produce beverages, with unique and pleasant flavors with additional health benefits. It is forecasted that the world's market of edible mushrooms will remarkably grow with a compounded annual growth rate (CAGR) of 7%, meaning that the value will grow from USD 45.3 billion in 2020 to 62.19 billion in 2023. It was also estimated that the medicinal mushroom market will grow up to USD 13.88 billion from 2018 to 2022 (Niego et al., 2021). Using different criteria, beverages can be sorted in several categories: dairy and non-dairy; fermented and non-fermented; alcoholic and non-alcoholic; sports and energy drinks; and all of them might be designated as functional foods. Functional foods and beverages used in diet might reduce health-care expenses for up to 20%, thus they are

Table 11.1. Mushrooms in various beverage production, quality, and bioactivity.

Mushroom	Beverages	Contribution of Mushroom	References
Agaricus bisporus	Yogurt	Antioxidative activity	Francisco et al., 2018
Cordyceps militaris	Fermented beverage Coffee	Antioxidative activity Antioxidative activity	Lao et al., 2020 Song, 2020
Flamulina velutipes	Green tea Apple juice Yogurt Beer Wine	Flavor Enzyme-catalyzed, food browning, inhibitor Prebiotic activity, ADH activity Anticoagulative activity, ADH activity, Anticoagulative and fibrinolytic activity	Rigling et al., 2021 Jang et al., 2002 Chou et al., 2013 Okamura et al., 2001a Okamura et al., 2001b
Ganoderma lucidum	Beer Distilled spirits Yakju Fermented soy milk Kombucha	Flavor Flavor, Antioxidative activity ACE activity inhibition, electron-donating ability, and superoxide dismutase activity Nutritional value improvement Antibacterial and antioxidative activity	Leskosek-Cukalovic et al., 2010 Pecic et al., 2016 Veljović et al., 2019 Kim et al., 2004 Yang and Zhang, 2009 Sknepnek et al., 2018
Grifola frondosa	Fermented soy milk	Nutritional value improvement	Yang et al., 2015
Helvella leucopus	Compound juice	Flavor	XuJie et al., 2008
Hericium erinaceus	Wine, vinegar *H. erinaceus* fermented beverage	Antioxidative activity, ɣ-aminobutyric acid (GABA) production	Li et al., 2014 Woraharn et al., 2015
Lentinus edodes	Wort fermentation Kombucha Yogurt	Flavor Immunomodulatory activity Prebiotic activity	Zhang et al., 2014 Sknepnek et al., 2021 Chou et al., 2013
Pleurotus eryngii	Yogurt	Prebiotic activity	Chou et al., 2013
Pleurotus ostreatus	Yogurt	Prebiotic activity	Pelaes Vital et al., 2015
Trametes versicolor	Wort fermentation Kombucha	Flavor Immunomodulatory activity	Zhang et al., 2014 Sknepnek et al., 2021
Tricholoma matsutake	Beer	ADH activity, Fibrinolytic activity	Okamura et al., 2001a

important for public health protection strategies. The definition of functional foods is not internationally accepted, but it is difficult to give the precise data regarding the worlds' market of such foods. Nevertheless, it is reported that the sector of functional foods is the fastest growing food industry, while the functional beverages occupied up to 59% of complete US functional food market (Corbo et al., 2014). Application of mushrooms in beverages represents the great opportunity for mushrooms and its producers to satisfy an increasing demand of consumers for the innovative and sensory-pleasant products, with a positive impact on the quality of life.

3.1 Macrofungi in Fermented Beverages

To provide an undisturbed fermentation process of beverages, microorganisms require sufficient amounts of macronutrients, trace elements, and growth factors (Walker and Stewart, 2016). In the production of fermented beverages, yeasts require nitrogen in assimilable organic or inorganic forms. Its role is important in the biosynthesis of structural proteins and enzymes as well as nucleic acids. Nitrogen in the fermentation media also plays an important role in the pleasant flavor development, through the growth of higher alcohols, and avoidance of unpleasant sulfur odors (Lin et al., 2010; Walker and Stewart, 2016). Many researches have proved that mushrooms are a good source of proteins; its content ranges from 20 to 40% on dry weight basis (Wasser, 2005; Sarangi et al., 2006). On the other hand, mushrooms do not contain sufficient amounts of fermentable sugars necessary for yeast fermentation processes, and thus, a carbon source is usually added in the broth before the fermentation (Lin et al., 2010; Li et al., 2014; Sknepnek et al., 2021). Broth prepared from deionized water, cane sugar, and *L. edodes* extract (60 g/L), produced from shiitake stipe, which is usually discarded or sold as a cheap byproduct, contained even higher amounts (287 mg N/L) of yeast assimilable nitrogen (YAN) than it is required (140 mg N/L) for optimal duration of the wine fermentation process. According to this finding, mushroom extracts could be used as a source of nitrogen for the development of novel and functional beverages with unique fragrances. The final ethanol content reached sufficient values required for wines, ranging between 10.9–14.1 mL/100 mL, and its content differed depending on the *Saccharomyces cerevisiae* strain used (Lin et al., 2010). *Hericium erinaceus* (lion's mane) is also a good source of nitrogen which recommends this mushroom as a good base for the application in the fermentation processes. The disadvantage in the application of lion's mane mushroom in the food matrices is its bitter taste, which is also characteristic for many other medicinal mushrooms. In the two-step fermentation process, *H. erinaceus* mushroom powder was first mixed with water and sugar to perform an alcoholic fermentation using *S. cerevisiae* and to produce the *H. erinaceus* wine (HW). The best organoleptic attributes of this wine, considering alcohol flavor, mushroom flavor and overall acceptance was reached when 5% of the mushroom was applied. In the second step, acetic acid fermentation using oxidative activity of the bacteria *Acetobacter aceti* was carried out, and the *H. erinaceus* vinegar (HV) was produced. During this process, similar acidity was developed (4.09%, after 9 days) as it earlier was reached in the cider vinegar, brown rice vinegar, persimmon vinegar, and it was similar to garlic vinegar, proving that this mushroom is a good substrate for the alcoholic and acetic acid fermentation processes (Li et al., 2014). Similarly, fruit bodies of the three different macrofungi (*C. versicolor*, *G. lucidum*, and *L. edodes*) were applied in the intertwined alcoholic and acetic acid fermentation process to produce fermented beverage kombucha, by a symbiotic activity of yeasts and acetic acid bacteria. Mushrooms' nutritive and bioactive components were extracted using hot water, and prior to the fermentation process sucrose was added to obtain liquid part. The souring activity started after the addition of a symbiotic consortium of bacteria and yeasts (SCOBY) from the traditionally produced kombucha. In comparison to some herbal teas that lack purine derivatives, and thus are not suitable for the kombucha fermentation (Velićanski et al., 2014), using medicinal mushroom as a substrate for successful kombucha production was proved. The monitoring of the processes showed that the substrates prepared from the medicinal mushrooms are suitable for the growth of yeasts and acetic acid bacteria and synthesis of higher acetic acid concentrations (22.8 g/L – 33.5 g/L) than in traditional kombucha. As being a decisive factor for the determination of the fermentation end point, higher total acids' content in kombucha beverages from mushrooms lead to the reduction of time for kombucha souring. Chemical analyses showed that mushroom kombucha beverages are a complex mixture of various components including polysaccharides, monosaccharides, ethanol, phenols, flavonoids, proteins, and lipids, whose content depended on the applied mushroom (Sknepnek et al., 2018; Sknepnek et al., 2021). Application of enzymatic hydrolysis with cellulase and pectinase, prior to the fermentation process, might be a good way to degrade macromolecules, as in *Cordyceps militaris* pulp, when this step led to an increased amount of the reducing sugars (Lao et al., 2020). For this purpose, the most effective application of cellulase and pectinase is in the ratio 2:3. In the first part of the fermentation processes, five strains of the lactic acid bacteria (LAB) were involved followed by *S. cerevisiae* fermentation. The content of a polysaccharide—cordycepin, a biologically active substance, increased after enzymatic hydrolysis, but during the fermentation process it was noted that cordycepin amounts were reduced. Nevertheless, the content of cordycepin still remained higher after the process than before. The cordycepin was protected from degradation most probably due to a higher amount of the other nutrients available to the microorganisms; those were released after the enzymatic treatment of the mushroom pulp (Lao et al., 2020).

3.2 Macrofungal Flavor in Beverages

Consumers' choice to include certain foods and beverages in diet is highly influenced by the sensory acceptance of the product (Ignat et al., 2020). Many researchers have focused their attention on flavor development and sensory acceptance of novel beverages using mushrooms. Increased research of food flavors and spices to mask the green flavor, usually appearing in green tea and green tea-containing beverages, was influenced by the rising sales trend of the tea in the European countries. Novel macrofungi beverages might be a source of the natural compounds that fulfill demands of actual trends in the beverage industry for the development of beverages without artificial flavours (Rigling et al., 2021). Basidiomycetes that produce natural flavor compounds can be used to develop novel beverages (Zhang, 2015). To influence a flavor of the beverages, mushrooms are applied in the form of dried fruit bodies, various types of extracts, or their mycelia as substrate prior to the fermentation activity.

The mycelia of various mushrooms (*L. edodes, F. velutipes, P. ostreatus, Kuehneromyces mutabilis, Pholiota nameko, Lepista nuda, T. versicolor, G. frondosa, P. eryngii*, and others) are used in the submerged fermentation processes of green tea infusions. The pleasant, innovative, and attractive aroma and flavors were developed by an action of diverse extracellular enzymes, and were perceived as chocolate-like, cinnamon-like, fruity, bitter almond-like, herbal, honey-like, vanilla-like, and others. Among all the mushrooms, *F. velutipes* (enokitake) was found to be the most favorable one and can be used as a novel tea-based drink. The flavor in this drink was converted from soapy, green, floral, and extremely bitter to savory nutty, without bitterness and with pleasant chocolate-like touch. For the first time, basidiomycetous mushroom biosynthesized 2-ethyl-3,5-dimethylpyrazine, the compound with a sensory impact, which is responsible for nutty and cocoa-like odor, while the bitterness and astringency disappearance might be a result of green tea polyphenols oxidation by enokitate culture (Rigling et al., 2021). In the fermentation process of the wort, 31 basidiomycete mushrooms were applied. The flavor of wort was formed in less than 48 h, because of the wort's complex composition, including fermentable components and flavor precursors. Pleasant odors from *L. edodes, Polyporus umbelaltus*, and *T. versicolor* were characterized as fruity, honey-like odor that originated from *Panellus serotinus* and *T. versicolor*, while addition of mycelium of some of the mushrooms resulted in the formation of an unpleasant odors (i.e., chlorine, medicinal, metallic, moldy, and smoky). Among all mushrooms, *L. edodes* stands out to be used in the application of the novel fermentation process, due to a fresh, fruity, plum-like, sweet and a bit of sour flavor that develops in the final product. As key-odor active compounds, 2-acetylpyrrole, ß-damascenone, (E)-2-nonenal, and 2-phenylethanol were detected. The most important pleasant flavor compound, 2-phenylethanol was transformed by the mushroom from L-phenylalanine, and was detected as a typical rose odor (Zhang et al., 2014). On the other hand, (E)-2-nonenal is the aldehyde that is responsible for the development of an oxidized odor or flavor and for the paper/cardboard character developed in aged beer (Santos et al., 2003). Additionally, in the overall flavor, other compounds were also involved: methyl 2-methylbutanoate, 2-phenylethanol acetate, and (E)-methyl cinnamate (Zhang et al., 2014). For the development of natural flavor compounds in wort-fermented by shiitake mycelium, critical process parameters in a stirred tank bioreactor that are required are volumetric power input (trough agitation rate) and concentration of the inoculum. Production of methyl 2-methylbutanoate, the compound with the highest odor activity, value and fruity flavor, was influenced by mixing the rate and the growth conditions, while higher concentrations of fruity esters were also produced when higher concentration of inoculum was applied (Özdemir et al., 2017). In the beverage made from *T. versicolor* mycelium, 27 aroma compounds were identified. Among them, the highest overall aroma impact that showed was by five compounds: 2-phenylacetaldehyde, ethyl 2-methylpropanoate, linalool, 2,3-butanedione, and methional. Two fruit aroma compounds belonging to ethyl-esters (ethyl 2-methylpropanoate and 2-methylbutanoate) were synthesized for the first time by *T. versicolor* and have been characterized as fruity-strawberry or fruity-green apple-like flavors, respectively. Considering their sensory characteristics, the *L. edodes* and *T. versicolor* are designated as the most promising species for the use in non-alcoholic beverage production (Zhang et al., 2014; Zhang et al., 2015a; Zhang et al., 2015b). *G. lucidum* ethanol extract applied in a Pilsner beer, contributed to the sensorial quality. Detected triterpenes with a bitter taste that originated from the mushroom extract, influenced the development of a more pleasant and harmonic bitterness than hops in a commercial beer, which served as control (Leskosek-Cukalovic et al., 2010). The *G. lucidum* fruit bodies were also added to various distillates (i.e., grain, plum, wine, and grape). After 60 days, influence of the mushroom addition through extraction of its components on the changes in volatile compounds profile and their influence on sensory quality was examined. Higher alcohols and aldehydes were the main detected compounds whose composition depends on a used distillate. Besides alcohols and aldehydes, the highest impact on the sensory quality of the spirits had the presence of ketones, esters, acids, terpens, and phenols. Higher content of esters and lower content of ketones influenced on obtaining higher scores for odor and flavor of the spirits. Simpler sensory profile of distillates was shown to be more suitable to achieve the best sensory characteristics after the addition of *G. lucidum*, while further enrichment of the spirits' volatile fractions was achieved after the addition of herbal ethanolic extracts (Veljović et al., 2019). *Cordyceps militaris* beverage, produced in the two-step fermentative process involving lactic acid bacteria and yeasts, had an increased content of volatile flavor components, and decreased bitterness, which was probably masked by the developed sourness (Lao et al., 2020). Consumers' acceptance of *L. edodes* wine was comparable with the commercial white wine from Taiwan, and was characterized as pleasant in flavor, color, mouth feel, and overall acceptance (Lin et al., 2010). In the development process of a functional *Cordyceps* coffee, green coffee beans of the Arabic variety were immersed in the medicinal mushroom extracts prepared as a complex of *C. militaris*, *Phellinus linteus*, and *Inonotus obliquus*, prior to the roasting process. The application of *C. militaris* alone led to the development of an unpleasant coffee aroma, which was solved by the application of *P. linetus* and *I. obliquus*, as it decreased the unpleasant scent of *C. militaris*. The application of the novel method for making functional *Cordyceps* coffee did not have a negative influence on flavor, maintaining the original coffee aroma (Song, 2020). In a compound beverage made from *Helvella leucopus* (BaChu mushroom) crude polysaccharides, hawthorn and apple juices, the highest impact on flavor expressed the hawthorn juice, followed by the mushroom juice. The good taste, odor and color were achieved when those compounds were combined in ratios: hawthorn juice 45.5%, mushroom polysaccharide juice 36.4%, and apple juice 18.2% (XuJie et al., 2008).

3.3 Macrofungal Nutrients in Beverages

Fermentation process is a good way to produce foods and beverages with improved nutritional value and health benefits, or to remove undesirable components. Microbial and enzymatic activities have an impact on the suppression of biogenic amines and phytic acid, and also reduce the amount of FODMAP components (i.e., fermentable oligosaccharides, disaccharides, monosaccharides, and polyol), improving overall the digestive human health (Lavefve et al., 2019). Various microorganisms are applied to overcome the limitations that arise from the presence of raw bean indigestible components, which belong to the oligosaccharides' family. Application of the *G. lucidum* and *G. frondosa* mycelium in the fermentation process of soy milk, decreased the content of oligosaccharides (e.g., stachyose and raffinose) and crude proteins, enhancing its acceptability. Additionally, application of these mushrooms resulted in the increase of free amino acids, polysaccharides, and B vitamins: niacin, riboflavin, and thiamin. In addition, soy milk is a good source of low-cost proteins suitable for vegetarians and people who suffer from lactose-intolerance and milk-allergies (Yang and Zhang, 2009; Yang et al., 2015). The application of *T. versicolor* mycelium in the wort fermentation for 38 hrs resulted in lowering of oxalic acid concentration, whose presence in beer can influence haze-forming and gushing. Oxalic acid is well known antinutritional factor, which can form insoluble salts in its reaction with calcium, magnesium, or iron. Oxalates have an adverse effect on human health and can induce abdominal pain, nausea, diarrhea or muscle pain, while calcium oxalate accumulates as kidney stones as well (Popova and Mihaylova, 2019). The *F. velutipes* showed remarkable potential to be applied as a natural food ingredient and a substitute for synthetic compounds in the prevention of enzyme-induced food browning. Enzyme-catalyzed browning of food is often prevented by bisulfite addition, which could be harmful for human health, mostly for asthmatic patients. The application of *F. velutipes* acetone or hot water mushroom extracts into apple slices, powder, or apple juice inhibited tyrosinase activity and protected the products from browning. Further investigation has to be focused on the detection of a compound responsible for the potent inhibitory potential of this mushroom on tyrosinase activity (Jang et al., 2002).

3.4 Fermented Alcoholic Beverages

Yeasts are mainly used in the industrial-scale alcohol and alcoholic beverage production from fermentable carbon sources, since they possess the alcohol dehydrogenase (ADH) enzyme, which is required for alcohol production from carbohydrates. The limitations of yeast-based alcohol production originate from inhibitors of yeast's activity, which could be found in a souring substrate, or can be developed during the fermentation process. As the main product of yeast's activity, ethanol in high concentrations (usually over 15%) is one of the most common yeast inhibitors (Mat Isham et al., 2019). Some mushrooms express potent ADH activity and can be used in the production of alcoholic beverages, as well (Okamura et al., 2001a; Okamura et al. 2001b; Matsui et al., 2009). The production of ethanol by some mushrooms is possible in both conditions, aerobic and anaerobic, while *S. cerevisiae* produce alcohol in the absence of oxygen. This indicates that the pathways of converting carbohydrates to pyruvate are not the same. During alcohol fermentation, mushrooms most probably use Embden-Mayerhof-Parans (EMP) and Entner-Doudoroff (ED) pathways, while *S. cerevisiae* uses the EMP pathway (Okamura-Matsui et al., 2003). Okamura et al. (2001a) tested the ADH activities of mushrooms and their potential to be used in beer brewing. *F. velutipes* expressed more pronounced ADH activity and consequently, higher ethanol concentration than *Tricholoma matsutake*. In a further research by Okamura et al. (2001b), the ADH activity was also found in other mushrooms; those were used in wine fermentation processes to test the characteristics of *F. velutipes*, *P. ostreatus*, and *A. blazei* wines. As being more complex organisms than yeasts, mushrooms tolerate higher contents of ethanol. This feature was used to develop an alternative ebullition process, which involves yeast (*S. cerevisiae*) and mushrooms (*P. pulmonarius* and *V. volvacea*), to reach an increased ethanol yield (Mat Isham et al., 2019). The Korean rice wine (yakju) is traditionally produced using *Aspergillus oryze* (koji mold), cooked rice, flour, and *S. cerevisiae*. Kim et al. (2005) studied the development of functionalized yakju with different amounts of *G. lucidum* fruit bodies (0.1 – 2.0%). It was established that the concentration of ethanol was lower in the yakju without application of the *G. lucidum* mushroom. Okamura-Matsui et al. (2003) modified the conventional method for sake brewing, which traditionally involves the action of two microorganisms: *A. oryze* and *S. cerevisiae*. They applied only mushrooms in sake brewing, since they express both the ADH and amylase activities.

3.5 Functional Potential of Macrofungi in Beverages

For the extraction of a bioactive components and macrofungal metabolites, various procedures can be applied, that affect their yield, structure, and the biological activities (Lu et al., 2020a; Zhang et al., 2020). To develop a compound beverage, *Helvella leucopus* (BaChu mushroom) crude polysaccharides were combined with hawthorn and apple juices. Extraction of the mushroom polysaccharides was optimized to maximize the yield, using the optimal temperature (95°C), during 120 min extraction time and a solvent to raw material ratio is 40:1 (XuJie et al., 2008). As previously mentioned, BaChu mushroom polysaccharides express antioxidative, life-prolonging, immunomodulatory, and antitumor activities (Zeng and Zhu, 2018; Zhang et al., 2020). The contribution of the applied polysaccharides in this novel beverage from hawthorn/mushroom/apple juices should be studied on the biological activities. The *A. bisporus* is the most consumed mushroom worldwide with great potential to be applied in food matrices. Polysaccharides and their purified fractions including high amounts of mycosterol (i.e., ergosterol) found in this mushroom influences its bioactivities: antioxidant, antimicrobial, anti-inflammatory, antitumor. and hypocholesterolemic. To overcome the low solubility of *A. bisporus* extracts in lipophilic and hydrophilic media, spray-drying microencapsulated technique was applied in a material with good water compatibility (Francisco et al., 2018; Lu, 2020a). Citric acid-maltodextrin cross-linked microspheres, as encapsulating material for the mushroom extract were shown to be the feasible carrier for the incorporation of the lipophilic *A. bisporus* extract in a hydrophilic formulation (e.g., yogurt). Encapsulation ensures protection and the gradual delivery of the extract in the medium during the storage period. Yogurt functionalized with *A. bisporus* thermally-treated microspheres, expressed the best antioxidant activity, which increased after seven days storage period, due to the thermal treatment coupled with spray drying influenced effective material crosslinking and the protection of the extract (Francisco et al., 2018). Application of the enzymatic treatment of *C. militaris* pulp prior the fermentation by LAB and yeasts, improved the antioxidant activity of the *C. militaris* fermented beverage, reaching the hydroxyl radical scavenging rate

28.76% and reducing power 6.91% which was higher in comparison with the control group (Lao et al., 2020).

3.5.1 Antioxidant Activities

Normal functioning of the biological organisms depends on the balance between the production of free radicals and their elimination and by the act of antioxidants, which prevent cellular lipids, proteins, and DNA damage (Badalyan et al., 2019). The antioxidant activity of mushroom beverages is dependent on the presence of antioxidative compounds in mushrooms, which are extracted from their fruit bodies, mycelium, or cultivation broth. These components are dominantly phenolic compounds, polysaccharides, and vitamins. Mushrooms phenolic compounds are crucial for the neutralization of free radicals. Although many other compounds which are found in mushrooms may also exhibit antioxidant potential, it was proved that phenols are the main carries of the antioxidant activities in *G. tsugae*, regardless of whether the basidiocarp, mycelium, or the culture broth filtrate was assessed (Mau et al., 2005a; Mau et al., 2005b). Like other mushroom bioactive compounds, the total phenolic content can vary not only between genera, but also between species within the same genera. Among the seven tested *Pleurotus* spp. (*P. citrinopileatus*, *P. djamor*, *P. eryngii*, *P. flabellatus*, *P. florida*, *P. ostreatus*, and *P. sajor-caju*), *P. eryngii* was found to have the highest content of phenolics, and it also had the highest DPPH radical scavenging activity (Mishra et al., 2013). In foods, polyphenols are usually present in the form of esters, polymers, or glycosides that are not possible to be absorbed in the small intestine (Kozarski et al., 2015). By the β-glucosidase activity of mushroom mycelia, it is possible to convert glycosides to aglycones during the fermentation process. Isoflavone aglycones content was increased during the fermentation process of soy milk by *G. lucidum* and the *G. frondosa* mycelia. Biological activity was found to be better for aglycones of isoflavones than glycosylated forms, resulting in better prevention of some cancer occurrence or cardiovascular disease (Yang and Zhang, 2009; Yang et al., 2015). Polyphenol content in *H. erinaceus* wine (HW) was higher than in commercial red and white grape wines, and the total polyphenol content in produced *H. erinaceus* vinegar (HV) was higher than in balsamic vinegars. The HW and HV expressed antioxidant activity comparable to other wines and vinegars (Li et al., 2014). Pecic et al. (2016) demonstrated that the total phenolic content increased with the rising concentration of *G. lucidum* during the maceration process in distilled grain brandy and significantly influenced the antioxidant capacities, while the extraction time over 21 days did not have any significant contribution. The pre-treatment of coffee beans by *C. militaris* extracts led to an increase of total polyphenol content in *Cordyceps* coffee compared to *C. militaris* mushroom and control coffee without mushroom, and to an increase of DPPH radicals scavenging activity (Song, 2020). Application of *G. lucidum* in the kombucha beverage resulted in the notably high DPPH scavenging ability (86.58%) with determined half-maximal effective concentration (EC_{50}) at 22.8 mg/mL. Additionally, reducing power of the kombucha beverage was also very high, reaching maximum absorbance of 2.01 and EC_{50} value at 10.61 mg/mL (Sknepnek et al., 2018). Besides, in coffee beans pretreated with *C. militaris* extract, glucans (α- and β-), as well as cordycepin were absorbed during the process, increasing the possibility of this coffee to express other biological activities (Song, 2020). Fermentation of soy milk, using mushroom mycelia, influenced the increase of exopolysaccharide concentration which could contribute to immunomodulatory, antitumor, antioxidative, anti-diabetic, or other biological activities of the product (Yang et al., 2015).

3.5.2 Immunomodulatory and Antimicrobial Activities

Mushroom bioactive components that are able to express immunomodulatory activities are usually commercially available in the form of nutraceuticals or supplements (Reis et al., 2017). Polysaccharide peptides from *C. versicolor* stimulate the production of interferons, interleukins and the tumor necrosis factor (Lee et al., 2006), while *L. edodes* polysaccharide, the lentinan, downregulates anti-inflammatory cytokines and upregulates proinflammatory cytokines (Wang et al., 2012). After the application of *C. versicolor* and *L. edodes*, in the kombucha fermentation processes, the polysaccharide extracts of the fruit body were analyzed for immunomodulatory and cytotoxic activity on human peripheral blood mononuclear cells (PBMC). Mushroom polysaccharides in kombucha expressed highly desirable immunomodulatory properties in human PBMCs without any cytotoxic effect. They inhibited Th2 cytokines (IL-4 and IL-5) and anti-inflammatory cytokine IL-10 indicating that these polysaccharides could be effective in the treatment of allergic reactions, asthma, or atopic dermatitis as typical Th2 mediated immunopathology. It was also assumed that stimulated proinflammatory cytokines (TNF-α and IL-6) and reduced IL-10 could act positively on an organism's defense against viruses or other external pathogens (Sknepnek et al., 2021). Different organic compounds act against microorganisms by various mechanisms, such as cell walls or proteins' synthesis prevention, plasmatic membrane permeability obstruction, or by their interference with chromosome replication (Matijašević et al., 2016). Bacterial infections of humans are a rising threat due to microbial resistance to available drugs. It is widely documented that mushrooms express antimicrobial activity (Matijašević et al., 2016; Soković et al., 2017) and beside their beneficial effect on human health, they might be used as food preservatives. Application of *G. lucidum* in the kombucha beverage showed remarkable inhibitory activity against all tested bacterial species and bactericidal activity against 14 out of 17 Gram-positive or Gram-negative bacteria. This activity is a consequence not only of synthesized acids, but also the mushroom's bioactive components and metabolites that developed during the ebullition process (Sknepnek et al., 2018).

3.5.3 Cardiovascular Disease Prevention

For leading cardiovascular disease (CVD) deaths worldwide, the main recognized risk factors are the high arterial pressure, high glucose, and high cholesterol levels in blood. Owing to the fibrin aggregation in blood, occurrence of thrombosis can also cause the CVD. Proteolytic enzymes from mushrooms can dissolve the fibrin by the action of fibrinolysis (Badalyan et al., 2019). Application of mycelium cultures from *T. matsutake* and *F. velutipes* in a beer production influenced the increase of thrombin time, suggesting that these types of beers could prevent thrombosis. The detected fibrinolytic activity of *T. matsutake* beer may also positively influence in the prevention of thrombosis (Okamura et al., 2001a). A similar effect with great potentiality to prevent thrombosis was reached in wine fermented with the action of *F. velutipes* mycelia (Okamura et al., 2001b). Drugs that are usually used for the inhibition of angiotensin I-converting enzyme (ACE), often express side effects and have limited efficacy. For several mushrooms, it was proved that they express ACE activity inhibition *in vitro* and in several *in vivo* studies (Badalyan et al., 2019). The ACE inhibitory activity of rice wine, yakju, with 0.1% of *G. lucidum* fruit body, increased to 63.4%, which was 10% higher than in yakju without mushroom. The substance with enhancing ACE inhibitory activity was found to be triterpene and ganoderic acid K. This beverage also expressed higher electron-donating ability and superoxide dismutase-like activity in comparison with the control group (Kim et al., 2004).

3.5.4 Macrofungi as Prebiotics

Lactic acid bacteria and bifidobacteria are responsible for the sustenance of the gastrointestinal microbiota, pathogen growth inhibition, and stimulation of normal growth of microbiota after antibiotic therapy. They support human metabolism and control the host immune system activity. Usually, these bacteria are used as test microorganisms for determining the prebiotic effect. They also produce vitamin B, reduce blood anemia, and high cholesterol levels. Lactobacilli are beneficial in the recovery process after *Salmonella* spp. infections, and normal bowel function in lactose-intolerant patients and infants (Aida et al., 2009). To support their growth, oligosaccharides and polysaccharides are mostly applied as prebiotics. Carbohydrates from mushrooms, mainly β-glucans, are resistant to digestive enzymes and acid hydrolysis in the stomach of mammals, for which they are recommended to be used as potential prebiotics (Singdevsachan et al., 2016). The *P. eriyngii* and *F. velutipes* bases, as well as those from *L. edodes* stipes, which are usually discarded as wastes, were used to prepare mushroom-water polysaccharide extracts. These polysaccharides enhanced the survival rate of *L. acidophilus*, *L. casei*, and *Bifidobacterium longum* and expressed a synergistic effect with yogurt culture peptides including amino acids. Consequently, during the 28 days of cold storage, the presence of mushrooms polysaccharides was shown to be the crucial factor for maintaining the number of probiotic cells above 10^7 CFU/mL. In the simulated gastric and bile juice conditions this synergistic effect was also found to act as a protectant on the probiotic cultures, with the best protective effect in the medium containing *L. edodes* polysaccharides. All aforementioned effects recommended mushroom wastes as an inexpensive source of prebiotics (Chou et al., 2013). Application of cellulase and pectinase, prior the fermentation of *C. militaris* pulp, influenced the increase of the reducing sugars and total polysaccharides contents from the mushroom cell wall, providing improved conditions for LAB growth in the *C. militaris* fermented beverage (Lao et al., 2020). Pelaes Vital et al. (2015) showed that water extract of *P. ostreatus* added to a reconstituted skim milk, powder (Elegê, Brazil) stimulated the growth *Streptococcus thermophilus* and *L. bulgaricus*, most probably due to the presence of dietary fibers (i.e., chitin, β- and α-glucans, xylans, mannans, and galactans). Similarly, addition of polysaccharides extracted from *P. eryngii* stimulated the bacterial growth during yogurt fermentation (Li and Shah, 2015).

Hericium erinaceus mushroom was used as a natural substrate for L-glutamic acid (GA) fermentation, and γ-aminobutyric acid (GABA) production. The GA is a precursor of GABA, which is produced by the glutamic acid decarboxylase (GAD) activity. The GABA is a highly desirable compound because of the physiological functions that it expresses, such as antidiabetic, antihypertensive, anticancer, and others (Woraharn et al., 2015). In a mammalian central nervous system, it is the main inhibitory neurotransmitter, and acts by upgrading the plasma concentration, growth hormones, and the brain proteins syntheses (Dhakal et al., 2012). Some lactic acid bacteria (LAB) are capable of producing GABA by the action GAD enzymes. For that purpose, *L. brevis* HP2 and *L. fermentum* HP3 were applied in the development of *H. erinaceus* fermented juice (FHJ) with high GABA amounts. Optimal conditions were set (pH 7 and 40°C) to produce the highest GA concentration (1.55 mg/mL) in the novel mushroom juice. It was noted that the highest influence on GA concentration expressed the cofactor (K_2HPO_4). Further, the production of the GABA was optimized during *H. erinaceus* fermentation by *L. fermentum* HP3 reaching 2.05 mg/mL and optimal conditions were set at pH 6.5, temperature 35°C. The results provided a pronounced impact for further production of the *H. erinaceus* LAB fermented probiotic/prebiotic drink/beverage having a high concentration of GABA (Woraharn et al., 2015).

4. Conclusion and Future Prospects

Economic value and the global production of edible and medicinal mushrooms is constantly increasing. They have been used for centuries in human diet, as a great source of nutrients, or in the traditional medicine owing to their bioactive components and health benefits. Numerous *in vitro* assays, animal and clinical studies have evaluated that mushroom-derived compounds may act as a preventive or in disease control or treatment. Mushrooms could be considered as the most diverse organisms for production of mushroom-based beverages to supplement desired nutrients and health-promoting compounds (nutraceuticals), in addition desired aroma and flavors. To reach a desirable effect, it is necessary to establish standard procedures and sustainable production of mushroom beverages under controlled conditions. It is also significant to determine the proper dosage of the compound of interest and to carry out appropriate *in vivo* studies to clarify the mechanisms and contribution of various bioactive components of a novel beverage. As discussed in this chapter, mushrooms usually have a positive impact on the sensorial profile of the beverage, but studies related to market testing are missing. Future research should be conducted in collaboration with the beverage producers on an industrial scale to develop systems for the mass production of mushroom beverages, which may dramatically increase the market value of beverages.

References

Afiukwa, C.A., Ebem, E.C. and Igwe, D.O. (2015). Characterization of the proximate and amino acid composition of edible wild mushroom species in Abakaliki, Nigeria. *AASCIT J. Biosci.*, 1(2): 20–25.

Aida, F.M.N.A., Shuhaimi, M., Yazid, M. and Maaruf, A.G. (2009). Mushroom as a potential source of prebiotics: A review. *Trends Food Sci. Technol.*, 20(11): 567–575. doi: 10.1016/j.tifs.2009.07.007.

Atila, F., Owaid, M.N. and Shariati, M.A. (2017). The nutritional and medical benefits of *Agaricus bisporus*: A review. *J. Microbiol. Biotechnol. Food Sci.*, 7(3): 281–286. doi: 10.15414/jmbfs.2017/18.7.3.281-286.

Atri, N.S., Sharma, S.K., Joshi, R., Gulati, Ashu and Gulati, A. (2013). Nutritional and neutraceutical composition of five wild culinary-medicinal species of genus *Pleurotus* (higher Basidiomycetes) from northwest India. *Int. J. Med. Mushrooms*, 15(1): 49–56. doi: 10.1615/intjmedmushr.v15.i1.60.

Badalyan, S.M., Barkhudaryan, A. and Rapior, S. (2019). Recent progress in research on the pharmacological potential of mushrooms and prospects for their clinical application. pp. 1–70. *In*: Agrawal, D. and Dhanasekaran, M. (Eds.). *Medicinal Mushrooms*. Singapore: Springer.

Bai, W.-F., Guo, X.-Y., Ma, L.-Q., Guo, L.-Q. and Lin, J.-F. (2013). Chemical composition and sensory evaluation of fermented tea with medicinal mushrooms. *Indian J. Microbiol.*, 53(1): 70–76. doi: 10.1007/s12088-012-0345-0.

Barros, L., Cruz, T., Baptista, P., Estevinho, L.M. and Ferreira, I.C.F.R. (2008). Wild and commercial mushrooms as source of nutrients and nutraceuticals. *Food Chem. Toxicol.*, 46(8): 2742–2747. doi: 10.1016/j.fct.2008.04.030.

Bellettini, M.B., Fiorda, F.A., Maieves, H.A., Teixeira, G.L., Ávila, S. et al. (2019). Factors affecting mushroom *Pleurotus* spp. *Saudi J. Biol. Sci.*, 26(4): 633–646. doi: 10.1016/j.sjbs.2016.12.005.

Bergendiova, K., Tibenska, E. and Majtan, J. (2011). Pleuran (β-glucan from *Pleurotus ostreatus*) supplementation, cellular immune response and respiratory tract infections in athletes. *Eur. J. Appl. Physiol.*, 111(9): 2033–2040. doi: 10.1007/s00421-011-1837-z.

Bouzgarrou, C., Amara, K., Reis, F.S., Barreira, J.C.M., Skhiri, F. et al. (2018). Incorporation of tocopherol-rich extracts from mushroom mycelia into yogurt. *Food Funct.*, 9(6): 3166–3172. doi: 10.1039/c8fo00482j.

Çaglarlrmak, N., Unal, K. and Otles, S. (2002). Nutritional value of edible wild mushrooms collected from the Black Sea region of Turkey. *Micol. Apl. Int.*, 14(1): 1–5.

Chang, S.T. and Miles, P.G. (1989). *Edible Mushrooms and Their Cultivation*. Boca Raton, Florida: CRC Press, p. 345.

Chang, S.T. and Buswell, J.A. (1996). Mushroom nutriceuticals. *World J. Microbiol. Biotechnol.*, 12(5): 473–476. doi: 10.1007/bf00419460.

Chang, S.T. and Miles, P.G. (2004). *Mushrooms : Cultivation, Nutritional Value, Medicinal Effect and Environmental Impact*. Boca Raton, Florida, USA: CRC Press, p. 451.

Chang, S.T. and Hayes, W.A. (2013). *The Biology and Cultivation of Edible Mushrooms*. London: Academic Press, p. 842.

Chen, L. and Cheung, P.C.K. (2014). Mushroom dietary fiber from the fruiting body of *Pleurotus tuber-regium*: fractionation and structural elucidation of nondigestible cell wall components. *J. Agric. Food Chem.*, 62(13): 2891–2899. doi: 10.1021/jf500112j.

Chou, W.T., Sheih, I.C. and Fang, T.J. (2013). The applications of polysaccharides from various mushroom wastes as prebiotics in different systems. *J. Food Sci.*, 78(7): M1041–M1048. doi: 10.1111/1750-3841.12160.

Corbo, M.R., Bevilacqua, A., Petruzzi, L., Casanova, F.P. and Sinigaglia, M. (2014). Functional beverages: the emerging side of functional foods: Commercial trends, research, and health implications. *Compr. Rev. Food Sci. Food Saf.*, 13(6): 1192–1206. doi: 10.1111/1541-4337.12109.

Cui, J. and Chisti, Y. (2003). Polysaccharopeptides of *Coriolus versicolor*: Physiological activity, uses, and production. *Biotechnol. Adv.*, 21(2): 109–122. doi: 10.1016/S0734-9750(03)00002-8.

Deepalakshm, K. and Sankaran, M. (2014). *Pleurotus ostreatus*: An oyster mushroom with nutritional and medicinal properties. *J. Biochem. Technol.*, 5(2): 718–726.

Dermiki, M., Phanphensophon, N., Mottram, D.S. and Methven, L. (2013). Contributions of non-volatile and volatile compounds to the umami taste and overall flavour of shiitake mushroom extracts and their application as flavour enhancers in cooked minced meat. *Food Chem.*, 141(1): 77–83. doi: 10.1016/j.foodchem.2013.03.018.

Dhakal, R., Bajpai, V.K. and Baek, K.H. (2012). Production of gaba (γ-aminobutyric acid) by microorganisms: a review. *Brazilian J. Microbiol.*, 43(4): 1230–1241. doi: 10.1590/s1517-83822012000400001.

Diamantopoulou, P. and Philippoussis, A. (2015). Cultivated mushrooms: Preservation and processing. pp. 495–525. *In*: Sinha, N., Hui, Y.H., Evranuz, E.Ö., Siddiq, M. and Ahmed, J. (Eds.). *Handbook of Vegetable Preservation and Processing*, John Wiley & Sons, Ltd.

Dimitrijevic, M.V., Mitic, V.D., Jovanovic, O.P., Stankov Jovanovic, V.P., Nikolic, J.S. et al. (2018). Comparative study of fatty acids profile in eleven wild mushrooms of Boletacea and Russulaceae families. *Chem. Biodivers.*, 15(1): e1700434. doi: 10.1002/cbdv.201700434.

dos Santos Bazanella, G.C., de Souza, D.F., Castoldi, R., Oliveira, R.F., Bracht, A. and Peralta, R.M. (2013). Production of laccase and manganese peroxidase by *Pleurotus pulmonarius* in solid-state cultures and application in dye decolorization. *Folia Microbiol. (Praha).*, 58(6): 641–647. doi: 10.1007/s12223-013-0253-7.

Duvnjak, D., Pantić, M., Pavlović, V., Nedović, V., Lević, S. et al. (2016). Advances in batch culture fermented *Coriolus versicolor* medicinal mushroom for the production of antibacterial compounds. *Innov. Food Sci. Emerg. Technol.*, 34: 1–8. doi: 10.1016/j.ifset.2015.12.028.

FAO. (2016). *Top 10 Country Production of Macrofungi and Truffles 2016*. Food Agric Organ., UN, Rome.

FAO. (2018). *Harvested Area and Production Quantity of Mushroom around World from 2009 to 2018*. Food Agric Organ., UN, Rome.

Farghaly, F.A. and Mostafa, E.M. (2015). Nutritional value and antioxidants in fruiting bodies of *Pleurotus ostreatus* mushroom. *J. Adv. Biol.*, 7(1): 1146–1152.

Farzana, T., Mohajan, S., Hossain, M. and Ahmed, M.M. (2017). Formulation of a protein and fibre enriched soy-mushroom health drink powder compared to locally available health drink powders. *Malays. J. Nutr.*, 23(1): 129–138.

Ferreira, I.C.F.R., Barros, L. and Abreu, R.M.V. (2009). Antioxidants in wild mushrooms. *Curr. Med. Chem.*, 16(12): 1543–1560. doi: 10.2174/092986709787909587.

Francisco, C.R.L., Heleno, S.A., Fernandes, I.P.M., Barreira, J.C.M., Calhelha, R.C. et al. (2018). Functionalization of yogurts with *Agaricus bisporus* extracts encapsulated in spray-dried maltodextrin crosslinked with citric acid. *Food Chem.*, 245: 845–853. doi: 10.1016/j.foodchem.2017.11.098.

Friedman, M. (2016). Mushroom polysaccharides: Chemistry and antiobesity, antidiabetes, anticancer, and antibiotic properties in cells, rodents, and humans. *Foods*, 5(4): 80. doi: 10.3390/foods5040080.

Giavasis, I. (2014). Bioactive fungal polysaccharides as potential functional ingredients in food and nutraceuticals. *Curr. Opin. Biotechnol.*, 26: 162–173. doi: 10.1016/j.copbio.2014.01.010.

Gibson, C., Park, Y.H., Myoung, K.H., Suh, M.K., McArthur, T. et al. (2006). *The bio-fortification of barley with selenium*. Proceedings of the Institute of Brewery & Distillating (Asia-Pacific Section), 19–24.

Govorushko, S., Rezaee, R., Dumanov, J. and Tsatsakis, A. (2019). Poisoning associated with the use of mushrooms: A review of the global pattern and main characteristics. *Food Chem. Toxicol.*, 128: 267–279. doi: 10.1016/j.fct.2019.04.016.

Győri, J. (2007). Study on the mushroom species *Agaricus blazei* (Murill). *Int. J. Hortic. Sci.*, 13(4): 45–48. doi: 10.31421/ijhs/13/4/772.

Hapuarachchi, K.K., Elkhateeb, W.A., Karunarathna, S.C., Cheng, C.R., Bandara, A.R. et al. (2018). Current status of global *Ganoderma* cultivation, products, industry and market. *Mycosphere*, 9(5): 1025–1052. doi: 10.5943/mycosphere/9/5/6.

Hawksworth, D.L. and Lücking, R. (2017). Fungal diversity revisited: 2.2 to 3.8 million species. pp. 79–95. *In*: Heitman, J., Howlett, B.J., Crous, P.W., Stukenbrock, E.H., James, T.Y. and Gow, N.A.R. (Eds.). *The Fungal Kingdom*. Washington, DC.: ASM Press.

Hobbs, C. (2000). Medicinal Value of *Lentinus edodes* (Berk.) Sing. (Agaricomycetideae): A Literature Review. *Int. J. Med. Mushrooms*, 2(4): 16. doi: 10.1615/intjmedmushr.v2.i4.90.

Hossain, M.S., Alam, N., Amin, S.M.R., Basunia, M.A. and Rahman, A. (2007). Essential fatty acids content of *Pleurotus ostreatus*, *Ganoderma lucidum* and *Agaricus bisporus*. *Bangladesh J. Mushroom*, 1(1): 1–7.

Houston, M.C. and Harper, K.J. (2008). Potassium, magnesium, and calcium: their role in both the cause and treatment of hypertension. *J. Clin. Hypertens.*, 10(7): 3–11. doi: 10.1111/j.1751-7176.2008.08575.x.

Hu, J., Yan, F., Zhang, Z. and Lin, J. (2012). Evaluation of antioxidant and anti-fatigue activities of *Ganoderma lucidum* polysaccharides. *J. Anim. Vet. Adv.*, 11(21): 4040–4044.

Ignat, M.V., Salanţă, L.C., Pop, O.L., Pop, C.R. and Tofană, M. (2020). Current functionality and potential improvements of non-alcoholic fermented cereal beverages. *Foods*, 9(8): 1031. doi: 10.3390/foods9081031.

Jang, M.S., Sanada, A., Ushio, H., Tanaka, M. and Ohshima, T. (2002). Inhibitory effects of 'Enokitake' mushroom extracts on polyphenol oxidase and prevention of apple browning. *LWT - Food Sci. Technol.*, 35(8): 697–702. doi: 10.1006/fstl.2002.0937.

Kakon, A.J., Choudhury, M.B.K. and Saha, S. (2012). Mushroom is an ideal food supplement. *J. Dhaka Natl. Med. Coll. Hosp.*, 18(1): 58–62. doi: 10.3329/jdnmch.v18i1.12243.

Kalač, P. (2012). A review of chemical composition and nutritional value of wild-growing and cultivated mushrooms. *J. Sci. Food Agric.*, 93(2): 209–218. doi: 10.1002/jsfa.5960.

Kertesz, M.A. and Thai, M. (2018). Compost bacteria and fungi that influence growth and development of *Agaricus bisporus* and other commercial mushrooms. *Appl. Microbiol. Biotechnol.*, 102(4): 1639–1650. doi: 10.1007/s00253-018-8777-z.

Khan, M.A. and Tania, M. (2012). Nutritional and medicinal importance of *Pleurotus mushrooms*: An overview. *Food Rev. Int.*, 28(3): 313–329. doi: 10.1080/87559129.2011.637267.

Kim, J.H., Lee, D.H., Lee, S.H., Choi, S.Y. and Lee, J.S. (2004). Effect of *Ganoderma lucidum* on the quality and functionality of Korean traditional rice wine, yakju. *J. Biosci. Bioeng.*, 97(1): 24–28. doi: 10.1016/s1389-1723(04)70160-7.

Kim, J.-H., Hubbard, N.E., Ziboh, V. and Erickson, K.L. (2005). Attenuation of breast tumor cell growth by conjugated linoleic acid via inhibition of 5-lipoxygenase activating protein. *Biochim. Biophys. Acta - Mol. Cell Biol. Lipids*, 1736(3): 244–250. doi: 10.1016/j.bbalip.2005.08.015.

Knop, D., Yarden, O. and Hadar, Y. (2015). The ligninolytic peroxidases in the genus *Pleurotus*: divergence in activities, expression, and potential applications. *Appl. Microbiol. Biotechnol.*, 99(3): 1025–1038. doi: 10.1007/s00253-014-6256-8.

Koutrotsios, G., Kalogeropoulos, N., Stathopoulos, P., Kaliora, A.C. and Zervakis, G.I. (2017). Bioactive compounds and antioxidant activity exhibit high intraspecific variability in *Pleurotus ostreatus* mushrooms and correlate well with cultivation performance parameters. *World J. Microbiol. Biotechnol.*, 33(5). doi: 10.1007/s11274-017-2262-1.

Koyyalamudi, S.R., Jeong, S.-C., Cho, K.Y. and Pang, G. (2009). Vitamin B12 is the active corrinoid produced in cultivated white button mushrooms (*Agaricus bisporus*). J. *Agric. Food Chem.*, 57(14): 6327–6333. doi: 10.1021/jf9010966.

Kozarski, M., Klaus, A., Jakovljevic, D., Todorovic, N., Vunduk, J. et al. (2015). Antioxidants of edible mushrooms. *Molecules*, 20(10): 19489–19525. doi: 10.3390/molecules201019489.

Lai, T., Gao, Y. and Zhou, S. (2004). Global marketing of medicinal Ling Zhi mushroom *Ganoderma lucidum* (W.Curt.:Fr.) Lloyd (Aphyllophoromycetideae) products and safety concerns. *Int. J. Med. Mushrooms*, 6(2): 189–194. doi: 10.1615/intjmedmushr.v6.i2.100.

Lao, Y., Zhang, M., Li, Z. and Bhandari, B. (2020). A novel combination of enzymatic hydrolysis and fermentation: Effects on the flavor and nutritional quality of fermented *Cordyceps militaris* beverage. *LWT-Food Sci. Technol.*, 120: 108934. doi: 10.1016/j.lwt.2019.108934.

Lavefve, L., Marasini, D. and Carbonero, F. (2019). Microbial ecology of fermented vegetables and non-alcoholic drinks and current knowledge on their impact on human health. *Adv. Food Nutr. Res.*, 87: 147–185. doi: 10.1016/bs.afnr.2018.09.001.

Lee, C.-L., Yang, X. and Wan, J.M.-F. (2006). The culture duration affects the immunomodulatory and anticancer effect of polysaccharopeptide derived from *Coriolus versicolor*. *Enzyme Microb. Technol.*, 38(1-2): 14–21. doi: 10.1016/j.enzmictec.2004.10.009.

Leskosek-Cukalovic, I., Despotovic, S., Lakic, N., Niksic, M., Nedovic, V. and Tesevic, V. (2010). *Ganoderma lucidum*: Medical mushroom as a raw material for beer with enhanced functional properties. *Food Res. Int.*, 43(9): 2262–2269. doi: 10.1016/j.foodres.2010.07.014.

Li, S. and Shah, N.P. (2015). Effects of *Pleurotus eryngii* polysaccharides on bacterial growth, texture properties, proteolytic capacity, and angiotensin-I-converting enzyme–inhibitory activities of fermented milk. *J. Dairy Sci.*, 98(5): 2949–2961. doi: 10.3168/jds.2014-9116.

Li, T., Lo, Y.M. and Moon, B. (2014). Feasibility of using *Hericium erinaceus* as the substrate for vinegar fermentation. *LWT - Food Sci. Technol.*, 55(1): 323–328. doi: 10.1016/j.lwt.2013.07.018.

Lin, F.-Y., Lai, Y.-K., Yu, H.-C., Chen, N.-Y., Chang, C.-Y. et al. (2008). Effects of *Lycium barbarum* extract on production and immunomodulatory activity of the extracellular polysaccharopeptides from submerged fermentation culture of *Coriolus versicolor*. *Food Chem.*, 110(2): 446–453. doi: 10.1016/j.foodchem.2008.02.023.

Lin, P.-H., Huang, S.-Y., Mau, J.-L., Liou, B.-K. and Fang, T.J. (2010). A novel alcoholic beverage developed from shiitake stipe extract and cane sugar with various *Saccharomyces strains*. *LWT - Food Sci. Technol.*, 43(6): 971–976. doi: 10.1016/j.lwt.2010.02.006.

Lisiecka, J., Sobieralski, K., Siwulski, M. and Jasinska, A. (2013). Almond mushroom *Agaricus brasiliensis* (Wasser et al.)–properties and culture conditions. *Acta Sci. Pol. Hortorum Cultus*, 12(1): 27–40.

Liu, Y., Fukuwatari, Y., Okumura, K., Takeda, K., Ishibashi, K. et al. (2008). Immunomodulating activity of *Agaricus brasiliensis* KA21 in mice and in human volunteers: Evidence-based complement. *Altern. Med.*, 5(2): 205–219. doi: 10.1093/ecam/nem016.

Llarena-Hernández, R.C., Largeteau, M.L., Farnet, A.-M., Foulongne-Oriol, M., Ferrer, N. et al. (2013). Potential of European wild strains of *Agaricus subrufescens* for productivity and quality on wheat straw based compost. *World J. Microbiol. Biotechnol.*, 29(7): 1243–1253. doi: 10.1007/s11274-013-1287-3.

Lu, C.-C., Hsu, Y.-J., Chang, C.-J., Lin, C.-S., Martel, J. et al. (2016). Immunomodulatory properties of medicinal mushrooms: differential effects of water and ethanol extracts on NK cell-mediated cytotoxicity. *Innate Immun.*, 22(7): 522–533. doi: 10.1177/1753425916661402.

Lu, H., Lou, H., Hu, J., Liu, Z. and Chen, Q. (2020a). Macrofungi: A review of cultivation strategies, bioactivity, and application of mushrooms. *Compr. Rev. Food Sci. Food Saf.*, 19(5): 2333–2356. doi: 10.1111/1541-4337.12602.

Lu, J., He, R., Sun, P., Zhang, F., Linhardt, R.J. and Zhang, A. (2020b). Molecular mechanisms of bioactive polysaccharides from *Ganoderma lucidum* (Lingzhi): A review. *Int. J. Biol. Macromol.*, 150: 765–774. doi: 10.1016/j.ijbiomac.2020.02.035.

Lu, X.X., Xie, C.H. and Gu, Z.X. (2009). Optimisation of fermentative parameters for GABA enrichment by *Lactococcus lactis*. *Czech J. Food Sci.*, 27(6): 433–442. doi: 10.17221/45/2009-cjfs.

Lübbert, A. (2003). Optimization of bioprocesses. pp. 365–381. *In*: Berovič, M. and Kieran, P. (Eds.). *Bioprocess Engineering*, University of Ljubljana, Faculty of Chemistry and Chemical Technology.

Maftoun, P., Johari, H.J., Soltani, M., Malik, R., Othman, N.Z. and El Enshasy, H.A. (2015). The edible mushroom *Pleurotus* spp.: I. Biodiversity and nutritional values. *Int. J. Biotechnol. Wellness Ind.*, 4(2): 67–83. doi: 10.6000/1927-3037.2015.04.02.4.

Maga, J.A. (1981). Mushroom flavor. *J. Agric. Food Chem.*, 29(1): 1–4. doi: 10.1021/jf00103a001.

Maggi, F., Papa, F., Cristalli, G., Sagratini, G. and Vittori, S. (2010). Characterisation of the mushroom-like flavour of *Melittis melissophyllum* L. subsp. *melissophyllum* by headspace solid-phase microextraction (HS-SPME) coupled with gas chromatography (GC–FID) and gas chromatography–mass spectrometry (GC–MS). *Food Chem.*, 123(4): 983–992. doi: 10.1016/j.foodchem.2010.05.049.

Mamiro, D.P., Royse, D.J. and Beelman, R.B. (2007). Yield, size, and mushroom solids content of *Agaricus bisporus* produced on non-composted substrate and spent mushroom compost. *World J. Microbiol. Biotechnol.*, 23(9): 1289–1296. doi: 10.1007/s11274-007-9364-0.

Mane, V.P., Patil, S.S., Syed, A.A. and Baig, M.M.V. (2007). Bioconversion of low quality lignocellulosic agricultural waste into edible protein by *Pleurotus sajor-caju* (Fr.) singer. *J. Zhejiang Univ. Sci. B*, 8(10): 745–751. doi: 10.1631/jzus.2007.b0745.

Mat Isham, N.K., Mokhtar, N., Fazry, S. and Lim, S.J. (2019). The development of an alternative fermentation model system for vinegar production. *LWT - Food Sci. Technol.*, 100: 322–327. doi: 10.1016/j.lwt.2018.10.065.

Matijašević, D., Pantić, M., Rašković, B., Pavlović, V., Duvnjak, D. et al. (2016). The Antibacterial activity of *Coriolus versicolor* methanol extract and its effect on ultrastructural changes of *Staphylococcus aureus* and *Salmonella* enteritidis. *Front. Microbiol.*, 7. doi: 10.3389/fmicb.2016.01226.

Matsui, T., Kagemori, T., Fukuda, S., Ohsugi, M. and Tabata, M. (2009). Characteristics of wine produced by mushroom fermentation using *Schizophyllum commune* NBRC4929. *Mushroom Sci. Biotechnol.*, 17(3): 107–111.

Mattila, P., Könkö, K., Eurola, M., Pihlava, J.-M., Astola, J. et al. (2001). Contents of vitamins, mineral elements, and some phenolic compounds in cultivated mushrooms. *J. Agric. Food Chem.*, 49(5): 2343–2348. doi: 10.1021/jf001525d.

Mau, J.-L., Chyau, C.-C., Li, J.-Y. and Tseng, Y.-H. (1997). Flavor compounds in straw mushrooms *Volvariella volvacea* harvested at different stages of maturity. *J. Agric. Food Chem.*, 45(12): 4726–4729. doi: 10.1021/jf9703314.

Mau, J.-L., Wu, K.-T., Wu, Y.-H. and Lin, Y.-P. (1998). Nonvolatile taste components of ear mushrooms. *J. Agric. Food Chem.*, 46(11): 4583–4586. doi: 10.1021/jf9805606.

Mau, J.-L., Lin, H.-C. and Chen, C.-C. (2001). Non-volatile components of several medicinal mushrooms. *Food Res. Int.*, 34(6): 521–526. doi: 10.1016/s0963-9969(01)00067-9.

Mau, J.-L., Tsai, S.-Y., Tseng, Y.-H. and Huang, S.-J. (2005a). Antioxidant properties of hot water extracts from *Ganoderma tsugae* Murrill. *LWT - Food Sci. Technol.*, 38(6): 589–597. doi: 10.1016/j.lwt.2004.08.010.

Mau, J.-L., Tsai, S.-Y., Tsheng, Y.-H. and Huang, S.-J. (2005b). Antioxidant properties of methanolic extracts from *Ganoderma tsugae*. *Food Chem.*, 93(4): 641–649. doi: 10.1016/j.foodchem.2004.10.043.

Meenu, M. and Xu, B. (2019). Application of vibrational spectroscopy for classification, authentication and quality analysis of mushroom: A concise review. *Food Chem.*, 289: 545–557. doi: 10.1016/j.foodchem.2019.03.091.

Meng, X., Liang, H. and Luo, L. (2016). Antitumor polysaccharides from mushrooms: A review on the structural characteristics, antitumor mechanisms and immunomodulating activities. *Carbohydr. Res.*, 424: 30–41. doi: 10.1016/j.carres.2016.02.008.

Mešić, A., Šamec, D., Jadan, M., Bahun, V. and Tkalčec, Z. (2020). Integrated morphological with molecular identification and bioactive compounds of 23 Croatian wild mushrooms samples. *Food Biosci.*, 37: 100720. doi: 10.1016/j.fbio.2020.100720.

Miletić, D., Turło, J., Podsadni, P., Pantić, M., Nedović, V. et al. (2019). Selenium-enriched *Coriolus versicolor* mushroom biomass: potential novel food supplement with improved selenium bioavailability. *J. Sci. Food Agric.*, doi: 10.1002/jsfa.9756.

Miletić, D., Turło, J., Podsadni, P., Sknepnek, A., Szczepańska, A. et al. (2020a). Turkey tail medicinal mushroom, *Trametes versicolor* (Agaricomycetes), crude exopolysaccharides with antioxidative activity. *Int. J. Med. Mushrooms*, 22(9): 885–895. doi: 10.1615/intjmedmushrooms.2020035877.

Miletić, D., Pantić, M., Sknepnek, A., Vasiljević, I., Lazović, M. and Nikšić, M. (2020b). Influence of selenium yeast on the growth, selenium uptake and mineral composition of *Coriolus versicolor* mushroom. *J. Basic Microbiol.*, 60(4): 331–340. doi: 10.1002/jobm.201900520.

Miletić, D., Turło, J., Podsadni, P., Sknepnek, A., Szczepańska, A. et al. (2021). Production of bioactive selenium enriched crude exopolysaccharides via selenourea and sodium selenite bioconversion using *Trametes versicolor*. *Food Biosci.*, 42: 101046. doi: 10.1016/j.fbio.2021.101046.

Mishra, K.K., Pal, R.S., ArunKumar, R., Chandrashekara, C., Jain, S.K. and Bhatt, J.C. (2013). Antioxidant properties of different edible mushroom species and increased bioconversion efficiency of *Pleurotus eryngii* using locally available casing materials. *Food Chem.*, 138(2-3): 1557–1563. doi: 10.1016/j.foodchem.2012.12.001.

Morris, H.J., Llauradó, G., Beltrán, Y., Lebeque, Y., Bermúdez, R.C. et al. (2017). The use of mushrooms in the development of functional foods, drugs, and nutraceuticals. pp.123–159. *In*: Ferreira, I.C.F.R., Morales, P. and Barros, L. (Eds.). *Wild Plants, Mushrooms and Nuts: Functional Food Properties and Application*. United Kingdom: John Wiley & Sons, Ltd.

Mukhopadhyay, R. and Guha, A.K. (2015). A comprehensive analysis of the nutritional quality of edible mushroom *Pleurotus sajor-caju* gron in deproteinized whey medium. *LWT - Food Sci. Technol.*, 61(2): 339–345. doi: 10.1016/j.lwt.2014.12.055.

Niego, A.G., Rapior, S., Thongklang, N., Raspé, O., Jaidee, W. et al. (2021). Macrofungi as a nutraceutical source: promising bioactive compounds and market value. *J. Fungi*, 7(5): 397. doi: 10.3390/jof7050397.

Obodai, M., Cleland-Okine, J. and Vowotor, K.A. (2003). Comparative study on the growth and yield of *Pleurotus ostreatus* mushroom on different lignocellulosic by-products. *J. Ind. Microbiol. Biotechnol.*, 30(3): 146–149. doi: 10.1007/s10295-002-0021-1.

Ohwada, S., Ogawa, T., Makita, F., Tanahashi, Y., Ohya, T. et al. (2006). Beneficial effects of protein-bound polysaccharide K plus tegafur/uracil in patients with stage II or III colorectal cancer: Analysis of immunological parameters. *Oncol. Rep.*, 15(4): 861–868. doi: 10.3892/or.15.4.861.

Okamura, T., Ogata, T., Minamimoto, N., Takeno, T., Noda, H. et al. (2001a). Characteristics of beer-like drink produced by mushroom fermentation. *Food Sci. Technol. Res.*, 7(1): 88–90. doi: 10.3136/fstr.7.88.

Okamura, T., Ogata, T., Minamoto, N., Takeno, T., Noda, H. et al. (2001b). Characteristics of wine produced by mushroom fermentation. *Biosci. Biotechnol. Biochem.*, 65(7): 1596–1600. doi: 10.1271/bbb.65.1596.

Okamura-Matsui, T., Tomoda, T., Fukuda, S. and Ohsugi, M. (2003). Discovery of alcohol dehydrogenase from mushrooms and application to alcoholic beverages. *J. Mol. Catal. B Enzym.*, 23(2–6): 133–144. doi: 10.1016/s1381-1177(03)00079-1.

Özdemir, S., Heerd, D., Quitmann, H., Zhang, Y., Fraatz, M.A. et al. (2017). Process parameters affecting the synthesis of natural flavors by Shiitake (*Lentinula edodes*) during the production of a non-alcoholic beverage. *Beverages*, 3(4): 20. doi: 10.3390/beverages3020020.

Papaspyridi, L.-M., Aligiannis, N., Topakas, E., Christakopoulos, P., Skaltsounis, A.-L. and Fokialakis, N. (2012). Submerged fermentation of the edible mushroom *Pleurotus ostreatus* in a batch stirred tank bioreactor as a promising alternative for the effective production of bioactive metabolites. *Molecules*, 17(3): 2714–2724. doi: 10.3390/molecules17032714.

Park, E.J., Lee, S.O. and Lee, S.P. (2017). Development of natural fermented seasoning with *Flammulina velutipes* powder fortified with γ-aminobutyric acid (GABA) by lactic acid fermentation. *Korean J. Food Preserv.*, 24(2): 237–245. doi: 10.11002/kjfp.2017.24.2.237.

Pecic, S., Nikicevic, N., Veljovic, M., Jardanin, M., Tesevic, V. et al. (2016). The influence of extraction parameters on physicochemical properties of special grain brandies with *Ganoderma lucidum*. *Chem. Ind. Chem. Eng. Q.*, 22(2): 181–189. doi: 10.2298/ciceq150426033p.

Pelaes Vital, A.C., Goto, P.A., Hanai, L.N., Gomes-da-Costa, S.M., de Abreu Filho, B.A. et al. (2015). Microbiological, functional and rheological properties of low fat yogurt supplemented with *Pleurotus ostreatus* aqueous extract. *LWT - Food Sci. Technol.*, 64(2): 1028–1035. doi: 10.1016/j.lwt.2015.07.003.

Pérez-Corona, M.T., Sánchez-Martínez, M., Valderrama, M.J., Rodríguez, M.E., Cámara, C. and Madrid, Y. (2011). Selenium biotransformation by *Saccharomyces cerevisiae* and *Saccharomyces bayanus* during white wine manufacture: Laboratory-scale experiments. *Food Chem.*, 124(3): 1050–1055. doi: 10.1016/j.foodchem.2010.07.073.

Phat, C., Moon, B. and Lee, C. (2016). Evaluation of umami taste in mushroom extracts by chemical analysis, sensory evaluation, and an electronic tongue system. *Food Chem.*, 192: 1068–1077. doi: 10.1016/j.foodchem.2015.07.113.

Poojary, M.M., Orlien, V., Passamonti, P. and Olsen, K. (2017). Improved extraction methods for simultaneous recovery of umami compounds from six different mushrooms. *J. Food Compos. Anal.*, 63: 171–183. doi: 10.1016/j.jfca.2017.08.004.

Popova, A. and Mihaylova, D. (2019). Antinutrients in plant-based foods: A review. *Open Biotechnol. J.*, 13(1): 68–76. doi: 10.2174/1874070701913010068.

Rajapakse, P.A.L.I.T.H.A. (2011). *New cultivation technology for paddy straw mushroom (Volvariella volvacea)*. Proceedings of the 7th International Conference on Mushroom Biology and Mushroom Products (ICMBMP7), pp. 446–451.

Rathore, H., Prasad, S. and Sharma, S. (2017). Mushroom nutraceuticals for improved nutrition and better human health: A review. *PharmaNutrition*, 5(2): 35–46. doi: doi: 10.1016/j.phanu.2017.02.001.

Rayman, M.P. (2000). The importance of selenium to human health. *The Lancet*, 356(9225): 233–241. doi: 10.1016/S0140-6736(00)02490-9.

Reis, F.S., Barros, L., Martins, A. and Ferreira, I.C.F.R. (2012). Chemical composition and nutritional value of the most widely appreciated cultivated mushrooms: An inter-species comparative study. *Food Chem. Toxicol.*, 50(2): 191–197. doi: 10.1016/j.fct.2011.10.056.

Reis, F.S., Martins, A., Vasconcelos, M.H., Morales, P. and Ferreira, I.C.F.R. (2017). Functional foods based on extracts or compounds derived from mushrooms. *Trends Food Sci. Technol.*, 66: 48–62. doi: 10.1016/j.tifs.2017.05.010.

Rigling, M., Yadav, M., Yagishita, M., Nedele, A.-K., Sun, J. and Zhang, Y. (2021). Biosynthesis of pleasant aroma by enokitake (*Flammulina velutipes*) with a potential use in a novel tea drink. *LWT - Food Sci. Technol.*, 140: 110646. doi: 10.1016/j.lwt.2020.110646.

Rop, O., Mlcek, J. and Jurikova, T. (2009). Beta-glucans in higher fungi and their health effects. *Nutr. Rev.*, 67(11): 624–631. doi: 10.1111/j.1753-4887.2009.00230.x.

Royse, D.J., Baars, J. and Tan, Q. (2017). Current overview of mushroom production in the world. pp. 5–13. *In*: Zied, D.C. and Padro-Giménez, A. (Eds.). *Edible and Medicinal Mushrooms Technology and Application*. United Kingdom: John Wiley & Sons Ltd.

Saiqa, S., Haq, N.B., Muhammad, A.H., Muhammad, A.A. and Ata, U.R. (2008). Studies on chemical composition and nutritive evaluation of wild edible mushrooms. *Iran. J. Chem. Chem. Eng.*, 27(3): 151–154.

Sánchez, C. (2010). Cultivation of *Pleurotus ostreatus* and other edible mushrooms. *Appl. Microbiol. Biotechnol.*, Springer, 1321–1337. doi: 10.1007/s00253-009-2343-7.

Sánchez-Martínez, M., da Silva, E.G.P., Pérez-Corona, T., Cámara, C., Ferreira, S.L.C. and Madrid, Y. (2012). Selenite biotransformation during brewing: Evaluation by HPLC–ICP-MS. *Talanta*, 88: 272–276. doi: 10.1016/j.talanta.2011.10.041.

Sanodiya, B.S., Thakur, G.S., Baghel, R.K., Prasad, G.B.K.S. and Bisen, P.S. (2009). *Ganoderma lucidum*: A potent pharmacological macrofungus. *Curr. Pharm. Biotechnol.*, 10(8): 717–742. doi: 10.2174/138920109789978757.

Santos, J.R., Carneiro, J.R., Guido, L.F., Almeida, P.J., Rodrigues, J.A. and Barros, A.A. (2003). Determination of E-2-nonenal by high-performance liquid chromatography with UV detection: Assay for the evaluation of beer ageing. *J. Chromatogr. A*, 985(1-2): 395–402. doi: 10.1016/s0021-9673(02)01396-1.

Sarangi, I., Ghosh, D., Bhutia, S.K., Mallick, S.K. and Maiti, T.K. (2006). Anti-tumor and immunomodulating effects of *Pleurotus ostreatus* mycelia-derived proteoglycans. *Int. Immunopharmacol.*, 6(8): 1287–1297. doi: 10.1016/j.intimp.2006.04.002.

El Sebaaly, Z., Assadi, F., Najib Sassine, Y. and Shaban, N. (2019). Substrate types effect on nutritional composition of button mushroom (*Agaricus bisporus*). *Agriculture For.*, 65(1): 73–80. doi: 10.17707/agricultforest.65.1.08.

El Sheikha, A.F. and Hu, D.-M. (2018). How to trace the geographic origin of mushrooms? *Trends Food Sci. Technol.*, 78: 292–303. doi: 10.1016/j.tifs.2018.06.008.

Shi, L. (2016). Bioactivities, isolation and purification methods of polysaccharides from natural products: A review. *Int. J. Biol. Macromol.*, 92: 37–48. doi: 10.1016/j.ijbiomac.2016.06.100.

Singdevsachan, S.K., Auroshree, P., Mishra, J., Baliyarsingh, B., Tayung, K. and Thatoi, H. (2016). Mushroom polysaccharides as potential prebiotics with their antitumor and immunomodulating properties: A review. *Bioact. Carbohydrates Diet. Fibre*, 7(1): 1–14. doi: 10.1016/j.bcdf.2015.11.001.

Sknepnek, A., Pantić, M., Matijašević, D., Miletić, D., Lević, S. et al. (2018). Novel kombucha beverage from lingzhi or reishi medicinal mushroom, *Ganoderma lucidum*, with antibacterial and antioxidant effects. *Int. J. Med. Mushrooms*, 20(3): 243–258. doi: 10.1615/intjmedmushrooms.2018025833.

Sknepnek, A., Tomić, S., Miletić, D., Lević, S., Čolić, M. et al. (2021). Fermentation characteristics of novel *Coriolus versicolor* and *Lentinus edodes* kombucha beverages and immunomodulatory potential of their polysaccharide extracts. *Food Chem.*, 342: 128344. doi: 10.1016/j.foodchem.2020.128344.

Smiderle, F.R., Alquini, G., Tadra-Sfeir, M.Z., Iacomini, M., Wichers, H.J. and Van Griensven, L.J.L.D. (2013). *Agaricus bisporus* and *Agaricus brasiliensis* (1→6)-β-d-glucans show immunostimulatory activity on human THP-1 derived macrophages. *Carbohydr. Polym.*, 94(1): 91–99. doi: 10.1016/j.carbpol.2012.12.073.

Smith, J.A., Rowan, N.J. and Sullivan, R. (2002). Medicinal mushrooms: a rapidly developing area of biotechnology for cancer therapy and other bioactivities. *Biotechnol. Lett.*, 24(22): 1839–1845.

Sohretoglu, D. and Huang, S. (2018). *Ganoderma lucidum* polysaccharides as an anti-cancer agent. *Anticancer. Agents Med. Chem.*, 18(5): 667–674. doi: 10.2174/1871520617666171113121246.

Soković, M., Ćirić, A., Glamočlija, J. and Stojković, D. (2017). The bioactive properties of mushrooms. pp. 83–122. *In*: Fereira, I.C.F.R., Morales, P. and Barros, L. (Eds.). *Wild Plants, Mushrooms and Nuts: Functional food properties and applications*, John Wiley & Sons, Ltd.

Song, H.-N. (2020). Functional *Cordyceps* coffee containing cordycepin and β-glucan. *Prev. Nutr. Food Sci.*, 25(2): 184–193. doi: 10.3746/pnf.2020.25.2.184.

Takahashi, M., Isogai, A., Utsunomiya, H., Nakano, S., Koizumi, T. and Totsuka, A. (2007). Change in the aroma of sake koji during koji-making. *J. Brew. Soc. Japan*, 102(5): 403–411. doi: 10.6013/jbrewsocjapan1988.102.403.

Takamura, K., Hoshino, H., Sugahara, T. and Amano, H. (1991). Determination of vitamin D2 in shiitake mushroom (*Lentinus edodes*) by high-performance liquid chromatography. *J. Chromatogr. A*, 545(1): 201–204. doi: 10.1016/s0021-9673(01)88709-4.

Tešanović, K., Pejin, B., Šibul, F., Matavulj, M., Rašeta, M. et al. (2017). A comparative overview of antioxidative properties and phenolic profiles of different fungal origins: fruiting bodies and submerged cultures of *Coprinus comatus* and *Coprinellus truncorum*. *J. Food Sci. Technol.*, 54(2): 430–438. doi: 10.1007/s13197-016-2479-2.

Thongklang, N. and Luangharn, T. (2016). Testing agricultural wastes for the production of *Pleurotus ostreatus*. *Mycosphere*, 7(6): 766–772. doi: 10.5943/mycosphere/7/6/6.

Tsai, S.-Y., Wu, T.-P., Huang, S.-J. and Mau, J.-L. (2007). Nonvolatile taste components of *Agaricus bisporus* harvested at different stages of maturity. *Food Chem.*, 103(4): 1457–1464. doi: 10.1016/j.foodchem.2006.10.073.

Turło, J., Gutkowska, B. and Malinowska, E. (2007). Relationship between the selenium, selenomethionine, and selenocysteine content of submerged cultivated mycelium of *Lentinula edodes* (Berk.). *Acta Chromatogr.*, 18: 36–48.

Turło, J. (2014). The biotechnology of higher fungi—current state and perspectives. *Folia Biol. Oecologica*, 10(1): 49–65. doi: 10.2478/fobio-2014-0010.

Ulziijargal, E. and Mau, J.-L. (2011). Nutrient compositions of culinary-medicinal mushroom fruiting bodies and mycelia. *Int. J. Med. Mushrooms*, 13(4): 343–349. doi: 10.1615/intjmedmushr.v13.i4.40.

Urbain, P., Singler, F., Ihorst, G., Biesalski, H.-K. and Bertz, H. (2011). Bioavailability of vitamin D2 from UV-B-irradiated button mushrooms in healthy adults deficient in serum 25-hydroxyvitamin D: A randomized controlled trial. *Eur. J. Clin. Nutr.*, 65(8): 965–971. doi: 10.1038/ejcn.2011.53.

Uryu, T., Ikushima, N., Katsuraya, K., Shoji, T., Takahashi, N. et al. (1992). Sulfated alkyl oligosaccharides with potent inhibitory effects on human immunodeficiency virus infection. *Biochem. Pharmacol.*, 43(11): 2385–2392. doi: 10.1016/0006-2952(92)90317-c.

US FDAs Center for Food Safety and Applied Nutrition. (2008). *US FDA/CFSAN – EAFUS List.*

Valverde, M.E., Hernández-Pérez, T. and Paredes-López, O. (2015). Edible mushrooms: improving human health and promoting quality life. *Int. J. Microbiol.*, 2015: 1–14. doi: 10.1155/2015/376387.

Velićanski, A.S., Cvetković, D.D., Tumbas Šaponjac, V.T. and Vulić, J.J. (2014). Antioxidant and antibacterial activity of the beverage obtained by fermentation of sweetened lemon balm (*Melissa officinalis* L.) tea with symbiotic consortium of bacteria and yeasts. *Food Technol. Biotechnol.*, 52(4): 420–429. doi: 10.17113/ftb.52.04.14.3611.

Veljović, S.P., Tomić, N.S., Belović, M.M., Nikićević, M.P. and Vukosavljević, P.V. (2019). Volatile composition, colour, and sensory quality of spirit-based beverages enriched with medicinal fungus *Ganoderma lucidum* and herbal extract. *Food Technol. Biotechnol.*, 57(3): 408–417. doi: 10.17113/ftb.57.03.19.6106.

Vetter, J. (2019). Biological values of cultivated mushrooms: A review. *Acta Aliment.*, 48(2): 229–240. doi: 10.1556/066.2019.48.2.11.

Walker, G.M. and Stewart, G.G. (2016). *Saccharomyces cerevisiae* in the production of fermented beverages. *Beverages*, 2(4): 30. 10.3390/beverages2040030.

Wang, H., Fu, Z. and Han, C. (2013). The medicinal values of culinary-medicinal royal sun mushroom (*Agaricus blazei Murrill*). *Evidence-based Complement. Altern. Med.*, 2013: 1–6. 10.1155/2013/842619.

Wang, J.-L., Zheng, B., Zou, J.-W. and Gu, X.-M. (2012). Combination therapy with lentinan improves outcomes in patients with esophageal carcinoma. *Mol. Med. Rep.*, 5(3): 745–748. 1 doi: 0.3892/mmr.2011.718.

Wani, B.A., Bodha, R.H. and Wani, A.H. (2010). Nutritional and medicinal importance of mushrooms. *J. Med. Plants Res.*, 4(24): 2598–2604. doi: 10.5897/jmpr09.565.

Wasser, S.P. (2005). Shiitake (*Lentinus edodes*). pp. 653–664. *In*: Coates, P.M., Blackman, M.R., Cragg, G., Levine, M., Moss, J., and White, J. (Eds.). *Encyclopedia of Dietary Supplements*. NY: Marcel Dekker.

Wilson, T. and Temple, N.J. (2016). How beverages impact health and nutrition. pp. 3–9. *In*: Wilson, T. and Temple, N.J. (Eds.). *Beverage Impacts on Health and Nutrition*. Cham.: Humana Press.

Woraharn, S., Lailerd, N., Sivamaruthi, B.S., Wangcharoen, W., Sirisattha, S. et al. (2015). Evaluation of factors that influence the L-glutamic and γ-aminobutyric acid production during

Hericium erinaceus fermentation by lactic acid bacteria. *CyTA - J. Food,* 14(1): 47–54. doi: 10.1080/19476337.2015.1042525.

XuJie, H., Na, Z., SuYing, X., ShuGang, L. and BaoQiu, Y. (2008). Extraction of BaChu mushroom polysaccharides and preparation of a compound beverage. *Carbohydr. Polym.*, 73(2): 289–294. doi: 10.1016/j.carbpol.2007.11.033.

Yamaguchi, S., Yoshikawa, T., Ikeda, S. and Ninomiya, T. (1971). Measurement of the relative taste intensity of some l-α-amino acid and 5'-nucleotides. *J. Food Sci.*, 36(6): 846–849. doi: 10.1111/j.1365-2621.1971.tb15541.x.

Yang, H. and Zhang, L. (2009). Changes in some components of soymilk during fermentation with the basidiomycete *Ganoderma lucidum*. *Food Chem.*, 112(1): 1–5. doi: 10.1016/j.foodchem.2008.05.024.

Yang, H., Zhang, L., Xiao, G., Feng, J., Zhou, H. and Huang, F. (2015). Changes in some nutritional components of soymilk during fermentation by the culinary and medicinal mushroom *Grifola frondosa*. *LWT - Food Sci. Technol.*, 62(1): 468–473. doi: 10.1016/j.lwt.2014.05.027.

Zeng, D. and Zhu, S. (2018). Purification, characterization, antioxidant and anticancer activities of novel polysaccharides extracted from Bachu mushroom. *Int. J. Biol. Macromol.*, 107: 1086–1092. doi: 10.1016/j.ijbiomac.2017.09.088.

Zhang, L., Li, X., Xu, X. and Zeng, F. (2005). Correlation between antitumor activity, molecular weight, and conformation of lentinan. *Carbohydr. Res.*, 340(8): 1515–1521. doi: 10.1016/j.carres.2005.02.032.

Zhang, M., Zhang, Y., Zhang, L. and Tian, Q. (2019). Mushroom polysaccharide lentinan for treating different types of cancers: A review of 12 years clinical studies in China. pp. 297–328. *In*: Zhang, L. (ed.). *Progress in Molecular Biology and Translational Science*. Cambridge: Academic Press, Chapter 13.

Zhang, T., Ye, J., Xue, C., Wang, Y., Liao, W. (2018). Structural characteristics and bioactive properties of a novel polysaccharide from *Flammulina velutipes*. *Carbohydr. Polym.*, 197: 147–156. doi: 10.1016/j.carbpol.2018.05.069.

Zhang, W.-N., Gong, L.-L., Liu, Y., Zhou, Z.-B., Wan, C.-X. et al. (2020). Immunoenhancement effect of crude polysaccharides of *Helvella leucopus* on cyclophosphamide-induced immunosuppressive mice. *J. Funct. Foods,* 69: 103942. doi: 10.1016/j.jff.2020.103942.

Zhang, Y., Fraatz, M.A., Horlamus, F., Quitmann, H. and Zorn, H. (2014). Identification of potent odorants in a novel nonalcoholic beverage produced by fermentation of wort with shiitake (*Lentinula edodes*). *J. Agric. Food Chem.*, 62(18): 4195–4203. doi: 10.1021/jf5005463.

Zhang, Y. (2015). *Development of Novel Fermentation Systems for the Production of Nonalcoholic Beverages with Basidiomycetes.* Cumulative dissertation, Justus Liebig University, Giessen, Germany, p. 61.

Zhang, Y., Fraatz, M.A., Müller, J., Schmitz, H.-J., Birk, F. et al. (2015a). Aroma characterization and safety assessment of a beverage fermented by *Trametes versicolor*. *J. Agric. Food Chem.*, 63(31): 6915–6921. doi: 10.1021/acs.jafc.5b02167.

Zhang, Y., Hartung, N.M., Fraatz, M.A. and Zorn, H. (2015b). Quantification of key odor-active compounds of a novel nonalcoholic beverage produced by fermentation of wort by shiitake (*Lentinula edodes*) and aroma genesis studies. *Food Res. Int.*, 70: 23–30. doi: 10.1016/j.foodres.2015.01.019.

Zhao, M.-H., Zhang, C.-S., Zeng, G.-M., Huang, D.-L. and Cheng, M. (2016). Toxicity and bioaccumulation of heavy metals in *Phanerochaete chrysosporium*. *Trans. Nonferrous Met. Soc. China,* 26(5): 1410–1418. 10.1016/s1003-6326(16)64245-0.

Zimmermanovà, K., Svoboda, L. and Kalac, P. (2001). Mercury, cadmium, lead and copper contents in fruiting bodies of selected edible mushrooms in contaminated Middle Spis region, Slovakia. *Int. J. Ecol. Probl. Biosph.*, 20(4): 440–466.

Zivanovic, S. (2006). *Identification of opportunities for production of ingredients based on further processed fresh mushrooms, off-grade mushrooms, bi-products, and waste material.* Mushroom Council, University of Tennessee, Department of Food Science and Technology, 33.

12

Overview on Major Mycotoxins Accumulated on Food and Feed

Tapani Yli-Mattila,[1,*] *Emre Yörük,*[2] *Asmaa Abbas*[1] and *Tuğba Teker*[3]

1. Introduction

Fungi could grow on indoor and outdoor environments throughout the world, and they produce a wide range of metabolites. Plants, animals, and humans are subjected to those metabolites which could be pharmaceutically useful compounds (e.g., penicillin), pigment molecules (e.g., aurofusarin) and/or harmful (e.g., nivalenol (NIV)) compounds. Mycotoxins are the most important secondary metabolites produced by fungi which could be present and accumulated on feed and food; these secondary metabolites present significant hazards to animal and human health via the food chain (Sudakin, 2003; McCormick et al., 2011). Mycotoxins are secondary metabolites with low molecular weight produced by various mycotoxigenic fungi. They represent great variations in terms of their chemistry and toxicology. *Alternaria, Aspergillus Claviceps*, and *Fusarium* are the most important mycotoxigenic fungi producing several important minor and major toxic metabolites. Mycotoxins could be specific to fungal species: *Fusarium sporotrichioides* produces T-2 toxin and *F. verticillioides* fumonisin, while *F. graminearum* produces deoxynivalenol (DON) and zearalenone (ZEN). The most frequently accumulated and seen mycotoxins at indoor and outdoor environments are aflatoxins, ochratoxins, trichothecenes, patulins, and fumonisins (Foroud and Eudes, 2009; Sabuncuoğlu et al., 2008; McCormick et al., 2011).

[1] Department of Life Technologies, Faculty of Technology, University of Turku, FI-20014, Turku, Finland.
[2] Department of Molecular Biology and Genetics, Faculty of Arts and Sciences, Istanbul Yeni Yuzyil University, 34010, Istanbul, Turkey.
[3] Institute of Graduate Studies in Sciences, Programme of Molecular Biotechnology and Genetics, Istanbul University, 34116, Istanbul, Turkey.
* Corresponding author: tymat@utu.fi

Mycotoxins are defined as natural products produced by filamentous fungi. Those fungi produce a toxic response when introduced in low concentration to higher vertebrates and other animals by a natural way (Bennett, 1987). For example, "Turkey x disease" caused by aflatoxins was discovered in 1960 by Goldblatt (1969), when more than 100,000 people died in England by symptoms of internal bleeding and liver necrosis. *Fusarium* toxicology started in 1961 (Brian et al., 1961) with the determination of the structure of the trichothecene diacetoxyscirpenol (DAS) and in 1966 with the determination of the structure of ZEN (Urry et al., 1966). The most active time of searching new *Fusarium* mycotoxins ended by finding out the chemical identity of the fumonisin B1 independently in South Africa (Bezuidehnhout et al., 1988; Gelderblom et al., 1988) and New Caledonia in 1989 (Laurent et al., 1989). To date, there are more than 300 known mycotoxins; however, only some of these mycotoxins reveal a real risk to food safety. The effects of some food-borne mycotoxins are acute with symptoms of severe illness. Some other mycotoxins occurring in food have longer term chronic or cumulative effects on health, such as cancers and immune deficiency (Tola and Kebede, 2016). The major mycotoxins causing serious contamination of foods and feeds are aflatoxins, fumonisins, ZEN, and trichothecenes, which are produced by *Aspergillus* and *Fusarium* (Jeswal and Kumar, 2015; Bhat et al., 2010). According to the Food and Agriculture Organization (FAO), a quarter of the world food crops is contaminated by mycotoxins (Wu, 2007; Pankaj et al., 2018). Fungi that are responsible for mycotoxin production and can invade food before harvest are called field fungi and those that invade food after harvest are known as storage fungi. *F. graminearum*, *F. verticillioides*, and *Aspergillus flavus* are some examples of field fungi which produce DON and NIV, fumonisins, aflatoxins, respectively. *Penicillium verrucosum*, which produces ochratoxins, is an example of storage fungi (Ayalew, 2010). Annually, 25% of the world's food crops are contaminated with mycotoxins, leading to huge economic losses in the billions of dollars (Marin et al., 2013). Wheat and barley losses caused by Fusarium head blight (FHB) epidemics in the United States (US) during the 1990s are estimated at close to $3 billion (Windels, 2000). In the US and the European Union (EU) countries, aflatoxins are mainly an economic concern (Mitchell et al., 2016); however, in the developing countries of Asia and Africa, aflatoxins cause hundreds of hepatocellular carcinoma cases each year (Liu and Wu, 2010). The estimated annual losses to the US industry sector because of aflatoxin contamination is $52.1 million (Mitchell et al., 2016). This chapter covers the updated information of four major toxins: aflatoxins, fumonisins, zearalenones, and trichothecenes. Understanding the basis of fungal secondary metabolites is the first step in minimizing and eliminating these metabolites and contributing to global food and feed safety by providing improvement of control strategies.

2. Chemistry

2.1 Aflatoxins

Aflatoxins are among the most natural carcinogens in humans and animals. They are difuranocoumarins derivatives, derived from the toxigenic *Aspergillus* species (Iqbal et al., 2015). The four major aflatoxins isolated from foods and feeds are aflatoxins

B1, B2, G1, and G2 (Fig. 12.1). These four aflatoxins are named based on their blue (B) or green (G) fluorescence under ultraviolet light, and their virtual mobility by thin layer chromatography on silica gel; therefore, using of fluorescence is important in identifying and differentiating between the B and G groups. These aflatoxin types exhibit molecular differences, e.g., aflatoxins B1 and B2 have a cyclopentane ring while aflatoxins G1 and G2 contain a lactone ring (Gourama and Bullerman, 1995). Aflatoxin M1 (AFMI) (Fig. 12.1) is a hydroxylated derivative metabolized from aflatoxin B1 by cows and secreted in milk (Van Egmond, 1989). Jolly et al. (2021) has investigated the presence of aflatoxin M1 in mother's breast milk and urine of infants in the Western Highlands of Guatemala. Results showed that AFM1 tested positive in 5% of breast milk and 15.7% of urine samples. In aflatoxins, the maximum limit

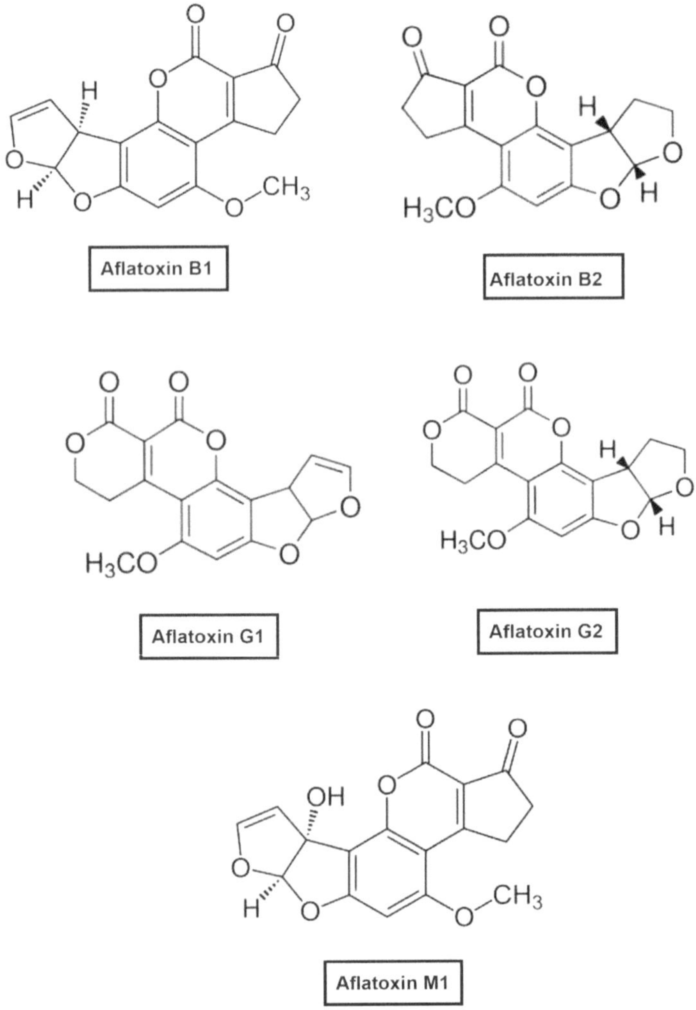

Fig. 12.1. Chemical structures of five aflatoxin types, aflatoxin B1 (AFB1), aflatoxin B2 (AFB2), aflatoxin G1 (AFG1), aflatoxin G2 (AFG2), and aflatoxin M1 (AFM1).

is 4 µg/kg for all AFs or 2 µg/kg for AFB1 only as published by the European Union (EC, 2010). Aflatoxin contamination is a serious problem especially in developing countries because in 2004, approximately 4.5 billion people were at risk for chronic, uncontrolled exposure to aflatoxins (Shepherd, 2003; Williams et al., 2004).

2.2 Fumonisins

The name fumonisin was formed based on the first letters of *Fusarium moniliforme*. Fumonisin B1 is a diester of propane-1,2,3, -tricarboxylic acid and 2-amino-12,16-dimethyl-3,510,14,15-pentahydroxyicosane (Fig. 12.2). The B series fumonisins B1, B2, and B3 account for the majority of fumonisins that occur in grain samples, which are contaminated by fumonisin-producing species. Fumonisin B4 is a minor metabolite. Fumonisins C1, C2, and C3 are the main fumonisins produced by *F. oxysporum*. In addition, there are P- and A-series fumonisins, which are minor metabolites (Scott, 2012). The regulatory limit for fumonisins in maize and by-products established by the European Union and Food and Drug Authority to prevent exposure of individuals to these fungal toxins is 200–4,000 µg kg^{-1} (van Egmond et al., 2007). According to the EU regulation, FUM content in unprocessed maize cannot exceed 4 ppm for human consumption (Commission Regulation, 2007).

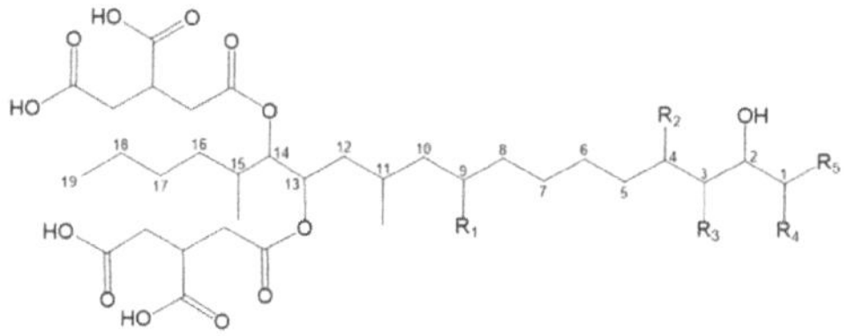

Fumonisin	R_1	R_2	R_3	R_4	R_5
A_1	OH	OH	H	$NHCOCH_3$	CH_3
A_2	OH	OH	H	$NHCOCH_3$	CH_3
A_3	OH	H	H	$NHCOCH_3$	CH_3
A_4	H	H	H	$NHCOCH_3$	CH_3
B_1	OH	OH	H	NH_2	CH_3
B_2	H	OH	H	NH_2	CH_3
B_3	OH	H	H	NH_2	CH_3
B_4	H	H	H	NH_2	CH_3
B_6	H	OH	OH	NH_2	CH_3
C_1	OH	OH	H	NH_2	H
C_2	H	OH	H	NH_2	H
C_3	OH	H	H	NH_2	H
C_4	H	H	H	NH_2	H
FP_1	OH	OH	H	3HP*	H
FP_2	H	OH	H	3HP*	H
FP_3	OH	H	H	3HP*	H

* 3-hydroxypyridinium

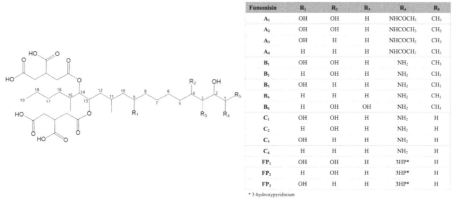

Fig. 12.2. Chemical structure of the most frequently occurring fumonisins (modified from Braun and Wink, 2018).

2.3 Zearalenone

Zearalenone (ZEN) (previously known as F-2 toxin) is the most well-known member of resorcylic acid lactones (RALs), which classified as fungal polyketide. It is also known as 6-(10-hydroxy-6-oxo-trans-1-undecenyl)-β-resorcyclic acid lactone ($C_{18}H_{22}O_5$, molecular weight (MW): 318.36 g/mol, Chemical Abstracts Service (CAS) Registry Number: 17924–92-4) (Fig. 12.3). The term 'zearalenone' comes from the combination of following words: maize (Zea mays), resorcylic acid lactones (RAL), 'en' (to have double bond) and 'one' (to have a ketone group) (Urry et al., 1966; Ropejko and Twarużek, 2021). Due to its structure, ZEN has demonstrated diverse biological activities and has versatile biological effects on the human and

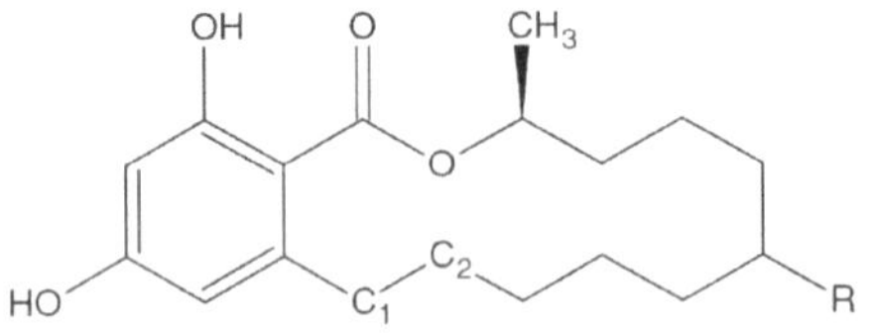

Compound	Formula	MW $(g{\cdot}mol^{-1})$	R	C_1-C_2
ZEN	$C_{18}H_{22}O_5$	318.36	=O	C=C
α-ZEL	$C_{18}H_{24}O_5$	320.38	····OH	C=C
β-ZEL	$C_{18}H_{24}O_5$	320.38	◀OH	C=C
α-ZAL	$C_{18}H_{26}O_5$	322.40	····OH	C—C
β-ZAL	$C_{18}H_{26}O_5$	322.40	◀OH	C—C
ZAN	$C_{18}H_{24}O_5$	320.38	=O	C—C

Fig. 12.3. Chemical structures and molecular weights of ZEN and its derivatives (adapted from Cavaliere et al., 2005).

animal health. It is commonly known for causing reproductive problems in farm animals, particularly in swine, and hyperoestrogenic disorders in humans (Kuiper-Goodman et al., 1987; Zinedine et al., 2007; EFSA, 2011; Zhang et al., 2018). The binding affinity of ZEN and its metabolites to estrogen receptors has led them to be termed as a 'non-steroidal estrogen', 'mycoestrogen' and 'phytoestrogen' (Hurd, 1977; Shier, 1998; Bennett and Klich, 2003).

The commonly elaborated derivatives, which occur during the Phase I metabolic processes of ZEN, are α-zearalenol (α-ZEL), β-zearalenol (β-ZEL), α-zearalanol (Zeranol) (α-ZAL), β-zearalanol (Taleranol) (β-ZAL), and zearalanone (ZAN) (Bottalico et al., 1985; Golinski et al., 1988; Mukherjee et al., 2014; EFSA, 2017) (Fig. 12.3). The estrogenic potency of the compounds is in the following order: α-ZEL> α-ZAL> ZEN~β-ZAL~ZAN> β-ZEL regarding the assessment "uterotrophic activity" in rodents (EFSA, 2017). Additionally, there are reports in which the relative estrogenic activity is ordered as shown in the Table 12.1.

Table 12.1. Estrogenicity potential of ZEN and its metabolites.

Estrogenic Potency	Methods for the Measurement of Estrogenicity Potential	Reference
α-ZEL> α-ZAL> β-ZAL> ZAN> ZEN> β-ZEL	MCF7 human breast cell proliferation assay	Shier et al. (2001)
α-ZEL> α-ZAL> β-ZAL> ZEN> β-ZEL	Stimulation of estrogen production by adrenocortical carcinoma cell	Minervini et al. (2005)
α-ZEL> α-ZAL> ZEN~β-ZAL~ZAN> β-ZEL	Based on their binding affinity to estrogenic receptors (method non-specified)	Mukherjee et al. (2014)
α-ZEL> α-ZAL> ZAN> ZEN> β-ZAL> β-ZEL	MCF-7 cell proliferation assay (E-screen assay)	Drzymala et al. (2015)

2.4 Trichothecenes

The first trichothecene 'trichothecin' was isolated from *Trichothecium roseum* as an antifungal metabolite (Freeman and Morrison, 1948). Trichothecenes are mycotoxins produced by *Fusarium*, *Memnoniella*, *Myrothecium*, *Stachybotrys*, *Trichoderma*, and *Trichothecium*, but the *Fusarium* species are the most important producers of trichothecenes in cereals and other foods and feeds. More than 200 toxins belonging to

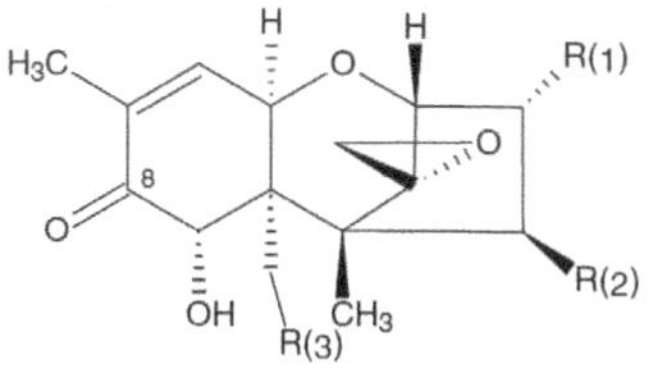

Compound	Formula	MW (g·mol⁻¹)	R(1)	R(2)	R(3)
NIV	$C_{15}H_{20}O_7$	312.31	OH	OH	OH
DON	$C_{15}H_{20}O_6$	296.32	OH	H	OH
3-ADON	$C_{17}H_{22}O_7$	338.35	OAc	H	OH
15-ADON	$C_{17}H_{22}O_7$	338.35	OH	H	OAc
FUS X	$C_{17}H_{22}O_8$	354.35	OH	OAc	OH

Fig. 12.4. Chemical structures and molecular weights of type B trichothecenes: NIV, DON, and their acetylated derivatives, 3-acetyldeoxynivalenol (3-ADON), 15-acetlydeoxynivalenol (15-ADON) (adapted from Cavaliere et al., 2005).

trichothecenes are reported from nature and DON is the most common trichothecene accumulated on food products (Sabuncuoğlu et al., 2008; Yazar and Omurtag, 2008; McCormick et al., 2011). Trichothecenes are non-volatile, amphipathic, small, and sesquiterpenoid mycotoxins. They are divided into four classes according to their chemical properties and producing fungi as A, B, C, and D. The most common types are A- and B-classes trichothecenes (Fig. 12.4) which contain at least 100 different types (Girgin et al., 2001; Sudakin, 2003; Desjardins and Proctor, 2007). Class A trichothecenes have a functional group other than a ketone at C8 position. Class B trichothecenes have a carbonyl function at C-8. Class C-trichothecenes have a C7-C8 epoxide, and class D-trichothecenes are the most structurally divergent with an additional ring linking the C-4 and C-15 positions (Yazar and Omurtag, 2008; McCormick et al., 2011).

3. Fungal Mycotoxins in Food and Feed

A total of 113 secondary metabolites was found in dairy cow feed samples collected from Thailand between August 2018 and March 2019. Both fungal metabolites and plant metabolites were investigated. Results showed that among the major mycotoxins, ZEN and fumonisins were mostly detected in the mixed feed samples, while DON and aflatoxin B1 were found in a percentage lower than 50% (Awapak et al., 2021). Due to the hazardous effects associated with mycotoxins, about 100 countries applied specific limits for the presence of mycotoxins in foodstuffs and feedstuffs by the end of 2003. These specific limits or recommendations have been established for numerous mycotoxins, including the naturally occurring aflatoxins and aflatoxin M1, DON, T-2 toxin, and HT-2 toxin, fumonisin B1, B2, and B3, EA, OTA, PAT, and ZEN (FAO, 2004).

3.1 Aflatoxins

Aflatoxins are carcinogenic, mutagenic, and teratogenic secondary metabolites produced by toxigenic *Aspergillus* spp. (Table 12.2). These fungi develop significantly in hot climates, humid conditions, or poor post-harvest handling. Hence, humans can be exposed to aflatoxins through ingesting of contaminated plant-derived and milk products. Their toxicity also links to their heat-resistant nature as they can survive at high temperatures of food processing (Medina et al., 2017). *A. flavus* can produce only aflatoxins B1 and B2. The lack of the ability to synthesize G aflatoxins by

Table 12.2. Collection of mycotoxins, fungal species, detection methods, and maximum limit permitted.

Mycotoxin Name	Fungal Species	Detection Methods	Maximum Limit	References
Aflatoxin B1	*Aspergillus flavus*, *A. parasiticus*, *A. nomius*, *A. emericella*	UFLC –MS/MS, UHPLC –QqLIT-MS, LC–MS/MS	2–10 μg/kg	(GB 2761-2017, 2017); (Kumar et al., 2017); (Li et al., 2016); (Xing et al., 2016); (Pallarés et al., 2017)
Fumonisin B1	*Fusarium verticillioides*, *F. proliferatum*, *F. nygamai*	UFLC –MS/MS, UHPLC –QqLIT-MS	1000–4000 μg/kg	(Rheeder et al., 2002); (Liu et al., 2020); (Li et al., 2016); (Xing et al., 2016)
ZEN	*Fusarium graminearum*, *F. culmorum*, *F. verticillioides*, *F. sporotrichioides*, *F. semitectum*, *F. equiseti*, *F. oxysporum*, *F. cerealis*, *F. incarnatum*	UFLC –MS/MS, UHPLC –QqLIT-MS, LC–MS/MS	20–200 μg/kg	(Pascari et al., 2018); (Li et al., 2016); (Xing et al., 2016); (Pallarés et al., 2017)
Trichothecenes (type A)	*F. sporotrichioides*, *F. poae*, *F. equiseti*	UHPLC –QqLIT-MS, GC-FID/ MS, LC–MS/MS	50–100 μg/kg	(Miró-Abella et al., 2017); (Xing et al., 2016); (Pallarés et al., 2017)
Trichothecenes (type B)	*F. graminearum*, *F. culmorum*, *F. pseudograminearum*, *F. equiseti*	UHPLC –QqLIT-MS, GC-FID/ MS, LC–MS/MS	500–1000 μg/kg	(Pascari et al., 2018); (Xing et al., 2016); (Pallarés et al., 2017)

A. flavus is due to deletion 0.8–1.5 kb region in the aflatoxin biosynthesis gene cluster (Ehrlich et al., 2004). *A. nomius* and *A. parasiticus* can produce all four aflatoxin types. There are two other aflatoxins produced in milk-generating animals, namely M1 and M2. They are hydroxylated metabolites of aflatoxin B1 and B2, respectively (Rahimi et al., 2010).

Aflatoxigenic fungi can contaminate a wide variety of main foods like cereals (maize, rice, barley, oats, and sorghum), nuts (peanuts, ground nuts, pistachio, almonds, walnuts), dried fruits and cottonseeds. Cows that are fed on contaminated diet with AFB1, their hepatic microsomal cytochrome P450 biotransform AFB1 to hydroxylated-AFB1 metabolite called AFM1 and can reach humans through milk or dairy products. AFM1 can be found in cheese with a higher concentration than that of the raw milk since AFM1 is heat stable, binds well to casein, and is not decayed by the cheese-making process (Alshannaq and Yu, 2017). The analysis of a recent study in Pakistan by Ajmal et al. (2021) to analyse AFB1 from 100 sesame seed samples, revealed that 92% fresh and 99% stored samples were contaminated with AFB1. From 2017 to 2020, 1,675 food samples with vegetal origin imported in Southern Italy from extra-EU countries were monitored for aflatoxin types. The results revealed that 295 samples (17.6%) were contaminated by aflatoxin B1 and 204 samples (12.2%) were contaminated by aflatoxins B/G (Gallo et al., 2021).

3.2 Fumonisin-producing Fusarium

Fumonisins are polyketides and they are produced only by the *Fusarium* species, while closely related compounds are also produced by the *Altenaria* species (Bottini and Gilchrist, 1981). High levels of fumonisin production have been found mainly in *F. proliferatum* and *F. verticillioides* strains of the *Gibberella fujikuroi* species complex. Species-specific polymerase chain reaction (PCR) is commonly used to clearly identify species inside fungus complexes such as *F. verticillioides* and *F. proliferatum* (Rahjoo et al., 2008; Waalwijk et al., 2008). *F. verticillioides*, the main fumonisin-producer, is a filamentous ascomycete that typically causes maize seedling blight and root, stalk and ear rots, but is also commonly associated with maize as an endophyte without any development of symptoms (Munkvold and Carlton, 1997; Logrieco et al., 2003; Bacon et al., 2008). Strains of *F. verticillioides* from banana do not have a FUM gene cluster and they do not produce fumonisins. These strains from banana also exhibit reduced virulence on maize seedlings (Glenn et al., 2008). Fumonisins can also be detected in symptomless infected maize kernels (Bacon and Hinton, 1996). Also, several other species of this species complex and some strains of *F. oxysporum* are producing fumonisins (Desjardins, 2006; Leslie and Summerell, 2006; Rossi et al., 2009).

The most common and important sources of fumonisin contamination for humans and animals are cereals (rice, wheat, barley, maize, rye, oat, and millet). The FB1 has also been reported to contaminate numerous food products like asparagus and garlic (Seefelder et al., 2002), barley foods (Park et al., 2002), beers (Kawashima et al., 2007), dried figs (Heperkan et al., 2012), and milk (Kamle et al., 2019). Maize and maize-based products are one of the most infected foods by FB1 (Desjardins, 2006). The occurrence of fumonisins in wheat and barley and asparagus in Europe has often been associated with infection by *F. proliferaturm* (Desjardins, 2006; Stępień et al., 2011; Guo et al., 2016), but recently *F. verticillioides* isolates were found even in wheat in Finland (Gagkaeva and Yli-Mattila, 2020).

3.3 Zearalenone

In Indiana, Minnesota and nearby areas of USA, *F. graminеareum* caused severe epidemics of maize rot during the 1950s and 1960s. Contaminated maize caused vomiting, feed refusal, and estrogenic symptoms in swines (Desjardins, 2006). Zearalenone (ZEN) was first reported in 1962 as the cause of a reproductive disorder in pigs which were fed spoiled grain infected with *Gibberella zeae* (the anamorph/asexual name *F. graminearum*, Stob et al., 1962). In 1966, this estrogenic compound was structurally characterized and named 'zearalenone' (Urry et al., 1966). It is mainly produced by *F. graminearum*, but also *F. culmorum*, *F. cerealis* (synonym, *F. crookwellense*), *F. equiseti*, *F. oxysporum*, and *F. semitectum* could create ZEN (Milano and López, 1991; Bennett and Klich, 2003; Zinedine et al., 2007; Yörük et al., 2015). ZEN generation by fungi takes place at a humidity level above 15% conditions (https://www.envirologix.com/; retrieved May 2021) and optimally

Table 12.3. Occurrence of ZEN in cereals.

Cereal	Country	Positive Samples (%) (Number of Samples)	Content (Average and Max.) [μg kg^{-1}]	References
Barley	Bulgaria	11.1% (18)	Mean: 29, Max: 36.6	Manova and Mladenova (2009)
Maize	Bulgaria	21.1% (19)	Mean: 80.6, Max: 148	Manova and Mladenova (2009)
Wheat	Bulgaria	1.9% (54)	Mean: 10, Max: 10	Manova and Mladenova (2009)
Corn	Germany	85% (41)	Mean: 48, Max: 860	Schollenberger et al. (2006)
Oats	Germany	24% (17)	Mean: 21, Max: not specified	Schollenberger et al. (2006)
Wheat	Germany	63% (41)	Mean: 15, Max: not specified	Schollenberger et al. (2006)
Wheat	Germany	80% (84)	Mean: 178, Max: 8038	Müller and Schwadorf, (1993)
Rye	Germany	16% (514)	Mean: 23, Max: 37	Meister (2009)
Wheat	Germany	41% (407)	Mean: 72, Max: 451	Meister (2009)
Maize	Italy	30% (46)	Mean: not specified, Max: 969	Cavaliere et al. (2005)
Wheat	Netherlands	62% (29)	Mean: 61, Max: 677	Tanaka et al. (1990)
Spring Barley	Poland	75% (8)	Mean: 13, Max: 31	Bryła et al. (2016)
Winter Barley	Poland	62.5% (16)	Mean: 7, Max: 19	Bryła et al. (2016)
Oats	Poland	100% (4)	Mean: 10, Max: 15	Bryła et al. (2016)

at approximately 15°C (Garcia et al., 2012; Wu et al., 2017). ZEN contamination in grains has been reported from several countries in temperate climate zones (Scientific Committee on Food 2000 (https://ec.europa.eu/food/index_en; retrieved May 2021) (Krska et al., 2003; Mostrom, 2016). It often coexists with other common mycotoxins produced by *Fusarium* such as trichothecenes including NIV and DON (Pittet, 1998; Krska et al., 2003). ZEN is most frequently present in corn (Zhang et al., 2018). It has also been detected in a variety of cereal grains such as wheat, barley and oats, rye at levels from few μg/kg to 8 mg/kg as shown in Table 12.3 (Tanaka et al., 1988; Placinta et al., 1999; Zinedine et al., 2007; EFSA, 2011; Ropejko and Twarużek, 2021). Although ZEN is primarily known as a field contaminant, it is also found in foodstuffs and animal feeds (Zinedine et al., 2007). Since ZEN is a heat-stable compound (up to 160°C) (Kuiper-Goodman et al., 1987), it cannot be completely eradicated during the processing of food and feed.

3.4 Fusarium Species Producing Trichothecenes

The most important trichothecenes; T-2, HT-2, DAS, DON and NIV accumulate on small grain cereals and reach animals and humans via food ingestion. Type A toxins, T-2 and HT-2 are co-produced by fungus whereas DON or NIV were co-present on cereals with their acetylated derivatives (Sabuncuoğlu et al., 2008; Yazar and Omurtag, 2008). Generally, *Fusarium* spp. produces either class A or B trichothecenes. However, several reports showed that class B producer *Fusarium* spp. could create NX-2 belonging to class A-trichothecenes (Varga et al., 2015; Kelly et al., 2016). Thus, chemotype assays on trichothecenes have become more focused and popular than in the past.

Fusarium acuminatum, *F. equiseti*, *F. langsethiae*, *F. poae*, *F. sibiricum*, and *F. sporotrichioides* are class A trichothecene producer species. *Tri5* gene was first discovered in *F. sporotrichioides* and the complete *tri5* gene cluster data was present only for *F. sporotrichioides* on GenBank (Brown et al., 2002; Chandler et al., 2003; Kimura et al., 2003, 2007; Yazar and Omurtag, 2008; Yli-Mattila et al., 2011). The *F. sporotrichioides* has only asexual stages. This fungal phytopathogen could be easily identified at species level on potato dextrose agar (PDA) and pentachloronitrobenzene agar (PCNB) media. It possesses two mating type idiomorphic alleles. The *F. sporotrichioides* is one of the major causal agents of head blight diseases of small grain cereals in tropical and temperate regions worldwide, while *F. langsethiae* is found mainly in Europe and *F. sibiricum* mainly in northern Asia. *F. graminearum* and *F. culmorum*, which are the most important class B-trichothecene manufacturers worldwide, are the major causal agents of head blight and crown rot diseases in temperate and cooler regions, respectively (Miedaner et al., 2008; Matny, 2015; Yörük and Yli-Mattila, 2019). Both species had completed genome projects and *tri5* gene sequences in the GenBank. *F. graminearum* has sexual and asexual stages while *F. culmorum* has no sexual stage. However, presence of both mating type alleles and para sexual stage were reported for both species. Both the species could be easily identified on the PDA medium. Moreover, sequence characterized amplified regions' markers have been widely used in identification (Schilling et al., 1996; Nicholson et al., 1998; Miedaner et al., 2001, 2008; Çepni et al., 2013; Albayrak et al., 2016). In addition, there are several other species of *F. graminearum* species complex in more restricted areas (Yli-Mattila et al., 2009; Yli-Mattila and Gagkaeva, 2016).

4. Mycotoxins-Biosynthetic Clusters

4.1 Biosynthesis of Aflatoxin B1

Aflatoxin Biosynthetic Cluster (ABC) involves at least 27 enzymatic reactions (Fig. 12.5) and contains several interconnecting networks. However, to date, the functions of some genes belonging to the cluster remain unclear. The first step in the aflatoxin pathway is the synthesis of the starter unit hexanoate from acetyl-CoA and malonyl-CoA catalyzed by *fas-1* and *fas-2* genes. This hexanoate produces norsolorinic acid anthrone (NAA) through extension and condensation reactions, catalyzed by the polyketide synthase *pksA*. *hypC* is a gene that encodes for a 17-kDa enzyme, responsible for the catalytic conversion of NAA into the first stable metabolite in the AF pathway, norsolorinic acid (NOR). The conversion from NOR to averantin (AVN) occurs through the hydroxylation reaction catalyzed by *nor-1* which encodes the ketoreductase. However, the contribution of two other

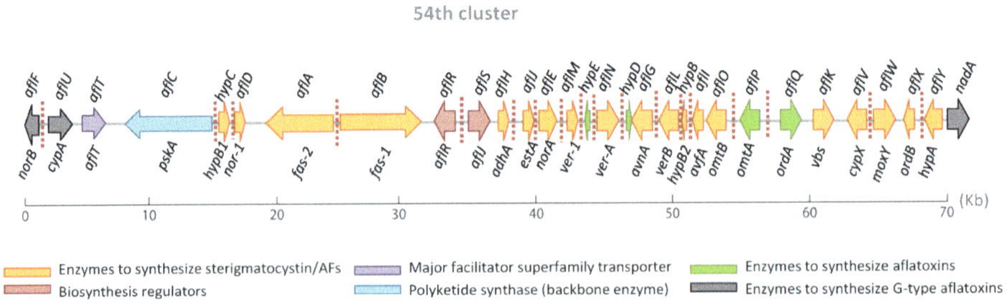

Fig. 12.5. Aflatoxin gene cluster including the old and new cluster gene names.

Note. Red dotted lines represent the binding sites of AflR in the above pathway (adapted from Caceres et al., 2020).

genes (*norA* and *norB*) was associated with this step (Zhou and Linz, 1999). *avnA* gene encodes a cytochrome P450 monooxygenase which converts AVN to 5'-hydroxyaverantin (HAVN) by a microsome and cytosol enzyme. The conversion process from HAVN to averufin (AVF) is catalyzed by the *adhA* gene and involved several intermediates. *avfA* gene is necessary for conversion of AVF to versiconal hemiacetal acetate (VHA) where the enzyme encoding this gene catalyzes the ring-closure step during the formation of hydroxyversicolorone (HVN) (Ehrlich, 2009). *estA* gene was considered in the transformation of VHA to versiconal (VAL). After that, VAL is transformed into versicolorin B (VERB) with the help of the *vbs* gene. The *verB* gene, which encodes P450 desaturase is responsible for alteration of VERB to versicolorin A (VERA). *ver-1*, *verA*, and *hypA* genes are in charge of the conversion of VERA to demethylsterigmatocystin (DMST) (Yu, 2012). Two O-methyltransferases (*omtA* and *omtB*) are reported to be involved in the aflatoxin pathway. *omtB* catalyzes the transfer of the methyl from S-adenosylmethionine (SAM) to the DMST to create sterigmatocystin (ST). *omtA* catalyzes the insertion of methyl group into ST and DHST to produce O-methylsterigmatocystin (OMST). The final process of conversion OMST to aflatoxin B1 (AFB1) occurs with the help of *ordA* which changes OMST into AFB1 by the oxidation of the A-ring of OMST.

4.2 Biosynthesis of Fumonisins

Fumonisin biosynthetic (FUM) gene clusters have been reported in *F. verticillioides*, *F. proliferatum*, and a single strain of *F. oxysporum* (Proctor et al., 2003, 2008; Waalwijk et al., 2004). It has also been reported that *Aspergillus niger* strains are able to produce FB2 (Frisvad et al., 2007; Aerts et al., 2018). The biosynthesis of FBs in *Fusarium* species requires a 16-gene cluster (Fig. 12.6, Table 12.4) as well as regulatory and transport proteins that can be conserved between several fungal genera (Desjardins, 2006; Proctor et al., 2013). The gene FUM1 encodes a polyketide synthase necessary to produce fumonisins, which catalyzes the initial steps in fumonisin biosynthesis (Bojja et al., 2004). This gene can be used to detect FB1 production by *F. verticillioides* (López-Errasquín et al., 2007).

Unlike the biosynthetic gene clusters of natural products from fungi, there is no pathway-specific regulatory gene for FB biosynthesis pathway in the FUM cluster of the *Fusarium* species (Woloshuk et al., 1994; Proctor et al., 2003; Flaherty and

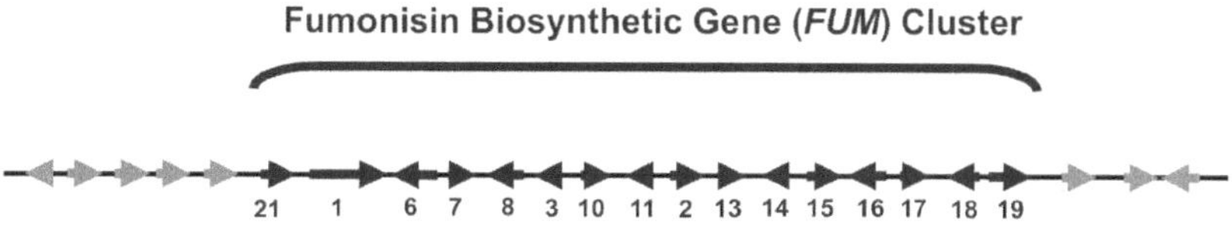

Fig. 12.6. Organization of the 16-gene *FUM* cluster described in *F. oxysporum*, *F. proliferatum*, and *F. verticillioides* (Proctor et al., 2013).

Table 12.4. List of genes in fumonisin biosynthetic gene cluster from *F. oxysporum* (Proctor et al., 2008).

Gene	Gene Product	Protein ID	Position (Strand)	Exon number
FUM21	FUM21	ACB12566.1	2798–5386 (+)	9
FUM1	FUM1	ACB12550.1	7418–15569 (+)	7
FUM6	FUM6	ACB12553.1	15974–19590 (–)	4
FUM7	FUM7	ACB12554.1	20143–21402 (+)	1
FUM8	FUM8	ACB12555.1	21605–24459 (–)	6
FUM3	FUM3	ACB12552.1	25435–26337 (–)	1
FUM10	FUM10	ACB12556.1	26806–28668 (+)	3
FUM11	FUM11	ACB12557.1	29064–30234 (–)	5
FUM2	FUM2	ACB12551.1	30746–32439 (+)	4
FUM13	FUM13	ACB12558.1	33317–34420 (+)	1
FUM14	FUM14	ACB12559.1	34791–36744 (–)	5
FUM15	FUM15	ACB12560.1	37203–39063 (+)	2
FUM16	FUM16	ACB12561.1	39224–41447 (–)	4
FUM17	FUM17	ACB12562.1	41916–43137 (+)	2
FUM18	FUM18	ACB12563.1	43248–44737 (–)	5
FUM19	FUM19	ACB12564.1	45063–49854 (+)	5
CPM1	CPM1	ACB12565.1	50651–52611 (–)	4

Woloshuk, 2004). The proposed biosynthetic pathway for FBs is shown in Fig. 12.7. The function of some FUM genes is known, e.g., FUM13 encodes a C-3 ketoreductase of FBs (Desjardins, 2006; Medina et al., 2013), FUM3 encodes a 2-ketoglutarate-dependent dioxygenase that catalyzes the conversion of FB B3 to B1 (Proctor et al., 2003; 2013), and FUM3P catalyzes the C-5 hydroxylation (Desjardins, 2006; Lazzaro et al., 2012). The known regulatory genes of FB biosynthesis are not linked to the FUM cluster and include *FCC1*, *FCK1*, *PAC1*, *ZFR1*, and *GBP1*.

4.3 Structural Organization of PKS Gene Cluster and Biosynthetic Pathway of Zearalenone

The biosynthetic pathway of zearalenone (ZEN) was discovered through forward and reverse genetic approaches in *F. graminearum* (Kim et al., 2005; Lysøe et al., 2006; Lysøe et al., 2009; Lee et al., 2011; Park et al., 2015). ZEN is synthesized by a polyketide pathway which a set of genes participate known as polyketide synthase (*PKS*) gene cluster. The size of PKS gene cluster is 49.993 bp and consists of 11 genes

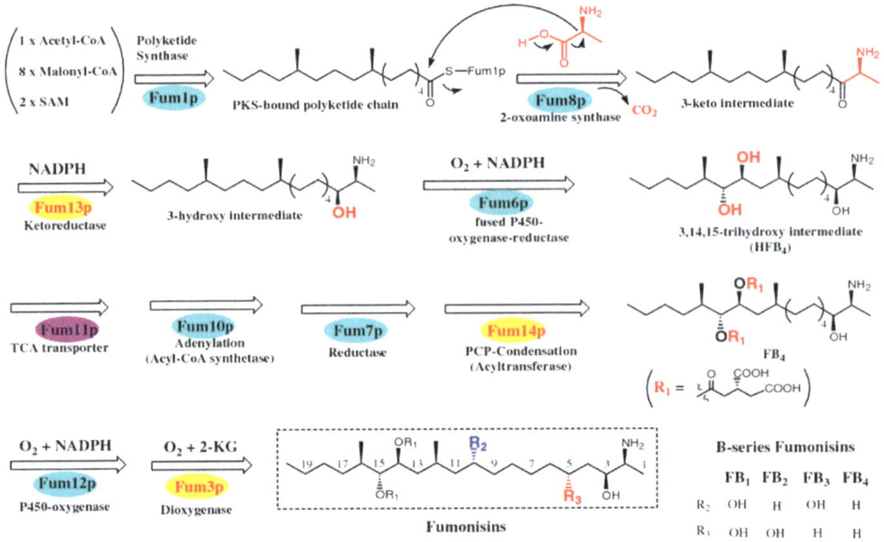

Fig. 12.7. A proposed biosynthetic pathway for fumonisins.

Note. *SAM S*-adenosyl methionine, *TCA* tricarballylic acid, *2-KG* 2-ketoglutarate (modified from Du et al., 2008 and Abodalam, 2019).

which located in four contigs (Table 12.5, Fig. 12. 8). Two fungal PKS encoding genes *PKS4* and *PKS13*, ZEN biosynthesis gene 1 (*ZEB1*), and gene 2 (*ZEB2*) are core genes of the *PKS* gene cluster. These four genes positioned together with genes encode the factors that are involved in and regulate ZEN biosynthesis (please see the NCBI GenBank accession number: DQ019316.1) (Kim et al., 2005; Gaffoor and Trail, 2006; Lysøe et al., 2006; Lysøe et al., 2009). The *PKS4* encodes a reducing polyketide synthase with functional domains ß-ketoacyl synthase (KS), ß-ketoacyl transferase (AT), dehydratase (DH), enoyl reductase (ER), ß-ketoacyl reductase (KR), and acyl carrier protein (ACP) (also known phosphopantetheine attachment site [PP]). The *PKS13* encodes a non-reducing polypeptide, PKS13, which possesses a starter unit acyltransferase (SAT), KS, AT, product template (PT), ACP, and thioesterase (TE) domains (Hansen et al., 2015). The KS, AT, and ACP constitute the conserved structure for PKSs, whereas the other domains may be absent in PKSs (Kroken et al., 2003). The 5' region of *PKS4* is towards the 5' region of *PKS13*. *ZEB1*, *ZEB2*, *GzSTK*, and *GzACA* are located downstream of the *PKS4* gene. The remaining genes of the PKS gene cluster *GzKAT*, GzMCT, *GzNPS*, *GzHET*, and *GzALD* are in the 3′ region of the *PKS13* (Table 12.5, Fig. 12.8). *ZEB1* encodes an isoamyl alcohol oxidase, which catalyzes an oxidation step for the conversion of β-zearalenol (β-ZOL) into ZEN (Kim et al., 2005). *ZEB2* encodes a transcription factor (TF). Two alternative forms of *ZEB2* transcripts (Zeb2L, Zeb2S) directly regulate the ZEN biosynthetic cluster genes in *F. graminearum* by forming an activator (ZEB2L-ZEB2L homooligomer) and an inhibitor (ZEB2L-ZEB2S heterodimer). Although ZEB2L carries a basic leucine zipper (bZIP) DNA-binding domain at its N-terminus, whereas ZEB2S is an N-terminally truncated form of ZEB2L that lacks the bZIP domain (Kim et al., 2005; Park et al., 2015; Park et al., 2016).

Table 12.5. List of genes involved in ZEN production.

Gene	**Gene Product**	**Protein ID**	**Contig [Position (Strand)]**	**Exon number**
GzALD [FG02392]	Aldehyde dehydrogenase	ABB90277.1	1.117 [2685–4268 (-)]	3
GzHET [FG02393]	Heterokaryon incompatibility protein	ABB90278.1	1.117 [4763–8515 (-)]	1
GzNPS [FG02394]	Non-ribosomal peptide synthetase	ABB90279.1	1.117 [9054–16031 (-)]	1
GzMCT [FG13438]	Monocarboxylase transporter	ABB90280.1	1.117 [16901–18249 (+)]	4
GzKAT [FG12015	K+ channel β-subunit	ABB90281.1	1.118 [18764–19912 (+)]	3
*PKS13** [FG02395]	Polyketide synthase	ABB90282.1	1.118 [20206–26529 (-)]	5
*PKS4** [FG12126]	Polyketide synthase	ABB90283.1	1.119 [27570-34937 (+)]	6
ZEB1 [FG12056]	Isoamyl alcohol oxidase	ABB90284.1	1.119 [35515–37387 (+)]	4
ZEB2 [FG02398]	bZIP domain-containing TF	ABB90285.1	1.119 [37591–38638 (-)]	2
GzSTK [FG02399]	Protein kinase Eg2-like	ABB90286.1	1.120 [40407–41496 (+)]	2
GzACA [FG02400]	Ca2+ transporting ATPase	ABB90287.1	1.1120 [42036-46050 (-)]	3

Notes. Brackets represent the gene designations by MIPS FGDB (Mewes et al., 2004; Güldener et al., 2006).

* *PKS13* and *PKS4* genes were also called *ZEA1* and *ZEA2*, respectively (Gaffoor and Trail, 2006).

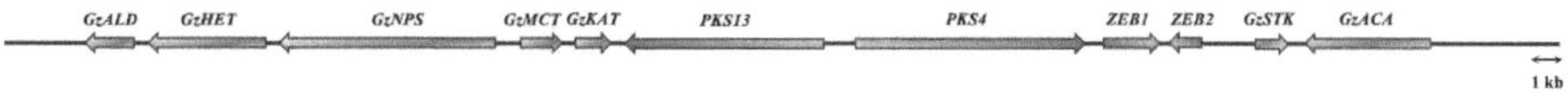

Fig. 12.8. Gene clusters of the ZEN biosynthesis pathway (11 genes, 49.993 bp) in *F. graminearum* (Kim et al., 2005).

Note. Position and direction of transcription of individual genes/ORFs indicated by the arrows. Orange arrows show the identified genes, which are essential in ZEN biosynthetic pathway.

ZEN is a product of enzymatic cascades in which PKS is considered as the key enzyme, in a similar way as other polyketide mycotoxins such as fumonisins, aflatoxins, etc. (Nahle et al., 2021). The biosynthetic pathway of ZEN is initiated by PKS4 which catalyzes the fusion of acetyl-CoA and five malonyl-CoA molecules into a hexaketide. The product of PKS4 functions as a starter unit for PKS13. The PKS13 triggers the further elongation of the polyketide chain by adding three malonyl-CoA molecules, resulting in the formation of a nonaketide. Then, the formation of an aromatic ring and a macrolide ring structure containing a lactone bond are produced because of two rounds of intramolecular cyclization reactions. As the final step,

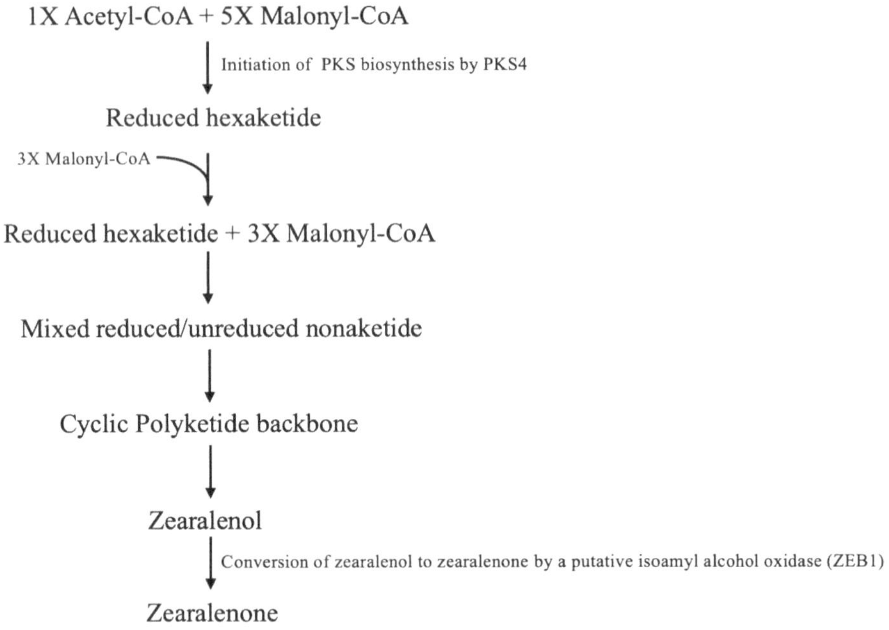

Fig. 12.9. ZEN biosynthetic pathway (as proposed by Kim et al., 2005; Gaffoor and Trail, 2006).

ZEB1 protein catalyzes an oxidation step for the conversion of produced β-zearalenol to ZEN. It transforms the macrolide bound hydroxyl group into a ketone group. ZEB2 transcripts have regulatory roles in the ZEN production pathway (Fig. 12.9) (Kim et al., 2005; Gaffoor and Trail, 2006; Park et al., 2015; Park et al., 2016).

4.4 Biosynthesis of Trichothecenes

At least ten different enzymes catalyze conformational alteration steps such as acetylation and deacetylation in the trichothecene biosynthetic pathway (Table 12.6, Fig. 12.10). There are important differences in different chemotypes of specific strains including class A- and class B-trichothecene chemotypes. Differences/variations related to expression of *tri3* gene located on core gene cluster, presence of a specific gene such as *tri10*, gene size differences such as *tri7* have been reported and these differences have been widely used in chemotype identification via generic PCR assays (Lee et al., 2001, 2002; Chandler et al., 2003; Brown et al., 2004; Jennings et al., 2004; Isebaert et al., 2009; Alexander et al., 2011). Trichothecene producing *Fusarium* spp. was identified via *tri5* gene amplification (Hue et al., 1999); disruption or deletion in *tri7* and *tri13* genes was targeted via PCR assays in differentiation of DON or NIV chemotypes (Chandler et al., 2003; Waalwijk et al., 2003); sub-chemotypes, 3-ADON and 15-ADON, were identified via *tri3* gene size differences by PCR assays (Jennings et al., 2004); and each chemotype was identified via specific PCR assay by amplification of *tri12* gene (Wang et al., 2008a).

In contrast to the data related to the differences of genes located on core *tri5* gene cluster, common expression profiles have been reported. *tri4*, *tri5* and *tri6* are

Table 12.6. Genes located in core *tri5* core gene cluster and their expression status related to the accession number AF359360.3 (*F. sporotrichioides*).

Gene	Product	Size (bp)	Exon/ intron	Expression Status
tri8	Trichothecene C-3 deacetylase	1344	1/0	+[1]
tri7	3-acetyltrichothecene 4-*O*-acetyltransferase	1379	2/1	±[2]
tri3	3-acetyltrichothecene 15-*O*-acetyltransferase	1783	5/4	–[3]
tri4	Multifunctional oxygenase	1743	4/3	+[1]
tri6	Positive regulatory protein	654	1/0	+[1]
tri5	Trichodiene synthase	1185	2/1	+[1]
tri10	Regulatory protein	1340	2/1	+[1]
tri9	*TRI9* with unknown function	132	1/0	–[3]
tri11	C-15 hydroxylase	1722	5/4	+[1]
tri12	Efflux pump	1902	3/2	+[1]
tri13	cytochrome P450	1833	2/1	±[2]
tri14	Hypothetical protein	1176	2/1	–[3]

Note. [1]+ means expressed gene; [2]± means pseudogene in precise chemotype or strain; [3]– means gene with no crucial expression.

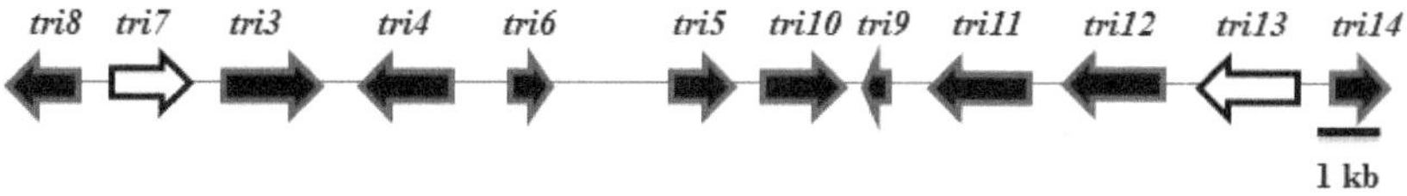

Fig. 12.10. Genes located on tri5 core gene cluster in *F. sporotrichioides*' genome (Yörük and Yli-Mattila, 2019).

some of the genes in which an expression is essential in each kind of chemotype including T-2, HT-2, DAS, 3-ADON, 15-ADON, and NIV. Figure 12.10 represents the *F. sporotrichioides tri5* core gene cluster not including three additional genes (a two-gene cluster: *tri1-tri16* and *tri101* locus). All genes are expressed in T-2 toxin producers. However, in the *F. graminearum* H-11 strain (DON producer) *tri5* core gene cluster, *tri9* and *tri10* genes are deleted and *tri7* and *tri13* genes are not expressing. Thus, knowledge of gene by gene in the core gene cluster is important for each strain with specific chemotype (Chandler et al., 2003; Kimura et al., 2007; McCormick et al., 2011).

The *tri5* is the first gene to express in the core gene cluster. Its product, TRI5 protein, converts farnesyl pyrophosphate to trichodiene via cyclization reaction. Trichodiene, is then converted to (iso)trichotriol via four different oxygenation series which are processed by TRI4 protein (cyctochrome P450 monooxygenase) coded by *tri4* gene. Trichotriol undergoes a non-enzymatic cyclization step forming isotrichodermol. Isotrichodermal is then converted to isotrichodermin by the expression of *tri101* gene. Another hydroxyl group was added via the *tri11* gene's product and 15-decalonectrin was formed. The last common step was carried out by TRI3 protein coded by *tri3* and 15-decalonectrin was converted to calonectrin. TRI1, TRI13, and TRI8 enzymes are responsible for the formation of DON,

3-ADON, and 15-ADON while TRI1, TRI13, TRI7 and TRI8 are responsible for the formation of NIV and its derivatives. Contrarily, in the biosynthetic pathway of class B-trichothecenes, TRI, TRI7, TRI1, TRI16, and TRI8 are involved in the production of the T-2 toxin (Kimura et al., 2007; McCormick et al., 2011).

5. Toxicity

5.1 Aflatoxins

Exposure to aflatoxins, fumonisins, ZEN, and trichothecenes through dietary products either directly from contaminated food or by ingesting animal-derived food previously fed on polluted feed, causes health problems because of their biological effects such as, carcinogenicity, immune toxicity, teratogenicity, neurotoxicity, nephrotoxicity, mutagenicity, and hepatotoxicity (Fig. 12.11). (Dalié et al., 2010).

Ingestion of aflatoxins has been associated with both acute toxicity and chronic carcinogenicity in humans and animals. In humans, symptoms of acute aflatoxicosis are vomiting, abdominal pain, pulmonary and cerebral edema, coma, and convulsions. Animals have symptoms of gastrointestinal dysfunction, reduced reproduction, reduced feed conversion and efficiency, depreciated milk and egg production (Alshannaq and Yu, 2017). Aflatoxin B1 has been involved in cancer development causing hepatocellular carcinoma (HCC). There are many risk factors

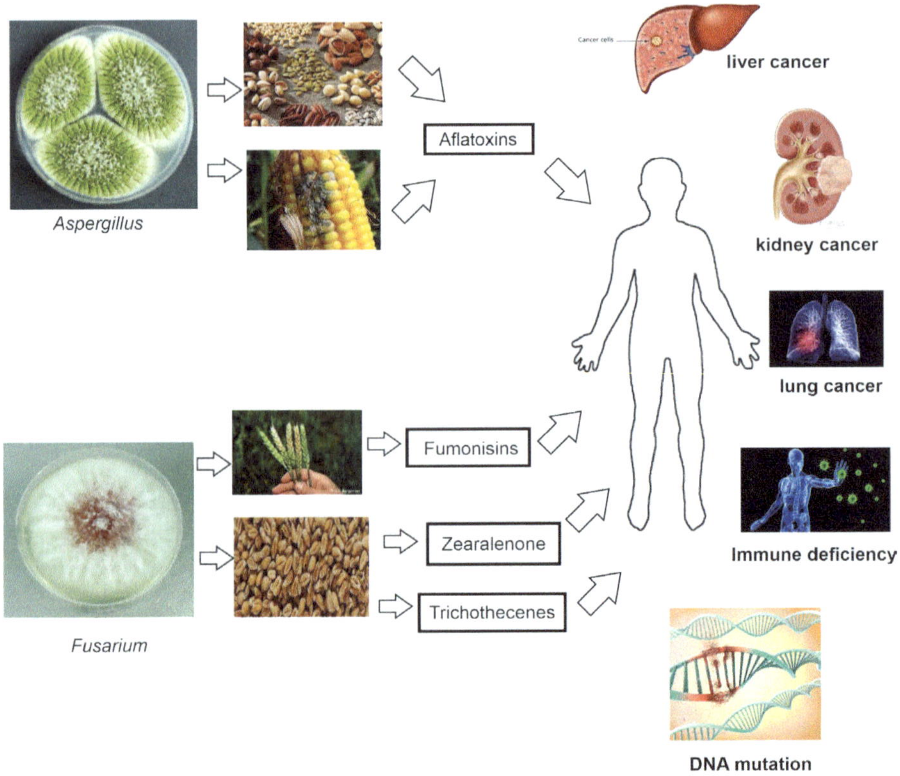

Fig. 12.11. Schematic diagram of mycotoxin contamination in food and toxic effects on human health.

for HCC development such as cirrhosis (chronic liver damage caused by fibrosis), infection with hepatitis B virus (HBV) or hepatitis C virus (HCV), and metabolic syndrome (EE, 2012). Aflatoxin B1 appears to be cooperative with HBV in causing HCC in Africa. The AFB1 metabolism plays a vital role in determining the toxicity of the toxin. AFB1 targets the liver of humans and animals, where it is metabolized into AFB1-8,9-Exo-epoxide (AFBO) by cytochrome 450. Epoxidation of AFB1 to the exo-8, 9-epoxide is a critical step in the genotoxic pathway of this carcinogenic mycotoxin. The exo-epoxide is highly unstable, and binds to DNA, particularly mitochondrial and nuclear nucleic acids, and nucleoproteins to form the predominant trans-8,9-dihydro-8-(N7-guanyl)-9-hydroxy-AFB1 (AFB1-N7-Gua) adduct, a pro-mutagenic DNA lesion (Wild and Turner, 2002). The resulted DNA adduct can induce mutation in the third base, No. 249 codon of the p53 genes, G–T, resulting in arginine to be replaced by tryptophan (Gouas et al., 2009). Because of the life-threatening concerns about AF contamination in food and feed and their dangerous health and economic impacts, AFs have been controlled by the FDA since 1969. Among all mycotoxins, AFs are the only ones regulated by established FDA action levels; others are subject only to advisory levels.

5.2 Fumonisins

Leudoencephalomalacia was first found in horses and other farm animals in USA more than one hundred years ago (Desjardins, 2006). A connection was found between the disease and mouldy maize grain, and it was possible to cause the disease in horses and pigs by feeding them with mouldy maize from disease outbreak areas. Later natural outbreaks of the disease were reported in South America, China, Africa, and southern Europe (Desjardins, 2006). Fumonisins have also been connected to pulmonary edema syndrome in pigs in USA (Osweiler et al., 1992), esophageal cancer in South Africa (Rheeder et al., 1992), and neural tube effects in USA and Mexico (Marasas et al., 2004; Desjardins, 2006). The disease was connected to *F. verticillioides* (which was later identified as *F. moniliforme*) in USA and subsequently in South Africa (Desjardins, 2006). Finally, the chemical identity of the isolated toxin in maize was characterized in South Africa and New Caledonia and named fumonisin B1 (Bezuidehnhout et al., 1988; Gelderblom et al., 1988; Laurent et al., 1989).

Fumonisins inhibit sphingolipid metabolism (Riley et al., 1996) and can cause eucoencephalomalacia, which can damage horse brains and kill them after only a few days of eating contaminated feed (Wilson et al., 1990; Desjardins, 2006). Fumonisins have been found to be carcinogenic to rodents (Lemmer et al., 1999; Gelderblom et al., 2001) and they have also been associated epidemiologically with human diseases (Marasas, 2001; Marasas et al., 2004). The International Agency for Research on Cancer classified fumonisin B1 in Group 2B (IARC, 2002). Recent surveys have raised concerns about the extent of fumonisin B1 contamination and its implications for animal health and productivity (Rodrigues and Naehrer, 2012; Boutigny et al., 2014; Cendoya et al., 2014; Abd-El Fatah et al., 2015). Consumption of fumonisin-contaminated maize leads to disruption of sphingolipid metabolism, associated with human esophageal cancer and several toxicoses in livestock animals, and increases the risk for neural tube defects in children (Marasas, 1995; Marasas et al., 2004; Missmer et al., 2006).

5.3 Zearalenone

Consumption of zearalenone-contaminated grains and the corresponding products are the main contributors of the dietary intake of ZEN. Although the specific food consumption patterns varied among different countries, in particular grain milling products, bread and fine bakery wares are considered as the main sources of human exposure to ZEN in European countries (EFSA, 2011). Dermal route and airborne ZEN inhaled by human and animals are another source of mycotoxin exposure that causes health problems in animals as well as in humans (Peraica et al., 1999; Wang et al., 2008b; EFSA, 2011). Humans also get exposed to ZEN and its metabolites by consuming foodstuffs of animal origin, which were fed on contaminated grains. However, contribution to human intake of ZEN residues via food products of animal origin is negligible compared to direct consumption of grain-based food, since the contaminant is diluted in these food products through various processes, including the formation of conjugated forms of mycotoxin (Goyarts et al., 2007; EFSA, 2011; Liu and Applegate, 2020). This modified form of ZEN and its reduced metabolites are generated from its Phase II conjugation reaction with monosaccharides and sulfates, e.g., zearalenone-14-glucoside (ZEN-14-G), α-zearalenol-14-glucoside (α-ZEL-14G), and β-zearalenol-14-glucoside (β-ZEL-14G), zearalenone-4-sulfate (ZEN-4S) (Mirocha et al., 1981; Olsen et al., 1986; Berthiller et al., 2013; Cirlini et al., 2016; Dellafiora, 2016; EFSA, 2017; Mahato et al., 2021).

The ZEN exhibits low acute toxicity after oral administration (oral LD50-4000 up to > 20000 mg per kg b. wt), however serious health problems could occur depending on the dose and long-term duration of exposure to this toxin (NTP, 1982; JECFA, 2001; Rai et al., 2020). The primary effect of ZEN is estrogenic activity in various animal species. ZEN and some of its metabolites compete with endogenous steroidal sex hormone 17β-estradiol to bind the oestrogen receptors. It leads to reproductive abnormalities; decreased fertility; increased embryonic resorptions; reduced litter size; changes in weight of adrenal, thyroid, and pituitary glands; and changes in serum levels of progesterone and 17β-oestradiol (Peraica et al., 1999; EFSA, 2011; Mostrom, 2012; Gil-Serna et al., 2014; Ropejko and Twarużek, 2021). It has a potency to be an etiological agent of the hyper estrogenic syndrome in humans (Massart and Saggese, 2010; Poór et al., 2015; Zhang et al., 2018).

In addition to estrogenic properties of ZEN, its effect on hamatological changes and hepatic disturbances was reported from different animals (rats, mice, piglets) (Maaroufi et al., 1996; Abbès et al., 2006; Jiang et al., 2010). In the context of *in vivo* and *in vitro* studies of genotoxicity, reduction of the mitotic activity, and chromosomal aberrations in various cell type were observed due to ZEN exposure (EFSA, 2011). Hepatocellular adenomas in female mice and pituitary adenomas in both male and female mice were determined in studies devoted to carcinogenic activity of ZEN in experimental animals (EFSA, 2011). Several reports reveal the possible carcinogenic activity of ZEN for human esophageal cancer and breast cancer; however, it has been listed by the International Agency for Research on Cancer (IARC) as group 3 carcinogen (not classifiable as to their carcinogenicity to human) (IARC, 1993).

5.4 Trichothecenes

The historical background of trichothecene problems and discoveries of different trichothecenes have been described by Ueno (1983), Desjardins (2006) and Yli-Mattila and Gagkaeva (2016). The most important trichothecenes are T-2, HT-2, DON, and NIV. The toxicity of trichothecenes from higher to lower is as follows T-2 toxin, diacetoscirpenol (DAS), DON, and NIV. However, particularly high levels of DON and NIV could be a reason for shock-like death (Sabuncuoğlu et al., 2008; Yazar and Omurtag, 2008). They are most associated with cereal crops including wheat, barley, maize, rye, and oat throughout the world. Trichothecenes have been associated with taumelge treide and alimentary toxic aleukia in Russia by the end of 19th century. Vomiting, headache, diarrhea, and vertigo are symptoms of trichothecene related diseases. At the molecular level, protein synthesis inhibition, DNA-RNA synthesis inhibition, and apoptosis are triggered by the intake of trichothecene related products (Yazar and Omurtag, 2008; McCormick et al., 2011).

6. Detection Methods

Accurate detection of aflatoxins provides a strong guarantee for certifying food safety. At the time of the aflatoxin discovery, a thin layer chromatography (TLC) was the most common method among the chromatographic methods. Since then, high performance liquid chromatography (HPLC), and liquid chromatography mass spectroscopy (LCMS) replaced TLC, besides the enzyme-linked immunosorbent assay (ELISA). Nevertheless, these traditional techniques have drawbacks, as they are time-consuming and require skilled personnel to operate. Chromatographic techniques used for major mycotoxins detection are summarized in Table 12.2.

The PCR and visual methods, which possess the merits of rapidity and sophisticated instrument-free present an excellent potential for the on-site detection of AFs such as fluorescence/near-infrared spectroscopy (FS/NIRS) and hyperspectral imaging (HSI) (Mahato et al., 2019). Visual sensors use antibodies, aptamers, or nanomaterials to recognize the AFs and then transform these recognizing events to visible signals through paper-based immunoassay, chromogenic reactions, or fluorescent emission (Wang et al., 2021). Zhongzhi and Limiao (2020), managed to use hyperspectral-imaging technology to detect the degree of aflatoxin contamination based on a support vector machine (SVM) combining band index and narrow band. They artificially infected peanut kernels with five different concentrations from aflatoxin. Next, hyperspectral images with 33 bands (400–720 nm) were taken using a hyperspectral imaging system under 365 nm UV light. Then four fluorescence indices, namely, Radiation Index (RI), Difference Radiation Indices (DRI), Ratio Radiation Index (RRI), and Normalized Difference Radiation Index (NDRI) were calculated. Finally, Fisher's method was used to optimize and obtain a narrowband spectrum, and the RBF-SVM model was used to recognize aflatoxin. Jia et al. (2021) have developed a simple and highly sensitive quantum-dot nanobeads-based (QBs) lateral flow fluorescent strip immunosensor for detection of AFB1 in edible and medicinal lotus seeds. Results revealed that the new test strip sensor could achieve rapid detection of AFB1 within 15 mins, with a limited detection (LOD) of 1 ng/mL (2 μg/kg) and a linear range of 1–19 ng/mL (2–38 μg/kg). Wang et al. (2019) introduced an aptamer that could hybridize with the G-rich DNA probes and form a G-quadruplex DNAzyme with hemin to recognize the targets. The presence of AFB1 would lead to the dissociation of aptamer/G-quadruplex DNAzyme probe and deprive its peroxidase-like catalytic activity. This method's detection limit could reach 0.02 μg/L with a spectrofluorometer for measuring the fluorescent signal. Aptamers specific to AFB1 could form folded structures upon exposure to the target. This folding behavior could resist against exonuclease I digestion to show a strong response to SYBR Gold. However, these visual methods for detection of AFs don´t require a sophisticated instrument and save time; they lack in quantifying the exact quantity of targets, possess a challenge to multiplex analysis capability, and extracting the target molecules from the food and feed matrix remains the obstacle to on-site detection (Wang et al., 2021).

Liquid chromatography with fluorescence detection is the most widely used method for the detection and quantification of fumonisins (Shephard et al. 1990; 2009). But other methods such as, MS/MS detection (Adejumo et al. 2007) should be used for detection of A- and P-series of fumonisins. Thin-layer chromatography, gas-liquid chromatography, and immunoassays are also used for identification of fumonisins. A fluorescence-based magnetic separation immunoassay for simultaneous detection of mycotoxins fumonisin B1 and ZEN) was recently established (Li et al., 2021). The method employed high fluorescent upconversion-nanoparticles (UCNPs) conjugated with biotinylated antigens as upconversion fluoroscent probes. Magnetic nanoparticles (MNPs) immobilized with monoclonal antibodies were used as immune-capture probes. Highly sensitive detection of FB1 and ZEN was achieved based on the luminescence properties of UCNPs and the separation effects of MNPs. There are also modified fumonisins, which are more difficult to detect (Braun and Wink, 2018).

The ZEN and its metabolites can be quantified by conventional analytical methods like high performance liquid chromatography (HPLC), gas chromatography (GC), liquid chromatography with tandem mass spectrometry (LC-MS/MS), gas chromatography-mass spectrometry (GC-MS), thin layer chromatography (TLC) (Liu and Applegate, 2020; Mahato et al., 2021). Enzymatic assays and immunoassay methods such as enzyme-linked immunosorbent assay (ELISA) and lateral flow immunoassay (LFA) are other technologies for the analysis of ZEN in food and feed (https://food.r-biopharm.com/; retrieved June 2021).

Different methods have been used in detection of trichothecenes. Analytical methods including qualitative and quantitative approaches, have been widely used in class A- and class B-trichothecenes detection. Thus, precise and reliable data have been obtained related to *Fusarium* chemotypes (Pasquali and Migheli, 2014; Yörük and Yli-Mattila, 2019). Thin layer chromatography, gas chromatography-mass spectroscopy, high-pressure liquid chromatography and ELISA methods have been widely used in T-2 toxin, HT-2, and NX-2 detection (McCormick et al., 2011; Varga et al., 2015). Similarly, chemotypes of class B-trichothecenes have been precisely detected and quantified via analytical methods (Pasquali and

Migheli, 2014; Pasquali et al., 2016; Yörük and Yli-Mattila, 2019). Even if analytical methods provide reliable and precise data related to chemotype knowledge of class B-trichothecene producer *Fusarium* spp., these methods include plenty of steps, and they are time-consuming. However, PCR provides fast and reliable data on chemotype of *Fusarium* spp. Investigations related to chemotype distribution studies including PCR and/or analytical methods yielded great scale data accumulation on *Fusarium* spp. chemotypes. These results showed that *F. graminearum* and 15-ADON chemotypes were predominant factors in fusarium head blight (FHB) in southern and central Europe. Similarly, in cooler regions of Europe, the 3-ADON chemotype of *F. gramineaum* and *F. culmorum* have been predominating factors in FHB (Yörük and Yli-Mattila, 2019). It must be mentioned, however, that the PCR tests are not always reliable as compared to the chemotype determined by sensitive chemical analysis (Pasquali et al. 2011; Villafana et al., 2020). Thus, the usage of both PCR and analytical methods is the best way to obtain reliable data on class B-trichothecene chemotyping.

7. Conclusion and Future Perspectives

Mycotoxin contaminations have great significance on human and livestock health. The occurrence of mycotoxin contamination can be seen during the cultivation in the field and contaminated food can occur at any stage of production, processing, distribution, storage, or preparation (Food Standards Agency, 2011; Abbas et al., 2017; Sobrova et al., 2010; Shikhaliyeva et al., 2020). Since mycotoxins are naturally occurring, their presence in foods cannot be completely avoided. However, appropriate control strategies have been considered to avoid the negative effects of mycotoxins on animal and human health. These control strategies mainly involve prevention of moulds in crops and other feedstuffs, application of efficient decontamination methods to mycotoxin contaminated foods and feeds, and continuous surveillance of mycotoxins in agricultural crops. The efficient and common disease management strategies (such as strategies used in head blight and crown rot diseases) have been widely and indirectly sued in mycotoxin contamination management worldwide. Disease resistant cultivar usage or development such as *Triticum aestivum* L. cv. Sumai-3, fungicide treatment, e.g., thiophanate methyl and tetraconazole, biocontrol agent usage such as *Trichoderma* spp., or crop rotations could be useful and important examples of mycotoxin levels controlling.

The novel strategies including the use of plant or bacteria derived secondary metabolites such as essential oils have been initiated for mycotoxin management in recent years (Arif et al., 2009; Özsoy et al., 2020). Plant extracts contain aromatic secondary metabolites, such as alkaloids, flavonoids, saponins, glycosides and tannins (Paranagama et al., 2003). Several lines of evidence demonstrated that some plant extracts have antifungal or anti-aflatoxigenic activities, because they are rich with antioxidants (El Khoury et al., 2017; Kohiyama et al., 2015; Zhao et al., 2018). In addition to these *in planta* strategies, *in vitro* approaches (e.g., H_2O_2 treatment, pH alterations or siRNA usage) could also be accepted as useful in the fight against mycotoxin controlling strategies (Ponts et al., 2009; Yörük and Albayrak, 2019).

However, these approaches need more detailed investigations on *in vivo* and *in planta* systems.

Conflicts of Interest

The authors declare no conflict of interest.

References

Abbas, H.K., Shier, W.T., Plasencia, J., Weaver, M.A., Bellaloui, N. et al. (2017). Mycotoxin contamination in corn smut (*Ustilago maydis*) galls in the field and in the commercial food products. *Food Control*, 71: 57–63. doi: 10.1016/j.foodcont.2016.06.006.

Abbès, S., Ouanes, Z., ben Salah-Abbès, J., Houas, Z., Oueslati, R. et al. (2006). The protective effect of hydrated sodium calcium aluminosilicate against haematological, biochemical, and pathological changes induced by zearalenone in mice. *Toxicon*, 47(5): 567–574. doi: 10.1016/j.toxicon.2006.01.016.

Abd-El-Fatah, S.I., Naguib, M.M., El-Hossiny, E.N., Sultan, Y.Y., Abodalam, T.H. et al. (2015). Molecular versus morphological identification of *Fusarium* spp. isolated from Egyptian corn. *Res. J. Pharm. Biol. Chem. Sci.*, 6(4): 1813–1822.

Abodalam, T. (2019). *Molecular Approaches for Mycotoxin Risk Reduction*. PhD thesis. University of Turku. Grano Oy - Turku, Finland.

Adejumo, T.O., Hettwer, U. and Karlovsky, P. (2007). Survey of maize from south-western Nigeria for zearalenone, α-and β-zearalenols, fumonisin B1 and enniatins produced by *Fusarium* species. *Food Addit. Contam.*, 24(9): 993–1000. doi: 10.1080/02652030701317285.

Aerts, D., Hauer, E.E., Ohm, R.A., Arentshorst, M., Teertstra, W.R. et al. (2018). The FlbA-regulated predicted transcription factor Fum21 of *Aspergillus niger* is involved in fumonisin production. *Antonie Leeuwenhoek*, 111(3): 311–322. doi: 10.1007/s10482-017-0952-1.

Ajmal, M., Akram, A., Hanif, N.Q., Mukhtar, T. and Arshad, M. (2021). Mycobiota isolation and aflatoxin b1 contamination in fresh and stored sesame seeds from rainfed and irrigated zones of Punjab, Pakistan. *J. Food Prot.*, 84(10): 1673–1682. doi: 10.4315/JFP-21-060.

Albayrak, G., Yörük, E., Gazdağli, A. and Sharifnabi, B. (2016). Genetic diversity among *Fusarium graminearum* and *F. culmorum* isolates based on ISSR markers. *Arch. Biol. Sci.*, 68(2): 333–343. doi: 10.2298/ABS150630025A.

Alexander, N.J., McCormick, S.P., Waalwijk, C., van der Lee, T. and Proctor, R.H. (2011). The genetic basis for 3-ADON and 15-ADON trichothecene chemotypes in *Fusarium*. *Fungal Genet. Biol.*, 48(5): 485–495. doi: 10.1016/j.fgb.2011.01.003.

Alshannaq, A. and Yu, J.H. (2017). Occurrence, toxicity, and analysis of major mycotoxins in food. *Int. J. Environ. Res. Public Health*, 14(6): 632. doi: 10.3390/ijerph14060632.

Arif, T., Bhosale, J.D., Kumar, N., Mandal, T.K., Bendre et al. (2009). Natural products–antifungal agents derived from plants. *J. Asian Nat. Prod. Res.*, 11(7): 621–638. doi: 10.1080/10286020902942350.

Awapak, D., Petchkongkaew, A., Sulyok, M. and Krska, R. (2021). Co-occurrence and toxicological relevance of secondary metabolites in dairy cow feed from Thailand. *Food Addit. Contam., A*, 1–15. doi: 10.1080/19440049.2021.1905186.

Ayalew, A. (2010). Mycotoxins and surface and internal fungi of maize from Ethiopia. *Afr. J. Food Agric. Nutr. Dev.*, 10(9): 4109–4123. doi: 10.4314/ajfand.v10i9.62890.

Bacon, C.W. and Hinton, D.M. (1996). Symptomless endophytic colonization of maize by *Fusarium moniliforme*. *Can. J. Bot.*, 74(8): 1195–1202. 10.1139/b96-144.

Bacon, C.W., Glenn, A.E. and Yates, I.E. (2008). *Fusarium verticillioides*: managing the endophytic association with maize for reduced fumonisins accumulation. *Toxin Reviews*, 27(3-4): 411–446. doi: 0.1080/15569540802497889.

Bennett, J. and Klich, M. (2003). Mycotoxins. C lin. *Microbiol. Rev.*, 16: 497–516. doi: 10.1128/CMR.16.3.497–516.2003.

Bennett, J.W. (1987). Mycotoxins, mycotoxicoses, mycotoxicology, and mycopathologia. *Mycopathologia*, 100: 3–5. doi: 10.1007/BF00769561.

Berthiller, F., Crews, C., Dall'Asta, C., Saeger, S.D., Haesaert et al. (2013). Masked mycotoxins: A review. *Mol. Nutr. Food Res.*, 57(1): 165–186. doi: 10.1002/mnfr.201100764.

Bezuidenhout, S.C., Gelderblom, W.C., Gorst-Allman, C.P., Horak, R.M., Marasas, W.F., Spiteller, G. and Vleggaar, R. (1988). Structure elucidation of the fumonisins, mycotoxins from *Fusarium moniliforme*. *J. Chem. Soc. Chem. Comm.*, (11): 743–745. doi: 10.1039/C39880000743.

Bhat, R., Rai, R.V. and Karim, A.A. (2010). Mycotoxins in food and feed: present status and future concerns. *CRFSFS*, 9(1): 57–81. doi: 10.1111/j.1541-4337.2009.00094.x.

Bojja, R.S., Cerny, R.L., Proctor, R.H. and Du, L. (2004). Determining the biosynthetic sequence in the early steps of the fumonisin pathway by use of three gene-disruption mutants of *Fusarium verticillioides*. *J. Agric. Food Chem.*, 52(10): 2855–2860. doi: 10.1021/jf035429z.

Bottalico, A., Visconti, A., Logrieco, A., Solfrizzo, M. and Mirocha, C.J. (1985). Occurrence of zearalenols (diastereomeric mixture) in corn stalk rot and their production by associated *Fusarium* species. *AEM*, 49(3): 547–551. doi: 10.1128/aem.49.3.547-551.1985.

Bottini, A.T. and Gilchrist, D.G. (1981). Phytotoxins. I. A 1-aminodimethylheptadecapentol from *Alternaria alternata* f. sp. lycopersici. *Tetrahedron. Lett.*, 22(29): 2719–2722. doi: 10.1016/S0040-4039(01)90534-9.

Boutigny, A.L., Ward, T.J., Ballois, N., Iancu, G. and Ioos, R. (2014). Diversity of the *Fusarium graminearum* species complex on French cereals. *Eur. J. Plant Pathol.*, 138(1): 133–148. doi: 10.1007/s10658–013-0312-6.

Braun, M.S. and Wink, M. (2018). Exposure, occurrence, and chemistry of fumonisins and their cryptic derivatives. *CRFSFS*, 17(3): 769–791. doi: 10.1111/1541-4337.12334.

Brian, P.W., Dawkins, A.W., Grove, J.F., Hemming, H.G. and Norris, G.L.F. (1961). Phytotoxic Compounds produced by *Fusarium equiseti*. *J. Exp. Bot.*, 12(1): 1–12. doi: 10.1093/jxb/12.1.1.

Brown, D.W., Mccormick, S.P., Alexander, N.J., Proctor, R.H. and Desjardins, A.E. (2002). Inactivation of a cytochrome P-450 is a determinant of trichothecene diversity in *Fusarium* species. *Fungal Genet. Biol.*, 36: 224–233. doi: 10.1016/S1087-1845(02)00021-X.

Brown, D.W., Dyer, R.B., McCormick, S.P., Kendra, D.F. and Plattner, R.D. (2004). Functional demarcation of the *Fusarium* core trichothecene gene cluster. *Fungal Genet. Biol.*, 41(4): 454–462. doi: 10.1016/j.fgb.2003.12.002.

Bryła, M., Waśkiewicz, A., Podolska, G., Szymczyk, K., Jędrzejczak, R. et al. (2016). Occurrence of 26 mycotoxins in the grain of cereals cultivated in Poland. *Toxins*, 8(6): 160. doi: 10.3390/toxins8060160.

Caceres, I., Al Khoury, A., El Khoury, R., Lorber, S., P Oswald, I. et al. (2020). Aflatoxin biosynthesis and genetic regulation: A review. *Toxins*, 12(3): 150. doi: 10.3390/toxins12030150.

Cavaliere, C., D'Ascenzo, G., Foglia, P., Pastorini, E., Samperi, R. et al. (2005). Determination of type B trichothecenes and macrocyclic lactone mycotoxins in field contaminated maize. *Food Chem.*, 92(3): 559–568. doi: 10.1016/j.foodchem.2004.10.008.

Cendoya, E., Monge, M.P., Palacios, S.A., Chiacchiera, S.M., Torres, A.M. et al. (2014). Fumonisin occurrence in naturally contaminated wheat grain harvested in Argentina. *Food Control*, 37: 56–61. doi: 10.1016/j.foodcont.2013.09.031.

Çepni, E., Tunalı, B. and Gürel, F. (2013). Genetic diversity and mating types of *Fusarium culmorum* and *Fusarium graminearum* originating from different agro-ecological regions in Turkey. *J. Basic Microbiol.*, 53(8): 686–694. doi: 10.1002/jobm.201200066.

Chandler, E.A., Simpson, D.R., Thomsett, M.A. and Nicholson, P. (2003). Development of PCR assays to Tri7 and Tri13 trichothecene biosynthetic genes, and characterisation of chemotypes of *Fusarium graminearum*, *Fusarium culmorum* and *Fusarium cerealis*. *Physiol. Mol. Plant. Path.*, 62(6): 355–367. doi: 10.1016/S0885-5765(03)00092-4.

Cirlini, M., Barilli, A., Galaverna, G., Michlmayr, H., Adam, G. et al. (2016). Study on the uptake and deglycosylation of the masked forms of zearalenone in human intestinal Caco-2 cells. *Food Chem. Toxicol.*, 98: 232–239. doi: 10.1016/j.fct.2016.11.003.

Commission Regulation (EC) No 1126/2007 of 28 September 2007 amending Regulation (EC) No 1881/2006 setting maximum levels for certain contaminants in foodstuffs as regards *Fusarium* toxins in maize and maize products. *OJEU*, 255: 14–17.

Dalié, D.K.D., Deschamps, A.M. and Richard-Forget, F. (2010). Lactic acid bacteria–Potential for control of mould growth and mycotoxins: A review. *Food Control*, 21(4): 370–380. doi: 10.1016/j.foodcont.2009.07.011.

Dellafiora, L., Perotti, A., Galaverna, G., Buschini, A. and Dall'Asta, C. (2016). On the masked mycotoxin zearalenone-14-glucoside. Does the mask truly hide? *Toxicon*, 111: 139–142. doi: 10.1016/j.toxicon.2016.01.053.

Desjardins, A.E. (2006). *Fusarium Mycotoxins: Chemistry, genetics, and biology*. St. Paul, MN, USA: American Phytopathological Society (APS Press).

Desjardins, A.E. and Proctor, R.H. (2007). Molecular biology of *Fusarium* mycotoxins. *Int. J. Food Microbiol.*, 119(1-2): 47–50. doi: 10.1016/j.ijfoodmicro.2007.07.024.

Drzymala, S.S., Binder, J., Brodehl, A., Penkert, M., Rosowski, M. et al. (2015). Estrogenicity of novel phase I and phase II metabolites of zearalenone and cis-zearalenone. *Toxicon*, 105: 10–12. doi: 10.1016/j.toxicon.2015.08.027.

Du, L., Zhu, X., Gerber, R., Huffman, J., Lou, L. et al. (2008). Biosynthesis of sphinganine-analog mycotoxins. *J. Ind. Microbiol. Biotechnol.*, 35(6): 455–464. doi: 10.1007/s10295-008-0316-y.

EASL-EORTC (EE) (2012). Clinical practice guidelines: Management of hepatocellular carcinoma. *J. Hepatol.*, 56: 908–43. doi: 10.1016/j.jhep.2011.12.001.

EC (2010). European Commission. Commission Regulation (EC) No 165/2010 of 26 February 2010 amending Regulation (EC) No 1881/2006 setting maximum levels for certain contaminants in foodstuffs as regards aflatoxins. *OJEU*, 50: 8–12.

EFSA Panel on Contaminants in the Food Chain (2011). Scientific Opinion on the risks for public health related to the presence of zearalenone in food. *EFSA J.*, 9(6): 2197. doi: 10.2903/j.efsa.2011.2197.

EFSA Panel on Contaminants in the Food Chain (CONTAM), Knutsen, H.K., Alexander, J., Barregård, L., Bignami, M., Brüschweiler, B. et al. (2017). Risks for animal health related to the presence of zearalenone and its modified forms in feed. *EFSA J.*, 15(7): e04851. doi: 10.2903/j.efsa.2017.4851.

Ehrlich, K.C., Chang, P.K., Yu, J. and Cotty, P.J. (2004). Aflatoxin biosynthesis cluster gene cypA is required for G aflatoxin formation. *AEM*, 70(11): 6518–6524. doi: 10.1128/AEM.70.11.6518-6524.2004.

Ehrlich, K.C. (2009). Predicted roles of the uncharacterized clustered genes in aflatoxin biosynthesis. *Toxins*, 1(1): 37–58. doi: 10.3390/toxins1010037.

El Khoury, R., Caceres, I., Puel, O., Bailly, S., Atoui, A. et al. (2017). Identification of the anti-aflatoxinogenic activity of *Micromeria graeca* and elucidation of its molecular mechanism in *Aspergillus flavus*. *Toxins*, 9(3): 87. doi: 10.3390/toxins9030087.

FAO. (2004). *Worldwide Regulations for Mycotoxins in Food and Feed in 2003*. Rome, Italy: Food and Agriculture Organization of the United Nations.

Flaherty, J.E. and Woloshuk, C.P. (2004). Regulation of fumonisin biosynthesis in *Fusarium verticillioides* by a zinc binuclear cluster-type gene, ZFR1. *AEM*, 70(5): 2653–2659. doi: 10.1128/AEM.70.5.2653-2659.2004.

Food Standards Agency (2011). Foodborne Disease Strategy. Scanning 4.

Foroud, N.A. and Eudes, F. (2009). Trichothecenes in cereal grains. *Int. J. Mol. Sci.*, 10(1): 147–173. doi: 10.3390/ijms10010147.

Freeman, G.G. and Morrison, R.I. (1948). Trichothecin: An antifungal metabolic product of *Trichothecium roseum* Link. *Nature*, 3; 162(4105): 30. doi: 10.1038/162030a0.

Frisvad, J.C., Smedsgaard, J., Samson, R.A., Larsen, T.O. and Thrane, U. (2007). Fumonisin B2 production by *Aspergillus niger*. *J. Agric. Food Chem.*, 55(23): 9727–9732. doi: 10.1021/jf0718906.

Gaffoor, I. and Trail, F. (2006). Characterization of two polyketide synthase genes involved in zearalenone biosynthesis in Gibberella zeae. *AEM*, 72(3): 1793–1799. doi: 10.1128/AEM.72.3.1793-1799.2006.

Gagkaeva, T.Y. and Yli-Mattila, T. (2020). Emergence of *Fusarium verticillioides* in Finland. *Eur. J. Plant Pathol.*, 158(4): 1051–1057. doi: 10.1007/s10658-020-02118-2.

Gallo, P., Imbimbo, S., Alvino, S., Castellano, V., Arace, O. et al. (2021). Contamination by aflatoxins B/G in food and commodities imported in Southern Italy from 2017 to 2020: A risk-based evaluation. *Toxins*, 13(6): 368. doi: 10.3390/toxins13060368.

Garcia, D., Barros, G., Chulze, S., Ramos, A.J., Sanchis, V. et al. (2012). Impact of cycling temperatures on *Fusarium verticillioides* and *Fusarium graminearum* growth and mycotoxins production in soybean. *J. Sci. Food Agric.*, 92(15): 2952–2959. doi: 10.1002/jsfa.5707.

GB 2761-2017. (2017). *National Food Safety Standard: Maximum levels of mycotoxins in foods*. National Health Commission & State Administration of Market Regulation.

Gelderblom, W.C., Jaskiewicz, K., Marasas, W.F., Thiel, P.G., Horak, R.M., Vleggaar, R. and Kriek, N.P. (1988). Fumonisins--novel mycotoxins with cancer-promoting activity produced by *Fusarium moniliforme*. *AEM*, 54(7): 1806–1811. doi: 10.1128/aem.54.7.1806-1811.1988.

Gelderblom, W.C., Abel, S., Smuts, C.M., Marnewick, J., Marasas, W.F. et al. (2001). Fumonisin-induced hepatocarcinogenesis: Mechanisms related to cancer initiation and promotion. *Environ. Health. Perspect.*, 109(suppl 2): 291–300. doi: 10.1289/ehp.01109s2291.

Gil-Serna, J., Vázquez, C., González-Jaén, M.T. and Patiño, B. (2014). Mycotoxins: Toxicology. pp. 887–892. *In*: Batt, C.A. and Tortorello, M.L. (Eds.). *Encyclopedia of Food Microbiology*; Cambridge, MA, USA: Academic Press.

Girgin, G., Başaran, N. and Şahin, G. (2001). Mycotoxins in Turkey and the world. *Turk. Bull. Hyg. Exp. Biol.*, 58(3): 97–118.

Glenn, A.E., Zitomer, N.C., Zimeri, A.M., Williams, L.D., Riley, R.T. et al. (2008). Transformation-mediated complementation of a FUM gene cluster deletion in *Fusarium verticillioides* restores both fumonisin production and pathogenicity on maize seedlings. *Mol. Plant. Microbe. Interact.*, 21(1): 87–97. doi: 10.1094/MPMI-21-1-0087.

Goldblatt, L.A. (1969). *Aflatoxin: Scientific background, control, and implications*. New York: Academic Press.

Golinski, P., Vesonder, R.F., Latus-Zietkiewicz, D. and Perkowski, J. (1988). Formation of fusarenone X, nivalenol, zearalenone, alpha-trans-zearalenol, beta-trans-zearalenol, and fusarin C by *Fusarium crookwellense*. *AEM*, 54(8): 2147–2148. doi: 10.1128/aem.54.8.2147-2148.1988.

Gouas, D., Shi, H. and Hainaut, P. (2009). The aflatoxin induced TP53 mutation at codon 249 (R249S): Biomarker of exposure, early detection, and target for therapy. *Cancer Lett.*, 286(1): 29–37. doi: 10.1016/j.canlet.2009.02.057.

Gourama, H. and Bullerman, L.B. (1995). Detection of molds in foods and feeds: Potential rapid and selective methods. *J. Food Prot.*, 58(12): 1389–1394. doi: 10.4315/0362-028X-58.12.1389.

Goyarts, T., Dänicke, S., Valenta, H. and Ueberschär, K.H. (2007). Carry-over of Fusarium toxins (deoxynivalenol and zearalenone) from naturally contaminated wheat to pigs. *Food Addit. Contam.*, 24(4): 369–380. doi: 10.1080/02652030600988038.

Güldener, U., Mannhaupt, G., Münsterkötter, M., Haase, D., Oesterheld, M. et al. (2006). FGDB: a comprehensive fungal genome resource on the plant pathogen *Fusarium graminearum*. *Nucleic Acids Res.*, 34: 456–458.

Guo, Z., Pfohl, K., Karlovsky, P., Dehne, H.W. and Altincicek, B. (2016). Fumonisin B1 and beauvericin accumulation in wheat kernels after seed-borne infection with *Fusarium proliferatum*. *Agric. Food Sci.*, 25(2): 138–145. doi: 10.23986/afsci.55539.

Hansen, F.T., Gardiner, D.M., Lysøe, E., Fuertes, P.R., Tudzynski, B. et al. (2015). An update to polyketide synthase and non-ribosomal synthetase genes and nomenclature in *Fusarium*. *Fungal Genet. Biol.*, 75: 20–29. doi: 10.1016/j.fgb.2014.12.004.

Heperkan, D., Güler, F.K. and Oktay, H.I. (2012). Mycoflora and natural occurrence of aflatoxin, cyclopiazonic acid, fumonisin and ochratoxin A in dried figs. *Food Addit. Contam. A.*, 29(2): 277–286. doi: 10.1080/19440049.2011.597037.

Hue, F.X., Huerre, M., Rouffault, M.A. and De Bievre, C. (1999). Specific detection of *Fusarium* species in blood and tissues by a PCR technique. *JCM*, 37(8): 2434–2438. doi: 10.1128/JCM.37.8.2434-2438.1999.

Hurd, R.N. (1977). Structure activity relationships in zearalenones. pp. 379–391. *In*: Rodricks, J.V., C.W. Hesseltine, and M.A. Mehlman (Eds.). *Mycotoxins in Human and Animal Health*, Park Forest South, Chicago, USA: Pathotox Publications, Inc.

IARC (International Agency for Research on Cancer) (1993). IARC monographs on the evaluation of the carcinogenic risk of chemicals to humans, some naturally occurring substances: Heterocyclic aromatic amines and mycotoxins. *Lyon*: 56: 39–444.

IARC Working Group (International Agency for Research on Cancer) (2002). *Some traditional herbal medicines, some mycotoxins, naphthalene, and styrene*. Vol. 82: 1–556.

Iqbal, S.Z., Jinap, S., Pirouz, A.A. and Faizal, A.A. (2015). Aflatoxin M1 in milk and dairy products, occurrence, and recent challenges: A review. *Trends Food Sci. Technol.*, 46(1): 110-119. doi: 10.1016/j.tifs.2015.08.005.

Isebaert, S., De Saeger, S., Devreese, R., Verhoeven, R., Maene, P. et al. (2009). Mycotoxin-producing *Fusarium* species occurring in winter wheat in Belgium (Flanders) during 2002–2005. *J. Phytopathol.*, 157(2): 108–116. doi: 10.1111/j.1439-0434.2008.01443.x.

JECFA (2001). *Evaluation of certain food additives and contaminants: Fifty-fifth report of the JOINT FAO WHO Expert Committee on Food Additives*. Geneva.

Jennings, P., Coates, M.E., Turner, J.A., Chandler, E.A. and Nicholson, P. (2004). Determination of deoxynivalenol and nivalenol chemotypes of *Fusarium culmorum* isolates from England and Wales by PCR assay. *Plant Pathol.*, 53(2): 182–190. doi: 10.1111/j.0032-0862.2004.00985.x.

Jeswal, P. and Kumar, D. (2015). Mycobiota and natural incidence of aflatoxins, ochratoxin A, and citrinin in Indian spices confirmed by LC-MS/MS. *Int. J. Microbiol.*, 2015. doi: 10.1155/2015/242486.

Jia, B., Liao, X., Sun, C., Fang, L., Zhou, L. et al. (2021). Development of a quantum dot nanobead-based fluorescent strip immunosensor for on-site detection of aflatoxin B1 in lotus seeds. *Food Chem.*, 356: 129614. doi: 10.1016/j.foodchem.2021.129614.

Jiang, S., Yang, Z., Yang, W., Gao, J., Liu, F. et al. (2010). Physiopathological effects of zearalenone in post-weaning female piglets with or without montmorillonite clay adsorbent. *Livest. Sci.*, 131(1): 130–136. doi: 10.1016/j.livsci.2010.02.022.

Jolly, P.E., Mazariegos, M., Contreras, H., Balas, N., Junkins, A. et al. (2021). Aflatoxin exposure among mothers and their ınfants from the Western Highlands of Guatemala. *Matern. Child Health J.*, 1–10. doi: 10.1007/s10995-021-03151-1.

Kamle, M., Mahato, D.K., Devi, S., Lee, K.E., Kang, S.G. et al. (2019). Fumonisins: Impact on agriculture, food, and human health and their management strategies. *Toxins*, 11(6): 328. doi: 10.3390/toxins11060328.

Kawashima, L.M., Vieira, A.P. and Soares, L.M.V. (2007). Fumonisin B1 and ochratoxin A in beers made in Brazil. *Food Sci. Technol.*, 27: 317–323. doi: 10.1590/S0101-20612007000200019.

Kelly, A., Proctor, R.H., Belzile, F., Chulze, S.N., Clear, R.M. et al. (2016). The geographic distribution and complex evolutionary history of the NX-2 trichothecene chemotype from *Fusarium graminearum*. *Fungal Genet. Biol.*, 95: 39–48. doi: 10.1016/j.fgb.2016.08.003.

Kim, Y.T., Lee, Y.R., Jin, J., Han, K.H., Kim, H. et al. (2005). Two different polyketide synthase genes are required for synthesis of zearalenone in *Gibberella zeae*. *Mol. Microbiol.*, 58(4): 1102–1113. doi: 10.1111/j.1365-2958.2005.04884.x.

Kimura, M., Tokai, T., O'Donnell, K., Ward, T.J., Fujimura, M. et al. (2003). The trichothecene biosynthesis gene cluster of *Fusarium graminearum* F15 contains a limited number of essential pathway genes and expressed non-essential genes. *FEBS Lett.*, 539(1-3): 105–110. doi: 10.1016/S0014-5793(03)00208-4.

Kimura, M., Tokai, T., Takahashi-Ando, N., Ohsato, S. and Fujimura, M. (2007). Molecular and genetic studies of *Fusarium* trichothecene biosynthesis: pathways, genes, and evolution. *Biosci. Biotechnol. Biochem.*, 71: 2105–2123. doi: 10.1271/bbb.70183.

Kohiyama, C.Y., Ribeiro, M.M.Y., Mossini, S.A.G., Bando, E., da Silva Bomfim, N. et al. (2015). Antifungal properties and inhibitory effects upon aflatoxin production of *Thymus vulgaris* L. by *Aspergillus flavus* Link. *Food Chem.*, 173: 1006–1010. doi: 10.1016/j.foodchem.2014.10.135.

Kroken, S., Glass, N.L., Taylor, J.W., Yoder, O.C. and Turgeon, B.G. (2003). Phylogenomic analysis of type I polyketide synthase genes in pathogenic and saprobic ascomycetes. *PNAS*, 100(26): 15670–15675. doi: 10.1073/pnas.2532165100.

Krska, R., Pettersson, H., Josephs, R.D., Lemmens, M., Mac Donald, S. et al. (2003). Zearalenone in maize: stability testing and matrix characterisation of a certified reference material. *Food Addit. Contam.*, 20(12): 1141–1152. doi: 10.1080/02652030310001615203.

Kuiper-Goodman, T., Scott, P. and Watanabe, H. (1987). Risk assessment of the mycotoxin zearalenone. *Regul. Toxicol. Pharmacol.*, 7(3): 253–306. doi: 10.1016/0273-2300(87)90037-7.

Kumar, P., Mahato, D.K., Kamle, M., Mohanta, T.K. and Kang, S.G. (2017). Aflatoxins: A global concern for food safety, human health, and their management. *Front. Microbiol.*, 7: 2170. doi: 10.3389/fmicb.2016.02170.

Laurent, D., Platzer, N., Kohler, F., Sauvant, M.P. and Pellegrin, F. (1989). Macrofusin and micromonilin: two new mycotoxins isolated from corn infested by *Fusarium moniliforme*. Sheld. *Microbiologie Aliment Nutrition*, 7: 9–16.

Lazzaro, I., Falavigna, C., DallAsta, C., Proctor, R.H., Galaverna, G. et al. (2012). Fumonisins B, A, and C production and masking in *Fusarium verticillioides* and *F. proliferatum* grown with different water activity regimes. *Int. J. Food Microbiol.*, 159: 93–100. doi: 10.1016/j.ijfoodmicro.2012.08.013.

Lee, S., Son, H., Lee, J. et al. (2011). A putative ABC transporter gene, *ZRA1*, is required for zearalenone production in *Gibberella zeae*. *Curr. Genet.*, 57: 343. doi: 10.1007/s00294-011-0352-4.

Lee, T., Oh, D.W., Kim, H.S., Lee, J., Kim, Y.H. et al. (2001). Identification of deoxynivalenol-and nivalenol-producing chemotypes of *Gibberella zeae* by using PCR. *AEM*, 67(7): 2966–2972. doi: 10.1128/AEM.67.7.2966-2972.2001.

Lee, T., Han, Y.K., Kim, K.H., Yun, S.H. and Lee, Y.W. (2002). Tri13 and Tri7 determine deoxynivalenol- and nivalenol-producing chemotypes of *Gibberella zeae*. *AEM*, 68(5): 2148–2154. doi: 10.1128/AEM.68.5.2148-2154.2002.

Lemmer, E.R., de la Motte Hall, P., Omori, N., Omori, M., Shephard, E.G. et al. (1999). Histopathology and gene expression changes in rat liver during feeding of fumonisin B1, a carcinogenic mycotoxin produced by *Fusarium moniliforme*. *Carcinogenesis*, 20(5): 817–824. doi: 10.1093/carcin/20.5.817.

Leslie J.F. and Summerell B.A. 2006. *The Fusarium Laboratory Manual*. Oxford, UK: Blackwell Publishing Ltd.

Li, J., Zhao, X., Wang, Y., Li, S., Qin, Y. et al. (2021). A highly sensitive immunofluorescence sensor based on bicolor upconversion and magnetic separation for simultaneous detection of fumonisin B1 and zearalenone. *Analyst*, 146(10): 3328–3335. doi: 10.1039/d1an00004g.

Li, M., Kong, W., Li, Y., Liu, H., Liu, Q. et al. (2016). High-throughput determination of multi-mycotoxins in Chinese yam and related products by ultra fast liquid chromatography coupled with tandem mass spectrometry after one-step extraction. *J. Chromatogr. B.*, 1022: 118–125. http://dx.doi.org/10.1016/j.jchromb.2016.04.014.

Liu, J. and Applegate, T. (2020). Zearalenone (ZEN) in Livestock and Poultry: Dose, toxicokinetics, toxicity and estrogenicity. *Toxins*, 12(6): 377. doi: 10.3390/toxins12060377.

Liu, Y. and Wu, F. (2010). Global burden of aflatoxin-induced hepatocellular carcinoma: A risk assessment. *Environ. Health Perspect.*, 118(6): 818–824. doi: 10.1289/ehp.0901388.

Liu, Y., Galani Yamdeu, J.H., Gong, Y.Y. and Orfila, C. (2020). A review of postharvest approaches to reduce fungal and mycotoxin contamination of foods. *CRFSFS*, 19(4): 1521–1560. doi: 10.1111/1541-4337.12562.

Logrieco, A., Bottalico, A., Mulé, G., Moretti, A. and Perrone, G. (2003). Epidemiology of toxigenic fungi and their associated mycotoxins for some Mediterranean crops. *Eur. J. Plant Pathol.*, 109(7): 645–667. doi: 10.1023/A:1026033021542.

López-Errasquín, E., Vázquez, C., Jiménez, M. and González-Jaén, M.T. (2007). Real-time RT-PCR assay to quantify the expression of FUM1 and FUM19 genes from the fumonisin-producing *Fusarium verticillioides*. *J. Microbiol. Methods*, 68(2): 312–317. doi: 10.1016/j.mimet.2006.09.007.

Lysøe, E., Klemsdal, S.S., Bone, K.R., Frandsen, R.J., Johansen, T. et al. (2006). The PKS4 gene of *Fusarium graminearum* is essential for zearalenone production. *AEM*, 72(6): 3924–3932. doi: 10.1128/AEM.00963-05.

Lysøe, E., Bone, K.R. and Klemsdal, S.S. (2009). Real-time quantitative expression studies of the zearalenone biosynthetic gene cluster in *Fusarium graminearum*. *Phytopathology*, 99(2): 176–184. doi: 10.1094/PHYTO-99-2-0176.

Maaroufi, K., Chekir, L., Creppy, E.E., Ellouz, F. and Bacha, H. (1996). Zearalenone induces modifications of haematological and biochemical parameters in rats. *Toxicon*, 34(5): 535–540. doi: 10.1016/0041-0101(96)00008-6.

Mahato, D.K., Lee, K.E., Kamle, M., Devi, S., Dewangan, K.N. et al. (2019). Aflatoxins in food and feed: An overview on prevalence, detection, and control strategies. *Front. Microbiol.*, 10: 2266. doi: 10.3389/fmicb.2019.02266.

Mahato, D.K., Devi, S., Pandhi, S., Sharma, B., Maurya, K.K. et al. (2021). Occurrence, impact on agriculture, human health, and management strategies of zearalenone in food and feed: A review. *Toxins*, 13(2): 92. doi: 10.3390/toxins13020092.

Manova, R. and Mladenova, R. (2009). Incidence of zearalenone and fumonisins in Bulgarian cereal production. *Food Control*, 20(4): 362–365. 10.1016/j.foodcont.2008.06.001.

Marasas, W.F. (1995). Fumonisins: Their implications for human and animal health. *Nat. Toxins*, 3(4): 193–198. doi: 10.1002/nt.2620030405.

Marasas, W.F. (2001). Discovery and occurrence of the fumonisins: A historical perspective. *Environ. Health Perspect*, 109(Suppl. 2): 239–243. doi: 10.1289/ehp.01109s2239.

Marasas, W.F., Riley, R.T., Hendricks, K.A., Stevens, V.L., Sadler, T.W. et al. (2004). Fumonisins disrupt sphingolipid metabolism, folate transport, and neural tube development in embryo culture and *in vivo*: A potential risk factor for human neural tube defects among populations consuming fumonisin-contaminated maize. *Nutr. J.*, 134(4): 711–716. doi: 10.1093/jn/134.4.711.

Marin, S., Ramos, A.J., Cano-Sancho, G. and Sanchis, V. (2013). Mycotoxins: Occurrence, toxicology, and exposure assessment. *Food Chem. Toxicol.*, 60: 218–237. doi: 10.1016/j.fct.2013.07.047.

Massart, F. and Saggese, G. (2010). Oestrogenic mycotoxin exposures and precocious pubertal development. *Int. J. Androl.*, 33(2): 369–376. doi: 10.1111/j.1365-2605.2009.01009.x.

Matny, O.N. (2015). *Fusarium* head blight and crown rot on wheat and barley: Losses and health risks. *APAR*, 2(1): 00039. doi: 10.15406/apar.2015.02.00039.

McCormick, S.P., Stanley, A.M., Stover, N.A. and Alexander, N.J. (2011). Trichothecenes: from simple to complex mycotoxins. *Toxins*, 3(7): 802–814. doi: 10.3390/toxins3070802.

Medina, A., Schmidt-Heydt, M., Cárdenas-Chávez, D.L., Parra, R., Geisen, R. et al. (2013). Integrating toxin gene expression, growth and fumonisin B1 and B2 production by a strain of *Fusarium verticillioides* under different environmental factors. *J. R. Soc. Interface*, 10(85): 20130320. doi: 10.1098/rsif.2013.0320.

Medina, A., Gilbert, M.K., Mack, B.M., OBrian, G.R., Rodriguez, A. et al. (2017). Interactions between water activity and temperature on the *Aspergillus flavus* transcriptome and aflatoxin B1 production. *Int. J. Food Microbiol.*, 256: 36–44. doi: 10.1016/j.ijfoodmicro.2017.05.020.

Meister, U. (2009). *Fusarium* toxins in cereals of integrated and organic cultivation from the Federal State of Brandenburg (Germany) harvested in the years 2000–2007. *Mycotoxin Res.*, 25(3): 133–139. doi: 10.1007/s12550-009-0017-z.

Mewes, H.W., Amid, C., Arnold, R., Frishman, D., Güldener, U. et al. (2004). MIPS: Analysis and annotation of proteins from whole genomes. *Nucleic Acids Res.*, 32: D41–D44.

Miedaner, T., Schilling, A.G. and Geiger, H.H. (2001). Molecular genetic diversity and variation for aggressiveness in populations of *Fusarium graminearum* and *Fusarium culmorum* sampled from wheat fields in different countries. *J. Phytopathol.*, 149(11-12): 641–648. doi: 10.1046/j.1439-0434.2001.00687.x.

Miedaner, T., Cumagun, C.J.R. and Chakraborty, S. (2008). Population genetics of three important head blight pathogens: *Fusarium graminearum*, *F. pseudograminearum*, and *F. culmorum*. *J. Phytopathol.*, 156(3): 129–139. doi: 10.1111/j.1439-0434.2007.01394.x.

Milano, G.D. and López, T.A. (1991). Influence of temperature on zearalenone production by regional strains of *Fusarium graminearum* and *Fusarium oxysporum* in culture. *Int. J. Food Microbiol.*, 13(4): 329–333. doi: 10.1016/0168-1605(91)90092-4.

Minervini, F., Giannoccaro, A., Cavallini, A. and Visconti, A. (2005). Investigations on cellular proliferation induced by zearalenone and its derivatives in relation to the estrogenic parameters. *Toxicol. Lett.*, 159(3): 272–283. doi: 10.1016/j.toxlet.2005.05.017.

Miró-Abella, E., Herrero, P., Canela, N., Arola, L., Borrull, F. et al. (2017). Determination of mycotoxins in plant-based beverages using QuEChERS and liquid chromatography–tandem mass spectrometry. *Food Chem.*, 229: 366–372. doi: 10.1016/j.foodchem.2017.02.078.

Mirocha, C.J., Pathre, S.V. and Robison, T.S. (1981). Comparative metabolism of zearalenone and transmission into bovine milk. *Food Chem. Toxicol.*, 19: 25–30. doi: 10.1016/0015-6264(81)90299-6.

Missmer, S.A., Suarez, L., Felkner, M., Wang, E., Merrill Jr., A.H. et al. (2006). Exposure to fumonisins and the occurrence of neural tube defects along the Texas–Mexico border. *Environ. Health Perspect.*, 114(2): 237–241. doi: 10.1289/ehp.8221.

Mitchell, N.J., Bowers, E., Hurburgh, C. and Wu, F. (2016). Potential economic losses to the US corn industry from aflatoxin contamination. *Food Addit. Contam.*, A, 33(3): 540–550. doi: 10.1080/19440049.2016.1138545.

Mostrom, M. (2016). Mycotoxins: Classification. pp. 29–34. *In*: Caballero, B. Finglas, P.M., Toldrá, F. (Eds.). *Encyclopedia of Food and Health*. Academic Press. doi: 10.1016/B978-0-12-384947-2.00478-5.

Mostrom, M.S. (2012). Zearalenone. pp. 1266–1271. *In*: Gupta, R. (Ed.). *Veterinary Toxicology; Basic and clinical principles*. Cambridge, MA: Academic Press, USA.

Mukherjee, D., Royce, S.G., Alexander, J.A., Buckley, B., Isukapalli, S.S. et al. (2014). Physiologically based toxicokinetic modeling of zearalenone and its metabolites: Application to the Jersey girl study. *PLOS One*, 9(12): e113632 doi: 10.1371/journal.pone.0113632.

Müller, H.M. and Schwadorf, K. (1993). A survey of the natural occurrence of *Fusarium* toxins in wheat grown in a southwestern area of Germany. *Mycopathologia*, 121(2): 115–121. doi: 10.1007/BF01103579.

Munkvold, G.P. and Carlton, W.M. (1997). Influence of inoculation method on systemic *Fusarium moniliforme* infection of maize plants grown from infected seeds. *Plant Dis.*, 81(2): 211–216. doi: 10.1094/PDIS.1997.81.2.211.

Nahle, S., El Khoury, A. and Atoui, A. (2021). A current status on the molecular biology of zearalenone: Its biosynthesis and molecular detection of zearalenone producing *Fusarium* species. *Eur. J. Plant Pathol.*, 159: 247–258. doi: 10.1007/s10658-020-02173-9.

Nicholson, P., Simpson, D.R., Weston, G., Rezanoor, H.N., Lees, A.K. et al. (1998). Detection and quantification of *Fusarium culmorum* and *Fusarium graminearum*in cereals using PCR assays. *Physiol. Mol. Plant. Path.*, 53(1): 17–37. doi: 10.1006/pmpp.1998.0170.

NTP. (1982). *Carcinogenicity Bioassay of Zearalenone in F344/N Rats and F6C3F1 Mice*. National Toxicology Program Technical Reports Series, UK, 235: 7–51.

Olsen, M., Mirocha, C.J., Abbas, H.K. and Johansson, B. (1986). Metabolism of high concentrations of dietary zearalenone by young male turkey poults. *Poult. Sci.*, 65(10): 1905–1910. doi: 10.3382/ps.0651905.

Osweiler, G.D., Ross, P.F., Wilson, T.M., Nelson, P.E., Witte, S.T. et al. (1992). Characterization of an epizootic of pulmonary edema in swine associated with fumonisin in corn screenings. *JVDI*, 4(1), 53–59. doi: 10.1177/104063879200400112.

Özsoy, E., Kesercan, B. and Yörük, E. (2020). Antifungal activity of epecific plant essential oils against *Fusarium graminearum*. *Phytopathology*, 90: 17–21. doi: 10.29328/journal.jpsp.1001052.

Pallarés, N., Font, G., Manes, J. and Ferrer, E. (2017). Multimycotoxin LC–MS/MS analysis in tea beverages after dispersive liquid–liquid microextraction (DLLME). *J. Agr. Food Chem.*, 65(47): 10282–10289. doi: 10.1021/acs.jafc.7b03507.

Pankaj, S.K., Shi, H. and Keener, K.M. (2018). A review of novel physical and chemical decontamination technologies for aflatoxin in food. *Trends Food Sci. Technol.*, 71: 73–83. doi: 10.1016/j.tifs.2017.11.007.

Paranagama, P.A., Abeysekera, K.H.T., Abeywickrama, K. and Nugaliyadde, L. (2003). Fungicidal and anti-aflatoxigenic effects of the essential oil of *Cymbopogon citratus* (DC.) Stapf. (lemon grass) against *Aspergillus flavus* Link. isolated from stored rice. *Lett. Appl. Microbiol.*, 37(1): 86–90. doi: 10.1046/j.1472-765X.2003.01351.x.

Park, A.R., Son, H., Min, K., Park, J., Goo, J.H. et al. (2015). Autoregulation of ZEB 2 expression for zearalenone production in *Fusarium graminearum*. *Mol. Microbiol.*, 97(5): 942–956. doi: 10.1111/mmi.13078.

Park, A.R., Fu, M., Shin, J.Y., Son, H. and Lee, Y.-W. (2016). The protein Kinase A pathway regulates zearalenone production by modulating alternative ZEB2 transcription. *J. Microbiol. Biotechnol.*, 26(5): 967–974. doi: 10.4014/jmb.1601.01032.

Park, J.W., Kim, E.K., Shon, D.H. and Kim, Y.B. (2002). Natural co-occurrence of aflatoxin B1, fumonisin B1 and ochratoxin A in barley and corn foods from Korea. *Food Addit. Contam.*, 19(11): 1073–1080. doi: 10.1080/02652030210151840.

Pascari, X., Ramos, A.J., Marín, S. and Sanchís, V. (2018). Mycotoxins and beer: Impact of beer production process on mycotoxin contamination: A review. *Food Res. Int.*, 103: 121–129. doi: 10.1016/j.foodres.2017.07.038.

Pasquali, M., Beyer, M., Bohn, T. and Hoffmann, L. (2011). Comparative analysis of genetic chemotyping methods for Fusarium: Tri13 polymorphism does not discriminate between 3- and 15-acetylated deoxynivalenol chemotypes in *Fusarium graminearum*. *J. Phytopathol.*, 159(10): 700–704. doi: 10.1111/j.1439-0434.2011.01824.x.

Pasquali, M. and Migheli, Q. (2014). Genetic approaches to chemotype determination in type B-trichothecene producing *Fusaria*. *Int. J. Food Microbiol.*, 189: 164–182. doi: 10.1016/j.ijfoodmicro.2014.08.011.

Pasquali, M., Beyer, M., Logrieco, A., Audenaert, K., Balmas, V. et al. (2016). A European database of *Fusarium graminearum* and *F. culmorum* trichothecene genotypes. *Front. Microbiol.*, 7: 406. doi: 10.3389/fmicb.2016.00406.

Peraica, M., Radić, B., Lucić, A. and Pavlović, M. (1999). Toxic effects of mycotoxins in humans. *B World Health Organ.*, 77(9): 754.

Pittet, A. (1998). Natural occurrence of mycotoxins in foods and feeds: An update review. *Rev. Med. Vet.*, 6: 479–492 (*France*).

Placinta, C.M., D'Mello, J.F. and Macdonald, A.M.C. (1999). A review of worldwide contamination of cereal grains and animal feed with *Fusarium* mycotoxins. *Anim. Feed Sci. Tech.*, 78(1-2): 21–37. doi: 10.1016/S0377-8401(98)00278-8.

Ponts, N., Couedelo, L., Pinson-Gadais, L., Verdal-Bonnin, M.N., Barreau, C. et al. (2009). *Fusarium* response to oxidative stress by H2O2 is trichothecene chemotype-dependent. *FEMS Microbiol. Lett.*, 293(2): 255–262. doi: 10.1111/j.1574-6968.2009.01521.x.

Poór, M., Kunsági-Máté, S., Sali, N., Kőszegi, T., Szente, L. et al. (2015). Interactions of zearalenone with native and chemically modified cyclodextrins and their potential utilization. *J. Photochem. Photobiol. B.*, 151: 63–68. doi: 10.1016/j.jphotobiol.2015.07.009.

Proctor, R.H., Brown, D.W., Plattner, R.D. and Desjardins, A.E. (2003). Co-expression of 15 contiguous genes delineates a fumonisin biosynthetic gene cluster in *Gibberella moniliformis*. *Fungal Genet. Biol.*, 38(2): 237–249. doi: 10.1016/S1087-1845(02)00525-X.

Proctor, R.H., Busman, M., Seo, J.A., Lee, Y.W. and Plattner, R.D. (2008). A fumonisin biosynthetic gene cluster in *Fusarium oxysporum* strain O-1890 and the genetic basis for B versus C fumonisin production. *Fungal Genet. Biol.*, 45(6): 1016–1026. doi: 10.1016/j.fgb.2008.02.004.

Proctor, R.H., Van Hove, F., Susca, A., Stea, G., Busman, M. et al. (2013). Birth, death, and horizontal transfer of the fumonisin biosynthetic gene cluster during the evolutionary diversification of *Fusarium*. *Mol. Microbiol.*, 90(2): 290–306. doi: 10.1111/mmi.12362.

Rahimi, E., Bonyadian, M., Rafei, M. and Kazemeini, H.R. (2010). Occurrence of aflatoxin M1 in raw milk of five dairy species in Ahvaz, Iran. *Food Chem. Toxicol.*, 48(1): 129–131. doi: 10.1016/j.fct.2009.09.028.

Rahjoo, V., Zad, J., Javan-Nikkhah, M., Gohari, A.M., Okhovvat, S.M. et al. (2008). Morphological and molecular identification of *Fusarium* isolated from maize ears in Iran. *Int. J. Plant Pathol.*, 90: 463–468.

Rai, A., Das, M. and Tripathi, A. (2020). Occurrence and toxicity of a *Fusarium* mycotoxin, zearalenone. *Crit. Rev. Food Sci. Nutr.*, 60(16): 2710–2729. doi: 10.1080/10408398.2019.1655388.

Rheeder, J.P., Marasas, W.F., Thiel, P.G., Sydenham, E.W., Shephard, G.S. et al. (1992). *Fusarium moniliforme* and fumonisins in corn in relation to human esophageal cancer in Transkei. *Phytopathology*, 82: 353–357.

Rheeder, J.P., Marasas, W.F. and Vismer, H.F. (2002). Production of fumonisin analogs by *Fusarium* species. *AEM*, 68(5): 2101–2105. doi: 10.1128/AEM.68.5.2101-2105.2002.

Riley, R.T., Wang, E., Schroeder, J.J., Smith, E.R., Plattner, R.D. et al. (1996). Evidence for disruption of sphingolipid metabolism as a contributing factor in the toxicity and carcinogenicity of fumonisins. *Natural Toxins*, 4(1): 3–15. doi: 10.1002/19960401NT2.

Rodrigues, I. and Naehrer, K. (2012). A three-year survey on the worldwide occurrence of mycotoxins in feedstuffs and feed. *Toxins*, 4(9): 663–675. doi: 10.3390/toxins4090663.

Ropejko, K. and Twarużek, M. (2021). Zearalenone and its metabolites: General overview, occurrence, and toxicity. *Toxins*, 13(1): 35. doi: 10.3390/toxins13010035.

Rossi, V., Scandolara, A. and Battilani, P. (2009). Effect of environmental conditions on spore production by *Fusarium verticillioides*, the causal agent of maize ear rot. *Eur. J. Plant Pathol.*, 123(2): 159–169. doi: 10.1007/s10658-008-9351-9.

Sabuncuoğlu, S.A., Baydar, T., Giray, B. and Şahin, G. (2008). Mikotoksinler: toksik etkileri, degredasyonları, oluşumlarının önlenmesi ve zararlı etkilerinin azaltılması. *HUJPHARM*, (1): 63–92. https://dergipark.org.tr/en/pub/hujpharm/issue/49847/639173.

Schilling, A.G., Moller, E.M. and Geiger, H.H. (1996). Polymerase chain reaction-based assays for species-specific detection of *Fusarium culmorum*, *F. graminearum*, and *F. avenaceum*. *Phytopathology*, 86(5): 515–522.

Schollenberger, M., Müller, H.M., Rüfle, M., Suchy, S., Plank, S. et al. (2006). Natural occurrence of 16 *Fusarium* toxins in grains and feedstuffs of plant origin from Germany. *Mycopathologia*, 161(1): 43–52. doi: 10.1007/s11046-005-0199-7.

Scientific Committee on Food (2000). *Opinion on Fusarium Toxins*—Part 2: Zearalenone (ZEA) (22 June 2000).

Scott, P.M. (2012). Recent research on fumonisins: A review. *Food Addit. Contam. A*, 29(2): 242–248. doi: 10.1080/19440049.2010.546000.

Seefelder, W., Gossmann, M. and Humpf, H.U. (2002). Analysis of fumonisin B1 in *Fusarium proliferatum*-infected asparagus spears and garlic bulbs from Germany by liquid chromatography–electrospray ionization mass spectrometry. *J. Agric. Food Chem.*, 50(10): 2778–2781. doi: 10.1021/jf0115037.

Shephard, G., Berthiller, F., Dorner, J., Krska, R., Lombaert, G. et al. (2009). Developments in mycotoxin analysis: An update for 2007–2008. *World Mycotoxin J.*, 2(1): 3–21. doi: 10.3920/WMJ2008.1095.

Shephard, G.S., Sydenham, E.W., Thiel, P.G. and Gelderblom, W.C.A. (1990). Quantitative determination of fumonisins B1 and B2 by high-performance liquid chromatography with fluorescence detection. *J. Liq. Chromatogr.*, 13(10): 2077–2087. doi: 10.1080/01483919008049014.

Shepherd, G.S. (2003). Aflatoxin and food safety: Recent Africa perspectives. *Toxin Rev.*, 22: 267–286.

Shier, W.T. (1998). Estrogenic mycotoxins. *Rev. Med. Vet.*, 149(6): 599–604.

Shier, W.T., Shier, A.C., Xie, W. and Mirocha, C.J. (2001). Structure-activity relationships for human estrogenic activity in zearalenone mycotoxins. *Toxicon*, 39(9): 1435–1438. doi: 10.1016/S0041-0101(00)00259-2.

Shikhaliyeva, I., Teker, T., Albayrak, G. (2020). Masked mycotoxins of deoxynivalenol and zearalenone – unpredicted toxicity. *BJSTR*, 29(2): doi: 10.26717/BJSTR.2020.29.004773.

Sobrova, P., Adam, V., Vasatkova, A., Beklova, M., Zeman, L. et al. (2010). Deoxynivalenol and its toxicity. *Interdiscip. Toxicol.*, 3: 94–99.

Stępień, Ł., Koczyk, G. and Waśkiewicz, A. (2011). Genetic and phenotypic variation of *Fusarium proliferatum* isolates from different host species. *J. Appl. Genet.*, 52(4): 487–496. doi: 10.1007/s13353-011-0059-8.

Stob, M., Baldwin, R.S., Tuite, J., Andrews, F.N. and Gillette, K.G. (1962). Isolation of an anabolic, uterotrophic compound from corn infected with *Gibberella zeae*. *Nature*, 196(4861): 1318–1318. doi: 10.1038/1961318a0.

Sudakin, D.L. (2003). Trichothecenes in the environment: relevance to human health. *Toxicol. Lett.*, 143(2): 97–107. doi: 10.1016/S0378-4274(03)00116-4.

Tanaka, T., Hasegawa, A., Yamamoto, S., Lee, U.S., Sugiura, Y. et al. (1988). Worldwide contamination of cereals by the *Fusarium* mycotoxins nivalenol, deoxynivalenol, and zearalenone. 1. Survey of 19 countries. *J. Agric. Food Chem.*, 36(5): 979–983.

Tanaka, T., Yamamoto, S., Hasegawa, A., Aoki, N., Besling, J.R. et al. (1990). A survey of the natural occurrence of *Fusarium* mycotoxins, deoxynivalenol, nivalenol, and zearalenone, in cereals harvested in the Netherlands. *Mycopathologia*, 110(1): 19–22.

Tola, M. and Kebede, B. (2016). Occurrence, importance, and control of mycotoxins: A review. *Cogent Food Agric.*, 2(1): 1191103. doi: 10.1080/23311932.2016.1191103.

Turner, N.W., Subrahmanyam, S. and Piletsky, S.A. (2009). Analytical methods for determination of mycotoxins: A review. *Anal. Chim. Acta*, 632(2): 168–180. doi: 10.1016/j.aca.2008.11.010.

Ueno, Y. (1983). General toxicology. pp. 135–146. *In*: Ueno, Y. (Ed.). *Trichothecenes-Chemical, Biological, and Toxicological Aspects* (Developments in Food Science, Vol. 4), Amsterdam, The Netherlands, Elsevier Science Ltd.

Urry, W.H., Wehrmeister, H.L., Hodge, E.B. and Hidy, P.H. (1966). The structure of zearalenone. *Tetrahedron Lett.*, 7(27): 3109–3114. doi: 10.1016/S0040-4039(01)99923-X.

Van Egmond, H.P. (1989). Current situation on regulations for mycotoxins; Overview of tolerances and status of standard methods of sampling and analysis. *Food Addit Contam.*, 6(2): 139–188. doi: 10.1080/02652038909373773.
van Egmond, H.P., Schothorst, R.C. and Jonker, M.A. (2007). Regulations relating to mycotoxins in food. *Anal. Bioanal. Chem.*, 389(1): 147–157. doi: 10.1007/s00216-007-1317-9.
Varga, E., Wiesenberger, G., Hametner, C., Ward, T.J., Dong, Y. et al. (2015). New tricks of an old enemy: Isolates of *Fusarium graminearum* produce a type A trichothecene mycotoxin. *Environ. Microbiol.*, 17(8): 2588–2600. doi: 10.1111/1462-2920.12718.
Villafana, R.T., Ramdass, A.C. and Rampersad, S.N. (2020). TRI Genotyping and chemotyping: A balance of power. *Toxins*, 12(2): 64. doi: 10.3390/toxins12020064.
Waalwijk, C., Kastelein, P., de Vries, I., Kerényi, Z., van der Lee, T. et al. (2003). Major changes in *Fusarium* spp. in wheat in the Netherlands. *Eur. J. Plant Pathol.*, 109(7): 743–754. doi: 10.1023/A:1026086510156.
Waalwijk, C., van der Heide, R., de Vries, I., van der Lee, T., Schoen, C. et al. (2004). Quantitative detection of *Fusarium* species in wheat using TaqMan. *Eur. J. Plant Pathol.*, 110(5): 481–494. doi: 10.1023/B: EJPP.0000032387.52385.13.
Waalwijk, C., Koch, S., Ncube, E., Allwood, J., Flett, B. et al. (2008). Quantitative detection of *Fusarium* spp. and its correlation with fumonisin content in maize from South African subsistence farmers. *World Mycotoxin J.*, 1(1): 39–47. doi: 10.3920/WMJ2008.x005.
Wang, J.H., Li, H.P., Qu, B., Zhang, J.B., Huang, T. et al. (2008a). Development of a generic PCR detection of 3-acetyldeoxynivalenol-, 15-acetyldeoxynivalenol-and nivalenol-chemotypes of *Fusarium graminearum* clade. *Int. J. Mol. Sci.*, 9(12): 2495–2504. doi: 10.3390/ijms9122495.
Wang, L., Zhu, F., Chen, M., Zhu, Y., Xiao, J. et al. (2019). Rapid and visual detection of aflatoxin B1 in foodstuffs using aptamer/G-quadruplex DNAzyme probe with low background noise. *Food Chem.*, 271: 581–587. doi: 10.1016/j.foodchem.2018.08.007.
Wang, L., He, K., Wang, X., Wang, Q., Quan, H. et al. (2021). Recent progress in visual methods for aflatoxin detection. *Crit. Rev. Food Sci. Nutr.*, 6: 1–18. doi: 10.1080/10408398.2021.1919595.
Wang, Y., Chai, T., Lu, G., Quan, C., Duan, H. et al. (2008b). Simultaneous detection of airborne Aflatoxin, Ochratoxin and Zearalenone in a poultry house by immunoaffinity clean-up and high-performance liquid chromatography. *Environ. Res.*, 107(2): 139–144. doi: 10.1016/j.envres.2008.01.008.
Wild, C.P. and Turner, P.C. (2002). The toxicology of aflatoxins as a basis for public health decisions. *Mutagenesis*, 17(6): 471–481. doi: 10.1093/mutage/17.6.471.
Williams, J.H., Phillips, T.D., Jolly, P.E., Stiles, J.K., Jolly, C.M. et al. (2004). Human aflatoxicosis in developing countries: A review of toxicology, exposure, potential health consequences, and interventions. *Am. J. Clin. Nutr.*, 80(5): 1106–1122. doi: 10.1093/ajcn/80.5.1106.
Wilson, T.M., Ross, P.F., Rice, L.G., Osweiler, G.D., Nelson, H.A. et al. (1990). Fumonisin B1 levels associated with an epizootic of equine leukoencephalomalacia. *JVDI*, 2(3): 213–216. doi: 10.1177/104063879000200311.
Windels, C.E. (2000). Economic and social impacts of Fusarium head blight: Changing farms and rural communities in the Northern Great Plains. *Phytopathology*, 90(1): 17–21. doi: 10.1094/PHYTO.2000.90.1.17.
Woloshuk, C.P., Foutz, K.R., Brewer, J.F., Bhatnagar, D., Cleveland, T.E. et al. (1994). Molecular characterization of aflR, a regulatory locus for aflatoxin biosynthesis. *AEM*, 60(7): 2408–2414. doi: 10.1128/aem.60.7.2408-2414.1994.
Wu, F. (2007). Measuring the economic impacts of *Fusarium* toxins in animal feeds. *Anim. Feed Sci. Tech.*, 137(3-4): 363–374. doi: 10.1016/j.anifeedsci.2007.06.010.
Wu, L., Qiu, L., Zhang, H., Sun, J., Hu, X. et al. (2017). Optimization for the production of deoxynivalenoland zearalenone by *Fusarium graminearum* using response surface methodology. *Toxins*, 9(2): 57. doi: 10.3390/toxins9020057.
Xing, Y., Meng, W., Sun, W., Li, D., Yu, Z. et al. (2016). Simultaneous qualitative and quantitative analysis of 21 mycotoxins in *Radix Paeoniae* Alba by ultra-high performance liquid chromatography quadrupole linear ion trap mass spectrometry and QuEChERS for sample preparation. *J. Chromatogr. B.*, 1031: 202–213. http://dx.doi.org/10.1016/j.jchromb.2016.07.008.
Yazar, S. and Omurtag, G.Z. (2008). Fumonisins, trichothecenes and zearalenone in cereals. *Int. J. Mol. Sci.*, 9(11): 2062–2090. doi: 10.3390/ijms9112062.

Yli-Mattila, T., Gagkaeva, T., Ward, T.J., Aoki, T., Kistler, H.C. et al. (2009). A novel Asian clade within the *Fusarium graminearum* species complex includes a newly discovered cereal head blight pathogen from the Russian Far East. *Mycologia*, 101(6): 841–852. doi: 10.3852/08-217.

Yli-Mattila, T., Ward, T.J., O'Donnell, K., Proctor, R.H., Burkin, A.A. et al. (2011). *Fusarium sibiricum* sp. nov, a novel type A trichothecene-producing *Fusarium* from northern Asia closely related to *F. sporotrichioides* and *F. langsethiae*. *Int. J. Food Microbiol.*, 147(1): 58–68. doi: 10.1016/j.ijfoodmicro.2011.03.007.

Yli-Mattila, T. and Gagkaeva, T. (2016). *Fusarium* toxins in cereals in northern Europe and Asia. pp. 293–317. *In*: Deshmukh, S.K., Mishra, J.K., Tewari, J.P. and Tapp, T. (Eds.). *Fungi: Applications and Management Strategies*. Boca Raton, USA: CRC Press.

Yörük, E., Karlik, E., Gazdagli, A., Kayis, M., Kaya, F. et al. (2015). Expression analysis of PKS13, FG08079. 1 and PKS10 genes in *Fusarium graminearum* and *Fusarium culmorum*. *Iran. J. Biotechnol.*, 13(2): 51–55. doi: 10.15171/ijb.1035.

Yörük, E. and Albayrak, G. (2019). siRNA quelling of tri4 and tri5 genes related to deoxynivalenol synthesis in *Fusarium graminearum* and *Fusarium culmorum*. *J. Environ. Biol.*, 40(3): 370–376. doi: 10.22438/jeb/40/3/MRN-942.

Yörük, E. and Yli-Mattila, T. (2019). Class B-trichothecene profiles of *Fusarium* species as causal agents of head blight. pp. 347–376. *In*: *Advancing Frontiers in Mycology & Mycotechnology*, . Singapore: Springer. doi: 10.1007/978-981-13-9349-5_14.

Yu, J. (2012). Current understanding on aflatoxin biosynthesis and future perspective in reducing aflatoxin contamination. *Toxins*, 4(11): 1024–1057. doi: 10.3390/toxins4111024.

Zhang, G.L., Feng, Y.L., Song, J.L. and Zhou, X.S. (2018). Zearalenone: A mycotoxin with different toxic effect in domestic and laboratory animals' granulosa cells. *Front. Genet.*, 9: 667. doi: 10.3389/fgene.2018.00667.

Zhao, X., Zhi, Q.Q., Li, J.Y., Keller, N.P. and He, Z.M. (2018). The antioxidant gallic acid inhibits aflatoxin formation in *Aspergillus flavus* by modulating transcription factors FarB and CreA. *Toxins*, 10(7): 270. doi: 10.3390/toxins10070270.

Zhongzhi, H. and Limiao, D. (2020). Aflatoxin contaminated degree detection by hyperspectral data using band index. *Food Chem. Toxicol.*, 137: 111159. doi: 10.1016/j.fct.2020.111159.

Zhou, R. and Linz, J.E. (1999). Enzymatic function of the Nor-1 protein in aflatoxin biosynthesis in *Aspergillus parasiticus*. *AEM*, 65(12): 5639–5641. doi: 10.1128/AEM.65.12.5639-5641.1999.

Zinedine, A., Soriano, J.M., Molto, J.C. and Manes, J. (2007). Review on the toxicity, occurrence, metabolism, detoxification, regulations, and intake of zearalenone: An estrogenic mycotoxin. *Food Chem. Toxicol.*, 45(1): 1–18. doi: 10.1016/j.fct.2006.07.030.

Biofuels

13

Pichia pastoris
Multifaced Fungal Cell Factory of Biochemicals for Biorefinery Applications

Bikash Kumar and *Pradeep Verma**

1. Introduction

Pichia pastoris also named *Komagataella phaffii* is considered as an emerging model organism in fundamental research applicable to biotechnological and pharmaceutical industries (Bernauer et al., 2021). It is one of the most sought-after eukaryotic expression systems followed by the leading prokaryotic expression system *Escherichia coli* (Zhu et al., 2019). The *P. pastoris* is also known as biotech yeast and is a distant cousin of *Saccharomyces cerevisiae*, which is to date the most studied model yeast (Żymańczyk-Duda et al., 2017; Heistinger et al., 2020). Several studies suggest that *P. pastoris* is widely used in the expression, production, characterization followed by structural analysis of proteins having application in the field of biopharmaceuticals and biotechnological industries. Several proteins such as epidermal growth factor, enzymes (e.g., lipase, phytase, mannanase, and xylanases), hepatitis B surface antigen, human serum albumin, and others have been produced on a commercial scale via *P. pastoris* as an expression system (Weinacker et al., 2013; Spohner et al., 2015). The ability of *P. pastoris* to efficiently assimilate methanol for their energy and carbon needs (Heistinger et al., 2020) made it a model organism for studying methanol assimilation (Riley et al., 2016), pexophagy (Nazarko et al., 2007) and peroxisome biogenesis (Agrawal and Subramani, 2016).

Apart from being a well-known protein expression system, *P. pastoris* is a potential cell factory to produce several chemicals (Schwarzhans et al., 2017; Peña et al., 2018). Earlier petroleum refineries encountered the needs of chemicals and energy for mankind (Kumar and Verma, 2021). However, with the increase in

Bioprocess and Bioenergy Laboratory, Department of Microbiology, Central University of Rajasthan, NH-8, Bandarsindri, Kishangarh, Ajmer 305817, Rajasthan, India.
* Corresponding author: vermaprad@yahoo.com, pradeepverma@curaj.ac.in

population, industrialization, and modernization, these chemical and energy needs have gone up. However, the petroleum-based fuels are depleting at a faster rate. Thus, there is a need to develop several systems that can fulfill our needs for chemicals and fuels. Microbe-based systems have the potential to meet this demand provided the production of different chemicals is be performed at a mass scale with efficient downstream production ability (Pham et al., 2019). The microbial biomass generated after the production of biochemicals can be exploited for biofuel production (Chisti, 2007; Talebi et al., 2016; Majidian et al., 2018). Biotechnological approaches for inducting several desired properties in the yeast genome for its application as the combined lignocellulolytic and ethanologenic ability has been at the forefront for biorefinery applications (Nevoigt, 2008; Mellitzer et al., 2012; Karbalaei et al., 2020). Thus, *P. pastoris* is one of the most relevant eukaryotic systems that can be used as a protein expression system, whole-cell microbial bio-factory, model organisms, and modern systems biology tool for biorefinery applications. The chapter gives a systematic preview of the *P. pastoris* as an efficient multifaceted bio-tool for biorefinery applications.

2. Model Organism for Industrial Applications

2.1 Protein Expression System

Since the last 30 years, *P. pastoris* has evolved as a highly efficient eukaryotic expression system (Hasslacher et al., 1997; Ahmad et al., 2014). Application of *P. pastoris* has several advantages (Zhu et al., 2019) as detailed below:

(a) Efficient secretory expression that helps in the reduction of efforts required for separating and purifying target proteins.
(b) Even after integration with a foreign gene, it shows a stable expression.
(c) It is suitable for a wide range of proteins expression.
(d) Good protein production levels were suitable for the preparatory purpose.

However, several limitations that act as roadblocks in the application of *P. pastoris* as the most efficient expression system (Zhu et al., 2019) are as follows:

(a) Variable protein secretion ability (mg L^{-1} to g L^{-1}) with only 5% of total protein produced intracellularly, i.e., very low as compared to other industrial-scale intracellular expression system.
(b) Needs strategies towards the enhancement in the expression level of extracellular heterologous proteins.

The protein expressions could be divided into several levels, i.e., transcription, translation, and secretion level. Several works have researched for overcoming the limitations for efficient overexpression of the protein.

2.1.1 Transcription

During the stage of transcription level, scientists focused on increasing the gene copy number (i.e., gene dosage) of a foreign gene. The episomal vectors for *P. pastoris* are found to be unstable (Cregg et al., 1985) and exist on low copy numbers

(Ito et al., 2018; Nakamura et al., 2018). Thus, several studies worked for the construction and screening of multicopy strains, where expression cassettes consisting of promoter, coding sequence, and terminator were integrated into yeast genome multiple times. This was achieved either by the *in vivo* or *in vitro* method (Romanos et al., 1998; Cereghino and Cregg, 2000) or by a combination of both the approaches (Sunga et al., 2008; Marx et al., 2009; Zhu et al., 2009a,b). Other factors that need to be considered at this level are promoter strength and messenger RNA (mRNA) secondary structures (Zhu et al., 2019).

2.1.2 Translation

The application of methanol as a substrate for *P. pastoris* can lead to a scarce supply of carbon sources but the increases in gene dosage can help overcoming this limitation (Cos et al., 2005; Zhu et al., 2009b). The increased gene dosage or multicopy transformants help in improvement in protein expression, but not in the proportion as expected (Zhu et al., 2019). The protein expression levels in *P. pastoris* are associated closely with its specific growth rate. The other factors which are critical for enhanced expression are the availability of a sufficient amount of amino acid precursors and energy supply (Curvers et al., 2002; Jungo et al., 2006; Jia et al., 2017). Also, an increased copy number can result in downregulation of alternative oxidase (AOX) expression (Zhu et al., 2011) as evident from the investigation of transcription level (Zhu et al., 2011; Cámara et al., 2017). This suggests a lower availability of carbon and energy when it is needed the most. These translation level bottlenecks can be overcome via:

(a) Solving rare codon usage, i.e., amplifying rare codon tRNA (Abad et al., 2010).
(b) Additional supply of carbon and energy source, i.e., co-feeding of methanol with auxiliary carbon sources (Looser et al., 2015) such as lactate (Xie et al., 2005), mannitol (Gu et al., 2015), and sorbitol (Thorpe et al., 1999), which do not repress AOX1 (Zhu et al., 2019).
(c) Lowering of induction temperature which can help to increase the energy efficiency leading to enhanced expression of foreign protein along with enhanced biomass yield (Cos et al., 2006; Looser et al., 2015).

2.1.3 Protein Secretion

An increase in gene dosage can have a positive impact on protein expression at the secretion level but several studies highlighted some limitations such as:

(a) Increased gene dosage could reduce expressions as the optimal copy number complementary to each type of foreign protein is experimentally proved (Hohenblum et al., 2004; Inan et al., 2006).
(b) An increase in copy number above an optimum level may cause an increase, decrease, or no changes in the final product yield (Cos et al., 2005; Zhu et al., 2011).
(c) Overexpression of a foreign protein will lead to a generation of a several new types of polypeptides that need to be folded and modified (Inan et al., 2006).

Table 13.1. Possible solutions and strategies to overcome limitations associated with the protein secretion stage.

Solution	Methodology	References
Co-expression of UPR regulator	• Co-expression of homologous protein from eukaryotic organisms (yeast or higher system).	Zhu et al., 2019
Utilization of existing knowledge for enhanced secretion of protein and	• Improvement of protein folding, modification, and trafficking via overexpression of chaperone genes responsible for these assignments. • Improved mechanism for unfolded protein recognition and interference of genes key to unfolded or misfolded protein degradation.	Idiris et al., 2010; Delic et al., 2014; Puxbaum et al., 2015
Application of advanced bioinformatics-based techniques along with reverse engineering approach	• Advanced 'omics' based approaches utilizing information available on different databases to develop diagnostic tools to identify potential targets (i.e., upregulated genes) for reverse engineering. • Overcoming the limitations of rational genetic engineering via reverse engineering of secretion pathway.	Gasser et al., 2007; Stadlmayr et al., 2010; Huangfu et al., 2015
Strategies to develop a balance between protein synthesis and secretion	Combinatorial application of co-feeding strategy with secretion pathway engineering.	Zhu et al., 2019

The over-saturation of cells with foreign protein may lead to failure of protein processing and quality inspection machinery leading to movement of unfolded or misfolded protein to degradation via the ubiquitin proteosome pathway and lysosomal proteolysis (Travers et al., 2000; Gasser et al., 2008; Pfeffer et al., 2011). Table 13.1 gives an overview of major solutions and strategies for the limitations faced during the protein secretion stage of foreign protein production via *P. pastoris* system.

3. A Model Organism for Fundamental Research

Several organisms such as *Caenorhabditis elegans*, *Drosophila melanogaster*, *Escherichia coli*, and *Mus musculus* have been evolved as suitable genetic model organisms due to their easy availability and amenability to genetic modification (Pal and Kasinski, 2017; Bernauer et al., 2021). Yeasts as unicellular eukaryotes pose several properties akin to prokaryotic *E. coli* that were considered to be the first model organism for molecular biology (Cooper et al., 2007). Some of the advantages of utilizing *P. pastoris* as a model organism relies on the fact that it shows speedy growth, low cost, and straightforward cultivation conditions with well-established and well-known genetic manipulation and modification strategies (Steensels et al., 2014). The application of unicellular organisms also provides ease of handling of a many organisms to study rare phenotypes and finding a crucial gene for specific biological processes (Alberts et al., 2002; Roberts et al., 2002). Variation in medium composition and growth conditions can be studied along with an application of high-throughput screening assays (Szymański et al., 2012; Larsson et al., 2020).

Yeast is one of the simplest eukaryotes that possesses properties (cytoskeletal organization) and processes (i.e., cell organelle biogenesis, DNA replication, cell cycle progression, and protein secretion) that are conserved from yeast to humans (Mullock and Luzio, 2005; Perocchi et al., 2006). The *S. cerevisiae* has dominated as a model organism for a long time, however, several advantages of *P. pastoris* are better osmo-tolerant and thermo-tolerance (Steensels et al., 2014; Karbalaei et al., 2020). In addition, high cell densities, respiratory growth, efficient protein secretion mechanism, capability to produce high-level expression of recombinant proteins due to the presence of strong constitutive and inducible promoters (Karbalaei et al., 2020). Several studies suggested that modification of glycosylation pathways can facilitate enhanced production of secretory mammalian protein (Hamilton et al., 2003; Hamilton and Gerngross, 2007; Jacobs et al., 2009). Besides, *P. pastoris* can produce membrane proteins which help in its application in different bioprocess applications (Jahic et al., 2006; Byrne, 2015). Based on the above discussion it can be suggested that *P. pastoris* can serve as a model organism for fundamental research with application in improving heterologous protein production for large-scale applications.

4. Application as Whole-Cell Biofactories

Besides protein expression and a model organism, *P. pastoris* can also act as the whole-cell bio-factories for the synthesis of several biochemicals via biotransformation and fermentation-based strategies (Karbalaei et al., 2020; De et al., 2021). Biotransformation involves the participation of the whole cell for conversion of a given substrate into products utilizing native or heterologous expressed enzymes (Lin and Tao, 2017). The application of whole-cell mechanism of *P. pastoris* for biotransformation was first demonstrated in 1989, where oxidative conversion of benzyl alcohol to benzaldehyde using high levels of alcohol oxidase was carried out (Duff and Murray, 1989). Later in 1995, *P. pastoris* with high expression of spinach glycolate oxidase for conversion of glycolic acid to glyoxylic acid was engineered (Payne et al., 1995). The *P. pastoris* has evolved as an efficient catalyst for whole-cell biosynthesis. It is often considered robust as compared to *E. coli* due to higher stress (temperature, pH) tolerance, low maintenance cost, low-cost non-food competing substrates (methanol), and can achieve high cell densities, high level of expression of foreign protein, and above all has the suitability for expression of eukaryotic proteins (Rosano and Ceccarelli, 2014; Tripathi and Shrivastava, 2019). Several chemicals are synthesized utilizing *P. pastoris* as whole-cell catalysts using a mechanism such as oxidation-reduction reaction, reduction reactions utilizing NADPH regeneration, ATP DNA synthesis, cell surface-based reactions, and chemical reactions without utilizing any cofactor (Zhu et al., 2019). Some of the examples of the different mechanisms by which the *P. pastoris* act as whole-cell biocatalysts are presented in Table 13.2.

5. *P. pastoris* as Cell Factories for Biorefinery Application

The *P. pastoris* has been utilized as whole-cell biocatalysts and protein expression systems for the generation of several proteins such as enzymes and other biochemicals (Ahmad et al., 2014; Karbalaei et al., 2020). Similarly, it can be widely applied for

Table 13.2. Classification of mechanism for application of *P. pastoris* as whole-cell biocatalysts for the synthesis of useful chemicals with suitable examples.

Mechanism		Examples	References
Oxidation-reduction reaction	Oxidation	*P. pastoris* NRRL Y-21001 utilized 2-Hydroxybutanoic acid for the generation of 2-Ketobutanoic acid with a maximum yield of 38.7%.	Das et al., 2010
		P. pastoris utilized D-Phenylalanine as a substrate for the generation of D-Phenylpyruvate with 99% efficiency and can be used up to 13 cycles.	Tan et al., 2007
	Reduction	*P. pastoris* GTS115 mediated conversion of 4-Androstene-3,17-dione to Testosterone with a yield of 11.6 g L^{-1}.	Shao et al., 2016
		P. pastoris CBS7435-BDH1 utilized Acetoin as a substrate for generation of 2,3-Butanediol.	Schroer et al., 2010
ATP dependent synthesis		*P. pastoris* based generation of S-Adenosyl-L-methionine via utilization of L- methionine, and ATP with the yield on 6.14–13.5 g L^{-1}.	Chu et al., 2013
Hydrolysis reaction		*P. pastoris* GS115 utilized 7-β-xylosyl-10-deacetyltaxanes for generation for 10-deacetyltaxanes and D-xylose with the conversion efficiency of 81%.	Yu et al., 2013
Carbon condensation		*P. pastoris* GS115 utilized β-hydroxy pyruvic acid and glycolaldehyde as a substrate for the generation of L-erythrulose with a maximum productivity of 46.58 g L^{-1} h^{-1}.	Wei et al., 2018

the generation of enzymes such as lignocellulolytic enzymes (Haon et al., 2015; Chukwuma et al., 2020). It can also act as an expression system for genes critical for providing essential properties (such as tolerance to extreme conditions or product-based negative feedback inhibition to yeast) for ethanol generation (Nevoigt, 2008; Robinson, 2015; Peña et al., 2018) (Fig. 13.1).

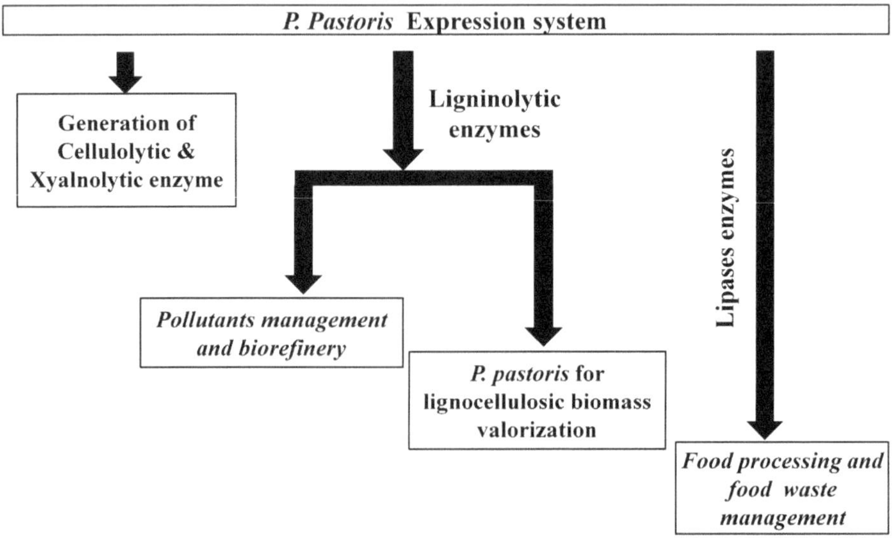

Fig. 13.1. Different application *P. pastoris* as cell factories for biorefinery application.

5.1 Generation of Cellulolytic and Xylanolytic Enzymes

Several genes coding for cellulolytic, xylanolytic, and lipolytic enzymes are cloned and overexpressed in *P. pastoris* for enhanced productions for biomass valorization and biofuel generation.

A cellulase is a group of enzymes that is key to breaking down the crystalline cellulose into an amorphous hydrolyzable form that can be further converted to hexoses and acts as a substrate for ethanol (Vermelho et al., 2012; Li et al., 2021). Several fungi and bacterial strains have shown a great ability for cellulase production however, the yield and activity are low (Li et al., 2021). Thus, modern recombinant DNA technology or genetic engineering approach along with metabolic and surface engineering has been applied to enhance cellulase production (Adrio and Demain, 2010; Acharya and Chaudhary, 2012). Zhang et al. (2018) successfully demonstrated heterologous expression of GHF9 endoglucanase from *Reticulitermes speratus* (termite) in *P. pastoris*. The biochemical characterization of the recombinant GHF9 endoglucanase showed optimum activity over pH 4–9 and stability over a wide range of pH, i.e., 4–11. The recombinant protein also showed high activity at 40°C, but above it the activity decreased. The Km and Vmax with carboxy methyl cellulose were reported to be 7.6 mg mL^{-1} and 5.4 µmol $min^{-1}mg^{-1}$. This protein has potential application in the field of biofuels and chemical generation and application in the laundry and textile industries. Some of the noticeable work for the generation of recombinant proteins for cellulase and hemicellulase enzyme expression in *P. pastoris* has been tabulated in Table 13.3. Several attempts have been made to minimize the cost of enzyme production. One such approach is the co-production of two enzymes in a single expression system such as *P. pastoris*. De Amorim Araújo et al. (2015) demonstrated how the endoglucanase II (*eglII*) and cellobiohydrolase II (*cbhII*) genes from *Trichoderma reesei* were fused in-frame. A list of cellulase and xylanase enzymes production using *P. pastoris* system has been presented in Table 13.3.

A self-processing 2A peptide from foot-and-mouth disease virus separated two different cellulases genes in the construct (de Amorim Araújo et al., 2015). The protein fusion was successful and showed that both enzymes are fully functional with the same catalytic ability based on analysis of methanol-induced yeast transformants. Thus, combining two or more cellulolytic enzymes in a single construct can be a cost-efficient alternative for future biorefinery (Geier et al., 2013). Such an approach can help in developing one-pot lignin removal and enzymatic hydrolysis of cellulose and hemicellulose content. The development of enzymes that are stable at high temperatures and over a wide range of pH is necessary to be able to work in the demanding environment at industrial units (Bhardwaj et al., 2019). Lu et al. (2016) demonstrated that a high-level expression thermo-alkaline stable xylanase from *Bacillus pumilus* HBP8, when expressed in *P. pastoris*, was found stable at 60°C for 30 min with 1.5-fold increase in enzyme activity. The study also demonstrated codon optimization, i.e., codon usage bias for *P. pastoris* suggesting an increase in xylanase activity by 39.5%. The study demonstrated that the Y16 strain can form a large halo with xylanase activity of 6403 U mL^{-1} and consists of multiple

Table 13.3. A list of cellulase and xylanase enzymes production using *P. pastoris* system.

Enzyme	**Gene Source**	***P. Pastoris* Strain**	**Yield and Properties of Recombinant Protein**	**References**
α-L-arabinofuranoside	*A. niger* ATCC120120	*P. pastoris* X-33	• 23 UmL^{-1} (after 5th day) • Maximum activity at pH 4 and 50°C. Km and Vmax on *p*-NPA were 0.93 mm and 17.86 $mmolmL^{-1}min^{-1}$, respectively	Alias et al., 2011
endo-β-1,4-xylanase	*T. reesei* Rut C-30	*P. pastoris* X-33	• After 3rd day of induction, expression of 4350 $nkatmL^{-1}$ • Maximum activity at pH 6 and 50°C. • Specific activity, Km and Kcat over birchwood xylan was 4687 $nkatmL^{-1}$, 2.1 $mgmL^{-1}$, and 219.2 s^{-1} respectively. • Ability to generate xylotriose, but lacks cellulase activity	He et al., 2009
endo-β-1,4-xylanase (xyn11A)	*Thermobifida fusca* YX	*P. pastoris* X-33	• Supernatant specific activities of rXyn11A of 149.4 Umg^{-1}, i.e., 4-fold higher than native enzyme (29.3 Umg^{-1}) • Purified protein specific activities 557.35 Umg^{-1} • Maximum activity at pH 8 and 80°C with more than 60% activity between pH 6–9 and 60–80°C	Zhao et al., 2015
β-xylosidase	*A. niger*	*P. pastoris*	• 100 mg L^{-1} of protein recovered • The purified enzyme showed its maximal activity at 55°C and pH 4 • Km and Vmax over 4-nitrophenyl-β-xylopyranoside of 1.0 mM and 250 mmol $min^{-1}mg^{-1}$, respectively	Kirikyali et al., 2014
Acetyl xylan esterase (AXE)	*Coprinopsis cinerea* Okayama 7 (#130)	*P. pastoris*	• Enzyme showed optimum activity at 40°C and pH 8 • Km and Vmax with 4-nitrophenyl acetate was 4.3 mM 2.15 U $mg^{-1}L^{-1}$, respectively • Km and Vmax with 4-nitrophenyl butyrate was 0.11 mM and 0.78 U mg L^{-1}, respectively • Highly stable against protease enzyme due to the presence of two additional amino acid residues at its native N terminus	Juturu et al., 2013

GH10 endo-xylanase	*Aspergillus niger*	*Pichia pastoris*	• Demonstrated double plasmid approach and showed an increase in yield by 33% as compared to the conventional single plasmid approach • The maximum yield of GH10 XynC of 1650 U mL^{-1} was observed with a recombinant system. • Optimum activity at pH 5.0 and stable over pH 4.5–7.0 • Optimum temperature was found to be 55°C • K_m and V_{max} values with beechwood xylan were 3.5 mg mL^{-1} and 2327 U mg^{-1}, respectively	Long et al., 2020
Endo-1,4-β-xylanase XylP	*Pyromyces finnis*	*Pichia pastoris*	• Specific xylanase activity of recombinant protein 4700 U mg^{-1} of protein • Optimum pH and temperature were observed at 5.0 and 50°C, respectively • KM and Vmax with birch xylan is 0.51 mg mL^{-1} and 7395.3 μmol min^{-1} mg^{-1}, respectively	Kalinina et al., 2020
Endoglucanase II	*T. reesei*	*P. pastoris*	• Highest enzyme activity of 2358 UmL^{-1} after 72 h • Optimum activity was observed at pH 4.8 and 75°C • Specific activity, Km, Kcat with low viscosity CMC was found to be 2620.9 Umg^{-1}, 2.83 g L^{-1} and 2.87 s^{-1} respectively	Akbarzadeh et al., 2014
Celluase (TaCel5A & AfCel12A)	*Thermoascus aurantiacus* & *A. fumigatus*	*Pichia pastoris*	• Total protein using glyceraldehyde-3-phosphate (GAP) promoter was observed to be 3.5 g L^{-1} • CMCase activities for AfCel12A:1200 nkat mL^{-1} • CMCase activities for TaCel5A 170 nkat mL^{-1} r	Várnai et al., 2014
Cellobiohydrolase 2 and β-mannase	*T. reesei*	*P. pastoris* CBS 7435 MutS	• Highest cellobiohydrolase 2 and β-mannase yield of 6.55 and 1.14 g L^{-1}, respectively	Mellitzer et al., 2012
Xylanase A	*T. lanuginosus*	*P. pastoris* CBS 7435 MutS	• The highest xylanase yield was 1.2 g L^{-1}	Mellitzer et al., 2012

copies of the xylanase gene observed by qPCR. The maximum xylanase activity of 48,241 U/ml was observed (5 L bioreactor) suggesting that the recombinant strain was capable of high-density fermentation.

5.2 Ligninolytic Enzyme in Management of Pollutants and Biorefinery

Laccase enzymes play a key role in overcoming the biomass recalcitrance by degradation of phenolic lignin from lignocellulosic biomass. Thus, as a key enzyme for the lignocellulosic biomass-based biorefinery, the laccase enzymes also play a key role in waste (pollutants) degradation such as dyes, chlorinated phenolic, pesticides and polycyclic aromatic hydrocarbons (PAH), and others. The properties that are key for the industrial application of laccases is their stability at adverse conditions and sufficient availability. However, the conventionally produced laccase enzyme has a yield, stability, and activity. Thus, the recombinant system using such as *P. pastoris* has been suggested as an efficient system for laccase production. (Colao et al., 2006) demonstrated the heterologous expression of the laccase (lcc1) gene from *Trametes trogii* in *P. pastoris*. The maximum yield of recombinant protein was obtained at 17 mg/l with the highest production level of 2520 UL^{-1} in a fed batch system. The specific productivity was calculated to be 31.5 Ug^{-1} biomass. The recombinant enzyme thus produced also showed tremendous ability for decolourization of azo, anthraquinonic, indigo carmine, and triarylmethane that could be further enhanced by addition of redox mediators such as viol uric acid and 1-hydroxybenzotriazole (Colao et al., 2006).

Bronikowski et al. (2017) demonstrated cloning and expression of laccase (Mrl2) from *Moniliophthora roreri* in *P. pastoris* and gave a high yield of 1.05 g L^{-1}. The recombinant laccase obtained showed high Kcat values of 316, 74, 36, and 20 s^{-1} towards 2,2′-azino-bis (3-ethylbenzthiazoline-6-sulfonic acid) (ABTS), 2,6-dimethoxyphenol (DMP), guaiacol, and syringaldazine, respectively. Also, it showed stability in an alkaline environment starting from pH 6. The laccase has demonstrated the ability to cause degradation of non-steroidal anti-inflammatory drugs (NSDAIs) and endocrine-disrupting chemicals (EDCs) at a much faster rate as compared to *T. versicolor*. For example, recombinant laccase was able to remove more than 90% of bisphenol A, 17α-ethinyl estradiol, 17ß-estradiol, and estriol after 30 minutes and 56% of drug diclofenac after 20 hrs (Bronikowski et al., 2017). Similarly, recombinant laccase demonstrated excellent degradation of insecticides such as chlorpyrifos (Xie et al., 2013).

Liu et al. (2020a) demonstrated heterologous expression of the laccase gene from *Lentinula edodes* in *P. pastoris*. The recombinant laccase showed maximum activity at pH 3.0 and 4.0 using ABTS and o-tolidine, respectively. The variation in optimum temperature was observed with change in substrate, i.e., the optimum temperature of 60°C and 50°C with ABTS and o-tolidine, respectively. The recombinant protein helped degradation of lignin component of rape straw and promoting cellulose hydrolysis for enhanced biofuel production. The recombinant enzyme had the potential remove soluble phenols and thus play a key role in phenol-rich biomass and pollutant degradation. In addition, the ability of the recombinant protein to tolerate 15% of ethanol and methanol makes it suitable for industrial-scale delignification, enzymatic hydrolysis, and ethanol generation simultaneously through simultaneous delignification, hydrolysis, and fermentation.

5.3 Expression of Constitutive Enzymes for Lignocellulosic Biomass Valorization

The laccase enzyme works on the different phenolic compounds but breaks down the complex lignin into a simpler compound. These compounds can be used as starting material for several other economical valuable compounds. Conacher (2018) demonstrated the constitutive expression of lignocellulolytic enzymes in *P. pastoris* for use in lignin valorization. Lignocellulolytic enzymes, i.e., cellobiose dehydrogenase, laccase, and glucuronoyl esterase from *Neurospora crassa*, *T. versicolor*, and *Hypocrea jecorina* were cloned and successfully expressed in *P. pastoris*. Similarly, Mellitzer et al. (2012) demonstrated the expression of lignocellulolytic enzymes (i.e., beta-mannanase, cellobiohydrolase 1, cellobiohydrolase 2 gene from *Trichoderma reesei*, and xylanase A gene from *Thermomyces lanuginosusi*) in *P. pastoris*. They reported that gene optimization and strain characterization are important parameters for improving secretion levels. The role of promoter strength and gene dosage required is important for efficient improvement of the secretory production of lignocellulolytic enzymes in *P. pastoris*.

5.4 Generation of Lipases

The utilization of waste biomass such as waste cooking oil could be one of the important components of future integrated biorefinery with the valorization of food and other waste biomass (Kumar and Verma, 2020). Native lipases capable of degumming the vegetable oils are available; however, they are not very efficient. Thus, an attempt was made to express the gene for lipases using different expression systems. Ciofalo et al. (2006) demonstrated expression of the gene encoding for Type C phospholipid-specific lipase (BD16449; EC. 3.1.4.3) from *Bacillus cereus* which expressed in the protease deficient *P. pastoris* strain SMD1168 by homologous site-specific recombination. The recombinant protein showed activity of 315 U PLC mg^{-1} with optimal pH and temperature of 7.5 and 60°C (Ciofalo et al., 2006). Similarly, Wang et al. (2014) demonstrated the expression of pro-form lipase from *Rhizopus oryzae* in the *P. pastoris* X-33 strain. The recombinant lipases showed maximum lipase activity of 21,000 U mL^{-1} after 168 hrs induction with methanol in a 50-L bioreactor. The optimum pH and temperature of pH 9.0 and 40°C, respectively, were obtained for the recombinant protein with pH and temperature stability in the range of pH 4.0–9.0 and 25–55°C, respectively. The recombinant lipase showed high enzyme activity towards triglyceride-Trilaurin (12:0) and Tripalmitin (C16:0) (Wang et al., 2014). Recently, Theron et al. (2020) demonstrated a comprehensive comparison between *P. pastoris* and *Yarrowia lipolytica* as an expression system. The study demonstrated lipase B gene (*Candida antarctica*) production and secretion in both the expression system. However, *Y. lipolytica* was observed to be an efficient system in terms of biomass yield, protein yield, and requirement of short cultivation time as compared to *P. pastoris*. But the ability of *P. pastoris* to express 7-fold higher levels of CalB mRNA as compared to *Y. lipolytica*, suggests that a multifaceted overview is required for the selection and development of the recombinant lipase production system.

6. Conclusion and Future Perspectives

The advancement made in the field of biotechnology and bioinformatic-based technologies has enhanced our knowledge of *P. pastoris* as a host for protein production (de Schutter et al., 2009). The genomic information on *P. pastoris*, widely available in databases and thus can be subjected to regulatory responses at the fluxomic, metabolomic, proteomic, and transcriptomic levels (Zahrl et al., 2017; Manzoni et al., 2018). In general, transcriptomics studies, well supported by whole-cell proteomics analysis, revealed that at the transcription level a high degree of gene regulation is observed (Zrimec et al., 2020). The accurate localization of several useful, key metabolic enzymes, e.g., methanol assimilation pathway in the peroxisomal region was also elucidated utilizing the targeted proteomics approach (van der Klei et al., 2006; Schrader and Fahimi, 2008). Metabolomic (metabolome quantification) and fluxomic studies have helped in fine-tuning several media components and physical parameters (temperature, pH, salt stress, and others) for regulating the key protein's enhanced productions (Prosser et al., 2014; Zahrl et al., 2017; Zhang et al., 2021).

The development of fast genome-sequencing technologies and genome annotations approaches resulted in the publication of metabolic models. It has benefited the development of advanced bioengineering strategies such as metabolic engineering and cell surface engineering via the prediction of rational cell engineering targets (Schwarzhans et al., 2017; Peña et al., 2018; Siripong et al., 2018). Several effective engineering targets have been identified in *P. pastoris* for enhancing different protein production such as regulating central carbon metabolism through conventional and advanced genomic, transcriptomics, and metabolomics approaches (Dong et al., 2013; Schwarzhans et al., 2017; Yang et al., 2017; Peña et al., 2018; Siripong et al., 2018).

The major limitations associated with utilizing the omics approaches are that most studies have been performed in chemostat cultures. As fed-batch process is considered for the industrial production process, omics studies need to be performed for the fed-batch system as well (Gmeiner et al., 2015; Liu et al., 2020b). To move ahead of methanol induction and adaptation phase in the early fed-batch process and understand the limitation of production methods, a need to understand cellular responses occurring during fed-batch cultivation is necessary. Other than the fed-batch system, continuous processing (Nieto-Taype et al., 2020), simultaneous lignocellulolytic enzyme production (Kumar and Verma, 2020), and combining the ethanologenic potential (Dong et al., 2020) in a single *P. pastoris* system can be future alternatives for the biorefinery application (Peebo and Neubauer, 2018).

Thus, understanding the genomic stability and potential of *P. pastoris* to adapt in a lengthy, continuous culture, apart from the ability to withstand stress and inhibitions during industrial processes is necessary (Vanz et al., 2012; Nieto-Taype et al., 2020). Thus, this modern systems biology approach integrating genetic engineering, molecular biology, omics, and synthetic biology can be instrumental for further development in *P. pastoris* studies.

Zahrl et al. (2017) presented a mini review on system biotechnology approaches for enhanced protein production via utilization of *P. pastoris*. They provided a systematic summary of systems biology applications in *P. pastoris* and suggested an improved understanding of cell physiology and recombinant protein production strategies in developing these microbial cell factories for mass production units for biofuel and biochemical (Fig. 13.2). Therefore, understanding physiology and modern cell engineering tools can enable scientists to develop *P. pastoris* as robust and efficient secretory cell factories via system biology-mediated cell reprogramming.

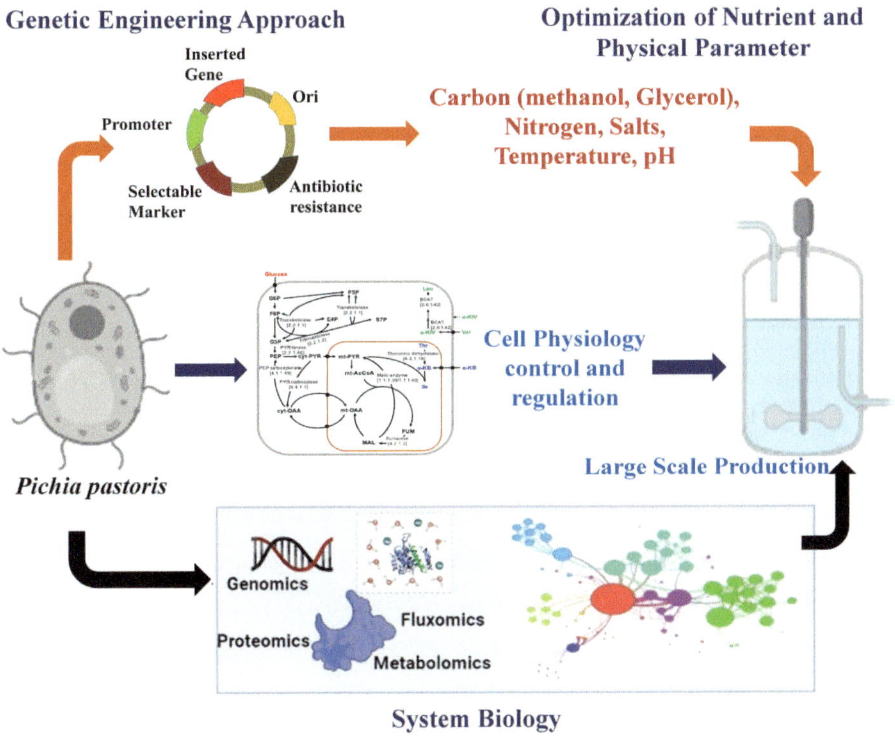

Fig. 13.2. Schematic representation of approaches for *P. pastoris* strain as microbial cell factories.

References

Abad, S., Nahalka, J., Bergler, G., Arnold, S.A., Speight, R., Fotheringham, I., Nidetzky, B. and Glieder, A. (2010). Stepwise engineering of a *Pichia pastoris* D-amino acid oxidase whole cell catalyst. *Microb. Cell Fact.*, 9(1): 1–12. doi: 10.1186/1475-2859-9-24.

Acharya, S. and Chaudhary, A. (2012). Bioprospecting thermophiles for cellulase production: a review. *Braz. J. Microbiol.*, 43: 844–856. doi: 10.1590/S1517-83822012000300001.

Adrio, J.-L. and Demain, A.L. (2010). Recombinant organisms for production of industrial products. *Bioeng. Bugs*, 1(2): 116–131. doi: 10.4161/bbug.1.2.10484.

Ahmad, M., Hirz, M., Pichler, H. and Schwab, H. (2014). Protein expression in *Pichia pastoris*: recent achievements and perspectives for heterologous production. *Appl. Microbiol. Biotechnol.*, 98(12): 5301–5317. doi: 10.1007/s00253-014-5732-5.

Akbarzadeh, A., Siadat, S.O.R., Motallebi, M., Zamani, M.R., Tashnizi, M.B. and Moshtaghi, S. (2014). Characterization and high-level expression of acidic endoglucanase in *Pichia pastoris*. *Appl. Biochem. Biotechnol.*, 172(4): 2253–2265. doi: 10.1007/s12010-013-0672-6.

Alberts, B., Johnson, A., Lewis, J., Raff, M., Roberts, K. and Walter, P. (2002). Studying gene expression and function. *In*: Alberts, B., Johnson, A., Lewis, J., Raff, M., Roberts, K. and Walter, P. (Eds.). *Molecular Biology of the Cell.* (4th edition.) New York City, USA: Arland Science.

Alias, N.I., Mahadi, N.M., Murad, A.M.A., Bakar, F.D.A., Rabu, A. and Illias, R.M. (2011). Expression optimisation of recombinant α-Larabinofuranosidase from *Aspergillus niger* ATCC 120120 in *Pichia pastoris* and its biochemical characterisation. *Afr. J. Biotechnol.*, 10(35): 6700–6710.

Bernauer, L., Radkohl, A., Lehmayer, L.G.K. and Emmerstorfer-Augustin, A. (2021). *Komagataella phaffii* as emerging model organism in fundamental research. *Front. Microbiol.*, 11: 3462. doi: 10.3389/fmicb.2020.607028.

Bhardwaj, N., Kumar, B., Agarwal, K., Chaturvedi, V. and Verma, P. (2019). Purification and characterization of a thermo-acid/alkali stable xylanases from *Aspergillus oryzae* LC1 and its application in Xylo-oligosaccharides production from lignocellulosic agricultural wastes. *Int. J. Biol. Macromol.*, 122: 1191–1202. doi: 10.1016/j.ijbiomac.2018.09.070.

Bronikowski, A., Hagedoorn, P.-L., Koschorreck, K. and Urlacher, V.B. (2017). Expression of a new laccase from *Moniliophthora roreri* at high levels in *Pichia pastoris* and its potential application in micropollutant degradation. *AMB Express*, 7(1): 1–13. doi: 10.1186/s13568-017-0368-3.

Byrne, B. (2015). *Pichia pastoris* as an expression host for membrane protein structural biology. *Curr Opin. Struct. Biol.*, 32: 9–17.doi: 10.1016/j.sbi.2015.01.005.

Cámara, E., Landes, N., Albiol, J., Gasser, B., Mattanovich, D. and Ferrer, P. (2017). Increased dosage of AOX1 promoter-regulated expression cassettes leads to transcription attenuation of the methanol metabolism in *Pichia pastoris*. *Sci. Rep.*, 7(1): 1–16. doi: 10.1038/srep44302.

Cereghino, J.L. and Cregg, J.M. (2000). Heterologous protein expression in the methylotrophic yeast *Pichia pastoris*. *FEMS Microbiol. Rev.*, 24(1): 45–66. doi: 10.1111/j.1574-6976.2000.tb00532.x.

Chisti, Y. (2007). Biodiesel from microalgae. *Biotechnol. Adv.*, 25(3): 294–306. doi: 10.1016/j.biotechadv.2007.02.001.

Chu, J., Qian, J., Zhuang, Y., Zhang, S. and Li, Y. (2013). Progress in the research of S-adenosyl-L-methionine production. *Appl. Microbiol. Biotechnol.*, 97(1): 41–49. doi: 10.1007/s00253-012-4536-8.

Chukwuma, O.B., Rafatullah, M., Tajarudin, H.A. and Ismail, N. (2020). Lignocellulolytic enzymes in biotechnological and industrial processes: A review. *Sustainability*, 12(18): 7282. doi: https://doi.org/10.3390/su12187282.

Ciofalo, V., Barton, N., Kreps, J., Coats, I. and Shanahan, D. (2006). Safety evaluation of a lipase enzyme preparation, expressed in *Pichia pastoris*, intended for use in the degumming of edible vegetable oil. *Regul. Toxicol. Pharmacol.*, 45(1): 1–8. doi: 10.1016/j.yrtph.2006.02.001.

Colao, M.C., Lupino, S., Garzillo, A.M., Buonocore, V. and Ruzzi, M. (2006). Heterologous expression of lcc1 gene from *Trametes trogii* in *Pichia pastoris* and characterization of the recombinant enzyme. *Microb. Cell Fact.*, 5(1): 1–11. doi: 10.1186/1475-2859-5-31.

Conacher, C.G. (2018). *Constitutive Expression of Enzymes in Pichia pastoris for use in Lignin Valorisation.* Stellenbosch: Stellenbosch University.

Cooper, G.M., Hausman, R.E. and Hausman, R.E. (2007). *The Cell: A molecular approach* (Vol. 4). Washington, DC.: ASM press.

Cos, O., Serrano, A., Montesinos, J.L., Ferrer, P., Cregg, J.M. and Valero, F. (2005). Combined effect of the methanol utilization (Mut) phenotype and gene dosage on recombinant protein production in *Pichia pastoris* fed-batch cultures. *J. Biotechnol.*, 116(4): 321–335. doi: 10.1016/j.jbiotec.2004.12.010.

Cos, O., Ramón, R., Montesinos, J.L. and Valero, F. (2006). Operational strategies, monitoring and control of heterologous protein production in the methylotrophic yeast *Pichia pastoris* under different promoters: A review. *Microb. Cell Fact.*, 5(1): 1–20. doi: 10.1186/1475-2859-5-17.

Cregg, J.M., Barringer, K.J., Hessler, A.Y. and Madden, K.R. (1985). *Pichia pastoris* as a host system for transformations. *Mol. Cell. Biol.*, 5(12): 3376–3385. doi: 10.1128/mcb.5.12.3376-3385.1985.

Curvers, S., Linnemann, J., Klauser, T., Wandrey, C. and Takors, R. (2002). Recombinant protein production with *Pichia pastoris* in continuous fermentation--kinetic analysis of growth and product formation. *Eng. Life Sci.*, 2(8): 229–235. doi: 10.1002/1618-2863(20020806).

Das, S., Glenn IV, J.H. and Subramanian, M. (2010). Enantioselective oxidation of 2-hydroxy carboxylic acids by glycolate oxidase and catalase co-expressed in methylotrophic *Pichia pastoris*. *Biotechnol. Prog.*, 26(3): 607–615. doi: 10.1002/btpr.363.

de Amorim Araújo, J., Ferreira, T.C., Rubini, M.R., Duran, A.G.G., De Marco, J.L., de Moraes, L.M.P. and Torres, F.A.G. (2015). Co-expression of cellulases in *Pichia pastoris* as a self-processing protein fusion. *AMB Express*, 5(1): 1–10. doi: 10.1186/s13568-015-0170-z.

De, S., Mattanovich, D., Ferrer, P. and Gasser, B. (2021). Established tools and emerging trends for the production of recombinant proteins and metabolites in *Pichia pastoris*. *Essays Biochem.*, 65(2): 293–307. doi: 10.1042/EBC20200138.

de Schutter, K., Lin, Y.-C., Tiels, P., van Hecke, A., Glinka, S., Weber-Lehmann, J., Rouzé, P., de Peer, Y. and Callewaert, N. (2009). Genome sequence of the recombinant protein production host *Pichia pastoris*. *Nat. Biotechnol.*, 27(6): 561–566. doi: 10.1038/nbt.1544.

Delic, M., Göngrich, R., Mattanovich, D. and Gasser, B. (2014). Engineering of protein folding and secretion—strategies to overcome bottlenecks for efficient production of recombinant proteins. *Antioxid Redox Signal.*, 21(3): 414–437. doi: 10.1089/ars.2014.5844.

Dong, C., Qiao, J., Wang, X., Sun, W., Chen, L., Li, S., Wu, K., Ma, L. and Liu, Y. (2020). Engineering *Pichia pastoris* with surface-display minicellulosomes for carboxymethyl cellulose hydrolysis and ethanol production. *Biotechnol. Biofuels*, 13(1): 1–9. doi: 10.1186/s13068-020-01749-1.

Dong, J.-X., Xie, X., He, Y.-S., Beier, R.C., Sun, Y.-M., Xu, Z.-L., Wu, W.-J., Shen, Y.-D., Xiao, Z.-L., Lai, L.-N. et al. (2013). Surface display and bioactivity of *Bombyx mori* acetylcholinesterase on *Pichia pastoris*. *PlOS One*, 8(8): e70451. doi:10.1371/journal.pone.0070451.

Duff, S.J.B. and Murray, W.D. (1989). Oxidation of benzyl alcohol by whole cells of *Pichia pastoris* and by alcohol oxidase in aqueous and nonaqueous reaction media. *Biotechnol. Bioeng.*, 34(2): 153–159. doi: 10.1002/bit.260340203.

Gasser, B., Saloheimo, M., Rinas, U., Dragosits, M., Rodriguez-Carmona, E., Baumann, K., Giuliani, M., Parrilli, E., Branduardi, P., Lang, C. et al. (2008). Protein folding and conformational stress in microbial cells producing recombinant proteins: a host comparative overview. *Microb. Cell Fact.*, 7(1): 1–18. doi: 10.1186/1475-2859-7-11.

Gasser, B., Sauer, M., Maurer, M., Stadlmayr, G. and Mattanovich, D. (2007). Transcriptomics-based identification of novel factors enhancing heterologous protein secretion in yeasts. *Appl. Environ. Microbiol.*, 73(20): 6499–6507. doi: 10.1128/AEM.01196-07.

Geier, M., Braun, A., Fladischer, P., Stepniak, P., Rudroff, F., Hametner, C., Mihovilovic, M.D. and Glieder, A. (2013). Double site saturation mutagenesis of the human cytochrome P450 2D6 results in regioselective steroid hydroxylation. *The FEBS Journal*, 280(13): 3094–3108.

Gmeiner, C., Saadati, A., Maresch, D., Krasteva, S., Frank, M., Altmann, F., Herwig, C. and Spadiut, O. (2015). Development of a fed-batch process for a recombinant Pichia pastoris $Δ$ och1 strain expressing a plant peroxidase. *Microb. Cell Fact.*, 14(1): 1–10.

Gu, L., Zhang, J., Liu, B., Du, G. and Chen, J. (2015). High-level extracellular production of glucose oxidase by recombinant *Pichia pastoris* using a combined strategy. *Appl. Biochem. Biotechnol.*, 175(3): 1429–1447. doi: 10.1007/s12010-014-1387-z.

Hamilton, S.R., Davidson, R.C., Sethuraman, N., Nett, J.H., Jiang, Y., Rios, S., Bobrowicz, P., Stadheim, T.A., Li, H., Choi, B.K. and Hopkins, D. (2006). Humanization of yeast to produce complex terminally sialylated glycoproteins. *Science*, 313(5792): 1441–1443. doi: 10.1126/science.1130256.

Hamilton, S.R. and Gerngross, T.U. (2007). Glycosylation engineering in yeast: the advent of fully humanized yeast. *Curr. Opin. Biotechnol.*, 18(5): 387–392. doi: 10.1016/j.copbio.2007.09.001.

Haon, M., Grisel, S., Navarro, D., Gruet, A., Berrin, J.-G. and Bignon, C. (2015). Recombinant protein production facility for fungal biomass-degrading enzymes using the yeast *Pichia pastoris*. *Front. Microbiol.*, 6: 1002. doi: 10.3389/fmicb.2015.01002.

Hasslacher, M., Schall, M., Hayn, M., Bona, R., Rumbold, K., Lückl, J., Griengl, H., Kohlwein, S.D. and Schwab, H. (1997). High-level intracellular expression of hydroxynitrile lyase from the tropical rubber Tree *Hevea brasiliensis* in microbial hosts. *Protein Expr. Purif.*, 11(1): 61–71. Doi: 10.1006/prep.1997.0765.

He, J., Yu, B., Zhang, K., Ding, X. and Chen, D. (2009). Expression of endo-1, 4-beta-xylanase from *Trichoderma reesei* in *Pichia pastoris* and functional characterization of the produced enzyme. *BMC Biotechnol.*, 9(1): 1–10. doi: 10.1186/1472-6750-9-56.

Heistinger, L., Gasser, B. and Mattanovich, D. (2020). Microbe Profile: *Komagataella phaffii*: a methanol devouring biotech yeast formerly known as *Pichia pastoris*. *Microbiology*, 166(7): 614–616. doi: 10.1099/mic.0.000958.

Hohenblum, H., Gasser, B., Maurer, M., Borth, N. and Mattanovich, D. (2004). Effects of gene dosage, promoters, and substrates on unfolded protein stress of recombinant *Pichia pastoris*. *Biotechnol. Bioeng.*, 85(4): 367–375. doi: 10.1002/bit.10904.

Huangfu, J., Qi, F., Liu, H., Zou, H., Ahmed, M.S. and Li, C. (2015). Novel helper factors influencing recombinant protein production in *Pichia pastoris* based on proteomic analysis under simulated microgravity. *Appl. Microbiol. Biotechnol.*, 99(2): 653–665. doi: 10.1007/s00253-014-6175-8.

Idiris, A., Tohda, H., Kumagai, H. and Takegawa, K. (2010). Engineering of protein secretion in yeast: strategies and impact on protein production. *Appl. Microbiol. Biotechnol.*, 86(2): 403–417. doi: 10.1007/s00253-010-2447-0.

Inan, M., Aryasomayajula, D., Sinha, J. and Meagher, M.M. (2006). Enhancement of protein secretion in *Pichia pastoris* by overexpression of protein disulfide isomerase. *Biotechnol. Bioeng.*, 93(4): 771–778. doi: 10.1002/bit.20762.

Ito, Y., Watanabe, T., Aikawa, S., Nishi, T., Nishiyama, T., Nakamura, Y., Hasunuma, T., Okubo, Y., Ishii, J. and Kondo, A. (2018). Deletion of DNA ligase IV homolog confers higher gene targeting efficiency on homologous recombination in *Komagataella phaffii*. *FEMS Yeast Res.*, 18(7): foy074. doi: 10.1093/femsyr/foy074.

Jacobs, P.P., Geysens, S., Vervecken, W., Contreras, R. and Callewaert, N. (2009). Engineering complex-type N-glycosylation in *Pichia pastoris* using GlycoSwitch technology. *Nat. Protoc.*, 4(1): 58–70. doi: 10.1038/nprot.2008.213.

Jahic, M., Veide, A., Charoenrat, T., Teeri, T. and Enfors, S.-O. (2006). Process technology for production and recovery of heterologous proteins with *Pichia pastoris*. *Biotechnol. Prog.*, 22(6): 1465–1473. doi: 10.1021/bp060171t.

Jia, L., Tu, T., Huai, Q., Sun, J., Chen, S., Li, X., Shi, Z. and Ding, J. (2017). Enhancing monellin production by *Pichia pastoris* at low cell induction concentration via effectively regulating methanol metabolism patterns and energy utilization efficiency. *PlOS One*, 12(10): e0184602. doi: 10.1371/journal.pone.0184602.

Jungo, C., Rérat, C., Marison, I.W. and von Stockar, U. (2006). Quantitative characterization of the regulation of the synthesis of alcohol oxidase and of the expression of recombinant avidin in a *Pichia pastoris* Mut+ strain. *Enzyme Microb. Technol.*, 39(4): 936–944. doi: 10.1016/j.enzmictec.2006.01.027.

Juturu, V., Aust, C. and Wu, J.C. (2013). Heterologous expression and biochemical characterization of acetyl xylan esterase from *Coprinopsis cinerea*. *World J. Microbiol. Biotechnol.*, 29(4): 597–605. doi: 10.1007/s11274-012-1215-y.

Kalinina, A.N., Borshchevskaya, L.N., Gordeeva, T.L. and Sineoky, S.P. (2020). Expression of the Xylanase Gene from *Pyromyces finnis* in *Pichia pastoris* and Characterization of the Recombinant Protein. *Appl. Biochem. Microbiol.*, 56(7): 787–793. doi: 10.1134/S0003683820070054.

Karbalaei, M., Rezaee, S.A. and Farsiani, H. (2020). *Pichia pastoris*: A highly successful expression system for optimal synthesis of heterologous proteins. *J. Cell. Physiol.*, 235(9): 5867–5881. doi:10.1002/jcp.29583.

Kirikyali, N., Wood, J. and Connerton, I.F. (2014). Characterisation of a recombinant β-xylosidase (xylA) from *Aspergillus oryzae* expressed in *Pichia pastoris*. *AMB Express*, 4(1): 1–7. doi: 10.1186/s13568-014-0068-1.

Kumar, B. and Verma, P. (2020). Enzyme mediated multi-product process: A concept of bio-based refinery. *Ind. Crops Prod.*, 154: 112607. doi: 10.1016/j.indcrop.2020.112607.

Kumar, B. and Verma, P. (2021). Biomass-based biorefineries: an important architype towards a circular economy. *Fuel*, 288: 119622. Doi: 10.1016/j.fuel.2020.119622.

Larsson, P., Engqvist, H., Biermann, J., Rönnerman, E.W., Forssell-Aronsson, E., Kovács, A., Karlsson, P., Helou, K. and Parris, T.Z. (2020). Optimization of cell viability assays to improve replicability and reproducibility of cancer drug sensitivity screens. *Sci. Rep.*, 10(1): 1–12. doi: 10.1038/s41598-020-62848-5.

Li, H., Dou, M., Wang, X., Guo, N., Kou, P., Jiao, J. and Fu, Y. (2021). Optimization of Cellulase Production by a Novel Endophytic Fungus *Penicillium oxalicum* R4 Isolated from *Taxus cuspidata*. *Sustainability*, 13(11): 6006. doi: 10.3390/su13116006.

Lin, B. and Tao, Y. (2017). Whole-cell biocatalysts by design. *Microb. Cell Fact.*, 16(1): 1–12. doi: 10.1186/s12934-017-0724-7.

Liu, C., Zhang, W., Qu, M., Pan, K. and Zhao, X. (2020a). Heterologous expression of laccase from *Lentinula edodes* in *Pichia pastoris* and its application in degrading rape straw. *Front. Microbiol.*, 11: 1086. doi: 10.3389/fmicb.2020.01086.

Liu, L., Wang, F., Pei, G., Cui, J., Diao, J., Lv, M., Chen, L. and Zhang, W. (2020b). Repeated fed-batch strategy and metabolomic analysis to achieve high docosahexaenoic acid productivity in *Crypthecodinium cohnii*. *Microb. Cell Fact.*, 19(1): 1–14. doi: 10.1186/s12934-020-01349-6.

Long, L., Zhang, Y., Ren, H., Sun, H., Sun, F.F. and Qin, W. (2020). Recombinant expression of *Aspergillus niger* GH10 endo-xylanase in *Pichia pastoris* by constructing a double-plasmid co-expression system. *J. Chem. Technol. Biotechnol.*, 95(3): 535–543.

Looser, V., Bruhlmann, B., Bumbak, F., Stenger, C., Costa, M., Camattari, A., Fotiadis, D. and Kovar, K. (2015). Cultivation strategies to enhance productivity of *Pichia pastoris*: a review. *Biotechnol. Adv.*, 33(6): 1177–1193. doi: 10.1016/j.biotechadv.2015.05.008.

Lu, Y., Fang, C., Wang, Q., Zhou, Y., Zhang, G. and Ma, Y. (2016). High-level expression of improved thermo-stable alkaline xylanase variant in *Pichia Pastoris* through codon optimization, multiple gene insertion and high-density fermentation. *Sci. Rep.*, 6(1): 1–9. doi: 10.1038/srep37869.

Majidian, P., Tabatabaei, M., Zeinolabedini, M., Naghshbandi, M.P. and Chisti, Y. (2018). Metabolic engineering of microorganisms for biofuel production. *Renew. Sustain. Energy Rev.*, 82: 3863–3885. doi: 10.1016/j.rser.2017.10.085.

Manzoni, C., Kia, D.A., Vandrovcova, J., Hardy, J., Wood, N.W., Lewis, P.A. and Ferrari, R. (2018). Genome, transcriptome, and proteome: the rise of omics data and their integration in biomedical sciences. *Brief. Bioinformatics*, 19(2): 286–302. doi: 10.1093/bib/bbw114.

Marx, H., Mecklenbräuker, A., Gasser, B., Sauer, M. and Mattanovich, D. (2009). Directed gene copy number amplification in *Pichia pastoris* by vector integration into the ribosomal DNA locus. *FEMS Yeast Res.*, 9(8): 1260–1270. doi: 10.1111/j.1567-1364.2009.00561.x.

Mellitzer, A., Weis, R., Glieder, A. and Flicker, K. (2012). Expression of lignocellulolytic enzymes in *Pichia pastoris*. *Microb. Cell Fact.*, 11(1): 1–11. doi: 10.1186/1475-2859-11-61.

Mullock, B.M. and Luzio, J.P. (2005). Theory of organelle biogenesis. pp. 1–18. *In*: *The Biogenesis of Cellular Organelles*. Boston, MA: Springer.

Nakamura, Y., Nishi, T., Noguchi, R., Ito, Y., Watanabe, T., Nishiyama, T., Aikawa, S., Hasunuma, T., Ishii, J., Okubo, Y. et al. (2018). A stable, autonomously replicating plasmid vector containing *Pichia pastoris* centromeric DNA. *Appl. Environ. Microbiol.*, 84(15): e02882--17. doi: 10.1128/AEM.02882-17.

Nazarko, T.Y., Polupanov, A.S., Manjithaya, R.R., Subramani, S. and Sibirny, A.A. (2007). The requirement of sterol glucoside for pexophagy in yeast is dependent on the species and nature of peroxisome inducers. *Mol Biol Cell.*, 18(1): 106–118. doi: 10.1091/mbc.e06-06-0554.

Nevoigt, E. (2008). Progress in metabolic engineering of *Saccharomyces cerevisiae*. *Microbiol. Mol. Biol. Rev.*, 72(3): 379–412. doi: 10.1128/MMBR.00025-07.

Nieto-Taype, M.A., Garcia-Ortega, X., Albiol, J., Montesinos-Segui, J.L. and Valero, F. (2020). Continuous cultivation as a tool toward the rational bioprocess development with *Pichia Pastoris* cell factory. *Front. Bioeng. Biotechnol.*, 8: 632. doi: 10.3389/fbioe.2020.00632.

Pal, A.S. and Kasinski, A.L. (2017). Animal models to study microRNA function. *Adv. Cancer Res.*, 135: 53–118. doi: 10.1016/bs.acr.2017.06.006.

Payne, M.S., Petrillo, K.L., Gavagan, J.E., Wagner, L.W., DiCosimo, R. and Anton, D.L. (1995). High-level production of spinach glycolate oxidase in the methylotrophic yeast *Pichia pastoris*: Engineering a biocatalyst. *Gene*, 167(1-2): 215–219. doi: 10.1016/0378-1119(95)00661-3.

Peebo, K. and Neubauer, P. (2018). Application of continuous culture methods to recombinant protein production in microorganisms. *Microorganisms*, 6(3): 56. doi: 10.3390/microorganisms6030056.

Peña, D.A., Gasser, B., Zanghellini, J., Steiger, M.G. and Mattanovich, D. (2018). Metabolic engineering of *Pichia pastoris*. *Metab Eng.*, 50: 2–15. doi: 10.1016/j.ymben.2018.04.017.

Perocchi, F., Jensen, L.J., Gagneur, J., Ahting, U., von Mering, C., Bork, P., Prokisch, H. and Steinmetz, L.M. (2006). Assessing systems properties of yeast mitochondria through an interaction map of the organelle. *PLOS Genet.*, 2(10): e170. doi: 10.1371/journal.pgen.0020170.

Pfeffer, M., Maurer, M., Köllensperger, G., Hann, S., Graf, A.B. and Mattanovich, D. (2011). Modeling and measuring intracellular fluxes of secreted recombinant protein in *Pichia pastoris* with a novel 34 S labeling procedure. *Microb. Cell Fact.*, 10(1): 1–11. doi: 10.1186/1475-2859-10-47.

Pham, J.V., Yilma, M.A., Feliz, A., Majid, M.T., Maffetone, N., Walker, J.R., Kim, E., Cho, H.J., Reynolds, J.M., Song, M.C. et al. (2019). A review of the microbial production of bioactive natural products and biologics. *Front. Microbiol.*, 10: 1404. doi: 10.3389/fmicb.2019.01404.

Prosser, G.A., Larrouy-Maumus, G. and de Carvalho, L.P.S. (2014). Metabolomic strategies for the identification of new enzyme functions and metabolic pathways. *EMBO Rep.*, 15(6): 657–669. doi: 10.15252/embr.201338283.

Puxbaum, V., Mattanovich, D. and Gasser, B. (2015). Quo vadis? The challenges of recombinant protein folding and secretion in *Pichia pastoris*. *Appl. Microbiol. Biotechnol.*, 99(7): 2925–2938. doi: 10.1007/s00253-015-6470-z.

Riley, R., Haridas, S., Wolfe, K.H., Lopes, M.R., Hittinger, C.T., Göker, M., Salamov, A.A., Wisecaver, J.H., Long, T.M., Calvey, C.H. et al. (2016). Comparative genomics of biotechnologically important yeasts. *Proc Natl. Acad. Sci.*, 113(35): 9882–9887. doi: 10.1073/pnas.1603941113.

Roberts, K., Alberts, B., Johnson, A., Walter, P. and Hunt, T. (2002). *Molecular Biology of the Cell*. New York: Garland Science.

Robinson, P.K. (2015). Enzymes: Principles and biotechnological applications. *Essays Biochem.*, 59: 1–41. doi: 10.1042/bse0590001.

Romanos, M., Scorer, C., Sreekrishna, K. and Clare, J. (1998). The generation of multicopy recombinant strains. pp. 55–72. *In*: *Pichia Protocols*. New Jersey, USA: Springer. Humana Press.

Rosano, G.L. and Ceccarelli, E.A. (2014). Recombinant protein expression in *Escherichia coli*: Advances and challenges. *Front. Microbiol.*, 5: 172. doi: 10.3389/fmicb.2014.00172.

Schrader, M. and Fahimi, H.D. (2008). The peroxisome: Still a mysterious organelle. *Histochem. Cell Biol.*, 129(4): 421–440. doi: 10.1007/s00418-008-0396-9.

Schroer, K., Luef, K.P., Hartner, F.S., Glieder, A. and Pscheidt, B. (2010). Engineering the *Pichia pastoris* methanol oxidation pathway for improved NADH regeneration during whole-cell biotransformation. *Metab. Eng.*, 12(1): 8–17. doi: 10.1016/j.ymben.2009.08.006.

Schwarzhans, J.-P., Luttermann, T., Geier, M., Kalinowski, J. and Friehs, K. (2017). Towards systems metabolic engineering in *Pichia pastoris*. *Biotechnol. Adv.*, 35(6): 681–710. doi: 10.1016/j.biotechadv.2017.07.009.

Shao, M., Zhang, X., Rao, Z., Xu, M., Yang, T., Li, H., Xu, Z. and Yang, S. (2016). Efficient testosterone production by engineered Pichia pastoris co-expressing human 17-β-hydroxysteroid dehydrogenase type 3 and *Saccharomyces cerevisiae* glucose 6-phosphate dehydrogenase with NADPH regeneration. *Green Chem.*, 18(6): 1774–1784. doi: 10.1039/C5GC02353J.

Siripong, W., Wolf, P., Kusumoputri, T.P., Downes, J.J., Kocharin, K., Tanapongpipat, S. and Runguphan, W. (2018). Metabolic engineering of *Pichia pastoris* for production of isobutanol and isobutyl acetate. *Biotechnol. Biofuels*, 11(1): 1–16.

Spohner, S.C., Müller, H., Quitmann, H. and Czermak, P. (2015). Expression of enzymes for the usage in food and feed industry with *Pichia pastoris*. *J. Biotechnol.*, 202: 118–134. doi: 10.1016/j.jbiotec.2015.01.027.

Stadlmayr, G., Benakovitsch, K., Gasser, B., Mattanovich, D. and Sauer, M. (2010). Genome-scale analysis of library sorting (GALibSo): Isolation of secretion enhancing factors for recombinant protein production in *Pichia pastoris*. *Biotechnol. Bioeng.*, 105(3): 543–555. doi: 10.1002/bit.22573.

Steensels, J., Snoek, T., Meersman, E., Nicolino, M.P., Voordeckers, K. and Verstrepen, K.J. (2014). Improving industrial yeast strains: exploiting natural and artificial diversity. *FEMS Microbiol. Rev.*, 38(5): 947–995. doi: 10.1111/1574-6976.12073.

Sunga, A.J., Tolstorukov, I. and Cregg, J.M. (2008). Post-transformational vector amplification in the yeast *Pichia pastoris*. *FEMS Yeast Res.*, 8(6): 870–876. Doi: 10.1111/j.1567-1364.2008.00410.x.

Szymański Paweland, Markowicz, M. and Mikiciuk-Olasik, E. (2012). Adaptation of high-throughput screening in drug discovery—toxicological screening tests. *Int. J. Mol. Sci.*, 13(1): 427–452. doi: 10.3390/ijms13010427.

Talebi, A.F., Dastgheib, S.M.M., Tirandaz, H., Ghafari, A., Alaie, E. and Tabatabaei, M. (2016). Enhanced algal-based treatment of petroleum produced water and biodiesel production. *RSC Advances*, 6(52): 47001–47009. doi:10.1039/C6RA06579A.

Tan, Q., Song, Q., Zhang, Y. and Wei, D. (2007). Characterization and application of D-amino acid oxidase and catalase within permeabilized *Pichia pastoris* cells in bioconversions. *Appl. Biochem. Biotechnol.*, 136(3): 279–289. doi: 10.1007/s12010-007-9026-6.

Theron, C.W., Vandermies, M., Telek, S., Steels, S. and Fickers, P. (2020). Comprehensive comparison of Yarrowia lipolytica and Pichia pastoris for production of *Candida antarctica* lipase B. *Sci. Rep.*, 10(1): 1–9. doi: 10.1038/s41598-020-58683-3.

Thorpe, E.D., d'Anjou, M.C. and Daugulis, A.J. (1999). Sorbitol as a non-repressing carbon source for fed-batch fermentation of recombinant *Pichia pastoris. Biotechnology Letters*, 21(8): 669–672. doi: 10.1023/A:1005585407601.

Travers, K.J., Patil, C.K., Wodicka, L., Lockhart, D.J., Weissman, J.S. and Walter, P. (2000). Functional and genomic analyses reveal an essential coordination between the unfolded protein response and ER-associated degradation. *Cell*, 101(3): 249–258. doi: 10.1016/s0092-8674(00)80835-1.

Tripathi, N.K. and Shrivastava, A. (2019). Recent developments in bioprocessing of recombinant proteins: expression hosts and process development. *Front. Bioeng. Biotechnol.*, 7: 420. doi: 10.3389/fbioe.2019.00420.

Van der Klei, I.J., Yurimoto, H., Sakai, Y. and Veenhuis, M. (2006). The significance of peroxisomes in methanol metabolism in methylotrophic yeast. *Biochim. Biophys. Acta.*, 1763(12): 1453–1462. doi: 10.1016/j.bbamcr.2006.07.016.

Vanz, A., Lünsdorf, H., Adnan, A., Nimtz, M., Gurramkonda, C., Khanna, N. and Rinas, U. (2012). Physiological response of *Pichia pastoris* GS115 to methanol-induced high-level production of the Hepatitis B surface antigen: Catabolic adaptation, stress responses and autophagic processes. *Microb. Cell Fact.*, 11(1): 1–11. doi: 10.1186/1475-2859-11-103.

Várnai, A., Tang, C., Bengtsson, O., Atterton, A., Mathiesen, G. and Eijsink, V.G.H. (2014). Expression of endoglucanases in *Pichia pastoris* under control of the GAP promoter. *Microb. Cell Fact.*, 13(1): 1–10. doi: 10.1186/1475-2859-13-57.

Vermelho, A.B., Supuran, C.T. and Guisan, J.M. (2012). Microbial enzyme: Applications in industry and in bioremediation. *Enzyme Res.*, 2012:980681. doi: 10.1155/2012/980681.

Wang, J.-R., Li, Y.-Y., Xu, S.-D., Li, P., Liu, J.-S. and Liu, D.-N. (2014). High-level expression of pro-form lipase from Rhizopus oryzae in *Pichia pastoris* and its purification and characterization. *International Int. J. Mol. Sci.*, 15(1): 203–217. doi: 10.3390/ijms15010203.

Wei, Y.-C., Braun-Galleani, S., Henriquez, M.J., Bandara, S. and Nesbeth, D. (2018). Biotransformation of β-hydroxypyruvate and glycolaldehyde to l-erythrulose by *Pichia pastoris* strain GS115 overexpressing native transketolase. *Biotechnol. Prog.*, 34(1): 99–106. doi: 10.1002/btpr.2577.

Weinacker, D., Rabert, C., Zepeda, A.B., Figueroa, C.A., Pessoa, A. and Farias, J.G. (2013). Applications of recombinant *Pichia pastoris* in the healthcare industry. *Braz. J. Microbiol.*, 44: 1043–1048. doi: 10.1590/s1517-83822013000400004.

Xie, H., Li, Q., Wang, M. and Zhao, L. (2013). Production of a recombinant laccase from *Pichia pastoris* and biodegradation of chlorpyrifos in a laccase/vanillin system. *J. Microbiol. Biotechnol.*, 23(6): 864–871. Doi: 10.4014/jmb.1212.12057.

Xie, J., Zhou, Q., Du, P., Gan, R. and Ye, Q. (2005). Use of different carbon sources in cultivation of recombinant *Pichia pastoris* for angiostatin production. *Enzyme Microb. Technol.*, 36(2-3): 210–216. doi: 10.1016/j.enzmictec.2004.06.010.

Yang, S., Lv, X., Wang, X., Wang, J., Wang, R. and Wang, T. (2017). Cell-surface displayed expression of trehalose synthase from *Pseudomonas putida* ATCC 47054 in *Pichia pastoris* using Pir1p as an anchor protein. *Front. Microbiol.*, 8: 2583. doi: 10.3389/fmicb.2017.02583.

Yu, W.-B., Liang, X. and Zhu, P. (2013). High-cell-density fermentation and pilot-scale biocatalytic studies of an engineered yeast expressing the heterologous glycoside hydrolase of 7-β-xylosyltaxanes. *J. Ind. Microbiol. Biotechnol.*, 40(1): 133–140.

Zahrl, R.J., Peña, D.A., Mattanovich, D. and Gasser, B. (2017). Systems biotechnology for protein production in *Pichia pastoris. FEMS Yeast Res.*, 17(7): 1–15. doi:10.1093/femsyr/fox068.

Zhang, M., Yu, Z., Zeng, D., Si, C., Zhao, C., Wang, H., Li, C., He, C. and Duan, J. (2021). Transcriptome and metabolome reveal salt-stress responses of leaf tissues from *Dendrobium officinale*. *Biomolecules*, 11(5): 736. doi: 10.3390/biom11050736.

Zhang, P., Yuan, X., Du, Y. and Li, J.-J. (2018). Heterologous expression and biochemical characterization of a GHF9 endoglucanase from the termite *Reticulitermes speratus* in *Pichia pastoris*. *BMC Biotechnol.*, 18(1): 1–9. doi: 10.1186/s12896-018-0432-3.

Zhao, L., Geng, J., Guo, Y., Liao, X., Liu, X., Wu, R., Zheng, Z. and Zhang, R. (2015). Expression of the *Thermobifida fusca* xylanase Xyn11A in Pichia pastoris and its characterization. *BMC Biotechnol.*, 15(1): 1–12. doi: 10.1186/s12896-015-0135-y.

Zhu, T., Guo, M., Sun, C., Qian, J., Zhuang, Y., Chu, J. and Zhang, S. (2009a). A systematical investigation on the genetic stability of multi-copy *Pichia pastoris* strains. *Biotechnol. Lett.*, 31(5): 679–684. doi: 10.1007/s10529-009-9917-4.

Zhu, T., Guo, M., Tang, Z., Zhang, M., Zhuang, Y., Chu, J. and Zhang, S. (2009b). Efficient generation of multi-copy strains for optimizing secretory expression of porcine insulin precursor in yeast *Pichia pastoris*. *J. Appl. Microbiol.*, 107(3): 954–963. doi: 10.1111/j.1365-2672.2009.04279.x.

Zhu, T., Guo, M., Zhuang, Y., Chu, J. and Zhang, S. (2011). Understanding the effect of foreign gene dosage on the physiology of *Pichia pastoris* by transcriptional analysis of key genes. *Appl. Microbiol. Biotechnol.*, 89(4): 1127–1135. doi: 10.1007/s00253-010-2944-1.

Zhu, T., Sun, H., Wang, M. and Li, Y. (2019). *Pichia pastoris* as a versatile cell factory for the production of industrial enzymes and chemicals: Current status and future perspectives. *Biotechno J.*, 14(6): 1800694. Doi: 10.1002/biot.201800694.

Zrimec, J., Börlin, C.S., Buric, F., Muhammad, A.S., Chen, R., Siewers, V., Verendel, V., Nielsen, J., Töpel, M. and Zelezniak, A. (2020). Deep learning suggests that gene expression is encoded in all parts of a co-evolving interacting gene regulatory structure. *Nat. Commun.*, 11(1): 1–16. Doi: 10.1038/s41467-020-19921-4.

Żymańczyk-Duda, E., Brzezińska-Rodak, M., Klimek-Ochab, M., Duda, M. and Zerka, A. (2017). Yeast as a versatile tool in biotechnology. *Yeast Ind Appl.*, 1: 3–40. doi:10.5772/intechopen.70130.

14

Wood Rot Fungi in the Advanced Biofuel Production

Chu Luong Tri,[1,2] *Le Duy Khuong*[2] and *Ichiro Kamei*[1,*]

1. Introduction

Climate change and its negative impacts on the biosphere is one of the most critical issues that living organisms are facing at present. As a commitment to build sustainable societies, in the Paris Agreement 2015, a historic deal was mutually agreed upon by 195 countries as an effort to limit the increase of global warming below 1.5°C. Contributing to that control, the reduction of greenhouse gases (GHGs) emitted from human activities has a significant role to play (Puricelli et al., 2020). More important is the use of renewable biofuels, including ethanol, butanol, or diesel as an alternative to traditional fossil fuels as a possible strategy to reduce GHGs emissions. Further, the use of biofuels is a positive impact on the efficiency and safety of combustion engines (Manoj Babu et al., 2021).

The United States, Brazil, Argentina, Germany, and China are the top countries using biofuel for the combustion engine as an alternative to traditional fossil fuels (Das, 2020). Globally, the use of biofuels in 2014 was up to 70.8 Megaton of oil equivalent (Mtoe) and by 2050 it is expected to rise up to 720 Mtoe (Hao et al., 2018).

Industrial biofuel production is currently using starch, sugar, food, and feed residue to produce bioethanol and biodiesel. In this process, with well-known fermentation technology, biofuel production could obtain a high yield and efficiency. However, it is raising the conflict between food security and biofuel. To the pursuit of reasonable production, the intensive studies are focusing on the utilization of lignocellulose as an alternative resource for biofuel.

[1] Faculty of Agriculture, University of Miyazaki, 1-1 Gakuen-kibanadai-nishi, Miyazaki 889-2192, Japan.
[2] Faculty of Environment, Ha Long University, 258 Bach Dang Street 02316, Uong Bi district, Quang Ninh province, Vietnam.
* Corresponding author: kamei@cc.miyazaki-u.ac.jp

Lignocellulose is the most abundant biomass resource on earth and could be classified as virgin biomass (woody and grasses), waste biomass from agriculture and forestry (sugarcane bagasse, corn stover, wheat straw, and others), or energy crops. It is reported that the global yearly production of yield of lignocellulose is around 200×10^9 tons, comparing to synthetic polymers is 1.5×10^8 tons (Mohanty et al., 2000). Contributing to this quantity, the annual production of lignocellulose from China, the United States, Canada, and India are up to 1 billion tons, 1.3 billion tons, > 200 million m^2, and 0.2 billion tons, respectively (Zhang, 2008). Such huge yields cause several environmental pollutions, waste accumulation, and reduction of agricultural efficiency (Green, 2019; Zou et al., 2020).

Lignocelluloses are mostly composed of three polymers: cellulose, hemicellulose, and lignin. Thus, the characteristics of lignocellulose are closely related to the properties of their components individually as well as in composed form. The highlighted features of lignocellulose are pointed out as under (Chen, 2015):

(1) Richness and Renewable: Until photosynthesis on the earth exists, lignocelluloses are produced as the main part of plant biomass. It is recognized that lignocelluloses are inexhaustible.
(2) Degradable: Physical, chemical, or biological treatments affect the degradation of lignocelluloses, therefore, the pollution caused by lignocelluloses possibly could be solved using environment-friendly methods.
(3) Transformable: Lignocelluloses are polymers, hence, they have general characteristics of an organic polymer, such as flammability and molecular weight distribution inhomogeneity. Besides, according to their functional groups, the targeted chemical reactions could be designed to achieve modified lignocelluloses.

Overall, polysaccharides (cellulose and hemicellulose) in lignocelluloses are ideal carbon sources to produce biofuels using the microbial fermentation process. Besides economic profit, lignocellulose utilization also reduces the negative impacts of lignocellulose waste on the environment and improves the efficiency of agricultural activities.

Even though lignocellulose is showing great potential as mentioned above, the current utilization of lignocellulose to produce biofuel is still at a low yield and mainly under the laboratory scale. The main challenge is to improve this low yield of the recalcitrant structure of lignocellulose against the microbial fermentation process (Fig. 14.1). The main contribution that should be counted in the recalcitrant structure of lignocellulose is the formation of cellulose fibrils. These fibrils, containing both crystalline and amorphous regions with β-(1→4) linked D-glucose and hydrogen bonding, making the high tensile strength of lignocellulose. The mechanical role of cellulose fibers in the lignocellulose is significantly responsible for its strong structural resistance, which can be compared to that of the reinforcement bars in concrete. In this concretely lignocellulose structure, the hemicellulose is likely the connecting wires of cellulose and lignin playing the role of the hardened cement paste between the polysaccharides. This special complex structure protects lignocellulose from the attacks of microbial, enzymes, and environmental factors. However, the

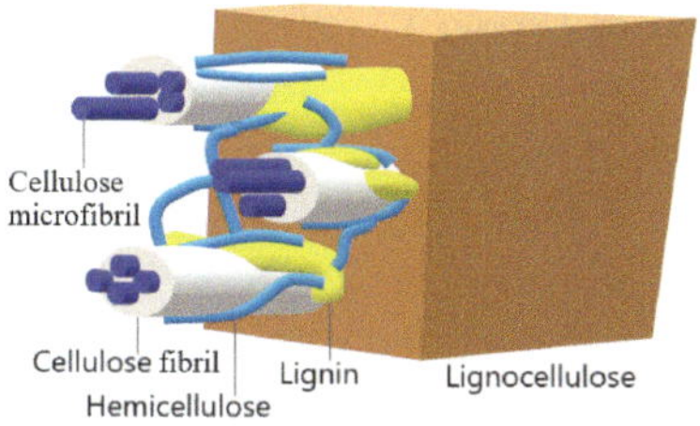

Fig. 14.1. The stimulation of lignocellulosic structure.

highly recalcitrant structure of lignocellulose is resistant to implementation on the industrial scale of the artificial biofuel production process.

To overcome this bottleneck, the current strategy in the conversion of lignocellulose to biofuels requires employing a series of three steps: (1) pretreatment; (2) hydrolysis of polysaccharides; (3) fermentation of sugars. The first step, pretreatment, is the most important and expensive: its goals are to modify the initial substrates, break the "lignin cement paste", and disrupt the crystalline structure of cellulose (Mosier et al., 2005). After pretreatment, the modified materials will be not only more biodegradable but also have a higher polysaccharide content. In the second step, the substrate with the high polysaccharide content is more accessible to the microbial enzymes that are supposed to hydrolyze to the pentose (C5) and hexose (C6) sugars. Finally, the microbial fermentation process metabolites these sugars into biofuels (ethanol and butanol).

Many studies have been focused on the pretreatment step as necessary to improve the yield of biofuel production. It is reported that chemical (alkaline, diluted acid, and organic solvents) or thermo-chemical/physical (steam exploitation, wet oxidization, supercritical CO_2, and others) pretreatments are effective on the modification and removal of a lignin portion in lignocellulose. More importantly, advanced pretreatment is required to degrade lignin with less negative impacts on the environment. Among these pretreatments, the biological methods that employ microbial agents to degrade lignin through enzymatic activities need to be considered. The use of biological pretreatment not only provides efficiency in environmental lignin degradation but also retains a significantly higher content of polysaccharides than those in thermo-chemical/physical pretreatments (Roy et al., 2020).

The pretreated lignocellulose containing polysaccharides (hemicelluloses and cellulose) prefers a suitable approach to saccharify them to fermentable sugars (hexoses and pentoses). Besides, the byproducts of pretreatment and saccharification need to show reduced negative impacts on microbial fermentation in the promising design.

With these expectations regarding ideally sustainable biofuel production, the use of white-rot fungi (WRF) is widely documented as an advantage (Singh and Singh, 2014; Wan and Li, 2012). The details about the successful application of WRF in lignin degradation as biological pretreatment and their simultaneous saccharification and co-fermentation of pentoses and hexose to produce biofuel are discussed in this chapter.

2. Biofuel Production from Lignocelluloses using White-Rot Fungi

2.1 Biological Characteristics of White-rot Fungi

The common name of a heterogeneous group of fungi that mostly belong to the basidiomycetes division is white-rot fungi. In nature, the common feature of these fungi, and only them, is that they can degrade three main components of lignocellulosic biomass (lignin, cellulose, and hemicelluloses) causing rotted woody biomass to feel moist, soft, spongy, or stringy, and appear white or yellow (Schwarze, 2007; Kamei, 2020). In the research on biofuel production, several popular WRF are nominated: *Phanerochaete chrysosporium*, *Pleurotus ostreatus*, *Trametes versicolor*, *Cyathus stercoreus*, etc. Many WRF are cultivating as edible mushrooms, providing nutrient food such as *Lentinula edodes* (Shiitake mushroom), *P. ostreatus* (oyster mushroom), *Armillaria mellea* (Honey mushroom), and others.

2.2 White-rot Fungi and Lignin Degradation

Several bacteria or soft-rot and brown-rot fungi involve in the hydrolysis of polysaccharides in lignocellulose, while the white-rot fungi remove the lignin and retain the higher polysaccharide content behind through activities of the oxidative and extracellular ligninolytic enzymes: lignin peroxidases (LiP, EC 1.11.1.14), manganese peroxidases (MnP, EC 1.11.1.13 versatile peroxidase (VP, EC 1.11.1.16), and laccases (EC 1.10.3.2) (Sánchez, 2009; Wan and Li, 2012). Among the different secreted enzymes involved in lignin degradation, LiP and MnP are the key enzymes to oxidize phenolic lignin units by consecutive one-electron oxidation steps with intermediate phenoxy radical formations. Besides, LiP with high redox potential is also capable to oxidize non-phenolic, veratryl alcohols, or methoxy-substituted lignin units through highly reactive aryl cation radicals. It was also reported that MnP metabolizes non-phenolic units in the lignin biodegradation process through thiol radicals and lipid peroxidation reactions. The versatile peroxidase (VP) showed both characteristics of MnP and LiP, where the Mn^{2+} will be oxidized to Mn^{3+} to metabolize phenolic lignin (likely MnP) or form aryl cation radicals (likely LiP) to convert non-phenolic lignin units (Jensen et al., 1996; Sánchez, 2009; Mäkelä et al., 2021). Laccase activity is the oxidation of the phenolic compounds with the concurrent reduction of molecular oxygen to water. With the low redox potential, the role of laccases in lignin biodegradation is dependent on mediator compounds or other ligninolytic enzymes (Viikari et al., 2009; Tuomela and Hatakka, 2011; Sindhu et al., 2016).

The review data for operation of lignin degradation by WRF incubation as biological pretreatment for lignocellulosic biofuels have indicated that the optimal temperature is 25–30°C and the suitable moisture content in biomass is 70–80% with the presence of oxygen (Rouches et al., 2016; Sindhu et al., 2016). Besides, WRF biological pretreatment for lignocellulose in submerged (less than 5.0% solid loading) or solid-state (up to 30.0% solid loading) has shown the advantage in low energy consumption, fewer reagent requirements, less or no inhibitors for the fermentation step. However, it requires weeks to months to partly degrade lignin in lignocellulose by WRF secreted enzymes (Zabed et al., 2019). It is a consideration that during the biological pretreatment, a significant portion of hemicellulose and cellulose could also be utilized by the fungal metabolism (Sánchez, 2009).

2.3 White-rot Fungi and Saccharification

Saccharification is a process that converts polysaccharides to monosaccharides. In the concept of sustainable second-generation biofuel production, the dulcification of cellulose and hemicellulose in lignocellulose material is supposed to be carried out by enzymatic hydrolysis. The cellulases including endoglucanase (EC 3.2.1.4), exoglucanase (EC 3.2.1.91), and β-glucosidase (EC 3.2.1.21) are responsible for the hydrolysis of cellulose to produce glucose (Fig. 14.2A). In nature, these cellulases are produced by various microorganisms; however, most commercial cellulases are produced using *Trichoderma reesei* (Meenu et al., 2014; Uzuner and Cekmecelioglu, 2019). Hemicellulose, the branched polysaccharide with diverse substitute groups, requires a complex hemicellulase enzyme for the hydrolysis

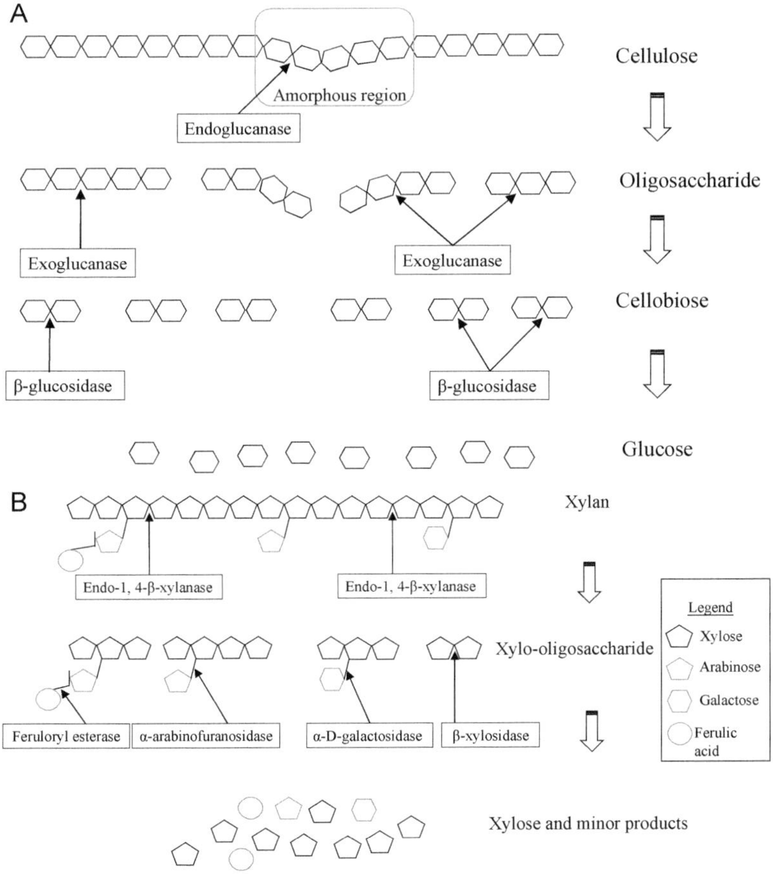

Figs. 14.2A,B. The schematic diagram of WRF enzymatic saccharification of cellulose (A) and xylanhemicellulose (B).

process. The hemicellulase includes endo-1, 4-β-xylanase (EC 3.2.1.8), β-xylosidase (EC 3.2.1.37), α-arabinofuranosidase (EC 3.2.1.55), and esterases (E.C. 3.1.1.72) (Binod et al., 2019; Yi, 2021). The role of hemicellulase in the hydrolysis of xylan (the main hemicellulose in lignocellulose) is described in Fig. 14.2B.

The above cellulase and hemicellulase are classified as glycoside hydrolase, which belongs to sequence-based carbohydrate-active enzymes (CAZy). Many studies have proved that the WRF induces simultaneous saccharification of cellulose and hemicellulose in lignocellulosic forms by the activities of cellulase and hemicellulase, however, the activity of cellulase is more significant (Qinnghe et al., 2004; Kobakhidze et al., 2016).

It is concluded that the growth of WRF and their induction of hydrolytic enzymes in the bioreactor depends on operating conditions, including incubation temperature, mineral compositions, and pH of the fermentation medium (Okal et al., 2020). Besides, the physico-chemical properties of lignocellulose also affect the saccharification yield (Yoon et al., 2014). During the enzymatic dulcification, the accumulation of glucose, fructose, and cellobiose probably causes feedback inhibition to cellulase activities (Agrawal et al., 2016; Okal et al., 2020). There are several options to enhance the efficiency of enzymatic dulcification of lignocellulose (Agrawal et al., 2021; Guo et al., 2018): (1) using cocktails enzyme with a proper ratio between cellulase and xylanase; (2) supplement of surfactants to improve the accessibility of enzyme to cellulose and hemicellulose that finally resulted in acceleration of saccharification rate; (3) modification of bioprocess conditions to induce higher enzyme production (Guo et al., 2018; Agrawal et al., 2021).

2.4 White-rot Fungi and Ethanol Fermentation

Current industrial biofuel production is widely using bacteria or yeast such as *Zymomonas mobilis* and *Saccharomyces cerevisiae* due to their high specific ethanol productivity and tolerance to high ethanol concentration. Unfortunately, these wild-type fermenters are usually unable to utilize pentose sugars. To pursuit economical biofuel production, the conversion of saccharified pentoses from lignocellulosic pentoses, such as xylose and arabinose, is needed to count.

The efficient conversion of pentose(s) and hexose(s) to ethanol could be processed by WRF under facultative anaerobic fermentation (Okamoto et al., 2014; Kamei, 2020; Kamei et al., 2020). As explained in Fig. 14.3, generally, in the ethanol fermentation step, one molecule of glucose produces two molecules of pyruvate in glycolysis. After that, under limited oxygen concentration, two molecules of pyruvate are converted to two ethanol molecules by enzymes alcohol dehydrogenase and pyruvate decarboxylase. In the case of pentose, such as xylose, the conversion is a little more complicated. At first, xylose is reduced to xylitol, then the xylitol is oxidized to xylulose, which is finally phosphorylated to xylulose-5-phosphate by the enzyme xylulokinase. The xylulose-5-phosphate then entering the pentose phosphate pathway to produce ethanol (Noor et al., 2010; Kudahettige et al., 2012; Yano, 2015). Even a co-fermentation of pentose and hexose by WRF has been reported; however, the hexose is more favorable than pentose as the result of carbon catabolite repression

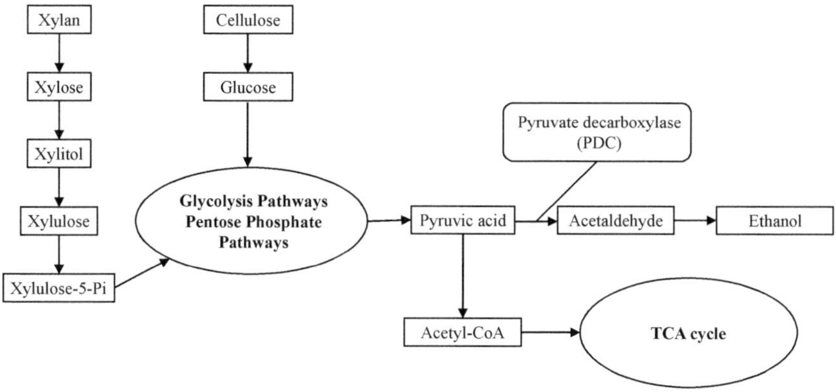

Fig. 14.3. The schematic diagram of WRF ethanol conversion from polysaccharides.

(Kamei et al., 2020; Robak and Balcerek, 2020). The biofuel production process still needs refinement in the conversion of hemicellulose to biofuel.

3. Advanced Biofuel Production using White-rot Fungi

3.1 Integrated Fungal Fermentation Process

Depending on the substrates, microorganisms, and experimental conditions, modification, or combination of delignification, saccharification, and fermentation the biofuel production from lignocellulose could be performed. As the description in Fig. 14.4, several strategies have been used for biofuel production from lignocellulose: (1) separate hydrolysis and fermentation (SHF); (2) simultaneous saccharification and fermentation (SSF); (3) simultaneous saccharification and co-fermentation (SSCF); (4) consolidated bioprocessing (CBP) (Khoo, 2015; Rastogi and Shrivastava, 2017; Amiri and Karimi, 2018). In the SHF, the optimum condition for hydrolysis and fermentation could be applied to achieve the highest yield in each step. However, the separation of biomass brings the risk of contamination leading to a high weight loss during the process. The application of SSF or SSCF in a single bioreactor could reduce the risks of contamination, decline the feedback inhibition of saccharified sugars, resulting in the enhancement of the overall yield (Robak and Balcerek, 2020). Recently, biofuel production has been developing CBP using the WRF to secrete their hydrolytic enzymes, operating saccharification and (co-) fermentation in a single bioreactor without the addition of commercial enzymes. Employing WRF *Phlebia* sp. MG-60, which was screened and isolated from the mangrove forest in Okinawa (Japan), the CBP production of ethanol was obtained from various biomass resources, including alkaline pretreated sugarcane bagasse and bamboo stem powder or biologically pretreated spent mushroom waste (Kamei et al., 2014; Khuong et al., 2014a; Tri et al., 2018).

In the aspect of an eco-friendly approach, the most updated bioethanol fermentation process named Integrated Fungal Fermentation Process (IFFP) was introduced (Kamei et al., 2012). Using hardwood powder as the carbon source in this IFFP, the integration of biological delignification in aerobic solid-state incubation

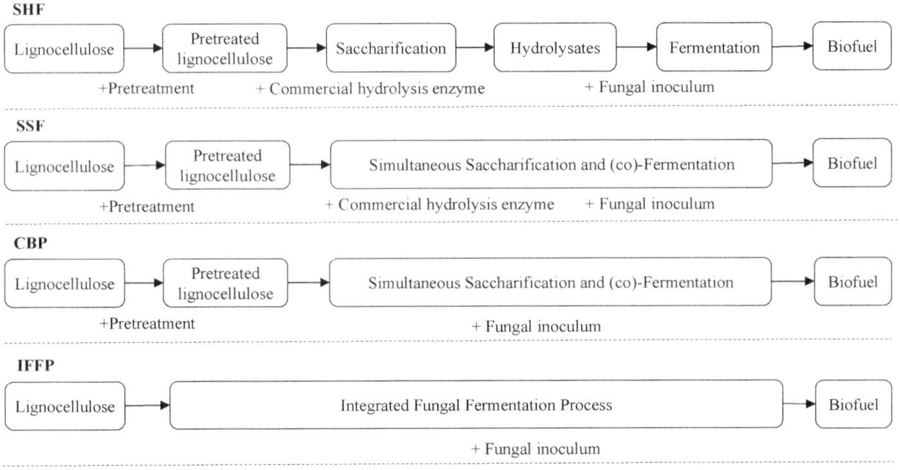

Fig. 14.4. The schematic diagram of biological processes for biofuel production employing WRF.

and simultaneous saccharification and co-fermentation in anaerobic liquid-state incubation was solely functioned by *Phlebia* sp. MG-60. Advantageously, after 56 days of solid-state incubation, the selective degradation of lignin content in hardwood was achieved (40.7% w/w) while the cellulose and hemicellulose contents were almost kept intact. Subsequently, in liquid-state incubation, *Phlebia* sp. MG-60 converted 43.9% of the theoretical maximum of polysaccharides in biological pretreated hardwood to ethanol. A further study revealed that the moisture content of substrates and inorganic nutrients also affect their biological delignification and final yield by IFFP (Khuong et al., 2014b).

It is remarked that lignin is an aromatic polymer which gets oxidatively depolymerized and degraded by WRF ligninolytic enzymes, thus probably the intermediates and or final products of the biological delignification in IFFP are valuable chemicals such as acetic and ferulic acid, syringyl alcohols, syringic and vanillic acid, and short-chain fatty acids. The recovery of these possible compounds is considered to improve the overall efficiency of the lignocellulosic utilization process (Koncsag et al., 2012; Chio et al., 2019).

3.2 Metabolic Engineering

Even though the WRF showed the direct conversion of lignocellulose into biofuels, but it is still necessary to enhance the efficiency of their lignin degradation, saccharification, and fermentation for cost-effective biofuel production in an industry. Using genetic/metabolic engineering approaches one could modify encoding genes resulting in the exhibition/inhibition of related WRF metabolism towards an advanced biofuel production process.

There are few studies on heterologous expressions of recombinant WRF ligninolytic enzymes involving lignin degradation as reported (Asemoloye et al., 2021). A functional expression system for MnP of *Phanerochaete chrysosporium* was developed using the *Escherichia coli* S30 coupled transcription/translation system, which has increased a 9-fold higher H_2O_2 stability of MnP as compared to those of the wild type (Miyazaki-Imamura et al., 2003). Using the industrial fungal host *Aspergillus niger*, the laccase (*lac1*) gene of WRF *P. cinnabarinus* was overexpressed to produce an 80-fold increased recombinant laccase under the control of the glyceraldehyde-3-phosphate dehydrogenase promoter and terminator (Record et al., 2002). Notably, it was successful to use WRF *Phlebia* sp. MG-60 to produce protoplast, and then insert plasmids containing their manganese peroxidase isozyme 2 gene (MGmnp2) into protoplast to induce transformants with higher delignification on wood powder (Yamasaki et al., 2014). With their fast growth, simple morphology, and high possibility for post-translational modifications, yeasts are considered as suitable hosts for large-scale production of WRF ligninolytic enzymes in industrial biofuel production from lignocellulose (Alcalde, 2015; Antošová and Sychrová, 2016). Many research groups also revealed that additional benefits include their application in waste water and organic pollutant treatments using WRF ligninolytic enzymes (Mir-Tutusaus et al., 2018; Zhuo and Fan, 2021).

Commercial lignocellulolytic enzymes (cellulase and hemicellulase) are widely used in the production of not only biofuel but also foods, beverages, pharmaceuticals, and functional foods (Tirado-González et al., 2016). It is reported that the induction of these enzymes is varied and dependent on WRF species, the composition of growth medium and fermentation methods (Elisashvili et al., 2008; Tirado-González et al., 2016; Okal et al., 2020). To actualize the reasonable biofuel conversion from lignocellulose using WRF, the robust improvement of lignocellulolytic enzyme activities through genetic manipulation has been considered as a significant advantage.

In the liquid culture, the transcription lignocellulolytic enzymes are stimulated by the presence of lignocellulose-derived oligosaccharides and inhibited by the accumulation of fermentable mono-sugars known as the carbon catabolite repression (CCR). Therefore, there is a possibility to enhance the expression of lignocellulolytic enzyme genes by the repression of β-glucosidases (BGs), which are responsible for the hydrolysis of oligosaccharides to fermentable sugars (Liu and Qu, 2019; Sukumaran et al., 2021).

To alter the substrate preference during the biological process, the knockout of regulator genes Cre1 (act as a repressor in the process of carbon catabolite repression) in WRF *P. ostreatus*, PC9 was conducted and finally resulted in higher secreted cellulolytic activity (Yoav et al., 2018). In addition, a study about the regulation of lignocellulolytic enzymes indicated that the xylanase regulator (Xyr1) activates both cellulase and xylanase in *T. reesei* (Stricker Astrid et al., 2006).

Metabolic engineering is also applied in the conversion of glucose to ethanol. The glycolytic enzyme pyruvate decarboxylase (PDC), that catalyzes the non-oxidative decarboxylation of pyruvate to produce acetaldehyde, plays an important role in WRF ethanol production. With the overexpression of the self PDC gene in transformant strain GP7, the ethanol fermentation yield was 1.4-fold higher than those of wildtype *P. sordida* YK-624 (Wang et al., 2016). In contrast, the transformant line KO77 with the suppression of the PDC gene, which was obtained by the transfection of the PDC knockout construct into the protoplast of *Phlebia* sp. MG-60-P2, had successfully inhibited ethanol fermentation and resulted in the accumulation of glucose from the cellulose medium (Tsuyama et al., 2017). Using the RNAi-mediated silencing method for *Phlebia* sp. MG-60-P2, the RNAi transformants have shown success in knockdown of the PDC gene, leading to inhibiting the ethanol conversion, while the extracellular peroxidase activity is slightly increased (Motoda et al., 2019). However, the above knockdown technique that applied to the PDC gene in *Phlebia* sp. MG-60-P2 induced various phenotypes of transformants with a wide range of efficiency in gene expression (low-moderate-severe). This situation promotes the development of a new strategy in biofuel production using the co-culture of WRF. Thus, WRF is expected to degrade lignin, convert polysaccharides in lignocellulose to fermentable sugars for other microbes in co-culture to stably produce biofuels as well as valuable chemicals.

3.3 Co-culturing

In the biofuel production process from lignocellulose using a single microorganism as monoculture, even metabolic engineering could be applied, it is very difficult to enhance the whole enzymatic production that is needed for all the pretreatment, saccharification, and fermentation. Advantageously, co-culture is an excellent strategy to obtain the synergistic metabolism between distinct genomes in the same culture, such as: (1) improvement of enzyme activities; (2) acceleration of bioconversion; (3) production of novel metabolites such as hydrogen, methane, butanol, bio-diesel (Zhao et al., 2018; Sperandio and Ferreira Filho, 2019).

In general, the concept of co-culture technique is the interaction of two or more organisms or different life stages of the same organism ultimately resulting in: (1) activation of silent/cryptic gene cluster and production of the unusual secondary metabolites; (2) induction of the specific regulatory pathway (Bhattarai et al., 2020).

Previously, a study about co-culturing among WRF showed that the pair of *Ceriporiopsis subvermispora* and *P. ostreatus* resulted in higher induction of lignin-degrading enzymes and or their isoform composition, leading to significantly stimulate wood decay as compared to monocultures. However, this effect is species-specific, which means the combinations of other fungi were only slightly stimulating or not stimulatory at all (Chi et al., 2007).

Co-culture between WRF *P. sanguineus* and yeast *S. cerevisiae* to produce bioethanol from alkaline pretreated bagasse was also reported. In this combination, the WRF showed the ability to induce cellulase for saccharification. The addition of yeast subsequently utilizes saccharified sugars to produce 4.5 g ethanol per 100 g bagasse (Yoon et al., 2019).

Recently, the co-culture between white-rot fungus and bacterium *Clostridium* has been developed to produce butanol from lignocellulose. Compared to ethanol, butanol is preferable for combustion engines with a higher heating value, while being safer to transport and handle (Trindade and Santos, 2017; Li et al., 2019). However, it is recognized that wildtype *Clostridium* produces butanol from monosaccharides and disaccharides, such as glucose, saccharose, or cellobiose, through the acetone-butanol-ethanol (ABE) fermentation process under anaerobic conditions (Li et al., 2019; Veza et al., 2021). To use this process in an industrial scale, it is desirable to utilize raw, inexpensive, and abundant materials such as cellulose or lignocelluloses.

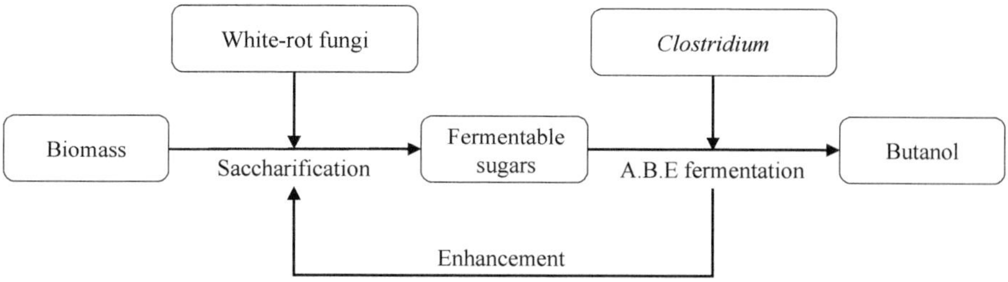

Fig. 14.5. The schematic diagram of co-culture between white-rot fungus *Phlebia* sp. MG-60-P2 and bacterium *Clostridium* to produce butanol from UHKP biomass.

Unfortunately, *Clostridium* is incapable of delignification and saccharification (Schwarz, 2001; Lynd et al., 2002; Jang et al., 2012; Godoy et al., 2018; Ibrahim et al., 2018).

The co-culture of two bacterial strains, cellulolytic *Clostridium thermocellum* NBRC 103400 and butanol-producing *Clostridium saccharoperbutylacetonicum* strain N1-4, produced 5.5 g/l of butanol from 40 g/L at 1% (w/v) NaOH pretreated rice straw (Kiyoshi et al., 2015). In a consolidated bioprocess using lignocellulosic unbleached hardwood kraft pulp (UHKP), the co-culture between WRF *Phlebia* sp. MG-60-P2 and *C. saccharoperbutylacetonicum* synergistically produced butanol and enhanced saccharification (as explained in Fig. 14.5). Besides, knockout of the pyruvate decarboxylase gene in the transformant line KO77 of WRF MG-60-P2 led to inhibition of ethanol fermentation and high accumulation of fungal saccharified cellobiose and glucose from UHKP. These dulcified sugars in co-culture were mostly dedicated to the metabolism of *C. saccharoperbutylacetonicum*, which resulted in enhanced butanol production up to 3.2 g/l from 20 g/l of UHKP substrate (Tri and Kamei, 2020). These novel results promote the construction of the co-culture between WRF and other microorganisms to produce not only butanol but also other valuable chemicals such as organic acids and exopolysaccharides from lignocellulose in future studies.

4. Conclusion and Future Perspectives

Utilization of lignocellulose to produce biofuel not only provides economic benefits but also reduces environmental pollutant issues. Microbial reactions, such as saccharification and fermentation of the polysaccharide, are the key technologies to produce biofuels. Integration of the several steps including delignification to produce biofuel from lignocellulose into a single microorganism has a great economic advantage to reduce the cost and environmental burden. Among various microorganisms, only white-rot fungi have an ability to degrade all the components of plant cell wall, cellulose, hemicellulose, and lignin. Therefore, white-rot fungi are the most significant candidate to construct a biological process for biofuel production from lignocellulose. Although bioethanol production from lignocellulose materials via biological delignification, saccharification, and fermentation by specific white-rot fungus was achieved on a laboratory scale, further study is needed to develop it on an economically viable scale process using white-rot fungi. Recently, three main approaches are in the process to obtain efficient conversion from lignocellulose to biofuel using white-rot fungi: (1) Integration and improvement of solid-state biological delignification and liquid-state saccharification and fermentation in a single biological procedure; (2) Application of metabolic engineering to enhance the enzymatic activities of white-rot fungi to improve the bioconversion step in the operation; (3) The combination of white-rot fungi and other microorganism/s in co-culture to induce distinct mechanisms from the properties of the original white-rot fungal genome, such as butanol production. For the integration of delignification and saccharification and fermentation of lignocellulose by white-rot fungi, elucidation of the mechanism of metabolic change between biological delignification and dulcification should be clarified. Normally, delignification by white-rot fungi is achieved on a solid-state condition whereas saccharification occurs in a liquid-state condition. The detailed mechanisms of change of these different metabolic phases are unknown. Artificial control of this metabolic phase will have the key to efficient integration of delignification and saccharification and fermentation. To achieve the metabolic engineering of white-rot fungi, development of an efficient transformant and gene targeting method is needed. Normally, the transformation and gene targeting of white-rot fungi is difficult, so the reports are limited. To alter the metabolic pathway of white-rot fungi, finding and development of a strong promoter for forced expression of suitable gene and gene-targeting method will be the key for metabolic engineering. Combination of white-rot fungi and other microorganism/s have tremendous potential to develop the unique biological process to convert lignocellulose. If we can choose and design the combination of white-rot fungi which outperformed to delignification and saccharification and of bacteria and suitable for the conversion of sugar to valuable chemicals, it will lead to the development of an environmental-friendly biological process to produce several chemicals from lignocellulose. Finally, the utilization of lignin should not be forgotten. Although lignin is an important aromatic resource, presently just removal of lignin is focused on the production of biofuel from polysaccharides. Recently, the production of low molecular weight aromatic fragment from lignin is concentrated on how to obtain the source of biomaterial such as resin. To obtain the lignin fragment from the process of white-rot fungi, further clarification of the delignification mechanism is required. It is considered that delignification is caused by the radical reaction induced by fungal peroxidases. However, detailed mechanism of the fragmentation of lignin is unknown. Elucidation and control of lignin degradation by white-rot fungi will play the key role to improve and enhance the economically viable process for production of biofuels.

References

Agrawal, R., Verma, A.K. and Satlewal, A. (2016). Application of nanoparticle-immobilized thermostable β-glucosidase for improving the sugarcane juice properties. *Inn. Food Sci. Emerg. Technol.*, 33: 472–482.

Agrawal, R., Verma, A., Singhania, R.R., Varjani, S., Di Dong, C. and Kumar Patel, A. (2021). Current understanding of the inhibition factors and their mechanism of action for the lignocellulosic biomass hydrolysis. *Biores. Technol.*, 332: 125042. doi: 10.1016/j.biortech.2021.125042.

Alcalde, M. (2015). Engineering the ligninolytic enzyme consortium. *Tr. Biotechnol.*, 33(3): 155–162.

Amiri, H. and Karimi, K. (2018). Pretreatment and hydrolysis of lignocellulosic wastes for butanol production: Challenges and perspectives. *Biores. Technol.*, 270: 702–721.

Antošová, Z. and Sychrová, H. (2016). Yeast hosts for the production of recombinant laccases: A review. *Mol. Biotechnol.*, 58(2): 93–116.

Asemoloye, M.D., Marchisio, M.A., Gupta, V.K. and Pecoraro, L. (2021). Genome-based engineering of ligninolytic enzymes in fungi. *Microb. Cell Fact.*, 20(1): 20. https://doi.org/10.1186/s12934-021-01510-9.

Bhattarai, K., Bastola, R. and Baral, B. (2020). Antibiotic drug discovery: Challenges and perspectives in the light of emerging antibiotic resistance. *Adv. Genet.*, 105: 229–292.

Binod, P., Gnansounou, E., Sindhu, R. and Pandey, A. (2019). Enzymes for second generation biofuels: Recent developments and future perspectives. *Biores. Technol. Rep.*, 5: 317–325.

Chen, H. (2015). Lignocellulose biorefinery engineering: An overview. *In*: Chen, H. (Ed.). *Lignocellulose Biorefinery Engineering*. Cambridge: Woodhead Publishing, pp. 1–17.

Chi, Y., Hatakka, A. and Maijala, P. (2007). Can co-culturing of two white-rot fungi increase lignin degradation and the production of lignin-degrading enzymes? *Int. Biodeter. Biodegr.*, 59(1): 32–39.

Chio, C., Sain, M. and Qin, W. (2019). Lignin utilization: A review of lignin depolymerization from various aspects. *Renew. Sust. Energy Rev.*, 107: 232–249.

Das, S. (2020). The National Policy of Biofuels of India: A perspective. *Energy Policy*, 143: 111595.

Elisashvili, V., Penninckx, M., Kachlishvili, E., Tsiklauri, N., Metreveli, E. et al. (2008). *Lentinus edodes* and *Pleurotus* species lignocellulolytic enzymes activity in submerged and solid-state fermentation of lignocellulosic wastes of different composition. *Biores. Technol.*, 99(3): 457–462.

Godoy, M.G., Amorim, G.M., Barreto, M.S. and Freire, D.M.G. (2018). Agricultural residues as animal feed: Protein enrichment and detoxification using solid-state fermentation. pp. 235–256. *In*: Pandey, C., Larroche, C.R. and Soccol, A. (Eds.). *Current Developments in Biotechnology and Bioengineering*. Elsevier.

Green, A. (2019). Agricultural waste and pollution. pp. 531–55.1 *In*: Letcher, T.M. and Vallero, D.A. (Eds.). *Waste* (2nd Edn). London: Academic Press.

Guo, H., Chang, Y. and Lee, D.-J. (2018). Enzymatic saccharification of lignocellulosic biorefinery: Research focuses. *Biores. Technol.*, 252: 198–215.

Hao, H., Liu, Z., Zhao, F., Ren, J., Chang, S. et al. (2018). Biofuel for vehicle use in China: Current status, future potential, and policy implications. *Renew. Sust. Energy Rev.*, 82: 645–653.

Ibrahim, M.F., Kim, S.W. and Abd-Aziz, S. (2018). Advanced bioprocessing strategies for biobutanol production from biomass. *Renew. Sust. Energy Rev.*, 91: 1192–1204.

Jang, Y.-S., Malaviya, A., Cho, C., Lee, J. and Lee, S.Y. (2012). Butanol production from renewable biomass by *Clostridia*. *Biores. Technol.*, 123, 653–663.

Jensen, K.A., Bao, W., Kawai, S., Srebotnik, E. and Hammel, K.E. (1996). Manganese-dependent cleavage of nonphenolic lignin structures by *Ceriporiopsis subvermispora* in the absence of lignin peroxidase. *Appl. Environ. Microbiol.*, 62(10): 3679–3686.

Kamei, I., Hirota, Y. and Meguro, S. (2012). Integrated delignification and simultaneous saccharification and fermentation of hard wood by a white-rot fungus, *Phlebia* sp. MG-60. *Biores. Technol.*, 126: 137–141.

Kamei, I., Nitta, T., Nagano, Y., Yamaguchi, M., Yamasaki, Y. and Meguro, S. (2014). Evaluation of spent mushroom waste from *Lentinula edodes* cultivation for consolidated bioprocessing fermentation by *Phlebia* sp. MG-60. *Int. Biodeter. Biodegr.*, 94: 57–62.

Kamei, I. (2020). Wood-rotting fungi for biofuel production. pp. 123–147. *In*: Salehi Jouzani, G., Tabatabaei, M. and Aghbashlo, M. (Eds.). *Fungi in Fuel Biotechnology*. Cham, Springer International Publishing.

Kamei, I., Uchida, K. and Ardianti, V. (2020). Conservation of xylose fermentability in *Phlebia* species and direct fermentation of xylan by selected fungi. *Appl. Biochem. Biotechnol.*, 192(3): 895–909.

Khoo, H.H. (2015). Review of bio-conversion pathways of lignocellulose-to-ethanol: Sustainability assessment based on land footprint projections. *Renew. Sust. Energy Rev.*, 46: 100–119.

Khuong, L.D., Kondo, R., De Leon, R., Kim Anh, T., Shimizu, K. and Kamei, I. (2014a). Bioethanol production from alkaline-pretreated sugarcane bagasse by consolidated bioprocessing using *Phlebia* sp. MG-60. *Int. Biodeter. Biodegr.*, 88: 62–68.

Khuong, L.D., Kondo, R., Leon, R.D., Anh, T.K., Meguro, S. et al. (2014b). Effect of chemical factors on integrated fungal fermentation of sugarcane bagasse for ethanol production by a white-rot fungus, *Phlebia* sp. MG-60. *Biores. Technol.*, 167: 33–40.

Kiyoshi, K., Furukawa, M., Seyama, T., Kadokura, T., Nakazato, A. and Nakayama, S. (2015). Butanol production from alkali-pretreated rice straw by co-culture of *Clostridium thermocellum* and *Clostridium saccharoperbutylacetonicum*. *Biores. Technol.*, 186: 325–328.

Kobakhidze, A., Asatiani, M., Kachlishvili, E. and Elisashvili, V. (2016). Induction and catabolite repression of cellulase and xylanase synthesis in the selected white-rot basidiomycetes. *Ann. Agrarian Sci.*, 14(3): 169–176.

Koncsag, C.I., Eastwood, D., Collis, A.E.C., Coles, S.R., Clark, A.J. et al. (2012). Extracting valuable compounds from straw degraded by *Pleurotus ostreatus*. *Resour. Conser. Recycl.*, 59: 14–22.

Kudahettige, R.L., Holmgren, M., Imerzeel, P. and Sellstedt, A. (2012). Characterization of bioethanol production from hexoses and xylose by the white-rot fungus *Trametes versicolor*. *BioEner. Res.*, 5(2): 277–285.

Li, Y., Tang, W., Chen, Y., Liu, J. and Lee, C.-f.F. (2019). Potential of acetone-butanol-ethanol (ABE) as a biofuel. *Fuel*, 242: 673–686.

Liu, G. and Qu, Y. (2019). Engineering of filamentous fungi for efficient conversion of lignocellulose: Tools, recent advances and prospects. *Biotechnol. Adv.*, 37(4), 519–529.

Lynd, L.R., Weimer, P.J., Van Zyl, W.H. and Pretorius, I.S. (2002). Microbial cellulose utilization: Fundamentals and biotechnology. *Microbiol. Mol. Biol. Rev.*, 66(3): 506–577.

Mäkelä, M.R., Hildén, K.S. and Kuuskeri, J. (2021). Fungal lignin-modifying peroxidases and H_2O_2-producing enzymes. pp. 247–259. *In*: Zaragosa, Ó. and Casddevall, A. (Eds.). *Encyclopedia of Mycology*. Volume 2, Elsevier. https://doi.org/10.1016/B978-0-12-809633-8.21127-8.

Manoj Babu, A., Saravanan, C.G., Vikneswaran, M., Edwin Geo, V., Sasikala, J. et al. (2021). Analysis of performance, emission, combustion and endoscopic visualization of micro-arc oxidation piston coated SI engine fuelled with low carbon biofuel blends. *Fuel*, 285: 119189. doi: 10.1016/j.fuel.2020.119189.

Meenu, K., Singh, G. and Vishwakarma, R.A. (2014). Molecular mechanism of cellulase production systems in *Trichoderma*. pp. 319–32. *In*: Gupta, V.K., Schmoll, M., Herrera-Estrella, A., Upadhyay, R.S., Druzhinina, I. and Tuohy, M.G. (Eds.). *Biotechnology and Biology of Trichoderma*. Amsterdam: Elsevier, Chapter 22.

Mir-Tutusaus, J.A., Baccar, R., Caminal, G. and Sarrà, M. (2018). Can white-rot fungi be a real wastewater treatment alternative for organic micropollutants removal? A review. *Wat. Res.*, 138: 137–151.

MiyazakiImamura, C., Oohira, K., Kitagawa, R., Nakano, H., Yamane, T. and Takahashi, H. (2003). Improvement of H_2O_2 stability of manganese peroxidase by combinatorial mutagenesis and high-throughput screening using *in vitro* expression with protein disulfide isomerase. *Prot. Eng. Des. Selec.*, 16(6): 423–428.

Mohanty, A.K., Misra, M. and Hinrichsen, G. (2000). Biofibres, biodegradable polymers, and biocomposites: An overview. *Macromol. Mat. Eng.*, 276-277(1): 1–24.

Mosier, N., Wyman, C., Dale, B., Elander, R., Lee, Y.Y. et al. (2005). Features of promising technologies for pretreatment of lignocellulosic biomass. *Biores. Technol.*, 96(6): 673–686.

Motoda, T., Yamaguchi, M., Tsuyama, T. and Kamei, I. (2019). Down-regulation of pyruvate decarboxylase gene of white-rot fungus *Phlebia* sp. MG-60 modify the metabolism of sugars and productivity of extracellular peroxidase activity. *J. Biosci. Bioeng.*, 127(1): 66–72.

Noor, E., Eden, E., Milo, R. and Alon, U. (2010). Central carbon metabolism as a minimal biochemical walk between precursors for biomass and energy. *Mol. Cell*, 39(5): 809–820.

Okal, E.J., Aslam, M.M., Karanja, J.K. and Nyimbo, W.J. (2020). Mini review: Advances in understanding regulation of cellulase enzyme in white-rot basidiomycetes. *Microb. Pathogen.*, 147: 104410. doi: 10.1016/j.micpath.2020.104410.

Okamoto, K., Uchii, A., Kanawaku, R. and Yanase, H. (2014). Bioconversion of xylose, hexoses and biomass to ethanol by a new isolate of the white-rot basidiomycete *Trametes versicolor*. *SpringerPlus*, 3(1): 121. doi: 10.1186/2193-1801-3-121.

Puricelli, S., Cardellini, G., Casadei, S., Faedo, D., van den Oever, A.E.M. and Grosso, M. (2020). A review on biofuels for light-duty vehicles in Europe. *Renew. Sust. Energy Rev.*, 110398. doi: 10.1016/j.rser.2020.110398.

Qinnghe, C., Xiaoyu, Y., Tiangui, N., Cheng, J. and Qiugang, M. (2004). The screening of culture condition and properties of xylanase by white-rot fungus *Pleurotus ostreatus*. *Proc. Biochem.*, 39(11): 1561–1566.

Rastogi, M. and Shrivastava, S. (2017). Recent advances in second generation bioethanol production: An insight to pretreatment, saccharification, and fermentation processes. *Renew. Sust. Energy Rev.*, 80: 330–340.

Record, E., Punt, P.J., Chamkha, M., Labat, M., van den Hondel, C.A.M.J.J. and Asther, M. (2002). Expression of the *Pycnoporus cinnabarinus* laccase gene in *Aspergillus niger* and characterization of the recombinant enzyme. *Eur. J. Biochem.*, 269(2): 602–609.

Robak, K. and Balcerek, M. (2020). Current state-of-the-art in ethanol production from lignocellulosic feedstocks. *Microbiol. Res.*, 240: 126534. doi: 10.1016/j.micres.2020.126534.

Rouches, E., Herpoël-Gimbert, I., Steyer, J.P. and Carrere, H. (2016). Improvement of anaerobic degradation by white-rot fungi pretreatment of lignocellulosic biomass: A review. *Renew. Sust. Energy Rev.*, 59: 179–198.

Roy, R., Rahman, M.S. and Raynie, D.E. (2020). Recent advances of greener pretreatment technologies of lignocellulose. *Curr. Res. Green Sust. Chem.*, 3: 100035. doi: 10.1016/j.crgsc.2020.100035.

Sánchez, C. (2009). Lignocellulosic residues: Biodegradation and bioconversion by fungi. *Biotechnol. Adv.*, 27(2): 185–194.

Schwarz, W. (2001). The cellulosome and cellulose degradation by anaerobic bacteria. *Appl. Microbiol. Biotechnol.*, 56(5): 634–649.

Schwarze, F.W.M.R. (2007). Wood decay under the microscope. *Fungal Biol. Rev.*, 21(4): 133–170.

Sindhu, R., Binod, P. and Pandey, A. (2016). Biological pretreatment of lignocellulosic biomass: An overview. *Biores. Technol.*, 199: 76–82.

Singh, A.P. and Singh, T. (2014). Biotechnological applications of wood-rotting fungi: A review. *Biomass Bioener.*, 62: 198–206.

Sperandio, G.B. and Ferreira Filho, E.X. (2019). Fungal co-cultures in the lignocellulosic biorefinery context: A review. *Int. Biodeter. Biodegr.*, 142: 109–123.

Stricker Astrid, R., Grosstessner-Hain, K., Würleitner, E. and Mach Robert, L. (2006). Xyr1 (Xylanase regulator 1) Regulates both the hydrolytic enzyme system and d-xylose metabolism in *Hypocrea jecorina*. *Eukaryotic Cell*, 5(12): 2128–2137.

Sukumaran, R.K., Christopher, M., Kooloth-Valappil, P., Sreeja-Raju, A., Mathew, R.M. et al. (2021). Addressing challenges in production of cellulases for biomass hydrolysis: Targeted interventions into the genetics of cellulase producing fungi. *Biores. Technol.*, 329: 124746. doi: 10.1016/j.biortech.2021.124746.

Tirado-González, D.N., Jáuregui-Rincón, J., Tirado-Estrada, G.G., Martínez-Hernández, P.A., Guevara-Lara, F. and Miranda-Romero, L.A. (2016). Production of cellulases and xylanases by white-rot fungi cultured in corn stover media for ruminant feed applications. *Anim. Feed Sci. Technol.*, 221: 147–156.

Tri, C.L., Khuong, L.D. and Kamei, I. (2018). The improvement of sodium hydroxide pretreatment in bioethanol production from Japanese bamboo *Phyllostachys edulis* using the white-rot fungus *Phlebia* sp. MG-60. *Int. Biodeter. Biodegr.*, 133: 86–92.

Tri, C.L. and Kamei, I. (2020). Butanol production from cellulosic material by anaerobic co-culture of white-rot fungus *Phlebia* and bacterium *Clostridium* in consolidated bioprocessing. *Biores. Technol.*, 305: 123065. doi: 10.1016/j.biortech.2020.123065.

Trindade, W.R.d.S. and Santos, R.G.d. (2017). Review on the characteristics of butanol, its production and use as fuel in internal combustion engines. *Renew. Sust. Energy Rev.*, 69: 642–651.

Tsuyama, T., Yamaguchi, M. and Kamei, I. (2017). Accumulation of sugar from pulp and xylitol from xylose by pyruvate decarboxylase-negative white-rot fungus *Phlebia* sp. MG-60. *Biores. Technol.*, 238: 241–247.

Tuomela, M. and Hatakka, A. (2011). Oxidative Fungal Enzymes for Bioremediation. pp. 183–196. *In*: Moo-Young, M. (Ed.). *Comprehensive Biotechnology* (2nd Edn). Burlington: Academic Press.

Uzuner, S. and Cekmecelioglu, D. (2019). Enzymes in the beverage industry. pp. 29–43. *In*: Kuddus, M. (Ed.). *Enzymes in Food Biotechnology*, Academic Press, Chapter 3.

Veza, I., Muhamad Said, M.F. and Latiff, Z.A. (2021). Recent advances in butanol production by acetone-butanol-ethanol (ABE) fermentation. *Biomass Bioener.*, 144: 105919. doi: 10.1016/j.biombioe.2020.105919.

Viikari, L., Suurnäkki, A., Grönqvist, S., Raaska, L. and Ragauskas, A. (2009). Forest products: Biotechnology in pulp and paper processing. pp. 80–94. *In*: Schaechter, M. (Ed.). *Encyclopedia of Microbiology* (3rd Edn). Oxford: Academic Press.

Wan, C. and Li, Y. (2012). Fungal pretreatment of lignocellulosic biomass. *Biotechnol. Adv.*, 30(6): 1447–1457.

Wang, J., Hirabayashi, S., Mori, T., Kawagishi, H. and Hirai, H. (2016). Improvement of ethanol production by recombinant expression of pyruvate decarboxylase in the white-rot fungus *Phanerochaete sordida* YK-624. *J. Biosci. Bioeng.*, 122(1): 17–21.

Yamasaki, Y., Yamaguchi, M., Yamagishi, K., Hirai, H., Kondo, R. et al. (2014). Expression of a manganese peroxidase isozyme 2 transgene in the ethanologenic white-rot fungus *Phlebia* sp. strain MG-60. *SpringerPlus*, 3(1): 699. doi: 10.1186/2193-1801-3-699.

Yano, S. (2015). Enzymatic saccharification and fermentation technology for ethanol production from woody biomass. *J. Jap. Petrol. Inst.*, 58(3): 128–134.

Yi, Y. (2021). Chapter 7—Tiny bugs play big role: Microorganisms' contribution to biofuel production. pp. 113–136. *In*: Lü, S. (Ed.). *Advances in 2nd Generation of Bioethanol Production*. Cambridge: Woodhead Publishing.

Yoav, S., Salame, T.M., Feldman, D., Levinson, D., Ioelovich, M. et al. (2018). Effects of cre1 modification in the white-rot fungus *Pleurotus ostreatus* PC9: Altering substrate preference during biological pretreatment. *Biotechnol. Biofuels*, 11: 212. doi: 10.1186/s13068-018-1209-6.

Yoon, L.W., Ang, T.N., Ngoh, G.C. and Chua, A.S.M. (2014). Fungal solid-state fermentation and various methods of enhancement in cellulase production. *Biomass Bioener.*, 67: 319–338.

Yoon, L.W., Ngoh, G.C., Chua, A.S.M., Abdul Patah, M.F. and Teoh, W.H. (2019). Process intensification of cellulase and bioethanol production from sugarcane bagasse via an integrated saccharification and fermentation process. *Chem. Eng. Proc. Proc. Intens.*, 142: 107528. doi: 10.1016/j.cep.2019.107528.

Zabed, H.M., Akter, S., Yun, J., Zhang, G., Awad, F.N. et al. (2019). Recent advances in biological pretreatment of microalgae and lignocellulosic biomass for biofuel production. *Renew. Sust. Energy Rev.*, 105: 105–128.

Zhang, Y.H.P. (2008). Reviving the carbohydrate economy via multi-product lignocellulose biorefineries. *J. Ind. Microbiol. Biotechnol.*, 35(5): 367–375.

Zhao, C., Deng, L. and Fang, H. (2018). Mixed culture of recombinant *Trichoderma reesei* and *Aspergillus niger* for cellulase production to increase the cellulose degrading capability. *Biomass Bioener.*, 112: 93–98.

Zhuo, R. and Fan, F. (2021). A comprehensive insight into the application of white-rot fungi and their lignocellulolytic enzymes in the removal of organic pollutants. *Sci. Tot. Environ.*, 778: 146132. doi: 10.1016/j.scitotenv.2021.146132.

Zou, L., Liu, Y., Wang, Y. and Hu, X. (2020). Assessment and analysis of agricultural non-point source pollution loads in China: 1978–2017. *J. Environ. Manag.*, 263: 110400. doi: 10.1016/j.jenvman.2020.110400.

15

An Insight into the Applications of Fungi in Ethanol Biorefinery Operations

Navnit Kumar Ramamoorthy,[1] *Puja Ghosh,*[2] *Renganathan S.*[1] and *Vemuri V. Sarma*[2,*]

1. Introduction

The second-generation bioethanol (2G bioethanol), one of the potential and major alternate fuels to evade the fossil fuel crisis, involves an expensive production methodology (Jin et al., 2015). The US Energy Information Administration (USEIA) defines biomass as the renewable organic content, which originates from plant and animal sources (Source: Biomass-renewable energy from plants and animals, USEIA. https://www.eia.gov/energyexplained/biomass/). Lignocellulosic biomass predominantly comprises plant cell walls, with the structural carbohydrates, i.e., celluloses (a ß-D-glucose monomer), hemicelluloses, and the heterogeneous phenolic polymer lignin. A biorefinery is a renewable analogue of a petroleum refinery. Crude oil is the raw material in a petro-refinery while renewable lignocellulosic biomass is the raw material in a biorefinery.

The 2G bioethanol refinery involves three major stages: pretreatment, saccharification, and fermentation, as shown in a flow diagram in Fig. 15.1. Table 15.1 shows that in fermentation operations involving yeasts, the bioethanol yield is 3-fold higher in comparison to bacterial fermentations.

[1] Centre for Biotechnology, Anna University, Chennai - 600025, India.
[2] Department of Biotechnology, Pondicherry University, Kalapet, Pondicherry - 605014, India.
* Corresponding author; sarmavv@yahoo.com
The first and second authors have equally contributed.

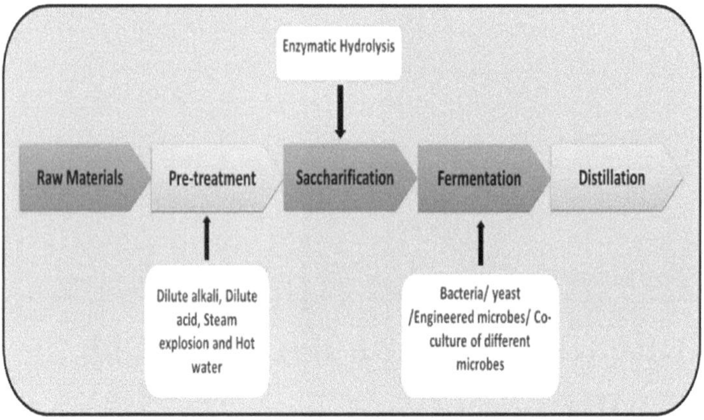

Fig. 15.1. The conventional biorefinery concept for 2G bioethanol production.

2. Composition and Architecture of Lignocellulosic Biomass

In lignocellulosic biomass, xylose and ß-D-glucose are the major sugar moieties (Parthasarathi et al., 2016). Cellulose, with repeating cellobiose (two units of six carbon ß-D-glucose) residues, is a linear glucan polysaccharide having ß-1,4-glycosidic bonds between its ß-D-glucose monomers (Fig. 15.2).

Intramolecular hydrogen bonds and intermolecular Van der Waals forces have been observed in the cellulose microfibrils architecture (Martínez-Sanz et al., 2017). Hemicelluloses are polysaccharides having ß-(1,4)-glycosidic linkages between the 5 carbon pentose sugars (Fig. 15.3) such as xylose and arabinose, which make up the backbone structure of hemicelluloses.

Additionally, hemicelluloses also bond with 6 carbon hexose sugars such as mannose and lactose having glucose as the repeating units. The similarity in the glycosidic linkages in hemicellulose and cellulose results in a strong non-covalent association of hemicelluloses with cellulose microfibrils (Berglund et al., 2016). Lignin is a non-sugar-based phenolic polymer consisting of phenylpropane units having carbon-carbon linkages and predominantly ß-O-4 aryl ether linkages (Ma et al., 2015a). Lignin biosynthesis commences when phenylalanine is deaminated to cinnamic acid; p-coumaryl, coniferyl, and sinapyl alcohols are the three major fundamental components of lignin architecture (Fig. 15.4) (Geng et al., 2020).

3. Strategies of Pre-Treatment

3.1 Pretreatment

Pretreatment, one of the most cost-intensive steps in the 2G bioethanol refinery, improves the accessibility of the cellulose to the cellulose hydrolyzing/saccharifying enzymes (cellulase) during the saccharification process (Li et al., 2019). Pretreatment reduces the recalcitrance of lignin or removes lignin to a considerable extent (Loow et al., 2015). At the end of the process, the cellulose component of the biomass is readily available to be hydrolyzed/saccharified and converted to fermentable

Table 15.1. A list of various lignocellulosic biomass, the fermentative microbes used, and the bioethanol yields obtained.

Biomass	Pretreatment Technique	Fermentative Microorganism	Ethanol Yield	References
Cyanobacteria: Spirulina	Mechanical grinding followed by dilute acid pretreatment	*Saccharomyces cerevisiae*	15.2 g/L	Hossain et al., 2015
Micro algae: *Chlamydomonas reinhardtii*	-	Photofermentation by supplementing with nutrients	19.24 g/L	Costa et al., 2015
Potato waste	Biomass recalcitrance was reduced using a mixture of pectinase and cellulase.	*Saccharomyces cerevisiae*	0.2 g/g sweet potato residue (SPR)	Wang et al., 2016
Water hyacinth	Combined acid pre-treatment and enzymatic hydrolysis	*Saccharomyces cerevisiae*	1.289 g/L	Zhang et al., 2016
Brown algae	Mechanical pulverization	*Defluviitalea phaphyphila* Alg1	0.44 g/gβ-D-glucose	Ji et al., 2016
Paper sludge	Direct SSF performed without a pre-treatment	*Saccharomyces cerevisiae*	22.7 g/L	Mendes et al., 2017
Pinecones, paper, and domestic wastes like corncob and corn	10% (v/v) sulphuric acid pretreatment	*Lactobacillus plantarum* M24	Corn cob: 1.236 mg/L	Soleimani et al., 2017
			Corn: 2.556 mg/L	
Carnauba straw	Alkali pretreatment	*Kluyveromyces marxianus* (ATCC-36907)	7.53 g/L	Silva et al., 2018
Castor bean cake	Direct enzyme hydrolysis	*Saccharomyces cerevisiae*	35 g/L	Abada et al., 2018
Waste surgical cotton-waste packaging Cardboard mixture	15% (v/v) ammonia pretreatment; Biological pre-treatment using *Paecilomyces inflatus*; cold non-thermal plasma pre-treatment.	*Saccharomyces cerevisiae* RW 143	Ammonia treated: 0.4 g/g of β-D-glucose	Ramamoorthy et al., 2018
			Biologically pretreated: 0.42 g/g of β-D-glucose	Ramamoorthy et al., 2020a
			Plasma pre-treated: 0.4 g/g of β-D-glucose	Ramamoorthy et al., 2020b
Corn stover	NaOH and ozone combined pre-treatment	*Saccharomyces cerevisiae*	101.50% w/v	Shi et al., 2019

Table 15.1 contd. ...

...Table 15.1 contd.

Biomass	Pretreatment Technique	Fermentative Microorganism	Ethanol Yield	References
Municipal solid wastes	Crushed to smaller pieces and acid hydrolyzed (7.5% v/v)	*Saccharomyces cerevisiae*	0.13 g/g of wastes	Thapa et al., 2019
Potato peels	Soaking assisted thermal pre-treatment (SATP)	*Saccharomyces cerevisiae*	0.32 g/g of sugar	Naseeha et al., 2020
Cassava peels	Dilute aid pre-treatment	*Saccharomyces cerevisiae*	16.80 g/L	Gabriel et al., 2020
Pineapple waste	Mechanical pre-treatment	*Saccharomyces cerevisiae*	5.4% (v/v) fermentation media	Gil et al., 2018
Napier grass	2 M NaOH pre-treatment	*Saccharomyces cerevisiae*	44.7 g/L	Kongkeitkajorn et al., 2020
Pre-treated dairy manure	Dilute NaOH pre-treatment	*Zymomonas mobilis*	71.91% (v/w) of the pre-treated biomass	You et al., 2017b
Food wastes	*Fusarium oxysporum* was used for both hydrolysis and fermentation	*Fusarium oxysporum*	16.3 g/L	Prasoulas et al., 2020

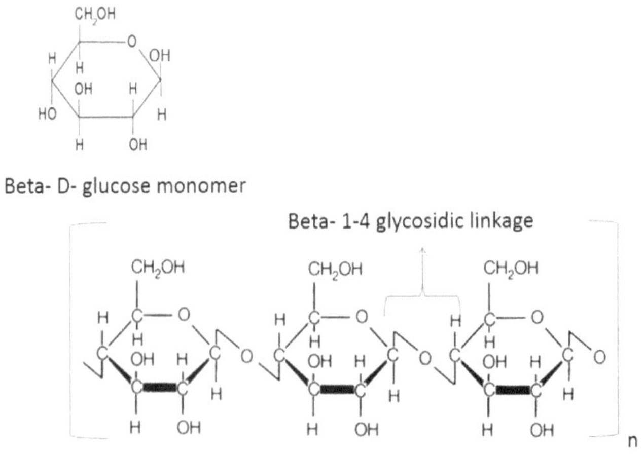

Fig. 15.2. The linkages in a cellulose molecule.

ß-D-glucose monomers. Some of the common pretreatment processes include the employment of chemical or physiochemical agents or biological techniques using various microbes (Yang et al., 2016). Pretreatment techniques at a low pH result in the hydrolysis of hemicelluloses (Chen, 2015), while techniques that employ a

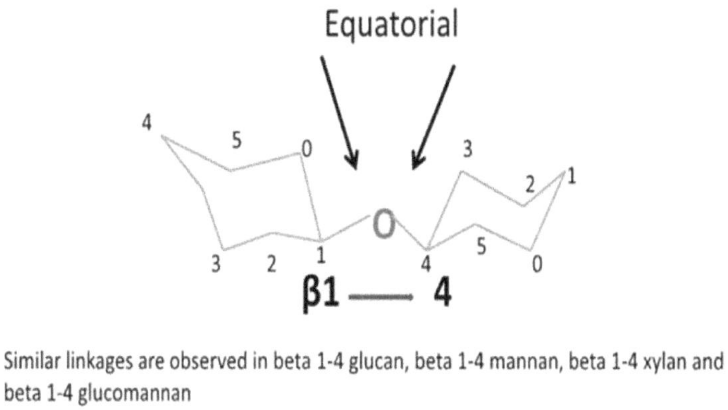

Equatorial
β1 — 4
Similar linkages are observed in beta 1-4 glucan, beta 1-4 mannan, beta 1-4 xylan and beta 1-4 glucomannan

Fig. 15.3. The linkages in hemicellulose.

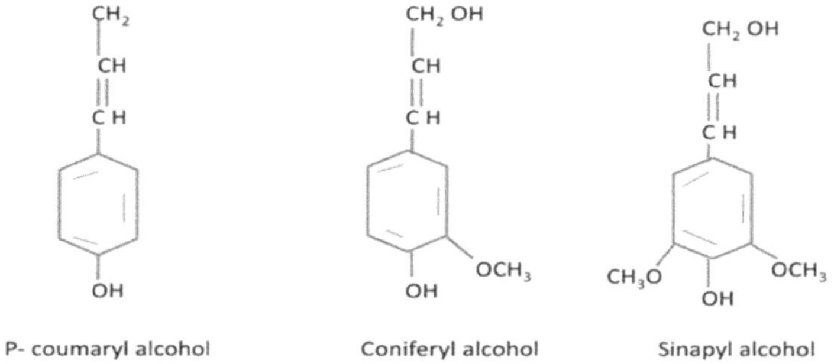

Fig. 15.4. The fundamental mono-lignol units of lignin.

higher pH depolymerize lignin (Fernández-Rodríguez et al., 2017) to a major extent; at neutral pH, there is partial hemicellulose hydrolysis (Kucharska et al., 2018).

Mechanical pretreatments such as chipping, milling, and grinding of the biomass result in a size reduction and a decrease in its crystallinity, thereby enabling access to the celluloses (Ani et al., 2016). Hydrothermal/steam pretreatment which efficiently depolymerizes lignin (0.5%–1% depolymerization), subjects the biomass to the action of saturated steam at high temperature (150–290°C) and pressure (5–50 atm) for a few minutes.

During ultrasonication-based pretreatment, ultrasonic cavitation creates shear forces which help in disintegrating the complex biomass architecture (Bundhoo and Mohee, 2018). In one of our previous works (Ramamoorthy et al., 2020a), when a step of ultrasonication (80% amplitude; 30s ON/10s OFF; total cycle time, 20 min) was used prior to biological pretreatment of a waste cotton-cardboard mixture, lignin removal was around 76% (w/w).

3.1.1 Biological Pretreatment

Biological pretreatment employs microbes, such as fungi, bacteria, and actinomycetes, which secrete certain saccharolytic/cellulolytic, hemicellulolytic, and ligninolytic enzymes to break down lignin, hemicelluloses, and celluloses (Capolupo and

Faraco, 2016). Selective delignification and no generation of inhibitory byproducts (Ramamoorthy et al., 2020a) are the salient features of this approach. Our research group performed a 14-day long biological pretreatment of a waste cotton-cardboard mixture using a fungus *Paecilomyces inflatus* ATCC® 32919™ (Ramamoorthy et al., 2020a) and it was observed that there were inconspicuous quantities/complete absence of inhibitors. Laccases, lignin peroxidases, manganese peroxidases, and versatile peroxidases are the significant ligninolytic enzymes, which efficiently degrade lignin in association with mediators and assisting enzymes, such as feruloyl esterases, quinone reductases, catechol 2,3 dioxygenases, among others. Laccases catalyze the conversion of 4 benzenediol + O_2 to 4 benzo semiquinone + $2H_2O$. Lignin peroxidise catalyzes a hydrogen peroxide-dependent oxidative depolymerization of lignin. Manganese peroxidase, in the manganese ion, effectively deconstructs lignin molecules (Kumar and Chandra, 2020). A comparison of some of the significant innovative pretreatment processes' efficiencies with a biological pretreatment's efficiency using a fungal source has been shown in Table 15.2.

Table 15.2. A comparison of various pretreatment techniques with a fungal-based pretreatment technique.

Pretreatment Technique	Biomass	Impact of the Pre-treatment	References
A hybrid, thermal-acid-organ Solv technique; 200°C; 1% (w/w) H_2SO_4; 60% (v/v) ethanol.	Birch	86.2% (w/w) delignification; 8.8% (w/w) hemicelluloses removal; 77% (w/w) cellulose recovery	Matsakas et al., 2018
Biological pretreatment using a fungus *Paecilomyces inflatus*	Waste surgical cotton-waste packaging cardboard mixture	76% (w/w) lignin removal; 97% (w/w) hemicelluloses hydrolysed; no inhibitor generation; bioethanol yield of 0.42 g/g of glucose.	Ramamoorthy et al., 2020a
Pretreatment performed using deep eutectic solvent (DES) and protonic acid	Hybrid *Pennisetum*	78% lignin removal; 93% hemicelluloses removal; 95% cellulose yield after saccharification	Wang et al., 2020
Cold non-thermal plasma-based pretreatment	Waste surgical cotton-waste packaging cardboard mixture	68% (w/w) lignin removal and 95% (w/w) hemicelluloses removal; no inhibitor generation; bioethanol yield 0.4 g/g of glucose.	Ramamoorthy et al., 2020b

3.2 Pretreatment Conditions

Pretreatment durations lesser than 4 hours and a temperature range lesser than 160°C need to be chosen to avoid the formation of inhibitory compounds, such as furfurals, acetic acid, *p*-coumaric acid, ferulic acid, and 5-hydroxy-methyl-furfural (5-HMF) (Wang et al., 2020). Non-optimal and extreme reaction conditions result in hemicellulose hydrolysis as the pentose sugar monomers dehydrate to form the inhibitor furfural. Furfurals are furan aldehydes (Mariscal et al., 2016), which, in rare cases, are fermented to furfuryl alcohol by certain fermentative organisms during the fermentation of the saccharified hydrolysate (Zhang et al., 2019). Yeasts can tolerate a furfural concentration of around 3 mg/L (Field et al., 2015). Furfurals create oxidative stress and decrease glycolytic activity and hamper the tri-carboxylic acid cycle (TCA) in yeasts (Cheng et al., 2018). Hexose sugars, such as glucose, degrade to form 5-HMF. During the fermentative step for ethanol production, furfurals and HMFs, apart from hampering cell growth and respiration, are also involved in inhibition of the action of the key fermentative enzymes: alcohol dehydrogenase, aldehyde dehydrogenase, and pyruvate decarboxylases. In combination, furfurals and HMF cause repression of the translation activity in *Saccharomyces cerevisiae* due to the formation of stress granules (SG's) and cytoplasmic mRNP granules (Iwaki et al., 2013). Furthermore, this limits the reusability of the fermentative microbial culture (Olofsson et al., 2008). In native strains of *S. cerevisiae* and *Pichia stipitis*, around 10–120 mM HMF can be tolerated (Liu et al., 2004).

3.2.1 Breakdown Products

During pretreatment steps, apart from minuscule quantities of other breakdown products, hemicelluloses are found to be the sources of: (i) water, (ii) acetic acid, (iii) formic acid, and (iv) xylose furfural (Jönsson and Martin, 2016). Acetic acid lowers the cell pH resulting in reduced cell activity of the fermentative microbes (Casey et al., 2010). Its intracellular concentrations close to 120 mM or more result in a 50% suppression of the activities of enolase and phosphoglyceromutase (Caspeta et al., 2015).

Lignin breakdown products are usually aromatic, polyaromatic, phenolic, and aldehyde compounds, which hamper cell growth and sugar assimilation by creating a loss of rigidity of the cell membranes of the fermenting organisms (Zeng et al., 2014). Lignin has the potential to non-specifically bind to the cellulase enzyme, thereby resulting in reduced efficiency of cellulose hydrolysis by several manifolds (Azar et al., 2020). Yeasts tolerate a maximum concentration of 1 g/L of furans and phenolic degradation products (Caspeta et al., 2015).

3.3 Biological Abatement

Biological abatement/bio-abatement is a method in which the removal of inhibitors is carried out using organisms like *Coniochaeta ligniaria.* Around 50% of HMF, furfural, and phenolic inhibitory compounds have been reported to have been removed using bio-abatement (Cao et al., 2013). Furthermore, researchers have been working on the development of engineered microbes whose bio-catalytic activities are not obstructed through these toxic compounds.

4. Saccharification and Fermentation

4.1 Enzymatic Saccharification

Saccharification is the process where a complex polysaccharide such as cellulose is broken down/hydrolyzed to its monomeric units (monosaccharide ß-D-glucose) by breaking the chemical bonds (Li et al., 2019). Cellulase belongs to the family of glycoside hydrolases and it comprises three enzymes: endoglucanases/CMCases (EC 3.2.1.4); exoglucanases/FPases, including cellobiohydrolases (majorly CBH) (EC 3.2.1.91); and β-glucosidase (BG) (EC 3.2.1.21) (Ramamoorthy et al., 2019a). The synergistic activity of the three major cellulolytic enzymes (Fig. 15.5), cellobiohydrolase-1 (CBH-I), cellobiohydrolase-II (CBH-II), and endoglucanases, catalyzes the complete breakdown of cellulose to its repeating homo-disaccharide (β-D-glucose) units, cellobiose (Lakhundi et al., 2015). An exo-endo synergism is a primary mechanism in enzymatic saccharification. Endoglucanase/CMCases randomly cleave the β-1,4 glycosidic linkages (depolymerization) of the lengthy cellulose chain. The resultant broken cellulose chains contribute to the presence of additional regions for exoglucanase activity; the synergism of exo-endo is enhanced to resulting in the simultaneous actions of CBH-I on the reducing ends and CBH-II

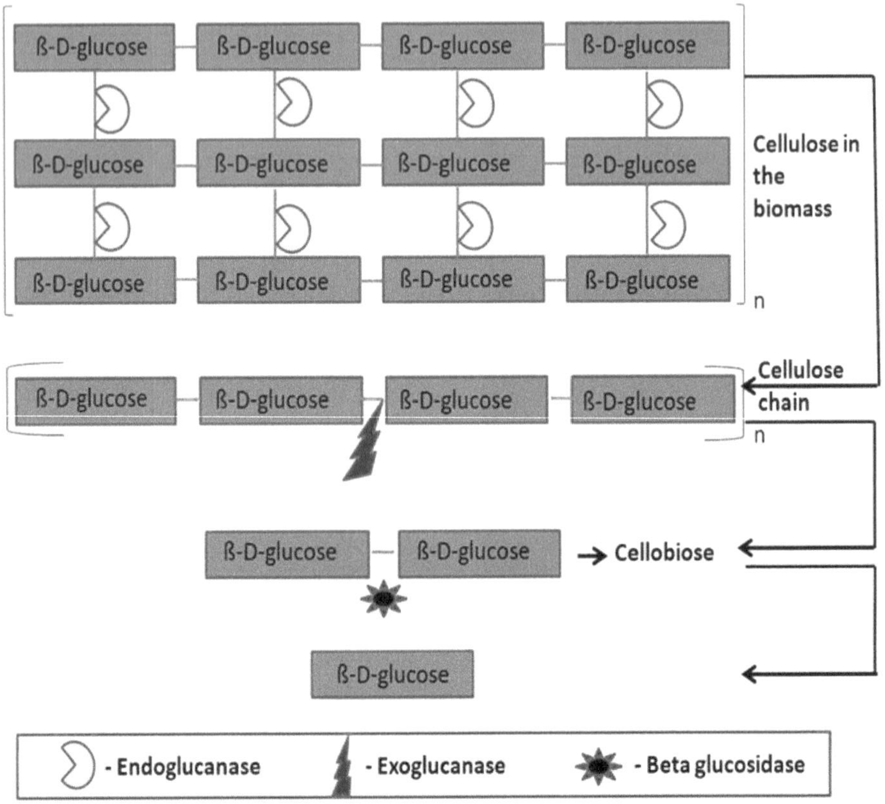

Fig. 15.5. The sequence of enzymatic saccharification of lignocellulosic biomass.

on the non-reducing ends of the depolymerized cellulose chain, further exposing additional regions for saccharification. In general, exoglucanase/FPases cleave glucose commencing from the reducing end of the broken chain. β-glucosidases cleave the resultant cellobioses to individual β-D-glucose molecules (Liu et al., 2020). Glycosyl hydrolases involve a retention or inversion hydrolytic mechanism initially proposed by Koshland (Elferink et al., 2020). In retention hydrolytic mechanism, a dual-staged, double-displacement is observed in an acid/base and a nucleophilic residue. In the initial glycosylation step, a proton transfer occurs to the glycosidic oxygen from the acid/base residue. Consequentially, an attack on the anomeric carbon of the carbohydrate (in the −1 binding site) by the nucleophile results in the formation of a glycosyl-enzyme intermediate (GEI). An enzyme catalyzed alteration of the −1 glycosyl residue (from the native chair conformation) occurs during the progress of glycosylation. Deglycosylation, the second step, progresses when a water molecule attacks the anomeric carbon and cleaves the GEI bond. This results in a proton transfer to the acid/base. Following this, the acid/base residue and the nucleophile are restored to further complete the catalysis (Ribeiro et al., 2019). Commercially available cellulases are expensive (Navnit Kumar et al., 2020). Hence, biofuel industries and laboratories producing ethanol have been relying more on in-house fungal cellulases (Sambavi et al., 2019). In two cellulase production processes carried out by us employing *Trichoderma* spp. and using vegetable-fruit peel-coir mixture and a waste cotton-cardboard mixture, it was found that the cost of the produced in-house cellulase batches were \$8.5 (per 100 mL of 159 CMCU) and \$8 (per 100 mL of 20 FPU), respectively (Sambavi et al., 2019; Ramamoorthy et al., 2019b).

Cellobioses are reported to cause a significant product inhibition in the saccharification processes (Chen, 2015). Cellobiose binds to the tryptophan residue, positioned close to the active site of CBH, thereby causing steric hindrance. This blocks the further diffusion of fragments of cellulose molecules entering the active site of the enzyme (CBH's). Additionally, cellobiose binding alters the molecular conformation of CBH, which results in non-productive adsorption of cellulose to CBH. Such an end-product inhibition makes it difficult for the microfibrils to be removed from the cellulose chain, consequentially halting further saccharification (Zhao et al., 2004).

4.2 Ethanol Fermentation

Biochemically, fermentation is a metabolic process where an organism metabolizes a carbohydrate, usually starch or sugar, and converts it to an acid or an alcohol. In yeasts, fermentation is a process by which they obtain energy by converting sugar to alcohol. *Saccharomyces cerevisiae* (ethanol yield per gram of glucose, 0.36 g/g to 0.4 g/g (Qiu and Jiang, 2017; Ramamoorthy et al., 2018)), the commonly employed fermentative yeast, and certain bacteria such as *E. coli* (ethanol yield: 0.3–0.4 g/g glucose (Wang et al., 2019), convert pyruvate obtained from glycolysis into carbon-dioxide and ethanol in a final fermentative step (Maicas, 2020). Together, the glycolytic pathway and the fermentative steps comprise the well-established Embden-Mayerhof-Parnas (EMP) pathway (Sánchez-Pascuala et al., 2017). The EMP pathway is also observed in the fermentative bacteria *E. coli*, during fermentation of glucose (Sánchez-Pascuala et al., 2017). In a fermentation process, the theoretical maximum ethanol titer achievable is 0.51 g/g of the glucose consumed (Ramamoorthy et al., 2018). The chemical equation for ethanol fermentation is shown in Equation 1. The pathway for the conversion of sugar to ethanol in yeasts (Fig. 15.6). Pyruvate produced at the end of the glycolysis pathway is non-oxidatively decarboxylated by pyruvate decarboxylase to acetaldehyde and carbon dioxide. Acetaldehyde is reduced to ethanol involving the enzyme alcohol dehydrogenase.

$$C_6H_{12}O_6 \rightarrow 2C_2H_5OH + 2CO_2 \quad (1)$$

Under anaerobic conditions (complete absence of oxygen) or micro-aerophilic conditions (requiring very little oxygen), acetaldehyde is converted to ethanol along with the generation of 2 moles of ATP. However, this process does not sustain as higher quantities of glucose need to be consumed by the cells to generate enough ATP. An excessive sugar concentration (more than 25% v/v) in the fermentation broth results in an extended lag phase during the cell multiplication, apart from also causing higher osmotic stress over the cells (de Silva et al., 2013). Fermentation results in the accumulation of ethanol (Maicas et al., 2020). At higher concentrations of ethanol (more than 115–200 g/L (Caspeta et al., 2015)), the phospholipid composition of yeast's plasma membrane is affected, thereby resulting in altered permeability and ineffective glucose uptake within the cells (de Silva et al., 2013). Its cellular enzymes are denatured, inhibited (de Silva et al., 2013) and the cell cycle and cell division stalled (Maicas et al., 2020). Ethanol dehydrates and results in osmosis, due to the removal of water from the cells (de Silva et al., 2013). Due to all the above-mentioned conditions, fermentation halts and does not proceed further (Maicas et al., 2020). *S. cerevisiae* with an ethanol tolerance of up to 12% (v/v) has been produced using diploidization (Burcu et al., 2017). In a recent demonstration by Sukwong et al. (2020), it was observed that both glucose and galactose present in the red seaweed repress the galactose genes (MIG2, MIG1, GLK1) and over-express phosphoglucomutase gene (*PGM2*), which controls galactose metabolism in *S. cerevisiae* (Sukwong et al., 2020). This process enhances galactose consumption rate (0.24 g/L/h). Certain yeast strains such as *Spathaspora passalidarum* (ethanol titer, 0.48 g/g of xylose) (Selim et al., 2020); *Pichia stipitis* (57 g/L bioethanol) (Selim et al., 2018); and *Candida shehatae* (Selim et al., 2018), effectively ferment xylose to bioethanol (Selim et al., 2020). They utilize the pentose phosphate pathway as an intermediate step (pentose phosphate shunt) before taking the course of the EMP pathway (see Fig. 15.6) (Selim et al., 2020).

Bacteria such as *Lactobacillus brevis*, under micro-aerophilic conditions, utilize a similar pathway (see Fig. 15.6) to metabolize xylose and later ferment it to ethanol (8.3 g/L). In this process, ethanol is one of the by-products along with lactic acid (0.53 g/L) and acetic acid (5.1 g/L) (Zhang and Vadlani et al., 2015). In a continuous culture technique using *Clavispora opuntiae*, around 92% (v/v) of the theoretical yield of ethanol was obtained by the fermentation of D-xylose (Nigam, 2015). *Escherichia coli* has been reported to be a potential candidate for co-fermentation, which can ferment both glucose and xylose in the saccharified hydrolysate. Though *E. coli*

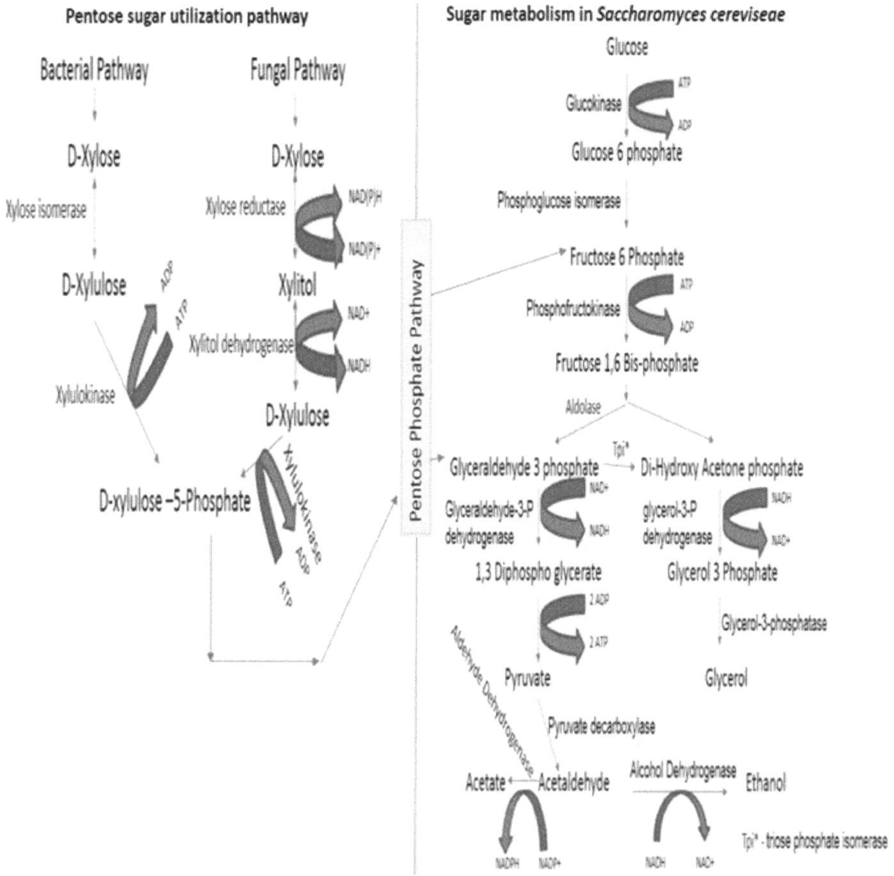

Fig. 15.6. The biochemical pathways of the fermentation of glucose and xylose to ethanol in fermenting microbes.

prefers to utilize glucose in the hydrolysate (carbon catabolite repression), it also is capable of consuming xylose post the complete consumption of glucose (Fernández-Sandoval et al., 2019). Apart from counteracting issues related to inhibitor presence (from pre-treatment) and ethanol tolerance, the usage of a higher cell density (of the fermentative microbe) eliminates the lag/adaptation phase and enhances the rate of fermentation (Li et al., 2019). Fermentation yield improved by two to three folds while this technique was employed in processes involving *S. cerevisiae* and *Z. mobilis* (Li et al., 2019). *Z. mobilis*, a co-fermenting, gram-negative, non-sporulating, facultative anaerobic bacterium, with an acetic acid tolerance of 8 g/L and furfural tolerance of 3 g/L (Wang et al., 2020), survive in a wider range of pH (3.5–7.5) and shows higher bioethanol productivity (ethanol yield: 0.4 g/g of glucose). A higher surface area of the microbe enables enhanced glucose assimilation in comparison to yeasts and its ability to utilize N_2 as the direct nitrogen source (instead of the usually expensive NH_4) makes *Z. mobilis* an economical choice too (Yang et al., 2016). A quicker completion of the fermentation process (50% faster than yeasts) is observed in this microbe as it takes the course of the Entner-Doudoroff (ED) pathway (Fig. 15.7), which completes fermentation with 50% lesser ATP and reduced bio-catalytic stages, in contrast to the EMP pathway (Yang et al., 2016). To overcome the production cost owing to excessive nitrogen sources and water supply, while using biogas and dairy manure as an alternative to nitrogen sources during fermentation, *Z. mobilis* strains yield 72.63% (v/v) (You et al., 2017a) and 71.9% (v/v) bioethanol (You et al., 2017b). A minuscule quantity of Na^+ ions (0.175 M) and ammonium ions, which arise from pre-treatments using an acid/alkali or during the neutralization steps (of the biomass), reduce the cell viability of strains of *Zymomonas* spp. (Gao et al., 2018). *Clostridium* spp., a gram-positive, thermophilic, anaerobic bacterium, comprises a cellulosome that degrades sugarcane bagasse directly (Qu et al., 2017). Singh and his co-workers reported 95.32% cellulose conversion to ethanol using *C. thermocellum* ATCC 31924 (Singh et al., 2017). The application of UV, introduction of random mutagenesis followed by protoplast fusion, were used for improving the acetic acid and ethanol tolerance of *C. ragsdalei* (DSM 15248). The mutant strain thus raised could produce around 14.92 ± 0.75 g/L of ethanol, while the wild strain could produce only 1.92 ± 0.52 g/L of ethanol (Patankar et al., 2021). Recent studies also showed that several thermotolerant yeasts such as *Candida galbrata*, *C. nivariensis*, *C. tropicalis*, *Kluyveromyces marxianus*, and *Pichia kudriavzevii* obtained from buffalo rumen, are also having high fermentative efficiency (Avchar et al., 2021). Though the fermentation yield of several microbes has been studied, the higher ethanol tolerance of *S. cerevisiae* (115 and 200 g/L) (Caspeta et al., 2015) in comparison to *E. coli* and *Z. mobilis* (60–130 g/L) (Yang et al., 2016) makes it the most preferred microbe for ethanol fermentation.

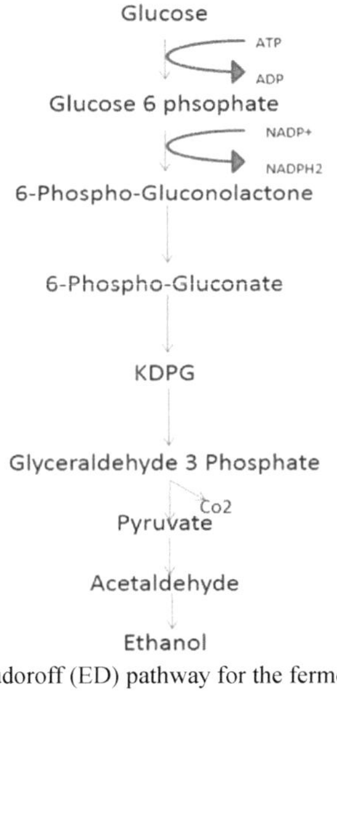

Fig. 15.7. The Entner-Doudoroff (ED) pathway for the fermentation of glucose to ethanol.

5. Combination of Biorefinery Stages

5.1 Simultaneous Saccharification and Fermentation

In a separate hydrolysis and fermentation process, an end-product inhibition reduces the yield of the fermentable ß-D-glucose units. This problem could be circumvented if the produced sugars causing the inhibition are removed on a timely basis (based on the rate of saccharification) from the saccharification reactor. In simultaneous saccharification and fermentation (SSF), concurrently, the saccharification of cellulose into β-d-glucose monomers and fermentation of the produced monomeric sugar units to bioethanol occur in a single reaction vessel (Jason Charles et al., 2018). Since the sugars are simultaneously fermented as they are produced after saccharification, they do not participate in end-product inhibition (Jason Charles et al., 2018). In an earlier work, we found that a co-culture of *Trichoderma harzianum* ATCC® 20846™ (15% v/v) and *S. cerevisiae* RW 143 (5% w/v) could enhance cellulase production (2.035 FPU/mL) and ethanol concentration (0.09 g/L) and we surmized that this process could counteract carbon catabolite repression (due to the accumulation of cellobiose/glucose by the action of cellulase) (Jason Charles et al., 2018).

The regulation of pH and temperature in such a combinatorial process is a major bottleneck. Enzymatic saccharification occurs at a higher temperature (close to 50°C) than the temperature required for fermentation (usually 25–30°C for yeasts) (Navnit Kumar et al., 2018b). The attainment of saccharification at an optimum temperature has been an ordeal and several researchers have reported temperatures close to 37°C to be optimal (Jason Charles et al., 2018). In a recent process innovation, an unconventional, fed-batch, non-isothermal-SSF, it has been reported that the temperature alteration every 6 hrs could elevate the ethanol production up to 30.1% (Wang et al., 2020). To reduce the inhibitor production, Mendes and his co-associates employed fed-batch/batch SSF for bioethanol production (22.7 g/L) using *S. cerevisiae* from the sludge (60% carbohydrates and unbleached pulp) coming out of paper and pulp manufacturing industries (Mendes et al., 2017). The SSF process can become more efficient and profitable if the fermentative microbes utilize both pentose and hexose sugars in the reactor.

5.2 Semi-simultaneous Saccharification and Co-Fermentation

In semi-simultaneous saccharification and co-fermentation (SScF), the co-fermentation of both hexose and pentose sugars is performed together in a single SSF reactor (Lee et al., 2017). The SScF is usually employed when significant quantities of pentose sugars are obtained after saccharification (Lee et al., 2017). This could either be based on the nature of the biomass or because of the pre-treatment strategy, with pentoses not being completely removed due to partial hydrolysis. Genetically modified *S. cerevisiae* XUSE (expressing genes from the pentose phosphate pathway) and an improved variant of the recombinant *S. cerevisiae* XUSEA, which express *xylA*3* and RPE_1 co-ferment xylose to ethanol (ethanol yield: 0.39 g/g xylose) (Hoang Nguyen Tran et al., 2020). Bondesson et al. (2016), used a xylose and glucose

Table 15.3. A list of some of the significant process improvements reported in the microbe-based combinatorial procedures of the 2G bioethanol refinery.

Process	**Biomass**	**Yield**	**References**
Simultaneous saccharification and fermentation (SSF) using *Bacillus coagulans* CC17	Bagasse sulphite pulp	Product yield 0.72 g/g of cellulose.	Zhou et al., 2016
Thermophilic simultaneous saccharification and fermentation; *Bacillus subtilis* IPE5-4, which simultaneously co-ferments xylose and glucose.	Corn cob	96.34% increase in cellulose utilization and an increase of 93.29% hemicellulose utilization.	Jia et al., 2017
Simultaneous saccharification and fermentation with delayed yeast extract feeding and *in situ* recovery.	Oil palm empty fruit bunch (OPEFB)	42% improvement in the bio-alcohol yield and 11% improvement while *in situ* stripping was performed.	Salleh et al., 2019
Coupled semi-continuous fermentation and simultaneous saccharification strategies.	Sugarcane bagasse	Enhanced bioethanol yield of 8%–10% (v/v).	Portero et al., 2020
Enzyme consortia employment (alpha-amylase; glucoamylase).	Chinese *Jiuqu*	A bioethanol production of 488–470 g/L.	Wang et al., 2020

co-fermenting strain *S. cerevisiae* (KE6-12) for producing bioethanol (ethanol yield: 0.32 g/g sugar) using SScF from wheat straw, which was impregnated with acetic acid and then steam hydrolyzed as part of the pretreatment (Bondesson et al., 2016). A list of certain significant, improved microbe-based combinatorial processes has been shown in Table 15.3.

5.3 Consolidated Bioprocessing

The consolidated bioprocessing (CBP) integrates three major steps of the 2G bioethanol refinery. These include: (i) production of cellulases, (ii) saccharification of the biomass by the produced cellulases, and finally (iii) microbial fermentation of the sugars produced in the saccharification process (Cunha et al., 2020). A CBP was performed by our research group by using chemically fused protoplasts ($CaCl_2$—Polyethylene glycol mediated) of *T. harzanium* ATCC® 20846™ and *S. cerevisiae* RW 143 and an ethanol concentration of 0.04 g/L was recorded. The work was performed with an aim to attenuate carbon catabolite repression during cellulase production (0.9 FPU/mL was recorded). Similarly, a solid-state CBP was performed by our research group in a self-designed aerated tray reactor employing cotton-cardboard mixture as the substrate. The ethanol concentration recorded was 11.25 g/L (Navnit Kumar et al., 2019).

5.4 Improvement of the Microbial Strains

Though researchers have been using synthetic biology tools for improving the co-fermentative nature of the strains, the achievement of an efficient co-fermentation yield is still being investigated. Ning et al. (2018) used UV irradiation to obtain two tannin tolerant mutant strains of *Pachysolen tannophilus*, which reduced the extraction cost of tannin (from acorn starch) by around five times and in turn reduces the ethanol production cost (Ning et al., 2018). In a previous work of ours, to produce exoglucanases, an enzyme that is least produced by cellulolytic microbes, a soluble heterologous expression was performed. The gene responsible for producing exoglucanases (*Cel 7A* from *T. resei*) was expressed in a recombinant host *E. coli* SHuffle, and an FPase yield of 2.5 IU/mL (0.7 g/L; 58 kDa) was recorded (Navnit Kumar et al., 2018a). Table 15.4 lists several significant microbial strain improvements/microbial culture techniques (with respect to the 2G bioethanol refinery) reported by researchers.

6. Recent Significant Biotechnological Advances

The entire genome of *Mucor circinelloides*, a dimorphic fungus, has been well studied. Apart from the several significant enzymes it produces, certain species of the fungi are known producers of cellulases, xylanases, and polygalacturonases. It grows in a wide range of temperatures (40°C – 60°C) and pH (3.5 – 5.5) making it a suitable candidate for a fungal-based biorefinery. The fungal biomass could serve as a nutritive feed to livestock (Rodrigues Reis et al., 2019), thereby counteracting the shortage of conventional lignocellulosic biomass, which has risen due to its huge demand in bio-fuel applications. The net present value of an ethanol biorefinery increases 8-fold approximately when the fungal biomass is marketed as feed (Bulkan et al., 2021). Mycelium composites, due to their high acoustic absorptivity, reduced thermal conductivity and fire safety, find wide usage in civil engineering applications, such as material insulation, door frame design, panel making, flooring and cabinet designs (Jones et al., 2020). Filamentous fungi produce certain organic acids with low molecular weights. Wood-decaying fungi, saprophytic fungi, and fungi belonging to the Basidiomycetes classes synthesize oxalic acid, which efficiently hydrolyzes the holocellulose content of lignocellulosic biomass. Such fungi could be grown on complex substrates to perform an efficient pretreatment (Javaid et al., 2019). In a recent work by Amoah et al. (2017), a CBP strain named *S. cerevisiae* MT8-1 was engineered to express five cellulase producing genes (BGL, XYNII, EGII, CBHI, and CBHII). This strain has been reported to perform significant biomass saccharification and fermentation in a single-pot process. Using this strain, and ionic liquid pretreated bagasse as the substrate, 0.9 g/L ethanol was produced. Similarly, while unbleached kraft pulp (UKP) was used as the substrate, around 0.7 g/L of ethanol was produced by employing the CBP strain (Amoah et al., 2017). In an attempt at performing a partial CBP, Althuri et al. (2017) used the biomass of *Ricinus communis*, *Saccharum officinarum*, and *Saccharum spontaneum*, and subjected them to a non-isothermal simultaneous pretreatment and dulcification process, which was supplemented with a blend of laccase (produced from *Pleurotus djamor*) and holocellulase (produced from *Trichoderma reseei* RUT C30). Later, a co-fermentation of the hydrolysate resulted in 62 g/L ethanol production (Althuri et al., 2017). Alan Grant et al. (2020) performed a consolidated bioprocessing (CBP) by co-culturing *Clostridium thermocellum*, *C. stercorarium*, and *Thermoanaerobacter*

Table 15.4. Microbial strain improvement techniques/microbial culture processes employed to enhance the biorefinery process.

Process	**Enhancement in the Process**	**References**
Zymomonas mobilis ZM4 grown as a biofilm.	Bioethanol yield of 0.37 g/g sugars from 13.4 g/L rice bran hydrolysate.	Todhanakasem et al., 2015
An acid tolerant mutant ZMA7-2 of *Zymomonas mobilis*	Maximum bioethanol yield of 0.5g/g of sugar.	Ma et al., 2015b
Protoplast fusion *Saccharomyces cerevisiae* and *Pichia stipitis*	Xylose fermentation was enhanced.	Shalsh et al., 2016
Sodium-proton antiporter gene *nhaA* (ZMO0119) introduction into the genome of *Zymomonas mobilis*	Improved sodium ion tolerance (as high as 150 mM) of *Zymomonas mobilis*. Ethanol fermentation time was reduced by 24 hrs.	Gao et al., 2018
Cel 7A from *T. resei* was expressed in a recombinant host *E. coli* SHuffle	An increased FPase activity of 2.5 IU/mL (0.7 g/L; 58 kDa).	Navnit Kumar et al., 2018a
PP_2680 gene (encoding NAD^+-dependent aldehyde. dehydrogenase) from *Pseudomonas putida* was introduced into *Zymomonas mobilis*	Engineered strain showed improved expression of ED pathway genes by 16.5 and 1.5 folds. Oxidation of the phenolic aldehydes was also improved by 4.6 and 2.8 folds.	Yi et al., 2019
Mutation in *adhE* gene (coding for alcohol dehydrogenase enzyme) of *C. thermocellum*	Improved ethanol production of 29.9 g/L.	Holwerda et al., 2020
A natural CBP isolate *Clostridium* spp. DBT-IOC-C19	94.6% degradation of 5g/L Avicel in 96 hrs.	Singh et al., 2017
CBP performed using fused protoplasts of *T. harzanium* ATCC® 20846™ and *Saccharomyces cerevisiae* RW 143	A bioethanol yield of 11.25 g/L was obtained from the waste cotton-cardboard mixture (1:1).	Navnit Kumar et al., 2019
A solid state CBP performed by co-culturing *Trichoderma harzanium* ATCC® 20846™ (15% v/v) and *Saccharomyces cerevisiae* RW 143(5% w/v)	An enhanced cellulase production (2.035 FPU/mL) and a bioethanol yield of 0.09 g/L from the enzyme-saccharified hydrolysate of 15% (v/v) ammonia pre-treated mixture (containing 1% w/v cellulose) of waste cotton-cardboard mixture (1:1).	Navnit Kumar et al., 2018b
A simultaneous saccharification and co-fermentation (SScF) using a co-culture of *Bacillus cereus* (cellulolytic and sugar fermenting) GBPS9 and *Bacillus thuringiensis* serovar Kurstaki HD1	A bioethanol yield of 19.08 g/L from 4% (w/v) steam exploded sugarcane bagasse.	Ire et al., 2016
Saccharomyces cerevisiae (W5) and *Candida shehatae* 20335	Ethanol yield of 0.54 g/g of xylose was achieved in 72 hrs.	Jingping et al., 2017

Table 15.4 contd. ...

...Table 15.4 contd.

Process	Enhancement in the Process	References
A co-culture-based co-fermentation employing *Zymomonas mobilis* and *Candida shehate*. Kans grass was the biomass	Bioethanol yield of 67.2 g/L was obtained from the fermentation broth containing 100 g/L glucose and 59 g/L xylose.	Mishra and Ghosh, 2020
A CBP strain of *Saccharomyces cerevisiae* has been created by introducing *cbhC16* gene from *Orpinomyces* spp. Y102	As the cellulase (due to *cbhC16* gene incorporation) hydrolyzes cellulose to cellobiose and glucose from pre-treated Avicel, *Saccharomyces cerevisiae* ferments it to 0.16 g/L ethanol in a single step.	Liu et al., 2020
Employing co-culture of *Zymomonas mobilis* and *Pichia stipitis*	Enhanced ethanol productivity of 0.705 g/L/hr.	Wirawan et al., 2020
Adaptive laboratory evolution (ALE) of *Clostridium thermocellum* with a disruption in *hfsB* gene and mutation of *adhE* gene	Bioethanol yield of 29.9 g/L from 120 g/L cellulose in the fermentation broth.	Holwerda et al., 2020
Different strains of *Clostridium ragsdalei* (DSM 15248)	Improved ethanol yield of 14.92 g/L.	Patankar et al., 2021

thermohydrosulfuricus. Post fermentation of the resultant hydrolysate, an ethanol titre of 0.298 g/L has been reported (Froese et al., 2020).

7. Conclusion and Future Prospects

Bioethanol blending to the commercially marketed fossil fuels has already been mandated by several nations, while the government of India has recommended a 10%–20% (v/v) bioethanol blending by the year 2022. Apart from producing a plethora of commercially significant products through complete valorization of the biomass, an ideal fungal-based biorefinery aims at bringing together the aspects of social and environmental sustainability, and immensely contributes to the circular economy and bioeconomy. As elucidated in this chapter, a paradigm shift in the nature of the bioprocesses (of a biorefinery) from being completely chemical in nature to being holistically biological is being addressed by the usage of biocatalysts/enzymes derived from fungi, majorly belonging to the classes, Ascomycetes, Zygomycetes, and Basidiomycetes. Future research with respect to the in-depth transcriptional profiling, proteomic profiling, and metabolomic analyses of the genes in fungi, which are responsible for producing cellulolytic and ligninolytic enzymes, are being performed as well. Techniques such as CRISPR and advances in systems biology could help in high-throughput screening and the engineering of multiple bio-catalytic machineries into a robust host, which may be capable of integrating the individual bio-processes of a biorefinery (pretreatment, saccharification, and fermentation). The application of fungi in a biorefinery has certain shortcomings too, such as the scalability of the process; the safety of the fungal species during large-scale applications; and significant batch-to-batch variation in the duration of the processes and their yields. This chapter would benefit the upcoming research scholars/students, biofuel research groups, and biofuel producing industry personnel alike by elucidating the application of fungi in a typical 2G bioethanol refinery process, its pitfalls, contemporary bio-process enhancements such as combinatorial processes, and the possible frontiers for future developments.

References

Abada, E., Al-Fifi, Z. and Osman, M. (2018). Bioethanol production with carboxymethyl cellulase of *Pseudomonas poae* using castor bean (*Ricinus communis* L.) cake. *Saudi J. Biol. Sci.*, 26(4): 866–871.

Althuri, A., Gujjala, L.K.S. and Banerjee, R. (2017). Partial consolidated bioprocessing of mixed lignocellulosic feedstocks for ethanol production. *Bioresour. Technol.*, 245: 530–539.

Amoah, J., Ishizue, N., Ishizaki, M., Yasuda, M., Takahashi, K., Ninomiya, K., Yamada, R., Kondo, A. and Ogino, C. (2017). Development and evaluation of consolidated bioprocessing yeast for ethanol production from ionic liquid-pretreated bagasse. *Bioresour. Technol.*, 245(Part B): 1413–1420.

Ani, F. (2016). Utilization of bioresources as fuels and energy generation. *Electric Renewable Energy Systems*, 140–155. https://doi.org/10.1016/B978-0-12-804448-3.00008-6.

Avchar, R., Lanjekar, V., Kshirsagar, P., Dhakaphalkar, P.K., Dagar, S.S. and Baghela, A. (2021). Buffalo rumen harbours diverse thermotolerant yeasts capable of producing second-generation bioethanol from lignocellulosic biomass. *Renewable Energy*, 173: 795–807.

Ázar, L.R., Bordignon-Junior, S.E., Laufer, C., Specht, J., Ferrier, D. and Kim, D. (2020). Effect of lignin content on cellulolytic saccharification of liquid hot water pretreated sugarcane bagasse. *Molecules*, 25(3): 623.

Berglund, J., Angles d'Ortoli, T., Vilaplana, F., Widmalm, G., Bergenstråhle-Wohlert, M., Lawoko, M., Henriksson, G., Lindström, M. and Wohlert, J. (2016). A molecular dynamics study of the effect of glycosidic linkage type in the hemicellulose backbone on the molecular chain flexibility. *Plant J.*, 88(1): 56–70.

Bondesson, P.M. and Galbe, M. (2016). Process design of SSCF for ethanol production from steam-pretreated, acetic-acid-impregnated wheat straw. *Biotechnology for Biofuels*, 9: 222. https://doi.org/10.1186/s13068-016-0635-6.

Bulkan, G., Ferreira, J.A. and Taherzadeh, M.J. (2021). Retrofitting analysis of a biorefinery: Integration of 1st and 2nd generation ethanol through organo Solv pretreatment of oat husks and fungal cultivation. *Bioresour. Technol. Reports*, 15. https://doi.org/10.1016/j.biteb.2021.100762.

Bundhoo, Z.M.A. and Mohee, R. (2018). Ultrasound-assisted biological conversion of biomass and waste materials to biofuels: A review. *Ultrasonics Sonochemistry*, 40 (Part A): 298–313. https://doi.org/10.1016/j.ultsonch.2017.07.025.

Burcu, T.Y., Benbadis, L., Alkim, C., Sezgin, T., Aksit, A., Gokce, A., Ozturk, Y., Baykal, A.T., Cakar, Z.P. and Francois, J.M. (2017). *In vivo* evolutionary engineering for ethanol-tolerance of *Saccharomyces cerevisiae* haploid cells triggers diploidization. *J. Biosci. Bioeng.*, 124(3): 309–318. https://doi.org/10.1016/j.jbiosc.2017.04.012.

Cao, G., Ximenes, E., Nichols, N.N., Zhang, L. and Ladisch, M. (2013). Biological abatement of cellulase inhibitors. *Bioresour. Technol.*, 146: 604–610. https://doi.org/10.1016/j.biortech.2013.07.112.

Capolupo, L. and Faraco, V. (2016). Green methods of lignocellulose pretreatment for biorefinery development. *Appl. Microbiol. Biotechnol.*, 100(22): 9451–9467. https://doi.org/10.1007/s00253-016-7884-y.

Casey, E., Sedlak, M., Ho, N.W. and Mosier, N.S. (2010). Effect of acetic acid and pH on the co-fermentation of glucose and xylose to ethanol by a genetically engineered strain of *Saccharomyces cerevisiae*. *FEMS Yeast Res.*, 10(4): 385–393.

Caspeta, L., Castillo, T. and Nielsen, J. (2015). Modifying yeast tolerance to inhibitory conditions of ethanol production processes. *Front. Bioeng. Biotechnol.*, 3: 184. https://doi.org/10.3389/fbioe.2015.00184.

Chen, H. (2015). Lignocellulose biorefinery feedstock engineering. *Lignocellulosic Biorefinery Engineering*, 37–86. https://doi.org/10.1016/B978-0-08-100135-6.00003-X.

Cheng, C., Tang, RQ., Xiong, L., Hector, R.E., Bai, F.W. and Zhao, X.Q. (2018). Association of improved oxidative stress tolerance and alleviation of glucose repression with superior xylose-utilization capability by a natural isolate of *Saccharomyces cerevisiae*. *Biotechnol. Biofuels*, 11(28). https://doi.org/10.1186/s13068-018-1018-y.

Costa, R.L., Oliveira, T.V., Ferreira, J., Cardoso, V.L. and Batista, F.R. (2015). Prospective technology on bioethanol production from photo fermentation. *Bioresour. Technol.*, 181: 330–337. https://doi.org/10.1016/j.biortech.2015.01.090.

Cunha, J.T., Romaní, A., Inokuma, K., Johansson, B., Hasunuma, T., Kondo, A. and Domingues, L. (2020). Consolidated bioprocessing of corn cob-derived hemicellulose: engineered industrial *Saccharomyces cerevisiae* as efficient whole cell biocatalysts. *Biotechnol. Biofuels.*, 13(138). https://doi.org/10.1186/s13068-020-01780-2.

de Silva, R.O., Batistote, M. and Cereda, M. (2013). Alcoholic fermentation by the wild yeasts under thermal, osmotic, and ethanol stress. *Braz. Arch. Biol. Technol.*, 56(2): 161169. https://doi.org/10.1590/S1516-89132013000200001.

Elferink, H., Bruekers, J., Veeneman, G.H. and Boltje, T.J. (2020). A comprehensive overview of substrate specificity of glycoside hydrolases and transporters in the small intestine: "A gut feeling". *Cellular and Molecular Life Sciences* (*CMLS*), 77(23): 4799–4826.

Fernández-Rodríguez, J., Erdocia, X., Sánchez, C., Alriols, M.G. and Labidi, J. (2017). Lignin depolymerization for phenolic monomers production by sustainable processes. *J. Energy Chem.*, 26(4): 622–631.

Fernández-Sandoval, M.T., Galíndez-Mayer, J., Bolívar, F., Gosset, G., Ramírez, O.T. and Martinez, A. (2019). Xylose-glucose co-fermentation to ethanol by *Escherichia coli* strain MS04 using single- and two-stage continuous cultures under micro-aerated conditions. *Microb. Cell. Fact.*, 18: 145. https://doi.org/10.1186/s12934-019-1191-0.

Field, S.J., Ryden, P., Wilson, D., James, S.A., Roberts, I.N., Richardson, D.J., Waldron, K.W. and Clarke, T.A. (2015). Identification of furfural resistant strains of *Saccharomyces cerevisiae* and *Saccharomyces paradoxus* from a collection of environmental and industrial isolates. *Biotechnol. Biofuels*, 8: 33. https://doi.org/10.1186/s13068-015-0217-z.

Froese, A.G., Nguyen, T.N., Ayele, B.T. and Sparling, R. (2020). Digestibility of wheat and cattail biomass using a coculture of thermophilic anaerobes for consolidated bioprocessing. *Bioenergy Res.*, 13: 325–333. https://doi.org/10.1007/s12155-020-10103-0.

Gabriel, S., Aruwajoye, Y., Sewsynker-Sukai, E.B. and Gueguim K. (2020). Valorisation of cassava peels through simultaneous saccharification and ethanol production: Effect of prehydrolysis time, kinetic assessment, and preliminary scale up. *Fuel*, 278.

Gao, X., Gao, Q. and Bao, J. (2018). Improving cellulosic ethanol fermentability of *Zymomonas mobilis* by over-expression of sodium ion tolerance gene ZMO0119. *J. Biotechnol.*, 282: 32–37. https://doi.org/10.1016/j.jbiotec.2018.05.013.

Geng, P., Zhang, S., Liu, J., Zhao, C., Wu, J., Cao, Y., Fu, C., Han, X., He, H. and Zhao, Q. (2020). MYB20, MYB42, MYB43, and MYB85 regulate phenylalanine and lignin biosynthesis during secondary cell wall formation. *Plant Physiol.*, 182(3): 1272–1283.

Gil, L. and Fito Maupoey, P. (2018). An integrated approach for pineapple waste valorisation. Bioethanol production and bromelain extraction from pineapple residues. *J. Clean. Prod.*, 172: 1224–1231. https://doi.org/10.1016/j.jclepro.2017.10.284.

Hoang Nguyen Tran, P., Ko, J.K., Gong, G., Um, Y. and Lee, S.M. (2020). Improved simultaneous co-fermentation of glucose and xylose by *Saccharomyces cerevisiae* for efficient lignocellulosic biorefinery. *Biotechnology for Biofuels*, 13(12). https://doi.org/10.1186/s13068-019-1641-2.

Holwerda, E.K., Olson, D.G., Ruppertsberger, N.M., Stevenson, D.M., Murphy, S., Maloney, M.I., Lanahan, A.A., Amador-Noguez, D. and Lynd, L.R. (2020). Metabolic and evolutionary responses of *Clostridium thermocellum* to genetic interventions aimed at improving ethanol production. *Biotechnol. Biofuels*, 13: 40. https://doi.org/10.1186/s13068-020-01680-5.

Hossain, M., Basu, J. and Mamun, M. (2015). The production of ethanol from micro-Algae *Spirulina*. *Procedia Eng.*, 105: 733–738. https://doi.org/10.1016/j.proeng.2015.05.064.

Ire, F.S., Ezebuiro, V. and Ogugbue, C.J. (2016). Production of bioethanol by bacterial co-culture from agro-waste-impacted soil through simultaneous saccharification and co-fermentation of steam-exploded bagasse. *Bioresour. Bioprocess.*, 3: 26. https://doi.org/10.1186/s40643-016-0104-x.

Iwaki, A., Kawai, T., Yamamoto, Y. and Izawa, S. (2013). Biomass conversion inhibitors furfural and 5-hydroxymethylfurfural induce formation of messenger RNP granules and attenuate translation activity in *Saccharomyces cerevisiae*. *Appl. Environ. Microbiol.*, 79(5): 1661–1667.

Jason Charles, S., Navnit Kumar, R., Sambavi, T.R. and Renganathan, S. (2018). Yeast co-culture with *Trichoderma harzanium* ATCC® 20846™ in submerged fermentation enhances cellulase production from a novel mixture of surgical waste cotton and waste cardboard. *International Journal of Modern Science and Technology*, 3(5): 117–125.

Javaid, R., Sabir, A., Sheikh, N. and Ferhan, M. (2019). Recent advances in applications of acidophilic fungi to produce chemicals. *Molecules*, 24(4). https://doi.org/10.3390/molecules24040786.

Ji, S.Q., Wang, B., Lu, M. and Li, F.L. (2016). Direct bioconversion of brown algae into ethanol by thermophilic bacterium *Defluviitalea phaphyphila*. *Biotechnol. Biofuels*, 9: 81. https://doi.org/10.1186/s13068-016-0494-1.

Jia, X., Peng, X., Liu, Y. and Han, Y. (2017). Conversion of cellulose and hemicellulose of biomass simultaneously to acetoin by thermophilic simultaneous saccharification and fermentation. *Biotechnol. Biofuels*, 10: 232. https://doi.org/10.1186/s13068-017-0924-8.

Jin, M., Sousa, L., Schwartz, C., He, Y., Sarks, C., Gunawan, C., Balan, V. and Dale, B. (2015). Toward lower cost cellulosic biofuel production using ammonia-based pretreatment technologies. *Green Chem.*, 18(4): 957–966. https://doi.org/10.1039/C5GC02433A.

Jingping, G., Renpeng, D., Gang, S., Yuhuan, Z. and Wenxiang, P. (2017). Metabolic pathway analysis of the xylose-metabolizing yeast protoplast fusant ZLYRHZ7. *J. Biosci. Bioeng.*, 124 (4): 386–391.

Jones, M., Mautner, A., Luenco, S., Bismarck, A. and John, S. (2020). Engineered mycelium composite construction materials from fungal biorefineries: A critical review. *Materials & Design*, 187: 108397. https://doi.org/10.1016/j.matdes.2019.108397.

Jönsson, L.J. and Martín, C. (2016). Pretreatment of lignocellulose: Formation of inhibitory by-products and strategies for minimizing their effects. *Bioresour. Technol.*, 199: 103–112. https://doi.org/10.1016/j.biortech.2015.10.009.

Kongkeitkajorn, M.B., Sae-Kuay, C. and Reungsang, A. (2020). Evaluation of Napier grass for bioethanol production through a fermentation process. *Processes*, 8(5): 567. https://doi.org/10.3390/pr8050567.

Kucharska, K., Rybarczyk, P., Hołowacz, I., Łukajtis, R., Glinka, M. and Kamiński, M. (2018). Pretreatment of lignocellulosic materials as substrates for fermentation processes. *Molecules*, 23(11): 2937. https://doi.org/10.3390/molecules23112937.

Kumar, A. and Chandra, R. (2020). Ligninolytic enzymes and its mechanisms for degradation of lignocellulosic waste in environment. *Heliyon*, 6(2). https://doi.org/10.1016/j.heliyon.2020.e03170.

Lakhundi, S., Siddiqui, R. and Khan, N.A. (2015). Cellulose degradation: A therapeutic strategy in the improved treatment of *Acanthamoeba* infections. *Parasites Vectors*, 8: 23. https://doi.org/10.1186/s13071-015-0642-7.

Lee, C.R., Sung, B., Lim, K.M., Kim, M.J., Sohn, M.J., Bae, J.H. and Sohn, J.H. (2017). Co-fermentation using recombinant *Saccharomyces cerevisiae* yeast strains hyper-secreting different cellulases for the production of cellulosic bioethanol. *Sci. Rep.*, 7: 4428. https://doi.org/10.1038/s41598-017-04815-1.

Li, Y., Zhai, R., Jiang, X., Chen, X., Yuan, X., Liu, Z. and Jin, M. (2019). Boosting ethanol productivity of *Zymomonas mobilis* 8b in enzymatic hydrolysate of dilute acid and ammonia pretreated corn stover through medium optimization, high cell density fermentation and cell recycling. *Front. Microbiol.*, 10: 2316. https://doi.org/10.3390/pr811145910.3389/fmicb.2019.02316.

Liu, J.C., Chang, W.J., Hsu, T.C., Chen, H.J. and Chen, Y.C. (2020). Direct fermentation of cellulose to ethanol by *Saccharomyces cerevisiae* displaying a bifunctional cellobiohydrolase gene from *Orpinomyces* sp. Y102, *Renew. Energy*, 159: 1029–1035. https://doi.org/10.1016/j.renene.2020.05.118.

Liu, Z.L., Slininger, P.J, Dien, B.S., Berhow, M.A., Kurtzman, C.P. and Gorsich, S.W. (2004). Adaptive response of yeasts to furfural and 5-hydroxymethylfurfural and new chemical evidence for HMF conversion to 2,5-bis-hydroxymethylfuran. *J. Ind. Microbiol. Biotechnol.*, 31(8): 345–352.

Loow, Y.L., Wu, T.Y., Tan, K.A., Lim, Y.S., Siow, L.F., Jahim, J.M., Mohammad, A.W. and Teoh, W.H. (2015). Recent advances in the application of inorganic salt pretreatment for transforming lignocellulosic biomass into reducing sugars. *J. Agric. Food Chem.*, 63(38): 8349–8363.

Ma, R., Xu, Y. and Zhang, X. (2015a). Catalytic oxidation of biorefinery lignin to value-added chemicals to support sustainable biofuel production. *ChemSusChem.*, 8(1): 24–51. https://doi.org/10.1002/cssc.201402503.

Ma, Kedong, Ruan, Z., Shui, Z., Wang, Y., Hu, G. and Mingxiong, H. (2015b). Open fermentative production of fuel ethanol from food waste by an acid-tolerant mutant strain of *Zymomonas mobilis*. *Bioresour. Technol.*, 203: 295–302. https://doi.org/10.1016/j.biortech.2015.12.054.

Maicas, S. (2020). The role of yeasts in fermentation processes. *Microorganisms*, 8(8): 1142. https://doi.org/10.3390/microorganisms8081142.

Mariscal, R., Maireles-Torres, P., Ojeda, M., Sadaba, I. and Granados, L.G.M. (2016). Furfural: A renewable and versatile platform molecule for the synthesis of chemicals and fuels. *Energy Environ. Sci.*, 9(4): 1144–1189.

Martínez-Sanz, M., Pettolino, F., Flanagan, B., Gidley, M.J. and Gilbert, E.P. (2017). Structure of cellulose microfibrils in mature cotton fibres. *Carbohydrate Polymers*, 175: 450–463. https://doi.org/10.1016/j.carbpol.2017.07.090.

Matsakas, L., Nitsos, C., Raghavendran, V., Yakimenko, O., Persson, G., Olsson, E., Rova, U., Olsson, L. and Christakopoulos, P. (2018). A novel hybrid organo Solv: Steam explosion method for the efficient fractionation and pretreatment of birch biomass. *Biotechnol. Biofuels*, 11: 160. https://doi.org/10.1186/s13068-018-1163-3.

Mendes, C., Rocha, J., de Menezes, F.F. and Carvalho, M. (2017). Batch and fed-batch simultaneous saccharification and fermentation of primary sludge from pulp and paper mills. *Environ. Technol.*, 38(12): 1498–1506.

Mishra, A. and Ghosh, S. (2020). Saccharification of Kans grass biomass by a novel fractional hydrolysis method followed by co-culture fermentation for bioethanol production. *Renew. Energy*, 146: 750–759. https://doi.org/10.1016/j.renene.2019.07.016.

Navnit Kumar, R., Jason Charles, S., Sambavi, T.R., Kabilan, S. and Renganathan, S. (2018a). Heterologous expression of exoglucanase from *Trichoderma resei* in *E. coli*. *International Journal of Modern Science and Technology*, 3(3): 65–71.

Navnit Kumar, R., Sambavi, T.R. and Renganathan, S. (2018b). Consolidated bioprocessing in solid state fermentation for the production of bioethanol from a novel mixture of surgical waste cotton and waste cardboard. *International Journal of Modern Science and Technology*, 3(8): 173–180.

Navnit Kumar, R., Sambavi, T.R. and Renganathan, S. (2019). A novel strain developed through protoplast fusion for consolidated bioprocessing of lignocellulosic waste mixture. *International Journal of Modern Science and Technology*, 4(5): 128–137.

Navnit Kumar, R., Sambavi, T.R., Baskar, G. and Renganathan, S. (2020). Experimental validation of optimization by statistical and CFD simulation methods for cellulase production from waste lignocellulosic mixture. *International Journal of Modern Science and Technology*, 5(2): 45–58.

Naseeha, A., Chohan, G.S., Aruwajoye, Y., Sewsynker-Sukai, E.B. and Gueguim, K. (2020). Valorisation of potato peel wastes for bioethanol production using simultaneous saccharification and fermentation: Process optimization and kinetic assessment. *Renew. Energy*, 146: 1031–1040. https://doi.org/10.1016/j.renene.2019.07.042.

Nigam, J.N. (2015). Continuous ethanol production from D-xylose: Using free cells of *Clavispora opuntiae*. *Energy Sources, Part A: Recovery,* Utilization, *and Environmental Effects*. 37(19): 2107–2113. https://doi.org/10.1080/15567036.2011.608105.

Ning, Z., Jian-Chun, J., Jing, Y., Min, W., Jian, Z., Hao, X., Jing-Cong, X. and Ya-Juan, T. (2018). Ethanol production from acorn starch by tannin tolerance mutant *Pachysolen tannophilus*. *Energy Sources A: Recovery Util. Environ. Eff.*, 40(5): 572–578.

Olofsson, K., Bertilsson, M. and Lidén, G. (2008). A short review on SSF: An interesting process option for ethanol production from lignocellulosic feedstocks. *Biotechnol. Biofuels*, 1(1): 7. https://doi.org/10.1186/1754-6834-1-7.

Parthasarathi, R., Sun, J., Dutta, T., Sun, N., Pattathil, S., Konda, N.V.S.N.M., Peralta, A.G., Simmons, B.A. and Singh, S. (2016). Activation of lignocellulosic biomass for higher sugar yields using aqueous ionic liquid at low severity process conditions. *Biotechnol. Biofuels*, 9: 160. https://doi.org/10.1186/s13068-016-0561-7.

Patankar, S., Dudhane, A., Paradh, A.D. and Patil, S. (2021). Improved bioethanol production using genome-shuffled *Clostridium ragsdalei* (DSM 15248) strains through syngas fermentation. *Biofuels*, 12(1): 81–89.

Portero, B.P., Bastidas Mayorga, B., Martín-Gil, J., Martín-Ramos, P. and Carvajal Barriga, E.J. (2020). Cellulosic ethanol: Improving cost efficiency by coupling semi-continuous fermentation and simultaneous saccharification strategies. *Processes*, 8(11): 1459. https://doi.org/10.3390/pr8111459.

Prasoulas, G., Gentikis, A., Konti, A., Kalantzi, S., Kekos, D. and Mamma, D. (2020). Bioethanol Production from food waste applying the multienzyme system produced on-site by *Fusarium oxysporum* F3 and mixed microbial cultures. *Fermentation*, 6(2): 39. https://doi.org/10.3390/fermentation6020039.

Qiu, Z. and Jiang, R. (2017). Improving *Saccharomyces cerevisiae* ethanol production and tolerance via RNA polymerase II subunit Rpb7. *Biotechnol. Biofuels*, 10: 125. https://doi.org/10.1186/s13068-017-0806-0.

Qu, X.S., Hu, B.B. and Zhu, M. (2017). Enhanced saccharification of cellulose and sugarcane bagasse by *Clostridium thermocellum* cultures with Triton X-100 and β-glucosidase/Cellic® CTec2 supplementation. *RSC Adv.*, 35(7): 21360–21365. https://doi.org/10.1039/C7RA02477K.

Ramamoorthy, N.K., Ravi, S. and Sahadevan, R. (2018). Production of bioethanol from an innovative mixture of surgical waste cotton and waste cardboard after ammonia pre-treatment. *Energy Sources, Part A: Recovery, Utilization, and Environmental Effects.*, 40(20): 2451–2457. https://doi.org/10.1080/15567036.2018.1502843.

Ramamoorthy, N.K., Sambavi, T.R. and Renganathan, S. (2019a). A study on cellulase production from a mixture of lignocellulosic wastes. *Process Biochem.*, 83: 148–158.

Ramamoorthy, N.K., Sambavi, T.R. and Sahadevan, R. (2019b). Assessment of fed-batch strategies for enhanced cellulase production from a waste lignocellulosic mixture. *Biochemical Engineering Journal*, 152. https://doi.org/10.1016/j.bej.2019.107387.

Ramamoorthy, N.K., Sambavi, T.R. and Renganathan, S. (2020a). Production of bioethanol by an innovative biological pretreatment of a novel mixture of surgical waste cotton and waste cardboard. *Energy Sources, Part A: Recovery, Utilization, and Environmental Effects.*, 42 (8): 942–953. https://doi.org/10.1080/15567036.2019.1602208.

Ramamoorthy, N.K., Nagarajan, R., Ravi, S. and Sahadevan, R. (2020b). An innovative plasma pre-treatment process for lignocellulosic bioethanol production. *Energy Sources, Part A: Recovery, Utilization, and Environmental Effects.* https://doi.org/10.1080/15567036.2020.1815900.

Ribeiro, A., Tyzack, J., Borkakoti, N., Holliday, G. and Thornton, J. (2019). A global analysis of function and conservation of catalytic residues in enzymes. *J. Biol. Chem.*, 295(2): 314–324.

Rodrigues Reis, C.E., Bento, H.B., Carvalho, A.K., Rajendran, A., Hu, B. and De Castro, H.F. (2019). Critical applications of *Mucor circinelloides* within a biorefinery context. *Crit. Rev. Biotechnol.*, 39(4): 555–570.

Salleh, M.S.M., Ibrahim, M.F., Roslan, A.M. and Abd-Aziz, S. (2019). Improved biobutanol production in 2-L simultaneous saccharification and fermentation with delayed yeast extract feeding and *in situ* recovery. *Sci. Rep.*, 9: 7443. https://doi.org/10.1038/s41598-019-43718-1.

Sambavi, T.R., Ramamoorthy, N.K. and Renganathan, S. (2019). Mixture of potato, sapodilla, kiwi peels, and coir as a substrate for the production of cellulases using trichoderma atroviride ATCC® 28043™ by a solid state cyclic fed-batch strategy and evaluation of its saccharification efficiency. *Int. J. Sci. Eng. Managem.*, 4(6): 130–133.

Selim, K.A., El-Ghwas D.E., Easa, S.M. and Abdelwahab Hassan, M.I. (2018). Bioethanol a microbial biofuel metabolite: New insights of yeasts metabolic engineering. *Fermentation*, 4(1): 16. https://doi.org/10.3390/fermentation4010016.

Selim, K.A., Easa, S.M. and El-Diwany, A.I. (2020). The xylose metabolizing yeast *Spathaspora passalidarum* is a promising genetic treasure for improving bioethanol production. *Fermentation*, 6(1): 33. https://doi.org/10.3390/fermentation6010033.

Shalsh, F., Ibrahim, A., Mohammed, A. and Meor Hussin, A.S. (2016). Optimization of the protoplast fusion conditions of *Saccharomyces cerevisiae* and *Pichia stipitis* for improvement of bioethanol production from biomass. *Asian J. Biol. Sci.*, 9(1-2): 10–18. https://doi.org/10.3923/ajbs.2016.10.18.

Shi, X., Liu, Y., Dai, J., Liu, X., Dou, S., Teng, L., Meng, Q., Lu, J., Ren, X. and Wang, R. (2019). A novel integrated process of high cell-density culture combined with simultaneous saccharification and fermentation for ethanol production. *Biomass and Bioenergy*, 121: 115–121. https://doi.org/10.1016/j.biombioe.2018.12.020.

Silva, F.L., Campos, A.O., Santos, D.A., Magalhaes, RB., Macedo, G.B. and Santos, E.S. (2018). Valorization of an agroextractive residue-carnauba straw-for the production of bioethanol by simultaneous saccharification and fermentation (SSF). *Renewable Energy*, 127: 661–669. https://doi.org/10.1016/j.renene.2018.05.025.

Singh, N., Mathur, A.S., Tuli, D.K., Gupta, R.P., Barrow, C.J. and Puri, M. (2017). Cellulosic ethanol production via consolidated bioprocessing by a novel thermophilic anaerobic bacterium isolated from a Himalayan hot spring. *Biotechnol. Biofuels*, 10(73). https://doi.org/10.1186/s13068-017-0756-6.

Soleimani, S.S., Adiguzel, A. and Nadaroglu, H. (2017). Production of bioethanol by facultative anaerobic bacteria. *J. Inst. Brew.*, 123(3): 402–406.

Sukwong, P., Sunwoo, I.Y., Jeong, D.Y., Kim, S.R., Jeong, G.T. and Kim, S.K. (2020). Improvement of bioethanol production by *Saccharomyces cerevisiae* through the deletion of GLK1, MIG1, MIG2 and overexpression of PGM2 using the red seaweed *Gracilaria verrucosa*. *Process. Biochem.*, 89: 134–145. https://doi.org/10.1016/j.procbio.2019.10.030.

Thapa, B., Patidar, S.K., Khatiwada, N.R., KC, A.K. and Ghimire, A. (2019). Production of ethanol from municipal solid waste of India and Nepal. *In*: *Waste Valorisation and Recycling*. Switzerland: Springer-Nature, 2: 47–58.

Todhanakasem, T., Narkmit, T., Areerat, K. and Thanonkeo, P. (2015). Fermentation of rice bran hydrolysate to ethanol using *Zymomonas mobilis* biofilm immobilization on DEAE-cellulose. *Electron. J. Biotechnol.*, 18(3): 196–201. https://doi.org/10.1016/j.ejbt.2015.03.007.

USEIA. Biomass-renewable energy from plants and animals, https://www.eia.gov/energyexplained/biomass/

Wang, B., Wu, Q., Xu, Y. and Sun, B. (2020). Synergistic effect of multiple saccharifying enzymes on alcoholic fermentation for Chinese baijiu production. *Appl. Environ. Microbiol.*, 86(8). https://doi.org/10.1128/AEM.00013-20.

Wang, F., Jiang, Y., Guo, W., Niu, K., Zhang, R., Hou, S., Wang, M., Yi, Y., Zhu, C., Jia, C. and Fang, X. (2016). An environmentally friendly and productive process for bioethanol production from potato waste. *Biotechnol. Biofuels*, 9: 50. https://doi.org/10.1186/s13068-016-0464-7.

Wang, L., York, S.W., Ingram, L.O. and Shanmugam, K.T. (2019). Simultaneous fermentation of biomass-derived sugars to ethanol by a co-culture of an engineered *Escherichia coli* and *Saccharomyces cerevisiae*. *Bioresour. Technol.*, 273: 269–276. https://doi.org/10.1016/j.biortech.2018.11.016.

Wang, L.Q., Cai, L.Y. and Ma, Y.L. (2020). Study on inhibitors from acid pretreatment of corn stalk on ethanol fermentation by alcohol yeast. *RSC Adv.*, https://doi.org/10(63): 38409–38415.

Wang, W., Dai, L., Wu, B., Qi, B.F., Huang, TF., Hu, G.Q. and He, M. (2020). Biochar-mediated enhanced ethanol fermentation (BMEEF) in *Zymomonas mobilis* under furfural and acetic acid stress. *Biotechnol. Biofuels*, 13: 28. https://doi.org/10.1186/s13068-020-1666-6.

Wang, Z., Ning, P., Hu, L., Nie, Q., Liu, Y., Zhou, Y. and Yang, J. (2020). Efficient ethanol production from paper mulberry pretreated at high solid loading in Fed-non-isothermal-simultaneous saccharification and fermentation. *Renewable Energy*, 160: 211–219. https://doi.org/10.1016/j.renene.2020.06.128.

Wang, Z.K., Li, H., Lin, X.C., Tang, L., Chen, J.J., Mo, J.W., Yu, R.S. and Shen, X.J. (2020). Novel recyclable deep eutectic solvent boost biomass pretreatment for enzymatic hydrolysis. *Bioresour. Technol.*, 307: 123237. https://doi.org/10.1016/j.biortech.2020.123237.

Wirawan, F., Cheng, C.L., Lo, Y.C., Chen, C.Y., Chang, J., Leu, S.Y. and Lee, D. (2020). Continuous cellulosic Bioethanol co-fermentation by immobilized *Zymomonas mobilis* and suspended *Pichia stipitis* in a two-stage process. *Applied Energy*, 266: 114871. https://doi.org/10.1016/j.apenergy.2020.114871.

Yang, S., Fei, Q., Zhang, Y., Contreras, L.M., Utturkar, S.M., Brown, S.D., Himmel, M.E. and Zhang, M. (2016). *Zymomonas mobilis* as a model system for production of biofuels and biochemicals. *Microb. Biotechnol.*, 9(6): 699–717.

Yang, S., Zhang, Y., Yue, W., Wang, W., Wang, Y.Y., Yuan, T.Q. and Sun, R.C. (2016). Valorization of lignin and cellulose in acid-steam-exploded corn stover by a moderate alkaline ethanol post-treatment based on an integrated biorefinery concept. *Biotechnol. Biofuels.*, 9: 238. https://doi.org/10.1186/s13068-016-0656-1.

Yi, X., Gao, Q. and Bao, J. (2019). Expressing an oxidative dehydrogenase gene in ethanologenic strain *Zymomonas mobilis* promotes the cellulosic ethanol fermentability. *J. Biotechnol.*, 303: 1–7. https://doi.org/10.1016/j.jbiotec.2019.07.005.

You, Y., Wu, B., Yang, Y.W., Wang, Y.W., Liu, S., Zhu, Q., Qin, H., Tan, F., Ruan, Z., Ma, K., Dai, L., Zhang, M., Hu, G. and He, M. (2017a). Replacing process water and nitrogen sources with biogas slurry during cellulosic ethanol production. *Biotechnol. Biofuels*, 10: 236. https://doi.org/10.1186/s13068-017-0921-y.

You, Y., Liu, S., Wu, B., Wang, Y.W., Zhu, Q., Qin, H., Tan, F.R., Ruan, Z.Y., Ma, K.D., Dai L.C., Zhang, M., Hu, G.Q. and He, M.X. (2017b). Bioethanol production by *Zymomonas mobilis* using pretreated dairy manure as a carbon and nitrogen source. *RSC Adv.*, 7: 3768–3779. https://doi.org/10.1039/C6RA26288K.

Zeng, Y., Zhao, S., Yang, S. and Ding, S.Y. (2014). Lignin plays a negative role in the biochemical process for producing lignocellulosic biofuels. *Curr. Opin. Biotechnol.*, (27): 38–45. https://doi.org/10.1016/j.copbio.2013.09.008.

Zhao, Y., Wu, B., Yan, B. and Gao, P. (2004). Mechanism of cellobiose inhibition in cellulose hydrolysis by cellobiohydrolase. *Sci. China Ser. C.-Life Sci.*, 47(1): 18–24. https://doi.org/10.1360/02yc0163.

Zhang, Q., Weng, C., Huang, H., Achal, V. and Wang, D. (2016). Optimization of bioethanol production using whole plant of water hyacinth as substrate in simultaneous saccharification and fermentation process. *Front. Microbiol.*, 6(1411). https://doi.org/10.3389/fmicb.2015.01411.

Zhang, X.Y., Xu, Z.H., Zong, M.H., Wang, C.F. and Li, N. (2019). Selective synthesis of furfuryl alcohol from biomass-derived furfural using immobilized yeast cells. *Catalysts*, 9(1): 70. https://doi.org/10.3390/catal9010070.

Zhang, Y. and Vadlani, P.V. (2015). Lactic acid production from biomass-derived sugars via co-fermentation of *Lactobacillus brevis* and *Lactobacillus plantarum*. *J. Biosci. Bioeng.*, 119(6): 694–699.

Zhou, J., Ouyang, J., Xu, Q. and Zheng, Z. (2016). Cost-effective simultaneous saccharification and fermentation of l-lactic acid from bagasse sulfite pulp by *Bacillus coagulans* CC17. *Bioresour. Technol.*, 222: 431–438.

Plant Protection

16

Nematicidal Potential of Nematophagous Fungi

Ewa B. Moliszewska, Małgorzata Nabrdalik* and *Paweł Kudrys*

1. Introduction

Nematodes are causal agents of several diseases with non-specific symptoms in many crops, thus in some cases they may be misdiagnosed. Symptoms usually range from wilting to reduced growth, including reduction of nutrients uptake, reduced root system, reduced flowering and fruiting, and in some cases even plant death. Symptoms of nematode infection, except those mentioned above, also appear as chloroses, necrosis, root galls (when created below the ground), nematode cysts and clumps. In general, nematode derived losses are higher in tropical than in moderate climates, but there are some important crops strongly threatened by them, like sugar beets, potatoes, and carrots (Chitwood, 2002). Nematodes cause economic losses also in livestock, as well as being important parasites of pets and humans. Thus, there is demand for pesticides for crop protection as well as anthelmintic treatments and nematicidal drugs. Specialized literature terms refer to design 'nematicides' as drugs used against plant-parasitic nematodes and 'anthelmintics' against nematode parasites of animals (Silvestre and Cabaret, 2004; Braga and de Araújo, 2014; Pineda-Alegrı´a et al., 2017; Comans-Pérez et al., 2021). Chemical control of these pathogens includes many side effects, including environmental hazards and being prone to developing resistance in nematodes, thus safer and more ecological methods are demanded in combating harmful nematodes. For that reason, fungi seem the most interesting source of potential drugs and preparations; however, their biology and modes of action as well as possibilities to create innovative preparations still need broad research.

University of Opole, Faculty of Natural Sciences and Technology Institute of Environmental Engineering and Biotechnology, 45-040 Opole, Poland.
* Corresponding author: ewamoli@uni.opole.pl

Soil, as the natural environment, exposes miscellaneous mycelia to unfavorable effects of various organisms inhabiting it. During the process of evolution, mycelia have developed defensive abilities, thanks to which they are able to resist the attack of mycophagic organisms, such as nematodes, and adapt to win the competition for nutrient uptake from the soil. Nematophagy was developed by members of almost all the main fungal taxa (*Ascomycota*, *Basidiomycota*, *Zygomycota*, *Chytridiomycota*) as well as in fungi-like *Oomycota*. The most important nematode trapping fungi belong to genera: *Arthrobotrys*, *Drechmeria*, *Fusarium*, *Harposporium*, *Hirsutella*, *Nematophthora*, *Monacrosporium*, *Paecilomyces*, *Verticillium*, and others, while *Coprinus*, *Climacodon*, *Lentinus*, and *Pleurotus* developed rather toxic defensive abilities. An oomycete *Nematophthora gynophila* produces biflagellate zoospores. It exists exclusively in soils of the British Islands and lives as an obligate parasite of cyst nematodes. *N. gynophila* infects females as soon as they emerge from parasitized roots and destroy them within 1 week (Moosavi and Zare, 2012; Saxena, 2018; Yang et al., 2020; Noweer, 2020).

The defensive properties of different mycelia against nematodes have been described for over 700 species of various genera. Briefly, fungi may directly create traps on their hyphae or secrete sticky and highly toxic substances against nematodes. Such fungal means of defense may protect mycelia against predators and provide an alternative way to obtain nutrients from digested nematode bodies. Broad ecological views on the mode of action of nematophagous fungi show that they may act directly as obligate parasites or facultative saprotrophs. These may be gained by capturing nematodes by differently formed traps: adhesive hyphal branches, adhesive-network traps, adhesive knobs, non-constricting rings, and constricting rings (Siddiqui and Mahmood, 1996; Moosavi and Zare, 2012; Li et al., 2014; Noweer, 2020). The methods of defense against nematodes include: (1) specific trap structures produced in the shape of a loop or lasso; (2) a densely woven network of transformed hyphae into appressoria on the mycelium surface; (3) production of digestive enzymes; (4) adhesive spores. These modes of fungal action against nematodes are usually divided into four groups: (1) nematode-trapping, which is characteristic to predatory fungi; (2) endoparasites, which need mycophagy on the nematode side; (3) egg- and female-parasites; (4) toxin-producing fungi (Fig. 16.1).

The most important differentiating feature for the mode of action of the nematophagous fungi is the nematode feeding style. Those carrying a stylet are unable to swallow fungal spores, whereas saprotrophic microbiophagous nematodes can take up spores orally (Moosavi and Zare, 2012; Li et al., 2014; Yang and Zhang, 2014; Abd-Elgawad and Askary, 2018; Saxena, 2018). The defensive ability by itself is not just a defense for some species of fungi, but nematodes are an ideal source of nutrients, including nitrogen, sulfur, and carbon compounds (Li et al., 2014; Saxena, 2018).

Such defensive features empower protection of the mycelium against harmful effects of nematodes and lead to the reduction of their number in the environment (Degenkolb and Vilcinskas, 2016b; Devi, 2018). All the natural abilities of fungi to protect themselves against nematodes show the excellent way for their practical use. They are the most frequently applied in plant protection against plant parasitic nematodes, e.g., *Heterodera* spp., *Globodera* spp., *Meloidogyne* spp., *Helicotylenchus* spp.

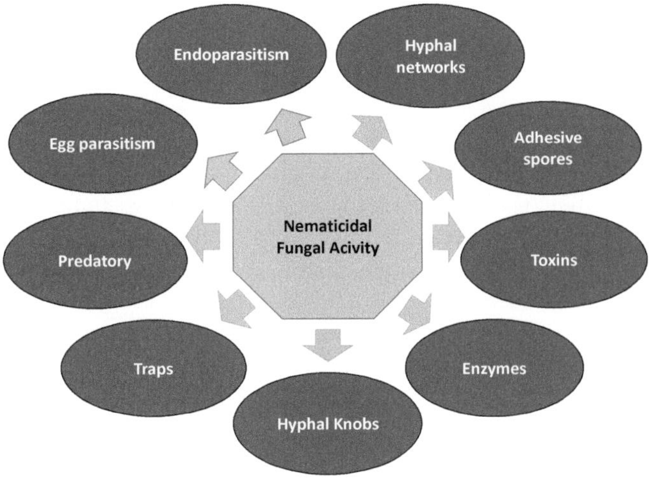

Fig. 16.1. Nematicidal fungal activity: Fungal modes of action.

(Agrios, 2005; Tileubayeva et al., 2021) and as animal drugs against parasites (Braga et al., 2015; Rodrigues et al., 2018).

2. Nematode Trapping Fungi

Nematode-trapping fungi are dated at 100 million years and have been found preserved in amber. Most nematode destroying fungi are found in soils with organic matter. In general, they used to live as saprotrophs in soil, but in the presence of host nematodes they switch to a parasitic phase (Su et al., 2017; Devi, 2018).

Traps are created against moving stages of soil inhabiting nematodes. They are mostly non-host-specific and of various shapes and sizes. Some fungi may use more than one type of trapping devices (Siddiqui and Mahmood, 1996). They include: (1) specific trap structures; (2) a densely woven network of hyphae transformed into appressoria on the mycelium surface. Vegetative hyphae of some species form lateral branches in the shape of rings or loops. The rings act as non-constricting ones, which choke entrapped nematodes, or constricting rings built of three parts that swell quickly and tightly clench nematode bodies. Some of those fungi require external stimulation, which is mainly the presence of nematodes or nematode secretions, and that usually concerns the slowly growing species. Some of them, which grow fast, may form their traps without external stimulation (Su et al., 2017). Yang et al. (2020) observed that strains, that were highly sensitive to the nematode, developed traps faster and the nature of trap formation correlated with prey killing. The correlation between trapping abilities and fungal nature was also found in genome composition. They found that deletion of the G-protein β-subunit encoding gene in *A. oligospora* genome almost inhibited trap formation and demonstrated that the trap-formation ability is a sum of fitness in the adaptation mode (Yang et al., 2020). Induction of trap formation for most of the fungal species is usually positively correlated with juvenile size and motility but traps may be also induced by specific peptides and amino acids, as well as abscisic acid, acarosides, and lectins. Nematode extract was also effective in the induction of trap formation as well as some bacteria, soil and soil extract, cow faeces, CO_2, and cellophane (Su et al., 2017).

Trapping hyphae may be also armed with sticky structures which can be formed as adhesive branches, simple loops or more complicated three- or di-dimensional complexes. Another way of trapping nematodes by fungi depends on forming adhesive spores or knobs (Siddiqui and Mahmood, 1996; Su et al., 2017). Sticky substances that they produce commonly contain proteins and/or carbohydrates. The recognition of both nematodes and fungi is possible due to carbohydrates that cover their surfaces (Moosavi and Zare, 2012). The most interesting traps are exclusively created by members of the genera *Arthrobotrys* and *Drechslerella*, which actively capture nematodes when they enter such a trap. Then the three cells which comprise the ring swell immediately, in a tenth of a second, and tightly catch the prey (Degenkolb and Vilcinskas, 2016b). A good representative of such a group of fungi is *Arthrobotrys oligosporus*, the organism that produces circular structures to trap nematodes. Nematodes searching for food are trapped during their movement through the soil as they enter such a loop of the hyphae. Then the loop immediately absorbs water from the soil, and as a result its volume increases, enabling it to tighten the nematode. Nematodes intensify a pressure in the loop cells, which is followed by activation of G-proteins and an increase of Ca^{2+} content in loop cells, activation of calmodulin that results in the water channels opening. Ca^{2+} inflow leads to calmodulin regulation because of an increase in Ca^{2+}, which reverses calmodulin inhibiting antagonists. As a consequence, the nematode is paralyzed and dies. The hypha then overgrows the nematode body, digests it, and uses it as an ideal source of nutrients (Moosavi and Zare, 2012). During saprotrophic growth in the soil, the fungus acquires mainly xylose and cellobiose as well as pectin, cellulose, and chitin, which are carbon sources, and nitrates and nitrites as sources of nitrogen. While feeding on nematodes, the mycelium also acquires cellobiose, glutamic acid, arginine, and asparagine, which are additional sources of carbon and nitrogen (Lee et al., 2020). *A. oligosporus* traps easily develop in low-nutrient soils or low-nutritive laboratory media. They are mainly composed of three cells, which present strong cytoplasmic movements and have dense structural bodies filled with peroxisomal organelles containing catalase and D-amino acid oxidase. Structures such as dense bodies can only be observed in traps during digesting nematodes; the victim is only a place for storing spores and creating conditions for them to grow and there are no dense bodies in the trapping hyphae. Traps also serve as a kind of spore because they can survive longer in the soil or in laboratory conditions than the hypha itself, and after catching the prey, they can develop further and form new spores inside the victim's body. This defensive method can also serve as an element in the spread of the fungus in the environment (Nordbring-Hertz, 2004).

For other *Arthrobotrys* species, traps typically may break away from the hypha, remaining on the nematode that can move around. After some time, the nematode's epidermis is damaged and the body is penetrated by the elements of the trap, which are not the clamping part of it. After that, new spores are created. Thanks to this way of invading nematodes, the fungus can 'travel' for long distances using them as a means of transportation (Nordbring-Hertz, 2004).

The interesting method of defending mycelium against nematodes is the ability to produce the specific defensive spinous cells—spiny balls—on the hyphae. An example of such cells are spiny balls produced by the *Coprinus comatus* hyphae. At first, they were thought to be terminal chlamydospores, but the lack of nuclei and quick formation make them incomplete cells, which disagrees with that conclusion. Isolated spiny balls show moderate, and mainly mechanical, nematocidal activity compared to the whole mycelium activity. So far, the devices and the appendages found on that, and other basidiomycetes are specialized cells producing toxins or adhesive substances (Luo et al., 2004; 2007; Soares et al., 2018). The spiny ball structures did not lose activity after treating them with organic solvents, but they were not able to act after being ground with liquid nitrogen. In addition, it is worth noting that the substances extracted from spiny balls had no effect on *Panagrellus redivivus*. However, in the case of that fungus, the defense mechanism consists also of the secreted substances from the hypha surface that aid in immobilization of the nematodes: 5-methylfuran-3-carboxylic acid and 5-hydroxy-3,5-dimethylfuran-2(5*H*)-one showed 90% toxicity at dose 200 μg/ml against both *M. incognita* and *P. redivivus*. Other substances: 5-hydroxy-3-(hydroxymethyl)-5-methylfuran-2(5*H*)-one, 4,6-dihydroxyisobenzofuran-1,3-dione, 4,6-dihydroxybenzofuran-3(2*H*)-one, 4,6-dimethoxyisobenzofuran-1(3*H*)-one, and 3-formyl-2,5-dihydroxybenzyl acetate showed far less toxicity, acting at higher doses from 400 to 800 μg/ml. As soon as the nematode receives the false and attracting signal generated by the hyphae, it begins to move towards them. Spiny balls attach to the nematode's surface and damage its epidermis causing leakage of inner substances. This allows the hypha to easily penetrate their bodies to digest them and obtain nutrients (Luo et al., 2007).

Another type of trapping nematodes presents *Harposporium harposporiferum*, an anamorph of *Podocrella harposporifera*, which does not produce specific cells but 'bites' the victims by producing spores which, in close contact with nematodes, are able to infect them. Infection goes through the area of the nematode's mouth or epidermis, resulting in new mycelium and the next generation of spores. Spores that have been eaten develop inside the nematode body absorbing nutrients to form hyphae and other spores. This predatory method of action also serves as a very useful way of spreading the fungus for long distances using the nematode body as the means of transportation before it dies (Chaverri et al., 2005; Dubiel, 2014).

3. Nematode Trapping Fungi and Endoparasites

The border between nematode trapping fungi and endoparasites is not clear because both lifestyles support each other. Typically, endoparasites are considered obligate parasites with a broad host range. Lebrigand et al. (2016) confirmed that opinion by the comparative genomic analysis of *Drechmeria coniospora* that produces non-motile spores which can stick to the nematode cuticle. Shortly after that the spore produces an appressorium and germinates, then pierces the cuticle and enters the body. A detailed study *in silico* of *D. coniospora* and 11 other fungal species demonstrated it as a highly specialized pathogen. Genomic analysis led to the discovery of gene families and genes potentially involved in nematode parasitism and virulence (Lebrigand et al., 2016).

Usually, they do not produce extensive mycelium, but they must develop an effective mode of trapping prey to be able to complete the life cycle. Their existence is concentrated as viable conidia which adhere to the cuticle surface of the host nematode, e.g., the *D. coniospora* conidia are armed with adhesive buds, or they may be ingested by nematodes, an example is the *Harposporium* spp. These fungi may attract nematodes by palatable conidia, which leads them to be eaten by mycophagous nematodes. Attracting of nematodes may be coincident with the specific signaling molecules (proteins, carbohydrates) released by the host. That was proved for plant parasitic nematodes, for which the signaling molecules are naturally occurring polyamines. Oota et al. (2020) found that *Meloidogyne incognita* may be specifically attracted by natural compounds that possess three to five methylene groups between two terminal amino groups.

Among organisms of that group the separate category consists of fungi that produce zoospores, which can actively attach to the host cuticle. After reaching the host nematode, its body is penetrated and digested by the predatory organism (Saxena, 2018). Adhesive conidia are found in genera like *Drechmeria*, *Haptocillium*, *Hirsutella*, *Pochonia*, *Purpureocillium* in *Hypocreales* of Ascomycota (Quandt et al., 2014), and *Meristacrum* of Zygomycota. The zoospores are produced by *Catenaria* (*Chytridiomycota*) and members of *Oomycota* including *Haptoglossa*, *Nematophthora*, and *Myzocytium* (Saxena, 2018).

The third group of endoparasite fungi concentrates on eggs. In some cases, the possibility to parasite on eggs by fungi is limited only to certain developmental stages, like for the *Hyalorbilia* genus. The fungus of that genus suppressed the hitching of 100% of the J2 forms taken from 3-week-old *Heterodera schachtii* females and 75% from 4-week-old females, unfortunately other stages were not affected (Becker et al., 2020).

Fungi of *Verticillium* and *Fusarium* genera have been often recognized as in majority eggs' parasites of *Heterodera avenae* and *H. schachtii* (Dackmanet al., 1989), while *Paecilomyces liliacinus* was a rarely isolated fungus from nematode eggs in Europe. On the other hand, *P. liliacinus* was well recognized as a nematodes' parasite and commercially used in biocontrol of cyst nematodes in Philippines and Peru. However, Dackman et al. (1989) observed that *Arthrobotrys oligospora*, *Cylindrocarpon destructans*, *Fusarium oxysporum*, *Paecilomyces lilacinus* (*Purpureocillium lilacinum*) and *Penicillium viridicatum* were unable to parasitize nematodes' eggs *in vitro*.

In some cases, the scenario of nematode-fungus associations may be more complicated and complex than a one-directed interaction, and a one mode of fungal nematicidal activity. A good example of broad relations among fungus, nematodes, wood, and its pest was described by Hajek and Morris (2014). In that relationship, nematode *Deladenus siricidicola* was used for pine biocontrol against a pine-killing wood wasp larva. The larvae were easily parasitized by nematode *D. siricidicola*, although the nematodes were attacked by a white-rot fungus *Amylostereum areolatum* which, on the other hand, is a food source for nematodes during their mycophagous stage of life. It was shown that *A. areolatum* can invade nematodes' vulva and eggs, respectively, by hyphal tips and cystidia (Degenkolb and Vilcinskas, 2016a).

4. Induction of Host-Plant Resistance

The association of arbuscular mycorrhizal fungi (AMF) and plant parasitic nematodes in the roots of different crops exert opposite effects on the host plant. Thus, the effect of interaction of these organisms on plant growth and yield should be carefully determined (Akhtar and Siddiqui, 2008). The AMF may cause the indirect suppression of plant parasitic nematodes by complex mechanisms of mutual interactions (Noweer, 2020). Udo et al. (2013) demonstrated strong inhibition of gall formation as well as egg-mass production by *M. incognita* race 1 on tomato by AMF. The mechanism of that reaction considered changing the overall susceptibility in tomato plants by AMF *Glomus etunicatum* and *G. deserticola*, which most effectively influenced the production of galls and eggs. In experiments by Udo et al. (2013), the ability was unequal among the different AMF species; some of them did not cause any effect. That type of interaction resulted in physiological and biochemical changes in plants, which resulted in increased production of phytoalexins, phenols, lignin, and enzymes like phenylamine and serine chitinase. Plants treated with AMF exuded fewer carbohydrates and presented improved nutrition. Changes were observed also in the morphology and histopathology roots. The protective effect of AMF fungi against nematodes was actively supported by *Paecilliomyces lilacinus* (Udo et al., 2013). Similar results were observed by da Silva Campos (2020) in the case of plants infected by *Meloidogyne* nematodes. However, the effects were highly variable, although they were positive, generally. AMF were also able to reduce nematode development. He pointed out, similarly as Udo et al. (2013) did, that interactions among AMF, plant and parasitic nematodes are complex and depend on the natural features of componential organisms (da Silva Campos, 2020; Poveda et al., 2020). Borowicz (2001) furnished the distinction between sedentary and migratory nematodes, pointing out that the results of complex interaction depend on the nematode mode of feeding. According to her analyses, AMF tended to worsen the harmful effects of nematodes on plants. AMF harmed the sedentary endoparasitic nematodes and improved the growth of migratory endoparasitic nematodes. However, there was no significance in her analyses (Borowicz, 2001). Confirmation for those suggestions for migratory nematodes was given by Elsen et al. (2008). In general, they observed that *Glomus intraradices* reduced *Pratylenchus coffeae* and *Radopholus similis* populations in *Musa* cv. Grand Naine from 72 to 84%, but these results did not differ significantly from the controls. Furthermore, among different experimental treatments, if *P. coffeae* or *R. similis* were present in the treatment, in neither of them any significance in necroses was observed. They observed the lack of plant growth promotion by AMF, so it was concluded that improved plant growth and nutrition had not influenced the protective mechanism, and they pointed out that bio-protection against nematodes was at least partially systemically induced (Elsen et al., 2008). Hol and Cook (2005) distinguished three groups of nematodes according to their mode of parasitism: (1) ectoparasites; (2) sedentary endoparasites; and (3) migratory endoparasites. They suggested that the AMF influence on plant and nematode interactions would depend on the nematode mode of parasitism. In their opinion, interactions between AMF and endoparasites were stronger than those between AMF and ectoparasites. They pointed out that AMF-infected plants were more severely damaged by ectoparasites than by endoparasites. They observed that the AMF influence on the sedentary endoparasites was greater than on migratory endoparasites, and migratory endoparasite nematodes could increase their number in co-existence in AMF-infected plants. However, in the case of migratory nematodes, relatively high levels of AMF infections and little nematode damage were observed compared to the other feeding types (Hol and Cook, 2005).

Akhtar and Siddiqui (2008) gave a broad list of AMFs tested and active against plant-parasitic nematodes. They traced literature starting from the 1970s and listed the parasitic nematodes and the effectiveness, and the mode of action of many AMFs. In general, AMF acts through production of strigolactones and induction of SAR, and ISR in plants followed by the transportation of defense components throughout the plant (Poveda et al., 2020).

5. Nematicidal Toxins

Chemical crop protection against nematodes seems to be a standard option in current agriculture and farm animal treatment; however, there is still a strong demand to get environmentally safe and beneficial alternatives. They should also be neutral to other non-target organisms and crop plants. The alternative way gives fungal defensive abilities, which include the synthesis and secretion of toxic substances by hyphae, thanks to which the nematode is completely paralyzed or even immediately killed. This process takes place before a physical contact, and is characteristic to *Pleurotus* and *Coprinus* genera, and to some ascomycete fungi (Degenkolb and Vilcinskas, 2016b).

A representative and good example of that kind of fungi is the oyster mushroom (*P. ostreatus*), a saprotrophic fungus inhabiting dead trees and highly humic soils. This fungus defends itself against nematodes by secreting a specific substance produced on the mycelial appendages (toxocysts). The substance, trans-2-decene dioic acid, paralyzes nematodes (Barron and Thorn, 1987). This acid is released only by the mycelium spread in the substrate to protect the fruiting body. After paralyzing the nematode with the toxin, the hyphae penetrate the nematode body interior, obtaining nutrients necessary for further growth. Most likely, the increase in the production and secretion of the toxin is induced by the availability of light: if more light is available, more toxin is synthesized and secreted to the surface of the hyphae. The toxocysts of *P. cystidiosus* develop well when grown on PDA on the light, but with the absence of nematodes. It was synchronized with the coremia formation (Truong et al., 2007). Additionally, an increase in the production of the toxin is followed by the mechanical damage imitating injuries caused by nematodes to the hyphae. The simple presence of a nematode has not been shown to induce the toxin production by *P. ostreatus* (Barron and Thorn, 1987). Hyphae of *P. ostreatus* produce (E)-2-decenedioic acid on the secretion knobs (Figs. 16.2, 16.3); that phenomenon was broadly described in the literature. It acts on a broad spectrum of nematodes; however, not equally on various species. At a concentration of 300 μg/ml it was able to inhibit the movement of 95% of a nematode *Panagrellus redivivus* population within 1 hour (Degenkolb and Vilcinskas, 2016a). Various *Pleurotus* species may kill with different effectiveness,varying from 20% to 96%, sugar beet cyst nematode

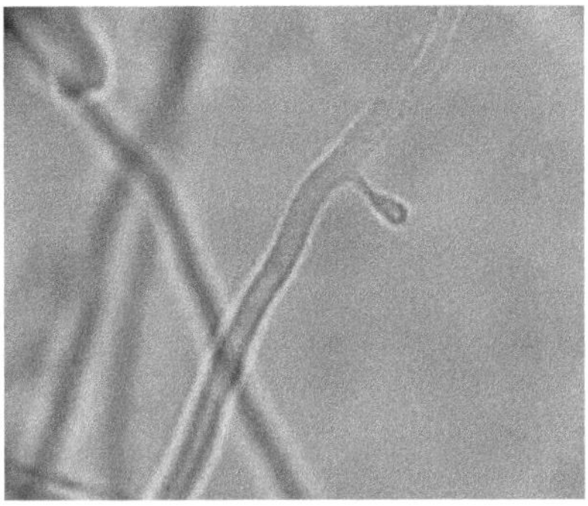

Fig. 16.2. Hyphal knob on young monokaryotic hyphae of *P. ostreatus*; no toxin secretion is visible, the hyphae was incubated on PDA medium (photo credit, E. Moliszewska).

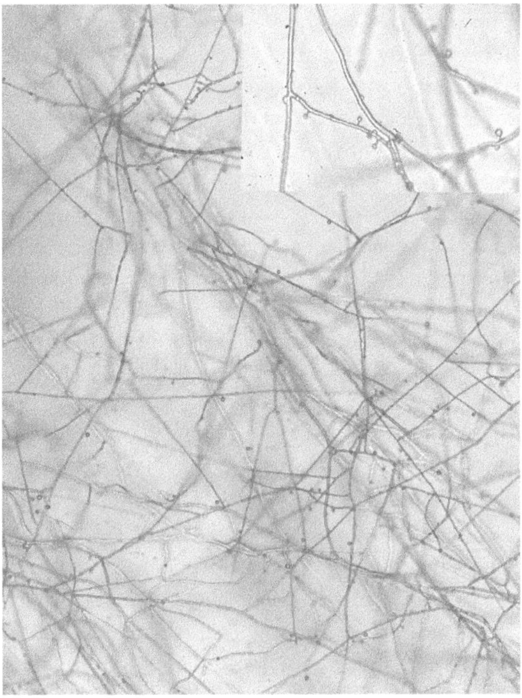

Fig. 16.3. Hyphal knobs with toxin droplets on the heterokaryotic hyphae of *P. ostreatus* (photo credit, E. Moliszewska).

(*Heterodera schachtii*) and *P. ostreatus* compost added to soil caused 85% reduction of cysts (Palizi et al., 2007). After killing nematodes with a toxin, the *P. ostreatus* hyphae start to regrow their bodies (Fig. 16.4).

Tanney and Hutchison (2012) observed the ability of two isolates of *Climacodon septentrionalis* to immobilize and after that kill a mycophagous nematode (*Aphelenchoides* sp.) *in vitro*. Both isolates produced toxic droplets formed at the apices of branched secretory cells. The secretory hyphae were measured as 700–1500 μm in height and 20–45 μm in diameter. The killing process started from

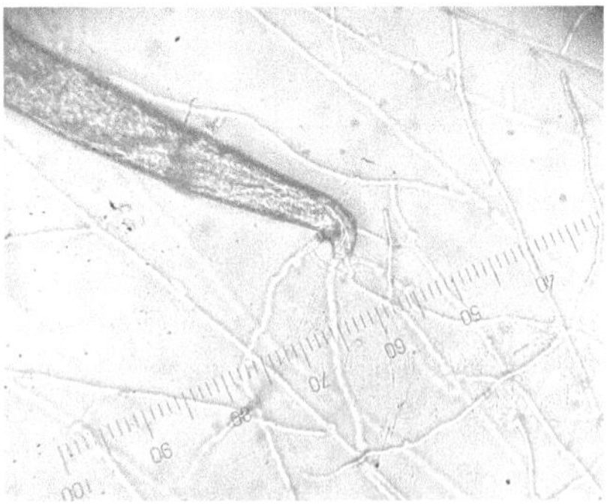

Fig. 16.4. A nematode killed and invaded by hyphae of *P. ostreatus* (photo credit, E. Moliszewska).

enveloping the nematode in the toxin droplet and the victim's death occurred after a few hours of being immobilized by the toxin. Decomposition and utilization of the nematode body lasted from several days to weeks.

Chemicals used by fungi to paralyze nematodes are detected in different ways, either as fungal culture filtrates or by direct observations. Culture filtrates used to treat nematodes showed numerous substances capable of poisoning them. The primary identified nematotoxic compound of *Arthrobotrys conoides* and *A. oligospora* was linoleic acid which is considered as a subcellular substance of trap cells. Its toxicity and other fatty acids, shown as LD_{50}, were measured at the level of approx. 5–10–50 µl/ml depending on the target organism. The amount of the produced linoleic acid was correlated positively with the number of fungal traps. It has likewise been reported in *A. bronchopagus* (syn. *Drechslerella brochopaga*) and *A. dactyloides* (syn. *D. dactyloides*). Metabolites of *A. oligospora* CBS 115.81, *A. bronchopagus* and *A. dactyloides* include the three other oils: oligosporon, oligosporol A, and oligosporol B (Table 16.1) (Stadler et al., 1993; Anke et al., 1995; Anderson et al., 1995; Chitwood, 2002; Degenkolb and Vilcinskas, 2016b).

Compounds that are able to influence the nematode lifespan belong to different chemical groups, like terpenoids (e.g., isovalleral), sesquiterpenoids (e.g., cheimonophyllal), chlorinated compounds (e.g., chloromycorrhizin), lactones (e.g., floccolactone, patulin), cyclic peptides (e.g., cyclosporin A, bursaphelocides A), anthraquinones (e.g., emodin), alkaloids (e.g., paraherquamides, marcfortines), macrolides (e.g., helmidiol, clonostachydiol), and fatty or organic acids (e.g., coriolic acid and linoleic acid, citric acid) (Table 16.1). Due to their different chemical compositions, they present varied modes of action on nematodes as targets (Stadler et al., 1995). The overview of the variability of fungal nematicidal compounds is given in Table 16.1.

Many nematicidal compounds actively reducing *C. elegans* or *M. incognita* were isolated and tested in the laboratory of Anke and Sterner and colleagues. They described the activity of several substances obtained from fungal culture filtrates of *Lachnum papyraceum* as following: mycorrhizins, lachnumol A and lachnumon, lachnumfuran, papryracons, and lachnumlactone A, mycorrhizin derivatives—chloromycorrhizinol A, (1–Z)-dechloromycorrhizin A, and the mycorrhizin-related analogs papyracons A, B, C, as well asmethylated and chlorinated derivatives of

Table 16.1. Nematicidal compounds produced by nematophagous fungi.

Compound	Group	Source	References
Lachnumol A	Mycorrhizins	*Lachnum papyraceum*	Stadler et al., 1993; Stadler et al., 1995; Degenkolb and Vilcinskas, 2016b
Lachnumon		*Lachnum papyraceum*	
Mycorrhizin A		*Lachnum papyraceum*, *Monotropa hypopitys*, and *Gilmaniella humicola*	
Chloromycorrhizin	Mycorrhizins, chlorinated compounds	*Lachnum papyraceum* and *Monotropa hypopitys*	Stadler et al., 1993
14-epicochlioquinone B		*Neobulgaria pura*	Anke et al., 1995
(1-Z)-dechloromycorrhizin A		Obtained during a total synthesis of mycorrhizin A	Stadler et al., 1993; Degenkolb and Vilcinskas, 2016b
Papryracons: papyracon A, B, C, and D; 6-O-methylpapyracon B and 6-O-methylpapyracon C	Mycorrhizins	*Lachnum papyraceum*	Shan et al., 1996; Stadler et al., 1995; Chitwood, 2002; Degenkolb and Vilcinskas, 2016b
Lachnumfuran A			
Lachnumlactone A			
Mycenone	Chlorinated benzoquinone compound	*Mycena* sp.	Chitwood, 2002; Li et al., 2007
Lethaloxin	Lactone	*Mycosphaerella lethalis*	
Floccolactone		*Strobilomyces floccopus*	
Patulin		*Penicillium* spp.	
1-methoxy-8-hydroxynaphthylene	Aromatic compound	*Daldinia concentrica*	
1,8-dimethoxynaphthylene			
14-epicochlioquinone B	Quinone	*Neobulgaria pura*	
Emodin	Anthraquinone	Various fungi	
Cheimonophyllal	Bisabolene sesquiterpenoids	*Cheimonophyllum candidissimum*	
Cheimonophyllons A-E, cheimonophyllal	Bisabolane sesquiterpenoids		
Dihydroxymintlactone	Terpenoid		
Isovalleral	Terpenoid	*Lactarius* spp.	
Marasmic acid	Terpenoid	*Marasmius conigeus*, *Lachnella villosa*, *Peniophora laeta*, and *Lachnella* sp.	
Phenoxazone	Heterocyclic compound	*Calocybe gambosa*	
Lactarorufin A and B, furantriol	Furan sesquiterpenoids	*Lactarius mitissimus*	Li et al., 2007

Table 16.1 contd. ...

...Table 16.1 contd.

Compound	Group	Source	References
Stereumin A, B, C, D, and E	Terpenoids	*Stereum* sp. culture	Li et al., 2007
Isovelleral	Terpenoid	*Lactarius vellereus*	Li et al., 2007
Ophiobolins	Sesterterpenes	*Aspergillus ustus*, *Cochliobolus heterostrophus*, *Helminthosporium* sp.	Li et al., 2007
Cannabiorcichromenic acid and its 8-chloro derivative	Terpenoids	*Cylindrocarpon olidum*	Li et al., 2007
Paeciloxazine	Terpenoid	*Paecilomyces* sp. BAUA3058	Li et al., 2007
p-anisaldehyde	Aldehyde	*Pleurotus pulmonarius*	Chitwood, 2002; Degenkolb and Vilcinskas, 2016a; Soares et al., 2018
Coriolic acid	Fatty acid		
Linoleic acid	Fatty acid	*Pleurotus pulmonarius*, *Chlorosplenium* sp., and *Arthrobotrys* spp.	Anke et al., 1995; Chitwood, 2002; Degenkolb and Vilcinskas, 2016a; Soares et al., 2018; Li et al., 2007
Oligosporon,oligosporol A and oligosporol B, 4',5'-dihydro-oligosporon	Terpenoids	*A. oligospora* (CBS 115.81), *A. Bronchopagus* and *A. dactyloides*	
Beauvericin	Cyclic depsipeptide	*Beauveria bassiana*	Chitwood, 2002; Li et al., 2007
Enniatin A	Cyclic depsihexapeptide	*Fusarium* sp.	
Illinitone A	Colorless oil – secoterpenoid	*Limacella illinita* (strain 99049)	Gruhn et al., 2007; Degenkolb and Vilcinskas, 2016b
Thermolides A, B, C, and D	Colorless oils	*Talaromyces thermophilus* (YM 3–4)	Degenkolb and Vilcinskas, 2016b
Omphalotins: omphalotin A – I	Cyclic dodecapeptide	*Omphalotus olearius* (TA 90170)	Mayer et al., 1997; 1999; Sterner et al., 1997; Büchel et al., 1998; Chitwood, 2002; Degenkolb and Vilcinskas, 2016b; Wang et al., 2017

Table 16.1 contd. ...

...Table 16.1 contd.

Compound	**Group**	**Source**	**References**
Ophiobolins	Sesterterpenoids	*Bipolaris* spp.	Degenkolb and Vilcinskas, 2016b
δ-lactone [6-n-pentyl-2H-pyran-2-one (6-PAP)]	Lactone	*Trichoderma* sp. (YMF 1.00416)	
Paraherquamides	Oxindole alkaloids	*Penicillium charlesii* (ATCC 20841)	
Caryopsomycins A–C	Resorcylic acid lactones	*Caryospora callicarpa* (YMF1.01026)	
Nafuredin	Epoxy-δ-lactone with olefinic side chain	*Aspergillus niger* (FT-0554)	
Nafuredin-γ	γ-lactone with olefinic side chain		
Clonostachydiol	Macrodiolide	*Clonostachys cylindrospora*	
2β,13- dihydroxyledol (= dichomitin B)	Sesquiterpene	*Dichomitus squalens*	
Pseudohalonectrin A and B	Azaphilones	*Pseudohalonectria adversaria* (YMF1.01019)	
Cochlioquinone A	Benzoquinone derivative	*Bipolaris sorokiniana*, *Cochliobolus miyabeanus*, *Helminthosporium leersii*, and *Helminthosporium sativum*	
Cyclosporine A	Cyclopeptide	*Tolypocladium inflatum*	Chitwood, 2002
Bursaphelocide A	Cyclodepsipeptides	*Mycelia sterilia* (strain D1084) (anamorphic fungus)	Kawazu et al., 1993; Chitwood, 2002; Degenkolb and Vilcinskas, 2016b
Bursaphelocide B			
5-pentyl-2-furaldehyde	Aldehyde	*Irpexlacteus* (dermateaceous Ascomycete)	Mayer et al., 1996
(E)-2-decenedioic acid	Carboxylic acid	*Pleurotus ostreatus*	Barron and Thorn 1987; Degenkolb and Vilcinskas, 2016a
p-anisyl alcohol	Alcohol	*Pleurotus pulmonarius*	Degenkolb and Vilcinskas, 2016a
1-(4-methoxyphenyl)-1,2-propanediol			
2-hydroxy(4'-methoxy)-propiophenone	Ketone		
Pleurotin, dihydropleurotinic acid, leucopleurotin	Aromatic compound	*Nematoctonus robustus*, *Nematoctonus concurrens*, *Hohenbuehelia* spp., and *Pleurotus* spp.	Anke et al., 1995; Degenkolb and Vilcinskas, 2016b; Soares et al., 2018

Table 16.1 contd. ...

...Table 16.1 contd.

Compound	**Group**	**Source**	**References**
Leucinostatins	Nanopeptides	*Paecilomyces liliacinus*, *P. marquandi*, and *Acremonium* sp.	Park et al., 2004
T2 toxin	Mycotoxin	*Fusarium solani*	Ciancio, 1995
Moniliformin			
Fusarenone			
Neosolaniol			
Oxalin	Mycotoxin	*Fusarium* spp., *Penicillium anatolicum*, *P. chrysogenum*, *P. oxalicum*, and *P. vermiculatum*	Devi, 2018
Secalonic acid D	Chiral dimeric tetrahydroxanthones		
Verrucarin A	Sesquiterpene		
5-methylfuran-3-carboxylic acid	Carboxylic acid	*Coprinus comatus*	Luo et al., 2004, 2007
5-hydroxy-3,5-dimethylfuran-2(5*H*)-one	Ketone		
Oxalic acid dihydrate	Dicarboxylic acid	*Syncephalastrum racemosum*	Sun et al., 2008
Tartaric acid	Alpha-hydroxy acid		
Flavipin	Antibiotic	*Chaetomium globosum*	Nitao et al., 2002
4,15-diacetylnivalenol and diacetoxyscirpenol	Trichothecenes	*Fusarium equiseti*	Nitao et al., 2001
Gymnoascole acetate	Indoloditerpenoid	*Gymnoascusreessii*	Liu et al., 2017
Citric acid, isomer of dimethyl citrate	Organic compound	*Aspergillus candidus*	Shemshura et al., 2016
Trichodermin, dermadin, and trichoviridin	Mycotoxins and Antibiotic	*Trichoderma viride*	Abd-Elgawad and Askary, 2018
Sesquiterpene heptalic acid			
Paraherquamide and its analogs paraherquamides B, C, D, E, F and G	Alkaloids	*Penicilliumm charlesii*	Li et al., 2007
Aspergillimide, 16-keto aspergillimide	Alkaloids	*Aspergillus* sp.	Li et al., 2007
Marcfortine A, B, and C	Alkaloids	*Penicillium roqueforti*	Li et al., 2007
Phenoxazone	Alkaloids	*Calocybe gambosa*, *Pycnoporus sanguineus*	Li et al., 2007
Radicicol	Macrolide	*Nectriaradicicola*, *Monosporiumbonorden*, and *Pencicillium luteoaurantium* and *Chaetomium chiversii*	Li et al., 2007

Table 16.1 contd. ...

...Table 16.1 contd.

Compound	Group	Source	References
Lethaloxin(9-lactide decane)	Macrolide	*Mycosphaerella lethalis*	Li et al., 2007
Clonostachydiol	Macrolide	*Clonostachys cylindrospora*	Li et al., 2007
βγdehydrocurvularin, αβ-dehydrocurvularin, 8-β-hydroxy-7-oxocurvularin, 7-oxocurvularin	Macrolide	*Aspergillus* sp., genera: *Curvularia*, *Penicillium*, *Cochliobolus*, and *Alternaria*	Li et al., 2007
Helmidiol	Macrolide	*Alternaria alternata*	Li et al., 2007

papyracons, such as 6-O-methylpapyracon B, 6-O-methylpapyracon C, etc. Other antinematicidal metabolites produced by the same fungus are lachnumfuran A and lachnumlactone A (Table 16.1). *L. papyraceum*, like other fungi, produces numerous secondary metabolites with different activity, although the activity against nematodes is present only in some of them, for instance, chloromycorrhizinol does not have any nematicidal activity, although it is a mycorrhizin derivative (Stadler et al., 1993; 1995; Anke et al., 1995; Shan et al., 1996).

The other interesting fungal sources of nematicidal chemicals are, for instance, *Daldinia concentrica* (1-methoxy-8-hydroxynaphthylene and 1,8-dimethoxynaphthylene), *Neobulgariapura*, which produces 14-epicochlioquinone B, or *Cheimonophyllum candidissimum*, which produces bisabolane sesquiterpenoids and terpenoids. Terpenoids are produced also by members of the genus *Lactarius*. Organic acids, including fatty acids, are produced by numerous fungi like members of the genera *Pleurotus* and *Chlorosplenium*. The separated group of toxic substances creates peptides, mostly cyclic ones, like beauvericin from *Beauveria bassiana*; enniatin A from *Fusarium* spp.; cyclosporine from *Tolypocladium inflatum*, and omphalotin A from *Omphalotisolearius*. The anthraquinone toxin—emodin—was found as a metabolite produced by various fungi. Various fungi, including edible produce p-anisaldehyde, which is typically known as one of the mushroom scent compounds. That component also shows the activity against nematodes (Table 16.1) (Chitwood, 2002; Li et al. 2007; Moliszewska et al., 2021). *Pleurotus ostreatus* produces well-known antinematicidal toxin trans-2-decene dioic acid, which is supported by ostreolysin. On the other hand, ostreolysin enhances fruiting of that mushroom (Berne et al., 2007).

Results of the Park et al. (2004) research suggest that leucinostatins produced by *Paecilomyces liliacinus* are indicators of its nematicidal activity; however, isolates of *P. lilacinus* differed in the efficacy of the toxin production. Leucinostatins belong to a group of nanopeptides produced also by *P. marquandi* and *Acremonium* spp. (Table 16.1). Their nematicidal activity was strong and depended on the age of nematodes, briefly—freshly hatched ones were killed 100%, while adult ones about 75%. The mode of action of leucinostatin-A leads to different phosphorylation pathways in ATP due the inhibition of the process of mitochondrial ATP synthesis (Park et al., 2004). The *P. lilacinus* isolates with a weaker nematicidal activity have been recognized as producers of unsaturated fatty acids: oleic, linoleic, and linolenic acid, and have shown to possess nematicidal activity against *C. elegans* (Park et al., 2004).

The chlorinated compound mycenone was obtained from *Mycena* sp., while from *Mycosphaerella lethalis*–lethaloxin and from *Strobilomyces floccopus* – floccolactone. Some mycotoxins like patulin, typically produced by *Penicillium* spp., as well as T2-toxin, moniliforrnin, verrucarin A, and cytochalasin B isolated from *Fusarium* spp. also showed nematicidal activity (Stadler et al., 1995; Ciancio et al., 1995). Ciancio et al. (1995) tested them *in vitro*, and the nematicidal activity was observed as a significant reduction of the viability of *Meloidogyne javanica* juveniles. The active concentrations were low and started from a absorption of 2 ppm. Enniatin B, the other tested mycotoxin, significantly decreased the population of *M. javanica* at 20 ppm. Some mycotoxins, such as T2-toxin, also showed nematicidal activity against *M. hapla* and *Pratylenchus neglectus* at low concentrations. However, that result was not repeated in pot experiments. In soil mycotoxins, T2-toxin and moniliformin did not affect *M. javanica* population on tomato plants, suggesting a complex interaction among nematodes, fungi, plant, and rhizosphere (Ciancio, 1995; Devi, 2018). Metabolites produced by *Trichoderma viride*–trichodermin, dermadin, trichoviridin, and sesquiterpene heptalic acid—are also involved in the suppression of nematodes (Table 16.1) (Abd-Elgawad and Askary, 2018).

Flavipin (Table 16.1), an antibiotic firstly isolated from *Epicoccum* sp., can inhibit *in vitro* egg hatching and juvenile mobility of root-knot nematode (*Meloidogyne incognita*); however, in soil the number of galls on roots increased with flavipin treatment and any effect on nematode populations was found at the harvest (Nitao et al., 2002).

6. Enzymes of Nematophagous Fungi

Some mushroom species have developed a way to eliminate their enemies while they are still in eggs. Some of them form appressoria, which secrete enzymes facilitating the enzymatic disruption of eggs' shells (Obrępalska-Stępiowska and Sosnowska, 2008; Lopez-Llorca et al., 2010; Gortari and Hours, 2008; Saxena, 2018). The enzyme production is previously initiated, as postulated by Stirling and Mankau (1979), by the precursory contact of hyphae with egg masses, which lead to forming appressoria followed by mechanical and chitinase disruption of eggs' shell. That mode of action is observed for *Dactylella oviparasitica* which parasitizes eggs of *Acrobeloides* sp., *Heterodera schachtii, Meloidogyne* spp., and *Tylenchulus semipenetrans* (Stirling and Mankau, 1979). The main enzymes implicated in digestion of nematoda body or eggs belong to chitinases, proteases, and collagenases; however, the latter group seems to be weakly examined (Obrępalska-Stępiowska and Sosnowska, 2008; Saxena, 2018). Other common examples which use enzymes in attacking nematodes are *Lecanicillium lecanii*, *Pochonia chlamydosporia*, *P. rubescens* and others, infecting the eggs and females by appressoria and zoospores in the case of *Catenaria auxiliaris* (Castillo and Lawrence, 2011; Saxena, 2018). This type of fungi also infects eggs via protease activity, which follows from three layers of the eggshell (Lopez-Llorca et al., 2010; Gortari and Hours, 2008; Saxena, 2018). The egg penetration facilitated by extracellular enzymes such as chitinases and proteases was postulated also by Yang et al. (2007). The major proteases identified from nematophagous fungi belong to the proteinase K family of subtilases (from peptidase S8 subtilase family) (Morton et al., 2004). However, the latest findings allow their identification as aspartyl proteases, cysteine proteases, metalloproteases, and serine proteases. Most of them belong to the large subtilisin family of endopeptidases MEROPS M8 found only in fungi and bacteria. Subtilisins play an important role in *A. oligospora* infection and immobilization of the pry (Herrera-Estrella et al., 2016).

Purpureocillium atypicola infects nematodes' eggs by synthesizing and secreting proteases when the hypha encounters nematodes on its way. Disruption of the eggshell allows the hypha to penetrate inside it after which it digests the contents of the egg, grows, and infects the remaining eggs, eliminating the nematode from the environment. Lipases have been implicated in the infection of *Heterodera schachtii* eggs but also other enzymes that are able to degrade inner lipid egg layers (Perry and Trett, 1986; Dackman et al., 1989; Park et al., 2004; Saxena, 2018). Becker et al. (2020) tested three fungal strains belonging to the genus *Hyalorbilia*, isolated from eggs of *H. schachtii*, *Meloidogyne incognita*, and *H. glycines*. All these fungi were able to produce extracellular hydrolytic enzymes, like acid phosphatase, α-glucosidase, β-glucosidase, and N-acetyl-β-glucosamidase on PDA, or when they were grown on *H. schachtii* females. The trypsin-like protease was additionally detected for the isolate obtained from *H. schachtii* (Becker et al., 2020).

Trichoderma harzianum secretes many lytic enzymes like chitinase, glucanases, and proteases after penetration of the eggs and juveniles by dissolving chitin cuticle. This ability facilitates the parasite *Meloidogyne* spp. and *Globodera* spp. eggs, then the hyphae of *T. harzianum* proliferates within the organism, and in the next step produces toxic metabolites. A similar way of interaction with nematodes is presented by *Paecilomyces liliacinus*, which produces enzymes such as protease and chitinase which are supported by the antibiotics leucinostatin and lilacin, as well as ammonia released from the decomposition of chitin, toxic to second-stage juveniles. This cocktail of such different substances destroys eggs and allows hyphae to grow inside rapidly. *Pochonia chlamydosporia* inhibits the hatching of eggs by exoenzymes that are supported by toxins and followed by hyphae penetrating as well (Park et al., 2004; Gortari and Hours, 2008; Abd-Elgawad and Askary, 2018).

Good sources of fungal enzymes may be spent mushroom composts. Crude extracts obtained from spent mushroom composts of *Flammulina velutipes* and *Hypsizygus marmoreus* demonstrated significant ability to reduce *Panagrellus* sp. population. Those crude extracts contained proteases; however, their activity was still present after boiling (denaturation of enzymes), which demonstrated undeniably that enzymes in those extracts were supported by other substances (toxins) (Ferreira et al., 2019; Soares et al., 2019).

Although enzymatic activity of nematophagous fungi is thought to be correlated with their pathogenicity, some experiments showed that this feature may be not obligatory. Park et al. (2004) and Dackman et al. (1989) demonstrated that high production of chitinase and protease by *P. lilacinus* was not always related to high parasitic activity, while high levels of parasitic activity were associated with noticeable chitinase activity. While in *Verticillium suchlasporium* and *V. chlamydosporium* chitinase activity expressed as N-acetyl-glucosamine, production was observed in amounts 3.7–14.6 μmol per mg crude protein per hour and, was correlated with parasitic activity (Dackman et al., 1989). On the other hand, protease production does not appear to be strongly correlated with *P. lilacinus* pathogenicity to the nematodes and may not be critically important for its nematophagous activities compared with chitinase activity. In *P. lilacinus* enzymatic activity supported by leucinostatin, production may be a useful indicator for the *P. lilacinus* nematophagous activity (Dackman et al., 1989; Park et al., 2004).

Enzymatic activity seems to be a mode of action for *Duddingtonia flagrans* (*Arthrobotrys flagrans*) against horse nematode complexes causing cyathostomin. The activity manifested as chitinase and proteases production leads to a significant reduction in the number of intact cyathostomin larvae L3. Enzymes prepared as crude extracts of such nematophagous fungi, among others, consisting of extracellular proteases, can reduce the number of larval stages of gastrointestinal nematodes by preventing their eggs' hatching (Braga and de Araújo, 2014).

7. Biological Control of Nematodes by Fungi

Chemical control of nematodes is very efficient; however, the pressure to use ecologically friendly products for plant protection management is widely demanded. This demand for broader use of biological pesticides leads to the development of different bioproducts (Ferreira et al., 2019). Many of them may be based on mycophagous fungi. That is possible because predacious and mycophagous fungi are natural members of microbiota, which would not be able to disturb the natural environment. They are broadly found in natural habitats, however, with age, a natural reduction in nematode-trapping efficiency is observed. This makes the complementation of the natural microbiota with actively growing nematicidal fungi necessary. As of date several attempts have been made to assess their potential efficacy in nematode controlling, and such a way of biological control with fungi is a viable and safe alternative for animals, humans, and the environment (Braga and de Araújo, 2014). For effective control, it is necessary that the nematode invasion of crops and other hosts must coincide with the period of the highest activity of nematode-trapping fungi. As Siddiqui and Mahmood (1996) suggest, maximum fungal activity is reached only in the period of 12–15 days after fungal inoculation. They also suggest that fungal activity might be stimulated by the addition of organic matter to the soil (Siddiqui and Mahmood, 1996). Braga and de Araújo (2014) pointed out that fungi are promising natural sources of new bioactive molecules or crude extracts derived from these organisms with the broad potential of use expanded to other parasites including ticks.

The first step in practical use is the proper isolation and identification of nematophagous fungi. Isolation sometimes needs sophisticated methods and classical fungal identification mostly depends not only on the morphology and other characteristics features, but also includes molecular methods. The next steps include testing of the nematicidal abilities and efficacy, preliminary preparation of the inoculum, and the method of its introduction into the soil or other environments.

Tests against the target nematode species as well as other species are also demanded (Kawazu et al., 1980; Li et al., 2014; Udo et al., 2014).

However, the practical use of fungi or their products against nematodes does not seem to be widely practiced. So far very few preparations have been developed based on fungi. The most actively used in the USA includes DiTera DF derived from the microorganism *Myrothecium verrucaria* and BioAct WG (MeloCon WG, Germany) (active organism: *Paecilomyces lilacinus* 251=*Purpureocillium lilacinus* strain 251), and some preparations based on killed *Burkholderia* bacteria (Desaeger and Watson, 2019). The DiTera, a crude fungal product from fungus *M. verrucaria*, is effective against a broad range of phytoparasitic nematodes. It controls nematodes by paralyzing the muscles that control feeding and locomotion and finally kills nematodes due to its synergistically acting low-molecular-weight, water soluble compounds. It also prevents the egg hatching of *Globodera rostochiensis* but not *Meloidogyne incognita*, inhibits nematode sensory organs, and causes disorientation. Its toxicity appears limited to plant-parasitic nematodes, as effects were observed neither to *Caenorhabditis elegans* nor to the rat intestinal parasites *Nippostrongylus brasiliensis* and *Panagrellus redivivus*. Toxicity of that preparation is due to synergistically acting low-molecular-weight, water soluble compounds (Chitwood, 2002). Outside Europe, across the world numerous preparations were developed; however the most used fungi in them belong to only a few species. This includes *P. lilacinus* as the most frequently applied, *Aspergillus niger*, *Pochonia chlamydosporia*, *Purpureocillium lilacinus*, *Trichoderma harzianum*, and *T. viride.* Bioformulation containing *P. lilacinus* as the active ingredient with trade name PL Gold was obtained from the Biological Control Products in South Africa and BIOCON in the Philippines. Preparations based on spore concentrates of *P. lilacinus* are widely produced in India. Other specimens were developed such as: Stanes Bio Nematon, Shakti Paecil, PAECILO, Paecilon, Nematofree, Gmaxbioguard, Yorker, Commander Fungicide; Miexianning, Pl Plus. In India, *Trichoderma harzianum* also serves as an ingredient of specimens ECOSOM and Commander Fungicide. That fungus serves as a bioactive component for Romulus in South Africa. In Colombia, the specimen Trichobiol is based on *T. harzianum* and Trifesol is based on *T. viride*. Another widely utilized fungus in India is *Aspergillus niger*, which serves as the main component for formulations: Kalisena, Beej Bandhu, PusaMrida, and Kalasipahi. Based on *Pochonia chlamydosporia*, specimens were created: KlamiC in the UK and Cuba, PcMR-1 strain (Portugal), IPP21 (Italy), and Xianchongbike (China). Inferno® preparation was produced in Turkey based on the *M. verrucaria* strain (AARC-0255 fermentation product), Bionematon containing 1.5% *P. lilacinus* strain PL1, and EndoRoots Soluble based on 23.5% live organism including spores of mycorrhizal fungi (*Glomus* spp.) (Obrępalska-Stępiowska and Sosnowska, 2008; Udo et al., 2014; Kepenekci et al., 2017; Abd-Elgawad and Askary, 2018; Devi, 2018).

In bio-protection against nematodes, all the mechanisms developed by nematophagous fungi are involved, although depending on the specimen species composition, and the way of formulation, usually only one or a few are involved in real protection mechanisms. From the group of enzymes secreting fungi, *Arthrobotrys dactyloides*, *Purpureocillium atypicola*, and *Trichoderma* spp. are examples of fungi that have gained application in agriculture as a biological method of eliminating the nematode *Meloidogyne incognita* (root-knot nematode) from soil (Chitwood, 2002; Devi, 2018; Poveda et al., 2020). Another example with great potential is *Aspergillus niger*, practically used in some of the given-above formulations. That fungus grows rapidly in contact with cysts or egg masses of nematodes colonizing them before larval formation has been completed (Gortari and Hours, 2008; Abd-Elgawad and Askary, 2018; Devi, 2018).

The effectiveness of bio-preparations depends not only on the main fungal component and target organism. The application of the specimen as well as its formulation are the next most important features influencing the results in practice. Typical examples of formulated fungal bio-preparations include liquids, capsules, granulate and water-dispersible granules, suspension concentrates, powder concentrates, wettable powders, and talc-based carriers (Abd-Elgawad and Askary, 2018). Direct application of the fresh mycelium is used only in experimental tests. So far, numerous tests for nematophagous fungi have been carried out, providing a base for further applications.

Eberlein et al. (2020) demonstrated that the fungus *Hyalorbilia* genus, strain DoUCR50, favored the establishment of soil suppressiveness in 4-year microplot field experiments (sugar-beet monoculture) infested with different genotypes of *H. schachtii* and planted with susceptible, resistant, and tolerant sugar-beet genotypes. Diseased eggs were observed starting from the second year and achieved the level 90% of infected eggs in the third cropping year, in the case of all three sugar-beet cultivars (susceptible, resistant, and tolerant). In all the years, the tolerant genotype produced the highest and most stable white sugar yields while yields of the other cultivars slowly improved during the monoculture. Results of this study suggested the presence of egg-infecting factors in this sugar-beet monoculture that dramatically increased the proportions of diseased eggs (Eberlein et al., 2020). These observations showed the practical use of fungal antagonists in protection of sugar beets. Similar effects were obtained by Olatinwo et al. (2006) for *Dactyllea oviparasitica* strain 50, used in the non-suppressive soil artificially infected with *H. schachtii*. The *D. oviparasitica* reduced eggs and juvenile densities to those levels observed in the suppressive soil. The levels of cysts were lower than in the suppressive soil after two nematode generations (Olatinwo et al., 2006).

Yan et al. (2011) tested 294 endophytic fungi obtained from cucumber seedlings for their potential as seed treatments against *Meloidogyne incognita*. Among them, 23 isolates significantly reduced galls formed by *M. incognita* in the greenhouse test. The most effective isolates belonged to the genera of *Acremonium*, *Chaetomium*, *Fusarium*, *Paecilomyces*, *Phyllosticta*, and *Trichoderma*. According to these results, the strain *Chaetomium* Ch1001 was considered the best candidate for seed treatments for *M. incognita* biocontrol (Yan et al., 2011).

Duddingtonia flagrans (correct name according to Index Fungorum: *Artrobotrys flagrans*) formulated on rice bran, showed predatory capability against swine parasite *Oesophagostomum* spp. Tests proved its effectiveness against infective larvae (L3) in *in vitro* tests in petri dishes, against eggs in faecal cultures, as well as the predative capability of the fungus after passing through the gastrointestinal tract of swine without loss of viability. This isolate may be considered as an alternative in the control of *Oesophagostomum* spp. in swine (Rodrigues et al., 2018).

Extracellular enzymes produced by *Duddingtoniaflagrans* (syn. *Arthrobotrys flagrans*) led to a significant reduction ($p < 0.01$) in the number of intact cyathostomin L3, when compared to the control (Braga et al., 2015). Cyanthostomin, a horse parasitic disease, is caused by at least ten nematode species creating specific cyathostomin communities, among which *Cylicocyclusnassatus*, *Cylicostephanus longibursatus*, and *Cyathostomum catinatum* seem to be the most prevalent (Bellaw and Nielsen, 2020). That study showed that chitinase had a stronger nematicidal effect against cyathostomin causal nematodes. Authors calculated the conditions for maximum chitinase activity: incubation time 2 days, and moisture 511% (Braga et al., 2015). Other interesting substances active against intestinal nematodes of ruminants are paraherquamides from *Penicillium charlesii*, clonostachydiol from *Clonostachys cylindrospora*, as well as nafuredin and nafuredin-γ from *Aspergillus niger* (Table 16.1) (Degenkolb and Vilcinskas, 2016b).

Comans-Pérez et al. (2021) demonstrated in petri dishes' tests that mycelia and extracts of *Pleurotus ostreatus*, *P. eryngii*, *P. cornucopiae*, *Coprinus comatus*, *Panus* sp., *Lentinula edodes*, and *L. boryanus* are candidates for future *in vivo* tests against *Haemonchus contortus* infective larvae. That experiment opens the possibility to curing animals by feeding them or to produce naturally originated drugs. Pineda-Alegrı´a et al. (2017) showed five major compounds (pentadecanoic, hexadecanoic, octadecadienoic and octadecanoic acids, and β-sitosterol) which were responsible for the antinematicidal activity of the extract of *Pleurotus djamor* and concluded that this edible mushroom could be used as an alternative anthelmintic treatment for livestock.

8. Conclusion and Future Perspectives

Serving a broad mode of nematicidal activity, fungi are important factors in natural nematode population regulation and show real promise in switching nematode control to bioprotective methods. The nematicidal potential of fungi seems to be huge and very interesting; however, they are only marginally used as bioprotectants. There is a demand to research their biological potential and conditions for practical use for broadening their application as bio-preparations. Nature serves as excellent solutions for agricultural problems with plant-parasitic nematodes and this is promising in the context of other nematode problems as, for example, animal parasites. Currently, the most of the scientific work with nematode-controlling organisms is directed towards searching for potentially useful organisms and assessing their efficiency. Some of the nematode-controlling fungi are now used commercially; however, only in a few countries with minimal quantities. Production of biological control preparations still has problems with mass production, standardization, formulation, storage, application, and safety (Noweer, 2020). Consequently, this demands the broadening of research methods and developing production cycles. The one and commonly forgotten problem is the popularization of biological methods in nematode control among farmers. Therefore, educational courses for farmers should be carried out in parallel.

References

Abd-Elgawad, M.M.M. and Askary, T.H. (2018). Fungal and bacterial nematicides in integrated nematode management strategies. *Egypt. J. Biol. Pest. Control*, 28: 74. https://doi.org/10.1186/s41938-018-0080-x.

Agrios, G.N. (2005). Plant diseases caused by nematodes. pp. 825–874. *In*: Agrios, G.N. (Ed.). *Plant Pathology* (5th Edn.). Academic Press. https://doi.org/10.1016/B978-0-08-047378-9.50021-X.

Akhtar, M.S. and Siddiqui, Z.A. (2008). Arbuscular mycorrhizal fungi as potential bioprotectants against plant pathogens. pp. 61–97. *In*: Siddiqui, Z.A., Akhtar, M.S., Futai, K. et al. (Eds.). *Mycorrhizae: Sustainable Agriculture and Forestry*. Springer Science+Business Media B.V.

Anderson, M.G., Jarman, T.B. and Rickards, R.W. (1995). Structures and absolute configurations of antibiotics of the oligosporon group from the nematode-trapping fungus *Arthrobotrys oligospora*. *J. Antibiot. (Tokyo)*, 48(5): 391–398. doi: 10.7164/antibiotics.48.391.

Anke, H., Stadler, M., Mayer, A. and Sterner, O. (1995). Secondary metabolites with nematicidal and antimicrobial activity from nematophagous fungi and Ascomycetes. *Can. J. Bot.*, 73(S1): 932–939. https://doi.org/10.1139/b95-341.

Barron, G.L. and Thorn, R.G. (1987). Destruction of nematodes by species of *Pleurotus*. *Can. J. Bot.*, 65: 774–778. doi: 10.1139/b87-103.

Becker, J.S., Borneman, J. and Becker, J.O. (2020). Effect of *Heterodera schachtii* female age on susceptibility to three fungal hyperparasites in the genus *Hyalorbilia*. *Journal of Nematology*, 52: 1–12. https://doi.org/10.21307/jofnem-2020-093.

Bellaw, J.L. and Nielsen, M.K. (2020). Meta-analysis of cyathostomin species-specific prevalence and relative abundance in domestic horses from 1975–2020: Emphasis on geographical region and specimen collection method. *Parasit. Vectors*, 13: 509. https://doi.org/10.1186/s13071-020-04396-5.

Berne, S., Pohleven, J., Vidic, I., Rebolj, K., Pohlevend, F., Turke, T., Maček, P., Sonnenberg, A. and Sepčić, K. (2007). Ostreolysin enhances fruiting initiation in theoyster mushroom (*Pleurotus ostreatus*). *Mycological Research* 111: 1431–1436. doi:10.1016/j.mycres.2007.09.005.

Borowicz, V.A. (2001). Do Arbuscular mycorrhizal fungi alter plant-pathogen relations? *Ecology*, 82(11): 3057–3068. https://doi.org/10.2307/2679834.

Braga, F.R. and Araújo, J.V. (2014). Nematophagous fungi for biological control of gastrointestinal nematodes in domestic animals. *Appl. Microbiol. Biotechnol.*, 98: 71–82. doi: 10.1007/s00253-013-5366-z.

Braga, F.R., Soares, F.E.F., Giuberti, T.Z., Lopes, A.D.C.G., Lacerdaa, T., de HollandaAyupea, T., Queirozc, P.V., de Souza Gouveiac, A., Pinheiroa, L., Araújoa, A.L., Queirozc, J.H. and Araújo, J.V. (2015). Nematocidal activity of extracellular enzymes produced by the nematophagous fungus *Duddingtonia flagrans* on cyathostomin infective larvae. *Vet. Parasitol.*, 212: 215–218. http://dx.doi.org/10.1016/j.vetpar.2015.08.018.

Büchel, E., Martini, U., Mayer, A., Anke, H. and Sterner, O. (1998). Omphalotins B, C, and D, nematicidal cyclopeptides from *Omphalotus olearius*: Absolute configuration of omphalotin A. *Tetrahedron*, 54(20): 5345–5352.

Castillo, J.D. and Lawrence, K.S. (2011). First report of *Catenaria auxiliaris* parasitizing the reniform nematode *Rotylenchulus reniformis* on cotton in Alabama. *Plant Dis.*, 95(4): 490. doi: 10.1094/PDIS-07-10-0524.

Chaverri, P., Samuels, G.J. and Hodge, K.T. (2005). The genus *Podocrella* and its nematode-killing anamorph *Harposporium*. *Mycologia*, 97(2): 433–443.

Chitwood, D. (2002). Phytochemical based strategies for nematode control. *Annu. Rev. Phytopathol.*, 40(1): 221–249. doi:10.1146/annurev.phyto.40.032602.130045.

Ciancio, A. (1995). Observations on the nematicidal properties of some mycotoxins. *Fundam. Appl. Nematol.*, 18(5): 451–454. 146.

Comans-Pérez, R., Sánchez, J.E., Al-Ani, L., González-Cortázar, M., Castañeda-Ramírez, G.S., Mendoza-de Gives, P., Sánchez-García, A.D., Millán-Orozco, J. and Aguilar-Marcelino, L. (2021). Biological control of sheep nematode *Haemonchuscontortus* using edible mushrooms. *Biol. Control*, 152: 104420. https://doi.org/10.1016/j.biocontrol.2020.104420.

da Silva Campos, M.A. (2020). Bioprotection by arbuscular mycorrhizal fungi in plants infected with *Meloidogyne* nematodes: A sustainable alternative. *Crop Prot.*, 135: 105203. https://doi.org/10.1016/j.cropro.2020.105203.

Dackman, C., Chet, I. and Nordbring-Hertz, B. (1989). Fungal parasitism of the cyst nematode *Heterodera schachtii*: Infection and enzymatic activity. *FEMS Microbiol. Ecol.*, 62: 201–208. https://doi.org/10.1111/j.1574-6968.1989.tb03694.x.

Degenkolb, T. and Vilcinskas, A. (2016a). Metabolites from nematophagous fungi and nematicidal natural products from fungi as alternatives for biological control. Part II: metabolites from nematophagous basidiomycetes and non-nematophagous fungi. *Appl. Microbiol. Biotechnol.*, 100: 3813–3824. doi: 10.1007/s00253-015-7234-5.

Degenkolb, T. and Vilcinskas, A. (2016b). Metabolites from nematophagous fungi and nematicidal natural products from fungi as alternatives for biological control. Part I: Metabolites from nematophagous ascomycetes. *Appl. Microbiol. Biotechnol.*, 100: 3799–3812. doi: 10.1007/s00253-015-7233-6.

Desaeger, J.A. and Watson, T.T. (2019). Evaluation of new chemical and biological nematicides for managing *Meloidogyne javanica* in tomato production and associated doublecrops in Florida. *Pest Manag. Sci.*, 75(12): 3363–3370. https://doi.org/10.1002/ps.5481.

Devi, G. (2018). Utilization of nematode destroying fungi for management of plant-parasitic nematodes: A review. *Biosci. Biotechnol. Res. Asia*, 15(2): 377–396.

Dubiel, G. (2014). Obserwacjegrzybów z rodzaju *Harposporium*–pasożytówsaprofagicznychnicieni (Observations of *Harposporium* fungi: Parasites of saprophagous nematodes). *Przegląd Przyrodniczy*, XXV(1): 37–41.

Eberlein, C., Heuer, H. and Westphal, A. (2020). Biological suppression of populations of *Heterodera schachtii* adapted to different host genotypes of sugar beet. *Frontiers in Plant Science*, 11: 812. doi: 10.3389/fpls.2020.00812.

Elsen, A., Gervacio, D., Swennen, R. and de Waele, D. (2008). AMF-induced biocontrol against plant parasitic nematodes in *Musa* sp.: A systemic effect. *Mycorrhiza*, 18: 251–256. doi: 10.1007/s00572-008-0173-6.

Ferreira, J.M., Carreira,1 D.N., Braga, F.R. and de Freitas Soares, F.E. (2019). First report of the nematicidal activity of *Flammulina velutipes*, its spent mushroom compost and metabolites. *3 Biotech*, 9: 410. https://doi.org/10.1007/s13205-019-1951-x.

Gortari, M.C. and Hours, R.A. (2008). Fungal chitinases and their biological role in the antagonism onto nematode eggs: A review. *Mycol. Prog.*, 7: 221–238.

Gruhn, N., Schoettler, S., Sterner, O. and Anke, T. (2007). Biologically active metabolites from the Basidiomycete *Limacella illinita* (Fr.) *Murr. Z. Naturforsch. C*, 62(11-12): 808–812. doi: 10.1515/znc-2007-11-1206.

Hajek, A.E. and Morris, E.E. (2014). Sirex woodwasp. pp. 331–346. *In*: Van Driesche, R.G. and Reardon, R. (Eds.). *The Use of Classical Biological Control to Preserve Forests in North America*. FHTET-2013-02. Morgantown, West Virginia: USDA Forest Service, Forest Health Technology Enterprise Team.

Herrera-Estrella, A., Casas-Flores, S. and Kubicek, C.P. (2016). Nematophagous fungi. pp. 247–267. *In*: Druzhinina, I.S. and Kubicek, C.P. (Eds.). *The Mycota Vol. IV: Environmental and Microbial Relationships* (3rd Edn.). Switzerland: Springer International Publishing.

Hol, G. and Cook, R. (2005). An overview of arbuscular mycorrhizal fungi–nematode interactions. *Basic Appl. Ecol.*, 6(6): 489–503. doi: 10.1016/j.baae.2005.04.001.

Kawazu, K., Nishii, Y., Ishii, K. and Tada, M. (1980). A convenient screening method for nematicidal activity. *Agric. Biol. Chem.*, 44(3): 631–635.

Kawazu, K., Murakami, T., Ono, Y., Kanzaki, H., Kobayashi, A., Mikawa, T. and Yoshikawa N. (1993). Isolation and characterization of two novel nematicidal depsipeptides from an imperfect fungus, Strain D1084. *Biosci. Biotechnol. Biochem.*, 57(1): 98–101. doi: 10.1271/bbb.57.98. PMID: 27316880.

Kepenekci, I., Dura, O. and Dura, S. (2017). Determination of nematicidal effects of some biopesticides against root-knot nematode (*Meloidogyne incognita*) on kiwifruit. *J. Agric. Sci. Technol. A*, 7: 546–551. doi: 10.17265/2161-6256/2017.08.004.

Lebrigand, K., He, L.D., Thakur, N., Arguel, M.-J., Polanowska, J., Henrissat, B., Record, E., Magdelenat, G., Barbe, V., Raffaele, S., Barbry, P. and Ewbank, J.J. (2016). Comparative genomic analysis of

Drechmeria coniospora reveals core and specific genetic requirements for fungal endoparasitism of nematodes. *PLOS Genet.*, 12(5): e1006017. doi: 10.1371/journal.pgen.1006017.
Lee, C.-H., Chang, H.-W., Yang, C.-T., Wali, N., Shie, J.-J. and Hsueh, Y.-P. (2020). Sensory cilia as the Achilles heel of nematodes when attacked by carnivorous mushrooms. *PNAS*, 117(11): 6014–6022. www.pnas.org/cgi/doi/10.1073/pnas.1918473117.
Li, G., Keqin Zhang, K., Xu, J., Dong, J. and Liu, Y. (2007). Nematicidal substances from fungi. *Recent Patents on Biotechnology*, 1: 000–000 1.
Li, J., Hyde, K.D. and Zhang, K-Q. (2014). Methodology for studying nematophagous fungi. pp. 13–40. *In*: Zhang, K.Q. and Hyde, K.D. (Eds.). *Nematode-Trapping Fungi*. Fungal Diversity Research Series # 23. doi: 10.1007/978-94-017-8730-7_2.
Liu, T., Meyer, S.L.F., Chitwood, D.J., Chauhan, K.R., Dong, D., Zhang, T., Li, J. and Liu, W. (2017). New nematotoxicindoloditerpenoid produced by *Gymno ascusreessii* za-130. *J. Agric. Food Chem.*, 65(15): 3127–3132.
Lopez-Llorca, L.V., Gómez-Vidal, S., Monfort, E., Larriba, E., Casado-Vela, J., Elortza, F., Jansson, H.-B., Salinas, J. and Martín-Nieto, J.(2010). Expression of serine proteases in egg-parasitic nematophagous fungi during barley root colonization. *Fungal Genetics and Biology*, 47(4): 342–351. https://doi.org/10.1016/j.fgb.2010.01.004.
Luo, H., Mo Mo, M., Huang, X., Li, X. and Zhang, K. (2004). *Coprinus comatus*: A basidiomycete fungus forms novel spiny structures and infects nematode, *Mycologia*, 96(6): 1218–1224. doi: 10.1080/15572536.2005.11832870.
Luo, H., Liu, Y., Fang, L., Li, X., Tang, N. and Zhang, K. (2007). *Coprinus comatus* damages nematode cuticles mechanically with spiny balls and produces potent toxins to immobilize nematodes. *Appl. Environ. Microbiol.*, 73(12): 3916–3923. https://doi.org/10.1128/AEM.02770-06.
Mayer, A., Köpke, B., Anke, H. and Sterner, O. (1996). Dermatolactone, a cytotoxic fungal sesquiterpene with a novel skeleton. *Phytochemistry*, 43(2): 375–376.
Mayer, A., Anke, H. and Sterner, O. (1997). Omphalotin, a new cyclic peptide with potent nematicidal activity from *Omphalotus olearius* I: Fermentation and biological activity. *Nat. Prod. Lett.*, 10: 25–32. https://doi.org/10.1080/10575639708043691.
Mayer, A., Kilian, M., Hoster, B., Sterner, O. and Anke, H. (1999). *In-vitro* and *in vivo* nematicidal activities of the cyclic dodecapeptide omphalotin A. *Pestic. Sci.*, 55: 27–30. doi: 10.1002/ps.2780550106.
Moliszewska, E., Nabrdalik, M. and Dickenson, J. (2021). Mushrooms as sources of flavours and scents. pp. 252–286. *In*: Sridhar, K.R. and Deshmukh, S.K. (Eds.). *Advances in Macrofungi: Pharmaceuticals and Cosmeceuticals*. USA:CRC Press, Taylor and Francis Group.
Moosavi, M.R. and Zare, R. (2012). Fungi as biological control agents of plant-parasitic nematodes. pp. 67–107. *In*: Mérillon, J.M. and Ramawat, K.G. (Eds.). *Plant Defence: Biological Control, Progress in Biological Contro 12*. Springer Science+Business Media B.V.
Morton, O., Hirsch, P. and Kerry, B. (2004). Infection of plant-parasitic nematodes by nematophagous fungi: A review of the application of molecular biology to understand infection processes and to improve biological control. *Nematology*, 6(2): 161–170. https://doi.org/10.1163/1568541041218004.
Nitao, J.K., Meyer, S.L.F., Schmidt, W.F., Fettinger, J.C. and Chitwood, D.J. (2001). Nematode-antagonistic trichothecenes from *Fusarium equiseti*. *J. Chem. Ecol.*, 27: 859–869.
Nitao, J.K., Meyer, S.L., Oliver, J.E., Schmidt, W.F. and Chitwood, D.J. (2002). Isolation of flavipin, a fungus compound antagonistic to plant-parasitic nematodes. *Nematology*, 4(1): 55–63.
Nordbring-Hertz, B. (2004). Morphogenesis in the nematode-trapping fungus *Arthrobotrys oligospora*: An extensive plasticity of infection structures. *Mycologist*, 18(3): 125–133.
Noweer, E.M.A. (2020). Production, formulation and application of fungi-antagonistic to plant nematodes. pp. 365–401. *In*: El-Wakeil, N. et al. (Eds.). *Cottage Industry of Biocontrol Agents and Their Applications*.Switzerland AG: Springer Nature. https://doi.org/10.1007/978-3-030-33161-0_12.
Obrępalska-Stępiowska, A. and Sosnowska, D. (2008). Czynniki chorobotwórcze w biologicznym zwalczaniu nicieni szkodników roślin (Microorganisms in biological control of plant-parasitic nematodes). *Biotechnologia* 2(81): 115–130.
Olatinwo, R., Yin, B., Becker, J.O. and Borneman, J. (2006). Suppression of the plantparasitic nematode *Heterodera schachtii* by the fungus *Dactylella oviparasitica*. *Phytopathology*, 96: 111–114.

Oota, M., Tsai, A.Y-L., Aoki, D., Matsushita, Y., Toyoda, S., Fukushima, K., Saeki, K., Toda, K., Perfus-Barbeoch, L., Favery, B., Ishikawa, H. and Sawa, S. (2020). Identification of naturally occurring polyamines as root-knot nematode attractants. *Mol. Plant.*, 13: 658–665.

Palizi, P., Goltapeh, E.M., Pourjam, E. and Safaie, N. (2007). Potential of oyster mushrooms for the biocontrol of sugar beet nematode (*Heterodera schachtii*). *Journal of Plant Protection Research*, 49(1): 27–33. doi: 10.2478/v10045-009-0004-6.

Park, J-O., Hargreaves, J.R., McConville, E.J., Stirling, G.R., Ghisalberti, E.L. and Sivasithamparam, K. (2004). Production of leucinostatins and nematicidal activity of Australian isolates of *Paecilomyces lilacinus* (Thom) Samson. *Lett. Appl. Microbiol.*, 38: 271–276. doi: 10.1111/j.1472-765X.2004.01488.x.

Perry, R.N. and Trett, M.W. (1986). Ultrastructure of the egg-shell of *Heterodera schachtii* and *H. glycines* (Nematoda: Tylenchida). *Rev Nématol*, 9: 399–403.

Pineda-Alegrı´a, J.A., Jose´ Ernesto Sa´nchez-Va´zquez, J.E., Gonza´lez-Cortazar, M., Zamilpa, A., Lo´pez-Arellano, M.E., Cuevas-Padilla, E.J., Mendoza-de-Gives, P. and Aguilar-Marcelino, L. (2017). The edible mushroom *Pleurotusdjamor* produces metabolites with lethal activity against the parasitic nematode *Haemonchuscontortus*. *J. Med. Food* 00, (0): 1–9. doi: 10.1089/jmf.2017.0031.

Poveda, J., Abril-Urias, P. and Escobar, C. (2020). Biological control of plant-parasitic nematodes by filamentous fungi inducers of resistance: *Trichoderma*, mycorrhizal and endophytic fungi. *Front. Microbiol.*, 11: 992. https://doi.org/10.3389/fmicb.2020.00992.

Quandt, C.A., Kepler, R.M., Gams, W. et al. (2014). Phylogenetic-based nomenclatural proposals for *Ophiocordycipitaceae* (*Hypocreales*) with new combinations in *Tolypocladium*. *IMA Fungus* 5: 121–134. https://doi.org/10.5598/imafungus.2014.05.01.12.

Rodrigues, J.V., Braga, F.R., Campos, A.K., de Carvalho, L.M., Araujo, J.M., Aguiar, A.R., Ferraz, C.M., da Silveira, W.F., Valadao, M.C., de Oliveira, T., de Freitas, S.G. and de Araújo, J.V. (2018). *Duddingtonia flagrans* formulated in rice bran in the control of *Oesophagostomum* spp. intestinal parasite of swine. *Exp. Parasitol.*, 184: 11–15. https://doi.org/10.1016/j.exppara.2017.11.001.

Saxena, G. (2018). Biological control of root-knot and cyst nematodes using nematophagous fungi. pp. 221–237. *In*: Giri, B., Prasad, R. and Varma, A. (Eds.). *Root Biology, Soil Biology* 52. Springer International Publishing AG. https://doi.org/10.1007/978-3-319-75910-4_8.

Shan, R., Stabler, M., Sterner, O. and Anke, H. (1996). New metabolites with nematicidal and antimicrobial activities from the Ascomycete *Lachnumpapyraceum* (Karst.) Karst VIII: Isolation, structure determination and biological activities of minor metabolites structurally elated to mycorrhizin A. *J. Antibiot.* (*Tokyo*), 49(5): 447–452. doi.org/10.7164/antibiotics.49.447.

Shemshura, O.N., Beckmakhanova, N.E., Mazunina, M.N., Meyer, S.L., Rice, C. and Masler, E.P. (2016). Isolation and identification of nematode-antagonistic compounds from the fungus *Aspergillus candidus*. *FEMS Microbiol. Lett.*, 363(5): 1–9.

Siddiqui, Z.A. and Mahmood, I. (1996). Biological control of plant parasitic nematodes by fungi: A review. *Bioresour. Technol.*, 58: 229–239.

Silvestre, A. and Cabaret, J. (2004). Nematode parasites of animals are more prone to develop xenobiotic resistance than nematode parasites of plants. *Parasite*, 11(2): 119–29. doi: 10.1051/parasite/2004112119. PMID: 15224572.

Soares, F.E.F., Sufiate, B.L. and de Queiroz, J.H. (2018). Nematophagous fungi: Far beyond the endoparasite, predator, and ovicidal groups. *Agriculture and Natural Resources*, 52(1): 1–8. https://doi.org/10.1016/j.anres.2018.05.010.

Soares, F.E.F., Nakajima, V.M., Sufiate, B.L., Satiro, L.A.S., Gomes, E.H., Fróes, F.V., Sena, F.P., Braga, F.B. and Queiroz, J.H. (2019). Proteolytic and nematicidal potential of the compost colonized by *Hypsizygusmarmoreus*. *Exp Parasitol.*, 197: 16–19. https://doi.org/10.1016/j.exppa ra.2018.12.006.

Stadler, M., Anke, H., Bergquist, K.E. and Sterner, O. (1993). Lachnumon and lachnumol a, new metabolites with nematicidal and antimicrobial activities from the ascomycete *Lachnumpapyraceum* (Karst.) Karst. II: Structural elucidation. *J. Antibiot.* (*Tokyo*), 46(6): 968–971. doi: 10.7164/antibiotics.46.968.

Stadler, M., Anke, H. and Sterner, O. (1995). Metabolites with nematicidal and antimicrobial activities from the ascomycete *Lachnumpapyraceum* (Karst.) Karst. V: Production, isolation, and biological activities of bromine-containing mycorrhizin and lachnumon derivatives and four additional new bioactive metabolites. *J. Antibiot.* (*Tokyo*), 48(2): 149–153. https://doi.org/10.7164/antibiotics.48.149.

Sterner, O., Etzel, W., Mayer, A. and Anke, H. (1997). Omphalotin, a new cyclic peptide with potent nematicidal activity from *Omphalotus olearius* II: Isolation and structure determination. *Nat. Prod. Lett.*, 10: 33–38. https://doi.org/10.1080/10575639708043692.

Stirling, G.R. and Mankau, R. (1979). Mode of parasitism of *Meloidogyne* and other nemaiode eggs by *Dactylella oviparasitica*. *J. Nematol.*, 11(3): 282–288.

Su, H., Zhao, Y., Zhou, J., Feng, H., Jiang, D., Zhang, K.-Q. and Yang, J. (2017). Trapping devices of nematode-trapping fungi: Formation, evolution, and genomic perspectives. *Biol. Rev.*, 92: 357–368. 357. doi: 10.1111/brv.12233.

Sun, J., Wang, H., Lu, F., Du, L. and Wang, G. (2008). The efficacy of nematicidal strain *Syncephalastrum racemosum*. *Ann. Microbiol.*, 58(3): 369–373.

Tanney, J.B. and Hutchison, L.J. (2012). The production of nematode-immobilizing secretory cells by *Climacodon septentrionalis*. *Mycoscience*, 53(1): 31–35.

Tileubayeva, Z., Avdeenko, A., Avdeenko, S., Stroiteleva, N. and Kondrashev, S. (2021). Plant-parasitic nematodes affecting vegetable crops in greenhouses. *Saudi J. Biol. Sci.*, 28: 5422–5433. https://doi.org/10.1016/j.sjbs.2021.05.075.

Truong, B-N., Okazaki, K., Fukiharu, T., Takeuchi, Y., Futai, K., Le X-T. and Suzuki A. (2007). Characterization of the nematocidal toxocyst in *Pleurotus* subgen. *Coremiopleurotus*. *Mycoscience*, 48: 222–230. doi:10.1007/s10267-007-0358-4.

Udo, I.A., Uguru, M.I. and Ogbuji, R.O. (2013). Pathogenicity of *Meloidogyne incognita* Race 1 on tomato as influenced by different arbuscular mycorrhizal fungi and bioformulated *Paecilomyces lilacinus* in a dystericcambisol soil. *J. Plant Prot. Res.*, 53(1): 71–78. doi: https://doi.org/10.2478/jppr-2013-0011.

Udo, I.A., Osai, E.O. and Ukeh, D.A. (2014). Management of root-knot disease on tomato with bioformulated *Paecilomyces lilacinus* and leaf extract of *Lantana camara*. *Braz. Arch. Biol. Technol.*, 57(4): 486–492. http://dx.doi.org/10.1590/S1516-89132014005000022.

Wang, X., Lin, M., Xu, D., Lai, D. and Zhou, L. (2017). Structural diversity and biological activities of fungal cyclic peptides, excluding cyclodipeptides. *Molecules*, 22: 2069. doi: 10.3390/molecules22122069.

Yan, X., Sikora, R.A. and Zheng, J. (2011). Potential use of cucumber (*Cucumis sativus* L.) endophytic fungi as seed treatment agents against root-knot nematode *Meloidogyne incognita*. *J. Zhejiang Univ. Sci. B*, 12(3): 219–225. doi:10.1631/jzus.B1000165.

Yang, C-T., de Ulzurrun, G.V., Gonçalves, A.P., Lin, H-C., Chang, C-W., Huang, T-Y., Chen, S-A., Lai, C-K., Tsai, I.J., Schroeder, F.C., Stajich, J.E. and Hsueh, Y.P. (2020). Natural diversity in the predatory behavior facilitates the establishment of a robust model strain for nematode-trapping fungi. *Proc. Natl. Acad. Sci.*, 117(12): 6762–6770. doi: 10.1073/pnas.1919726117.

Yang, J. and Zhang, K.Q. (2014). Biological control of plant-parasitic nematodes by nematophagous fungi. pp. 231–262. *In*: Zhang, K.Q. and Hyde, K.D. (Eds.). *Nematode-Trapping Fungi*. Fungal Diversity Research Series # 23, Mushroom Research Foundation. doi 10.1007/978-94-017-8730-7_5.

Yang, J., Tian, B., Liang, L. and Zhang, K.Q. (2007). Extracellular enzymes and the pathogenesis of nematophagous fungi. *Appl Microbiol Biotechnol.*, 75(1): 21–31. doi: 10.1007/s00253-007-0881-4.

Index

About the Editors

Sunil K. Deshmukh

Sunil Kumar Deshmukh received his Ph.D. in Mycology from Dr. H.S. Gour University, Sagar (M.P.) in 1983. Veteran industrial mycologist who spent a substantial part of his career at Hoechst Marion Roussel Limited [now Sanofi India Ltd.], Mumbai, and Piramal Enterprises Limited, Mumbai, in drug discovery. He has to his credit 8 patents, 135 publications, and 15 books on various aspects of fungi and natural products of microbial origin. He was the president of the Mycological Society of India (MSI). He is a fellow of MSI, the Association of Biotechnology and Pharmacy, the Society for Applied Biotechnology, and Maharashtra Academy of Science. He was Fellow at Nano Biotechnology Centre, TERI, New Delhi, and Adjunct Associate Professor in Deakin University, Australia, till Jan 2019 who had been working towards the development of natural food colours, antioxidants, and biostimulants through nanotechnology intervention. He is also an advisor to Agpharm Bioinnovations LLP, Patiala, Punjab, India and Greenvention Biotech, Uruli-Kanchan, Pune, India.

Kandikere R. Sridhar

Dr. Kandikere R. Sridhar is an adjunct professor in the Department of Biosciences, Mangalore University and Yenepoya (Deemed to be University). His main areas of research are "Diversity and Ecology of Fungi of the Western Ghats, Mangroves and Marine Habitats". He was NSERC postdoctoral fellow/visiting professor in Mount Allison University, Canada; Helmholtz Centre for Environmental Research-UFZ and Martin Luther University, Germany; Center of Biology, University of Minho, Portugal. He was recipient of The Shome Memorial Award (2004), became Vice-President (2013), President (2018) and Lifetime Achievement Awardee (2019) by the Mycological Society of India. He was recipient of The Fellow of Indian Mycological Society, Kolkata (2014), Distinguished Asian Mycologist (2015) and Outstanding Leader in Education and Research, Association of Agricultural Technology of Southeast Asia (2016). He was awarded UGC-BSR Faculty Fellowship (2014–2017). He is one among the world top 2% scientists in the field of mycology (2020–21). He has over 450 publications and edited eight books.

Susanna M. Badalyan

Prof. Badalyan finished her university studies in 1980 in Biology. In 1981–1982 she completed post-graduate studies in Mycology at the Lomonosov Moscow State University, Russia. She holds her PhD (1988) and DSc (1998) degrees in Botany/Mycology in Armenia. Prof. Badalyan has 40 years of research experience in Fungal Biology and Biotechnology, particularly related to studies of life cycle and asexual sporulation in Agaricomycetes fungi. From 1993–2005, she was a group leader to study medicinal mushrooms at the Yerevan State University (YSU). Since 2005, she is a full Professor and Principal Researcher, head of the Laboratory of Fungal Biology and Biotechnology, YSU.

Prof. Badalyan has supervised several national and international projects and was awarded with research fellowships and grants from NATO, DAAD and other organisations. She has carried out research and was an invited lecturer in universities of Germany, Italy, France and other countries. She has authored 270 publications and contributed to 80 international scientific meetings. She is an editorial board member, referee and reviewer of several international journals, PhD theses and research projects. Prof. Badalyan is a lecturer of Bachelor and Master Courses in Biology, Botany, Fungal Genetics and Cytology, Fungal Taxonomy and Phylogeny. She has supervised 70 PhD, Master and Bachelor students.